RADIOACTIVE WASTE MANAGEMENT AND DISPOSAL

RADIOACTIVE WASTE MANAGEMENT AND DISPOSAL

Proceedings of the Second European Community
Conference, Luxembourg, April 22–26, 1985

Edited by R. SIMON
Commission of the European Communities
Directorate-General Science, Research and Development, Brussels

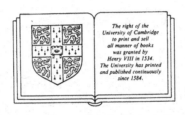

The right of the
University of Cambridge
to print and sell
all manner of books
was granted by
Henry VIII in 1534.
The University has printed
and published continuously
since 1584.

CAMBRIDGE UNIVERSITY PRESS
for the Commission of the European Communities

CAMBRIDGE UNIVERSITY PRESS
Cambridge, New York, Melbourne, Madrid, Cape Town, Singapore, São Paulo, Delhi

Cambridge University Press
The Edinburgh Building, Cambridge CB2 8RU, UK

Published in the United States of America by Cambridge University Press, New York

www.cambridge.org
Information on this title: www.cambridge.org/9780521115209

© ECSC, EEC, EAEC, Brussels and Luxembourg, 1986

First published 1986
This digitally printed version 2009

A catalogue record for this publication is available from the British Library

ISBN 978-0-521-32580-6 hardback
ISBN 978-0-521-11520-9 paperback

CONTENTS

This conference was organized by the Commission of the European Communities,
Directorate-General for Science, Research and Development,
Directorate for Nuclear Research and Development, Fuel cycle division,
Brussels

Programme Committee

Chairman: S. ORLOWSKI (CEC)

Members: O. CAHUZAC (CEN)
                    L. BAETSLE (CEN/SCK)
                    K. KüHN (GSF)
                    K. BRODERSEN (NRL)
                    H.D.K. CODEE (Ministerie van Volksgezondheid)
                    N.B.H. HARRYMAN (UKDOE)
                    F. GIRARDI (CEC Ispra)
                    A. DONATO (ENEA)

Publication arrangements:D. NICOLAY (CEC)

Local organization: R. LINSTER (CEC)

Scientific Secretary: R. SIMON (CEC)

Abbreviations

Aberdeen U.	University of Aberdeen (UK)	GSF	Gesellschaft für Strahlen- und Umweltforschung mbH (D)
Agip	Agip nucleare spa, Milano (I)	HMI	Hahn-Meitner-Institut für Kernforschung GmbH, Berlin (D)
ALKEM	Alkem GmbH, Hanau am Main (D)		
ANDRA	Agence nationale pour la gestion des déchets radioactifs, Paris (F)	ISMES	Instituto Sperimentale Modelli e Strutture, Roma (I)
Atkins	Atkins Research and Development, Surrey (UK)	JRC/CCR/GFS	Joint Research Centre/Centre commun de recherche/Gemeinsame Forschungsstelle
Battelle Ffm	Battelle Institut eV, Frankfurt am Main (D)	KFA	Kernforschungsanlage Jülich GmbH (D)
BGR	Bundesanstalt für Geowissenschaften und Rohstoffe, Hannover (D)	KfK	Kernforschungszentrum Karlsruhe GmbH (D)
		KUL	Katholieke Universiteit Leuven (B)
BGS	British Geological Survey, Nottinghamshire (UK)	LvG	Laboratorium voor Grondmechanica, Delft (NL)
BRE	Building Research Establishment, Herts (UK)	MHA	Mott, Hay & Anderson, Croydon (UK)
BRGM	Bureau de recherches géologiques et minières, Orléans (F)	NRC	Nuclear Research Centre, Greek Atomic Energy Commission, Aghia Paraskevi Attiki (GR)
CEA	Commissariat à l'énergie atomique (F)		
CEC/CCE/KEG	Commission of the European Communities/ Commission des Communautés européennes/ Kommission der Europäischen Gemeinschaften	NRL	National Research Laboratory, Risø (DK)
		NRPB	National Radiological Protection Board, Chilton (UK)
		NUKEM	NUKEM GmbH, Hanau (D)
CEN/SCK	Centre d'étude de l'énergie nucléaire/Studiecentrum voor Kernenergie, Mol (B)	ONDRAF/NIRAS	Organisme national des déchets radioactifs et des matières fissiles/Nationale Instelling voor het Beheer van radioaktive afval en splijstoffen, Brussel (B)
CNRS	Centre national de recherche scientifique, Vitry-sur-Seine (F)		
		Ove Arup	Ove Arup, London (UK)
DBE	Deutsche Gesellschaft zum Bau und Betrieb von Endlagern für Abfallstoffe mbH, Peine (D)	RNL	Risø National Laboratory, Roskilde (DK)
		RWTH	Rheinisch-westfälische Technische Hochschule, Aachen (D)
ECN	Energieonderzoek Centrum Nederland, Petten (NL)		
EMP	École des mines de Paris, Fontainebleau (F)	THD	Technische Hogeschool Delft (NL)
ENEA	Comitato nazionale per la ricerca e per lo sviluppo dell'Energia Nucleare e delle Energie Alternative, Roma (I)	TUM	Technische Universität München (D)
		TWC	Taylor Woodrow Construction Ltd, Southall (UK)
EP	École polytechnique, Palaiseau (F)	UCL	Université catholique de Louvain, Louvain-la-Neuve (B)
FhG	Fraunhofer Institut für Silikatforschung, Würzburg (D)	UKAEA	United Kingdom Atomic Energy Authority, London (UK)
FORAKY	FORAKY, Bruxelles (B)	UKDOE	Department of the Environment, London (UK)
FUB	Freie Universität München, Garching bei München (D)	U. Utrecht	Universität Utrecht (NL)

FOREWORD

The European Community's research programmes on "Management and disposal of radioactive waste" have the prime objective of finding effective means for ensuring the safety of man and his environment against the potential hazards arising from such wastes.

The EC's first R&D programme devoted specifically to radioactive waste was launched in 1973 at the Joint Research Centre, Ispra. In 1975, the Community's effort was substantially stepped up by the adoption of a five-year programme under the shared cost action scheme carried out at major laboratories in the Member States under CEC management. Since then, both research actions have been operating hand in hand.

In May 1980, the Commission of the European Communities held its first major meeting on this subject in Luxembourg. The second European conference, which once again follows the completion of a five-year R&D programme, presented and analysed the latest results achieved by sustained collaboration of leading laboratories in Europe, within the programme of the European Community.

The main topics addressed at the conference were:

(i)   **Treatment and conditioning technology** : solid, liquid and gaseous waste from reactor operation, spent fuel reprocessing and MOX fuel fabrication. Quality assurance in waste processing.

(ii) **Testing and evaluation of waste forms and packages** : characterization of medium active waste forms based on cement, bitumen and polymer resins, testing of vitrified HLW, waste container corrosion studies.

(iii) **Geologic disposal in salt, granite and clay formations** : mechanical and hydrogeologic characterization by deep drilling, analysis of cores and **in situ** investigations in experimental caverns (Mol, Asse); studies of backfilling and sealing for repositories.

(iv) **Migration (near-field and far-field)** : investigation of mechanisms and modelling (MIRAGE project).

(v)   **Performance analysis of geological isolation systems** (PAGIS project).

As the development of the technological elements for treatment and disposal of radioactive waste progressed, increased effort has been placed on examining industrial waste products as well as analysing management strategies and barrier systems. The broad scope of such evaluations has required close collaboration between CEC scientists and experts from the Member States. The extent of integration achieved is clearly demonstrated by the joint authorship of the conference papers presented in this volume of proceedings.

SESSION I

OPENING SESSION

Welcoming address
    J. LAHURE, State Secretary for Economic Affairs of
    the Grand Duchy of Luxembourg

Opening address
    N. MOSAR, Member of the Commission of the European
    Communities

The development of nuclear power and the research
effort in the Community
    D.H. DAVIES, Deputy Director-General for Research
    and Development, Commission of the European
    Communities, Brussels

Waste disposal in Europe - Looking ahead
    B. VERKERK, Chairman of the Advisory Committee on
    Programme Management

The role and results of the European Community's R & D
work on radioactive waste management
    S. ORLOWSKI (CEC); F. GIRARDI (JRC)

WELCOMING ADDRESS

Johny LAHURE
State Secretary for Economic Affairs of the
Grand Duchy of Luxembourg

Mr Commissioner, Mr Chairman, Ladies and Gentlemen,

It gives me great pleasure to welcome you, on behalf of the Luxembourg Government and myself, to the Second Conference on Radioactive Waste Management and Disposal. The Conference will concentrate on five main themes and, given the importance and interdependent nature of the subjects to be covered, will last five days.

In giving the welcoming address I have the advantage of not needing to go into technical detail. I must, however, exercise a certain minimum amount of caution.

Otherwise, I might find all too quickly that I am out of my depth addressing a very technical conference attended by a good number of eminent experts actively involved in the field.

I hope nevertheless that you will allow me, as the government spokesman of the host country, to make a few brief comments before the conference proper gets underway.

*

*   *

From the Community viewpoint, the conference will, by considering the various questions and their ramifications, establish meaningful links between subjects which, although seemingly dry and technical, in reality have a direct bearing on various strands of Community policy, namely energy policy and its technological and economic aspects, and environmental policy, which necessarily includes ecology.

I should like to draw your attention to four points that are fundamental to a rational and effective Community approach.

On 17 January, the European Parliament adopted a Commission proposal for a new, shared-cost (over the medium term) multiannual research programme in the field of radioactive waste.

Thus this conference would appear to have come at the right time.

Secondly, it must be stressed that the subjects covered form part of an interrelated complex involving energy, technology, the vast research programme and environmental protection as a whole.

Your work - to demonstrate the interrelationships and the basic need for action work and to emphasize the community nature of the right approach - is of fundamental importance to the Community.

International cooperation is of the essence, since analysis of scientific problems of such magnitude requires the finest brains.

Lastly, I should point out that this is a logical continuation of a process the Community started years ago.

The number of Commission documents in the field is enormous but I will only mention the document of 9 February 1982, outlining a Community energy strategy in the nuclear field which provides an excellent description of various problems, including those relating to radioactive wastes produced throughout the nuclear cycle, and covers the subject in depth.

*

*   *

Mr Chairman, ladies and gentlemen,

having briefly outlined the European and international aspect, I should like to underline the importance of ecology in this field.

My government intends - and I am pleased to note that this is also one of the questions with which you are concerned - to implement a programme of priority environmental measures.

Obviously, it would be inappropriate to go into details at this stage but I can nevertheless say that particular stress will be placed on waste management, prevention policy and environmental protection.

I should now like to look briefly at the question from a philosophical point of view.

The existence of the ecosystem, as a living and organized unit, is now a well accepted basic principle.

Several thinkers, particularly Edgar Morin, a sociologist and philosopher, teach that various elements are striving to establish a delicate balance : i.e. objective scientific knowledge and ultimate goals on the one hand, and the position of man and the requirements of nature on the other.

It is my belief that your work will throw more light on the numerous interrelationships involved which are essential to our overall understanding of the world in which we live.

*

*   *

Ladies and gentlemen,

To conclude my address, I would like to say three things :

- I thank you sincerely for having chosen to hold your Second Conference in the capital of the Grand Duchy, with its long history of Community involvement.

- I wish you every success in your work, which I consider fundamental, and hope that it will produce the desired response.

- Lastly, I will be happy if at the end of the Conference, all concerned can look back with general feelings of satisfaction on the five days of technical and economic discussions.

Thank you

OPENING ADDRESS

N. MOSAR
Member of the Commission of the European Communities

One of the basic aspects of our industrial society is the ability to convert available natural resources into useful energy. Since the invention of the steam engine, we have developed and exploited new and more economical sources of energy. We thus have a wide range of different primary energy sources today, mainly coal, oil, natural gas and nuclear energy, which meet most of Europe's needs.

The utilization of these primary energy sources and the extent to which they cover overall requirements depend primarily on price, on local or regional availability and on the technical feasibility of exploiting them - but the level of useful reserves, socio-economic conditions and environmental considerations play an increasingly important part. When planning for our medium and long-term energy supplies, we must therefore use our energy resources economically and rationally, and we cannot abandon any source of energy. This also holds true for nuclear energy which, together with coal, provides our electricity needs.

The trade and industry which provide a living for the vast majority of the people of Europe have always produced harmful waste. Some dangerous substances produced by the chemical industry, the textile industry, the iron and steel industry and conventional power stations are still today released into the environment, especially by old installations, in the form of waste gases or water. Nuclear energy also produces dangerous waste.

What is the difference between nuclear waste and other dangerous substances?

1. Unlike toxic metals such as cadmium, lead and mercury, certain radioactive components of nuclear waste do not occur in nature, although sources of similar natural radiation, although with very low radioactivity levels, are present everywhere in the environment, and even in the human body.
2. All radioactive substances ultimately become non-radioactive, so that in one generation most of the radiation from nuclear waste will have disappeared.

Nuclear waste also differs from the waste produced by other industries in that the quantities, their properties and the effects on the environment are perfectly well known, and since nuclear energy was first used techniques have been developed to remove nuclear waste from the environment and from society.

As early as 1957, the Treaty establishing the European Atomic Energy Community laid down basic standards for ionizing radiation and for limiting the release of radioactive substances into the environment, even though the Community's first nuclear power station did not begin operating until four years later.

Why, then, is non-nuclear waste treated differently?

Dangerous nuclear waste can be stored because:

- Firstly, nuclear power stations use very little fuel to produce large amounts of energy. Understandably, therefore, they produce only very small amounts of waste, the safe isolation of which has proved perfectly feasible.
Conventional power stations, on the other hand, produce such large amounts of gas and ash that it is virtually impossible to remove all dangerous substances.

- The second reason why radioactive waste is treated differently is that the effects of radioactivity on man and his environment were known even before nuclear energy was used. It was therefore possible to develop effective safety measures and incorporate them in the design of the first reactors. Since then there has been further substantial progress in the field of reactor safety.

If our health and environment are so well protected, why do we continue pouring money into waste management and research in that field?

Over a long period, waste gradually accumulates in nuclear installations and in temporary stores. Although most of the radioactivity disappears quite quickly, it takes thousands of years before waste with the longest half-life can be considered safe.

We therefore have to find a way of isolating it from the environment with no need for human intervention. Such conditions, providing safe storage for thousands of years without the need for human control, can be achieved only by disposal underground. Geological formations which are very unlikely to be disturbed over the next thousand years by earthquakes or other natural causes, or by human activity are found only at great depths below the surface of the earth.

Nevertheless, not content with trusting geological strata, we shall also encase nuclear waste in very tough containers.

The design of such a system of barriers and the careful examination of all matters relating to safety are the primary objectives of our research programme. In order to work as efficiently and as realistically as possible, the Commission will develop and exploit, together with certain Member States, large experimental installations in deep geological formations. For the first time experiments will be carried out on a small scale on the storage of highly radioactive "vitrified" waste. The prerequisite is that it must be possible, after the experiments, to remove the waste from the installations in completely sealed containers. The experience and know-how thus gained, as well as the other results of these research programmes, will be available to all Member States.

The effectiveness of the Community's activities in this field is known outside Europe and was emphasized by the Council when the third programme of research was adopted at the beginning of the year.

Over the next few days you will gain an overall impression of the results obtained, and will be able to discuss them. I would ask you to examine carefully the results and conclusions reached before any further steps are taken towards the permanent removal of radioactive waste.

We wish to continue in this direction, basing our actions fully on the knowledge acquired and on the reliability of the new techniques developed. The available scientific know-how, our technical resources and the time devoted to this subject can totally exclude the possibility that waste management becomes a game of chance with dire consequences for future generations.

*

*   *

Ladies and gentlemen, the technological and scientific problems which remain will require a great deal more work. It is all the more heartening, therefore, that we do not have to work in an uncoordinated way. The programme for this conference clearly shows that the researchers and research institutes of the Member States and of the Joint Research Centre are cooperating closely. In adopting the third programme of research, the Council recently confirmed that this cooperation is fruitful.

Let us not forget that national frontiers, customs duties and passport checks, which have still not yet been completely abolished after 28 years of the Community's existence, do not represent a barrier to the transport of dangerous substances by air and sea.

It therefore makes sense to establish a common front of science and technology to protect man and his environment in Europe. The spirit of cooperation which inspires research should also be present when final storage installations are designed and used in the future.

We regard such integration as one of the Commission's most pressing tasks, but we are aware that this will require patience, hard work and intuition on the part of our partners.

I would therefore like to thank very sincerely all those who have contributed towards the success of the Community programmes.

THE DEVELOPMENT OF NUCLEAR POWER AND THE RESEARCH EFFORT IN THE COMMUNITY

D.H. DAVIES, Commission of the European Communities, Brussels

Summary

The development of nuclear power in the Community is analysed at the light of the oil crisis which hit the world in 1973. Before 1973, nuclear energy was rapidly penetrating the market all over the world : nuclear power plants were being ordered in large numbers and the development of advanced nuclear reactors and of their fuel cycle was vigorously pursued in almost all industrialized countries.
In all logic the 1973 oil crisis should have quickned the pace of nuclear energy development; in reality the expected rapid expansion of nuclear energy in the most industrialized countries did not materialize. Despite the setbacks to the global pace of nuclear development, the nuclear energy's share of electricity production in the Community increased from 5.4% in 1973 to 22.4% in 1983.
To-day the installed nuclear electricity generating capacity is about 55 GWe and the nuclear energy's share exceeds 25%. In 1990 these figures should be about 100 GWe and 35% respectively.
The improvement of management of energy resources and the reduction of energy imports are some of the major goals to which Community research is directed. In this context, the further development of nuclear fission energy is considered as one of the main ways of reducing dependence on energy imports. The Community research strategy therefore provides for the consolidation and intensification of the research activities in the nuclear energy field. Among these research activities, the research effort deployed in the Community in the field of radioactive waste management is reviewed in more detail. Some achievements of the twelve year Plan of Action and of the multiannual R&D programmes are presented.

1. General Nuclear Power Situation

Dependence upon energy imports of the European Community has grown subtantially since 1950 in spite of the exploitation of North Sea Oil and gas : virtually all the increased dependence concerns oil.

Presently the European Community annually uses an amount of energy equivalent to approximately nine hundred millions tons of oil; of this 45 % is imported. The rest, that is approximately 55 %, is produced indigeneously. Of the energy which is imported, over 75 % is in the form of oil.

These data refer to the energy balance of the European Community as a whole. The situation is certainly not ideal. Individual Community Countries display significant variations with respect to the average, which makes the situation of some of them extremely dependent upon oil.

The predominant position gained by oil in the last few decades in the energy balance of the most industrialized Countries is a consequence of its flexibility of utilization, ease of transport and of storage. Furthermore oil was, until twelve years ago, a very cheap source of energy thanks to large and intensively exploited reservoirs. The 20th century has truly seen the triumph of oil.

The most developed Countries, shaped by the "oil civilization" which gave them unprecedented prosperity, have been caught in a dependence upon a source of energy which is no longer under their control and yet continues to be the vital bloodstream of their economies.

Over the past twelve years remarkable developments perturbed the world energy scene and nuclear energy has experienced its fair share of upheavals. Twelve years ago the oil crisis hit the world. While it came as a great shock, it should hardly have been a surprise.

It is useful to recall that the early motivations for the development of nuclear energy included conservation of fossil fuel resources, particularly oil, and reduction of energy dependence.

The early motivations for the development of nuclear energy are as valid to-day as they were then. The search for other alternative sources of energy (solar, geothermal, wave, etc...) which has been undertaken in the meantime has the same motivations and is complementary to but not substitute for nuclear energy. The experience of the last decade has indeed confirmed the importance of diversifying the energy sources. This is particularly true for the European Community.

In this context, the development of policies in the nuclear energy field, over the last decade, appears paradoxical.

Before the 1973 oil crisis, nuclear plants for electricity production were being ordered in large numbers, all over the world, on economically competitive terms. Nearly all of the nuclear plants now in service (i.e. approx. 65 GW in the US, 55 GW in the European Community and 18 GW in Japan) were ordered before 1973 (*).

Not only was nuclear energy rapidly penetrating the market, but also the development of advanced nuclear reactors and of their fuel cycle was vigorously pursued in almost all industrialized countries. This was in recognition of the fact that uranium resources were limited, and hence had to be rationally utilized, and that sooner or later nuclear energy would have to be used for non-electrical applications (e.g. process heat, metallurgical processes, production of synthetic fuels) in order to conserve oil and gas currently used for these applications. The successful development of fast breeder reactors and of high temperature gas cooled reactors could, thus, contribute to solving the problems of uranium supply and of hydrocarbon conservation. In parallel, reprocessing and radioactive waste disposal technologies were also being developed and gradually applied.

The 1973 oil crisis brought home the tangible message that oil supplies could not be guaranteed and that too strong a dependence upon oil and, more generally, energy imports was a most undesirable situation for any country.

In all logic the 1973 oil crisis was a factor that should have quickened the pace of nuclear energy development. The strategic case for nuclear energy was in fact reinforced and the economic competitivity of nuclear energy with respect to fossil energy seemed bound to improve.

In reality, not all the expectations placed upon nuclear energy could be fullfilled and the then expected rapid expansion of nuclear energy in the most industrialized countries did not materialize.

It is not intended to suggest reasons for this. For the present purpose, it is sufficient to review, briefly, the current situation :
- Nuclear power plant construction rate has not kept up to expectations of the seventies in any of the Countries of the western world except

(*) With the notable exception of France, nuclear power stations now take longer than 10 years from decision to build to completion.

France. In the United States, for instance, so many orders have been cancelled in last ten years that the total gigawatts ordered to date are less than ten years ago. Cancellations amount to about 110 GW for a cost which is estimated at some 10 billion dollars.
- In the European Community the rate of ordering was more modest and the massive cancellations experienced by the United States have not occurred. The total capacity which has been reached at the end of the last year is approximately 55 GW (before 1973 the target was about 100 GW).
- The advanced reactor development programmes have been adversely affected in the last twelve years. High Temperature Gas Cooled Reactor projects have suffered serious setbacks : the 8 GW ordered in the United States between 1971 and 1973 were all cancelled between 1974 and 1976. The joint European effort for the development of this system (the Dragon Project) terminated in 1976. Only a few scattered efforts to continue the development of this most interesting reactor system are still pursued in Germany, the United States and the Japan.

The Fast Breeder Reactor system, on the other hand, has continued to be vigorously and purposefully developed within the European Community. Even so, it would be difficult to deny that its prospects of commercial introduction have receded considerably; (ten years ago, 1990 was considered a realistic date for its commercial introduction; to-day 2005 is mentioned more and more frequently). Signs of hesitation are not rare even among the supporters of the developments.

Despite the setbacks to the global pace of nuclear development, the nuclear contribution in the Community has grown steadily. From 1973 to 1983 the nuclear contribution towards meeting the Community's total demand for energy increased from less than 2 % to about 10 %. The nuclear energy's share of electricity production in the Community increased from 5.4% to 22.4% in this period. To-day the nuclear share in electricity production in the Community exceeds 25 %. Furthermore, the Community's nuclear power capacity accounts for about one third of the world capacity. These figures show that nuclear power has become an essential part of the Community energy strategy.

The European Community has strong incentives to continue the peaceful development of nuclear power and has been able to maintain steady policies to this effect in the difficult last ten years : the development of fast reactors, the reprocessing of irradiated fuels and radioactive waste disposal are still being actively pursued.

The Community strategy, approved by the Council of Ministers in February 1980, aims at "closing" the fuel cycle, by reprocessing the spent fuel with the following objectives :
a) to extract plutonium as an energy source and thus pave the way for its recycling, particularly in fast breeder reactors, the advantages of which in terms of availability of supply are well known (of course, the enriched uranium still contained in irradiated fuels is also recoverable).
b) to separate out the highly radioactive fission products and to condition them with a view to final dispoal compatible with safety and environmental requirements.

The strong incentives to continue the development of nuclear power and the investment programmes in the Member Countries should lead, by 1990, to an installed nuclear electricity generating capacity of about 100 GWe. At that time, this capacity will account for about 35 % of electricity production in the Community and meet about 14 % of the Community overall demand for energy.

Concerning the outlook for the nineties, the Commission has recently proposed the adoption of the following objectives for development of nuclear power :
(i) to produce about 40 % of Community electricity in 1995, and
(ii) subsequently, to increase its share in electricity production considerably after the turn of the century.

The 1995 target would require commissioning of some 25 GWe of new capacity between 1991 and 1995. This figure takes into account a loss of 3 to 4 GWe resulting from the decommissioning of old nuclear power plants which is likely to take place in the first half of the nineties. Therefore, in 1995, the total installed nuclear electricity generating capacity should be of 120 GWe.

As far as the development of fast breeder reactors is concerned, the Commission recognizes that these reactors are not yet economically competitive with thermal reactors. Nevertheless, it thinks that the advantages of having fast breeder reactors available after the year 2000 will be such that it would be injudicious to wait until difficulties with uranium supplies arise before preparing for their commercial introduction.

The Commission therefore welcomes the cooperative agreement for fast reactor development signed by five Community Member States (Belgium, France, Italy, Germany and United Kingdom) on 10 January 1984 and suggests, as for a target, that these reactors should aim to reach comparable generation costs to thermal nuclear plants by 2005.

It is clear that the above objectives will only be achieved through a substantial effort of research, development and demonstration which will clearly benefit from international cooperation.

2. Research Effort in the Community

In 1983, the Council of Ministers adopted a proposal by the Commission for a European Scientific and Technical Strategy, expressed in the first framework programme 1984-1987 in which are set out the major goals to which Community research should be directed .

Among these goals, the improvement of management of energy resources and the reduction of energy dependence are considered as essential if the Community is to face up to the energy challenge. In order to attain these major goals, the Commission recommended an approach centred on specific objectives with a view to :
- facilitating the implementation of the research specifically desired by the Member States;
- facilitating the subsequent adoption of action programmes for implementation by identifying and putting into order the priority needs of the Community and thus the relative weighting to be given to the corresponding scientific and technical objectives.

As far as improving the management of energy resources and reducing energy dependence is concerned, the Commission proposed to concentrate its effort during the period 1984-1987 on the following four scientific and technological objectives : the rational use of energy, the development of renewable energy sources, controlled thermonuclear fusion and the development of nuclear fission energy.

Concerning the nuclear fission energy, the Commission considered again its development as one of the main ways of reducing, through the diversification of energy resources, the Community's dependence on oil. The continuation of a resolute nuclear programme is therefore an essential aspect of European energy policy. The Community strategy provides for the consolidation and intensification of research activities, in fields such as reactor safety, management of radioactive waste, fissile material safeguards and decommissioning of nuclear installations through the corresponding research action programme. Only the research activities related to the management of radioactive waste are considered here.

In some Community Countries research on radioactive waste management started in the early 1950s. The aim was to develop treatment and conditioning methods for the waste existing at that time, i.e. mainly low and medium level waste from nuclear research. Alongside this research effort, later expanded to cover waste from the first nuclear power stations to be commissioned, studies on the disposal of these categories of waste were started.

In this way treatment and conditioning processes and also the disposal routes for low and medium level waste have gradually been developed since the 1960s. To-day these are industrial operations.

With the entry into operation of the first spent fuel reprocessing plants (towards the beginning of the 1960s), research on the treatment and conditioning of high level waste was started, especially in France and the United Kingdom. The research work, which in the United Kingdom was suspended for some years, resulted in the commissioning in France, in 1969, of the first facility for waste solidification in glass.

Later several Member States embarked on studies and research on waste disposal in suitable geological formations at depths of up to several hundred metres. A major R&D effort was also launched on alpha-contaminated waste.

From the beginning of the 1970s, with the launching of these nume-rous national programmes or projects, the large number of possible options, and the need in the medium term to select and optimize methods and processes, Community cooperation became essential.

In 1973, the Council of Ministers of the European Communities established the principle of Commission competence for environmental protection and in particular nuclear and non-nuclear waste.

That same year a modest initial multiannual R&D programme on radio-active waste was started at the JRC. In 1975 it was supplemeted by a wider five-year programme comprising shared-cost projects with laborato-ries in the Member States.

This programme, the main results of which were described five years ago at the first European Community Conference on radioactive waste, covered activities designed to solve certain technological problems arising from the treatment, storage and disposal of waste of all cate-gories and to help towards the definition of a legal, administrative and financial framework for waste management activities.

This first programme yielded positive results and opened up encoura-ging prospects. The Council adopted, in 1980, a second five-year program-me (1980-1984) which was a logical follow-up to the first one and again fully coordinated with the direct action programme (1980-1983) carried out by the Joint Research Centre.The most important achievements of this second five year programme are :
- the improvement of the performances of treatment and conditioning processes for low- and medium level wastes with respect to the reduc-tion of the volume of produced waste and to the cost or quality or finished products;
- the identification of conditioning processes for spent fuel claddings which are sufficiently promising, account being taken of the results obtained on a laboratory scale, to merit a subsequent development on a

semi-pilot scale with radioactive materials;
- the accomplishement of R&D work in the field of the treatment of combustible solid waste contaminated with plutonium; this has made available, thanks to the operation of the corresponding pilot plants, treatment processes by acid digestion and high temperature combustion and the identification of the limitations of these processes for industrial applications;
- the information obtained on the long-term behaviour of the various waste categories in conditions representative of disposal; this information has already allowed the improvement of the conditioning of certain waste;
- the very important progress achieved as regards disposal of vitrified high level waste in deep geological formations, mainly clay, granite, salt and deep-sea sediments. The construction or operation of underground laboratories such as Asse (salt) or Mol (clay) has contributed to confirming the feasibility of geological disposal.

In addition, the understanding of the physical/chemical mechanisms governing the long term behaviour of waste in disposal conditions and the studies of prospective geology and hydrogeology have allowed the progress of safety studies and initiation of the Community project PAGIS (Performance Assessment of geological Isolation System).

Moreover, interesting results have been obtained with regard to the feasibility of disposal in deep-sea sediments.

In addition to these scientific achievements, the programme has allowed an important intra- and extra Community cooperation.

The joint study of the topics of the programme by teams from the various Member States has resulted in more specific ways of cooperation between the specialized bodies in certain Member States; one could mention the bilateral cooperation between the Dutch ECN* and the German GSF**, the latter having placed a cavern in the salt mine at Asse at the disposal of the former for a Dutch experimental drilling operation. The trilateral cooperation between some Italian and French bodies and the Belgian CEN/SCK*** is another example - the latter having placed a part of its experimental underground installation in clay at the disposal of the others.

Agreements for specific cooperation were reached with Atomic Energy of Canada Ltd in October 1980, with the United States Department of Energy in November 1982 and last year with the Swiss National Cooperative for the Storage of Radioactive Waste - NAGRA.

These cooperations result in fruitful exchanges (annual technical meetings, visits, joint seminars, projects) in the fields of common concern, in particular where the disposal of waste is concerned.

Relations with international nuclear bodies, particularly with the Nuclear Energy Agency (NEA) of the OECD and the International Atomic Energy Agency (IAEA) in Vienna have developed very favourable during the programme.

But, research and cooperation on research alone are not sufficient to resolve the problems raised by the management of radioactive waste. These problems constitute a combination of questions involving the perfecting of existing technologies and questions of a legal, administrative, financial and social nature which must be resolved in one and the same context.

---

*   Energieonderzoek Centrum Nederland
**   Gesellschaft für Strahlen und Umweltforschung mbH
***   Centre d'Etudes Nucléaires/Studiecentrum voor Kernenergie.

Conscious) of this necessity, the Council of Ministers, approved in 1980 a twelve year (1980-1992) Community Plan of Action in the field of radioactive waste. The implementation of this Plan involves the Commission in :

a) reporting periodically to the Community institutions and Member States on the situation prevailing in radioactive waste management in the Community;

b) promoting concerted action in the field of waste disposal and geological disposal in particular;

c) setting up Community consultation machinery on the criteria for the acceptance of conditioned waste for disposal;

d) promoting R&D and encouraging cooperation between laboratories in the Community;

e) ensuring that radioactive waste management is conducted in compliance with the standards and provisions of the Euratom Treaty relating to radiological protection and with the regulations covering the environment.

It was in this framework that the Commission issued in 1983 a first report on the analysis of the present situation and perspectives in the field of the radioactive waste management in the Community.

From this analysis it emerged that the disposal, which is the ultimate long term stage of management, has not yet been successfully demonstrated for long-lived and high level waste categories.

A solution to this problem is regarded in many countries as an important factor for the public acceptance of nuclear development and that is why the Commission considers that a convincing demonstration must be given, even if :

- disposal does not readily lend itself to direct demonstration since it needs safety assessments covering the very long term, which of necessity have to be based either on accelerated tests or on indirect evaluations;

- the need to allow highly active waste to "cool" for several decades rules out any industrial disposal before the end of the century.

Having this in mind, the Commission submitted in 1984 to the Council of Ministers and to the European Parliament a proposal for a third R&D programme on the management and disposal of radioactive waste for the period 1985-1989.

In March of this year, the Council adopted this programme. The funds estimated as necessary for the execution of the programme amount to 62 MECU.

To a large extent this programme is aimed at validating the basic information acquired so far in the field of waste conditioning and disposal, through to experiments approximating to real conditions and construction and/or operation of experimental underground disposal facilities.

Concerning the experimental underground disposal facilities, it should be pointed out that the R&D has now reached a stage where it is necessary to verify the feasibility by means of large facilities which reproduce at full scale and in real geological conditions the essential parts of a large and industrial underground disposal plant of the future. This is the case of the three projects included in part B of the programme and which are related to :

- a pilot underground facility in the Asse salt mine (Germany);

- a pilot underground facility in the argillaceous layer located under the Mol nuclear site (Belgium);

- an experimental underground facility in France in a geological medium of complementary nature.

All these projects have been opened to Community cooperation by the responsible bodies of the Member States on whose territory the facilities will be built. They will make it possible to determine in-situ the options, projects, and numerical values of the parameters to be taken into account in the construction of industrial disposal facilities and to develop the technologies for the disposal of radioactive waste.

The results to be expected from these experimental facilities and from the other research activities will be of very great interest. They would be reviewed in 1990 at the third Conference.

WASTE DISPOSAL IN EUROPE - LOOKING AHEAD

B. VERKERK

Chairman of the Advisory
Committee on Programme Management

Summary

In this introductory paper a short outline is given of the Commission's programme on management and disposal of radioactive waste, followed by a discussion of the programme structure. This leads to the very important aspect of evaluation of results obtained and the communication of the achievements to the outer world. The important role of the media in this respect is stressed.

Looking ahead, an important part of the Third Five year programme, the development of demonstration facilities, is projected against the problem of acceptability.

Thinking about the consequences of entering the demonstration stage with respect to future research it turns out that a broad field of work opens up, when the achievements reached in the radioactive waste area, could be transferred to problems of other toxic wastes and fusion wastes.

1. Introduction

It is a true honour to me that the Programme Committee of this Conference has asked me to present the introductory speech. The organisation of a large conference, where work of many people has to be reported is not an easy task and I wish to congratulate the Committee and the people of the Commission for their able preparation.

Many lecturers after me will present their share in the achievements of the Second Five Year Programme of the European Commission in more or less technical details and I think that my task is of a rather different character. Since an overview of the programme will be given by Mr. Orlowski, some remarks of a more contemplative nature inspired by the work performed in the second programme, the tasks set by the third and possible consequences thereof will be the main theme of my presentation.

2. Organisation of the Commission's programme

For delegates coming from outside the European Community a very short introduction may be needed to explain the broad lines of the programme in order for them to follow easier the course of my talk.

The Community programme on management and disposal of radioactive

waste has two main parts, supervised by the Brussels headquarters. First there are the so called Direct Action programmes, one running from 1979--1983, the present from 1984-1987, executed by the Ispra Centre with contributions from Karlsruhe Centre and comprising at present the following main components:
- Waste Management and the Fuel Cycle;
- Safety of Waste Disposal in Continental Geological Formations;
- Feasibility and Safety of Waste Disposal in Deep Oceanic Sediments.
Of course these projects are further subdivided and subsequent speakers will present more details.

The Indirect, or Shared Cost Action programmes (1980-1984 resp. 1985-1989), managed from Brussels are executed on the basis of research and development contracts, the cost of which is partly borne by the Commission, the remainder being paid by the Contractants, national institutions and or Governments. The second programme was set up on the basis of 8 main items, later expanded to 9, the so called sheets, covering respectively:
1. Characterisation of low- and medium level and alpha bearing waste forms.
2. Fuel cladding and dissolution residues.
3. Treatment and conditioning processes for medium level liquid wastes.
4. Treatment of alpha wastes.
5. Testing and evaluation of solidified high level waste forms.
6. Burial of low activity solid waste at shallow land depth.
7. Storage and disposal in geological formations.
8. Immobilisation and storage of gaseous waste.
9. Performance and safety evaluation of radioactive waste disposal in geological formations.

From offers for contracts relevant ones were selected by employees of the Commission assisted by technical expert groups for each sheet. The expert groups also had a task in the management of the sheets by regular progress meetings and workshops. The selected contracts had to be approved by an Advisory Committee on Programme Management, formed by experts from member states, this Committee being the main advisory body of the Commission in the field. The ACPM on radioactive waste management also acted as advisors to the Direct Action Programme. Over the years the ACPM and the expert groups have become valuable sources of ideas and focal points of international collaboration in the field. It will be clear from the programme listing given above that a close relation is needed and indeed is very well realized at the technical level between the direct and indirect actions.

3. The programme structure

Looking back to the Second shared-cost programme one is tempted to consider critically the goal the Commission was striving at and the way it was tried to achieve it. This means looking at the design of the programme, the management structure set up, the scrutiny of research proposals, the progress of the work, the very important aspect of evaluation of the results later in time and finally the task of communicating the outcome to the outer world, public and decision makers. In the following some of these aspects will be looked upon.

The second five-year programme, as written in 1980, in retrospect shows an extensive set of activities to be undertaken, or continued from the first programme. Forgetting for the moment part B of the programme, concerned with strategies, legal aspects etc., we see the splitting in the 9 sheets, according to which the programme management was also structured. And, although this structure promoted in an excellent way the thorough scientific and technical collaboration between experts in a certain area, it also prevented to some extent the cross-fertilisation between the various fields of radioactive waste research. Fortunately the programme was not static and there was a gradual evolution of priorities that led to the PAGIS (Performance Assessment of Geologic Isolation Systems), MIRAGE (MIgration of RAdionuclides in the GEosphere) and waste characterisation initiatives that succeeded in laying the necessary bridges. Still it must be seen as a significant improvement that the third programme is structured according to tasks, wherein a multi-disciplinary approach is evident. In this respect there remains an uncertainty; how easily will scientists and technicians, experts within different work areas, see the relative import-ance of their own work in the light of the general objective of the task of which their work is only part. Specialists have to become generalists to some extent and their own research may need to be modified according to the task objective. In the development of the new management structure, much organisational detailing and coordination has to be done and this certainly will also lay a burden on the employees of the Commission whose managerial tasks will become of a rather different nature.

These remarks lead me to the question:
Who is an expert? A crucial question indeed when we project it against the task of judging research and development proposals. In the past, when very much had to be learnt still, the reviewers, people of the Commission, as-sisted by the various expert groups indicated above, could be liberal to some extent, within the budget frame, to let interesting research items pass, when the proposals answered the criterion of community interest. Had they not been proposed by the real experts in the particular area? This li-beral approach has led to a certain divergence of the programme into a wide spectrum of subjects. This divergence cannot be allowed to continue now a wealth of knowledge has been accumulated in many areas. Therefore I highly welcome the task in the third programme of systems analysis and its subrou-tines sensitivity and uncertainty analysis. With its help one can better define what parts of the system are the crucial ones and where our know-ledge still needs to be improved. But it can also identify the items of less importance, so that responsible decisions can be made to finish cert-ain research and development tasks. This can lead to an objective conver-gence of the programme towards the ultimate goal, of great importance since the money does no longer flow so easily. Of course much has been done al-ready in the second programme by setting up concerted actions such as the characterisation of waste products and the PAGIS and MIRAGE projects, but also in these areas systems analysis will be of help in making the work converge.

4. Evaluation and communication of results

More or less opposite to this analysis is the necessary synthesis of research achievements, the critical judgement of the results.
The amount of knowledge accumulated is overwhelming and can hardly be digested even by people who devote a lot of time to it. And if one tries to cover also achievements outside the European Community even the most ardent reader will be at a loss.
How can we make the necessary synthesis of all the work done? A tre-mendous task but one that must be done, first by the workers in the field, by the Commission to see whether there has been a reasonable return of in-vestment and finally it must be done to inform the decision makers about the essentials of results obtained. The different interests of the three groups require three different approaches as regards the level of informa-tion and the language to be used. Workers in the field can if they wish

take the original publications and evaluate them in much detail, as much as required by their own research needs. This is part of their duties and needs no further discussion.

The evaluation of results by and for the Commission is already more complicated, since on the one hand there are the employees directly involved with the programme management and they can more or less be classified among the workers, if they had the time to do the evaluation and on the other hand the higher decision levels within the Commission. The latter group needs more condensed information in the sense of "executive summaries" and to provide this is in the first place a matter of internal organisation. But the Commission may also want, as it did before, to have a peer review of programme results by external consultants. Such a review still remains scientific and technical and can use the technical language. Input can be given by both the Commission employees and the contractants. Such a review and the publication of its report may be of some help in the acceptability area, but that would be an indirect approach.

The essential point is how we can transmit our results directly to the political sector, because we only have reached the ultimate goal of all our efforts when we can convince the political decision makers that radioactive waste of all sorts can be safely handled, treated and disposed off.

The media have the capability of translation for the general public but of necessity the information is transmitted in a rather superficial manner. The decision makers need more. A possible instrument is a status report such as published last year on geologic disposal by CEC and NEA and on sub-seabed disposal by NEA, but I fear that also this type of publication just misses its goal. What then? Is acceptance a matter of time only? We can truthfully tell that radioactive waste is not a dirty mess, that it is handled with great care, that we are really concerned about its future effects and that we have found means to prevent these. But there always comes the question, but what if?

5. The Third Five Year Programme

In the light of what was just said about communication of results and the right type of information for decision makers, it is of great importance that the Commission in the next programme gives priority to three demonstration projects in different geological media, where in pilot plant type of underground installations the possibility of safe handling and disposal of highly active materials will be demonstrated and where laboratory measurements and calculations will be checked in reality.

In the case of radioactive waste disposal, the demonstration pilot plants are not only intended to learn the engineers the solution of scaling-up problems, as is usual in many cases of big plant development. The point of demonstration to the outer world clearly stands out in our case. Indeed, if the planned demonstrations succesfully show the capability, that all the research and development efforts have indicated now, of safely placing high level waste in underground storage and to master the effects of heat and radiation in reality, an important step towards acceptability will be made. We then can show the day to day safety of the operations, but still the question but what if? is not answered. The PAGIS exercise hopes to provide that answer but there again the problem of long term safety will be solved in technical and scientific terms. Will people believe in the outcome of intricate calculations, in the scenario's and the radiological risks predicted? A great challenge lies ahead for the scientists involved; how can their arguments become convincing. Again the help of the press and the other communications media is indispensable. They are an important group with respect to translation of our scientific and technical presentations into the language of the public.

Coming back to the underground laboratories and demonstration plants in use or planned within the European Community and also in other countries, Canada, Sweden, the U.S. and Switzerland, far reaching implications of this development must be noted.

This shift in emphasis to technical realisation has consequences for further research as well as the fact that so much knowledge has been accumulated under the present research programmes. In other words, some of the waste research will come to an end in the next decade. Does this mean unemployment for a group of scientists and technicians? My personal impression is no, if we start soon enough to look for other kinds of relevant research. And especially for many people now working on radioactive waste I see a prospect of great significance, namely the application of their present expertise within another related domain, the disposal of chemical and other toxic wastes. I am aware that I touch an immense problem when mentioning "conventional" toxic wastes, but I also think that the achievements of radioactive waste research to some extent may be of significant help in the conventional waste domain.

There is of course one formidable obstacle in transferring our experience and that is cost.

The billions of kilowatthours produced in nuclear stations allow large amounts of money to be spent on safe disposal of nuclear wastes without impairing the economy of nuclear energy, but for chemical wastes costs of treatment and disposal, in view of their large amounts, must be strictly limited in order not to disturb the economics of many chemical productions. Nevertheless the problems of chemical wastes should be tackled as rigourously as we have done for long-lived radioactive wastes, otherwise we will gradually poison the environment forever. Fortunately Governments see the magnitude of the problems and measures are being taken, but I think that very long term risks of toxic wastes still are greatly underestimated, when compared with the utmost care and anxiousness as regards long-lived radioactivity. If we look at the limited amounts of alfa-wastes that may be put in shallow-land burial in rather insoluble form, and compare this with the thousands of tons of conventional toxic wastes, in rather soluble form, that rest in surface-dumps, a feeling of serious inconsistency easily creeps up.

If we would apply our safety analysis methodology to dumps as they exist now and other disposal techniques in use for conventional waste of various kinds, I think such exercises could be of help to authorities in making them aware of the magnitude of the problem and the urgency of improvement.

Other study areas, to my impression directly transferable to the other domain are migration of chemicals in soil, both on the basis of experimental and modelling studies; ways of decreasing leachability: study of the long term safety of geologic disposal of the most toxic wastes; improvement and expansion of chemical toxicology, in order to establish permissible doses to man.

In short: our efforts and methods to solve the radioactive waste problem can have a much wider applicability and it would be a pity if that possibility would not be fully explored and exploited in the coming time. The wider applicability also relates to a subject nearer to the present radioactive wastes. I wish to point here to the radioactive residues of fusion reactors. These will be wastes of a different character, they will arise in substantial quantities and activities and it would be wise to start giving attention to them.

Many in the field are still needed to provide the necessary scientific background to the development of the demonstration plants and real repositories and it must be a challenging task for the employees of the Commission and the contractors to make the Third Programme converge towards that goal.

For many people this Conference will be an interesting look-back. But looking ahead is also a must and I sincerely hope that our work of the coming years will bear fruit.

THE ROLE AND RESULTS OF THE EUROPEAN COMMUNITY'S R&D WORK ON RADIOACTIVE WASTE MANAGEMENT

S. ORLOWSKI, Commission of the European Communities, Brussels
F. GIRARDI, Commission of the European Communities, JRC Ispra

Summary

The titles of R&D programmes generally relate to a scientific discipline, a technology or a project: biotechnology, nuclear fission, etc. This is not so in the case of radioactive wastes, where R&D is focused on the management aspect. The role of R&D in general, and the contribution made by the Community programme in particular, are described and discussed with this in mind.
Community R&D in the field of radioactive waste emerges as a powerful tool for establishing a broad consensus on delicate scientific questions such as the feasibility and long-term safety of the final storage of high activity wastes. Such a consensus is based on the many results obtained jointly by Community research teams over the last ten years. The implementation of three projects concerning experimental underground facilities in the context of the Community's new five-year (1985-89) programme will provide the additional information that is needed before the large industrial disposal facilities of the future can be built.

1. Radioactive wastes: a management problem

When one looks at the titles of major R&D programmes, and those of the conferences which illustrate them, one notices that they mention a scientific discipline, a technology or a project. One thus speaks of R&D programmes relating to biotechnology, solar energy, wind power or nuclear fusion, to quote but a few examples taken from among the activities that go to make up the framework programme for European Community research.

That may be a rather trite observation, but it does serve to underline the unusual nature of the title of our conference this week, and of the programme with which we are dealing, namely: "radioactive waste management". Similar titles are, furthermore, given to the many R&D programmes relating to radioactive wastes throughout the world.

The unusual thing about this programme is the word "management".

What do we mean by that word?

In what ways does it make the R&D programmes specific and original?

Does it mean, in particular, that scientific and technical questions must be relegated to the background or regarded as solved? If not, what have we done and what remains to be done, particularly in the European Community context, to ensure that the R&D effort is properly channelled as support for the management of a system?

2. The meaning of the word "management"

Radioactive waste management consists in carrying out various industrial operations on a variety of substances whose only common denominator is the fact that they carry the labels "USELESS" and "RADIOACTIVE", with a view to disposing of them permanently with minimum

risk, or at any rate, while keeping the risks curred within reasonable
limits.

Let us look more closely at the constituent elements of this definition:

- The raw wastes are produced by nuclear installations, hospitals,
  industries, etc. and display a wide variety of physical and chemical
  properties.
- The common label "USELESS" means that the final aim of the undertaking is not to dispose of products on a market, but on the contrary
  to store them for an indefinite period.
- The industrial operations involved cover such widely differing
  activities as processing, transport and storage. The raw waste
  processing operations constitute essentially a chemical-engineering
  activity; storage and final disposal are operations which fall within
  the sphere of civil and mining engineering. These are well-
  established engineering sciences, but the existing know-how must be
  adapted for the specific purposes.
- During the next 15 years, between 50.000 and 100.000 m^3 of low-
  activity wastes, between 10000 and 20.000 m^3 of medium-activity
  wastes and up to about 1000 m^3 of high-activity wastes will undergo
  the above-described sequence of operations each year in the European
  Community. This should be compared with the present Community production of non-nuclear wastes, which amounts to several tens of millions
  of cubic metres annually (Figs. 1 and 2) (1). What does this represent at national level? Some 5000 tonnes of radioactive wastes arrive
  each year at the big La Manche storage centre in France, as against
  100.000-200.000 tonnes of toxic and hazardous wastes at a large
  incineration centre. In the United Kingdom, there are today more than
  150 sites where toxic wastes are handled or processed, whereas the
  number of existing and future sites for the storage of radioactive
  wastes is unlikely to exceed ten according to a recent communication
  (2)

A radioactive waste-management enterprise is therefore an industrial
enterprise on a modest scale, which is, however, highly diversified;
the problem of ensuring consistant, if not centralized, management of
such a system is easy to appreciate. This also helps to explain why
national management agencies or bodies have been set up in numerous
countries during the last few years.

This management problem - since the problem is indeed one of management - would be purely economic and commercial in nature and easy to
master if the enterprise were a "normal" one.

But the substances all carry the label "radioactive". This common
label covers specific levels of radioactivity (measured in Bq/m^3) which
vary over more than 12 orders of magnitude, half-lives ranging over six
orders of magnitude (at least in the case of the most significant ones)
and radiotoxicities (annual limits for uptake through ingestion or
inhalation) which vary to a similar extent. In other words, the substances are toxic to very widely varying degrees, and their toxicity can
in some cases persist for a very long time.

These factors have the following repercussions on management:

- The criterion of economic viability is replaced by that of "safety
  within reasonable limits": this is the ALARA (as low as reasonably
  achievable) principle laid down by the ICRP. The nuclear industry is,
  sofar at least, the only industry to have adopted this safety principle.
- It is necessary to make management projections into a fairly

distant future and, at all events, over a much longer timescale than
for other industrial undertakings, with the attendant consequences.
The major traditional industries plan for a period of three to five
years, the mining industry invests for between 20 and 30 years, and
civil engineers build structures (bridges, dams, etc.) which can last
for decades and even centuries, because regular checks on their
safety can be made thanks to easy and constant access. Here, the
time-scale ranges from a few centuries (in the case of the surface
storage of low-activity wastes) to a few thousand or a few tens of
thousands of years (in the case of the geological storage of alpha-
emitting high-activity wastes). The latter case rules out the possibility of permanently monitoring the storage facilities and of
retrieving the substances stored.
- It is necessary to take account of the problem facing society as a
   result of the development of industry in general and of nuclear
   energy in particular.

   These three considerations determine the particular context in which
radioactive wastes have to be managed.

3. The role of R&D

   What role can be played by R&D in making possible or facilitating
optimized management?
   Since its inception in mid-1970, the Community's R&D programme has
followed two lines of action in order to enable radioactive waste management to attain the safety objective it is pursuing:

The first line of action

   On the basis of today's nuclear installations and of their raw waste
production, to propose appropriate processes and technologies whereby the
feasibility and safety of each waste management operation are ensured. To
evaluate the corresponding strategies from an economic and radiological
standpoint and to establish very concrete reference strategies.

The second line of action

   In the context of broader studies based on the assumption that
modifications are made to nuclear fuel cycle installations, to develop
alternative strategies which can appear more promising than the above-
mentioned reference strategies from the standpoint of safety optimization. Clearly, such strategies are bound to be less precisely defined
than the reference strategies.

4. The first line of action; the conditioning and characterization of wastes

   In the context of the first approach, Community R&D from 1975 to
1980 concentrated on the task of identifying the advantages and disadvantages of existing processes with a view to developing new ones where
necessary. The various waste-conditioning processes and the corresponding
encapsulation matrices were studied: cement, bitumen and resins in the
case of low- and medium- activity wastes, and glasses in the case of high
activity wastes.
   During the period 1980-85, with which this conference is concerned,
process development was confined to areas where such processes were
lacking; the process concerned are essentially those for dealing with
certain types of waste which existed only in small quantities a few years
ago, for example cladding hulls and dissolution residues. The papers read
during session II and II will describe to you the results achieved.

Encapsulation matrices - and particularly those intended for low-and-medium-activity wastes - were the subject of an intense characterization effort carried out, as Dr. Verkerk has stressed, by a group of Community laboratories working according to a specific programme that was drawn up between them. Some ten "reference conditioned wastes" representative of the types of greatest interest to all the Member States, and another ten or so "additional conditioned wastes" of more specific interest, were selected for the 1980-84 programme (Table 1).

Among other results, this work demonstrated the swelling of certain wastes encapsulated in bitumen and the satisfactory retention of actinides by cement. Some ten reference materials, chiefly borosilicate glasses, were likewise the subject of a study coordinated between seven Community laboratories, which confirmed the satisfactory behaviour of the glasses. The work as a whole generated so much information that it appeared desirable to collect it together in a data bank, which is currently being set up. The papers read during session IV will throw light on this topic.

During the third period (1985-1989), which is now beginning, our ambition is, in particular, to develop what is an essential aspect of any system of safe management, namely quality control of the finished products. Such control already exists, of course, for the commonest categories of waste; it needs, however, to be perfected and extended to all categories, and its satisfactory performance on an industrial scale must be verified by means of a joint effort on the part of all the competent laboratories in the Community, including the JRC.

5. The first line of action; the final storage waste

As far as the final storage waste is concerned, the Community's work is proceeding according to a three-phase timetable similar to the preceeding one.

From 1975 to 1980, the questions raised by final storage were pinpointed and the key parameters identified. I am thinking here, for example, of the thermo-mechanical effects generated in the host rock during the storage of high activity wastes in deep geological formations. In addition, thanks to the joint effort of the national geological institutes, it was confirmed that formations geologically suitable for receiving radioactive wastes were present in the substrata of the European Community, and a catalogue of such formations was compiled

At the same time, studies were undertaken on the behaviour of solidified wastes in storage conditions and the transport of radionuclides in geological media.

Considerable simplifications were made, but the results were nevertheless extremely valuable. For example, the behaviour of glass was studied in water, the effect of the surrounding materials being ignored. The importance of surface and solubility phenomena, which in turn depend either on the chemical nature of the different radionuclides concerned or on the nature of the medium, was demonstrated.

As regards the migration of radioelements, especially actinides, the work conducted at the time revealed the shortcomings of the "distribution coefficient" concept, which had previously been used in safety studies to describe the behaviour of actinides in solution; the predominant processes are in fact filtration and redissolution, and the formation of anionic complexes by natural binding agents plays a fundamental role here. It then appeared necessary to tackle the problem more rigorously and to include research topics of a more academic nature, such as the determination of stability constants.

Lastly, during the same period the JRC launched the first studies on the risk inherent in storage. This was at the time a pioneering activity and was the subject of the first seminar in the world in that field, which heralded the setting-up of the PAGIS (Performance Assessment of Geological Isolation Systems) project, about which I will speak shortly.

During the period 1980-85, which is covered by the present conference, an impressive body of knowledge has been built up concerning the multiple-barrier geological confinement concept.

Both engineered barriers -especially canister and backfill materials - and the geological barrier itself have been studied in circumstances that are as close as possible to real conditions, account being taken, in particular, of the interactions between the various components of the system.

Without entering into the details of the work that will be presented during session V, it is fair to say that the earlier conservative assumptions can today gradually be replaced by more specific and more detailed laws which render more realistic the mathematical models describing the evolution of materials in storage conditions. It is heartening to note that all the data collected serve to confirm the safety of storage.

Still during the period 1980-85, the previously mentioned exploratory activities carried out in the seventies on actinide migration and risk analysis gave way to activities organized as Community projects, such as PAGIS and MIRAGE (Migration of Radionuclides in the Geosphere). These projects are concrete examples both of the close cooperation between the various laboratories concerned in the Community and of the importance of the Community effort. Both projects overlap the multiannual programmes and are entering a second more concrete phase.

The whole range of questions relating to the design of an underground facility for geological storage in clay, salt and granite has also been studied: rock mechanics, thermal effects, hydrogeology, the fracturing of crystalline rocks, etc.

Fig. 3 illustrates more effectively than a long speech the large number of locations where in situ experiments have been conducted.

The argillaceous sediments of the ocean floors have been added to the preceeding continental geological formations: the safety of storage under the seabed is based on the same criteria as continental storage, i.e. the multi-barrier confinement approach. It should therfore not be confused with the dumping of wastes at sea, the safety of which is based on the dilution and dispersion of the radioactivity in the masses of water that go to make up the oceans. Storage in marine sediments is a more futuristic and less highly developed option than the continental ones; the Community programme is making a significant contribution to the study of this option in the context of a vast international programme for which the OECD's Nuclear Energy Agency is acting as secretariat.

Sessions VI and VII describe all the research on geological storage to which I have briefly referred.

In light of this work programme, together with other results obtained throughout the world, the theoretical feasibility of geological storage can be said to have been confirmed.

The Community is therefore aiming to demonstrate during the period 1985-89 the practical feasibility of the final storage operation. To this end, the new R&D programme on radioactive wastes, which was approved by the Council of Ministers of the European Community in March 1985, provides for the Community's financial, technical and human participation in three projects for experimental underground facilities:

- Project 1: a pilot underground facility in the Asse salt mine;
- Project 2: a pilot underground facility in the clay stratum located under the Mol nuclear centre;
- Project 3: an experimental underground facility in France in a geological medium of a complementary nature.

Fig. 4 sets out the provisional timetable for this major activity.

6. R&D : the second line of action

The second approach consists in investigating whether or not the raw materials - that is to say radioactive wastes in the condition in which they arise at nuclear installations - should undergo special treatment to modify their specific radioactivity before going through the whole sequence of management operations leading to final storage.

The possibility of modifying nuclear installations themselves (for example, reprocessing plants) so that they give rise to different waste products can also be envisaged. The aim is clearly to obtain in all cases a product whose subsequent management, and particularly final storage, will be simpler and safer.

An initial ambitious strategy was studied by the JRC between 1975 and 1980 and was the subject of two international seminars held at Ispra with the cooperation of the NEA. The strategy involved the chemical separation, followed by in-pile recycling of certain highly toxic and long-lived substances contained in some wastes, namely the "minor actinides" and their transmutation into short-lived substances of low radio-toxicity. The conclusions reached internationally at the time demonstrated that the separation of actinide from wastes is possible, though certain reservations are called for owing to the fact that the studies were carried out chiefly on a laboratory scale. On the other hand, the recycling of such actinides in fission reactors, although theoretically possible, would be liable to increase in the near future the radiation dose to the workers concerned in return for a limited hypothetical benefit in terms of the risks to the general public in 10.000 or 100.000 years' time. Major developments in robotics and the widespread use of fast breeder reactors at the beginning of the next century appeared to be the only factors capable of modifying these conclusions. The strategy is nevertheless attractive, at least intellectually; it was recently recommended by the group of French experts chaired by Professor Castaing.

During the period 1980-85, a less ambitious strategy was studied and will be the subject of papers read during session II.

The idea is to apply chemical separation techniques to wastes contaminated by alpha-emitters. The advantage would be that the majority of the alpha-wastes would be reduced considerably, if not completely eliminated, by separating such wastes into two categories: the first category, containing no alpha-emitters, non longer poses any industrial management problem today, while the second category, which would be rich in alpha-emitters, would have to be managed together with high activity wastes, using the means designed for the latter.

The advantage of such a strategy remains to be proven; it would perhaps be smaller than that expected from the actinide-recycling strategy, but could undoubtedly be exploited sooner.

The Commission of the European Communities therefore considered it necessary to equip the JRC with a major tool that should enable it to make a more realistic evaluation of the proposed strategy than the laboratory studies in progress. This tool is called PETRA (Project for Evaluating the Treatment of Radioactive Wastes and Actinides) (Fig. 5).

It will take the form of a pre-industrial plant capable of producing the same range of waste streams as a reprocessing plant on a scale of a ten kilograms, thanks to the introduction into the production programme of several variants including, of course, actinide separation.

The same optimization approach can naturally be applied not only to the production and conditioning, but also to the storage of wastes. The two aspects are closely related and are rather two parts of one and the same system which should be evaluated and optimized as an indivisible whole.

During the period 1975-84, relatively little was done in this area; nevertheless, a remarkable amount of data has been built up, and this must be processed and supplemented. PAGIS, for example, will throw a great deal of light on the relationships between the safety of a repository and both its contents and the site parameters.

This information can constitute a valid point of departure for a more integrated study of the system as a whole, to be carried out in the years ahead.

7. Management projections into the future

I am now turning to another aspect which is quite specific to radioactive-waste management: where long-lived wastes are concerned, the storage solutions adopted must remain valid for extremely long periods. In other words, the solutions adopted today must continue to ensure that the radioactivity in the waste is still adequately confined 100, 1.000, 10.000 or more years from now, depending on the case in point.

This is a new problem facing mankind, and R&D has an essential role to play here; a fresh subject, namely "forward study", must be added to the numerous scientific disciplines concerned.

The Community's R&D programme adopts the following approaches:

8. Forward study and materials science

As regards materials science, the aim is to study all the types of ageing that affect waste encapsulation matrices, such as cement, bitumen and glass, in circumstances that are most representative of storage. The most obvious procedure consists in carrying out protracted experiments (lasting one or more years) in the laboratory or in situ, in order to establish the laws governing the phenomena incomed and then to extrapolate into the future. Clearly, in certain cases - such as that of irradiation effects - the duration of the experiments would be considerable, and it is necessary to resort to such artifices as doping with alpha-emitting radionuclides with shorter half-lives; irradiation damage can also be provoked by other means, for example by using fission fragments (internal irradiation) or accelerated particles (surface irradiation). The three methods, each of which has advantages, limitations and areas of optimum application, have made it possible to draw a fairly complete and reassuring picture of the consequences of irradiation-damage phenomena up to the time of almost complete decay of the radionuclides contained in wastes (one million years).

However, only archeological and geological parallels can give us information "in real time", ranging, in the case of the former, over a few hundred or a few thousand years and, in the case of the latter, possibly over several million years. The Community programme has focused, and will continue to focus, on the study of such parallels as regards the corrosion of waste canister metals, the behaviour of cements, the long-term integrity of glasses, etc.

Let us take the example of cements: under a contract entitled "A history of cements from 5.600 BC to the present day", cooperation is being established with a view to gathering and analysing samples of cements of widely differing ages. Very early results show that the hydration products of ancient hydraulic cements are similar to those of Portland cements, even though the reaction mechanisms may be different. Present-day cements would therefore appear to contain extremely durable substances. Furthermore, attention was very recently focused on a calcium silicate called Larnite, which was formed when basalt magma penetrated into a bed of limestone in Northern Ireland a few million years ago. This mineral has since aged in rock crevices without significant deterioration. It can therefore again be concluded, by analogy, that the main components of modern cements would remain stable over extremely long periods.

Clearly, all this will require further work and confirmation, and our new programmes will contribute thereto.

Session IV is devoted to these topics.

9. Forward study and earth sciences: the study of confinement

As far as the earth sciences are concerned, the first task is to develop methods that make it possible to evaluate, over a time-scale extending into the distant future, the probable evolution of the geological media and sites that are intended to accommodate underground waste-storage facilities. Such sites and media must not evolve in a way that significantly reduces their capacity to confine the radioactivity of the waste products stored in them. The Community programme is advancing in two directions:

-   the development of an operational methodology for the forward study of the confinement characteristics of a site, for which the term "geospective" has been coined. This is an original approach which has been developed in the wake of the forward studies carried out by B. de Jouvenel, among others. Its applications are operational and it combines:
    . an automatic simulation of geological evolution scenarios for a given site;
    . allowance for the interaction between the repository itself and its environment (the release of heat and disturbance of groundwater flows).
    When applied to a specific site and a well-defined storage project, this method of study makes it possible, by working backwards, to take the most appropriate measures as regards the choice of site and the design of the storage facility;
-   the use of artifices for studying certain phenomena in accelerated time. For example, the study of the long-term safety of the storage of heat-emitting wastes in a stratified salt formation demonstrates that the risk of dome formation in the very long term, with all the disturbances that would involve for the storage galeries, cannot be ruled out automatically. Tests using a centrifuge capable of spinning a one-tonne load at 100 g are being prepared with equivalent materials that are suitable from the standpoint of their cohesion, viscosity, etc. A TV camera makes it possible to observe the movements of the material during centrifugation. Fig. 6 shows the intrusion (in an experiment using a press) of a clay with low viscosity into a clay of higher viscosity, thus representing the formation of the salt dome in the overburden, starting from the initial stratum.

10. Forward study and the earth sciences: study of the transport of radioelements

It is also necessary, however, to obtain information, based on geological evidence, relating to the possible transport of radioelements through the underground media. Such information should confirm the validity of our experiments and predictions concerning the scale and distribution in the future of any return to man of the radioactivity contained in the stored wastes. Although no radioactive waste stores exist in nature, the latter harbours systems which have functioned for thousands or millions of years and present similarities with the components of storage facilities. A search for such natural parallels has been undertaken and will be described in particular in paper N°3 during session V. You will be told, for example, how a set of sediments in Loch Lomond and adjacent freshwater lochs in Scotland constitutes a system that is similar to a radioactivity gradient concentrated around a waste canister buried in an argillaceous formation (Fig. 7). This analogy has made it possible to compare the diffusion profiles of 27 elements, determined over several thousand years, with the diffusion profiles of the radioelements contained in wastes, as predicted by laboratory models.

This research will be continued during the new 1985-89 programme, and we are planning to hold a symposium on the topic in the spring of 1986.

All the work and results I have just mentioned make up the data base used in the studies on the long-term safety of radioactive waste storage which form part of the European Community's PAGIS project. The last session in this conference, session VIII, will take stock of Community work in this field.

11. Financial management and the long term

The need constantly to assess the future consequences of today's management choices concerns not only the technologies used and their safety, but also investments.

Design studies demonstrate that the design, construction, operation and closure of an industrial disposal system comprising interim storage facilities and a geological storage facility, capable of confining the high activity and alpha-emitting wastes produced over a period of 30 years by a nuclear power programme representing a total installed capacity of several tens of GWe, will extend over a period of 50 to 100 years. Fig. 8 gives an idea of the phasing of the expenditure. In this example, the total expenditure, including fixed and variable costs, amounts to some 2.000 million ECU over 60 years. In order to put things into perspective, I note that in the United States investment over the next ten years in plant for the incineration of household refuse is estimated at more than 20.000 million ECU (3).

By virtue of the "polluter pays" principle, the electricity producers must bear this expenditure, the absolute value of which should not frighten us, since it corresponds to a surcharge of no more than a few per cent on the kWh cost of the electricity produced.

The financial management of such an undertaking presents the financial and legal experts with interesting problems: such management must be effectively ensured for many decades, even if some of the principal parties to the venture were to disappear in the meantime.

As paper N° VII-5 will reveal, the skill of the financial and legal experts is perfectly adequate to enable them to set up the necessary mechanisms for financing projects submitted to them by researchers and engineers.

12. The socio-political aspects of waste management

It is very often asserted that radioactive waste management is impossible because there are no safe waste storage solutions, in the same way as people say that a firm is not viable because it does not have a market for its products.

First of all, it should not be forgotten that the management of low activity wastes, which represent nearly 80% by volume of total waste production, has the benefit of 30 years'practical experience. Clearly, the management of this category of wastes should also derive constant benefit from technological progress.

The preceding statement therefore concerns, above all, high activity and long-lived wastes.

This belief is so widespread among the public that certain governments have made the continued lawful development of nuclear energy conditional upon demonstration of the feasibility and long-term safety of the storage of radioactive wastes and spent fuels.

It is therefore necessary to convince people, if not to prove, that radioactive waste management is a viable undertaking which has outlets for the storage of all its products in compliance with national and international regulations.

Who has to be convinced? And again, what role should be played by R&D, and particularly Community R&D?

It is first of all necessary to convince the safety authorities, who will one day have to issue the construction permits and operating licences for the major underground storage facilities of the future. The role of Community R&D here is crystal-clear: it has to establish the broadest possible scientific consensus on the most important research results and on their use for safety assessments. The functioning of specialized working parties, which over the years have become genuine European research teams, and the MIRAGE and PAGIS projects, among other things, indicate that the Community is travelling in the right direction.

But it is also necessary to convince the general public; here, too, the broadest possible consensus on the scientific foundations for waste management has to be reached, although a direct demonstration, intended to prove that the system can be built, operated and closed down in total safety and at acceptable cost with the use of available mining and civil engineering experience, would also be useful. Such a demonstration could not fail to benefit from concrete, close and lasting cooperation between the national bodies responsible, and more generally between the Member States. This is the philosophy underlying the Community's 12-year plan of action for radioactive wastes, which the European Council adopted in 1980. This is what the European Community has been doing for the last ten years through its R&D programmes, culminating in the three projects concerning experimental underground facilities in the Federal Republic of Germany, Belgium and France, projects in which all the Member States can cooperate and which form an integral part of the Community's new 1985-89 R&D programme. It is through research and major development projects that relationships of solidarity and common interests are forged. These relationships and these interests will in the longer term assure all the Member States of the Community, including the smallest ones, of nuclear development in a climate of reason and public acceptance, thanks to the setting-up of machinery for cooperation in radioactive waste storage which will go hand-in-hand with abandonment of the disastrous "not in my back yard" principle.

13. Conclusions

In this paper I have attempted to show how the role of R&D in the field of radioactive waste is one of providing support for the solution of a management problem and is thus entirely different from the central role played by R&D in nuclear fusion, for example.

This role is nevertheless decisive in extending the frontiers of conventional science and technology in the fields of chemical, civil and mining engineering by providing the additional know-how that is necessary for dealing with the specific aspects of radioactive waste management, namely the radiotoxicity and long life of the substances incomed.

The Community's R&D programme, and all those who are taking part in it, are endeavouring to play this role as effectively as possible.

But the Community's action must contribute something more: a multi-national scientific consensus – all the papers read at this conference are multinational – and the emergence of European solidarity in the management of radioactive wastes.

It is for you to judge, as you listen to the papers presented during this conference, whether we have been successful to date and how far we have still to go in order to attain these objectives.

REFERENCES

1. "L'état de l'élimination des déchets toxiques et dangereux dans les pays de la CCE", rapport EUR 8439 (1983).
2. BROMLEY, J. "The comparison of control and public reactions to radioactive and toxic waste disposal". OYEZ. Conference on Radioactive Waste Management : "Technical hazards and public acceptance", 5th/6th March 1985, London (1).
3. LEINSTER, C. "The sweet smell of profits from trash", Fortune, 1 April 1984/5.

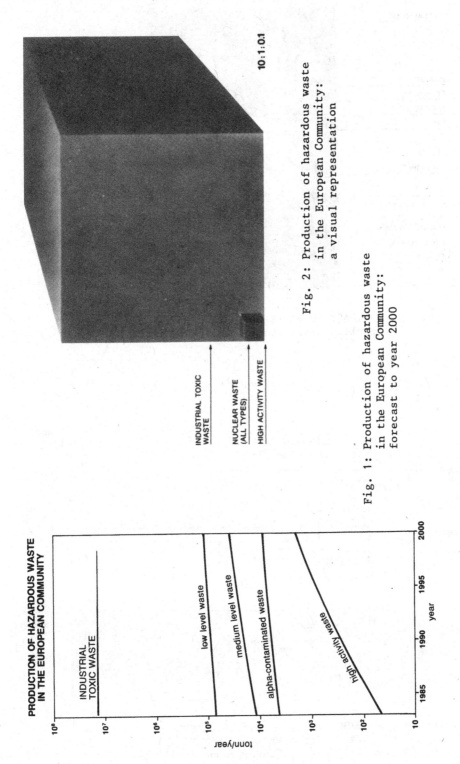

Fig. 2: Production of hazardous waste
in the European Community:
a visual representation

Fig. 1: Production of hazardous waste
in the European Community:
forecast to year 2000

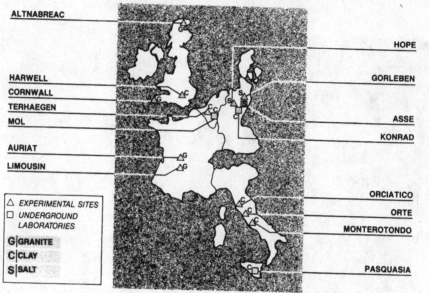

Fig. 3   RADIOACTIVE WASTE DISPOSAL
         Specific sites studied under CEC programme

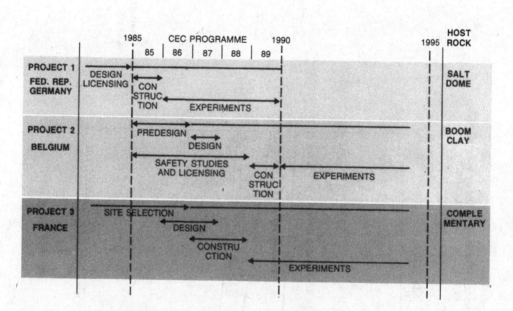

Fig. 4   EUROPEAN COMMUNITY PROGRAMME
         ON RADIOACTIVE WASTE MANAGEMENT
         Experimental underground facilities

Fig. 5: Mock-up of the PETRA
facility

Fig. 6: Intrusion of a low-viscosity
clay (dark) into a clay of
higher viscosity (light)

geological situation

analogy with
waste disposal

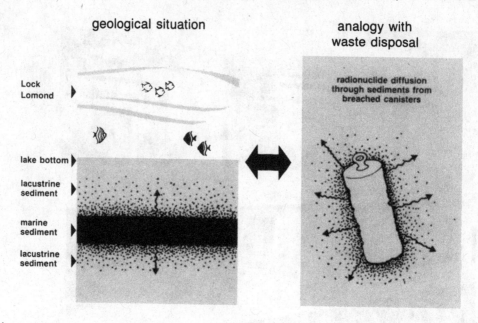

Fig. 7 - The Lock Lomond "natural analogue"

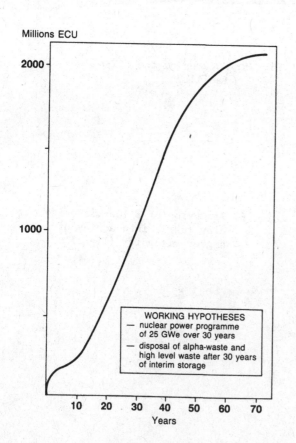

WORKING HYPOTHESES
— nuclear power programme
  of 25 GWe over 30 years
— disposal of alpha-waste and
  high level waste after 30 years
  of interim storage

Fig. 8 - Typical total
cumulated costs of
geological disposal

THE REFERENCE WASTE FORMS

No.	Reference Waste Form	Lead Laboratory
1	BWR evaporator concentrates/cement or pozzolana cement	Casaccia
2	PWR evaporator concentrates/cement or pozzolana cement	Cadarache
3	PWR evaporator concentrates/polyester or epoxide	Cadarache
4	Ion exchange resin/polystyrene	Cadarache
5	Ion exchange resin/polyester or epoxide	Cadarache
6	Magnox fuel pond water sludge/cement	Harwell
7	Reprocessing concentrates/bitumen	Mol
8	Reprocessing concentrates/cement	Karlsruhe
9	Reprocessing sludges/bitumen	Cadarache
10	Incinerator slags	Mol

Table I

SESSION II

TREATMENT AND CONDITIONING PROCESSES - PART 1

Chemical precipitation processes for the treatment of low and medium level liquid waste

Electrochemical and ion-exchange processes

PETRA, a hot cell facility for waste management studies

Characterization of spent fuel hulls and dissolution residues

Methods of conditioning waste fuel decladding hulls and dissolver residues

Conditioning of alpha waste

CHEMICAL PRECIPITATION PROCESSES FOR THE
TREATMENT OF LOW AND MEDIUM LEVEL LIQUID WASTE

L. CECILLE   CEC Brussels
K. GOMPPER   KFK Karlsruhe
R. GUTMAN   AERE Harwell
S. HALASZOVICH   KFA Jülich
F. MOUSTY   JRC Ispra
J.F. PETTEAU   SENA Chooz

Summary

New applications of chemical precipitation processes for the treatment of various radioactive low and medium level liquid waste have been investigated. For reducing the overall management cost and improving the long-term safety of disposal, partitioning of the reprocessing concentrate into different streams for separate conditioning, packaging and disposal has been studied through chemical precipitation of the whole activity (actinides + main gamma emitters) or the actinides only. Results achieved on testing of real sample of reprocessing concentrate (lab-scale) are presented and discussed.
In order to comply with the ALARA principle, an industrial flocculator prototype has been constructed and successfully operated for the treatment of utility liquid waste arising at the Chooz PWR site.
Combination of chemical precipitation with ultrafiltration seems quite promising for improving both decontamination and volume reduction factors for the treatment of various radwastes. On the basis of experimental tests performed successively on lab and technical scales, a pilot plant has been designed, constructed and commissioned for the treatment of Harwell low and medium level liquid wastes. First active runs confirm the merits of the process.

1. Introduction

Chemical precipitation methods are commonly used for the treatment of liquid effluents at research establishments as well as at some reprocessing plants. Their main advantages over other treatment processes (e.g. ion-exchange) are: simple implementation based on proven technology, low process cost, relatively good decontamination factors irrespective of the chemical form of the radionuclides to be removed and possible application to highly salt-laden effluents. In addition, combination of chemical precipitation with other processes, such as evaporation, ultrafiltration or electro-osmosis, can render it still more attractive when high decontamination factors are aimed at. For all these reasons, it appeared worthwhile to investigate the merits of chemical precipitation methods for some new applications in the general frame of optimisation of radioactive liquid waste management at various nuclear sites (e.g. reprocessing plants, power reactors and research establishments).

2. Application to reprocessing concentrate

In some reprocessing plants (e.g. the WAK plant in West Germany), low and medium level liquid wastes are mixed altogether prior to concentration by evaporation. The resulting waste, termed "reprocessing concentrate", is then directly incorporated into cement prior to interim storage and disposal in geological formations. It is expected that reprocessing of one tonne of spent fuel will lead to arisings of 1.4 m^3 of reprocessing concentrate and, finally, to 4.5 tonnes of cemented waste product[1]. In fact, to cope with regulations regarding the transportation of radioactive waste and due to the presence of gamma emitters (mainly Cs-134/137, Ru-106 and Sb-125) in the waste, the cemented products have to be packaged with shielded containers, hence increasing by a factor of 11 the overall weight of waste[1] to be handled and transported. Therefore, on the basis of system studies[1], important savings in the overall management cost of the reprocessing concentrate (reduction by a factor of about 2) are expected if the main gamma emitters responsible for shielding can be removed and subsequently added to the high level liquid waste for vitrification.

On the other hand, actinide separation from reprocessing concentrate may present certain advantages if for example, various disposal routes are available (shallow land burial and geological formation) or if a more sophisticated matrix for embedding alpha-bearing waste (e.g. ceramics) proved necessary for long term disposal[2]. To cope with all these different management options, processes based on chemical precipitation methods were developed at KFK-Karlsruhe, KFA-Jülich and the JRC-Ispra.

2.1. Removal of the main gamma emitters

For handling and transporting cemented reprocessing concentrate in unshielded container casks, it has been shown that decontamination factors of at least 280, 12 and 6 should be aimed at for Cs-134/137, Sb-125 and Ru-106 respectively.

For this purpose, two similar processes, consisting mainly of three steps (lowering of the waste acidity, chemicals addition for selective precipitation and liquid/solid phases separation) have been developed at KFK-Karlsruhe and KFA-Jülich.

Lowering of the waste acidity from 1M HNO_3 to pH 1-4 is performed by means of denitration with HCHO or HCOOH. Whatever, the reducing agent selected and the procedure adopted, denitration of reprocessing concentrate proceeds quickly and gives rise to similar off-gas composition (table 1). As an alternative to denitration, neutralisation with sodium hydroxide is also possible but entails an increase of the sodium nitrate content by about 25%.

Addition of selective precipitants, mainly for Cs, is then carried out while the pH of the waste is brought alkaline.

Two kinds of specific precipitants for Cs have been tested: sodium tetraphenylborate $NaB(C_6H_5)_4$ and cobalt or nickel hexaferrocyanide (Ni_2FeCy). In terms of decontamination factors, $NaB(C_6H_5)_4$ appeared, at first sight, much more efficient than Ni_2FeCy since their respective D.F.'s differed by about one order of magnitude (at pH 11, D.F.'s higher than 10.000 were recorded for $NaB(C_6H_5)_4$). Unfortunately, during the subsequent steps of the process, the behaviour of $NaB(C_6H_5)_4$ containing sludges proved problematic since these were shown to be difficult to centrifuge and, likewise, started to decompose in highly volatile and flammable organic products from 115°C onwards. Therefore, application of $NaB(C_6H_5)_4$ to partitioning of reprocessing concentrate has been given up.

The caesium removal through Ni_2FeCy precipitation is a well-known procedure applied in nearly all the radioactive liquid waste treatment facilities. Moreover, it is worth noting that resulting sludges settle reasonably well and that no noxious gas (e.g. HCN) has been detected during this thermal treatment. Another advantage over $NaB(C_6H_5)_4$ process is that addition of ferrous ion to Ni_2FeCy can remove, at the same pH range (8-9), noticeable amounts of Sb and Ru, hence allowing the D.F.'s target to be reached in only one precipitation step.

Procedure for	HNO_3 reduction into (%-vol)		
denitration	NO_2	NO	N_2O
HCHO feed into boiling waste	3	91	6
HCOOH feed into boiling waste	5	91	4
waste feed into boiling HCHO	1	75	24
waste feed into boiling HCOOH	2	73	25

Table 1: Balance of HNO_3 reduction into various volatile nitric oxides through denitration of reprocessing concentrate (simulated) under different operating conditions.

As an illustration of the process performance, typical D.F.'s achieved on samples of real reprocessing concentrate are quoted in table 2. On the whole, these D.F.'s are quite beyond the target for Cs and Sb. However, for Ru removal, the decontamination factor is hardly higher than the limit showing that some improvement in this area would be desirable, perhaps by combining chemical precipitation with ultrafiltration as will be shown further.

Radio-nuclide	Cs-134/137	Ru-106	Sb-125	Ce-144	Eu-154	Co-60	Am-241	Total alpha
D.F.'s	1000-2500	~ 7	20-60	>1000	>400	>54	>300	140-200

Table 2: decontamination factors achieved on real reprocessing concentrate (lab-scale) under following operating conditions: 6×10^{-3} M $K_4Fe(CN)_6$, 6×10^{-3} M $Ni(NO_3)_2$, 10^{-2} M $FeSO_4$ and pH 8.5

Scaling-up of the process will obviously require the use of industrial liquid/solid separation hardware. Therefore, various equipment available on the market was tested and evaluated on the basis of its separation performance, ease of maintenance and suitability for operation under remote conditions. It resulted that the decanting centrifuge type seemed to be the most suitable equipment, especially when combined with a polishing treatment (e.g. bag filter).

The resulting centrifuged sludge volume does not exceed 5% of the initial waste volume but unavoidably carries along some sodium nitrate. Due to the fact that these sludges are intended to be subsequently mixed with high level liquid waste for vitrification, washing the sodium nitrate out of the precipitate is not necessary since this could easily replace part (max. 15%) of the Na_2O needed as a glass-making additive.

Preliminary vitrification tests of precipitation sludges (alone or with HLLW calcinate) according to the GP 98/12 glass formulation set-up by KFK-Karlsruhe for HLLW have been successful, demonstrating the good compatibility between glass components and chemicals added for precipitation.

Most of the foregoing results and statements have been achieved at KFK-Karlsruhe at the lab-scale for active experiments and up to the industrial scale for testing liquid/solid separation hardware.

However, in order to get more representative data on the technical feasibility of the process, active runs, each involving 60 litres of real reprocessing concentrate (arising from operation of the WAK pilot reprocessing plant) are about to start at KFA-Jülich making use of the FIPS-II hot-cell equipment. More reliable conclusions about the feasibility of the whole process should be drawn when these active runs are completed.

It has to be pointed out that in addition to possible savings in cost of waste transportation, this new management option also warrants a safer disposal of radioactive waste since the bulk of the activity (actinides + main gamma emitters) containing reprocessing concentrate is conditioned into borosilicate glass, the long-term stability of which is well known.

2.2. Actinide separation

Actinide separation from radioactive waste can be envisaged for various purposes; extensive Pu-recycling or improvement of the disposal safety by separate conditioning and disposal of the short-lived and long-lived radionuclides present in the waste.

As a first approach, a threshold of 100 nCi α/g has been defined for enabling the waste products to be either conditioned into plain matrix (e.g. cement or bitumen) or disposed of in shallow land burial. Considering the rather low alpha content of the reprocessing concentrate and taking into account that cementation has been chosen for conditioning; this means that an alpha D.F. of about 20 is required for waste partitioning.

To this end, two similar processes, based on co-precipitation of actinides with lanthanides oxalate salts at pH about 1, have been successively developed both at the JRC-Ispra and the KFK-Karlsruhe. As previously, these processes involve an acidity lowering step of the reprocessing concentrate carried out equally well by denitration or by partial neutralisation with sodium hydroxide. Since the lanthanide concentration in the waste is too low to have the actinides precipitated, some carrier has to be added along with the oxalic acid.

Decontamination factors achieved on simulated reprocessing concentrate[3] looked promising (table 3).

Element or compounds	Pu	Ce	Fe	Sr	Mo	H_2MBP	overall phosphates
D.F.'s	250	100	5	1.2	1.4	1.3	1.3

Table 3: Alpha decontamination of reprocessing concentrate (lab-scale) under following operating conditions: 5×10^{-3} M $Ce(NO_3)_3$, 2.5×10^{-2} M $H_2C_2O_4$, pH 1.2 - 1.8

It is worth noting that depending on the final pH, some precipitation of iron as well as phosphates may occur. Moreover, the higher the pH, the more important the disturbing effect is of the complexing agents susceptible to be present in the waste. This could argue in favour of partial neutralisation with sodium hydroxide instead of denitration for lowering the waste acidity.

Since separation between oxalate precipitate and supernate is never complete, some radionuclides (e.g. Cs-134/137, Sb-125 and Ru-106) may contaminate the actinide/lanthanide fraction.

However, since the precipitate is crystalline (in absence of complexing agents), these contaminants can easily be washed out.

Comparative tests performed on a real sample of reprocessing concentrate showed a distinct decrease in the decontamination factors so that the amount of lanthanide carrier had to be increased by nearly one order of magnitude (table 4).

radionuclide	overall alpha	Am-241	Ce-144
D.F.'s 10^{-2} M Ce^{3+}	4	5	2
D.F.'s 5×10^{-2} M Ce^{3+}	70	>300	>1000

Table 4: Variation of the D.F.'s as a function of the amount of Ce carrier.
Operating conditions: real reprocessing concentrate adjusted to pH 2-3 by NaOH addition
0.03 and 0.15 M $H_2C_2O_4$

Conditioning of the resulting alpha bearing precipitate into ceramic matrices is tackled in detail in Dr. Halaszovich's paper[4]. However, it is worth mentioning that the specific alpha activity of the ceramic product is expected to be $\sim$ 50 Ci/m^3. With respect to waste arisings, this process is anticipated to generate less than 75 kg of ceramic waste products per tonne of irradiated fuel in addition to the 4.5 tonnes of cemented reprocessing concentrate.

This rather low additional waste arising can further be decreased if an extensive use of oxalic acid and cerium is performed in waste treatment facilities as suggested by the JRC-Ispra[2] (see fig. 1).

3. Utility liquid waste

Utility liquid wastes generated during PWR operation are very heterogeneous in terms of radioactivity and chemical composition since these encompass a wide range of low active streams (floor drains, shower water, laundry water, decontamination effluents ...). Usually, their radioactivity level is sufficiently low to enable their discharge into the environment (sea or river) after simple filtration. However, it has to be pointed out that utility liquid waste contributes to approximately 95% of the overall activity released as liquid effluents.

Therefore, at the Chooz PWR site, in order to comply with the ALARA principle, it has been decided to arbitrarily reduce the overall radio-activity discharged into the Meuse river (with the exception of tritium) to only 5-10% of the authorized limit by treating, in a specific way, utility liquid wastes. Formerly, these were treated by evaporation, but due to large variations in their chemical composition (salt, detergents, ...) this treatment process proved quite troublesome (foaming problems) and finally very expensive (energy consuming). Therefore, as an alter-native, treatment of utility liquid waste through chemical precipitation has been investigated and an industrial flocculator prototype was con-structed (5 m^3/h capacity). Basically, the prototype which operates continuously consists of two mixers (static + tubular) and a lamellar settler. To prevent possible carrying over of flocs, an hydroanthracite filter has been placed downstream. Addition of precipitants and pH adjustment is performed automatically. According to the detergent content in the waste, chemical precipitation of Cs-134/137, Mn-54 and Co-60 (main radionuclides containing utility liquid waste) is carried out at pH 7 (high content) or 8.5 (low content) and involves addition of copper ferrocyanide and ferric chloride as precipitants.

After natural settling, the sludges are further concentrated by centrifugation (horizontal decanting centrifuge type) prior to being embedded into cement or even epoxy resins.

Industrial testing of the flocculator (see table 5) proved success-ful since on average, the residual bêta-activity of treated utility liquid waste achieved was around 370 Bq/1 (10^{-8} Ci/1) irrespective of the stream origin.

Moreover, if the sludges are intended to be cemented, the floccu-lation process enables a volume reduction factor of about 800!

Likewise, in comparison with evaporation, this process appeared both less expensive (by a factor of 2) and more efficient.

4. Combination of ultrafiltration with chemical precipitation

Due to the difficulty of completely separating the activity bearing suspended solids and sludges from the liquid phase, relatively moderate decontamination factors are usually achieved through chemical precipi-tation methods. Therefore, a polishing treatment of the supernate is often desirable.

By permitting a thorough rejection of suspended and colloidal solids and even macromolecules of 2 nm in size, it is obvious that utilisation of ultrafiltration membranes for solid/liquid separation would improve much the overall decontamination factors. On the other hand, because membrane separation is very efficient, the amount of precipitate needed should be considerably less than that required for gravity floc sedimentation, hence leading to less waste volume to be

waste type	waste volume (m³)	breakdown of the activity before treatment (Bq/l)					breakdown of the activity after treatment (Bq/l)				
		overall β-activity	Mn-54	Co-60	Cs-134	Cs-137	overall β-activity	Mn-54	Co-60	Cs-134	Cs-137
various drains	45	4140	2100	4340	480	1670	340	1100	340	-	-
various drains	40	3280	1040	1030	575	1760	460	530	315	250	620
various drains	30	5040	1490	7620	375	2200	430	450	1000	-	155
laundry waste	42	560	200	1130	-	135	125	120	390	-	-
various drains	25	29610	7560	24900	965	12200	230	400	900	-	-
various drains	47	7000	2150	5970	420	2430	330	750	400	245	580
various drains		5510	1490	3300	380	1520	345	145	860	-	65

Table 5 : Decontamination of various arisings of utility liquid waste by means of the flocculator prototype (operating conditions: pH range 6.4-9, 6×10^{-5} M $K_4Fe(CN)_6$, 10^{-4} M $CuSO_4$, 10^{-3} M $FeCl_3$, 1ppm Drew Floc.

subsequently disposed of. Therefore, in order to assess the merits of ultrafiltration combined with chemical precipitation over classical floc process, an extensive research programme was launched in AERE-Harwell starting with lab-scale experiments which gradually scaled-up to operation of an active UF pilot plant.

4.1. Feasibility experiments

Potential application of ultrafiltration for the treatment of a variety of radioactive liquid waste$^{5-6}$ (e.g. plutonium evaporator overheads, plutonium oxalate production effluents and general nuclear site effluents) has been successfully demonstrated on the basis of lab-scale experiments using a small UF test circuit with flat discs of commercially available membrane. It was thus shown that, by appropriate choice of chemical pretreatment conditions, most encouraging D.F.'s (greater than 10^2 in most cases and sometimes up to 10^5) could be achieved for the most important radionuclides. However, more emphasis has been placed on the decontamination of simulated magnox pond waters. In this regard, two stages of ultrafiltration carried out successively at pH 11.5 and 8 after addition of $TiCl_4$ and $Cu_2Fe(CN)_6$ (4 g/m^3 each) enabled the achievement of D.F.'s higher than 100 for caesium and strontium$^{7-8}$.

Moreover, experiments with real radwastes, such as the Harwell site medium level and low level wastes, which contained α and β/γ activity levels ranging from 10^{-5} to 10^{-2} Ci/m^3 were shown to be equally effective to those performed on simulated wastes. These experiments also confirmed the merits of UF over conventional floc process in terms of D.F.'s (see table 6).

Process	α D.F.	β D.F.
Conventional floc sedimentation Process	33	3
Direct ultrafiltration (pH $\sim$ 9-10)	330–3300	4–33
Addition of Fe^{3+}, $Fe(CN)_6^{4-}$, Co^{2+} at pH 4.5 then UF	>10000	7

Table 6: Comparison between conventional floc with UF membrane processes for the decontamination of real samples of Harwell MLLW (α, β and γ activities $\sim$ 10^{-2} - 10^{-3} Ci/m^3)

4.2. Process scale-up

Decontamination UF tests on magnox pond waters were scaled-up to the processing of 100 litre batches of simulated effluent using two quite different types of commercially available modules: plastic hollow fibres and large bore tubular inorganic membranes. Both systems were able to reproduce, at membrane areas of 0.05 - 0.1 m^2, high D.F.'s similar to those achieved with the flat sheet UF test unit. However, these experiments also demonstrated that resistance of the inorganic membrane system to fouling was superior to hollow fibres which were rapidly blocked by thick accumulations of precipitate despite the use of high cross-flow velocities. For this reason and also because plastic membranes cannot tolerate high exposure levels[9] (> 1 Mrad), the inorganic membrane system has been selected for further development.

4.3. Pilot plant construction and operation

A small scale, automated unit, containing a single inorganic ultrafiltration module and 0.05 m^2 of membrane area, was constructed and used to remove Sr activity from simulated pond water after the formation of a titanium hydroxide precipitate. With this unit, shown schematically in Figure 2, it was possible to demonstrate for the first time the continuous processing of an effluent, in this case at a rate of 50-100 litres per day, for a period of several months. It was confirmed that stable performance could be maintained while operating at high volume factors of about 250. The ultrafiltration concentrate stream was a dilute sludge, suitable for dewatering prior to encapsulation.

A membrane flux of over 1 m^3/d and a Sr decontamination factor of 100 were maintained during the operation of the unit. Techniques were developed for cleaning chemically the ultrafiltration membranes at intervals of several weeks and for dealing with the secondary waste arisings.

After the extended test programme with a simulated effluent, the same automated unit was used to process a real radwaste, i.e. the Harwell site low level waste. This effluent contained considerable amounts of coarse suspended solids, and the problems encountered during its processing served to highlight the limitations of the design of the automated unit, particularly with regard to both the selection of pumps and valves for the processing of solids laden streams, and the necessary control strategy to protect the plant in the event of malfunction.

The experience gained during operation of the small unit was subsequently utilised during the design of and equipment specification for a much larger, more sophisticated pilot plant, which can process over 1000 litres of effluent per day. This pilot plant has been constructed and installed recently within the Harwell effluent treatment facility.

The plant is located within a ventilated enclosure, and the ultrafiltration module, which contains 37 membrane tubes and a total area of 0.85 m^2 is mounted behind a lead shielding wall. The plant operates under the control of a microcomputer system that also logs plant data and controls ancillary operations such as the cleaning cycle.

The results of the first active decontamination campaign (about 40 days operation) on real Harwell low level liquid waste appeared quite encouraging, especially for the alpha removal. Thus, with respect to conventional chemical precipitation, the overall alpha decontamination factors increased by a factor of 2 in the case of direct UF treatment and up to 7 in the presence of 10 ppm of titanium hydroxide. However, for the bêta-gamma emitters, only moderate D.F.'s have been achieved in these preliminary experiments. With respect to sludges arising, this process led to a volume reduction factor of about 200 with a solids density of 1%. Operation of the pilot plant is still on-going and more definite answers concerning the performance of the UF technique will be given in the next few months, when the process has been optimised.

5.   Conclusions

On the basis of experimental tests performed on real waste at the lab-scale, application of chemical precipitation methods for implementing new management schemes for reprocessing concentrate proved quite feasible.

If handling and transportation of cemented reprocessing concentrate in unshielded waste container casks proved to be economically profitable, removal of the main gamma emitters containing waste by means of precipitation with ferrocyanide/ferrous iron mixture seemed quite efficient, especially in the case of caesium and antimony.

Moreover, by mixing the resulting sludges with the high level liquid waste for vitrification, good glass product can be achieved without any significant increase in the overall glass volume. Accordingly, besides its economical interest, this management option seems particularly recommended if a safer conditioning of the radionuclides containing reprocessing concentrate is desired.

A further improvement in the conditioning of long-lived radio-nuclides containing reprocessing concentrate is also attainable by removing actinides through coprecipitation with lanthanides oxalate followed by incorporation into ceramics. In terms of waste product arisings, this management option only gives rise to limited amount of ceramics with respect to cemented reprocessing concentrate for which, as for the preceding option, disposal in shallow land burial can be envisaged (alpha specific activity < 100 nCi/g).

In another area of radwaste management, the construction and operation of an industrial flocculator prototype for the treatment of utility liquid waste generated at the PWR-Chooz site proved quite successful since this enables a strong reduction of the radioactivity discharged as liquid effluent into the Meuse river to 5-10% of the authorized limit.

Concerning the liquid/solid separation hardware, which, actually, governs the efficiency of most chemical precipitation processes, it was demonstrated in all the foregoing R and D activities that centrifugation, especially the horizontal, decanting centrifuge type, was the most suitable technique for active operation.

Improvement of chemical precipitation process when combined with ultrafiltration has clearly been demonstrated from lab-scale experiments for various waste types up to pilot plant operation for the Harwell low level liquid waste. Over conventional flocculation, higher decontamination and volume reduction factors have been achieved without encountering any serious fouling problems on the tested membranes especially for the inorganic type. Thus, decontamination performance of chemical precipitation methods can be levelled to conventional ion exchange processes.

REFERENCES

1. W. BAEHR - "Entsorgung in Wiederaufarbeitungsanlagen"
    EUR 9342 DE - Ecomed publishers (1984)

2. H. DWORSCHAK et al. - "High level liquid waste splitting: feasibility and incentive for a separate actinides conditioning and disposal". Proceedings 2nd technical meeting on the nuclear transmutation of actinides. Ispra, Italy, April 21-24, 1980
    EUR 6929 EN/FR (1980)

3. F. MOUSTY et al. - "L'acide oxalique et le traitement des déchets provenant du traitement du combustible irradié". European Appl. Res. Repr. Nucl. Sci. Technol. vol. 4, N° 4 (1982) pp. 835-878 -
    Harwood Academic Publishers

4. S. HALASZOVICH et al. "Conditioning of alpha waste". Proceedings of this conference

5. P. BIDDLE, R. GUTMAN - "The use of ultrafiltration for the clean-up of alkaline plutonium containing effluents" - AERE R10942 (June 1983)

6. R. KNIBBS - "The decontamination of alpha bearing waste streams using coprecipitation with ferric hydroxide in conjunction with ultrafiltration". AERE R10269 (1984)

7. G. WILLIAMS, R. GUTMAN - "Strontium removal from magnox pond waters by ultrafiltration" - AERE R11103 (May 1983)

8. I. REED - "Caesium removal from magnox pond waters by ultrafiltration" - AERE R11096 (May 1984)

9. R. KNIBBS - "The effects of radiation on ultrafiltration membranes"
    AERE R10368 (October 1984)

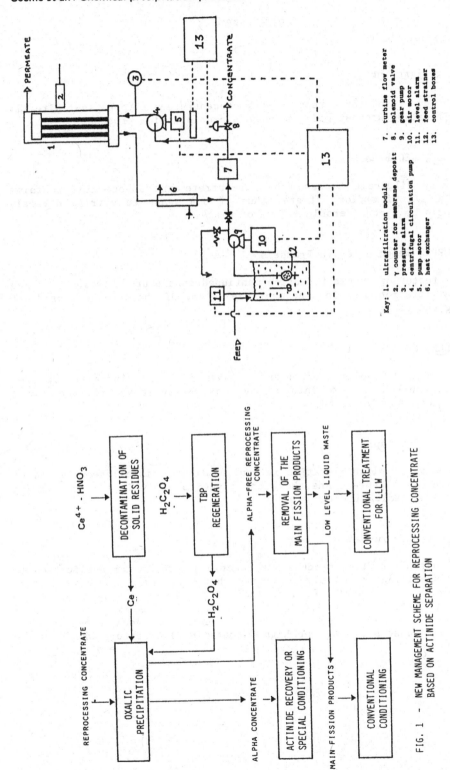

Key: 1. ultrafiltration module
     2. γ counter for membrane deposit
     3. pressure alarm
     4. centrifugal circulation pump
     5. pump motor
     6. heat exchanger
     7. turbine flow meter
     8. solenoid valve
     9. gear pump
     10. air motor
     11. level alarm
     12. feed strainer
     13. control boxes

FIG. 2 - AUTOMATED ULTRAFILTRATION TEST UNIT

FIG. 1 - NEW MANAGEMENT SCHEME FOR REPROCESSING CONCENTRATE
          BASED ON ACTINIDE SEPARATION

DISCUSSION

UNKNOWN SPEAKER, UKAEA

Is there any reason for a reprocessing concentrate not to be sent directly
to vitrification without going through a precipitation process?

L. CECILLE, CEC Brussels

It is very difficult because the concentrate of reprocessing contains
about 300 g/l of sodium nitrate, thus if one wants to vitrify directly
there would be a considerable volume of glass.

M. VANDORPE, Belgonucléaire Brussels

Does the factor of volume reduction mentioned in the presentation obtained
by ultrafiltration take into account the volume of the detergent used for
cleaning the cartridge?

L. CECILLE, CEC Brussels

No. The factor of volume reduction is taken from the sludge volume produced. If one had to take into account the detergent volume, the factor
would no longer be about 60.

F. GIRARDI, CEC Ispra

The procedure of ultrafiltration is applied to the miscellaneous waste of
Harwell Research Center. Are there any plans to apply it to reprocessing
waste such as medium level waste?

L. CECILLE, CEC Brussels

Yes, the ultrafiltration process was described here in its application to
the wastes arising from Research Centres, but this study took into account
a broader field of application including the waste arising from reprocessing operations. It was applied to research wastes simply because these
wastes were available. However, this process has good prospects for
reprocessing wastes due to the high decontamination factors which can be
reached.

ELECTROCHEMICAL AND ION-EXCHANGE PROCESSES

A.D. TURNER*, J.F. DOZOL** and P. GERONTOPOULOS[‡]
*UKAEA, AERE Harwell, England.
**CEA, CEN Cadarache, France.
[‡]Agip/ENEA, JRC Karlsruhe, West Germany.

Summary

The potential role of ion-exchange and electrochemical processes in the decontamination of liquid wastes has been studied experimentally. Inorganic exchangers not only can have a high specificity for active species, but by simple thermal treatment can be converted into a stable waste form suitable for long term storage. Batch experiments on commercially available materials recommend ammmonium phosphotungstate, titanium and zirconium phosphates at pH < 3, and zeolites at pH 3-12 for Cs removal; polyantimonic acid at pH < 4, hydrous titania at pH 4-10, and sodium titanate or zirconate at pH > 10 for Sr, Co, Np, Pu, Am removal. Gel-precipitated hydrous titania and titanium phosphate xerogel spheres are particularly suited to column use due to their size and strength, and also their excellent kinetics. Electrochemical control of exchangers can cause cation adsorption under cathodic potentials with DF's of ~ 2000 up to a 75% loading, and elution into water under anodic potentials - thus concentrating the waste at only $\frac{1}{2}$% the energy needed for evaporation. The system has already been cycled > 700 times. Electro-kinetic dewatering of gravity-settled flocs can concentrate them to ~ 30% solids with 99.99% solids retention at only ~ 5% the evaporation energy. Scaled-up cells have been successfully demonstrated with genuine wastes.

1. Introduction

The use of water as a process medium in nuclear plants of all descriptions inevitably gives rise to the production of medium and low level activity liquid waste streams. Immobilizing these without further treatment would give a large volume of solid waste for disposal, while evaporation of such volumes of water would be very energy intensive. Both of these strategies would thus be very expensive to operate. Alternative energy efficient processes, therefore, are needed that can selectively re-concentrate active species into a small volume of higher level waste in such a way that the bulk of inactive materials can safely be discharged to the environment. This will minimize the volume, and hence the cost of immobilized waste requiring controlled disposal. Both ion-exchange and electrical processes have been demonstrated to have a potential role to play in this conditioning of low and medium level liquid wastes.

2.   Inorganic Ion-exchange

2.1   Introduction

Medium and low activity liquid wastes arise not only from the reprocessing of spent fuel and the operation of reactors, but also from fuel fabrication plants and Research Establishments.

In France, liquid wastes have traditionally been treated by chemical precipitation to remove the main radionuclides before release into the environment. The sludges arising from this treatment have subsequently been immobilized in bitumen, but because of their α, ^{90}Sr and ^{137}Cs activity, they need to be stored in geological formation. Although this route has some advantages, the current trend is to use bitumen only for the immobilization of low and medium activity wastes containing short lived radio-elements. For wastes containing long lived α and $\beta\gamma$ emitters, the main disposal options being considered by the CEA are incorporation into a glass or ceramic host matrix before storage in a continental geological formation. These trends are consistent with the recommendations made by the Castaing Commission. As a result of this, the study of liquid waste treatment methods different from those used until now was initiated.

2.2   Objective of this Study

One of the possible methods is the direct removal from the unconcentrated effluents of the majority of long lived radionuclides on ion-exchangers. This process gives rise on the one hand to a solution where most of the long lived actinides and fission products have been removed (which would then be conditioned as before by chemical precipitation and immobilization of the resulting sludge in bitumen before disposal by shallow land burial), and on the other hand to one or more mixed exchangers containing the majority of the activity of the initial effluent, but in a smaller volume.

Inorganic ion-exchangers are of interest as not only do some of them display a selectivity for the adsorption of active species, but they can also be converted into stable matrices suitable for long term storage by simple thermal treatment. The purpose of this study has been to test on simulated and real effluents the effectiveness of a range of inorganic exchange media in fixing the actinides and the long-lived $\beta\gamma$ emitters.

The treatment of a liquid waste from La Hague is a good example of a potential application of this process, and will be the subject of future work. This effluent is at present treated by chemical precipitation – essentially removing almost all of the radionuclides before discharge to the environment. The sludges obtained by this precipitation are later immobilized in bitumen. The volume reduction factor – VRF (ratio of the volume of waste to the volume of sludge after immobilization) – is typically about 100 for medium activity liquid wastes. Therefore, as the α activity of the original liquid waste is 0.11 Ci/m^3, and that due to ^{90}Sr and ^{137}Cs is respectively 1.05 and 3.1 Ci/m^3, the resulting activities of the main long lived radionuclides in the solid waste after conditioning are: α activity, 11 Ci/m^3; ^{90}Sr, 105 Ci/m^3; ^{137}Cs, 310 Ci/m^3. However, currently at the La Manche Storage Center, solid waste can be stored in shallow land burial only if the specific activity is less than 1 Ci/m^3 for actinides, 4 Ci/m^3 for ^{90}Sr and 200 Ci/m^3 for ^{137}Cs. Therefore, in order to reach this requirement for the shallow land burial of the bitumen immobilized sludges arising from the final treatment of the effluent before discharge, the liquid waste must receive a pre-treatment with decontamination factors (DF) of at least 11 for the α activity, 25 for ^{90}Sr and 1.5 for ^{137}Cs. These numeric values are merely given by way of example and are not representative of the La Hague average liquid waste composition.

2.3 Choice of Inorganic Ion-exchangers

The spectrum of materials studied has spanned the most widely used and easily available products. These included zirconium phosphate (PHOZIR), titanium phosphate (PHOTI D), ammonium phosphotungstate in zirconium phosphate (PHOMIX) and polyantimonic acid from Applied Research; sodium titanate, sodium zirconate, sodium tantalate, sodium niobate produced by CERAC Inc. (a supplier of the SANDIA Laboratories where these products were initially studied); synthetic zeolites IE 96, A 51 and 13 X (manufactured by Union Carbide) and Zeolon 500, Zeolon 900 from Norton. Two varieties of clinoptilolite: Zeolon 400 from Norton and ZBS 15 from Oxymin were also examined. The amorphous silico-aluminate Decalso Y was supplied by Doulite International, while the titanium oxide tested came from two sources: Karlsruhe (Agip ENEA), and Cadarache, where it is used for uranium recovery.

2.4 Measurements of Distribution Coefficients

2.4.1 Sorption of cobalt, strontium and caesium

The distribution coefficients of these elements on the adsorbers have been determined under a variety of solution composition conditions using a small scale batch technique. One millilitre of an ion-exchanger was equilibrated for 15 h by shaking with 100 millilitres of pre-filtered (5 µm) effluent spiked with long lived fission products: ^{60}Co, ^{85}Sr as a tracer model for ^{90}Sr and ^{137}Cs. Before being tested, the ion-exchangers were pre-treated by shaking with water at the desired pH value.

Measurements have been made at different sodium (0.23 - 1 - 10 g/ℓ), calcium (0.1 - 0.3 - 1 g/ℓ) and magnesium (0.1 g/ℓ) concentrations and at various pH. The solutions contain only Na$^+$ (or Ca^{2+} or Mg^{2+}) and NO$_3^-$ ions; the desired values of pH and concentrations are obtained by mixing only sodium (or calcium or magnesium) hydroxide, sodium (or calcium or magnesium) nitrate and nitric acid.

After contacting with the exchanger, the solution was filtered (5 µm) again before its residual activity was measured with a Ge-Li crystal connected to a multi-channel analyser.

The distribution coefficient K_d was calculated from:

$$K_d = 100 \, \frac{a_o - a}{a} \, , \text{ where}$$

a_o = activity of liquid before contacting
a = activity of liquid after contacting.

2.4.2 Sorption of Np, Pu, Am

The ion-exchangers studied were the same as described previously, though as the zeolites are only efficient at high pH (7-12), K_d's were only measured for the products that gave the best results for γ emitters: ZBS 15 clinoptilolite and Zeolon 900, IE 96, A 51, 13 X zeolites. In any case A51 is unstable in acidic media.

The neptunium, plutonium and americium simulated wastes were prepared from nitric acid solutions, diluted in order to obtain a pH of 2 and the various desired concentrations of sodium, calcium and magnesium.

The experimental procedure was the same as the one used for γ emitters, though the activity in the solution was measured instead by a liquid scintillation counter.

2.5 Results

2.5.1 ^{60}Co retention

In all the solutions tested, the great majority of the ion-exchangers show a maximum effectiveness at pH 7. Hydrolysis of cobalt causes a decrease of retention factors under more alkaline conditions:

$$Co^{++} \xrightarrow{\ OH^-\ } Co\,(OH)^+ \xrightarrow{\ OH^-\ } Co(OH)_2$$

In synthetic effluents containing sodium, the best adsorbers for the removal of cobalt at pH > seven are sodium titanate, sodium zirconate and also to a lesser degree zeolites 13 X and IE 96. For high sodium concentrations, Agip ENEA titanium oxide and titanium phosphate xerogel particles both give good results.

At low calcium concentration (0.1 g/ℓ), the highest retention factors were obtained with the same ion-exchangers that performed best in sodium containing media - sodium titanate, sodium zirconate, Agip titanium oxide, zeolite A 51. However, at higher calcium concentrations (1 g/ℓ), the effectiveness of sodium titanate and sodium zirconate decrease sharply, though titanium oxide and zeolite Zeolon 500 are less seriously affected at high pH values.

For the majority of these ion-exchangers the values of distribution coefficients are, for the same concentrations, very close.

It should be noted, however, that in acidic media polyantimonic acid is the only exchanger for which the distribution coefficient exceeded 100.

2.5.2 ^{85}Sr retention

As a general rule, the distribution coefficients are higher for strontium than for cobalt, especially at elevated pH. The ion-exchangers that were successful in the removal of cobalt also proved effective for strontium retention. At pH 12, distribution coefficients larger than 10^5 (for sodium concentration 0.23 g/ℓ) and equal to 5.10^4 (for sodium concentration 10 g/ℓ) have been found for sodium titanate and sodium zirconate. Titanium oxide behaves similarly at low pH but distribution coefficients are less high than for the previous compounds with increasing pH. While Zeolites A 51 and 13 X are less effective than sodium titanate, their distribution coefficients are still high. The high affinity of polyantimonic acid for strontium over the whole pH range is notable, even at high sodium concentrations (10 g/ℓ).

At low calcium levels (0.1 g/ℓ), distribution coefficients higher than 5.10^4 have been found for sodium titanate, sodium zirconate, titanium oxide, zirconium phosphate and zeolites A 51 and 13 X. However, when the calcium concentration reaches 1 g/ℓ, distribution coefficients decrease sharply - to 10^3 or below. Polyantimonic acid, though, is remarkable in its consistency of adsorption, with distribution coefficients relatively insensitive to calcium concentrations – exceeding 10^3 even in acid media.

2.5.3 ^{137} Cs retention

Two categories of product are effective for caesium retention. The first comprises zirconium phosphate, titanium phosphate, and ammonium phosphotungstate mixed with zirconium phosphate, which have very high distribution coefficients in acid media, declining with increasing pH. The second category includes zeolites – in particular IE 96 and Zeolon 500 – which have a consistently high distribution coefficient in the pH range 3 to 12.

Of these materials, caesium retention by ammonium phosphotungstate is the least sensitive to high sodium concentrations (10 g/ℓ), while the zirconium phosphate distribution coefficient decreases even more sharply than that for titanium phosphate with increasing sodium concentration. The zeolite K_d's are also sensitive to concentration changes – decreasing by a factor of 10 with a sodium concentration increase of 50. Although the distribution coefficients for ammonium phosphotungstate and zirconium phosphate reach 10^5 at pH 3 in 0.1 g/ℓ Ca media, the K_d values obtained for zeolites, while lower in the pH range 3 to 12, are less sensitive to the calcium increase from 0.1 to 1 g/ℓ.

2.5.4 Np retention

NpO_2^+ is the dominant soluble species at pH < 10. At pH 2, only polyantimonic acid shows acceptable K_d values (800), which are relatively insensitive to salt content.

For ^{237}Np removal at pH 4, the best results were achieved using polyantimonic acid (K_d > 2500), though titanium oxide also performed well (K_d = 1000).

While titanium oxide and polyantimonic acid are **very** effective (K_d = 10^4) for neptunium removal at pH = 7, titanium phosphate, zirconium phosphate, sodium titanate, sodium tantalate and sodium zirconate also show reasonable K_d values.

2.5.5 Pu retention

Preliminary experiments performed in the laboratory showed that plutonium is not precipitated at pH < 8. As for neptunium, only polyantimonic acid shows a significant affinity for plutonium at pH 2, with K_d ~ 2000, independent of sodium, calcium or magnesium concentration.

The most complete removal of plutonium at pH 4 is achieved with polyantimonic acid, manganese dioxide and titanium oxide with K_d ~ 3000.

At pH 7, polyantimonic acid performs better (K_d = 5000) than manganese dioxide, sodium titanate, sodium zirconate or sodium tantalate (K_d = 2000), though not as well as titanium oxide (K_d > 10,000).

2.5.6 Am retention

As the solubility of Am^{3+}, the only valency state found in solution, decreases sharply at pH 7, K_d measurements were carried only at pH 2 and 4. For low salinity effluents polyantimonic acid is very effective, (K_d > 10^4) though the distribution coefficient falls with increasing salinity (K_d = 1000); titanium phosphate and zirconium phosphate show acceptable K_d

values.

In sodium, calcium or magnesium containing solutions at pH 4, the best results are achieved with sodium titanate ($K_d > 10,000$), polyantimonic acid and sodium tantalate. High distribution coefficients are also obtained in sodium solutions with sodium niobate and in calcium and magnesium solutions with zeolite A 51, manganese dioxide and titanium oxide ($K_d > 10,000$).

2.6 Conclusion

The measurement of distribution coefficients has enabled the identification of the most effective ion-exchangers. For caesium removal in neutral or alkaline media, zeolites, in particular Union Carbide IE 96, show excellent results. In acidic media, ammonium phosphotungstate is very effective, though its poor mechanical stability makes it less suitable for column use. Titanium and zirconium phosphates, however, are also effective in acidic medium, but are more physically robust; γ zirconium phosphate in particular is very promising, and is worthy of further study.

The other main long lived elements – strontium, cobalt, neptunium, plutonium and americium are removed from liquid wastes by the same exchangers – polyantimonic acid in acidic media, titanium oxide in the pH range 4-10 and sodium titanate or sodium zirconate above pH 10. It can be seen therefore that ion-exchange treatments for the complex wastes in the nuclear industry will need to use more than one specific adsorber in series in order to remove all the various active species sufficiently.

As the retention of an ion on an exchanger depends on its structure (amorphous, glassy, crystalline, etc.), further improvements in radionuclide removal may be made by modifying the microstructure of these media, or by producing insoluble co-compounds like silicoantimonic acid or phosphoantimonic acid.

At present, it is too early to know if the goals of the study have been achieved – that is the sufficiently complete removal from solution of the long lived radio-elements to allow the storage in shallow land burial of the flocs arising from the subsequent treatment of wastes. However, these preliminary results look promising. The ultimate plant design and final choice of adsorbers will be determined by the required throughput, the kinetics of adsorption and degree of selectivity of each particular exchanger – as well as other such factors as ease of subsequent immobilization, or mechanical strength which may make some media less suitable for column use than others.

3.   Xerogel Ion-exchanger

3.1 Introduction

Although employment of the inorganic ion-exchanger (IX) media discussed above appears to be very attractive for conditioning different waste streams, some of these materials have disadvantages in their currently supplied form for large-scale applications:-

- in order to assure satisfactory IX properties, commercially available products are usually prepared in the form of fine powders that cannot easily be used in fixed bed columns, and are unattractive for handling in an active plant;
- the use of expensive raw chemicals and complex preparative techniques often results in prohibitively high production costs.

However, significant advantages may accrue from the choice of a preparation route to produce a macroscopic structure more suited to column

use. One such process is gel precipitation, that was originally developed
for nuclear fuel manufacture as the "SNAM Process" - extensively
investigated in Italy and the UK (1,2). This programme has demonstrated
that this process can easily be used to prepare large, strong particles of
high surface area and good IX properties. Thus far, hydrous titania (HTiO)
and titanium and zirconium phosphates (TiP, ZrP) have been prepared, with
the promise of other materials. After these spheres have become loaded
with active materials, they can be easily transformed into a dense ceramic
waste form.

3.2 Preparation

The preparation of HTiO and TiP by gel precipitation, described below,
is in principle also applicable to the preparation of other hydrous metal
oxides and derived insoluble compounds of known IX properties such as
$ZrO_2.xH_2O$, ZrP (8), $Fe_2O_3.xH_2O$ etc.
In preparing a 1 kg batch of HTiO spheroids, 1.036 ℓ titanium
tetrachloride ($TiCl_4$) is first poured under continuous stirring in a cooled
bath of 5.180 ℓ tetrahydrofurfuryl alcohol (THFA). The strongly exothermic
reaction and the considerable viscosity increase of the reacting mixture
suggests the formation of condensed titanium species of the type
n ($Ti(OR)_4$). The resulting liquor is further thickened by adding 3.1 ℓ of
a 3.5% METHOCEL A4C solution in H_2O. The resulting viscosity of this feed
solution is about 120 cps (25°C), with a shelf life of more than two
weeks.
Atomization of the feed solution to controlled diameter liquid
droplets (e.g. 300 $\pm$ 50 μm after drying) is achieved by simple rotary cone
devices - suitable for industrial scale operation due to their high
production rates and clog free operation.
Conversion of the liquid droplets into spherical gel precipitated
particles takes place in a static 6 M NH_4OH bath, and is completed within
about 10 minutes. Effective removal of residual chloride impurities to
below 100 ppm is obtained by repeated H_2O washing (six times with 3 volumes
of water for 1 volume of gel material).
In the next step, residual water is removed from the gel material by
azeotropic distillation in a MARCUSSON type apparatus. This technique,
while only marginally more complex than other possible alternatives (such
as air drying, water extraction with a water soluble organic solvent etc.),
is preferred because it yields products with higher specific surface
areas.
Titanium phosphate particles are prepared by contacting the dried HTiO
material with phosphoric acid (24 hours in a 0.5-2.0 M H_3PO_4 bath) before
water washing and air drying at 110°C.
Although the above methods are still under development on the
laboratory scale (2 kg/day), it is anticipated that only moderate
industrial production costs will result - not substantially different from
those corresponding to the organic or zeolite ion-exchangers already widely
used in industrial radiochemical separative processes. This expectation is
based on the low cost of the raw chemicals and the simple process chemistry
used.

3.3 Physical and IX Properties

The ion-exchange material is produced in the form of compact, glassy
xerogel spheroids (Figure 1), that due to their peculiar internal sponge
like microstructure (Figure 2) have exceptionally high specific surface

areas and thus good IX properties.

The IX affinity of HTiO gel precipitated material for a wide range of waste elements has been previously described in detail (3-6) – often proving to be superior to the performance of fine powder products of the same chemical composition (including sodium titanate). The distribution coefficients for actinides, rare earths, strontium, antimony, cobalt and caesium are generally high for aqueous, low salinity solutions of pH greater than 2, thus permitting efficient uptake by percolation through fixed bed ion-exchange columns.

The potential for practical application of this type of material is illustrated by a combined precipitation-IX decontamination test on a waste stream generated by the French reprocessing plant at La Hague (7). The initial chemical composition of this effluent is basically 1.1 N HNO_3 containing mg/ℓ quantities of U, Ca, Fe as well as 0.2 g/ℓ phosphate. As the first treatment step, the pH value of the solution was raised to 10.5 before nickel hexacyano ferrate addition to remove caesium. The resulting supernatant was adjusted to pH 11.5, then passed through a 9 mm diameter 80 mm height column of gel precipitated HTiO spheroids (0.1-1.5 mm) at a rate of 50 mℓ/h (10 CV/h). The radiochemical composition of the solution before and after IX decontamination is given in Table I – showing excellent decontamination from ^{90}Sr and the actinides and moderate decontamination from ^{125}Sb. Further caesium decontamination falls drastically after only 35 CV while ^{60}Co is not removed at all.

Although less extensively investigated, titanium phosphate gel spheroids appear to have a promising IX affinity for caesium, even at low pH. For example, decontamination from ^{137}Cs of simulated MAW concentrates (containing 1 M HNO_3, 300 g/ℓ $NaNO_3$, a wide range of cations, and some organics arising from solvent washing) is given in Figure 3. In this case the solution, after adjustment of its pH value to 1.0, was passed through a 9 mm diameter, 40 mm high TiP IX column at a rate of 4-5 CV/hr. DF's of over 1000 are obtained up to 200 column volumes.

3.4 Conversion to Ceramics

The use of inorganic IX products in decontaminating liquid waste streams present the possibility that after being loaded with the waste elements, the IX material can be converted to leach-resistant waste forms by conventional ceramic powder techniques.

Utilization of gel precipitated IX spheroids offers the possibility of using simple ceramization routes such as cold pressing and sintering, and/or dispersion of sintered IX particles in a metal or glass host matrix.

It has been demonstrated that a cold pressing and sintering technique, previously developed for preparation of $(U,Pu)O_2$ nuclear fuel pellets from calcined gel precipitated spheroids (9), can be applied to the conversion of spent HTiO and TiP IX media to high density ceramics at temperatures of only 850°C.

3.5 Conclusions and Future Work

Dust-free gel precipitated IX spheroids of hydrous titania and/or titanium phosphate may be used in treating MAW streams instead of fine powder commercial products, with the following advantages:-
  - their preparation is simple, versatile and economic
  - they are suitable for use in fixed bed IX columns, and have a low potential for contamination diffusion

- they have good IX capacity and adsorption kinetics
- they can be converted directly to dense, leach-resistant ceramics by firing at acceptably low temperatures.

Future work should include (i) design, construction and operation of a pilot plant having a production capacity of about 50 kg/day, (ii) definition and testing of complete decontamination-immobilisation schemes working with real waste streams - possibly at a pilot plant scale, and (iii) search for new IX compounds, in particular those identified in Section 2, that can be prepared in xerogel form.

4.   Electrical Processes

While there is a widespread familiarity in the nuclear industry with the concept of ion-exchange as a separation process, electrical processes are generally less well known. They do, however, have an intrinsic attractiveness in that they can be controlled remotely and automatically by the extra reaction variable of applied voltage, which is available in addition to the normal process variables of temperature, pressure, concentration and mass transport. Also the use of electrons in place of inactive chemical reagent additions to perform redox and pH changing reactions is of considerable importance in helping to reduce the amount of material needing controlled disposal. An initial survey evaluated 10 candidate processes used in industry and laboratory, including: solid/liquid separation based on electrokinetic effects (electro-osmosis, electro-filtration, electro-flotation, electro-aggregation, electro-decantation and electro-flocculation); the deposition of dissolved ionic material either as a solid metal or an amalgam, or as an insoluble compound; or the adsorption of dissolved ionic material by ion-exchange under electrical control (10). On the basis of high DF and VRF, suitability for active use and an identifiable advantage over conventional technology, electro-osmotic dewatering and electrochemical ion-exchange were selected as being the most promising for immediate further investigation for application to the treatment of medium-active liquid wastes. Metal/ compound deposition, while still of interest, were at an even less advanced stage of development and need further investigation to evaluate their potential.

4.1 Electrokinetic Dewatering

Floc treatment of liquid wastes can essentially be viewed as co-precipitation combined with ion-exchange at a finely divided form of an exchanger produced in-situ in the waste stream. Though flocs present a high surface area with rapid kinetics, the effectiveness of such a process not only depends on the completeness of adsorption of active species on the precipitate, but also on the efficiency of the subsequent solid/liquid separation, so that the liquid is sufficiently decontaminated in order to permit its discharge. Slurries resulting from gravity settling may still contain ~ 95% water and be difficult to satisfactorily dewater further. However, concentration to 30% solids will lead to an additional 6 x reduction in the immobilized waste volume resulting from the addition of cement powder which makes use of the residual water content for the concrete hydration reaction. For wastes of low electrical conductivity $(0.01-1 \text{ Sm}^{-1})$, Electrokinetic Dewatering can meet the requirements of high solids retention factor (99.99%) in order to give a good overall DF, is able to handle slurries up to 30% solids, is able to achieve VRF's of $\geq$ 6, is energy efficient (typically using between 0.03 kWh/L-0.13 kWh/ℓ; this is

$^1/40$–$^1/15$ of the energy required for an evaporation process like bituminization) and has a high processing rate (typically 150–1500 $\ell/h/m^2$ of membrane area at current densities of ~ 1.5 kA m^{-2}) in order to keep the installed plant size, and hence overall cost, low. The clarity of the permeate is illustrated in Figure 4 for Mg(OH)$_2$ dewatering to 35% solids. It is essentially a direct electrical pumping of liquid from a solid suspension through a microporous membrane under a potential gradient with minimal externally applied hydrostatic pressure (11). The pores of the membrane become coated with the particulates being separated, thus imparting a small surface charge which is matched by ions of equal and opposite charge dispersed into the adjacent solution. This is characterized by the zeta potential across the mobile part of the diffuse double layer at the solid/liquid interface. As these particles are immobilized on the membrane, when an electric field is applied parallel to the pore walls (i.e. across the membrane) by external electrodes, the solution containing this local excess of counter-ions is transported along the pore. By normal solution viscosity effects, the complete fluid content of the pore is dragged through as well by the movement of only a very small electrical charge. This resulting flow rate is proportional to the current, zeta potential, and inversely proportional to the solution conductivity. In addition to this electro-osmotic effect, the dispersed particles in the feed are being electro-phoretically transported away from the membrane. Not only does this inhibit membrane fouling, thus maintaining the extraction flux, but it also dramatically reduces the fraction of very fine particles passing through. This system can also be combined with conventional UF, where the treatment rate is enhanced by reducing membrane fouling, and also by increasing permeation with simultaneous electro-osmosis. In both of these systems, in order to keep the feed sufficiently fluid with increasing solids content, and to prevent any build-up of deposits on the counter electrode at the back of the feed chamber, a cross-flow velocity of > 1.2 m/s is used. The electrochemical reactions that occur at the electrodes on either side of the membrane to convert the ionic/electro-kinetic currents back into electrons normally involve creating a small change in local pH. As at large VRF's this may become sufficiently great to cause the zeta potential, and hence the extraction rate to fall, the feed may be dosed with concentrated alkaki at a rate proportional to the current in order to maintain the pH.

A variety of membrane supports have been investigated for their performance and radiation resistance – including cellulose acetate/nitrate (100 kGy), cotton (1 MGy), stainless steel (microporous sinter, and mesh), zirconia cloth and zirconia coated microporous graphite. These latter obviously have superior radiation resistance. Small batch-scale laboratory trials have demonstrated the flexibility of the process in being able to treat a variety of feeds: ferric hydroxide, alumino-ferric hydroxide, calcium phosphate, hydrous zirconia, titania, Magnox sludge.

Scaled-up crossflow cells of both planar (Figure 5) and cylindrical geometry have been used to dewater simulant and genuine active waste generated on site, arising from a ferric hydroxide floc treatment. Solids contents have been raised from 5% to 25%, with the possibility of further volume reduction. Virtually quantitative retention of ^{60}Co and α emitters was found, and although the DF's were marginally lower in samples where significant quantities of detergents were present, these were still sufficiently high to permit discharge of the permeate. The results were a significant improvement over conventional vacuum filtration.

4.2 Electrochemical Ion-exchange

While the processes described above have essentially combined the concepts of immobilization with adsorption by operating a "once through" system, ion-exchange can also be used as a concentration process by elution of the exchanger into a small volume. Ideally this should be rapid, and generate as small a volume as possible that is compatible with any subsequent immobilization process – e.g. vitrification. Unfortunately, difficulties encountered in the chemical elution of some IX media have made this a less favoured choice. However, the use of electrical control of conventional ion-exchange processes can overcome these problems, thus making this a viable possibility.

Electrochemical ion-exchange (EIX) is basically the incorporation of ion-exchange material into an electrode structure based on an expanded metal mesh current feeder, so that it has many of the advantages of electrodialysis and ion-exchange, but few of the disadvantages – such as weeping seals. Because the exchanger is essentially mechanically supported by the current feeder, those materials that have insufficient strength to be used in a conventional ion-exchange column may be reconsidered – together with other finely divided media. For cation exchange, under a cathodic applied potential (with respect to a counter electrode), cations are rapidly adsorbed into the electrode structure (Figure 6). This is encouraged not only by an alkaline environment generated locally inside the electrode, which activates weak acid groups to cation adsorption, but cations are also transported into the bulk of the structure from surface sites by ion migration under the local field gradient. The application of an electric field extends the capabilities of the ion-exchanger by enhancing its kinetics, extending the pH range of its operation as well as enabling it to adsorb a significant fraction of its theoretical capacity. Studies to date have concentrated mainly on the behaviour of Cs, Na and Sr.

Corresponding anion responsive electrodes can also be fabricated using anion exchangers at anodic potentials. When these are used in conjunction with cationic electrodes, the feed pH is maintained during treatment, as an equal number of cation and anion equivalents is being replaced with H^+ and OH^- which combine to form water. This in turn maintains the cation adsorption kinetics and DF by inhibiting the reverse reaction normally arising from the fall in pH associated with the displacement of protons by adsorbing cations. On polarity reversal, these ions can be eluted into a limited volume of water (Figure 6), thus generating a concentrated ($\sim$ 0.2 M to date) product suitable for incorporation with high active waste for immobilization.

Even under an arduous accelerated test programme, electrodes have been cycled over 700 times with no detectable deterioration in performance. When such electrodes have been used in a low hydraulic resistance flow cell (Figure 7), Cs decontamination factors of over 2000 have been achieved on a single pass at flow rates of $\sim$ 6 bed volumes/hour (corresponding to 60 bed void volumes/hour). Typically this is maintained up to a loading of $\sim$ 75% of the theoretical electrode capacity before break-through is observed (Figure 6). With recent advances in cell design, the total energy requirement for volume reduction including pumping has been reduced to $^1/400$ of that needed for evaporation. This new design also permits the use of higher current densities (and hence more rapid kinetics), as well as higher exchanger loadings, which increases the cell capacity – thus reducing the size and cost of a plant for a given throughput.

In order to reduce even further the amount of inactive material being added to the highly active waste, EIX modules displaying selectivity only for active species are being developed. Work to date on organic and inorganic exchangers that are selective under the chemical conditions described above are showing promising results. Weak acid organic resins have capacities $\sim$ 10 meq/g and display K_d's $\sim$ 17,000 with a selectivity for Cs over Na of 3-4.5 in batch experiments. Inorganic exchangers, though, have the advantage of superior radiation stability. Zirconium phosphate has been demonstrated to have a Cs capacity of 5-7 meq/g with a selectivity of Cs over Na adsorption of 6.5/equilibrium stage in a batch experiment. The use of flow cells (like an IX column) and multiple units will not only greatly improve on this, but also permit the more effective use of electrical power.

Experiments with simulant fuel storage pond water containing 100 ppm Na at pH 10 have demonstrated the reduction of Cs content from 60 μCi/ℓ to background with both organic and zirconium phosphate electrodes. Treatment of genuine site wastes containing ^{60}Co, ^{137}Cs, ^{85}Sr, detergents, and other inactive salts at pH 8 had 99.6% of the activity removed at a zirconium phosphate electrode. It was most effective for Cs removal (Table II), as also found under the purely chemical adsorption conditions described elsewhere in this paper. As with any ion-exchange system, other adsorbers would be needed for the more specific removal of Sr and Co, possibly in a second unit. Thus EIX can be seen to have been developed into an efficient, robust process that can be easily externally controlled by electrical means. Versatility is built into the system by being able to use particular adsorbers for specific applications, though further work is needed in this area. By making multiple and complete use of the ion-exchange capacity of the EIX electrodes, partition of the waste into a small volume at a higher activity level while leaving the bulk at essentially background levels, will dramatically reduce the volume of solids requiring controlled disposal.

5. Conclusion

For all the processes described, which at present show real promise for the treatment of medium and low level liquid wastes, the next stage is the demonstration of these techniques with genuine feeds under typical operational conditions. This will also enable the development of process control, and the optimization of performance on a more realistic scale. From this information, an evaluation can be made for potential plant use.

REFERENCES

1. BRAMBILLA, G., GERONTOPOULOS, P., FACCHINI, A. and NERI, D. The SNAM process for production of ceramic nuclear fuel microspheres. Symposium on Sol Gel Processes and Reactor Fuel Cycles, Gatlinburg, Tennessee, USA. May 4-7, 1970. (CONF 700502).
2. HARDY, C.J. and LANE, E.S. Gel process development in the United Kingdom, Ibid.
3. GERONTOPOULOS, P., ARCANGELI, G., CAO, S., CRISPINO, E., FORNO, M. and MULLER, W. Gel precipitation for radwaste ceramics, (a) 185th ACS National Meeting, Seattle, Washington. March 20-25, 1983, (b) Radiochimica Acta 36, 69-74, (1984).
4. DOZOL, J.F. and EYMARD, S. Retention du ^{60}Co, ^{85}Sr, ^{137}Cs sur echangeurs mineraux. Rapport d'avancement 1er semestre 1983. Contract CEA-CCE WAS-316-53-31 F.

5. DOZOL, J.F., EYMARD, S. and LA ROSA, G. Etude de l' application d' echangeurs mineraux à la decontamination d' effluents radioactifs FA et MA. Rapport d' avancement 2ème semestre 1983. Contract CEA-CCE WAS-316-53-31 F.

6. DOZOL, J.F., EYMARD, S. and GAMBADE, R. Etude de l' application d' echangeurs mineraux à la decontamination d' effluents radioactfs FA et MA. Rapport d' avancement 1er semestre 1984. Contract CEA-CCE WAS-316-53-31 F.

7. CHAUVET, P. Private communication to J.F. Dozol.

8. CALETKA, R. and TYMPL, M. Sorption properties of zirconium phosphate prepared by the sol gel method. J. Inorg. Nucl. Chem. 39, 669 (1977).

9. COGLIATI, G., GERONTOPOULOS, P. and RICHTER, K., European Nuclear Conference, Hamburg, Germany, May 6-11, 1979, Trans. Am. Nucl. Soc. 31, (1979), 175.

10. TURNER, A.D., BOWEN, W.R., BRIDGER, N.J. and HARRISON, K.T. Electrochemical processes for the treatment of medium-active liquid wastes: A laboratory-scale evaluation (1984), EUR 9522 EN.

11. BOWEN, W.R. and TURNER, A.D. Electrical separation processes in the treatment of radioactive waste, in "Solid Liquid Separation", SCI-Ellis Horwood, Chichester (1984), Ed. J. Gregory.

Acknowledgement

The UK part of this work, concerning Electrical Processes, has been commissioned by the Department of the Environment as part of its radioactive waste management research programme. The results will be used in the formulation of government policy, but at this stage they do not necessarily represent such policy. The contribution to this work by W.R. Bowen, N.J. Bridger and D.R. Cox is gratefully acknowledged.

The Italian part of the work concerning xerogel ion-exchangers has been performed in the frame of a collaboration agreement on sol-gel technology between ENEA, Agip and the Commission of the European Community at the Institute for Transuranic Elements at Karlsruhe. The support of H. Bokelund, responsible for the laboratory facilities where the work has been done, is gratefully acknowledged.

Table I

Decontamination of MAW La Hague effluents by precipitation
and IX decontamination using gel precipitated HTiO spheroids

Radioelement	Effluent Activity, mCi/m^3					
	Feed	Superna-tant	Number of Column Beds Passed			
			35	70	140	240
^{60}Co	5.6	1.8	1.8	1.8	1.8	1.8
^{90}Sr	225	94	0.14	0.2	0.15	0.12
^{106}Ru	2100	620	32	42	140	207
^{125}Sb	112	60	nd	nd	nd	17
^{137}Cs	394	2	0.046	2	2	1
^{234}U + ^{237}Np	18	0.32	nd	nd	nd	nd
^{239}Pu	56	2.6	nd	nd	nd	nd
^{238}Pu + ^{241}Am	73	6.2	nd	nd	nd	nd

nd = not detected.

Table II

Decontamination of a Harwell Liquid Waste Stream by Electrochemical Ion-exchange at a Single Zirconium Phosphate Electrode

Radioelement	Activity Content mCi/m^3		DF
	Feed	Effluent	
^{60}Co	25	0.36	69
^{137}Cs	60	0.12	482
^{85}Sr	2	0.01	143

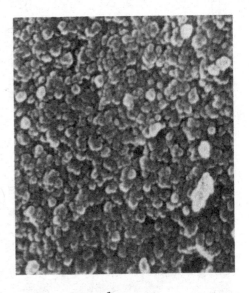

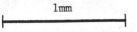

1mm

1µm

Fig 1. Hydrous titania xerogel
spheroids.

Fig 2. SEM image of the internal
structure of a fractured hydrous
titania particle dried at 180°C
in air.

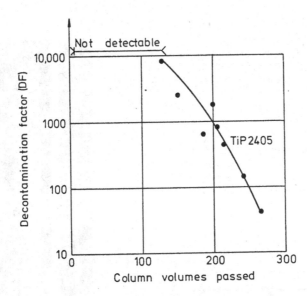

Fig 3. Caesium-137 Decontamination of a
simulated MAW concentrate by percolation
through a Titanium Phosphate ion-exchange
column.

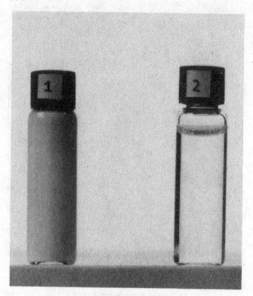

Fig 4. Electro-kinetic dewatering of Magnox sludge simulant: (1) original settled sludge, (2) permeate.

Fig 5. General view of a planar electro-extraction module of 100 cm^2 membrane area.

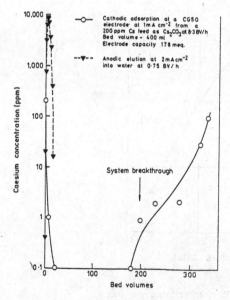

Fig 6. Caesium concentration of the effluent from an electro-chemical ion-exchange module during adsorption and elution.

Fig 7. Electrochemical ion-exchange flow cell.

DISCUSSION

R. DE BATIST, CEN/SCK Mol

It was mentioned in the respect to the sol-gel technique that it was possible to transform finished products into ceramics at low temperature. What is this reasonable temperature?

A.D. TURNER, UKAEA Harwell

This temperature is about 850°C.

C. SOMBRET, CEA Bagnols-sur-Cèze

The question refers to the process using inorganic exchange resins. It was the intention to transform the used ion exchanges into ceramics. What could be the final volume of the wastes which could be quite variable and even higher than normal if additives have to be added?

J.F. DOZOL, CEA Saint-Paul-lez-Durance

It is difficult to answer this question because the ceramisation aspect was not the aim of the study. However, starting from polyantimionic acids it is possible to obtain an antimony oxide which is insoluble and as a consequence the additives are not needed.

A. DONATO, ENEA Casaccia

What are the capabilities of ion exchange of some of the materials mentioned previously, such as the ammonium phospho-tungstate or, Zr phosphate, with respect to the behaviour of Sb in acid ambient. This is very important because most of the wastes arisings are acid. What about the state of the ion exchangers?

A.D. TURNER, UKAEA Harwell

The ion exchangers were powders, mixed by stirring, at the level of 1 ml of ion exchangers for 100 ml of solution. The polyantimionic acids were not very effective for elements like Cs.

A. DONATO, ENEA Casaccia

What is the behaviour of the ammonium phospho-tungstate with respect to the actinides in acid conditions?

A.D. TURNER, UKAEA Harwell

Probably it does not absorb at all.

J.F. DOZOL, CEA Saint-Paul-lez-Durance

The mentioned compound has practically no effect on other elements than Cs
in acid conditions.

H. KRAUSE, KFK Karlsruhe

Is it reasonable to apply ion exchange in alkaline pH conditions after
precipitation?

A.D. TURNER, UKAEA Harwell

The precipitation techniques are really ion exchange in a dispersed form
and then one has to separate the solids from liquids to obtain a good
decontamination. In ion exchange, the ion exchanger is in a form which
does not mix up with the liquid and so you have already done the separation.

SEBILLON, CEA

What is the capacity of the inorganic ion exchangers and of xerogel ion
exchangers and is this capacity constant with pH?

A.D. TURNER, UKAEA Harwell

Typically, inorganic exchangers have capacities between 2 and 7
milliequivalents but are dependent to a great extent on pH.

PETRA, A HOT CELL FACILITY FOR WASTE MANAGEMENT STUDIES

H. DWORSCHAK and F. GIRARDI
Commission of the European Communities
Join Research Centre - Ispra Establishment
I-21020 Ispra (Va), Italy

Summary

The PETRA plant is an experimental facility which will operate
at a pre-industrial, fully active scale, where the treatment of
LWR fuel material at high burn-up (33000 MWd/t) for the extraction
of uranium and plutonium and the successive treatment of high active
waste (HAW) and other typical waste streams generated during the
Purex type operations will be undertaken with the fundamental aim
to implement various treatment and conditioning processes on such
waste streams.
The plant will be installed in three of the hot cells available
in the ADECO complex, located in the ESSOR reactor, two of which
have not been used in "hot" conditions. A fourth cell will house
the hot analytical equipment.
The lay-out of the chemical process units allows to treat batches
containing about 6 kg U with a maximum annual throughput of 10
such batches. Maintenance and changing of the equipment can be
performed to a large extent remotely by the application of telemani-
pulative metal to metal couplings between the single units. In
view of the anticipated high variability in operation, the degree
of design safety applied is unusually high for the scale of this
facility.
The proposed research programme emphasizes, in conformity with
the general mandate of the JRC research objectives, safety aspects
related to the management of long-lived fractions of radioactive
waste, like the transuranium actinides or technetium. In-depth
studies of both the pattern of individual isotopes (e.g. Np, Tc)
over the complete process scheme and the behaviour of composite
streams under particular process conditions (e.g. evolution of
concentrated HAW composition) are scheduled. The facility is proposed
in the frame of international collaboration agreements for fully
active verifications of processes developed by other laboratories
at simulated or tracer level.

1 Introduction

Justification, individual dose limitations and optimization are key-words of the principles of radiation protection which have been recommended by the International Commission of Radiological Protection (ICRP) in its publication nr. 26. Directive nr. 80/836/EURATOM of 1980 of the Commission of European Communities has recommended those principles to E.C. countries for incorporation in their national legislation. The application of ICRP recommendations to waste management, however, is not straightforward and indeed ICRP itself has felt the need of further elaborating the recommendation for the specific application to waste management, particularly accounting for the probabilistic aspects of exposures when very long-time periods are involved like for the actinides bearing waste products.

For what concerns the first principle, justification, there is a general agreement that such a requirement refers to an entire practice (electrical power generation, in this case) rather than to each part of it. It is therefore not necessary to justify waste management in isolation.

For what concerns the second principle, individual dose limitation, protection against the hazards from radioactive waste is thought to be best achieved by their segregation from the biosphere over the time period required for their transmutation by decay. A multi-barrier disposal concept composed of up to six barriers equally divided into so-called technological (i.e. man-made) and natural (i.e. geological) barriers are considered adequate to provide such safe segregation (Fig.1).

The evaluation of the performances of such a barrier system is being done using probabilistic risk analysis methods which accounts not only for the consequences of a failure in the barrier system, but also of the probability of happening of the event and of the high degree of uncertainty of the data base. It should therefore be possible to comply with the individual dose limitation principle, at least from a scientific point of view, and research in that direction is well advanced.

The demonstration of compliance with the third principle, optimization, poses several difficulties, and relevant studies are perhaps less advanced than those dealing with the second principle. One difficulty is the definition of the "target" for optimization (should it refer to the entire practice, like justification, or to waste management in general, or to an individual project of waste disposal or even parts of it?).

Another difficulty is that although techniques such as cost-benefit analysis and multi-attribute analysis are in principle applicable to the waste management case, the data base needed as input to those decision-aiding methodologies is still rather limited, particularly at industrial scale. It is frequently difficult to judge from small scale laboratory experiments whether an alternative option which may look attractive from a radiological point of view can be implemented at an industrial scale while conserving the same cost-benefit balance.

In order to overcome some of these difficulties a pre-industrial facility, named PETRA, is presently under construction at the Joint Research Centre of the European Communities.

2 Operation, scope and mode of PETRA

The PETRA plant is an experimental facility which will operate at a pre-industrial, fully active scale, where the treatment of LWR fuel material at high burn-up (33000 MWd/t) for the extraction of uranium and plutonium and the successive treatment of high active waste (HAW) and other waste streams generated during this operation will be undertaken. The HAW, or indeed fractions of it, will be made available for conditioning in glass or ceramic matrices. The fundamental basic aim of the experimental facility is to implement treatment and conditioning processes on the various waste streams generated during Purex type operations.

Several operations on the back-end of the fuel cycle can be studied by means of PETRA. It will, in fact, be possible to:

- characterize, chemically and physically, waste streams arising from PUREX type operations, and eventually follow their evolution during interim storage;

- explore the feasibility of variations in the reprocessing scheme, aiming at a reduction of waste generation and optimization of waste categories;

- study the performance of different waste treatment and conditioning techniques and processes, inclusive of particular matrix materials for waste fractions such as actinides;

- prepare conditioned waste with the anticipated specific activity levels for characterization and behaviour testing, also in conditions relevant to interim storage and geological disposal.

It is necessary to underline at this point that the items put forward above, refer essentially to verifications of process and product performances. Taking into account the size of the operation units and the boundary conditions determined by the existing hot cells where PETRA will be installed, only proven technology will be considered. There are no plans to consider the development and hot testing of novel technological components in order to make PETRA operational. On the other side however, PETRA will afford the possibility as a test base for instrumentation, potentially applicable and advantageous in the back-end fuel cycle field.

The lay-out of the chemical process units allows the possibility to process LWR fuel material batches corresponding to about 6 kg of U. As a nominal maximum annual capacity the treatment of 10 such batches has been scheduled.

The material of each batch is processed in a sequential manner through the single unit operations which are essentially the following:

- dissolution and feed clarification (by filtration);

- HA co-decontamination cycle for U+Pu separation;

- exhaust solvent regeneration;

- HAW concentration;

- MAW concentration;

- denitration (and oxalates precipitation);

- purification of actinides (RE fraction by extraction);

- vitrification;

- off-gas treatment.

The purpose of such a sequential operation mode is to:

- limit the required man-power;

- minimize requirements especially for interim storage vessels and off-gas lay-out;

- limit the overall fissile and radioactive material inventory in the facility without curtailing its size.

3 <u>Description of the facility</u>

The process units are to be installed on a surface area of $\sim 43 \text{ m}^2$ in three existing hot cells of ADECO, (4305, 4306, 4307) a part of the ESSOR reactor complex, as shown in Fig. 2.
Four windows equiped with 8 heavy duty telemanipulators are available as working places. Another shielded area is available for the installation of the hot analytical support. Mechanical operations on the fuel pins or on waste products are performed in two other adjacent hot cells already in operation in the ADECO facility.
These cells could be available also for setting up experimental systems for conditioned waste "near field" interaction studies and leach tests.
For a more detailed characterisation of the matrix structures, use can also be made of the existing LMA cells at Ispra as well as the facilities available at the Transuranium Institute of the JRC at Karlsruhe.

The single units of the in-cell equipment are interconnected with metal
to metal couplings which can be handled remotely. Accordingly, to an
order of priority established on operational requirements (e.g. filters
and crucibles), on forecasted maintenance frequencies (e.g. dosimetric
pumps) and eventually on process scheme variations, the units have been
placed in positions accessible to the remote handling equipment, in order
to assure a high degree of flexibility whilst at the same time minimizing
exposure to operators.

An artistic view of the components lay-out in the cells is shown
in Fig. 3. It has been taken from a 1:5 model prepared for the purpose
of optimizing the lay-out. Due to this remotely operated connecting system
of the process equipment, the static and dynamic (i.e. ventilation)
containment of the hot cells play a predominant role in the safety assessment. It has been assured that in stand-by conditions, no active liquid
is in contact with the connections. Filters are drainable before being
removed. At the points where such "dirty" pieces must be handled, special
drainable driptrays are installed. All transfer systems, either air-lifts
or dosimetric pumps are installed above the static liquid head of the
vessels and are drainable back to the latter. The transfer procedure
is supported by underpressure provided from air ejectors. In stand-by
conditions there are therefore no dead-liquid volumes in the pipe work.
The composition of the in-cell equipment, schematically represented in
Fig. 4, can roughly be divided into:

- 18 vessels of cylindrical and slab shapes with volumes ranging from
  10 to 120 litres;

- 4 heatable units to be used as reactors or evaporators (with up to
  10 l/h evaporation rates);

- 2 furnaces for high-temperature processes (up to 1600°C);

- 6 mixer-settler banks for counter-current extraction operations, the
  largest units having a maximum throughput capacity of 2 kg U/24h;

- auxiliary equipment like filters, condensers, NO_x wash columns, vacuum
  air-jets, dosimetric pumps, etc.

In order to minimize the risk of fire, all units in which organic
solvent is involved are placed in one cell, from which on the contrary
any unit requiring heating is excluded. The HAW raffinate stream and
the U/Pu product stream, before being transferred to the evaporators
or reactors, are treated at the outlet from the mixer settler batteries
on columns filled with special sorbing resin, for the separation of dissolved
and entrained solvent.
This system has been kindly made available by the "Institut für Heisse
Chemic" of KFK Karlsruhe.

The general level of instrumentation will be similar to that applied in other facilities of this kind. There is one important exception, however, concerning the measurement of liquid levels. A system based on time domain reflectometry (TDR) is foreseen, based on the principle of reflection of a high frequency pulse in a coaxial probe. Advantages are that it is a tight "static" system (no dip-tubes with purge air) and that it is capable of detecting the presence of two immiscible liquid phases in a vessel, which will allow to control efficiently any unexpected presence of solvent for example.

The sequential operating mode and the possible utilisation of the same process unit for different operations in the timely dimension of one campaign coupled with the foreseen flexibility of the facility in general, imply that virtually no kind of routine operation standard will be achieved. In order to reduce start-up and shut-down procedures between the single process sequencies and to minimize at the same time the possibility of errors, a computerized process control system will be set up, including also the verification of positioning of all "manual" valves which have to be equiped with appropriate position feelers. This system will perform of course, also the acquisition of the process data. For the analytical control of the process, liquid samples can be withdrawn from all vessels and reactors into penicillin bottles (5ml) through capillaries, the bottles being then transferred to the hot analytical box by a pneumatic system. All radiometric determinations will be performed in the Radiochemistry laboratory, to which the samples are transferred, after dilution or extraction in the hot analytical box, by a second, independent pneumatic system.

4. Overview on specific research items

The proposed research programme emphasizes, in conformity with the general mandate of the JRC research objectives, safety related to the management of waste bearing long-lived components like transuranium actinides or Tc.

The PETRA facility offers the opportunity and basic hardware for pursuing these research objectives on LWR fuel material with increasing high burn-up rates to be considered typical for the future. As a consequence PETRA envelops a wide area of up-graded potential research items and topics for process analysis from waste generation, minimization and treatment optimization, through to waste product characterisation, allowing a "base line" waste management concept to be verified. Alternative concepts to be studied in PETRA ought to be referred to this "base-line" concept in terms of improvement and validation.

The scientific area of research is seen to be large and in order to maximize to its fullest potential the amount of information from each single campaign, the screening of the different possible research items will be conducted in such a way so as to establish an optimal number of verifications with a maximum output of information.

In order to illustrate the areas potentially suitable for research,

a simplified flow diagram has been prepared (see Fig. 5) in which only
those units are shown where any kind of transformation occur such as
evaporation, filtration, extraction, solidification, etc. Also indicated
in Fig. 5 are the input-streams and some basic data of the typical output
products as far as is defined by us at present.

Overall the areas of most interest are:

a) characterization of the typical Purex waste types
    - cladding (FP activity, activation products, etc.)
    - dissolver residues (FP activity, composition, etc.)
    - HAW concentration behaviour (precipitation of solids)
    - aqueous solvent wash liquors (FP, actinides, etc.).

b) characterization of conditioned wastes, (physical, chemical and radioactive properties)
    - HAW in glass matrix
    - Actinides in special matrices
    - MAW in glass
    - MAW merging with HAW in glass matrix
    - Dissolver residues in glass (or other) matrix
    - Cladding material conditioning

C) process patterns of actinides (Np) and Tc in the various streams namely
    - feed stream
    - U/Pu product stream
    - aqueous solvent wash waste stream
    - dissolver residues
    - HAW concentrate stream
    - vitrification (or other high temperature process) off-gas stream

d) evolution of phenomena like
    - solvent damage
    - composition of gases during
        . concentrated HAW storage
        . denitration process
        . conditioning processes (at high temperature)
        . dissolution
        . off-gas treatment (HNO_3, FP's in off-gas scrubber)
        . monitoring of gaseous effluent.

The points where such verifications could be performed are indicated
in Fig. 5 by numbers referring to the legend given in Table I.
It is recognised however that the research areas pinpointed above might
be neither complete nor exhaustive, for the possibilities offered by
this facility. Furthermore it should be recalled that PETRA is a flexible
facility allowing for components replacement and even addition in order
to perform also in the future, other unit operations if desired. Just
for the record the application of inorganic ion exchanger material for

waste streams treatment and radioactive products immobilisation shall
be included. In principle hot testing of on-line analytical process control
instrumentation is also applicable.

The PETRA plant is expected to become operational at a fully active
scale during 1987. The facility itself will be open to international
cooperation so that management schemes developed in the laboratories
of the European Community and other interested countries may be jointly
tested at a fully active pre-industrial scale without major investments.
Such collaboration is already a reality. For example a small group from
ENEA, the Italian nuclear organisation participated here at Ispra in
the design phase of the project. KFA Julich also have collaborated, with
their drum dryer concept being adopted for sludge drying prior to vitrification.

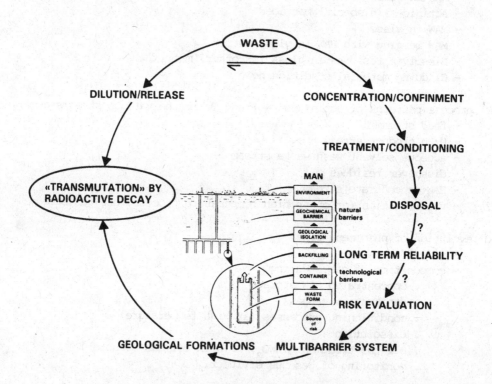

Fig. 1   Illustration of the approaches in nuclear
waste management highlighting the multi-
barrier disposal concept.

A.D.E.C.O.

UPPER LEVEL

FLOOR LEVEL

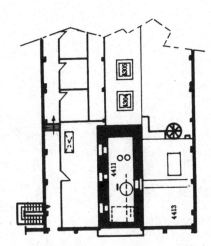

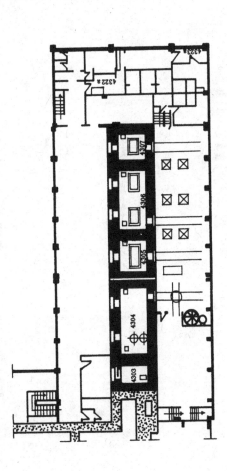

Fig. 2 The ADECO cells of the ESSOR reactor complex where PETRA will be installed (cells, 4305, 4306, 4307)

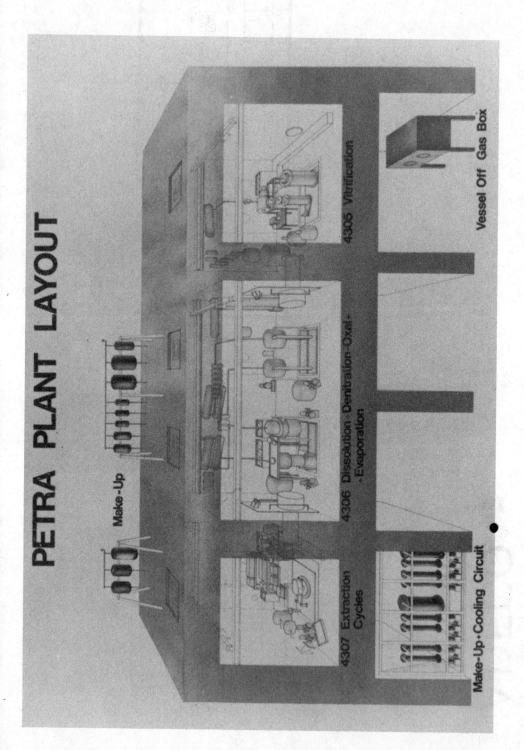

Fig. 3 An artistic view of the lay-out of the components in the ADECO cells.

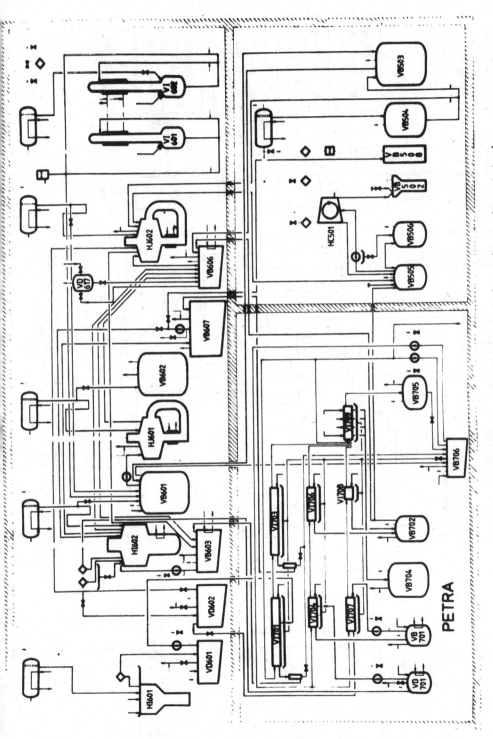

Fig. 4 A schematic representation of the main equipment to be installed in the three cells of ADECO.

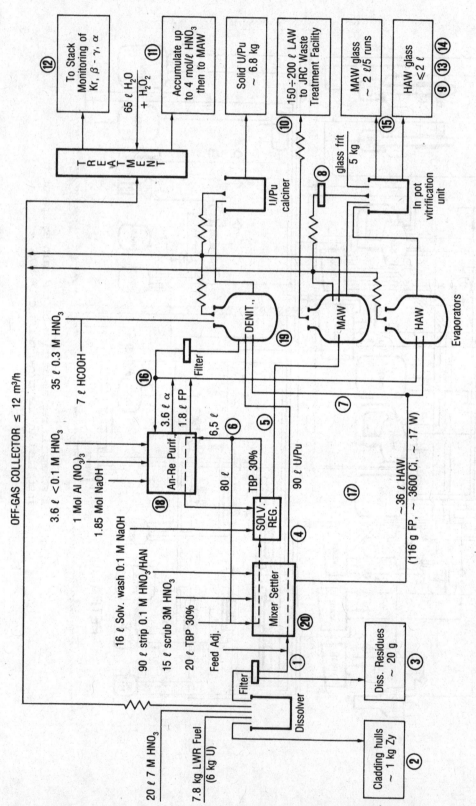

Fig. 5 A simplified flow diagram showing the main unit operations of PETRA. (To be read in conjuction with Table I)

Table I: TOPICAL LIST OF VERIFICATIONS IN PETRA*

Nr.	Stream	Verification	Special Items
1	Feed	characterisation	burn-up, Np, Am, Tc
2	Cladding	fuel residue	activation products
3	Dissolver Fines	composition/conditioning	
4	U/Pu prod.	DF from FP (Ru)	Np, Tc fraction
5	Solvent Wash Liquor	composition	Np, Ru, An, solvent degradion products
6	Solvent	characterisation	evolution of perman. radiation damage
7	HAW	composition/concentration	Np, Tc fraction/rad. gas, acidity, ppt.
8	Vitrif. Off-gas	volatile FP	
9	HAW-glass	various tests for charact.	Np, Tc leach behav.
10	LA liqu. waste	activity	evap. DF of Ru, Sb, influence of H^+
11	Off-gas Scrub	acidity, activity	NO_x absorp. effic.
12	Off-gas	contin. monit.	Kr, N_2O, I-129
13	Like 9		alpha-free HAW frac.
14	Like 9		alpha fraction of HAW
15	HAW/MAW glass	like 9	
16	OXAL Process	separation factors of alpha	off-gas compos.
17	1,4,5,7,8,9, etc.	distribution pattern	Np, Tc, Rn, etc.
18	HAW	heat generating FP sepn.	Cs,Sr
19	HAW	valuable material sepn.	noble metales
20	HAW/MAW	exhaustive actinide extr. extraction	"bidentates", new extractants

* To be read in conjunction with Fig. 5.

DISCUSSION

C. SOMBRET, CEA Bagnols-sur-Cèze

One of the cells of PETRA has been designed with a neutron shielding. What
neutron flux can be dealt with if examination of neutron generating
material is needed?

H. DWORSCHAK, JRC Ispra

If an examination is needed it could be done in more specialized cells
located in other facilities of the JRC, like the LMA or at the JRC Karlsruhe.

CHARACTERIZATION OF SPENT FUEL HULLS
AND DISSOLUTION RESIDUES

J.P. Gué, CEA, FONTENAY AUX ROSES, France
J.R. Findlay, UKAEA, HARWELL, United Kingdom
H. Andriessen, KFK, KARLSRUHE, Federal Republic of Germany

Summary

The main results obtained within the framework of CEC programmes, by KFK, UKAEA and CEA, are reviewed concerning the characterization of dissolution wastes. The contents were determined of the main radioactive emitters contained in the hulls originating in a whole fuel assembly sampled at the La Hague plant, or from Dounreay PFR fuels. Radiochemical characterizations were carried out by different methods including neutron emission measurement, alpha and beta-gamma spectrometry, and mass spectrometry. Decontamination of the hulls by using rinsings and supplementary treatment were also dealt with. The ignition and explosion risks associated with the zircaloy fines formed during the shearing of LWR fuels were examined, and the ignition properties of irradiated and unirradiated zircaloy powders were determined and compared. The physical properties and compositions of the dissolution residues of PFR fuels were defined, in order to conduct tests on the immobilization of these wastes in cement.

1   INTRODUCTION

Structural wastes of fuel elements and dissolution residues make up most of the high-level wastes that are recovered at the head end of reprocessing plants, and must be immobilized for subsequent disposal. Different immobilization methods are currently being investigated worldwide. Before finalizing these methods, however, a major effort is needed to characterize the wastes, both for their alpha/beta/gamma emitter contamination, as well as their stability. These data are actually indispensable for a definition of:
.   the safety of their handling and transfer,
.   monitoring of the activity levels,
.   rinsings and supplementary treatment,
.   the ideal packaging.
The characterization of these wastes is being investigated in European countries, and this document reviews the work carried out on this topic by UKAEA, KFK and CEA, under CEC programmes.

2   CHARACTERIZATION OF HULLS FROM IRRADIATED ASSEMBLIES

2.1   LWR FUEL HULLS

2.1.1   Radiochemical characterization and testing of hulls
              (Coquenstock programme developed at the CEA)

Operation Coquenstock consists of characterizing large volumes of LWR
hulls, taken from whole fuel assemblies reprocessed in the La Hague plant,
and hence representative of a waste obtained on the industrial scale. This
programme deals with the following points:
. examination of hulls and sampling by quartering,
. neutron measurements to estimate the residual amount of fuel in the
   hulls,
. gamma spectrometry measurements on containers measuring several
   litres,
. after dissolution of some 15 samples, determination of the uranium
   and transuranics contents, alpha/beta/gamma activities (activities of
   fission products, strontium 90 and activation products),
. measurement of tritium retention.
This exceptional operation was difficult to implement and was considerably
delayed.   So far characterizations have only begun on the first assembly
(Coquenstock 1).

All the hulls (about 100 litres) from an assembly taken from the West
German Obrigheim PWR reactor irradiated to 30,136 MWd·t^{-1} and discharged in
June 1979, were recovered in the La Hague plant.   After shearing of the
assembly in clusters, followed by dissolution and rinsing, the hulls were
transferred to Saclay and then partly to Fontenay aux Roses.   The photo-
graph in Figure 1 shows that these industrial hulls bear no resemblance to
those obtained in the laboratory or in a pilot plant (pin by pin shearing).
These hulls displayed a bulk density of 1.08 g·cm^{-3}, and we noted that their
appearance was quite different from that of hulls normally used in simulation
experiments.   The shape and length of each of the sections varied
considerably, and many of them were shredded, broken up and even crushed.

The main results of the radiochemical characterizations that were
performed are summarized and compared in Table I with the previous results
published by the CEC <1>.   For tritium, the different results show good
agreement (mean activity about 1.15 Ci·kg^{-1}) and they confirm the measurements taken at KFK on the same fuels <3>.   The fraction of tritium fixed
in the zircaloy represents 65% of the tritium formed in the reactor.   A
small part of this element (≤ 0.1%) was found in the form of tritiated
methane.   Apart from tritium, the zircaloy occludes small amounts of
krypton 85, up to 55 mCi·kg^{-1} on the average, or 0.1% of the total quantity
formed.   Spectrometry measurements by gamma scanning on large masses of
hulls (total 20 kg) indicated that the beta/gamma activity essentially
originated in the activation products contained in the structural parts of
the assembly, or in the pins (Co 60, Mn 54, Sb 125, grids, springs, spacers
and clads), together with fission products such as Ru 106/Rh 106, Cs 134,
Cs 137 and Sr 90, insolubilized or deposited during dissolution.

After five years of cooling, Co 60 predominates, but the distribution
of its activity in the containers, related to the disparity of the
structural materials, remains quite heterogeneous as shown in Figure 2.
Cesium, however, is distributed more or less uniformly.   In terms of
protection and packaging, it would therefore be irrelevant to define a mean
activity for cobalt 60.   The mean activity of the fission products
associated with the hulls represented about 0.2% of the quantity formed in

FIG. 1.– GENERAL APPEARANCE OF « OBRIGHEIM » HULLS RECOVERED AT H.A.O. PLANT IN LA HAGUE.

Table I

Characterization of Obrigheim PWR hulls

Comparison of preliminary results with those
obtained previously in the CEC programme

	Coquenstock 1		previous CEC work on PWR hulls [1]
CONCENTRATIONS (in mg/kg hulls)			
• uranium	687 to 1470 (a)		-
• neptunium	0.16 to 0.36 (a)		-
PLUTONIUM			
• by neutron emission measurement of curium 244	total measurement: • 26.2 (b) 10.6 to 11.6 (b)		- -
• by mass spectrometry	12.2 to 15.7 (a)		-
• by neutron activation	-		2.1 to 9.1 (b)
ALPHA ACTIVITIES (in mCi/kg hulls after 5 years of cooling)			
• total activity	6 to 8.7 (a)		2.9 to 3.75 (a)
• Am 241 activity	1 to 1.9		-
BETA-GAMMA ACTIVITIES (at discharge, in $Ci \cdot kg^{-1}$)	(a)	(b)	(a)
• total	29.7 to 35.7	-	20.4 to 46
• Ru 106 + Rh 106	12.5 to 15.3	51.3	15.6 to 30
• Ce 144 + Pr 144	11.7 to 14.8	-	3.1 to 13.4
• Cs 134	0.8 to 1	0.68	0.17 to 1.36
• Cs 137	0.57 to 0.7	0.47	0.27 to 0.78
• Co 60	0.056 to 0.086	8.83	0.02 to 0.406
• Mn 54	0.21 to 0.38	4	0.077 to 0.234
• Sb 125	2.8 to 3.4	1.73	0.83 to 3.16
• Sr 90	0.3 to 0.5	-	0.175 to 0.381
• tritium	0.93 to 1.29	-	0.486 to 1.485
• krypton 85	0.049 to 0.062	-	

(a) measurements after dissolution of all-zircaloy hulls
(b) direct measurements on hulls

CONTAINER № 9

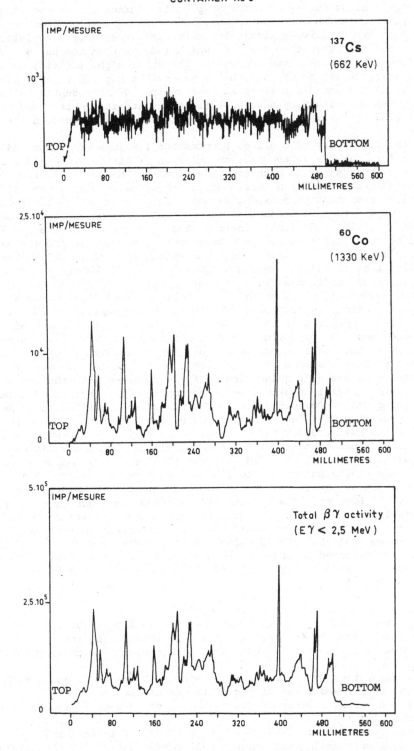

FIG. 2._ EXEMPLE OF AXIAL DISTRIBUTION OF βγ EMITTERS
IN HULLS CONTAINER.

the reactor (values confirmed after sample dissolution).   It was higher
for the pair Ru 106/Rh 106 (0.4 to 2%), but it is well-known that these
noble metals are relatively insoluble (main constituents of dissolution
residues) and that they form colloids which redeposit on the hulls during
dissolution of the irradiated oxide.   The total alpha activity measured
after dissolution of the zircaloy hulls was rather high (6 to 8.7 mCi·kg^{-1}).
Americium 241 accounted for 17 to 22% and curium 244 represented 14 to 20%.
Neptunium was determined by neutron activation, and its content was between
0.16 and 0.36 mg·kg^{-1}, or 0.02 to 0.04% of the quantity formed in the
reactor.

After four years of cooling, spontaneous fissions of curium 244
accounted for over 92% of the neutron emission of the fuel.   Assuming
that this element exhibits behaviour identical to that of plutonium, both
during irradiation and during dissolution (these two elements can diffuse
in the clad, migrate in the oxide, be adsorbed in solution on zirconia in
different ways), the measurement of its concentration in the hulls should,
in principle, help to determine their plutonium content.   The results
obtained by this neutron emission measurement technique, on 100 litres of
hulls, yielded an initial total plutonium estimate of 26 mg·kg^{-1} of hulls.
Out of 12 litres of hulls, sampled by quartering from the entire batch,
the measured values (12 measurements) were lower, between 10 and 12 mg·kg^{-1}.
Comparable values (12 to 16 mg·kg^{-1}, see Table I) were found by mass
spectrometry, after dissolution of the hulls of the same sample.

Hence it appears that the cross-check between these two methods is
fairly good, and that plutonium and curium display closely comparable
behaviour.   This measurement technique, easily adaptable to the
industrial scale, thus appears to be highly promising for the measurement
and detection of residual quantities of fuels combined with the hulls.
However, these initial results need to be confirmed on the second assembly
planned in operation Coquenstock.   The plutonium concentrations, as well
as the alpha activities detected, were much higher than those hitherto
defined in PWR hulls, under the CEC programmes [1,2].   However, a large
share of this alpha contamination can be eliminated by simple additional
rinsing with cold 3 N nitric acid.   The final residual contamination
threshold for plutonium appears to lie between 1 and 3 mg·kg^{-1}, and
corresponds approximately to the minimum contamination level reported so
far [1,2].   However, this rinsing only succeeds in reducing the beta/
gamma activity by about 10%.   To limit the alpha activity of these
wastes, it is clear that one should proceed with effective rinsings in the
plants, and we feel that the technique of dissolution and continuous
rinsings, as performed in UP3 [4] achieves much more effective countercurrent
rinsing than those currently performed by soaking.   Intensive
decontamination of the hulls by means of specific reagents is certainly
feasible.   However, it is limited (residual alpha activity produced by
U and Th contained as impurities in the zircaloy) and it often raises a
problem of recycling of the effluents produced in the reprocessing plant.

In conclusion, it is difficult to go further in the interpretation of
the still partial results that have been obtained.   We can nevertheless
point out that:
- the quantities of plutonium associated with these Obrigheim hulls lie
  between 0.04 and 0.08% of the plutonium of the fuel,
- neutron emission measurements appear to offer a valid method to test
  the hulls,
- alpha contamination is rather high but most of it is labile.
This observation recalls that, in the hull contamination process (diffusion
during irradiation, fission recoils, impregnation of corrosion layers), the

reprocessing conditions (shearing, dissolution, rinsings) certainly remain the most important parameters. .

2.1.2   Pyrophoric behaviour of irradiated zircaloy chips and fines
             (KFK investigation)

To promote the safe handling of Zry-fines arising in the shearing of fuel assemblies, chopping of fuel pins or compaction of leached hulls, an experimental programme was conducted to assess the ignition and explosion hazards.   To ensure that the results will meet the requirements of the licensing authorities, expert advice was obtained from Bundesanstalt für Materialforschung in Berlin (BAM).   This Institute, the principal establishment for testing dusts for explosion and fire properties in Germany, investigated non-radioactive Zry-fines under their standard procedures.   (A detailed account of the test methods used at BAM has been published in VDI Report No.304, 1978, pp.29-38). Based on these results, the methods for active tests at the hot cells of KWU-Karlstein were defined. Tests were performed on Zry-fines artificially produced from hulls of irradiated LWR fuel.   To evaluate the effect of irradiation on the pyrophoric properties, irradiated and unirradiated Zry-fines, generated by the same tool, were investigated and compared.   The investigations have not yet been completed.   The main findings obtained are described. All results refer to tests carried out on a < 100 μ sieve fraction.   · Information about practical applications of the results is given in VDI Guidelines 2263 and 3673.

Ignition properties of dust deposits

Zry-fines exposed to various ignition sources, such as flashlight or sparks from auermetall, turned out to be very easy to ignite, and subsequent burning was fast and self-sustaining.   It was also observed that sparks caused by rubbing or impact of Zry-hulls ignited Zry-dust deposits.

The minimum ignition temperatures of dust layers were determined by two methods.   In the first method, Zry-fines were deposited on an electrically-heated plate.   The lowest temperature at which ignition still occurred was measured and found to be 235 °C for the active fines and 295 °C for the non-radioactive Zry-fines.

In the second method (Gliwitzky), samples of different volumes were heated in a hot enclosure and the external surfaces of the dust deposit exposed to air at elevated temperatures.   Again a substantially lower ignition temperature was observed for active fines.   The results are plotted in Figure 3.   The linear dependence of the logarithmic volume-to-surface ratio on reciprocal ignition temperature allows a simple extrapolation to be made to the volumes in a practical case.   For a pile of 10 ℓ of irradiated Zry-fines, the minimum ignition temperature is 102 °C. Fines characterization of active and non-radioactive Zry-fines of the same sieve fraction, and prepared by the same tool, showed a finer grain and a bulk density higher by a factor of 3 of the active test material (2.4 g/cm³).   These discrepancies may well have contributed to the lower ignition temperatures of the active fines.   Supplementary tests will be performed to obtain conclusive results.

Ignition properties of raised fines

The minimum ignition temperature of a dust suspension was measured in the Godbert-Greenwald furnace.   The Zry-fines were blown by air in a

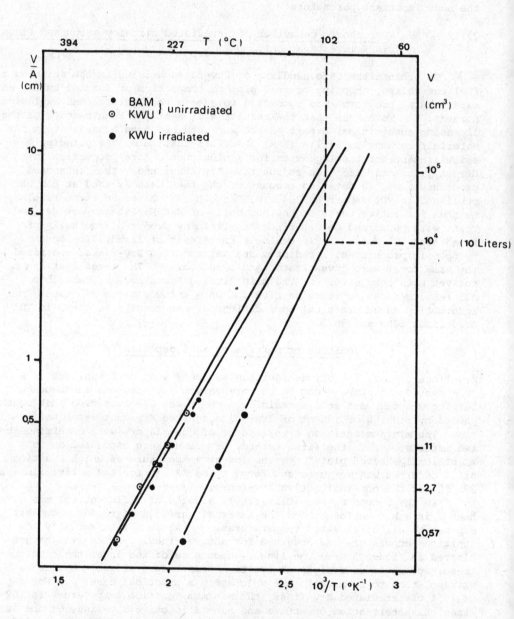

Figure 3 Dependence of the ignition temperature of zircaloy fines
         of a < 100 μm sieve fraction on the volume/surface ratio
         (V/A)

         (BAM, Berlin National Testing Laboratory)

suspension towards the preheated furnace.   The minimum ignition
temperature for the active fines was measured and found to be 365 °C,
which was only negligibly less than for the non-radioactive Zry-fines.

Dust explosibility

Explosion properties were determined in the 20 ℓ spherical explosion
apparatus.   The results are applicable in designing explosion protection
on an industrial scale.   The maximum pressure of unirradiated Zry-fines
reached a medium level (7.8 bar absolute) as did the maximum explosion rate
of rise (K_{st} value 200 bar m/s, dust explosion Class II).   The maximum
permissible oxygen concentration was also measured in the 20 ℓ sphere with
argon as the inert gas.   A dust explosion could still be triggered in an
atmosphere containing 1% (by volume) of oxygen.   Only in pure argon was
this no longer possible.   The explosion data of irradiated Zry-fines
still need to be measured.   These tests will presumably be completed by
the end of 1985.

2.2   FAST BREEDER REACTOR HULLS
         (characterization carried out by UKAEA)

A substantial programme has been undertaken at the Harwell Laboratory
on the examination of stainless steel hulls arising from irradiations in the
Prototype Fast Reactor (PFR) at Dounreay.
The fuel element hulls were obtained from the reprocessing operations
at Dounreay and were transported to Harwell for characterization and studies
on possible decontamination procedures.   Some preliminary work on methods
of immobilization has been undertaken.   The behaviour of all active
constituents has been examined.   Attention has been given to the behaviou:
of long-lived radionuclides because of their significance to disposal.
The fuel element hulls used came from four different irradiations in PFR,
and had achieved burnups between 3.9 and 7.3% fissions of the initial heavy
metal atoms in the fuelled section.   A further sample from an unirradiate(
fuel pin which had been sheared and put through the active dissolver system
was included, representing the zero burnup situation.   The hulls differed
in appearance, being either black or shiny.   The black hulls were
identified as coming from the central fuelled section of the pin where
corrosion and related deposition effects would be at a maximum.   The
shiny hulls originated from the axial breeder positions, where conditions
are less severe.   Examples of both types are shown in Figure 4.   The
hulls as received were 25 to 75 mm in length, and were cut into 1 mm rings
for ease of handling during treatment and analysis.   This technique
allowed determination of the activity profiles of individual fuel element
hulls.   Comparative work on complete hulls has confirmed that the
sectioning process does not invalidate the results obtained.   A wide
range of measurements were made within the programme.   The major gamma-
emitting species were measured directly using gamma spectrometry.   The
main fission product nuclides determined were Ru 106, Cs 137 and Sb 125.
The activation products seen were Mn 54 and Co 60.   Beta-emitting
nuclides, Sr 90 and Tc 99, were measured by chemical separation and beta
counting.   Carbon 14 was determined by complete dissolution of the hulls,
oxidation to CO_2, conversion to carbonate, and counting the final solution
by liquid scintillation methods.   The total fissile content of the hulls
was determined by using neutron interrogation methods.   A total alpha
count was made on solutions obtained from the complete dissolution of the
hulls.   This was followed by chemical separation and alpha spectrometry
to determine the individual actinides (Pu, Am and Cm).

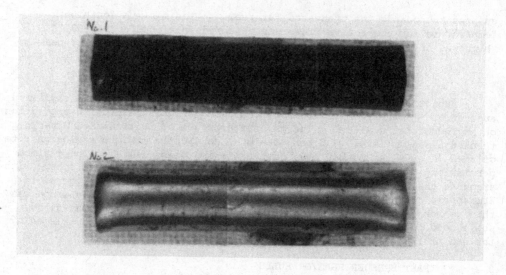

Fig. 4 - BLACK AND SHINY HULLS FROM PROTOTYPE FAST REACTOR FUEL

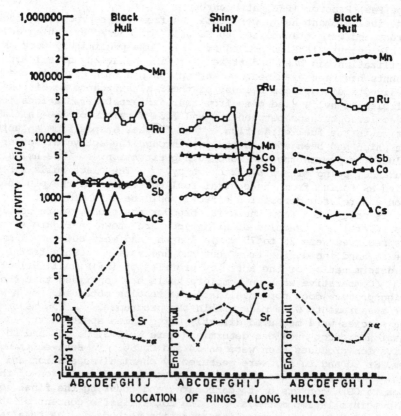

Fig. 5 - PROFILES OF ACTIVITIES ACROSS 7.3 % BURN UP HULLS

Results

The levels of fission product activity found on the fuel element
hulls were remarkably similar with the dominant activity being Ru 106
(10^{-2} to 10^{-1} $Ci \cdot g^{-1}$ of hull).   Next in importance was Sb 125
($\cong 10^{-3}$ $Ci \cdot g^{-1}$).   The results for Cs 137 were variable (10^{-5} to
10^{-3} $Ci \cdot g^{-1}$).   Sr 90 was present at 10^{-5} to 10^{-4} $Ci \cdot g^{-1}$.   Our first
measurements for Tc 99 suggest a level of about 10^{-6} $Ci \cdot g^{-1}$.   Typical
results for both black and shiny hulls from an irradiation at 7.3% burnup
are shown in Figure 5.   The levels of fission product activity are
generally similar.   The levels of cesium are lower on the shiny hulls
from the breeder section, possibly reflecting the lower mobility of cesium
in this region.   Activity measurements along a hull show that generally
the activity levels are the same within the 1 mm resolution provided by
the sectioning technique.   Some variation is however evident for Ru 106,
which may be associated with the presence or absence of alloy particles.
For the activation products, the levels of Co 60 were similar in the range
10^{-3} to 10^{-4} $Ci \cdot g^{-1}$.   The levels of Mn 54 were however substantially
higher in the black hulls, and it is this that identifies them as coming
from the core region of the reactor, since Mn 54 is formed by a fast
neutron capture reaction, and its formation is much suppressed in the
breeder sections of the fuel.   Levels of C 14 up to 5 $\mu Ci \cdot g^{-1}$ were found.
Measurements of the total fissile content of the hulls showed that,
throughout, the carry-over of fuel was < 0.1%.   The alpha activity level
on the hulls was in the range 10^{-4} to 10^{-6} $Ci \cdot g^{-1}$. It is notable that the
levels of both actinide and fission product activity on the hulls from
fuel that was not irradiated but passed through two solutions obtained
from the dissolution of irradiated fuel, were similar to those on the
hulls from irradiated fuel. This indicates that deposition of activity
on the hulls is important.

Treatment of the hulls

Experiments on decontamination methods for these hulls were undertaken successfully using several aqueous reagents. The favoured decontamination method is to reflux the hulls three times in succession in
9 M nitric acid. The use of nitric acid should allow the decontaminating
liquor to be fed to the fuel dissolver, and would thus be a preferred
process for reprocessing plants.

Methods of encapsulating undecontaminated hulls in cement were studied
as an alternative to decontamination.   Fully-active specimens immobilized
in sand/cement and cement/blast furnace slag matrices were prepared.   Leach
testing experiments are in progress using the ISO 6961 test method, and
cesium 137 and Sr 90 have been detected in the leachates.   Actinide
levels in the leach solutions are extremely low at the limits of detection
for the methods used.

3   CHARACTERIZATION OF DISSOLVER RESIDUES OF FAST BREEDER REACTOR FUELS
        (UKAEA)

A work programme has been undertaken in the Harwell Laboratories to
characterize the insoluble residues remaining after the dissolution of fast
breeder reactor fuel in nitric acid.   The work established the nature of
this particular waste arising, and provided information on the handling of

96

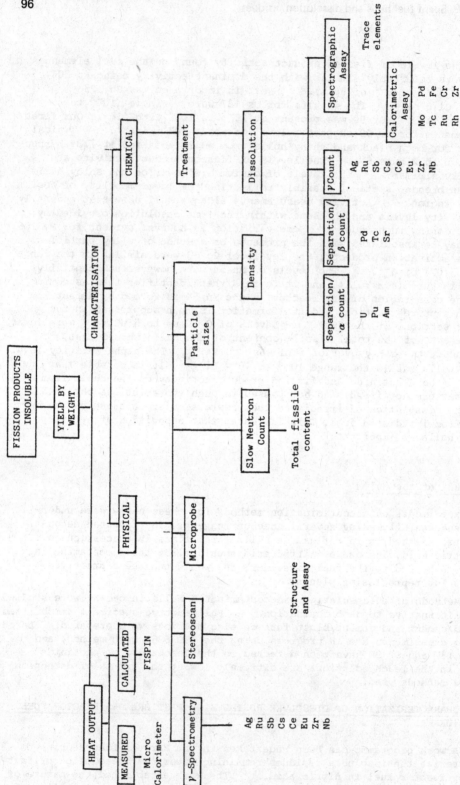

FIG .6. CHARACTERISATION OF INSOLUBLE DISSOLVEUR RESIDUES
FROM FAST REACTOR REPROCESSING.

these materials.   Six samples from differing irradiations and
dissolutions were examined, using both chemical and physical methods of
characterization.   These materials were immobilized in cement media and
subjected to leach testing to establish possible routes for their disposal.
The material examined originated from irradiations in PFR from co-
precipitated fuel at both inner and outer zone fuel compositions.
Irradiations ranged between 2.7 and 7.8% burnup of heavy metal atoms in the
fuelled section of the pin.   The residues were separated from the
dissolver liquor by centrifugation.   Samples for analysis were taken
from the resulting centrifuge bowls.

3.1   CHARACTERIZATION METHODS

The residues consist of several different materials and a range of
methods is needed to obtain full information.   The principal constituents
of the residues are the alloys of noble metal fission products, which are
formed as particles or inclusions in the fuel during irradiation and which
resist nitric acid treatment.   The main components are Ru, Rh, Pd, Mo
and Tc.   The dominant radioactive component is Ru 106, which has a strong
beta emission and, through its short-lived daughter (Rh 106), provides the
major source of heat emission from the residues.

To establish the proportion of the fission product yield appearing in
the residue for each of the constituent elements, the full fuel and fission
product inventory for the fuel pins, from which the residues were obtained,
was calculated using the FISPIN V computer program.   Decay corrections
were applied appropriate to the time of analysis, and heat outputs from
radioactive decay were calculated for comparison with experimental values.

The characterization programme and the methods used are summarized in
Figure 6.   A microcalorimeter was used for the heat output measurements,
which was capable of detecting emissions of 0.01 W, or lower in some
circumstances.   A complete dissolution procedure was used for chemical
analysis.   Samples were chlorinated in a sealed system, and then brought
into solution by acid treatment.   Physical characterization was
undertaken by scanning electron microscopy on mounted samples.   Supporting
chemical information was obtained using electron probe microanalysis (EPMA)
on polished sections of the mounted residues.   Neutron interrogation
techniques were used to determine fissile contents and were supported by
chemical analyses and alpha spectrometry.

The matrices used for experiments on immobilization methods were
either sand and cement, or cement with blast furnace slag (BFS), or
pulverized fuel ash (PFA) as an additive.   Specimens were prepared using
a shear mixer followed by vibro-packing into a mould.   The ISO 6961 test
was used for the leaching experiments.   Some control leaching experiments
were undertaken using un-immobilized material.

3.2   RESULTS

The results of the chemical analyses of the residues are shown in
Table 2, where the compositions of the metal alloys are given and are
compared with the yields expected theoretically from the computer
calculation.   Agreement is generally good, indicating that elements are
incorporated in the alloys approximately in the ratios of their yield,
although the experimental results for Pd are consistently low.   More
detailed examination of these residues by EPMA suggest there are variations
in composition with radial position which may be systematic.   The overall
yield of noble metal fission product alloy found experimentally from these

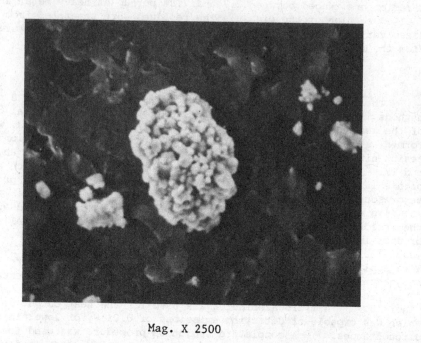

Mag. X 2500

|← 4 μm →|

FIG . 7. NOBLE METAL INCLUSIONS FROM REPROCESSED FAST REACTOR FUEL.

dissolutions is thought to account for approximately 50 % of the theoretical yield. The amounts of associated fissile material are small, although, as seen from Table II,the residues are rich in plutonium compared with the composition of the original fuel.

Measurements on heat content gave values between 0.08 and 0.13 $W.g^{-1}$ which are consistent with the yields of ruthenium after appropriate decay corrections of 2 to 4.5 years. The physical form of the residues are shown typically in Figure 7. The figure shows an agglomerated lump which, on examination, is seen to be composed of fine particles less than 2 μm in diameter. No pieces of undissolved fuel or pieces of cladding were observed. Density measurements have given values in the region 5.6 to 7.3 $g.cm^{-3}$ and are comparable with values reported in the USA.

Experiments on the leaching of residues immobilized in cement matrices showed that Cs 137 and Sr 90 were readily detected in the leach solutions. These materials may arise from residual dissolver solution retained on the samples. The presence of actinides or other long-lived activities could not be detected in the leach solutions. By contrast, most of the major activities, including the actinides, were seen in the experiments where the un-immobilized residues were subjected to leach testing.

4 DISCUSSION RELATED TO CEC PROGRAMMES

The investigations discussed above have served to identify the main characteristics of the hulls and dissolution residues, as well as the points that still require clarification to define their disposal method.

In general, the alpha contamination of the hulls is much higher than the values required for shallow ground disposal. The studies nevertheless hint at the possibility of considerably reducing this contamination, by rinsing and by specific treatment to be developed, simultaneously with sensitive testing methods. As for problems related to the pyrophoric properties of zircaloy, tests will have to be intensified in order to determine whether the ignition of zircaloy fines can propagate to large volumes of hulls. It will also be necessary to examine whether a synergistic effect exists in mixtures of zircaloy fines with insoluble fission products.

Efforts to characterize these high-level wastes should be continued, especially in line with further developments in the fuels of the two reactors (high burnups, MOX fuels for LWR, developments in claddings and oxide fabrication techniques in FBR). This is especially true of dissolution residues, whose characteristics are known to be highly sensitive to fuel fabrication and irradiation parameters. Preliminary encouraging results have been obtained during cementation tests on wastes from FBR fuels, but this complex area of immobilization and disposal is still largely open to investigation.

Acknowledgements

The work in this paper was commissioned by CEA, KfK Karlsruhe and the Department of the Environment. The results of the UK work will be used in the formulation of Government policy but, at this stage, they do not necessarily represent such policy.

Table II

Comparison of measured and computed results
PFR insolubles

sample	composition of noble metal alloy (%)					source	composition of insoluble fuel (%)	
	Mo	Tc	Ru	Rh	Pd		U	Pu
PFR 1	32	6	37	11	14	A	53	47
	33	9	30	9	19	C		
PFR 2	27	8	41	12	12	A	63	37
	32	9	32	9	18	C		
PFR 5	33	8	30	12	17	A	NA	NA
	31	8	29	10	22	C		
PFR 6	29	9	31	16	15	A	NA	NA
	31	8	29	10	22	C		

A   chemical assay
C   computed assay (FISPIN V)   NA   not available

REFERENCES

<1>   I.L. Jenkins et al, EUR, The characterization of activities associated with irradiated fuel element cladding, EUR 7671 EN (1982)

<2>   J.P. Gué, M. Isaac and P. Miquel, Measurement of the activities bound to LWR fuel hulls, Meeting on Packaging and Storage of Spent Fuel Element Hulls, EUR 9250, pp.271-290

<3>   A. Bleier, R. Kroebel et al, Tritium inventories and behaviour in zircaloy cladding of spent LWR fuel rods, ANS International Meeting on Fuel Reprocessing and Wastes, 26/29 August 1984, Jackson, Wyoming

<4>   P. Auchapt, L. Patarin and M. Tarnero, Development of a continuous dissolution process for the new reprocessing plant at La Hague, ANS International Meeting on Fuel Reprocessing and Wastes, 26/29 August 1984, Jackson, Wyoming.

DISCUSSION

F. MANNONE, JRC Ispra

It was mentioned that the Pu-contamination of the hulls decreases after washing with HNO_3. Is this due to chemical dissolution or to physical wash out?

J.P. GUE, CEA Fontenay-aux-Roses

It is due to a physical wash out.

R. KROEBEL, KFK Karlsruhe

As the oxyde layer is somehow an ion exchanger the Pu is on the outer surface as well as in the depth of the layer, thus by adding fresh acid it is possible to wash a certain part out nevertheless elements like Cs, implanted in depth by backscattering, remain.

J.P. GUE, CEA Fontenay-aux-Roses

In fact there are different levels of contamination and the first one is probably due to a defect of washing, but a second level is really fixed in the inner layers. Most probably phenomena of adsorption regulate the level of fixation, or the behaviour, of the Pu.

R. ATABEK, CEA Fontenay-aux-Roses

For the encapsulation of hulls in a cement matrix it is unimportant to have a good knowledge of the conditions of the external layers after successive washings. Have microscopic tests been performed in this sense?

J.P. GUE, CEA Fontenay-aux-Roses

Not up to now because the considered program was essentially oriented to the identification of the elements which were present in these hulls. However tests are foreseen in the near future because these aspects are becoming more and more important.

METHODS OF CONDITIONING WASTE FUEL DECLADDING HULLS AND DISSOLVER RESIDUES

P. DE REGGE, CEN/SCK, Mol, Belgium
A. LOIDA, KfK
T. SCHMIDT-HANSBERG, NUKEM
C. SOMBRET, CEA, Centre d'Études Nucléaires de VALRHO

Summary

Several methods for conditioning spent fuel decladding hulls or dissolver residues have been considered in various countries of the European Community. Five of these methods used embedding technique with or without prior compaction : they are based on incorporation in metallic alloys, glass, ceramics, cements and metals or graphite compounds. A sixth one consists in melting the decladding materials. The corresponding research programmes have been pursued to varying states of progress with regard to demonstrating their feasibility on an industrial scale and the use of genuine wastes in bench scale experiments.
The properties of the conditioned wastes have been investigated. Special attention has been paid to the corrosion resistance to various aqueous media as tap water, brine or clayey water. Although no categorical conclusion can be drawn from the initial results, the available findings provide a basis for assessing the different processes.

1. INTRODUCTION

The waste formed by pieces of irradiated fuel hulls contains radioactive materials of various kinds, in particular fissile materials, other alpha-emitters, fission products and, in the case of hulls from light-water reactors, a substantial amount of tritium. The dissolver residues of irradiated fuels are often considered as waste of the same type, owing to their metallic properties, despite their different sources and specific radioactivity.

Because of the radioactive and, indeed, pyrophoric hazards associated with the storage of such waste, studies have been conducted in several Community countries on methods of conditioning which permit storage under satisfactory conditions of safety.

2. CONDITIONING BY ENCAPSULATION

The purpose of these techniques is to isolate the waste from the environment by means of an inert material which prevents or delays any contact with natural agents resulting from incidents during storage. Various types of encapsulation materials are considered, all of mineral origin. In certain processes the waste is compacted prior to encapsulation.

2.1 Compaction and encapsulation in a metal with a low melting point

This process has for several years been the subject of studies at the SCK, CEN, Mol and is applied to hull waste. The waste is first compacted into cakes (briquettes) to permit easy handling with no risk of spontaneous ingnition. Pressures in excess of 200 MPa permit substantial volume reduction factors of up to 5 for non-irradiated hulls, corresponding to 75% of the theoretical density at 300 MPa. The compactability of irradiated hulls appears to be slightly lower (see Fig. 1). Compacting is followed by embedding in lead or a lead alloy with a low melting point and of sufficient viscosity to fill the remaining air pockets.

An evaluation of the wettability of zircaloy 4 by lead alloys without preliminary treatment or with treatment designed to produce a coating of zirconium oxide was made following contact times of 2 to 5 hours at 350, 450 and 530°C in an autoclave. Although interaction between the non-oxidized zircaloy and the alloy PbSb was confirmed by microprobe examination, irradiated zircaloy shows little or no wetting behaviour with respect to lead-based alloys, with the result that the contraction occuring when the alloy cools can cause cracks exposing the surface of the coating to the waste.

Corrosion tests were conducted under realistic storage conditions, i.e. using various aqueous phases and clay, since it is planned in Belgium to use clay formations for the storage of radioactive waste. To make allowance for the storage policies of other countries, brine was also used in the tests. Lead and lead alloy samples were subject to the action of various types of environment for 16 months at temperatures of 25 and 50°C in the case of the aqueous phases and 13°C in the case of clay.

The results showed that the samples in contact with the clay increased in thickness by up to 2.2 μm owing to the formation of an adherent layer, removal of which revealed corrosion depths of 1 to 6 μm. As far as the aqueous phases are concerned, brine is more corrosive than the interstitial water of clay. The effect of temperature is important, since corrosion is approximately 3 times greater at 50° than at 25°C.

The embedding technique was devised on the basis of several experiments on various volumes of hull waste with different geometrical features. The alloys (Pb 1.5 Sb and Pb 4 Sn 12 Sb) were exposed to temperature ranging from 350 to 450°C.

Leaching tests lasting 120 days were carried out on the basis of the ISO standard using interstitial clay water. The leach rate expressed as a fraction of the activity released per day is on balance between 10^{-8} and 10^{-7}. See Fig. 2 for the actinides.

Identical tests on cladding, both with and without a complete coating of the alloy, showed similar results, which raises doubts concerning the protective function of the matrix, despite the relatively low leach rates. The leach rates of fission products under the same conditions range from 10^{-10} and 10^{-3} depending on the alloy, the temperature of the coating and the surface integrity of the coated products (see Fig. 3).

This is probably due to the destruction by the molten alloy of the protective layer of oxides formed at the surface of the cladding during

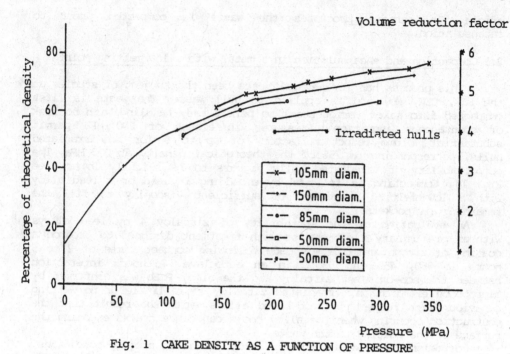

Fig. 1   CAKE DENSITY AS A FUNCTION OF PRESSURE

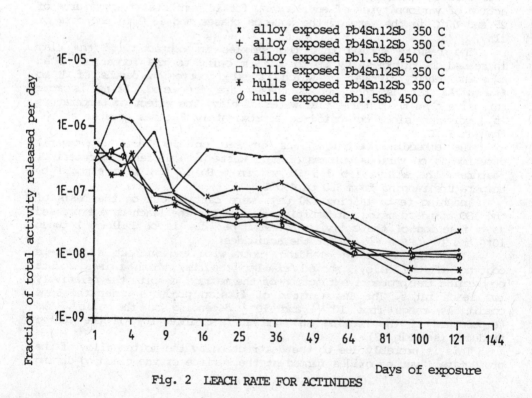

Fig. 2   LEACH RATE FOR ACTINIDES

Ceramic raw materials (by weight)	47% reactive corundum 33% kaolin	52% reactive corundum 14% kaolin 14% bentonite
Waste content	20% dissolver residues (light water type)*	20% dissolver residues (neutron type)**
Specific activity	21 mCi.g^{-1}	113 mCi.g^{-1}
Heat treatment	1300°C/10 mn	1300°C/10 mn
Nature of phases matrix waste	corundum (Al_2O_3), mullite ($Al_6Si_2O_{13}$) (U, Pu, Am)O_2, RuO_2, Ru, Pd, $CaMoO_4$	corundum (Al_2O_3) mullite ($Al_6Si_2O_{13}$) (U, Pu, Am)O_2, RuO_2, Ru, Pd, $CaMoO_4$
Density	2.35 g.cm^{-3}	3.20 g.cm^{-3}
Resistance to compression	120 MPa	90 MPa
Open porosity	1.85 vol. %	1.85 vol. %
Leach rate of Pu in water and brine at 25°C	10^{-8} – 10^{-9} g.cm^{-2}.j^{-1}	10^{-8} – 10^{-9} g.cm^{-2}.j^{-1}
Corrosion rate of matrix subject to water at 200°C	10^{-5} g.cm^{-2}.j^{-1}	10^{-5} g.cm^{-2}.j^{-1}

* fuel UO_2, irradiation 31,000 MWj.t^{-1}, reprocessing after 3.5 years of cooling
** rapid neutron fuel, irradiation 33,800 MWj.t^{-1}, reprocessing after two years of cooling

TABLE I – PROPERTIES OF CERAMIC MATERIALS CONTAINING 20% OF WASTE FROM VARIOUS SOURCES

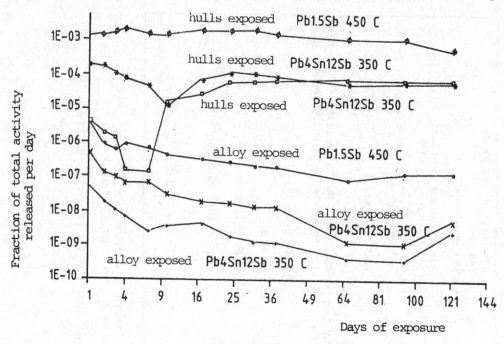

Fig. 3   LEACH RATES FOR FISSION PRODUCTS

Fig. 4   CONDITIONING INSTALLATION AT MOL

irradiation and reprocessing. Moreover, non-wettability by the alloy and its contraction during solidification create a network of tiny cracks. This would suggest that the alloy, which offers excellent resistance to corrosion, could serve as a component for a protective coating for a given unit volume of compacted waste. The design of a sealed container has been studied as part of this configuration, and a pilot installation will be constructed. The equipment used (Fig. 4) will permit compaction of 40 k charges to cakes 105 mm in diameter, which will be embedded in an alloy layer 10 to 20 mm thick. The installation is due to start operating in 1985, and plans for an industrial-scale unit producing cakes 350 mm in diameter will subsequently be drawn up.

2.2 Encapsulation in glass

This method involves coating the hulls at high temperature in glass to form a compact composite block with no air pockets and with a vitreous appearance over its entire surface.

An earlier study carried out at CEN/VALRHO using non-radioactive materials suggested that this process would be feasible on an industrial scale, i.e. using 100 to 200 kg of cladding waste. The study was supplemented by the production, in an armour-plated cell, of coatings for actual waste in order to examine the finished product. The zircaloy 4 cladding fragments came from the WURGASSEN reactor, the fuel of which was irradiated to 13 100 MW.j.t^{-1} and was reprocessed at the La Hague plant.

The coating is produced by heating a mixture of glass and cladding fragments in a vitreous carbon crucible for 3 hours at 115°C in an argon atmosphere. Several types of glass were used; Approximately 50 g of waste was used per 200 g of glass. Examination of the products obtained, which were good in appearance (Fig. 5), revealed in some cases that the vitreous medium had been reduced by the zircaloy.

This phenomenon is particularly pronounced when the glass used contains oxides of iron, nickel, chromium and molybdenum. As a result of reduction, inclusions from these metals are formed in the vitreous mass.

The following leach rates (expressed per 10^{-5} g.cm^{-2}.j^{-1}) were obtained from leaching tests using industrial water at ambient temperature in static mode and with daily renewal :

$$9.0 \quad \text{for Sr-90}$$
$$0.7 \quad \text{for Cs-137}$$
$$1.3 \quad \text{for Ru-106}$$
$$5.0 \quad \text{for Ce-144}$$
$$0.05 \text{ for Sb-95}$$

2.3 Encapsulation in an aluminosilicate ceramic material

This technique is used specifically for conditioning the residues resulting from the dissolution of irradiated fuels, mainly fission products such as Ru, Rh, Pd, Mo and Tc in the metallic state or as oxides, actinides in the form of oxides (U, Pu, Am) and zircaloy shearing fines in various states of oxidation.

Whereas embedding techniques using glass or cement allow only a relatively small amount of waste to be coated, coating with a ceramic material offers, among other advantages, the possibility of coating larger volumes.

The stages in the process studied by KfK-INE are as follows (see Fig. 6) :

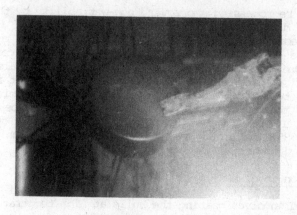

Fig. 5   RADIOACTIVE BAR COATED WITH A MIXTURE OF GLASS AND CLADDING WASTE

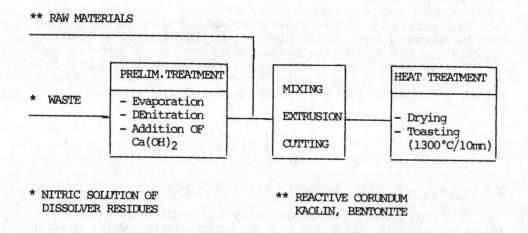

Fig. 6   OUTLINE OF THE PROCESS FOR EMBEDDING DISSOLVER
RESIDUES IN A CERAMIC MATRIX

- preliminary treatment of the solution in which residues are in suspension (concentration by evaporation and denitration);
- mixing and homogenization with suitable raw materials (reactive corundum and clay minerals such as kaolin and bentonite);
- shaping by extrusion and cutting into cylindrical blocks;
- drying and roasting for 10 minutes at a temperature not exceeding 1 300°C.

This technique was developed on the basis of experiments using actual waste from light-water and fast neutron reactors.

Heat treatment induces the formation of mullite ($Al_6Si_2O_{13}$) and the recrystallization of corundum (α Al_2O_3), the waste being incorporated in the microstructure of the material.

The volatility of ruthenium during the operation, carried out under normal atmospheric conditions, was less than 2%. Under the same conditions the molybdenum release rate is kept below 1.5% by the addition in stoichiometric proportion of $CaOH_2$ to form $CaMoO_4$ (powellite).

Because of the chemical composition of the waste, the crystalline phases of the fission product are (U, Pu, AmO_2, RuO_2, Ru, Pd, ZrO_2 ($ZrSiO_4$), $CaMoO_4$.

Fig. 7 shows the overall microstructure of the ceramic material. Maximum content is 40% by weight. Leaching tests on the basis of the draft ISO standard revealed leach rates for plutonium at ambient temperature under the influence of distilled water and brine of between 10^{-9} and 10^{-8} g.cm^{-2}.j^{-1}.

The mechanical resistance of between 50 and 150 MPa is acceptable.

Stability under the influence of radiation was assessed by doping by Pu-238 or Am-241. A dose of 2.5×10^{11} rads was assimilated in 20 months, which corresponds to storage for 10 000 years on the basis of a waste content of 20% by weight. No increase in the rate of leaching or energy accumulation was detected.

Table I gives some of the properties of the finished product. Because of the tendency of this type of ceramic material to assimilate various types of waste containing transuranian elements, the process will be demonstrated on a semi-industrial scale at the INE.

2.4 Embedding in cement

This process applied to cladding waste is based on a simple technique which has already been used at the WAK reprocessing plant at Karlsruhe using Portland cement.

Certain observations have been made, however, which cast doubt on the safety of the permanent storage of such waste :
- formation of hydrogen due to the radiolysis of the water;
- relatively low hydrolytic stability at high temperature.

For this reason experiments with ceramic cements with a low water content were carried out at KfK (IT and INE).

These cements comprise mainly sodium and potassium silicates, metallic salts and quartz. At 20°C, with a water/cement ratio of 0.15, a 400 l drum is filled in 60 minutes. The drum is 100% filled when raw waste is used. When the waste is compacted beforehand, an air pocket amounting to 13 to 22% was noted.

Average resistance to compression is approximately 25 MPa. Using an appropriate drying process can reduce the 100 l of water in a 400 l drum to 1l. This takes from 10 to 20 days.

Analysis of the product indicates the following :
- as expected, hydrogen formation by radiolysis was much less pronounced than with Portland cement;
- the type of cement used has practically no effect on the tritium release rate;
- the hydrolytic stability of materials coated in ceramic cement is much higher than that of waste embedded in Portland cement.

The present objectives of the study are to establish optimum drying conditions and to investigate the possibility of applying the process on an industrial scale.

2.5 Encapsulation in graphite or aluminium based materials

A method of coating cladding wastes and dissolver residues by applying pressure at high temperatures has been examined by the firm Nukem at Hanau.

Five materials were tested :
- graphite/sulphur composite (80/20);
- graphite/nickel sulphide composite (graphite 43.7, Ni 41.3, S 15);
- aluminium (purity 99.99%);
- alloy Al Si 12 (Al 85, Si 12);
- alloy Al Mg_3Si (Al 94, Mg 3, Si 1).

Five manufacturing processes were developed using non-irradiated fragments of reactor cladding and dissolver residues in two industrial-scale installations (see Fig. 8) whereby a diameter of 190 mm is obtained as follows : the matrix is first filled with the raw materials in powder form, followed by fragments of cladding waste 25 cm long and the dissolver residues to give a waste-containing core and a 30 cm thick waste-free shell. After moulding at low pressure (3 MPa) the block obtained is then heated to 130°C, subjected to pressure at 20 MPa, heated to 150°C for 10 minutes and is finally ejected when, after cooling, its temperature reaches 80°C. See Fig. 9.

The graphite/nickel sulphide matrices require greater pressure (40 MPa) and higher temperatures (410°C). Ejection occurs at 350°C. The aluminium or aluminium-based matrices are heated to 430°C without pressure and are then crushed at 50 MPa. After heating at 430°C for 15 minutes, the blocks are cooled and ejected at 400°C, then kept for 2 hours at between 300 and 350°C.

Maximum waste content by weight is :
58% in the graphite/sulphur composite;
20% in the graphite/nickel sulphide composite;
70% in the aluminium and alloy Al Si 12 (see Fig. 10);
48% in the alloy Al Mg_3 Si.

The physical properties of the coating materials listed in Table II are such that they can quite readily be used for permanent storage.

Tests to determine leaching and corrosion by brine at 90 and 100°C were carried out. Leach rates established for periods ranging from 40 to 350 days were determined by measuring the cesium released by the coated materials to which CsCl had already been added. The cesium concentrations in the leached product were in all cases below the detection limit, which resulted in the following values for 90°C (per 10^{-5} g.cm^{-2}.j^{-1}) :

< 7   for graphite/sulphur
< 5   for the graphite/nickel sulphide composite
< 1   for AlSi12
< 3   for AlMg$_3$Si
< 0.5 for Al

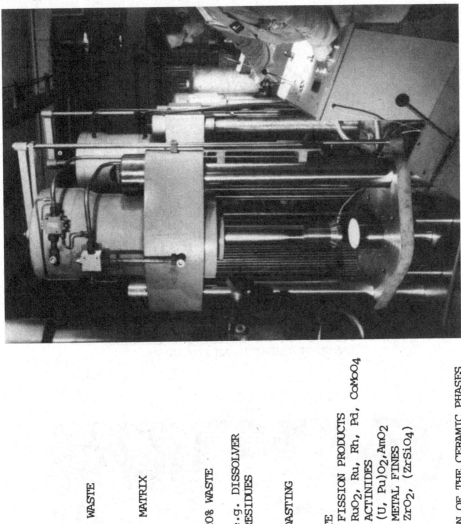

Fig. 8   NUKEM MOULDING PRESS

60–80% RAW MATERIALS

REACTIVE CORUNDUM
KAOLIN
BENTONITE

40–20% WASTE

e.g. DISSOLVER
RESIDUES

COMPOSITION OF PHASES AFTER ROASTING
(1300°C)

MATRIX
- RECRISTALLIZED CORUNDUM
  (α Al$_2$O$_3$)
- MULLITE
  (Al$_6$Si$_2$O$_{13}$)
- VITREOUS PHASE

WASTE
- FISSION PRODUCTS
  RuO$_2$, Ru, Rh, Pd, CoMoO$_4$
- ACTINIDES
  (U, Pu)O$_2$,AmO$_2$
- METAL FINES
  ZrO$_2$, (ZrSiO$_4$)

Fig. 7   MICROSTRUCTURE AND COMPOSITION OF THE CERAMIC PHASES
CONTAINING α WASTE

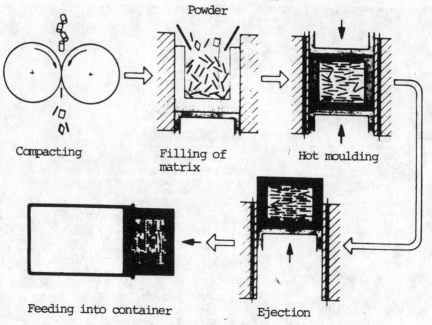

Fig. 9   PROCESS FOR THE EMBEDDING OF CLADDING
WASTE IN GRAPHITE AND ALUMINIUM

Fig. 10   SECTION OF A BLOCK (EMBEDDED IN THE ALLOY AlSi$_{12}$)

PROPERTIES OF MATERIALS	80% graphite 20% sulphur (by weight)	43.7% graphite 41.3% nickel 15.0% sulphur (by weight)	Al 99.99	AlSi12	AlMg3Si
Density g.cm^{-3}	2.17	3.28	2.61	2.59	2.64
Density increase (as % of theoretical density)	97.7	94.8	96.7	97.7	99.2
Thermal conductivity (W/cm K) * **	0.72 0.21	0.92 0.29	1.41 1.27	1.11 1.10	1.00 1.17
Dilatation (μm/m K) * **		4.92 11.82	24.78 23.15	19.32 19.39	22.79 21.91
Resistance to compression (MPa) * **	54.2 59.1	74.5 98.9	210.0 227.0	205.0 398.0	412.0 423.0
Resistance to breaking (MPa) * **		20.8 11.5	30.1 35.0	54.2 102.6	77.8 131.9
Resistance to bending (MPa) * **	42.5 23.3	69.5 23.3	55.2 60.3	115.8 188.7	159.2 263.9
Young's module (10$_2$ MPa) * **	58.7 22.2	76.8 17.7	63.7 62.5	73.2 77.4	66.5 71.1

* radial   ** axial

TABLE II – PROPERTIES OF GRAPHITE AND ALUMINIUM BASED ENCAPSULATION MATERIALS

Resistance to corrosion was determined by measuring weight loss. The following corrosion rates were obtained (per 10^{-5} g.cm^{-2}.j^{-1}) :

at 90°C after 660 days : < 0.4 for the graphite/sulphur composite;
at 90°C after 360 days : 1.6 for aluminium
                          5.8 for AlMg$_3$Si
                          5.6 for AlSi12;
at 100°C after 250 days : 0.9 for the graphite/nickel sulphide composite.

On the basis of these results, of possible waste content, i.e. of the volume reduction factor obtained, and of production requirements, the graphite/sulphur composite and aluminium were considered the most useful and were chosen for use in research and development using actual waste.

3. MELTING

The melting of cladding waste or dissolver residues has the advantage of creating maximum density. Unfortunately, the temperatures needed are very high, especially in the case of zircaloy.

The melting technique applied to cladding waste studied at CEN/VALRHO involves the use of a metal additive which lowers the melting point by forming an appropriate eutectic mixture. This process, which can be applied to both stainless steel and zircaloy hulls, had been examined in a feasibility study using non-radioactive materials, the outcome of which was positive. The study was continued using zircaloy fuel cladding waste irradiated to 32 500 MWj.t^{-1} in the Borssele reactor. The additive used is copper, representing 21.5% of the end product. Samples of various weights (approximately 150 to 400 g) have been produced in an armour-plated cell by melting a mixture of cladding fragments and copper chips in a crucible heated in a non-oxidizing atmosphere. Melting occurs at around 1 150°C. Allowance has also been made for the possibility of decontamination following slag formation, and this has resulted in a number of samples being taken. The slag results from the introduction into the crucible, before heating and after the waste and the copper have been introduced, of a preformed blend of CaF2 and MgF2, with CaF2 accounting for 55% by weight. This blend represents one quarter of the total weight of the materials used.

The samples obtained without slag are well formed, as are those produced with slag. After melting and cooling, the slag is found at the upper part of the ingot and is easily removed from the metallic part. During treatment a loss of activity due to the volatility of the cesium and tritium was noted. The volatility of the cesium was found to vary by 10% in the presence of slag and by 20% without slag. Tritium loss may amount to 0.4 mCi per gramme, i.e. it is very likely to be removed by virtually 100%, although the initial content could not be determined.

Radiochemical analyses of the ingots and their corresponding slags showed the decontaminating effect of the slags to be insufficient. Only 70 to 90% of all alpha emitters are removed by the slag. The percentages are even lower for beta emitters - approximately 45% for Ce-144 + Pr and practically 0 for Sb-95 and Ru-106 + Rh.

Leaching tests were carried out using industrial water at ambient temperature in static mode with daily renewal.

The leach rates, expressed in terms of 10^{-5} g.cm^{-2}.j^{-1}, are on average as follows :

10 to 15	for Sr-90
1 to 5	for Cs-137
0.5 to 2	for Ru-106
0.5 to 1	for Ce-144
0.04 to 0.08	for Sb-125
< 2	for all alpha emitters.

The release rate for alpha emitters is thus on balance ten times lower than for cladding leached in bulk.

Samples taken from ingots are currently undergoing metallographic analysis.

4. CONCLUSION

All the techniques considered have certain built-in advantages and drawbacks. However, these can only be established with certainty on the basis of experimentation with actual waste. The characteristics of the package produced, ease of handling and, to some degree, its cost, are highly important factors to be considered when assessing the feasibility of a particular process. Work on some of the processes considered is not yet sufficiently advanced for any definite conclusions to be drawn. However, the results obtained indicate that the incorporation of dissolver residues in a ceramic material is a promising technique capable of yielding a product of acceptable quality and volume.

With regard to cladding waste :
- Pressure compacting followed by packaging in lead alloy containers has the advantage of greatly reducing waste volume and of not requiring heat treatment, thus avoiding tritium loss and the production of secondary waste.
- Embedding in glass is a method which is very simple and easy to apply in a radioactive environment but results in a slight increase in the volume of waste to be stored.
- The drawbacks inherent in cementation may be avoided by using special hydraulic binding agents. However, work on this simple technique has not yet advanced far enough to allow any definitive assessment.
- Embedding in the graphite/nickel sulphide composite and in aluminium is a process requiring relatively low pressures and temperatures, thus avoiding tritium loss in the case of the graphite composite. Given a coating with a peripheral thickness of 30 mm, extrapolation of the corrosion rate of the protective layer indicates that the waste will be totally isolated from the external environment for over 1 000 years. However, the feasibility of this technique will have to be confirmed on the basis of experiments with actual waste.

The characteristics of the product obtained by coating with glass will also have to be confirmed, since they will be the main factor determining whether this process will be adopted (its simplicity counterbalancing the increase in volume).

Melting is an attractive process since it greatly reduces volume. Here again, we shall have to wait for a full evaluation of the properties of the end product to determine whether the reduction in volume compensates for the relative complexity of this method.

DISCUSSION

<u>J. LEWI</u>, CEA Fontenay-aux-Roses

Does a relation between the pyrophoric properties of shearing fines and the performances of the matrices exist?

<u>C. SOMBRET</u>, CEA Bagnols-sur-Cèze

Nobody has up to now studied the risk of pyrophoricity of these products after conditioning. However, due to the great dispersion of these products in the matrix, there are few risks of ignition.

<u>LAZARET</u>, KFK Karlsruhe

What is the main role of Al in the metallic matrix with respect to corrosion resistance and mechanical stability?

<u>C. SOMBRET</u>, CEA Bagnols-sur-Cèze

Various matrices are used for both reasons namely of enhancing the mechanical stability and improving the corrosion properties.

CONDITIONING OF ALPHA WASTE

S. HALASZOVICH (KFA), P. GERONTOPOULOS (Agip)
D. HENNART (CEN/SCK), F.W. LEDEBRINK (ALKEM)
A. LOIDA (KfK), D.C. PHILLIPS (UKAEA)
N. VAN DE VOORDE (CEN/SCK)

Summary

The long life and high radiotoxicity of the alpha-emitting transuranics in radioactive waste provide an incentive for the constant
improvement of existing processes and waste forms or the development of new alternatives, to isolate them safely from the biosphere. In the following, five processes at differing stages of
development are outlined, the products ranging between cement,
glass and ceramics :
- a process developed by ALKEM for the cementation of waste from
  fuel element manufacture,
- a process to improve the quality of cement products containing
  Magnox hulls, under development at AERE Harwell,
- high-temperature slagging incineration, developed at SCK/CEN,
- embedding of waste in an alumosilicate-based ceramic, being
  developed at KfK,
- embedding of waste in a titanium dioxide-based ceramic, proposed by Agip.

1. INTRODUCTION

On account of their long half-lives and radiotoxicity, the transuranics (TRUs) occupy a special position among radioactive waste. The
need to keep concentrations in the biosphere below certain levels has
led to a variety of approaches, ranging from secure storage to recycling
through a reactor. Amongst other things, this provides an incentive for
the further improvement of existing conditioning methods and repository
waste forms or the development of new alternatives. The processes described below are examples.

2. ALKEM PROCESS

ALKEM has developed a process for cementing waste resulting from
the manufacture of mixed-oxide fuel elements in glove boxes; It has been
supported by the German Federal Ministry for Research and Technology.
The wastes to be conditioned are liquids such as filtrates from plutonium conversion plant and analytical laboratories and solid waste such
as process wastes and discarded pieces of equipment. Statistical studies
over the last four years show that around 60% of solid waste consists of
organic materials such as PVC foil, neoprene gloves, cleaning rags,
cellulose and various plastic objects, with PVC accounting for the major

part with 50-70%. Inorganic materials such as glass, ceramics, metal objects and metal chips account for 40% of solid waste.

One way of reducing the volume of the combustible content of solid waste would be to incinerate it before final conditioning. This approach was rejected, however, since, in addition to ash, large quantities of secondary waste would be produced (from waste-gas scrubbing) due to the high PVC and neoprene content, so the advantage of reducing the final volume of the waste would be lost. Taking into account the composition of the waste considered here, it was decided to immobilize the raw waste directly in a cement matrix.

Cementation is considered to be suitable for solidifying both types of solid waste together with liquid waste in one waste form. One advantage of this process is that it is simple to use. It is cheaper than other methods, and there is considerable experience with its use in conventional processes. All steps in the process function naturally at low temperatures, hence ruling out the risk of fire. The alkaline environment in the cement form aids plutonium retention, because the plutonium – if not already present as insoluble PuO_2 – is converted into $Pu(OH)_n$, which is of extremely low solubility. The problem of disassociation of the waste form before hardening in cases where organic material such as foils and plastic objects are cemented is solved by shredding these materials before cementation.

2.1 DESCRIPTION OF PROCESS

The wastes do not contain any high γ-emitters and can therefore be handled directly using conventional glove-box techniques.

On entering the plant, they are sorted according to type and state (Fig. 1). Solid wastes from which it is worth recovering the plutonium (proportion of 20%) are first washed (1) and then integrated with the main flow of solid wastes. The soft materials are shredded to pieces of less than 5 mm diameter in a conventional shredder modified for use in glove boxes, which processes approximately 80 kg waste per hour. The shredded material is transported pneumatically to a cyclone above the cement mixture, where the waste is homogenized.

In the cementation section (Fig. 2), whose main component is a conventional, continous-operation screw conveyor, the shredded material is mixed with cement and the liquid waste, from which the plutonium has previously been extracted. All three material flows can be controlled separately to maintain the required process parameters. This type of process control forms part of the quality assurance arrangements for the waste form.

The mixture of cement, liquid waste and shredded solid waste is poured into a drum containing the hard solid waste. A vibrating cylinder ensures voidfree embedding of the solid waste pieces. A liquifier and a stabilizer are added to the cement mixture to obtain good flow behaviour even at low water-to-cement ratios.

2.2 Form properties

The end product of the process outlined above is a package consisting of a 200 l drum and a homogenous cement form enclosing the solid waste pieces. The drum is approximately 95% full.

Alpha activity in the drum due to the plutonium content averages 2.7×10^{11} Bq. The properties of the form are set out in Table I.

TABLE I : COMPOSITION AND PROPERTIES OF WASTE FORM

Composition	
Blast-furnace cement, water/cement :	0.30 - 0.55
Organic waste/waste form :	0 - 0.20
Salt/waste form :	0 - 0.25
Properties	
Compressive strength	35 - 10 N/mm^2
Porosity	1.3 - 5%
Leaching rate for Pu (in salt solution at 55°C after one year)	10^{-6}-10^{-8} g/cm^{2}d
Production of radiolysis gas :	
Pu added as PuO$_2$	1.6 dm^3 H$_2$/3.7·10^{10}Bq a
Pu added in solution	2.2 dm^3 H$_2$/3.7·10^{10}Bq a

Compressive strength and porosity depend on the composition of the waste, in particular the proportion of organic material. Blast-furnace cement is used because the cement form exhibits a better leaching resistance in saturated salt solutions than forms consisting of Portland cement. The leaching rates measured for plutonium agree closely with other published data (2). The differing results for radiolysis gas production are accounted for by self-shielding. Whereas PuO$_2$ occurs in the form of small particles ensuring a certain degree of self-shielding, the plutonium added in solution is characterized by a molecular distribution.

2.3 Stage of development

The cementation plant is in full-scale operation using 200 l drums. Processing of over 200 drums of raw waste in recent months has shown that the process functions effectively and satisfactorily.

3. POLYMER-MODIFIED CEMENT

Although cements are suitable for the solidification of most kinds of waste, problems may occur in some cases, particularly if the enclosed waste changes its volume by reacting with the matrix material. Progressive corrosion of cemented magnox hulls, for example, causes an increase in volume that may lead to cracking and even desintegration of the waste form.

AERE Harwell is working on a solution to this problem by increasing the breaking elongation of cement forms by adding polymers during mixing.

Following initial exploratory experiments (3), the effect of four different additives on the properties of Portland cement (OPC) and its suitability for waste solidification was studied in detail. The cement additives used were epoxy and polyurethane resins, a stryrene-butadiene copolymer and a bitumen/water emulsion.

3.1 Form properties

Table II shows the effect of the polymer/cement and water/cement ratios on the breaking elongation of cement forms after a 28-day setting period. The bending strength was measured.

The data show a considerable variance on account of the differing porosities of the individual cement samples, though some tendencies can be detected. The breaking elongation of normal cement is low at 5.10^{-4}. Adding bitumen does not result in any noticeable improvement, but the styrene-butadiene copolymer and epoxy resin at a polymer/cement ratio (by weight) higher than approximately 0.2 more than doubles this figure. Polyurethane at high polymer/cement ratios yields an improvement of more than one order of magnitude. The bending strength of the modified cement forms is either equal to that of non-modified cement or higher, whereas the compressive strength is lower, though still more than adequate.

Although irradiation may reduce deformability, provisional results show that some modified cement forms exhibit a greater degree of deformability than non-modified forms even after exposure to high doses (4).

TABLE II
BREAKING ELONGATION (10^{-4}) OF POLYMER-MODIFIED PORTLAND CEMENT
AFTER 28 DAYS

Additive	Polymer to cement	Water to cement				
		0.2	0.25	0.3	0.4	0.5
No additive	0	–	–	5+2	4+2	8+1
Styrene-butadiene	0.05	9+2	–	3+1	–	–
	0.10	4+1	–	8+1	5	–
	0.20	33+3	–	13+2	–	–
Bitumen	0.03	–	–	6+1	6+1	7+1
	0.06	–	–	6+2	7+2	4+2
	0.11	–	–	–	8+2	6+1
	0.22	–	–	–	–	6+1
Epoxy resin	0.05	–	9	4+1	9+1	–
	0.10	–	7+1	6+1	11+3	–
	0.15	–	10+1	8+1	10+1	–
	0.25	–	–	13+2	–	–
Polyurethane		Water to cement = 0.22				
	0.5	135				
	1.0	400				

3.2 Stage of development

Magnox metal has been successfully solidified in the laboratory using the above polymer-cement systems. In industrial-scale operation, the rapid hardening of polyurethane systems and the reaction heat this would release would cause difficulties, though this problem does not occur with styrene butadiene and epoxy resin systems. Since the polymer-modified cement forms possess a greater degree of deformability, they are expected to last longer before any significant cracking is caused by magnox corrosion because they are better able to handle theincreasing stresses. Initial accelerated corrosion tests at high temperatures appear to confirm this assumption, although the significance of the results so far obtained for actual storage and disposal operations must first be demonstrated by further investigations and model studies.

4. HIGH-TEMPERATURE SLAGGING INCINERATION

The 'High-Temperature Slagging Incineration' process (HTSI) has been under development by SCK/CEN since 1974. It is an integrated conditioning process in which all types of waste are converted in one step by incineration and fusion into a stable mineral product suitable for disposal.

The plutonium is oxidized to PuO_2 at high temperatures. However, although the high stability of PuO_2 improves safety, it may have economic disadvantages in that the plutonium can no longer be recovered from the waste form. This latter aspect is however becoming less important, since the licensing authorities are tending more and more to lay down upper limits for the plutonium content of materials to be discarded as low-level waste. Even if plutonium recovery is desired, the waste sent for conditioning will no longer contain economically recoverable quantities of plutonium. In this case, the HTSI process is seen as both environmentally safe and cost-effective.

4.1 Description of process

Combustible and non-combustible wastes are mixed together in suitable proportions before being fed into the furnace. The optimum mixture consists of 20% non-combustible and 65% combustible material. The non-combustible proportion may vary between 15% and 35%, however. The water content of the waste should be between 15% and 20%.

The operation of the furnace is shown in Fig. 3. The waste is fed into a annular gap around the burner and slides down as a cone into the combustion chamber. The surface of the waste exposed to the flame melts, covering the waste with a thin molten layer. The molten material drops out of the combustion chamber and is quenched in water.

There are three options for manufacturing the disposal packages :
- Encapsulating the glass granulate with cement in a waste drum,
- Sealing of the glass granulate in a waste drum with no additional filling,
- Producing a block from the glass granulate by remelting or hot pressing.

4.2 Properties of the form

On quenching, the molten potassium–magnesium–iron–alumosilicate mass turns into a form of glass similar to the natural obsidians.

The leaching rate for plutonium measured with the Soxhlet method is between 2.10^{-5} and 4.10^{-4} g/cm²d.

4.3 Stage of development

After several years of development, a pilot plant with a throughput of 40 kg/h entered service in Mol in 1981. Until mid-1983, it was used to condition low-active Pu waste from power reactors. The operating parameters were recorded and both the end product and the secondary waste were examined.

To demonstrate the suitability of this process for the solidification of plutonium waste, trials were carried out with 10 t of waste. A plutonium concentration of 1.5 ppm was obtained in the waste by adding plutonium. After incineration and slagging, 4.5 t glass granulate remained. The waste volume had been reduced by a factor of 24.

The secondary waste flows (dust from the bag filters and precipitation from the waste-gas scrubbers) and the exhaust air from the stack were checked for alpha activity. The exhaust air gave results only slightly above the background level.

The lower levels for plutonium recovery are 22.10^6 for the plant as a whole and 5.10^5 for the waste-gas scrubbing section consisting of bag filters, the soda scrubber and the HEPA filters.

Experience to date indicates that the process is highly reliable and environmentally safe.

5. IMMOBILIZATION OF ALPHA WASTE IN ALUMOSILICATE-BASED CERAMICS

The Institut für Nukleare Entsorgungstechniken at the Kernforschungszentrum Karlsruhe is investigating the solidification of alpha waste in alumosilicate-based ceramics. The use of such a ceramic as a matrix material is seen as a solution for problems that occur when using glass or cement for waste solidification.
- The formation of hydrogen by radiolysis is ruled out.
- The chemical composition of the waste has no significant impact on the quality of the waste form.
- A high waste content (up to 40%) is possible.
- Resistance to irradiation and leaching is high.

The ceramic raw material is a mixture of reactive corundum and clay minerals (kaoline and bentonite), similar to materials currently used in the ceramics industry.

5.1 Description of process

The simplified block diagram in Fig. 4 shows the process in schematic form. In general, the waste solutions require pretreatment before they can be mixed with the ceramic raw material. This consists of denitration of the nitric acid solutions, or neutralization with $Ca(OH)_2$ or $Ba(OH)_2$ if other acids are present, and concentration by means of evaporation. The waste solution and the ceramic raw material are then homogenized in a conventional extruder. The water content of the mixture should be $23\pm4\%$ during extrusion. The extruded cord is cut into cylindrical pellets, which are dried and then sintered for ten minutes at 1 300°C.

At this temperature, the ceramic raw material is converted by the following reaction :

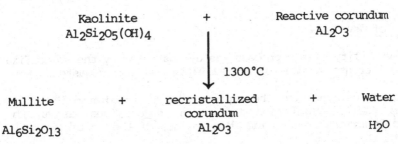

Kaolinite   +   Reactive corundum
$Al_2Si_2O_5(OH)_4$ Al_2O_3

1300°C

Mullite   +   recristallized   +   Water
                          corundum
$Al_6Si_2O_13$ Al_2O_3 H_2O

Before disposal, the ceramic pellets are placed in a special-steel or ceramic container and sealed in glass or dry cement.

5.2 Form properties

During sintering, the waste particles are enclosed between the interfaces of the newly formed matrix phases. The actinides are present as $(U, Pu, Am)O_2$ mixed crystals, which may form a solid solution with ZrO_2 and CeO_2 if these are also present. The actinides are not incorporated into the lattice of the matrix phases because the ion radii of Al(0.50 A) and Si(0.41 A) differ considerably from those of uranium, plutonium and americium (1.00 + 1.01 A).

If, depending on the type of neutralization, significant quantities of sodium, potassium, calcium or barium are present in the waste, the formation of corresponding alumosilicates as an additional crystal phase may also be observed. Table III sets out the composition and properties of the waste form.

The leaching rate for plutonium was determined in accordance with the ISO test in salt solutions and distilled water at room temperature. It is independent of the leaching medium.

TABLE III : COMPOSITION AND PROPERTIES OF THE WASTE FORM

Composition
Matrix : alumosilicate-based ceramic Waste content : 40% (by weight) maximum
Properties
Compressive strength : approx. 100 N/mm^2 Porosity : approx. 5% Leaching rate for Pu : $10^{-8} - 10^{-9}$ g/cm^{2}d

The resistance of the ceramic waste forms to irradiation was demonstrated by accelerated experiments, in which Am-241 and/or Pu-238 were used as additional internal irradiation sources. Neither an increase in the leaching rate nor a storage of energy could be observed as a result of the increased alpha dose (amounting to $2.5.10^{11}$ rad within 20 months, a dose which a standard ceramic form would be exposed to in 10 000 years).

5.3 Stage of development

The viability of the process was demonstrated by the solidification of five different examples of actual alpha-containing waste in the laboratory :

- dissolver residues from the reprocessing of irradiated LWR and fast-breeder fuel (a suspension of fission products such as Ru, Rh, Pd, Mo, Tc in metallic and oxide form, oxides of the actinides U, Pu, Am and zircaloy chips),
- actinide sludges from medium-active waste solutions,
- ashes from the incineration of alpha-containing combustible waste (mainly sodium, potassium, calcium, silicon and aluminium oxides with a PuO_2 content of approximately 15% (by weight)),
- residues from the acid digestion of alpha-containing combustible waste (sulphuric and nitric acid process solutions containing Am and Pu and filter cake containing Am after Pu extraction),
- actinide concentrates from fuel element production (nitric, phosphoric, sulphuric and fluoric acids with considerable quantities of U, Pu and Am).

The aim of current work is to develop a pilot process capable of solidifying all alpha waste from the fuel cycle in a ceramic matrix.

6. THE EDXP PROCESS

The EDXP process (Evaporative Deposition on Xerogel Precursors) is being studied by ENEA/Agip to examine its suitability for solidification of high-level waste in SYNROC ceramics (5). It is also proposed for the solidification of actinide concentrates. The matrix material takes the form of vitreous xerogel particles with a diameter of 0.1 to 0.3 mm. As a result of its sponge-like internal microstructure, its high specific surface (400 m^2/g after calcination at 450°C) and its low bulk density (0.5–0.6 g/cm^3), it is capable of absorbing many times its own weight in liquid. This property is used for the process proposed here.

6.1 Description of process

Fig. 5 shows the principle of the process. The SYNROC precursor is first manufactured by a sol-gel system closely resembling the manufacture of inorganic titanate-based ion exchangers (6). A bed of xerogel particles is placed in a rotating evaporator and heated to 120°C. The waste solution, which may contain up to 2 mol free nitric acid, is continuously sprayed onto the surface of the agitated bed. The particles are saturated with the waste solution, the water evaporates and the salt is deposited on the walls of the sponge structure, resulting in submicroscopic mixing of the waste with the matrix material. The evaporated water is condensed and taken off as low-active waste. When the precursor material has been charged with the predetermined quantity of nitrate salts, the bed in the evaporator is replaced.

The charged particles are heated to 800°C in a reducing atmosphere to break down the nitrates thermally. In a process analogous to the manufacture of fuel pellets (7) they are then cold-pressed at a pressure of 20 kN/cm^2, followed by sintering of the pressed forms at 1250–1350°C in an argon atmosphere with 5% hydrogen.

6.2 Form properties

The following table shows the composition and properties of the waste form.

TABLE IV : COMPOSITION AND PROPERTIES OF THE WASTE FORM

Composition
Matrix : 94.4% TiO_2, 3.22% ZrO_2, 2.36% CaO Waste content : 5–10% (by weight)
Properties
Density : 4.2–4.5 g/cm^3 after cold pressing and sintering Leaching rate (weight loss) : approx. 2 x 10^{-6} g/cm^{2}d

The leaching rate was measured at 14 bar and 200°C after 21 days leaching in water.

6.3 Stage of development

The feasibility of the process was demonstrated in laboratory tests in which pellets of 1 cm in diameter and 1 cm in height were manufactured from inactive material.

It is assumed that the process can be succesfully used for the solidification of alpha waste extracted from the MAW process by means of the OXAL method (8, 9). Assuming that 1.4 kg Ce is added as a precursor for every m^3 MAW (9), and that the titanate ceramic is charged by up to 20% (by weight) with CeO_2, around 2 dm^3 waste product would result per m^3 MAW.

7. OUTLOOK

A variety of different processes for solidifying alpha waste have been developed or are under development. Some of them are intended for conditioning specific waste flows, while others are claimed to be capable of processing all alpha wastes. The resulting waste forms cover the entire known range from cement to ceramics. This explains the diversity of the various processes.

A reduction in this diversity is unlikely, and with a view to achieving improvements perhaps undesirable, until conditions for approval have been fixed for repositories and until process concepts for disposal plant need to be defined. In reaching a decision, the plant operators, waste storage agencies and approval authorities will then assess these processes and their products mainly on the basis of their stage of development, reliability and safety concepts. The importance attached to the stability of the products will depend on whether a multi-barrier system is chosen to isolate the radionuclides from the biosphere or whether the repository alone is to perform this function.

REFERENCES

1. LEDEBRINK, F.W., GASPARINI, G., WIECZORECK, H., WILKINS, J.D., Dekontaminierung von Pu-Abfällen und Spaltstoffrückgewinnung, presented at this conference.
2. NEILSON, R., COLOMBO, P., BRADLEY, D., Plutonium leachability from alternative transuranic incinerator ash waste forms, IAEA-SM-246/45(1981)
3. BURNAY, S.O. and DYSON, J.R., A preliminary assessment of polymer-modified cements for use in immobilisation of intermediate level radioactive waste, AERE-R10599 (1982)
4. PHILLIPS, D.C., DE ANGELIS, G., KOESTER, R., Radiation, thermal and mechanical effects in low and medium active waste, presented at this conference.
5. RINGWOOD, A.E., KESSON, S.E., WARE, N.G., HIMBERSON, W.O., MAYOR, A., Nature 278 (1979) 219
6. TURNER, A.D., GERONTOPOULOS, P., DOZOL, J.F., presented at this conference
7. COGLIATI, G., GERONTOPOULOS, P., RICHTER, K., European Nuclear Conference, Hamburg, Germany, May 6-11, 1979, Trans. Am. Nucl. Soc. 32 (1979) 227
8. MOUSTY, F., BARBERO, P., TANET, G., EUR 7975 FR (1982)
9. GOMPPER, K., KUNZE, S., EDEN, G., LÖSCH, G., ZEMSKI, C., CHATTAS, N., KfK Karlsruhe, 1. Halbjahresbericht zum EG-Vertrag Nr. 288-83-31 WAS-D (B) (1984)

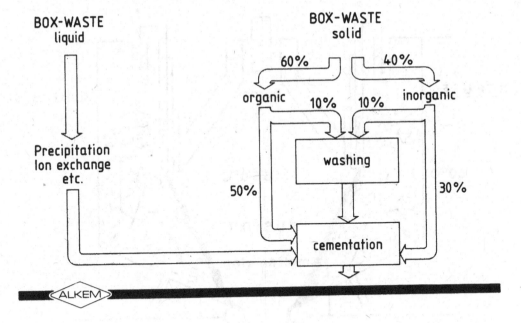

FIG. 1   SCHEMATIC OF   ALKEM   PROCESS

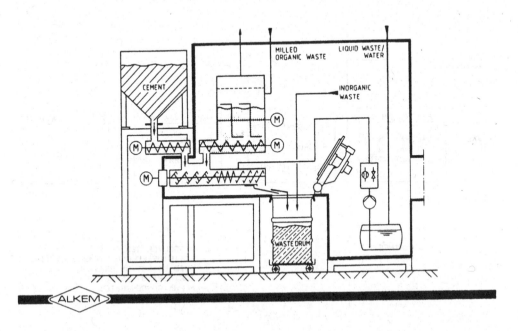

FIG. 2   CONTINUOUS CEMENTATION OF   TRU   WASTE

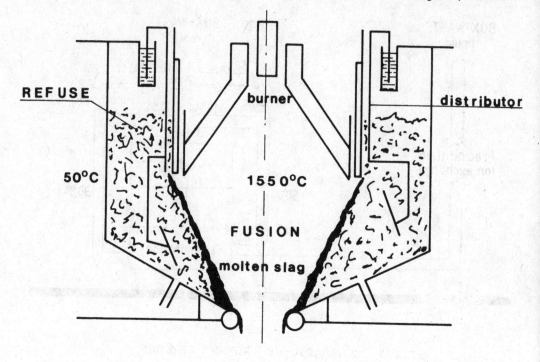

FIG. 3   SCHEMATIC OF HTSI FURNACE

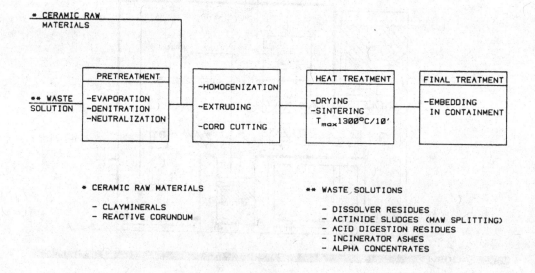

FIG. 4   BLOCK DIAGRAM OF THE SOLIDIFICATION OF TRU
WASTE IN A CERAMIC MATRIX

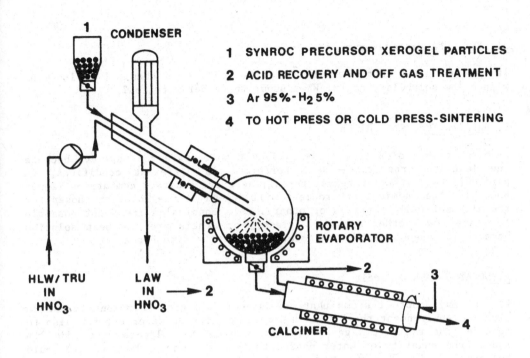

FIG. 5   PRINCIPLE OF   EDXP   PROCESS

D ISCUSSION

J. LEWI, CEA Fontenay-aux-Roses

It was mentioned that the cement matrix leads to a chemical environment with a low solubility of Pu. What about the other matrices?

S. HALASZOVICH, KFA Jülich

Concerning the solubility it is difficult to compare the matrices because the leaching experiments were performed in different conditions, in particular at different temperatures. Therefore a direct comparison is not possible. The chemical environment will depend on the solution chosen for the disposal (multibarriers or other). As a conclusion it is not possible to state now whether ceramics, glasses or others are the best solution because it depends on the characteristics of the site chosen.

A. DONATO, ENEA Casaccia

Without any doubt it seems that the cement is a good solution because in alcaline environment the Pu has a low solubility. However one has also to consider the radiolytic gas production. Therefore polymers were added to reduce the quantity of water needed to form the paste. Moreover it would be interesting to have further information on mechanical properties.

S. HALASZOVICH, KFA Jülich

Some data concerning compression are available and were presented. As to the production of radiolytic gas, additives are foreseen to reduce this phenomena in the Alkem process. Harwell has also foreseen to include polymers into the cement for the same reasons.

F. MANNONE, JRC Ispra

An integrated system applied by Alkem for concentration of alpha-waste from a fuel fabrication plant was described. Can this system also be applied to Pu wastes contaminated by fission products?

S. HALASZOVICH, KFA Jülich

The Alkem system was specially developped for the wastes arising from MOX fuel fabrication plants and not for wastes arising from other processes such as alpha-waste from sludges. The presence of fission products should be limited to avoid too high generation of heat.

H. KRAUSE, KFK Karlsruhe

About this point, it has to be noted that incorporation of fission products in cement will need remote operation and thus reduce the simplicity of the cementation.

SESSION III

TREATMENT AND CONDITIONING PROCESSES - PART 2

Plutonium recovery from α - wastes

Alpha monitoring of solid waste

Management of nuclear airborne wastes and tritium retention

Capture and immobilization of krypton-85

PLUTONIUM RECOVERY FROM α-WASTES

F.W. LEDEBRINK, ALKEM; J.D. WILKING, AERE, Harwell;
H. WIECZOREK, KfK; G. GASPARINI, ENEA, Roma

Summary

The recovery of plutonium from both primary wastes and the ash from inceneration procedures has been investigated and in certain cases demonstrated on an operational scale.

In view of the different composition of these wastes, 3 washing procedures have been tested. For soft materials a drum washing apparatus was used, and for hard materials an ultrasonic bath. PuO_2 contaminated wastes can be cleaned lucratively by using aqueous cleaning solutions or triflorotrichlorethane. Dilute nitric acid can be used successfully with plutonium nitrate contaminated wastes. Apart from the washing procedure, the nature of the surface of the waste, the contact time with the α-emitters and the degree of contamination proved to be the main variables in the recovery of the Pu.

Investigations with ash from incineration and pyrolysis procedures have shown that the highest recovery of Pu by means of leaching was reached with pyrolysis char which was oxidised at < 700°C. A solution of HNO_3/HF seems to be essential for higher leaching rates of > 99 %.

The feasibility of the acid digestion procedure on an operational scale was demonstrated with the recovery of 5 kg of plutonium from a throughput of 600 kg of waste. The process is characterised by high decontamination factors in off-gas treatment, and an overall recovery factor of ≥ 0.93.

The utilization of a neoalkylhydroxamic acid for the decontamination of an alkaline solution resulting from the regeneration of the solvent in a processing plant has been investigated. Tributylacetohydroxamic acid (TBAH) has been proposed as a separating agent for Pu, Am and Zr with liquid-liquid extraction as well as with extraction chromatography techniques.

1. INTRODUCTION

Plutonium-bearing solid wastes are generated during the processing of plutonium in mixed oxide fuel elements. As they are free from strong γ-emitters, they are also termed α-wastes, or "plutonium contaminated materials" (PCM). Both for ecological as well as economic reasons it is rational to recover the plutonium from these wastes.

Before considering how this can be done, it is important to identify the primary wastes. As a general principle, all materials which get into the glove-boxes (see Fig. 1) during the processing of plutonium at some time or another develop into α-wastes. These are mainly PVC lock linings, neoprene gloves, polyethylene bottles, filters, wipes for cleaning purposes, various plastic materials, as well as metal scrap and glass parts, which are contaminated with plutonium oxide powder or plutonium nitrate solution according to where they were utilized.

The statistical record of the accumulated wastes at ALKEM over several years has shown that organic materials of approx. 60 wt-% predominate slightly.(1) Further, extensive measurements on bundles of waste have shown that the plutonium is distributed highly unevenly in the waste (Fig. 2). The bulk of the waste contains only a small proportion, whereas in some few materials such as used air filters, high concentrations of plutonium are found. In determining the distribution, it can be seen that for instance about 80 % of the total Pu is found in only 20 wt-% of the waste. It is therefore recommendable in recovery to profit from this already identified concentration and only to treat waste which contains the bulk of the plutonium.

2. PROCEDURES FOR PLUTONIUM RECOVERY FROM SOLID α-WASTES

Within the framework of EEC programmes, processes have been developed based on these localised accumulations in waste, and others which initially bring about a concentration of the Pu. Washing procedures, which have been investigated by ALKEM and the Atomic Energy Research Establishment (AERE) in the last 5 years, are utilized for direct treatment of primary wastes. Concentration of the Pu before recovery is stimulated by both the acid digestion procedure, which was developed by the Karlsruhe Nuclear Research Centre (KfK), and pyrolysis or incineration.

2.1. Washing procedures

In view of the non-uniform nature of the primary wastes, it was obvious that optimal plutonium recovery cannot be achieved with only one washing procedure. According to the type of contamination, PuO_2 or plutonium nitrate, and the composition of the material - soft or hard - different washing agents and apparatus were required. Conventionally proven technology was reverted to (see Fig. 3).

For the cleaning of PuO_2 contaminated wastes, aqueous solutions - pure water with detergents - were selected, and 1.1.2 triflorotrichlorethane ($C_2F_3Cl_3$) obtainable under trade names such as Freon TF, Frigen-113 TR-T or Arklone-X, for dry cleaning. This organic cleaning fluid has the advantage that it cannot be burned, suffuses well, boils at a low temperature (47°C) and leaves the material dry.

Plutonium nitrate contaminated materials were washed in dilute nitric acid, to avoid the precipitation of $Pu(OH)_4$.

Whashing experiments whith soft wastes such as wipes and gloves were carried out in a small drum washing machine with an operating capacity of 4 l. To clean hard waste materials, e.g. metals, as conventionally proven to be very reliable, ultrasonic baths were utilised.

These washing procedures have the common advantage that they produce no secondary waste, as the washing solutions can be distilled and re-used.

Of major influence on the results of washing, from the point of view of material, the following main variables were established:
- degree of contamination,
- nature of the surface,
- aging of the contaminated material by α-radiation.

All washing trials were carried out with genuine wastes from the MOX production department, which inevitably had greatly varying α-contact times of between a few days and several years, and different levels of contamination. The results of the washing were correspondingly widely varied, so that with such a quantity of individual results in plutonium recovery, only ranges can be given here.

When washing PuO_2 contaminated soft waste with an aqueous solution containing EDTA at raised temperatures, recovery of between 60-95 % was attained (Table 1). High plutonium content in the waste and short contact times correspond with high recovery rates. If the individual items of waste contain less than 250 mg per kg of waste, the results from washing are substantially reduced, so that the investment for the amounts of plutonium recovered is hardly justified.

Really good results have also been achieved using 0.5 M HNO_3 on plutonium nitrate contaminated waste (Table 2). These results also show that non-porous surfaces, such as glass, considerably improve the recovery rate (> 95 %).

Dry-cleaning with Freon was tested on soft materials and metals, with additional agitation and the application of ultrasonic energy (Table 3). In comparison with water, the washing of soft materials with Freon did not produce better results. One positive aspect was the cavitation effect of ultrasound, noticeable during the treatment of hard PuO_2 contaminated materials. Cleaning with Freon and ultrasound produced higher recovery rates for waste with smooth surfaces.

These results coincide to a large extent with those achieved in the Atomic Energy Research Establishment (AERE) in comparable cleaning of metals using ultrasound. The cleaning effect of aqueous solutions in an ultrasonic bath was also investigated. Table 4 shows some typical results for the cleaning of highgrade steel. For PuO_2 contaminated parts, appreciable recovery rates were attained with all cleaning methods, in connection with which 1 M NaOH and water with 2 % "Decon 90" turned out to be particularly effective. Plutonium nitrate contaminated samples treated in Freon TF (Arklone X) led to only poor results (44 %), which were improved (to 85 %) by using Arklone W, which contains 6 % water.

The nature of the surface plays a key role not only in the pick-up of the contamination but also in later decontamination. As shown in Table 5, highly polished steel retains only a low amount of contamination and can be deconatminated to < 1 μg Pu/cm^2. Considerably higher residual contamination remains on the surface of organic materials (≤ 10 g Pu/cm^2). Consequently it is clear that by means of the correct choice of suitable working materials and their surface treatment, the pick-up of contamination on the surfaces can be reduced, and the subsequent decontamination considerably facilitated.

In conclusion it can be established that with the correct coordination of washing procedures with waste composition, the bulk of the plutonium can be recovered with very little expenditure. Additional important advantages of the washing procedures lie in their simple technology and safe working principles. They function without pressure, at relatively low temperatures, with non-burnable fluids, and do not generate off gas or liquid secondary waste, whose final disposal would cause further expense.

2.2 Plutonium leaching of ashes and char

The AERE has been dealing with the question of conditions under which pyrolysis or incineration should take place, in order to obtain optimal pre-conditions for the leaching of plutonium from ashes and char.(2) The ashes produced in an experimental furnace at operating temperatures between 500 - 900°C were leached under standard conditions, to facilitate a comparison between the various trial parameters.

The leaching of pyrolysis char turned out to be rather difficult, as the carbon protected the plutonium from the effects of the leaching agent (Fig. 4). After oxidation of the carbon, leaching improved.

A detailed comparison between incineration and pyrolysis followed by char oxidation at the same temperature has been carried out (Fig. 5). It can be seen that the leachability of the plutonium present falls with increasing temperature, with the best results being obtained at 700°C and below. In principle, however, the plutonium is most easily leached from the material treated by pyrolysis than from incineration ash, so that as far as recovery is concerned, pyrolysis followed by oxidation is preferable to incineration.

From among all types of contamination in primary waste, after thermal treatment only (U, Pu)O_2 showed significant solubility in pure nitric acid, so that in practise the use of an acid mixture of HNO_3/HF seems to be unavoidable for the leaching of residues from both incineration and pyrolysis/oxidation. Very high recoveries (> 99.5 %) can be obtained by means of repeated leaching with fresh batches of nitric acid/fluoride solution (HNO_3/HF).

As an alternative to conventional floc precipitation, the possibility of co-extracting the plutonium and americium with a mixture of tributyl phosphate/odourless kerosene has been investigated for the further treatment of solutions arising from the leaching of ash. Co-extraction is possible if the solution can be conditioned to reduced acidity (approx. 0.1 M H^+, 5 M NO_3^-). It has been determined that prior to neutralisation, ferric nitrate must be added, in order to prevent the formation of precipitates containing α -activity as the acidity is reduced. Additionally it has been found necessary to oxidise the plutonium (IV) to Pu(VI) using sodium bismuthate ($NaBiO_3$) to prevent the formation of non-extractable hydrolysed Pu(IV) species. Very good extraction results were obtained by this method (< 5 x 10_4 Ipm/ml).

2.3 Acid digestion procedure

In a joint development programme, KfK Karlsruhe and the Eurochemic Company in Mol, Belgium, have realised the acid digestion procedure of the waste and a process for Pu recovery in active operation. The entire procedure comprises the following main steps (Fig. 6) :
- waste pre-treatment (segregation, shredding, Pu assay),
- acid digestion (decarbonising, oxidising, Pu conversion, filtering),
- off-gas treatment (oxidation, washing, filtering),
- acid re-cycling (distillation, re-cycling),
- Pu recovery (leaching, extraction).

Acid digestion is based on the chemical conversion of organic wastes, which are firstly carbonised by dehydration with concentrated sulphuric acid and then oxidated and decomposed at approx. 250°C with nitric acid. The originally existing PuO_2 is thereby converted into soluble Pu(SO_4)$_2$ and reaches the filter cake as residue together with inorganic slurry after separation from the reagent acid. This is the concentrate from which the Pu is recovered.

The Pu(SO$_4$)$_2$ is dissolved from the filter cake with dilute sulphuric acid and separated according to the PUROSOLVEX-II process by extraction with an 0.65-molar Primene JMT solution in octanol/kerosene. After re-processing in 3-molar nitric acid and subsequent precipitation with oxalic acid the Pu is available in pure form for re-cycling. The main aim of the programme was to study the process with genuine α-wastes from Eurochemic and to prove its basic feasibility. The plant was erected on the Eurochemic site and has been in active operation since March 1983. Up till now approx. 600 kg of waste has been converted to ash and 5 kg of Pu recovered.(3) After a start-up phase, in which some technical improvements were made, the reactor reached a throughput of 7 kg of waste per day, which approaches the designed capacity of 10 kg/day.

As a result of using real waste, unexpected deposits of plutonium sulphate precipitations occurred in the reactor. To avoid the accumulation of Pu, the reactor was rinsed from time to time with dilute sulphuric acid.

In order to keep the liquid secondary waste from the acid digestion process low, as far as is technically and reasonably feasible, the unused acids were recycled. Thereby the actual losses in H$_2$SO$_4$ are only the adsorbed quantity in the filter cake material (0.1 l/kg waste). HNO$_3$ can only be recovered up to 40 %, as losses occur through the formation of nitrous oxide and nitrogen, and a further part becomes secondary waste as an equimolar mixture with the hydrochloric acid which arises. The total quantity of liquid waste is principally determined by the proportion of chlorine-bearing substances in the waste, such as PVC and neoprene.

Dissolution of the Pu by leaching the filter cake with dilute sulphuric acid raised no problems whatever, and high recoveries were attained in the subsequent extraction according to the PUROSOLVEX-II procedure (4) and oxalate precipitation (Fig. 7). The resulting factors are based on the measurements made on the secondary waste streams and do not make allowance for losses incurred from sampling and solid waste which arise during the operation of the plant. These Pu losses are estimated as not surpassing 3 % of the quantities recovered, so that a total recovery factor of ≥ 0.93 can be reckoned with.

Fig. 8 shows the decontamination factors which were measured in the treatment of off-gases. Residual activity of 10^{-13} α-Ci/Nm^3 was ascertained in the cleaned off-gas before absolute filtering, which corresponds to a decontamination factor DF of 7.10^{10}. The DF for liquid secondary waste amounts to 10^6.

Summarising, it can be maintained that the technological feasibility of acid digestion has been demonstrated in the operation of the plant up till now, and that the process features high Pu recovery rates both in leaching the residues and in the subsequent separation of the Pu by extraction.

3. DECONTAMINATION OF ALKALINE SOLVENT WASH WASTE

In addition to the processes for Pu recovery from alpha bearing waste, the utilization of a neo alkyl hydroxamic acid for the decontamination of an alkaline solution resulting from the regeneration of the solvent in a reprocessing plant, has been investigated at the italian ENEA laboratories. The Tri Butyl Aceto Hydroxamic acid (TBAH) has been proposed as separating agent for Pu, Am and Zr (5) (6) liquid-liquid extraction as well as with extraction chromatography techniques.

The best operative conditions have been checked with an alkaline solution simulating the aqueous solution resulting from the carbonate regeneration of the solvent in a first cycle classic Purex flow sheet. Complete extraction-reextraction runs with laboratory scale mixer settlers and simulated solutions have demonstrated the promising possibilities of liquid-liquid extraction techniques using hydroxamic acids as selective extractants.

The volume reduction is over 10 and a further volume reduction can be reached by calcination of the oxalic stripped solution. (Fig. 9).

The decontamination factors obtained in batch experiments using a real solution from a reprocessing plant is over 100 in a single extraction stage and 1400 with two consecutive stages.

4. NOTE

These investigations and trial runs were carried out as part of the second CEC Research and Development Programme for the Management and Storage of Radioactive Waste, and were partly or in some cases totally financed by the EC.

5. REFERENCES

1. W. Stoll, V. Schneider, F.W. Ledebrink – Alpha Waste Arising in Fuel Fabrication; EUR 8609
2. J.S. Isaacs et al. – An experimental Study of the Critical Factors in the Recovery of Plutonium Wastes by Incineration and Pyrolysis, EUR 7337
3. R. Swennen, H. Wieczorek – Alpha Waste at Eurochemic using Acid Digestion and a Plutonium Recovery Process, Radioactive Waste Management, Vol. 2, International Atomic Energy Agency, Vienna (1984) P. 335
4. R. Swennen, H. Cuyvers, J. Van Geel – The Treatment of Alpha Waste at Eurochemic by Acid Digestion – Feed Pretreatment and Plutonium Recovery; EUR 8609
5. ITALIAN PATENT 48741 A/71
6. Proceeding of the International Solvent Extraction Conference at Toronto (ISEC 77) Sept 1977.

Materials	
Organic 60 ± 10 w/o	
– Polyvinylchloride	Foils, Tubes
– Neoprene	Gloves
– Polyethylene	Bottles, Foils
– Cellulose	Wipes, Filters, Paper
– Other synthetics	PTFE, PA, PP, PMMA, etc.
Inorganic 40 ± 10 w/o	
– Metal scrap	
– Glass	
Elements (as Oxides or Nitrate)	
– Uranium	
– Plutonium	
– Americium	

Fig. 1   **Composition of Solid α-Waste**

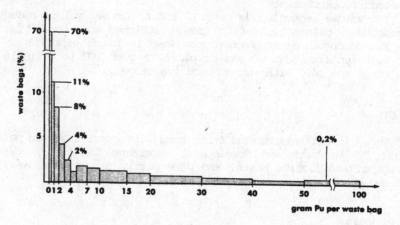

Distribution of Plutonium in the Primary Solid
α -Waste
Fig. 2

Conta minant	Washing Media		Waste Materials	Washing Apparatus
Pu-Dioxide	Freon / Aqueous Solutions		Soft	Wasching Machine
			Hard	Ultrasonic Bath
Pu-Nitrate	Nitric Acid		Soft	Washing Machine

Fig. 3   **Washing of Solid α -Waste**

Materials contaminated with PuO$_2$	Pu Recovered %
Wipes (Polypropylene)	70 - 95
Gloves (Neoprene)	60 - 85
Foils (PVC)	70 - 95

Tab.1 **Rates of Plutonium Recoverable by Washing with Aqueous Solutions**

Materials contaminated with Pu-Nitrate	Pu Recovered %
Gloves (Neoprene)	80 - 95
Metal (SS, AL)	90 - 95
Glass	>95

Tab.2 **Rates of Plutonium Recoverable by Washing with 0.5 M HNO$_3$**

Materials contaminated with PuO$_2$	Pu Recovered %
Wipes	50 - 83
Air Exhaust Filters	65 - 75
Al-Boxes	80 - 95
Stainless Steel	85 - 95
Glass	>95

Tab.3 **PuO$_2$ Recovered by Washing in an Ultrasonic Agitated Freon-Bath**

Contaminant	Wash Reagent	Pu Recovered %
PuO_2	H_2O	77
	$H_2O/2\%$ Decon 90 [1]	94
	1 M NaOH	95
	Arklone X [2]	70
	Arklone W [3]	71
$Pu(NO_3)_4$	1 M HNO_3	99
	Arklone X	44
	Arklone W [3]	85

1) Decon 90 is a proprietary cleaning reagent.
2) Arklone X = Freon TF = $C_2F_3Cl_3$
3) Arklone W contains 6% water and 2.5% surfactant

Tab. 4 **Ultrasonic Treatment of Stainless Steel**

Materials contaminated with PuO_2[1]	$\mu g\ Pu/cm^2$ before	$\mu g\ Pu/cm^2$ after
316L stainless steel mechanically polished	0.1	0.09
316L stainless steel polished rolled annealed at 900°C	4.7	3.5
Epoxy paint	12	5
Vinyl gloss paint	22	10

1) The materials were decontaminated in an ultrasonic bath with Arklone X for 30 min.

Tab. 5 **Pick Up of PuO_2 Contamination on Surfaces**

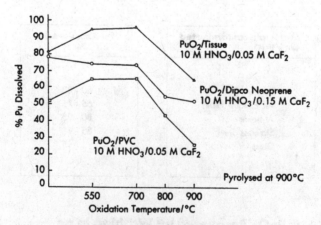

Fig. 4 **Leaching of Pyrolysis Char Oxidised at Various Temperatures**

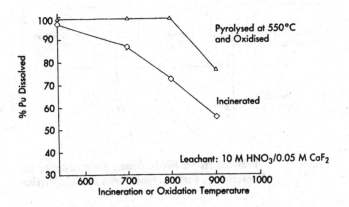

Fig.5   **Leaching of Incinerated or Pyrolysed and Oxidised PuO₂-Mixed Wastes**

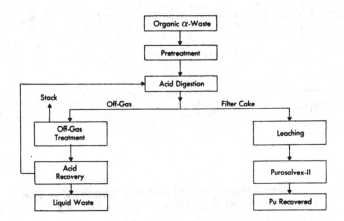

Fig. 6   **Pu-Recovery by Acid Digestion and Purosolvex-II**

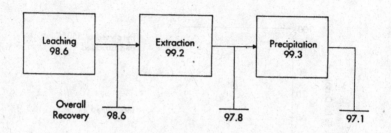

Fig. 7   **Percentage Plutonium Recoveries in Leaching, Extraction and Precipitation**

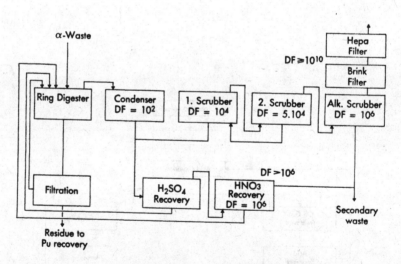

Fig. 8   **Decontamination Factors Reached in the Acid Digestion Process Facility**

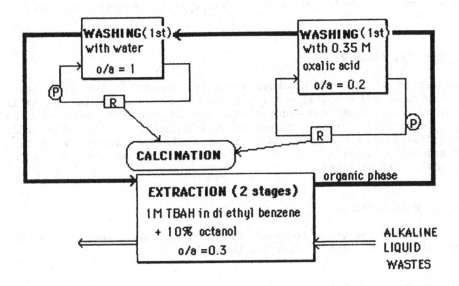

Fig 9 Proposed flow sheet for the elimination of alpha emitters
from alkaline liquid effluents by solvent extraction

DISCUSSION

H. KRAUSE, KfK Karlsruhe

In comparison to the acid digestion process, is there any threshold for
the Pu-content of the PCM waste to be assigned to pyrolysis/char oxydation?

F.W. LEDEBRINK, Alkem Hanau

The definition of such a threshold actually depends on two parameters;
plant throughput and disposal mode. If the plant throughput is low, it is
obvious that only the most active fraction of the PCM waste will be
treated. Accordingly, the Pu threshold will be high. On the other hand,
if geological disposal of PCM waste proved to be very expensive, a high
volume reduction factor will be the aim of treatment. In this case, a low
Pu threshold will be fixed.

G. GROSSI, ENEA Casaccia

Which management route is followed by the complexing reagents in the
washing solutions for decontaminating the PCM waste?

F.W. LEDEBRINK, Alkem Hanau

If complexing reagents are needed, these are subsequently destroyed by
boiling with concentrated nitric acid. Therefore, the further Pu purification step is not affected by the presence of these compounds and the
implementation of the washing procedure for PCM treatment does not give
rise to any secondary liquid waste.

ALPHA MONITORING OF SOLID WASTE

R. BOSSER (CEA), L. BONDAR (CEC), L. BORRMAN (KfK), G. GROSSI (ENEA),
J.W. LEAKE (UKAEA), J. LEY (CEC), H. WURTZ (KfK)

Summary

At the initiative of the CEC and with its support, a programme for
cooperation on non-destructive monitoring of solid wastes contaminated by alpha emitters was launched in 1980. This programme
centres around the intercomparison of assays carried out on a
single set of waste drums in eight European Laboratories. The four-
teen 220, 100 and 25 l drums include five model drums for multi-
parameter studies. A total of 83 configurations were selected for
comparison purposes. Of the 17 sets of assay equipment used, most
employ passive methods, half using neutron detection and the other
half gamma detection, while four installations use three different
techniques based on active methods. The assay programme started in
August 1982 and is scheduled to end in May 1985, more than half the
time being taken up by transport operations. A general report is
planned as soon as the laboratories have all submitted reports on
their results. The intensive exchange of information on the
methods, techniques and procedures developed in the individual laboratories is a further constructive aspect of this cooperative venture which it ought to be continued. Over the same period, the R &
D programmes proper and industrial trials with the assay equipment
tested have extended ideas concerning the possibilities and limitations of monitoring methods and concerning waste management
strategies.

1. INTRODUCTION

In 1980, a working party consisting of representatives of five
Community countries and the Commission drew up a programme for cooperation on non-destructive monitoring of alpha waste. The studies and work
carried out up to then did not yield a consensus on the methods or techniques needed to solve this problem, which is extremely complex due to
the variety of ways in which it can be approached. It was thought that
an inventory of the various approaches adopted would be a good idea, and
a proposal for an intercomparison of the results obtained in the various
European Laboratories from monitoring the same set of waste drums was
generally welcomed as a basis for cooperation with CEC support. It was
clear from the outset that a comparison limited to actual waste drums
could only demonstrate the existence of discrepancies between the different assays without permitting an analysis of their cause or enabling
steps to be taken to reduce error factors in a rational manner. The
approach adopted was a parameter study which was to be as extensive as
possible but with a limited number of drums. One difficulty which caused
some concern from the outset has unfortunately proved to be only too
real, namely the transport of drums containing plutonium.

The regulations involved and even more so their application have resulted in delays estimated at more than a year and a half in a programme started in 1982, and hence in obtaining results - which explains the limited contents of this paper. An analysis of the results of each laboratory in the form of a report presented according to a jointly defined model is due to appear in June 1985 after the final assays have been concluded. A general report will then be published by the working party. We shall limit ourselves here to outlining the assay configurations and describing the methods and techniques used by the participants in the exercise.

2. DESCRIPTION OF WASTE DRUMS

The intercomparison exercise has been limited to drums of light, low-activity, non-irradiant waste contaminated by plutonium, equivalent to the waste produced by plant for reprocessing irradiated fuel or for manufacturing plutonium fuel elements. This waste generally consists of polyvinyl film, latex gloves, cotton and rags, with an apparent density usually between 0.1 and 0.3 g . cm^3. However, to study the effects of conditioning waste in heavy matrices (concrete, bitumen, resin) on monitoring, we included a concrete drum in the exercise.

An essential condition for deriving clear conclusions from the assay results is a precise knowledge of the quality and quantity of the waste and in particular the contaminants. According to the experts consulted, destructive methods using actual waste - apart from being extremely expensive - cannot achieve the precision required for the small quantities of contaminants examined in our comparisons. All of the drums are thus entirely made up of precisely identified and quantified elements so as to include as many as possible of the parameters considered as important by the participants in the exercise, whose methods and objectives cover quite a wide range. These parameters include drum size, the nature and density of the waste, waste simulant or conditioning matrix, and the physicochemical form, isotopic composition, other associated radionuclides, quantity, geometrical form and distribution of the plutonium in the waste. Three types of drum were chosen with the above aims in mind.

The first type comprises non-sealed drums in which the waste is replaced by a compact polyethylene foam to simulate light waste and by ordinary concrete for the concrete matrix. The simulant cylinder, which completely fills the drum, contains narrow logitudinal holes into which contaminant samples can be introduced in sealed steel capsules, together with any other neutron or gamma sources that can fit into the holes. These drums allow detailed studies to be carried out of variations in the performance of the assay equipment for a variety of parameters using just a few containers. A total of five drums (two of 220 1, one of 100 1 and one of 25 1) with a simulant density varying between 0.1 and 0.3 g.cm^{-3} and one 220 1 concrete drum were used to prepare 74 configurations for the intercomparison exercise. Three types of plutonium in oxide form are employed, the first corresponding to plutonium recently produced by reprocessing fuel irradiated at 33 000 MWJ/t in a light water reactor, the second to old plutonium in which a large part of the 241 Pu has been converted to 241 Am, and the third to Pu from a GGR (natural uranium, gas, graphite reactor) reactor mixed with uranium, representing waste from the manufacturing of fuel for fast neutron reactors. A total of 43 samples 10 mm in diameter and of varying thicknesses, each containing a precisely weighed amount of Pu (between 40 and 800 mg) were available for setting up spatial configurations ranging from the extremely heterogeneous, using just one capsule, to more homogeneous arrangements with 12 capsules. The effects of varying the amounts Pu present can be shown even when very small quantities are involved.

In the second group, consisting of two 200 l drums and two 100 l drums, the waste is likewise simulated by means of a compact polyethylene foam cylinder with a density varying between 0.2 and 0.3 $g.cm^{-3}$, but in this case the contaminants in the steel capsule are arranged in fixed positions and the drums are subsequently sealed. The aim was to test the ability of the assay techniques to determine contaminant quantities in order to secure an unambiguous analysis of any discrepancies.

The third type of drum is most representative of actual waste. The content is a mixture of various common components of light waste : cotton, rags latex gloves, polyvinyl film etc., to which have been added precisely known quantities of plutonium and uranium in oxide or nitrate (etc.) form. The distribution of these contaminants is totally random. Three 220 l and two 100 l drums with densities varying between 0.1 and 0.2 $g.cm^{-3}$ were used.

For the last two types of drum, the quantity and composition of the contaminants, and their positions in the case of the second type, will be revealed after comparison with the assay results in order to analyse any discrepancies noted. In order to eliminate any weighing errors or to show up any heterogeneities in the sealed-capsule samples, the latter, which are identified by numbers, were subjected immediately after manufacture to intercomparisons based on the detection of neutron and gamma emissions. As far as development through time is concerned, especially for actual waste, a second assay is planned with two drums in the laboratory that carried out the first measurements (nearly three years before).

3. ASSAY METHODS

Non-destructive monitoring of alpha waste in all of the laboratories taking part in the intercomparison exercise is based on the detection of gamma rays and/or neutrons emitted by contaminants either naturally or as a result of fission induced in fissile radionuclides. Even though they use the same basic methods, the installations devised differ in certain respects and it is interesting to compare the effects of these differences on performance. Since they are easier to set up and use, passive detection installations make up the majority.

3.1 Passive methods

3.1.1 The gamma spectometry obtained with a high-resolution detector enables the characteristic lines to be identified for gamma-emitting radionuclides, thus providing a standard for quantifying these nuclides.

In the case of plutonium, each isotope (apart from 242 Pu) possesses characteristic lines. The energy range used for monitoring wastes contaminated by Pu, Am and uranium is generally between 100 to 450 KeV. The difficulties involved in precisely determining the contaminant quantity by this method are due to the fact that neither the waste nor the contaminant is distributed evenly throughout the drum and the fact that the correlation between the number of gammas detected and the number emitted is upset by essentially random distance and absorption effects. A number of remedies have been developed to reduce this interference, and the result is differences in the equipment and operating methods used. In the eight laboratories taking part, seven different types of gamma spectrometry device are used. In all cases, the drum rotates on a table turning about its vertical axis. Four installations operate with the detector in a fixed position, one permitting measurements of the contaminant quantity in a 25 l drum, while the others are only used for measuring the isotopic composition of the plutonium and the Pu/U or Am/Pu ratios. Since the isotopic abundance ratios are determined from the intensity ratios of neighbouring lines, or by interpolation, they are fairly insensitive to attenuation and propagation effects, provided that the isotopes compared are located together in the drum.

In the three other installations, the detector is located behind a rectangular-slit lead collimator and only scans part of the drum at a time with the drum being moved relative to the detector to enable the entire drum to be scanned parallel to its axis. Two installations operate with segmented scanning : the detector is in a fixed axial position relative to the (rotating) drum and the quantity of 239 Pu in the scanned section of the drum is determined from the gammas detected in a single channel tuned to the 414 KeV line. At the same time, transmission across this section is measured by detecting 400 KeV gammas from a selenium source located diametrically opposite the detector, enabling the figure for the mean attenuation of 239 Pu gammas to be adjusted. A fast pneumatic travel mechanism then enables the next section to be measured, with the travel steps corresponding to the height of the collimator slit. The third installation operates with continuous scanning over the entire height of the (rotating) drum. The spectrum for the sum of the gammas detected over the entire course of travel enables both the abundance ratio and the overall quantity of Pu contained in the drum to be derived. The axial profile of the overall spectrum for high-energy Pu gammas enables heterogeneous sources to be located and the effects of axial positions on these assays and neutron assays to be partly corrected. In order to take greater account of end effects, a system with a variable axial travel rate has been added to the continuous travel mechanism.

3.1.2 The second passive method for monitoring alpha waste is based on the detection of neutrons spontaneously emitted by the contaminants. These neutrons are created by two processes : spontaneous fission of even-numbered isotopes such as 238 Pu, 240 Pu, 242 Pu, with simultaneous emission of several neutrons for each fission reaction, and (α , n) reactions between alpha particles from the contaminants and target nuclei of low atomic mass, with a single neutron being emitted in each reaction. Only the spontaneous fission neutrons can be correlated in all cases with the quantity of contaminant-provided that the latter's isotopic composition is known. A signal proportional to the rates of emission of spontaneous fission neutrons therefore needs to be extracted from the neutrons detected.

TABLE I
FEATURES OF PASSIVE NEUTRON ASSAY EQUIPMENT

Installation No.	1	2	3	4	5	6
1-Cavity wall	paraffin	water	polyethyl +concrete	polyethyl	paraffin +compound	polyeth
$\emptyset$ internal (mm)	680	800	800	700	700	500
H internal (mm)	1072	900	1160	1100	1100	1200
2-Counters	BF3	BF3	^{3}He	^{3}He	^{3}He	^{3}He
diameter (mm)	51	51	25	25	25	25
active length (mm)	1070	1500	1000	1000	1000	1000
number	13	36	36x2+9x2=90	36+12x2=60	24	18
Number of rings	1	1	2	1	1	1
4 π	no	no	yes	yes	no	no
3-Overall effic. Source at centre %	5.2	12.4	18.	17.	7.	8.
4-Separation FS/(α,n)	TM	TM+ RD	RD	TM RD PTP PFA	TM	TM
5-Detection limit 240Pu mass (mg) in 1000 s.	15	27	1	2.5	4	20

TM : dead time   RD : shift register
PTP : pulse to pulse   PFA : pulse fluctuation analysis

This is done by using the fact that neutrons simultaneously emitted during a fission reaction are correlated in time on detection. Following a great deal of work, correlation systems have been developed to perform this extraction. Of the six neutron detection installations discussed here, five are equipped with a dead time system, three with a shift-register system and one with a time correlation analyser based on the analysis of the pulse train using computer algorithms. The last system was developed at Ispra and has recently been adapted to analogue operation. The neutrons are detected in a well-type geometry with BF3 or ^{3}He counters arranged in rings parallel to the drum sources.

Two of the installations also possess detectors perpendicular to the drum axis to ensure near 4 π detection, one of them having two rings of detectors in order to detect the maximum number of neutrons and to adjust for the slowing down of neutrons in the waste. One of the installations allows for simultaneous assaying of neutrons and gammas, with the diode travelling in a lateral slit housed in the neutron detector block. Paraffin, polyethylene or water are used as thermalizers and reflectors and for ambient protection. The main features of the six installations compared are given in Table I. The efficiency of spontaneous fission neutron detection is a function of the square of the efficiency of overall neutron detection is a function of the square of the efficiency of overall neutron detection, hence the importance of the latter parameter. The lower limit of detection is a function of the installation background and the efficiency of detection.

3.2 Active methods

Active methods may be used for alpha monitoring when the waste contains fissile isotopes, which is generally the case with plutonium and uranium contaminants. They involve the stimulation of fission reactions in these isotopes by means of an external high-energy gamma-ray or neutron source and the detection of the gammas or neutrons emitted by these induced reactions. The intensity of the activation source may be selected in accordance with the sensivity required, which offers more scope for manoeuvre compared with passive methods. As is the case with passive methods based on neutron detection, the isotopic composition of the contaminant needs to be known in order to take account of mixtures of fissile isotopes and associated non-fissile alpha-emitting radionuclides so that the total alpha activity of the waste can be determined. Several combinations of activation and detection methods are possible. Three different techniques have been developed in the laboratories participating in the intercomparison exercise.

One is based on activation using a 124 Sb–Be (γ, n) source emitting 24 KeV neutrons for the detection of high-energy neutrons emitted during induced fission. Five BF3 detectors placed diametrically opposite the source in relation to the drum are adjusted in such a way as to limit the detection of neutrons from the source and to maximize the detection of induced fission neutrons. The half-life of the 124 Sb source is only 60 days, which is a handicap as far as industrial use is concerned, although the cost of such a source is quite low. A 30 Ci source producing 10^7 n.s^{-1} will achieve a detection limit of 50 to 100 mg of Pu per drum in 1 000 s.

The second technique uses a 252 Cf source moved by a pneumatic system between a storage position and an activation position. The drum is irradiated for time t, after which the source is swiftly transferred (1.4 s) to the storage flask. This is followed by neutron detection for time t'. The delayed neutrons from induced fission reactions are detected in a well device identical to that used for the detection of spontaneous neutrons. The times t and t' each lasting 10 seconds ensure maximum overall efficiency of detection. The irradiation, transfer and assay cycle is repeated many times to obtain adequate statistics. One prototype operates with three sources irradiating the drum simultaneously at three different heights in order to ensure a more homogenous activation throughout the drum. For an installation with a neutron detection efficiency of 7% and a 252 Cf source emitting 10^9 n.s^{-1} under 4π, the detection limit in 1 000 s is estimated at 100 mg 235 U and 0.5 to 1 g of Pu in a 220 l drum containing light waste.

The low sensitivity of this technique is mainly due to the relatively low emission rate for delayed neutrons and to background caused by spontaneous neutrons in the case of Pu.

The third technique, which has been developed in recent years at Los Alamos, has been recently applied in our laboratories, with two installations in operation for assaying some of the current batch of drums. Fission reactions are induced by neutron pulses produced by a pulse neutron generator. The neutrons are emitted by (D/T) reactions at energies of 14 MeV and are then slowed down and thermalized in the graphite and polyethylene walls of a cavity containing the waste drum. After the generator has stopped emitting neutrons, the thermalized neutrons induce fission reactions for a certain time depending on their life in the drum and the drum surroundings. During this time, it is possible to detect prompt induced-fission neutrons emitted in the drum as fast neutrons. To do this, the cavity wall can be fitted with a neutron detector enclosed in cadmium with low sensitivity to thermalized neutrons and optimum sensitivity to fast neutrons. The BF3 or 3 He detectors are identical to those used for the passive method. The pulse-detection cycle is repeated at a rate of 10 to 100 Hz for a sufficient time to allow good statistics to be obtained. The pulse duration is in the order of some tens of μ s, while the detection period after each pulse is some ms. One of the neutron generators is a plasmatron operating with a tritiated target, which needs to be replaced after 80 h of continuous use at 10^9 n.s^{-1} on average, i.e. at a rate of 10^7 neutrons per pulse at a frequency of 100 Hz. In this case, the detection limit is estimated at 2 mg fissile Pu/180 s with 12 3He counters each 1 m long. The other generator, a sealed tube, produces 410^7 n per pulse at a frequency of 10 Hz with an estimated detection limit of 2 to 3 mg of Pu/100 s, the detection system being similar to the first. The high sensitivity of this technique is mainly due to its ability, using the pulsed neutron generator, to detect prompt neutrons, which have an emission factor of 156 for 235, 472 for 239 Pu and 195 for 241 Pu compared with delayed neutrons in the case of fission reactions induced by thermal neutrons. By extending the detection of neutrons to several seconds after each pulse, the delayed neutrons can be assayed and their ratio to prompt neutrons derived, while in certain simple cases the nature of the fissile isotope can be identified. The correlation of the neutron decay constant in the drum plus its surroundings permits the identification of the type of waste in the drum and hence a better calibration of the equipment. As with all techniques using radioactive sources, the neutrons spontaneously emitted by the contaminants contribute to background, which reduces performance in terms of detection limits.

Generally speaking, active methods are most suited when the contaminant mainly contains fissile isotopes such as highly enriched uranium or plutonium from slightly irradiated fuel. In cases where contaminants of this type are mixed with activates or fission products and/or the waste consists of heavy but not very hydrogenated material, active methods are the only ones that can be used. However, it should be stressed that although active methods can achieve extremely low detection limits useful for classifying wastes under the alpha activity threshold, in certain cases the error in the contaminant assays may be greater than with passive methods due to heterogeneous nature of the activation process.

4. CONCLUSION

The cooperation in the last five years within a working party supported by the CEC on non-destructive monitoring of solid alpha waste has led to a circuit for the intercomparison of waste drums and an exchange of detailed technical information. On the eve of the publication of the report on this exercise, some key points may be noted.

The range of monitoring equipment used to assay the same batch of drums has allowed a very large field to be investigated, a field that would never have been covered by a single laboratory. The various research and development efforts have yielded a number of different equipment designs and assay procedures which it is interesting to compare.

Even though intended to represent only light, non-irradiant, low-active waste contaminated by plutonium and with concrete conditioning, the batch of drums required a great deal of effort to prepare and has been designed for a fairly extensive multi-parameter study. The central aim behind setting up the configurations to be compared was to understand the phenomena causing different responses from different installations, this being the only way to achieve improvements or to make well-founded choices. Each assay technique has its own advantages and limitations, and no one method could replace the other.

The choice of a technique and the design of the equipment and its mode of operation largely depends on the waste management policy adopted. First of all, the conditioning of the waste by the producer before monitoring determines the various options and constraints. Indeed, the method(s) to be applied differ(s) depending on whether the waste comes in 25 l drums completely filled with light, non-irradiant waste contaminated by plutonium of a known isotopic composition or in 220 l drums partially filled with irradiant waste consisting of a mixture of metal objects and hydrogenated materials of various densities in unknown proportions, the alpha contaminant being a mixture of Pu and Cm of unknown isotopic composition, and conditioned in a heavy matrix; In the same way, the choice of monitoring methods will differ depending on whether the aim of monitoring is to provide a simple classification according to an alpha activity threshold or a method for assaying fissile material or assaying alpha or total activity.

In general, the passive gamma and neutron methods appear to be more accurate than active methods, whereas the latter can achieve lower detection limits, although these considerations are qualified by the nature and conditioning of the waste. Where they can be applied, the passive techniques are thus to be preferred for assaying large quantities of contaminant, whereas the active techniques, in particular the method using a pulsed neutron generator, are to be reserved for low-threshold assays and cases where passive methods will not work. Combining several assaying techniques will often enable additional information to be acquired, and reduce errors in determining the contaminant.

Gamma spectrometry is suitable for slightly irradiant or non-irradiant low-density waste, having the advantage of enabling the gamma-emitting isotopes to be identified and their quantities and abundance ratios derived and then correlated to determine the isotopic composition of Pu in certain cases.

Spontaneous fission neutron detection is suitable for irradiant waste, but it is essential to know the isotopic composition of the Pu or the alpha emitters. This technique is especially useful where the waste has a low hydrogen density and where the mass neutron emission rate due to spontaneous fission is high.

Activation techniques, in particular the method using a pulsed neutron generator, need to be studied in greater depth. However, although encouraging with respect to sensitivity and low detection limits, initial results with the latter method have been fairly unsatisfactory as regards the assay error, which continues to be substantial even if large quantities of contaminant are present.

Over the last five years, the work carried out by the individual laboratories in addition to this comparison work has enabled industrial experience to be acquired with the methods tested. Finally, it is clear that close cooperation with producers and waste managers is the only way of achieving rational monitoring procedures.

REFERENCES

1. K.P. LAMBERT, J.W.LEAKE, A.I. WEBB and F.J.G. ROGERS. A passive neutron well counter using shift register coincidence electronics. AERE-R9936, 1982.
2. K.P. LAMBERT and J.W. LEAKE.A comparison of the VDC and shift register neutron coincidence systems for Pu-240 assay. Nuclear materials management, Vol 7, N° 4, Winter 1978-'79.
3. K.P. LAMBERT, J.W. LEAKE, A.I. WEBB and F.J.G. ROGERS.Experience with a shift register neutron coincidence counter for passive plutonium assay. Proc. 3RD ESARDA Symposium on safeguards and nuclear material management, Karlsruhe, 1981.
4. L. BONDAR. Passive neutron assay IAEA-SM-260/54. International symposium on recent advances in nuclear materials safeguards. Vienna, Austria, 8-12 November 1982.
5. L. BONDAR and Al. Experimental and theoretical observations on the use of the Euratom time correlation analyser for the assay of up to 1 kg of plutonium. Proceedings of the 6th ESARDA Symposium on safeguards and nuclear management. Venice, Italy, 14-18 May 1984.
6. J. BOUCHARD, R. BERNE, R. BOSSER, C. FICHE. Le contrôle des déchets de faible activité contaminés alpha IAEA-SM-246/57. Proceedings of the symposium on the management of alpha-contaminated wastes. Vienna 2-6 June 1980.
7. R. BOSSER. Alpha contaminated waste control European interlaboratories comparison programme Benchmark configurations (coordination document) 1983.
8. W. EYRICH et al. Neutron well counter for plutonium assay in 200 liter-waste barrels. EUR 6692 EN 1979
9. M. EYRICH et al. Zerstoerungsfreie Plutoniumbestimmung in Abfallgebieten der Eurochemic in Mol KFK 3369, July 1982.
10. H. WUERZ. Plutonium-bestimmung in Abfallgebinden. KFK 3740, pp. 321-345, 1984.

DISCUSSION

P. JOURDE, CEA Fontenay-aux-Roses

What is the energy of the neutrons, the intensity and the duration of the pulse?

R. BOSSER, CEA Cadarache

The energy of the neutrons is 14 MeV. For both Pu-monitoring systems set-up in the UK and in the F.R.G., the intensity of the neutrons is 4×10^8 and 10^9 neutrons/second respectively. Every pulse lasts a few tenths of a microsecond and involves 4×10^7 and 10^7 neutrons. The pulse frequency is 10^{-1} and 10^{-2} sec^{-1} respectively.

MANAGEMENT OF NUCLEAR AIRBORNE WASTES AND TRITIUM RETENTION

W. HEBEL (CEC), A. BRUGGEMAN (CEN-SCK), A. DONATO (ENEA),
J. FURRER (KfK), I.F. WHITE (NRPB)

Summary

An overall review is presented of the outcome of the research activities that have been supported by the Community programme (shared cost action) for the last years in the field of management of nuclear airborne wastes. The kinds of waste dealt with originate essentially from the reprocessing of spent nuclear fuel and from certain treatment and conditioning processes of radioactive waste. Consideration is therefore given to the management of the volatile radionuclides Krypton-85, Iodine-129, Carbon-14 and tritium followed by that of nitric oxides and radioactive aerosols.

1. Introduction

Gaseous effluents from nuclear installations can contain radionuclides or radioactive matter which pass easily over into the airborne phase during certain treatment processes of radioactive material. The reprocessing of spent nuclear fuel and the associated waste treatment processes constitute major sources for such nuclear airborne waste.

Most prominent gaseous or easily volatilized radionuclides, contained in spent nuclear fuel are Krypton-85, Iodine-129, Carbon-14 and tritium. Their approximate contents in one tonne of used LWR fuel after the normal burn-up of 33 GWd/t are given in the following table (1; 2; 4; 8) :

	Half-life (a)	Approximate content (g/t)	Activity Ci/t
Krypton-85	10.8	29	11000
Iodine-129	16.10^6	170	0.03
Carbon-14	5730	0,6	2.2
Tritium	12.3	0,06	570

They are liberated from the spent fuel during its dissolution in nitric acid for being reprocessed. At this stage, also nitric oxides and aerosols are generated which make the second group of airborne waste species present in the dissolver off-gas and being of concern here.

Because of their small quantities, the former radionuclides are strongly diluted in general by the large gas flow (air) through the dissolver or are contained in low concentrations in the liquid effluents of the process. A fact which makes their trapping from the effluents expensive and technically difficult due to the large quantities of gaseous or liquid carrier ballast to be handled simultaneously.

Next to the spent fuel reprocessing, waste treatment processes such as high-level waste vitrification and low or intermediate-level waste incineration, are other important sources of airborne radioactivity.

Here, in addition, semi-volatile radionuclides play a major role like Cesium-137 and Ruthenium-106.

The Community funds committed to R&D activities in the field of airborne waste arrive at 4.4 million ECU in total, which have been spent between 1976 and 1984 in support of various contractual works conducted by research organizations in the Member States. The partitioning of the funds among the fore-mentioned subjects of concern is shown on Figure 1. It can be noted that almost half of the total went into research on Krypton-85, followed by that on tritium (27%) and aerosols (10%). Smaller parts went into the characterization of the off-gas from spent fuel dissolution (10%) and into the drawing up of status reports on the management modes for Iodine-129 and Carbon-14 (2; 3; 4). Also for Krypton-85, such a state of the art report was presented in 1982 (1), next to an earlier study on tritium immobilization (8).

The major effort on Krypton-85 was given to the development of krypton immobilization techniques followed by that of krypton trapping. It constitutes the subject of a separate paper at this conference. Therefore, brief mention is to be made here, only of some general aspects of krypton management.

2. Krypton-85

The CEA at Fontenay-aux-Roses and the University of Pisa drew up together, in the scope of the Community programme, a comparative study of management strategies for Krypton-85. This study was presented and discussed on the occasion of a specialists' seminar held in Brussels, in June 1982. (1) A few points of the outcome may be highlighted here, with regard to the situation of krypton management on the whole.

Mainly two techniques of trapping Krypton-85 from the fuel dissolution off-gas are being developed : the cryogenic rectification process and the fluorocarbon absorption process. The former has reached the stage of inactive pilot plant testing in several countries (e.g., U.S.A., Japan, Germany, Belgium, France). Apart from an older plant at the Idaho National Engineering Laboratory (U.S.A.), operation under real radioactive and industrial conditions has not been performed and would need testing and demonstration. Problems of operational safety remain to be solved for the cryogenic rectification process. They are due mainly to the specific composition and the relatively high air flow-rate of the dissolver off-gas.

The fluorocarbon absorption process, being considered as an alternate or back-up technique, has been developed to the inactive pilot plant scale in the U.S.A. and continues to get attention in Germany, but is less advanced towards industrial application.

Techniques for the immobilization of trapped krypton are also under development. They concern the incorporation in metal matrices (9) and the enclosure into the porous structure of zeolites (10; 12). The embedment of krypton in a metallic matrix by ion sputtering is a preferred process which has reached the state of half scale technical testing. Hot demonstration under representative radioactive conditions would need to be done here as well.

Krypton, being a noble gas, causes comparatively small radiological effects to men and its release into the atmosphere can be tolerated to a certain extent. However, as reprocessing capacities grow, the releases from individual plants might have to be restricted. Not only to keep the radiation exposure down as reasonably achievable in the local environment, but also to avoid an increased accumulation of Krypton-85 activity in the regional and global atmosphere.

In the case of unrestricted release of Krypton-85 emerging from a 1 500 t/a reprocessing plant, the most exposed public individuals in the vicinity of such a plant could receive radiation dose rates of the order of a few mrem/a. This radiation exposure is essentially due to the gamma rays which are associated with the predominant beta decay of Krypton-85. The latter radiation would mainly hit the skin or epidermis of the individuals by causing relatively little biological effects. The frequency of beta emission, however, is more than 100 times higher than that of the gamma rays and may therefore not be completely neglected as to its ionizing effects on the (less projected) biosphere as a whole.

The costs and technical problems of Krypton-85 retention are far from being negligible, and do not seem to be clearly outweighed at present by the radiological benefit of its abatement. But the question should merit a continuous development effort, in particular with regard to growing reprocessing capacities.

3. Iodine-129

Iodine-129 has the extremely long half-life of 16 million years. Even if Iodine-129 is trapped from the fuel dissolution off-gas, its isolation from the biosphere cannot be guaranteed for the timescale of its radioactive decay. It will eventually disperse on a regional scale and will then enter into global circulation, acting as a long-term source of low level radiation. The achievable objectives of waste management for Iodine-129 are thus limited to controlling the time, place and manner of its initial dispersion into the biosphere.

In a previous study (2) in the scope of the Community programme, the technological aspects of Iodine-129 management were examined by the United Kingdom Atomic Energy Authority (UKAEA) and the Commissariat à l'Energie Atomique (CEA), and the radiological aspects were examined by the UK National Radiological Protection Board (NRPB). A further contract was then given to NRPB to complete the radiological assessment and to use the results, together with cost data supplied by CEA under the previous contract, in an example of cost benefit analysis of management modes for Iodine-129.

The management modes considered in this generic study (3) are shown in Figure 2, which also indicates the environmental pathways that were modelled.

Both individual and collective doses were calculated for each of the management modes, normalized for 2 TBq (54 Ci) of Iodine-129 arisings, which represent one year's assumed throughput of one large reprocessing plant (1 500 t/a). The annual individual thyroid dose to members of the public from unlimited atmospheric discharge of that quantity of Iodine-129, calculated by using pessimistic assumptions as is appropriate for the protection of individual members of critical groups, was 700 mrem which is 14% of the annual dose limit of the thyroid or about 20 mrem/a, in terms of dose rate equivalent. In none of the other cases would potential individual doses approach the dose limit. Even if several plant-years' throughput of Iodine-129 were trapped and disposed of as a solid at one disposal site, none of the predicted potential individual doses would exceed the dose limit.

For all the Iodine-129 management modes considered, the predicted collective effective dose commitment to the world population tends towards almost the same value, about 2 000 man.Sv at times beyond 100 million years, largely determined by the global circulation of the Iodine-129. However, effluent discharge modes would give some collective doses immediately, while the less pessimistic of the assumptions for geologic disposal predict zero doses until 100 000 years, as can be noted from Figure 3.

It has already been shown earlier that Iodine-129 should be trapped from the off-gas of a large reprocessing plant, in order to control doses to individual members of the critical group. The present generic cost-benefit analysis comes to the conclusion that the radiological benefits from the reduction of the initial collective doses can outweigh the costs of trapping the Iodine-129 and of its appropriate disposal.

The greatest benefit is achieved by removing the Iodine-129 from the dissolver off-gas, which usually contains most of it, and that should result in a decontamination factor of about 100. Further stages of treatment to improve the decontamination factor do not appear worthwhile. This is because the further potential savings in radiological detriment costs are small, and any further stages of treatment will be more expensive because they must deal with larger and more dilute gas streams. The technical methods for the removal of Iodine-129 from the fuel dissolution off-gas exist and have been brought into use, such as caustic scrubbers or silver impregnated porous silica type filters ("AC 6120"). However, more development work is still needed to establish suitable iodine immobilization methods and matrices as well as accepted waste disposal routes.

4. Carbon-14

The situation with regard to the waste management of Carbon-14 was reviewed, in 1982, by a contractual study conducted jointly by UKAEA and NRPB (4) at Harwell (U.K.). The following findings may be briefly noted.

A reliable picture of the production and release of Carbon-14 from various reactor systems has been built up for the purpose of this study. It is based on a critical analysis of reported calculations and measurements and some new calculations in the case of AGRs and Magnox reactors. Generally, a good agreement exists between various sources of data.

The key problem in carbon-14 management is its retention in off-gas streams, particularly in the dissolver off-gas stream at reprocessing plants. In this stream, the nuclide is present as carbon dioxide, and is extensively isotopically diluted by the carbon dioxide content of the air. The size of plant required for retention, and the quantities of solid waste produced, are therefore determined by the air flow-rate through the dissolver.

Three alternative trapping processes that convert carbon dioxide into insoluble carbonates have been suggested. Any of them could be used as a preliminary to krypton removal, or as part of the product purification system of a fluorocarbon separation process for krypton, or to treat gas regenerated from a molecular sieve bed or cold trap. They are :

- The double alkali process. Carbon dioxide is absorbed here by sodium hydroxide solution. The resulting liquor is treated with calcium

hydroxide to form calcium carbonate and to regenerate sodium hydroxide for re-cycle.
- The direct process. Carbon dioxide reacts with an aqueous slurry of calcium hydroxide to give calcium carbonate directly.
- The barium hydroxide octahydrate process. Carbon dioxide reacts with a fixed bed of solid barium hydroxide octahydrate to give barium carbonate.

Any of these processes is capable of giving a high decontamination factor for Carbon-14.

It is probable that calcium or barium carbonates, produced in the above processes, could be incorporated into cement or bitumen matrices to provide immobilized waste forms. However, the stability of such waste forms to prolonged irradiation and to leaching needs to be investigated.

A number of disposal options for solid Carbon-14 wastes have been identified. The acceptability of the various options will be determined by the ALARA principle, i.e., that all exposures should be as low as reasonably achievable, economic and social factors being taken into account.

The maximum individual doses arising from the atmospheric discharge of Carbon-14 from a reference PWR reprocessing plant (1 200 t/a spent uranium) are estimated at some tenth of a mrem per year.

A conversion to a solid waste followed by disposal in a geologic repository could substantially reduce the collective dose commitment, relative to that arising from the unlimited discharge with effluents.

5. Tritium

Unlike the foregoing subjects, the management of tritium containing wastes has not been assessed by a recent study in the scope of the Community programme.

For several years, however, the development of a process is being supported which aims at the removal of tritium from the aqueous effluents of a reprocessing plant (5). The work is conducted by SCK/CEN at the nuclear research center in Mol (Belgium). It concerns a process (called ELEX) which uses water electrolysis combined with catalytic exchange between the gas and liquid phase, in order to concentrate by isotopic separation the tritium in a small liquid volume while the bulk of the aqueous effluent may be released with a strongly reduced tritium content. The effect of both volume concentration and tritium decontamination comes to a factor of about 100.

The process was first studied on laboratory scale which showed the technological feasibility of the approach. The equipment included a 1.5 KW electrolyser to generate hydrogen gas and a reflux isotopic exchange column of 2 cm inner diameter, filled with hydrophobic catalyst to amplify the tritium enrichment in the liquid phase, already partially realized by the water electrolysis. Essential part of the process is the water-repellent catalyst which contains platinum on a carrier of charcoal incorporated in a teflon matrix. Up to the end of 1982, more than 1 m^3 of tritiated water was processed by this bench scale equipment whereof the feed contained approximately 0.1 Ci/l of tritium; the latter being representative for the tritium content in the aqueous effluents from nuclear reprocessing plants.

Successively, and encouraged by the positive results, a technical scale pilot installation was designed and constructed in Mol. It comprises a 80 KW electrolytic gas generator, a 10 cm diameter isotopic exchange column and has the nominal throughput of about 5 l/h of tritiated water. At present, the pilot installation undergoes final commissioning tests prior to its demonstration runs (<u>Figure 4</u>). It will then contain an inventory of 1 000 Ci or 0.1 g of tritium, respectively.

6. <u>Nitric oxides</u>

A well known and straightforward method for the removal of nitrogen oxides from process off-gas, is the scrubbing by nitric acid or caustic liquor. Certain draw-backs of this method, however, have favoured the research and development into an alternative removal process, that is the catalytic reduction of the nitric oxides by ammonia gas (NH_3).

In the scope of the Community programme, two specific research works were conducted on the subject, one by ENEA in Italy (7) and the other by SCK-CEN in Belgium. In the former case, it is with regard to the off-gas cleaning related to the bituminization or vitrification of wastes charged with appreciable amounts of nitrates, and in the second case, it is with regard to the removal of the nitric oxides emerging from spent fuel dissolution, a necessary process step prior to Krypton removal.

Although the specific waste gas conditions are quite different in both cases, the basic characteristics chosen for the nitric oxide removal process, are similar. They both use a mordenite bed as catalyst, which operates at temperatures in the range of 400 to 450°C.

It was found, at ENEA, that the content of water vapour in the feed gas should be lower than 60 v/o, in order to achieve removal efficiencies of 99% or higher, for the nitric oxides. The control of the temperature of the catalyst bed showed to be important in both cases for a satisfactory NOx reduction. SCK-CEN observed in this respect, that the exact dosage of ammonia supply to the waste gas plays a major role. There, the residual NOx concentrations found in the exhaust gas of the catalyst bed were around 1 ppm compared with a content in the feed gas of somewhat less than 1% (10 000 ppm).

7. <u>Aerosols</u>

Work done by the Kernforschungszentrum Karlsruhe, concerned the development and testing of pre-filters composed of packed glass fibers (20 μm diameter) for the removal of aerosol droplets from the dissolution off-gas of spent fuel. The objective is to clean those filter elements again by spray-washing, in order to re-use them repeatedly for the purpose of reducing the aerosol load on the downstream HEPA-filters, thus increasing their operating life (6).

Development and testing of those prefilters was undertaken by using the so-called PASSAT facility at KfK which is the mock-up of a dissolver off-gas cleaning system, including the removal of nitric oxides, aerosols and Iodine-129. The effectiveness of the whashable pre-filters, in other words of the packed fiber mist eliminators, was studied there in connection with an up-stream droplet separator (wave plate) and a down-stream HEPA-filter (<u>Figure 5</u>).

At gas flow-rates through the system of the order of 100 m³/h and
with measured aerosol droplets in the range of 1 to 23 μm diameter, the
overall removal efficiency of the packed fiber filter was found to be
around and above 99.99% (Figure 6). This corresponds with an aerosol
decontamination factor between 500 and 10^6 depending, on the size of the
droplets.

Tests with radioactive tracer solution (Ba-139) sprayed into the
off-gas flow, showed that the packed fiber filter can be washed afterwards from its aerosol load by spray-flushing, during more or less 15
minutes. About 75 1 of water was found sufficient to bring the pressure
drop through the filter down again to normal values, while washing out
the bulk of activity at the same time. In the case of persisting
important activity load of the filter elements, they can be exchanged
remotely due to the special handling technique developed herefore in the
PASSAT facility.

The combined aerosol filter system, including droplet separator,
packed fiber mist eliminator and HEPA filter, showed a total aerosol
removal factor of about 2.10^7 for droplets of 4 μm average diameter.

In conclusion, it may be noted that the re-cleanable pre-filter can
extend the HEPA filter operating life about 10 times.

REFERENCES

1. W. HEBEL, G. COTTONE (ed.) :
    "Methods of Krypton-85 Management".
    EUR 8464 (1983) Harwood Academic Publishers, Radioactive Waste
    Management, Volume 10.
2. W. HEBEL, G. COTTONE (ed.) :
    "Management modes for Iodine-129".
    EUR 7953 (1982) Harwood Academic Publishers, Radioactive Waste
    management, Volume 7.
3. I.F. WHITE, G.M. SMITH :
    "Management modes for iodine-129".
    EUR 9267 (1984) Nuclear Science and Technology.
4. R.P. BUSH, I.F. WHITE, G.M. SMITH :
    "Carbon-14 Waste Management"
    EUR 8749 (1984) Nuclear Science and Technology.
5. A. BRUGGEMAN, R. LEYSEN, L. MEYENENDONCKX, C. PARMENTIER :
    "Separation of tritium from aqueous effluents".
    EUR 9107 (1984) Nuclear Science and Technology.
6. J. FURRER, A. LINEK :
    "Abscheidung von Tropfen- und Feststoff- Aerosolen in einem Tief-
    bett-Glasfaserpaket-Abscheider".
    EUR 9105 (1984) Kernforschung und Technologie.
7. A. DONATO, G. RICCI :
    "Denitrazione di rifiuti radioattivi liquidi a media attività con
    distruzione catalitica degli ossidi di azoto".
    EUR 9594 (1984), Scienze e Tecniche Nucleari.
8. H.A.C. McKAY :
    "Tritium Immobilisation".
    EUR 6270 (1979), Nuclear Science and Technology.
9. D.S. WHITMELL, R.S. NELSON, R. WILLIAMSON, M.J.S. SMITH, G.J. BAUER:
    "Immobilization of krypton by incorporation into a metallic matrix
    by combined ion implantation and sputtering".
    EUR 8711 (1983) European Applied Research Reports.

10. R.D. PENZHORN, H.E. NOPPEL, A. DOREA, K. GUNTHER, H. LEITZIG, P.
    SCHUSTER :
    "Long-term storage of Kr-85 in amorphous Zeolite 5A".
    EUR 9106 (1984) Nuclear Science and Technology.
11. M.G. NICHOLAS? P. TREVENA :
    "Screening of materials for embrittlement by rubidium".
    EUR 8184 (1983) European Applied Research Reports.
12. E.F. VANSANT, P. DE BIEVRE, G. PEETERS, A. THIJS. I. VERHAERT :
    "Occlusion and storage of Krypton in solids".
    EUR 9353 (1984) European Applied Research Reports.
13. M. KLEIN, W.R.A. GOOSSENS :
    "Aerosol filtration"
    EUR 8951 (1984) Nuclear science and technology.

4.4 M.ECU

7 %	IODINE-129+CARBON-14 (4)
10 %	DISSOLVER OFF-GAS (2)
10 %	AEROSOLS (3)

27 %	TRITIUM (6)
	  IMMOBILISATION
	  REMOVAL

46 %	KRYPTON-85 (15)
	  IMMOBILISATION
	  REMOVAL
	  MANAGEMENT MODES

E.C. FUNDS

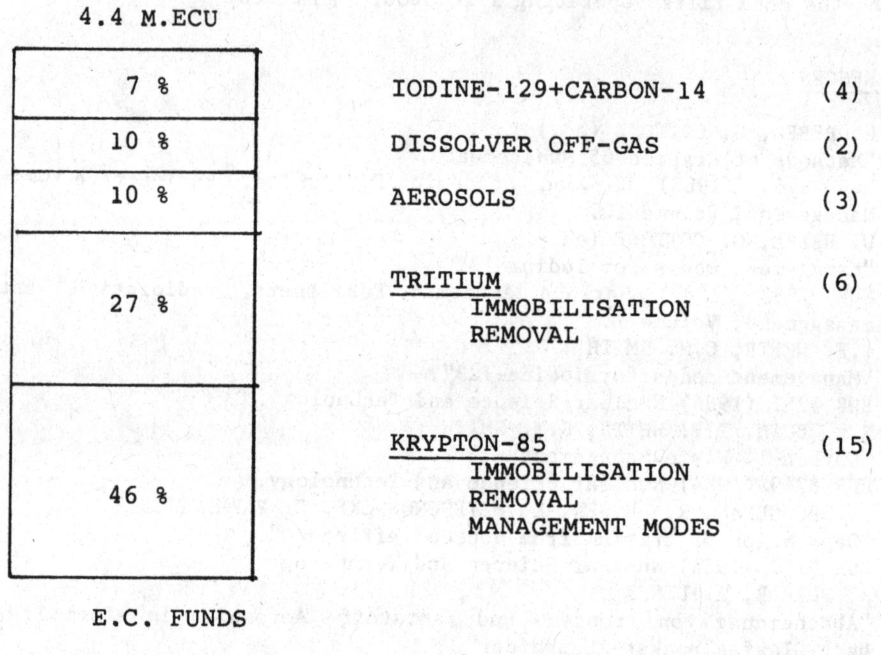

Fig. 1   E.C. SPONSORED R & D (1976-84)
ON NUCLEAR AIRBORNE WASTE

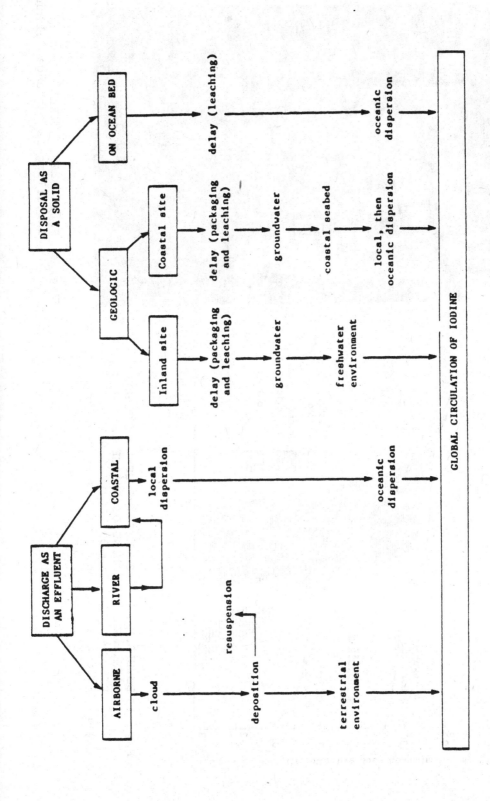

Fig. 2   OPTIONS FOR THE DISCHARGE AND DISPOSAL OF ^{129}I SHOWING SIMPLIFIED ENVIRONMENTAL PATHWAYS

Fig. 4   PILOT PLANT "ELEX"
Installing of exchange Column

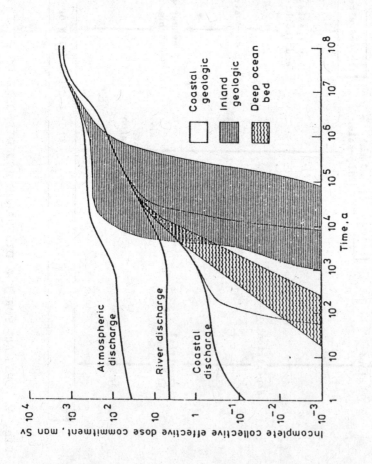

Fig. 3   DEVELOPMENT OF COLLECTIVE EFFECTIVE DOSE
COMMITMENT FOLLOWING DISCHARGE OR
DISPOSAL OF 2 TBq OF IODINE-129
(Note : Shaded areas denote ranges of results
for solid disposal)

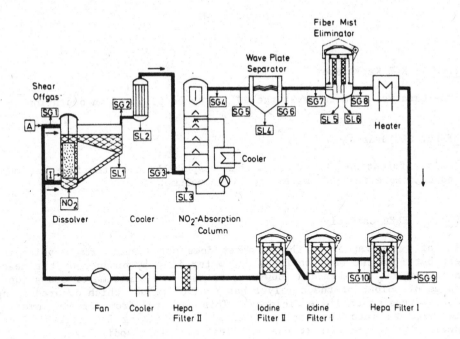

Fig. 5   TEST LOOP (PASSAT)
Off-gas cleaning of spent fuel
dissolution

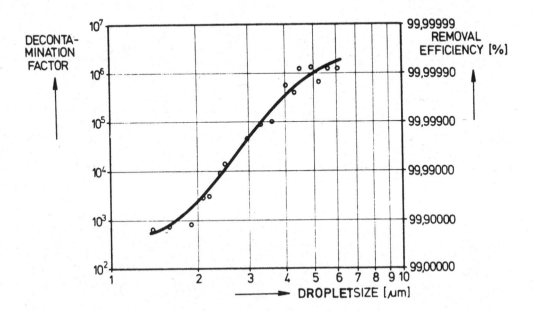

Fig. 6   AEROSOL REMOVAL EFFICIENCY OF THE
WASHABLE PACKED FIBER PREFILTER

DISCUSSION

H. KRAUSE, KfK Karlsruhe

Can the catalyst used to treat the NO_x gases be poisoned in operation?

W. HEBEL, CEC Brussels

The only information available in this respect is that the flue gas must not contain more than 60% water vapour in volume.

A. DONATO, ENEA Casaccia

Action of water vapour on the catalyst does not result from a poisoning effect but from the mass action since water is a by-product of the reaction between NO_x and ammonia. However, after a certain time of operation a change of colour of the catalyst has been observed, which occured along with a decreasing of its efficiency. This resulted from carrying over of waste droplets into the mordenite catalyst entailing a progressive exchange of its acidic form to the much less efficient sodic form.

CAPTURE AND IMMOBILIZATION OF KRYPTON-85

D.S. WHITMELL*, L. GEENS[+], R.D. PENZHORN[‡] and M.J.S. SMITH*
*UKAEA, AERE Harwell, England
[+]SCK/CEN, Mol, Belgium
[‡]KfK, Karlsruhe, Germany

Summary
It may become necessary to contain the krypton-85 released from nuclear fuel during reprocessing in order to reduce the exposure to the local population and the radioactive background throughout the world.

A brief description will be given of studies being carried out in the Indirect Action Programme. The separation of krypton from other off-gases by cryogenic distillation in the presence of oxygen is being studied at SCK/CEN Mol, together with the behaviour of ozone in the distillation column.

Two processes for the immobilization of krypton in solid forms have been successfully developed and demonstrated. At KfK Karlsruhe, krypton is encapsulated in vitrified zeolites; at AERE Harwell, krypton is immobilized within a metallic matrix. These processes offer excellent gas retention and either could be adopted for a reprocessing plant.

1. Introduction

At some time it may become necessary to contain the krypton-85 arising from the reprocessing of nuclear fuel in order to limit the exposure to the local population and the background radiation around the world. Krypton-85 is produced as a fission product at a rate of approximately 10^4 TBq (or 4 10^5 Curies) per GW(e) year and normally remains in the fuel until released during reprocessing when it forms part of the gaseous waste stream together with larger quantities of xenon and stable krypton isotopes. Since it is a beta and gamma emitter with a half life of 10.8 years it needs to be stored for at least 100 years. Containment poses particular problems since it is an inert gas, not forming stable solid compounds, and its decay product rubidium may cause corrosion of the container during storage. Storage as a gas in a pressurized cylinder is not ideal, since rupture of the cylinder during storage would result in the release of a large quantity of activity.

The envisaged method of treating krypton is therefore to separate it from other gases and then to immobilize it in a solid form. Separation reduces the volume of gas to be stored since krypton represents only about 60 ppm of the waste gas stream. Cryogenic distillation methods are normally used to separate the krypton [1-3]. In the past the elimination of oxygen was considered to be a necessary pretreatment step (in addition to the removal of other constituents such as iodine, NO_x, CO_2 and water) in order to prevent the formation of ozone in the cryogenic unit by radiolysis induced by the beta irradiation from the krypton-85. Oxygen can be eliminated by catalytic reduction with hydrogen, but hydrogen creates a

serious safety problem and the cost of the plant is increased
considerably.

A new operation mode is being investigated at SCK/CEN Mol in which the
oxygen is allowed to remain in the feed gas, since recent work[4] has
indicated that the accumulation of ozone in the cryodistillation unit can
be reduced or even prevented under the correct conditions provided the NO_x
is removed first. The work on the removal of the nitric oxides by
catalytic reduction with ammonia is described in another paper[5]. The
experience with oxygen and ozone in the feed gas to the cryogenic
distillation column is discussed below.

Two methods for the containment of krypton are being studied in
programmes supported by the Indirect Action programme: at KfK Karlsruhe the
krypton is encapsulated in vitrified zeolites, while at AERE Harwell it is
immobilized in a metallic matrix. Both promise to offer suitable, economic
methods for containing radioactive krypton.

2. Separation of Krypton from Feed Gas in the Presence of Oxygen

The system developed at Mol to investigate the separation of krypton
from feed gas in the presence of oxygen is shown schematically in Figure 1.

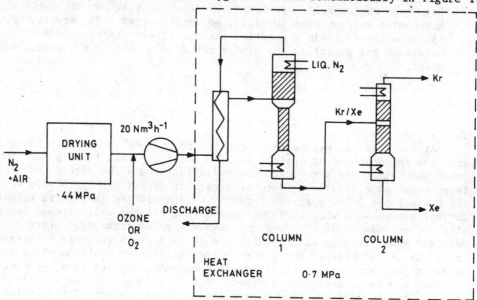

Fig. 1 Scheme of experimental cryogenic distillation equipment.

Air or a mixture of air and nitrogen is fed at a pressure of 0.42 MPa and a
flow rate of 20 $Nm^3.h^{-1}$ into a drying unit, which consists of 3 columns
packed with an acid resistant zeolite, in order to prepurify the gas stream
by removing water vapour and carbon dioxide before entering the cryogenic
distillation unit. The pressure of the feed gas is then increased to 0.7
MPa by a membrane compressor and the gas stream loaded with krypton, xenon,
and oxygen (or ozone).

The distillation unit consists essentially of two parallel heat
exchangers, where the feed gas is cooled; a first continuously operating
rectification column in which the krypton and xenon are separated from the

feed gas; and a second batch operated distillation column where the krypton and xenon are separated. The unit is shown in more detail in Figure 2.

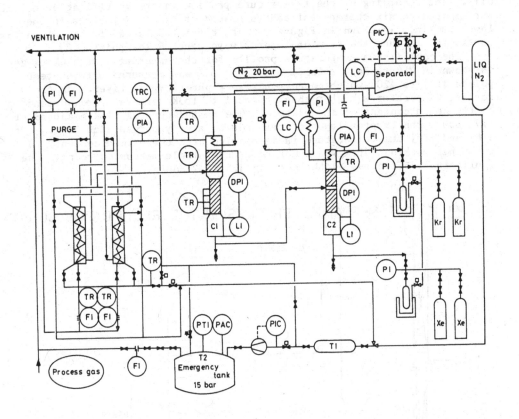

Fig. 2   Cryogenic distillation unit.

The main operating characteristics of the first column[3,6] are given in Table 1.

Table 1 : Main Characteristics of the Rectification Column

Gas flow rate	:	$25 \text{ NM}^3.\text{h}^{-1}$
Operating pressure	:	0.70 MPa
Inlet temperature	:	124 K
Reflux ratio	:	0.3-1.0
Packing	:	3 x 3 x 0.4 mm spring Raschig rings
Upper section	:	h = 0.80 m, d = 0.10 m
Lower section	:	h = 0.72 m, d = 0.07 m

2.1 Experimental Results with Oxygen

Experiments were carried out with three different oxygen concentrations in the feed gas: 6%, 12%, and 21% volume, (the latter being air). The recording of the temperature profile in the rectification column was limited to six thermo-resistances spaced at 7.5 cm intervals in the lower packing, as shown in Figure 3. The thermal balance of the column was controlled so that the krypton layer always reached the point [18].

The equilibrium temperature profile for the experiment with 6% oxygen is shown in Figure 3a. A temperature of 153K was measured between test points [18] to [20] which corresponds to a pure krypton layer. In the kettle, temperatures higher than or equal to 153K were recorded, depending on the xenon content of the mixture. The temperature of 116K measured at [17] above the krypton layer corresponds to a very oxygen-rich layer. At [16] and [15] the temperature is decreasing with the oxygen content. In the upper packing, 103K was recorded at the single measuring point. The equlibrium profile was established in 1.5 to 2 hours.

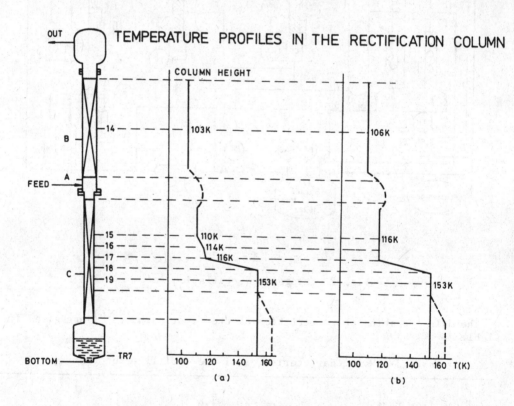

TEMPERATURE PROFILES IN THE RECTIFICATION COLUMN

Fig. 3   Temperature profiles in the rectification column.
(a) 6% vol. O_2 in feed.
(b) 12% or 21% vol. O_2 in feed.

Figure 3b shows the temperature profiles for oxygen concentrations of 12% and 21% which gave similar distributions. The temperature profile is

identical below [18]. However, a constant temperature of 116K was now
measured at the three points above [18]. When the krypton layer was
lowered it was observed that the oxygen-rich layer, which contained 85%
oxygen, extended over all six measuring points, so that in normal operation
it extended from [17] almost to the inlet section of the column. In the
upper packing 105K was recorded. The profile was established in about 60
minutes with 12% oxygen and in 30 minutes with 21% oxygen.

The oxygen analyses were limited to a few sampling points, summarized
in Table 2 where the gas phase compositions are given. In the bottom
product the oxygen concentrations were lower than the detection limit of
the oxygen monitor, (0.1 ppm volume). The table also gives the duration of
the three experiments.

Table 2 : Oxygen Concentrations in the Rectification Column

Experiment number	1	2	3
Feed	6.0	12.0	20.9% volume
Discharge	6.2	12.1	20.9% volume
A	7.0	13.0	22.8% volume
B	7.9	15.6	25.7% volume
C	0.1	0.1	0.1 ppm volume
Bottom	0.1	0.1	0.1 ppm volume
Duration	312	528	720 hours

According to the temperature profiles recorded, even with air, no
changes in any set point of the differential control and alarm systems had
to be made with different oxygen feeds. Furthermore, the very low oxygen
concentration of 0.1 ppm volume in the bottom product means that, if ozone
accumulation can be avoided, no oxygen will be transferred to the second
batch distillation column together with the 9:1 xenon-krypton mixture.

2.2 Experimental Results with Ozone

A preliminary experiment has been made to study the behaviour of ozone
in the cryogenic distillation column. When ozone was fed in concentrations
ranging from 50 to 100 ppm volume, no ozone was detected in the upper part
of the column or in the outlet stream. The ozone accumulated in the bottom
product of the rectification column at a concentration of 1 vol. %. The
temperature remained unchanged at this point. This ozone concentration did
not affect the transfer to or operation of the second batch distillation
column.

3. Encapsulation of Krypton in Zeolites

Recently it has been shown at KfK Karlsruhe that noble gases can be
trapped efficiently and stably in a variety of type A zeolites as well as
in H or Na zeolons, when the original crystalline structure is
hydrothermally destroyed in the presence of a densified gas[7,8]. This
laboratory observation led to the development of a technology for the
long-term storage of radioactive krypton released during reprocessing of
irradiated fuel from commercial nuclear reactors[9].

Briefly, the fixation process envisaged for disposal of Kr-85 from a reprocessing plant involves a one-way autoclave filled with zeolite 5A pellets. For gas fixation the zeolite is first dehydrated under vacuum. The amount of gas needed for the immobilization is then introduced into the autoclave by adsorption on the zeolite at a temperature a few degrees below 0°C (under these conditions the zeolite has a high sorption capacity for krypton). Next, the autoclave is closed by a valve and heated up to the fixation temperature. This causes gas desorption with a concomitant rise in pressure. At the fixation temperature the zeolite crystal structure vitrifies and noble gas is trapped. After completion of the vitrification the autoclave is cooled and the non-immobilized gas expanded into a large vessel. Finally, the autoclave is closed by pressing and welding the access tube. It is now ready for transport and interim or long-term disposal. The equipment is shown schematically in Figure 4.

3.1 Fixation Mechanisms

The mechanism of gas fixation has been examined systematically by electron and X-ray diffraction analysis[10]. With the latter method it was observed that with slow isobaric water desorption, crystalline zeolite 5 A transforms irreversibly at about 850°C into nepheline and anorthite. When amorphous zeolites, which had been loaded with noble gas at either 430 or 650°C, were slowly heated by the same procedure, recrystallization into the same products took place at the same temperature, regardless of the vitrification conditions. This observation, as well as electron microscopical pictures are in excellent agreement with results from gas leakage experiments. Thoroughly vitrified samples only liberate the trapped gas at the phase transformation temperature (850°C), and below this temperature the leakage rate is small. However, incompletely vitrified samples are characterized by a significant leakage at temperatures below the transformation point.

The crystalline/amorphous transition in zeolites can also be followed by high-resolution solid-state ^{29}Si NMR with magic angle spinning (MAS-NMR): a technique primarily sensitive to short-range order[12]. When the phase transition occurring during gas fixation in zeolites is examined, a spectrum is obtained which is very different from that of the parent zeolite. The chemical shift is now characteristic of Si surrounded tetrahedrically by four Al-centred tetrahedra. The large change in chemical shift shows that the structural identity of zeolite A has been lost, even at the unit cell level. This observation is substantiatied by IR spectroscopy which reveals a spectrum analogous to amorphous alumino silicates such as glasses.

Zeolite samples loaded with noble gases have been characterized by a number of physical methods[8]. Samples loaded with Kr as well as Ar/Xe and Kr/Xe mixtures were examined by microprobe analysis[12]. Comparison of the secondary electron images with corresponding X-ray images of unloaded, loaded and deloaded 2 mm pellets indicate that the density of the noble gas follows exactly the density of the aluminosilicate framework. Thus most of the noble gas is present in small units the size of zeolite crystals (about 2 μm or less), and not as bubbles in a sintered pellet.

It is concluded that hydrothermally immobilized gas is not trapped in the intercrystalline space inside a sintered pellet but trapped in a thermally very stable way in a truly amorphous solid. This is not only demonstrated by the microprobe analysis, but also by the fact that essentially no gas is released when a loaded pellet is ground.

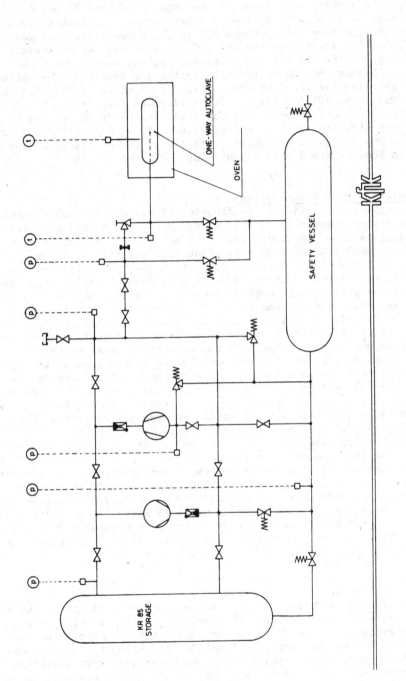

Fig. 4 Scheme of demonstration facility for the encapsulation of Kr in zeolite 5A.

3.2 Thermal Release Measurements

The rate of release of noble gases such as Ar, Kr and Xe out of vitrified type A zeolite containing the cations Mg, Ca and Sr has been carefully measured in a large number of experiments. The leakage data, obtained within the temperature range 450-700°C, demonstrate that gas evolution is well described by the $\sqrt{t}$ law. In general, the activation energies for gas diffusion through vitrified zeolite were found to be substantially higher than those through crystalline zeolite. The estimated activation energies for the diffusion of Kr out of vitreous MgA, CaA and SrA were found to be 191.6 $\pm$ 17.2, 210.3 $\pm$ 22.2 and 202.2 $\pm$ 7.6 kJ/mol respectively. The activation energy obtained for the Kr/CaA system is in good agreement with previously reported values of 197 kJ/mol[13] and 209 kJ/mol[14]. The estimated diffusion coefficients for Kr diffusion out of vitrified 5A zeolite are 3 x 10^{-20} cm^2/s at 500°C and 5 x 10^{-19} cm^2/s at 600°C.

3.3 Demonstration with Active Gases

The equipment employed for the fixation of radioactive noble gases in zeolites consists essentially of a 40 Ci Kr-85 vessel, a lead shielded 2 ℓ storage vessel, a 4 cm^3 pressurization trap and a 1 cm^3 autoclave containing the zeolite. The pressurization trap together with the autoclave constitute the actual fixation unit. The pressurization trap can be either heated externally up to 350°C with an oven or cooled with liquid N$_2$, fed into the glove box through a vacuum isolated fitting to trap the krypton. The autoclave is connected to the pressurization trap via a metal to metal seal provided with a stainless steel gasket. This arrangement permits easy recovery of the radioactive zeolite after completion of each run. The pressure is registered with two externally water cooled transducer sensors. All heating and cooling steps were controlled from outside the glove box with a temperature programmer unit. Information concerning the gas density inside the fixation unit could also be derived from a radioactivity measurement. For this purpose a scintillation detector provided with a plastic scintillator, was placed partially inside the lead shielding around the autoclave.

In a typical run approximately 1080 GBq (30 Ci) of gaseous Kr/Kr-85 was kept in the fixation unit for several hours at pressures up to 300 bar. During vitrification the temperature of the pressurization trap, the metal to metal seal and the autoclave reached 350, 150 and 520°C respectively. Since under these conditions no radioactivity was detected either in the stack or in the glove box, the employed metal to meal seal provides adequate gas containment. Several experiments with several hundred curies of Xe-133 activity substantiate this conclusion. After completion of each run the autoclave was opened and its radioactive contents transferred into small glass bulbs. This operation, which was carried out without any special precautions, occurred without measurable contamination of the glove box by radioactive aerosols.

Krypton from a reprocessing plant has a specific activity of the order of 3700 GBq/ℓ, which, when trapped in a zeolite, will give rise to a specific activity of about 93 GBq/g. The hottest Kr-85 samples prepared so far have about 20% of that activity (some Xe-133/5A samples are perhaps more than 5 times as hot). They may thus be considered representative for long-term characterization, which will be carried out in closed loops with mass spectrometric analysis. So far all Kr-85 and Xe-133 samples have been kept under a lab hood for a period of up to two years, without morphological changes or liberation of radioactive gas becoming apparent. The radiation dose of the highly radioactive Kr-85/5A samples (1 cm^3 volume), measured behind 1.3 mm glass, has remained unchanged at about 1 rem/h for a period of up to 9 months.

3.4 Trapping of Other Gases

When gaseous mixtures such as Xe:Ar = 1:1 or Xe:Kr = 1:1 were trapped in zeolite 5 A an enrichment of the species with highest sorption affinity was observed. Impurities such as N_2, O_2, noble gases, CO_2, H_2, etc. are trapped together with Kr.

An appreciable fraction of the C-14 produced in nuclear reactors is converted to gaseous effluents during fuel reprocessing. The typical composition of the off-gas of a HTGR fuel reprocessing plant is 95% C-14 contaminated CO_2, with ppm concentrations of Xe and Kr-85. In a 350 tU/year LWR reprocessing plant, handling fuel with burn-up of 33000 Mwd/t, an annual $^{14}CO_2$ arising of about 210 Ci is expected. If the dissolver atmosphere is air, a production of about 187 kg of $CO_2/^{14}CO_2$ can be estimated. Since $^{14}CO_2$ has been identified as a potential biohazard because of its long half-life and the ease with which it is incorporated into living matter, processes based on irreversible solid/gas reactions have been proposed to remove the gas from the off-gas[15]-[18]. However, arising problems have not been adequately solved.

Experimentally it has been shown that CO_2 and CH_4 can be incorporated in high yield in hydrothermally vitrified 3A, 4A and 5A zeolites, as well as in Na zeolon, H zeolon and clinoptilolite, employing relatively mild fixation conditions. For instance, at 490°C and 200 bar CO_2 a loading of 35 cm^3 STP/g was obtained in zeolite 4A employing a 5 ℓ one-way autoclave. At an initial CO_2 pressure of 1 bar sufficient gas was adsorbed by the zeolite to yield a pressure of 200 bar upon desorption when the autoclave was isolated and heated up to 490°C (sorbopressurization). Further optimization of the fixation parameters appears possible. Short and long-term leakage tests carried out with loaded samples containing CO_2 or CH_4 have demonstrated the high thermal stability of the encapsulates. In view of the above it appears possible to dispose radioactive CO_2 and CH_4 in vitrified zeolite together with Kr-85 in one single process, without necessitating a hydrocarbon oxidation step.

3.5 Demonstration Facility

A facility has now been designed, constructed and demonstrated which includes several 5 ℓ thick-walled externally heated one-way autoclaves (length 1000 mm and i.d. 80 mm) provided with a quick connector, a 7 kw oven, several pumps, a compressor, temperature and pressure registration devices, two 50 ℓ storage cylinders, etc. Two kinds of safety devices guarantee the integrity of the autoclave: on one hand three thermocouples touching the outer surface of the autoclave, which shut off the power supply when a preset temperature of 550°C is exceeded; on the other, two pressure relief valves, which expand the gas into a 50 ℓ vessel when a pressure of 300 bar is exceeded.

From measured temperature profiles it is concluded that after about two and a half hours a homogeneous temperature distribution in the autoclave is attained. Altogether 14 runs, completed so far with zeolite 4A and 5A and the gases Ar, Kr and CO_2, have demonstrated that previously obtained laboratory results can be reproduced on a technical scale.

The main advantage of the one-way autoclave concept is seen in the
simplicity of the process. Fixation is carried out without a high pressure
compressor, complex electronics, employing exclusively commercially
available technology. No transfer of radioactive zeolite pellets is
needed. Since the one-way autoclave is used only once, metal fatigue is
unlikely. The one-way autoclave can be considered a true second barrier
for transport and final storage, because of its inherent resistance towards
corrosion, high pressures, high temperatures and mechanical impact.
Equipment and operation costs are estimated to be comparatively low.

4. Immobilization in a Metal Matrix

4.1 Process

The process[19,20] developed at Harwell is shown schematically in
Figure 5. A glow discharge is generated between the cylindrical electrodes

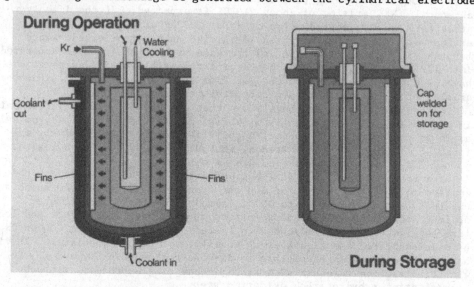

Fig. 5 Scheme of process to immobilize krypton in a metallic matrix.

by applying a negative potential of 3–5 kV in the presence of krypton gas
at a pressure of about 10 Pa. Ions produced in the discharge bombard the
negative electrode causing both ion implantation and sputtering of the
electrode. When the negative potential is applied to the outer electrode,
gas is implanted into its surface. The polarity of the discharge is then
reversed so that the implanted gas is coated with a fresh layer of metal
sputtered from the central electrode. The process is repeated typically a
few times per second so that a thick layer of matrix containing gas at a
concentration of 170 ℓ (at STP) per litre of metal is built up. When a
layer 20 mm thick has been laid down, the vessel is sealed and removed
without dismantling to provide a secondary container.

The process has been demonstrated using a half-scale, 50 kW inactive
pilot plant to prepare a copper matrix 22 mm thick, weighing 23 kg and
containing over 300 litres of krypton in a vessel 0.3 m long and 0.26 m
diameter[21,22]. The gas incorporation rate exceeded the design figure of

0.3 litres/hour, the vessel was operated at powers up to 30 kW, which corresponds to the power density envisaged for an industrial plant, and the power consumption was less than 100 kWh/litre. The efficiency remained unchanged at full power even when the matrix reached the maximum thickness. The process was tested using a gas mixture containing 15% of argon, nitrogen, xenon and hydrogen which were incorporated at the same rate. A vessel 1 m long has also been tested at low power densities and an automatic process controller developed.

Samples of copper, nickel, iron, aluminium, stainless steel, incaloy, monel, and the glassy alloy, nickel-10% lanthanum have been prepared and tested to determine the long term retention of the krypton by the matrix and its leach, corrosion and temperature resistance. Nickel showed the highest process efficiency: about 70% greater than copper, which represents a significant saving in running costs for an industrial plant[23].

4.2 Properties of Metal Matrix

Corrosion tests were carried out on copper and nickel in water and brine[24]. The weight losses of the copper and nickel in distilled water at 100°C were found to be only 0.003% and 0.03% per year respectively. The corrosion rates in brine at 90°C in the presence of oxygen were 1% and 0.1% per year respectively.

The release of gas from thick samples of copper and nickel has been measured for periods over 1 year. The gas release follows a diffusion controlled dependence after an initial incubation period, with the total amount increasing with the square root of time. The activation for copper is about 2.2 eV (212 kJ/mole). The amount predicted to be released from copper during storage for 100 years at a temperature of 150°C is less than 0.05%. Even at 300°C the amount released will be less than 1%. The thermal stability of krypton in nickel is even greater and the equivalent temperature for a given amount of release will be at least 100°C higher.

Metallurgical examination[24,25] showed that the matrix contains a very high density of minute bubbles (1 to 2 nm diameter) in grains of typical size 0.5 μm. When the matrix is heated the gas bubbles in the grain boundary grow, agglomerate and form long pipes of gas which finally intersect the surface. Recently it has been shown[26] that the pressure in the bubbles is so high that the krypton is solid and gives a characteristic electron diffraction spot pattern. When heated beyond 600K the krypton turns to gas and the diffraction spots disappear, only to return when the matrix is cooled.

No gas was released when the matrix was irradiated with 1 MeV beta or by gamma irradiation to doses up to those expected during storage for 100 years. The matrix is therefore resistant to the effects of radiation emitted by Kr-85.

The physical properties of the matrix have been studied[24] since the matrix material is unique. The densities of gas-filled matrices are typically 95% of the bulk parent material; X-ray diffraction gave broadened peaks and lattice parameters about 0.5% greater than the bulk material. The matrices were initially very hard (copper 330 VHN and nickel 700 VHN) but the hardness decreased on annealing. When gas was not implanted the hardness of the nickel was only 140 VHN. The thermal conductivity of the copper was about 20% of that for pure copper, and nickel about 30%, which are similar to the thermal conductivities found in conventional alloys. The thermal expansion coefficient was measured as a function of annealing temperature since it is possible that changes in

bubble size and internal pressure might alter the coefficient
significantly. Some variations were seen but there were also random
fluctuations in the measurements. Differential scanning calometric
measurements, neutron small angle scattering and positron annihilation
experiments have also been carried out using the matrix material which has
provided the first suitable gas-filled material for these studies.

Nickel was selected as the best material for the matrix since it
offers the optimum combination of process efficiency and gas retention
characteristics. A complete vessel containing a nickel matrix is now being
stored and the gas release rates measured as a function of temperature.

4.3 Tests with Active Gas

Tests have been carried out to demonstrate that the process works
equally well in the radiation levels which will develop in a fully active
plant. Calculations showed that the beta radiation level inside the vessel
rises rapidly as the Kr-85 is incorporated in the matrix but saturates at
80,000 R/hour when the layer is about 0.1 mm thick. The gamma dose
continues to increase to give a maximum of about 4000 R/hour internally
when the matrix is 20 mm thick.

Since the supply of Kr-85 is limited, 100 curies of Kr-85 were used,
enabling the internal beta levels to reach about 95% of the saturation
level. The rig, shown schematically in Figure 6, consists of a small

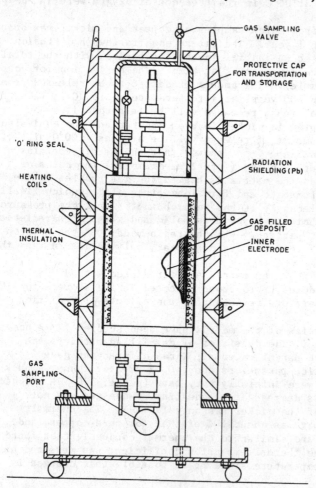

GAS SAMPLING VALVE

PROTECTIVE CAP FOR TRANSPORTATION AND STORAGE

'O' RING SEAL

HEATING COILS

THERMAL INSULATION

RADIATION SHIELDING (Pb)

GAS FILLED DEPOSIT

INNER ELECTRODE

GAS SAMPLING PORT

Fig. 6 Scheme of vessel
to immobilize radioactive
matrix.

vessel, (0.2 m long, 0.2 m diameter), within a 50 mm lead shield and mounted upon a trolley. A heater is wound around the vessel so that the vessel can be stored afterwards at a temperature of 150°C. A layer of nickel 3 mm thick containing inert gas was deposited and then the gas changed to the fully active gas (6% Kr-85) and an active layer 0.1 mm thick deposited. A surface closure layer was then laid down by changing back to inert gas. The experiment was completely successful, no effluent was produced, and there was no detectable exposure to the operators, demonstrating that the active gas could be safely handled in an open laboratory without any additional safety precautions such as glove boxes.

The vessel has been sealed, a cap placed over the insulators to provide additional protection, and the vessel moved to a store where it will stay for up to 5 years at 150°C. Samples of the gas taken from the vessel do not show any leakage of gas from the matrix. Subsequently samples of the nickel containing the active gas and its decay product, rubidium can be examined to determine the effect of decay.

Tests into the effects of gamma have also been carried out by placing gamma sources around the vessel to give a field of 7000 R/h. No effects were found.

4.4 The Design of a Full-Scale Plant

Previous conceptual designs for a plant have considered process vessels 1 m long, 0.26 m diameter with a nominal volume of 50 litres so that comparisons can be made with alternative storage methods. A vessel of this size has been tested at powers up to 50 kW in order to determine its electrical characteristics. Since its performance was satisfactory even larger vessels are likely to be feasible so reducing the number of vessels required to be on line at a large reprocessing plant. A 800 tonne/year plant would need only 3 vessels, 0.4 m diameter, 1.6 m long, on line. These would be changed every 60 days. The costs of the process are expected to be about US $ 0.5/kg U due mainly to capital and energy costs; labour costs are low since the process can be controlled automatically and handling requirements are minimal.

The metal matrix process offers an ideal method for the immobilization of radioactive krypton. Operation of the pilot plant has shown that the process is suitable for active gas, thick matrices can be prepared, impurities are incorporated and no effluent is produced in a sub-atmospheric process operated at ambient temperatures. The matrix can be formed from a wide range of materials to give a product which retains the gas well under normal storage and accident conditions and the process vessel itself provides additional protection. The cost of the process is small compared with other waste management costs.

5. Discussion and Conclusions

The aim of the CEC programme is to encourage the development of krypton waste management technologies so that they are available when required. Although separation of krypton by cryogenic distillation has already been carried out, steps to avoid the prior removal of oxygen would make it easier to meet safety standards, lead to simplified plant design and lower costs.

Within the programme, two methods for immobilization of krypton have been taken successfully from laboratory to pilot plant scale. Both have been shown to be suitable for active gas and they have been developed further towards industrial use than alternative processes elsewhere. The products possess excellent gas retention properties both in storage and potential accident conditions.

Although detailed costings have not yet been prepared, the costs will be relatively similar and small compared with other waste management costs. The cost is likely to be significantly smaller than that of storing krypton as a gas in cylinders within a special store designed to protect the cylinders. The choice of process will depend on the plant operator and regulatory requirements: the zeolite process is a high throughput batch process carried out at an elevated temperature and pressure, whereas the metal matrix process requires several units operating continuously in parallel at near ambient temperature and at sub-atmospheric pressure.

Both processes have been demonstrated at near full scale and with limited quantities of active gas. The next stage will be operation at an active krypton separation plant.

Acknowledgements

Many colleagues have contributed to the programmes and their help is gratefully acknowledged.

REFERENCES

1. BENDIXSEN, C.L. and OFFUT, G.F. (1969). Rare gas recovery facility at the Idaho Chemical Processing Plant. IN-1221.
2. von AMMON, R. et al. (1977). Auslegung der Tieftemperatur-Rektifikationsanlage KRETA, einschliesslich der vorreinigungsanlagen ADAMO und REDUKTION und erste Betriebserfahrung. Seminar "Fuel Reprocessing Plant Effluents", Karlsruhe.
3. GEENS, L. et al. (1981). Experience gained in the cryodistillation unit for krypton removal. BLG 547.
4. HUYSKENS, P. et al. Personal communication.
5. Hebel, W. et al. (1985). Management of airborne wastes and tritium retention. Second Europen Community Conference on "Radioactive Waste Management and Disposal", Luxembourg.
6. COLLARD, G. et al. (1982). Développements récents dans le domaine de la rétention du krypton-85 par distillation cryogénique. CEC Specialists' Meeting on "Methods of Krypton-85 Management", Brussels.
7. PENZHORN, R.-D. et al. (1980). Long-term storage of Kr-85 in zeolites. International Symp. on Management of Gaseous Waste from Nuclear Facilities, Vienna, Feb. 18-20.
8. PENZHORN, R.-D. et al. (1982). Noble gas immobilization in zeolites. Ber. Bunsenges. Phys. Chem. 86 1077.
9. PENZHORN, R.-D. et al. (1983). The long-term storage of radioactive krypton by fixation in zeolite 5A. Proceedings of the 17th DOE Nuclear Air Cleaning Conf., Denver, Colorado, CONF 820833 1 358.
10. PENZHORN, R.-D. and MERTIN, W. (1984). To appear in J. of Solid State Chem.
11. KLINOWSKI, J. et al. In press.
12. VANSANT, E.F. et al. (1984). Microprobe analysis of noble gas encapsulates in zeolites. Zeolites 4 35.
13. CHRISTENSEN, A.B. et al. (1981). The immobilisation of Kr-85 in zeolite 5A and porous glass. ENICO-1102.
14. PENZHORN, R.-D. (1980). Long-term storage of Kr-85 in zeolite 5A. Proceedings of the 16th DOE Air Cleaning Conf., CONF-801038 2 1047.

15. NOTZ, K.J. et al. (1980). Processes for the control of $^{14}CO_2$ during reprocessing. IAEA Symp. on Management of Gaseous Wastes from Nuclear Facilities, Vienna, IAEA-SM 245129, 191.
16. SCHMIDT, P.C. (1979). JUL-1567.
17. HAAG, G.L. (1983). Application of the carbon dioxide-barium hydroxide gas-solid reaction for the treatment of dilute carbon dioxide-bearing gas streams. ORNL-5887.
18. HAAG, G.L. et al. (1983). Carbon-14 immobilisation via the $Ba(OH)_2.8H_2O$ process. ORNL-5926.
19. WHITMELL, D.S., WILLIAMSON, R. and NELSON, R.S. (1977). Proceedings 'IPAT 77', CEP Consultants, Edinburgh 202.
20. WHITMELL, D.S., NELSON, R.S., WILLIAMSON, R. and SMITH, M.J.S. (1979). Nuclear Energy 18 349.
21. WHITMELL, D.S. (1982). Nuclear Energy 21 181.
22. WHITMELL, D.S., NELSON, R.S., WILLIAMSON, R., SMITH, M.J.S. and BAUER, G.J. (1983). European Applied Research Reports 5 (3), 513 (EUR-8711).
23. WHITMELL, D.S. (1984). 18th DOE Nuclear Airborne Waste Management and Air Cleaning Conference.
24. WILLIAMSON, R. (1984). AERE-R11273.
25. ELDRUP, M. and EVANS, J.H. (1982). J. Phys. F: Met. Phys. 12 265.
26. EVANS, J.H. and MAZEY, D.J. (1985). J. Phys. F: Met. Phys. 15 L1-6.

DISCUSSION

H. KRAUSE, KfK Karlsruhe

I was very much surprised about your statement that the implantation of
krypton in metal has a comparable management cost — or is even cheaper —
to storage in bottles. Could you comment on that?

D.S. WHITMELL, UKAEA Harwell

This results from the safety requirements imposed on the storage facilities for the disposal of pressure vessels in shallow land burial. This
has been extensively discussed in the CEC report (EUR 8464) on management
of Krypton-85.

PANEL SESSION

TREATMENT AND CONDITIONING OPTIONS FOR LOW AND INTERMEDIATE LEVEL WASTE

Chairman: R. Kroebel (KfK)
Secretary: G. Bertozzi (CEC Ispra)
Panel Members: A. Duncan (UKDOE)
                 W. Goossens (CEN/SCK)
                 G. Baudin (CEA)
                 R. Simon (CEC)
                 G. Grossi (ENEA)

R. KROEBEL, KfK Karlsruhe

We have decided to structurate the discussion so that my colleagues and myself can talk about medium and low level waste. We will talk about the EEC programme on this score. We will endeavour, independently of our countries, to discuss things in a European way and we will each tackle a particular field. Then, depending on the way time goes, we will discuss and we will have contributions from you.
Now, for what concerns categorizing waste in types, we have just mentioned some of them, but it is not an exhaustive definition for medium and low level. There is not such a bible yet, even if the IAEA has already suggested a categorization.
Now, let me give you the following examples. Nuclear wastes always pile up from fuel rod reprocessing plants, except for the first cycle raffinates; for example structural parts of fuel rods and scraps. Then, the category of solid non-burnable and the possible burnable residues; then, other process solutions and aqueous contaminants, alpha, beta gamma waste. Then, the organic extractants, organic liquids, organic combustible solids, ion exchangers, off-gases and off-gas components, such as filters and liquefied or bottled-up gas and tritium waste, tritiated water, for example.
For all these varieties there are operational procedures, tried and tested now for 20 years in some cases, or longer. More than 20 procedures exist for hulls and could be described. Now, it is impossible to define which one of these is really the best; this depends on many parameters, for example whether you should have a near-field effect or not, in any connection with their costs or adequate safety, and with the amount of room, shielding, transport, gas-release on the long term and so on. On top of that you have possible damage during transport from the intermediate to the final storage. An expensive sophisticated reprocessing may in the end be cheaper than a simple cheap one, because of transport and repository costs.

G. BAUDIN, CEA Fontenay-aux-Roses

Today's papers have shown us that there is never a single solution to a problem, and that researchers have lots of ideas to help find better solutions. Today we heard about what performances have to be obtained to come up with a solution which is satisfactory in a given storage situation;

what the impact of one or the other process or procedure is, given that
economics should not prevail over safety. The researcher who is a perfec-
tionist per sé, cannot always answer all these questions in a field where
he is very committed and sometimes looses sight of the end-objectives. We
have had the opportunity in France of being judged on a basis of our choice
of objectives and the quality of our research, by a committee set up by the
Minister of Industry and chaired by Prof. CASTAING. This committee included
scientists, civil servants and trade unionists. They were to give an opinion on the well-foundedness of the various treatment and management
procedures; I think it is important that I give you an idea of the opinion
of someone who is not so directly involved in research. The first comment
made by this committee was that it took note of the industrial mastering of
the situation, including the management of waste, outside geological repositories. This committee also clearly recommended that progress be made in
three main sectors. First of all: to increase the knowledge into the medium and long-term behaviour of conditioned waste forms. In order to assess
the reliability of the source term, you have to carry out this kind of
studies, unless you want to end up with unsatisfactory solutions in the
long run. Further, to have better knowledge of the behaviour of the alpha-
emitters; it has been suggested that we look into the possibilities of
separation and specific confinement, or transmutation; but in these fields
we are led to examine, in particular for the high level waste, whether
there is really better safety both in the long and the short term.
Third, the recommendation was made to reduce the volume of waste to be
stored, but this should not have an impact on safety at all. Now, at the
present state of research, the procedures - as our chairman said - have
reached such an industrial stage that we should still make progress with
the actual processes or procedures which we are already using.

A.G. DUNCAN, UK DOE London

I would like to support Mr. VERKERK's suggestion of yesterday that we
should take this opportunity to look forward as well as back over the last
five years. And I particularly want to encourage the new systems analysis
task, because I think this will have a significant influence on the studies of treatment processes and on how the results are applied.
Mr. DWORSCHAK reminded us this morning that the "optimization" principle,
recommended by ICRP, has been embraced by the CEC and, of course, we are
encouraged in all discussions to optimize over the complete waste management systems, including disposal.
I think that looking at proposals for waste treatment in the context of a
total waste management system will raise questions which may be already
overdue, and I will give you a few examples.

1) Mr. CECILLE described a chemical precipitation process for treating
reprocessing concentrates, with a view to incorporating the precipitated
residue in high-level waste. In answer to my question about direct calcination and vitrification, he said that this would increase the volume of
the glass. The important question which follows and which can only be
answered by system analysis, is: How much extra glass is it worth generating in order to avoid introduction of an extra process step with its additional complexity, its secondary waste, its decommissioning waste and its
operator exposure, etc?

2) Another example follows from the suggestion that we remove actinides
from alpha-contaminated wastes and incorporate them into HLW. What is the

benefit of this in the context of a total waste management system? It may
be of environmental benefit, but there are disposal scenarios for which
it may be of no benefit, and it may even be a positive disadvantage.
Again, this emerges only in the context of system analysis.

3) Then we have the question of selecting a matrix for encapsulation of
hulls, where - as Mr. KROEBEL has said - there are many possible choices.
I believe we ought to look to system analysis in order to justify selection
of any one. But we must be under no illusions that system analysis is an
easy task. To be effective, it requires evaluation of total systems against
some sets of properties or attributes. However, the question is: what
properties and what attributes? And who will advise on which shall be
included?

I wonder if help might be sought from the Art. 37 group of experts which
advises the Commission on waste management plans which may impact on the
water-, soil- or airspace of other member states. This Art. 37 expert group
might be even a customer for the results of the work.

Finally, I have spoken about analysis in the context of the total waste
management system, including disposal, but there remains one critical
question which I leave you to consider: How do we carry out such analysis and
justify choice of treatment processes if we do not know what the disposal
routes are going to be?

W. GOOSSENS, SCK/CEN Mol

Treating a more specific topic, I would first like to mention the general
characteristics of the gaseous waste. These properties are: 1) the gaseous
waste occurs as airborne waste in any type of nuclear installation; 2) the
treatment of the gaseous waste has to be done on the spot in line with the
process, since transport is not feasible; 3) technical measures against
gaseous release have to be taken immediately, since buffer storage is in
practice very limited; 4) the total hazard of the gaseous waste is often
conditioned by the process, the process equipment in use, and by the
working conditions.

In this way, the gaseous waste is a matter of concern mainly during the
design stage and the safety procedure. During the operation, the gaseous
waste is transformed into a secondary liquid or solid waste. Accident
conditions disturb the performance of on-line treatment systems for airborne waste. The operation of the gas-cleaning equipment available becomes
crucial under accidental circumstances in nuclear installations as well as
in any industrial activity handling hazardous compounds.

An accident provokes a challenge which is important since the potentially
poor performance of the gas-cleaning equipment at the moment of even a
simple accident implies the survival chances of the process, which in
itself is good; it might even imply the degree of public confidence in the
industrial community and more specifically in the nuclear business. The
technical possibilities of existing and new systems for treating gaseous
effluents from nuclear installations under routine operation conditions,
and under accidental conditions, will be discussed during the European
Conference on Gaseous Effluent Treatment in Nuclear Installations, next
October here in Luxembourg.

As a first specific subject on gaseous waste, I would like to mention the
aerosol-trapping item. Aerosols, either in a liquid or solid state, are
entrained by carrier gases in any type of nuclear installation. Although
the HEPA-filters in use perform well, a lot remains to be done to extend
their operation time, by installing proper pre-filters.

Furthermore, the aerosol trapping of condensating isotopes after a high-
temperature process, or after a high temperature waste conditioning facility
is open for further experimental investigation, even up to the point of
identification and qualification of the aerosols.
For the retention of iodine isotopes further research is needed on the
deterioration of stand-by active-carbon filters, on the optimal iodine
retention step in reprocessing plants and on the confirmation of the disposal route of the long-living iodine isotopes.
The tritium problem is still not totally solved for in-land reprocessing
plants which are foreseen in Europe. Alternative tritium trapping or concentration processes are still to be investigated on a technical scale,
permitting a scaling-up to the industrial stage.
Furthermore, the issue of noble gases and more specifically of krypton,
also needs more technical development with real off-gases.
Finally, there is a general need for a data base on the treatment and
disposal of gaseous waste, compiling the knowledge acquired in this field.
Moreover, the technical feasibility of different gas-treating systems
should be evaluated using technical process analyses, process engineering
and systems analysis, from trapping to final disposal.
In addition, an environmental and economic assessment within the European
programme is most desirable on the different routes possible for the
gaseous waste trapping and disposal. In this way it should perhaps become
possible to identify the best gas-treatment system that could potentially
fulfill the ALARA principle under technical, economical and ecological
conditions.

G. GROSSI, ENEA Casaccia

Radioactive wastes with low and medium activity, although in qualitative
terms are not the most harmful, are the most abundant. Most of these low
and medium activity types of waste are produced in nuclear power stations.
Their management and in particular the procedures for their treatment and
conditioning have become industrially mature. I think, however, that it
is necessary to recall the IAEA conclusion at the end of the international
conference on the management of radioactive waste, held in Seattle in 1983.
The resolution says: "The management of low and medium activity waste must
be the object of continuous efforts and attention. The criteria for the
acceptability of conditioned wastes and their economic factors urge us to
improve our treatment and conditioning processes in view of temporary or
final storage."
What improvements have to be made to the management of this waste? I would
now like to mention two types of improvements, legislative improvements
and technical improvements. As for the legislation side of it, it is necessary to clearly define the following points: First of all, the limit
below which a waste is no longer to be considered as being radioactive,
in order to avoid collection and management of materials which now - in
the absence of clear legislation - are considered as radioactive, and are
treated as such. Secondly, the limit below which waste containing long-
lived radionuclides - I am thinking of the transuranium elements - should
not be considered an alpha-waste. Thirdly, the characteristics that a final
product of our processes should respect in order to be considered acceptable.
Now, the situation in this legislative sector at European and world level
is, if not lacking, very different and very heterogeneous. So, I can recommend that a special effort be made to fill the gaps, but also to make

legislation as uniform as possible, although, of course, in the latter
case there are obstacles not only of a political nature, but also of a
technical nature.
Now, coming to the technical managerial improvements, I think that the already quoted quantitative aspects of the low and medium activity waste
should lead us to try to reduce its volume. Amongst the various possibilities that have been proposed and analysed, the ones which try to isolate
the truly radioactive component in the waste (which is usually quite small)
from matrices which may be of large volume, are - in my opinion - the most
interesting aspects. It does not seem rational, even though at the first
glance it may seem more economical, to produce products to immobilise
inactive matrices, rather than to immobilise the radionuclides we want to
isolate. An example of this are the borate solutions which are produced
as radioactive waste from power stations.
In this context, the development of systems for the selective separation
of the radioactive elements from the waste is indeed the path to be embarked upon very carefully.
Finally, I recall the need already mentioned in the IAEA recommendation,
that is: to take into account a greater severity in acceptability criteria
for the final product. I want to stress that it will be necessary to
introduce improvements also in the processes and techniques for the conditioning of the materials, both by introducing new waste-matrix combinations which will offer better guarantees from the point of view of the
physical, chemical and mechanical properties, and by developing more advanced matrices, for example cements of better quality or polymer-containing
concretes. Also in this case, the quality control for the final product
is very important. Research has to be focused on the development of non-
destructive methods to measure the quantity of the alpha-, beta- and
gamma-radioactivity present in both the waste before treatment and in the
waste after treatment.

R. SIMON, CEC Brussels

Mr. GROSSI has already given a short introduction to the theme I wanted
to speak about here and which is the quality of waste products.
In the discussion about the safety of disposal, we frequently refer to the
favourable characteristics of the one or other waste form or the long-
term integrity of the waste package. Indeed, the risk analyses, carried
out in Europe and in the USA, agree that the disposal of medium and low
level waste will present at most an insignificant risk on present or
future society if carried out according to some underlying hypotheses.
This confidence is based on the 20-30 year experience accumulated with
LLW disposal and confirmed by experiments. To implement our disposal concept we must make sure that the wastes and barriers will broadly follow
the models laid down in the risk analysis. This is the objective of quality
assurance.
Normally, the first approach to confirm that a product meets a requirement
is to test its performance in the "as delivered" state. This, however, is
obviously not the right method, if the task is to confirm that a 2-ton
block of reinforced concrete satisfies the requirements such as:
- It shall not contain more than 100 mCi alpha activity per m^3;
- It shall not swell when saturated with water;
- It shall not contain free liquids;
- It shall not release more than 1/100 of its nuclide inventory per year.
The only way to check all these characteristics on the existing block
would be by extensive destructive testing - clearly an absurd solution.

We must, therefore, find reliable non-destructive methods to implement our quality criteria. This can be achieved if the producer of the waste blocks has:

a) a complete understanding of the manufacturing process;
b) full knowledge and control of the process input;
c) control of all process operating conditions affecting product properties; and
d) meaningful output controls and non-destructive tests.

All this will be performed by the producer of a waste package if things are done following the specifications. But some external post-manufacturing tests are at least desirable if not necessary. We should only think of our present situation where many waste packages are being produced and most of them are probably quite satisfactory, but they will be stored perhaps 10 or 12 years before they will ever see an acceptance test at the entrance to a repository. In the meanwhile they could be dropped on their way from the railway station to the lorry and so it is quite desirable to introduce some external controls to the ones imposed on the producer at the outlet of his production line.

On the other hand, the good quality of a product is one of the liabilities of the producer to the public; the representative of the public is in this case the licensing authority or the nuclear inspectorate. These people have the right and the duty to verify and to make sure that the producer conforms in an adequate way to his liability. So we should have a system in which there are satisfactory controls on behalf of the producer; we should provide all the means to the producer to measure and to certify in good confidence that his product is in compliance with the criteria which are set by either the licensing authorities or by the operator of the disposal facilities. But there will be independent controls from the outside, be it by audits through the inspectorates or by simple inspections of the packages when they arrive at the site of the disposal facility. I think this is the only way to make sure that our confinement system, as it is presently conceived and designed, is really not only so good on the paper. It must be as perfect and as safe in reality. The translation from our concept to the reality has to pass through a severe quality control.

R. KROEBEL, KfK Karlsruhe

I am going to sum up in three sentences. The most important is that we have lots of procedures in Europe and beyond Europe. And now we must tie them all up in accordance with standards, and test them out. The procedures have to be looked at. Another important issue is that quality guarantees must be defined and implemented, putting the best and most promising procedures into practice. This applies to power stations and reprocessors. Do not forget that to date our present technology is in infants shoes! And if things go ahead as we think, we will have to don our running shoes, and we will have to condition our waste accordingly. I will close now and I would ask you to put your questions and ask for answers.

H. KRAUSE, KfK Karlsruhe

Let me add: I do think now we are at a crossroad. We do not need just technical demonstrations of new products. We need more; we must streamline what we already have, and I think here the EEC in its new programme could contribute by setting goals and standards and defining them. We should define our targets and then move towards them firmly. This is not easy. Mr. DUNCAN already raised the question of sytems analysis and certainly it is an important component of such thinking. We must decide what we want, what we need and how much money we are going to spend. Then, at the right stage we can get to the target, do something and dissipate the fog which exists, and the doubs over a broad front.

T. MARZULLO, ENEL Rome

This afternoon in this panel, I have received two flashes of information which are rather satisfactory. The frist came from Mr. GROSSI, who gave us a way of talking about the future five-year programme of the Community; and I also agree with Dr. KRAUSE, who actually took the words out of my mouth, because this panel should be the point in which we try to work out what to do in the future. I think this is high time to materialise and assemble all our experiences and pool them so that we can come up with single solutions. Italy needs to have guidance or orientation from the European Community, and I am talking on behalf of the energy-producing body in Italy. We use nuclear power and we will be using even more between now and 1990. Our national programme envisages a considerable increase, so these needs and requirements are Italian ones, but I should imagine that they are shared by many other countries, which are in the same conditions as Italy.
These proposals for legislation are welcome. I think this is a logical final step; as GROSSI was saying, the IAEA resolution would be materialised if we were to do this. The European Community should give us guidance as to temporary and final repositories, which are much simpler than the so-called back-end of the fuel.
The last five-year programme has been focused on this back-end, but now the fuel is irradiated because it has been used in a nuclear power station. Therefore, it produced not only power but also waste, and we cannot forget it.
Everything we have heard today, the medium and low activity question, seems to be a settled matter, and this is not the case.
Finally, I think I understood that in the repository of high activity waste, somebody is envisaging to store also low and medium activity waste. I did not quite understand here wether there is disposal which would cover both. I think that it would be ludicrous to put such products together in one single repository.

G. GROSSI, ENEA Casaccia

We must recall that the EC programme covers R&D activities, not into processes that have already been developed at industrial level, but rather on areas that should be at least an optimisation of processes and techniques that already exist, if not actually through innovation. I think that in any case the treating and conditioning of the power station waste has already been worked on; these types of radioactive waste may have been slightly neglected by researchers. But I think that the resolution of the IAEA at the end of the Seattle conference should be taken into account; already in the next European Community programme there are clear guidelines which will allow us to meet the concerns you mentioned.

K. KUEHN, GSF Braunschweig

I agree entirely with what Mr. KRAUSE said. But there is another question, that is in the opposite direction. Mr. DUNCAN said that we need systems analysis and Mr. BAUDIN also mentioned results of the CASTAING report. In both statements there was a note of uncertainty about whether our range of instruments was complete; in the CASTAING report there were shades of criticism about consolidation procedures. I would like to ask the brains trust now, whether is satisfied with what has been done so far. If I may simplify, there are three major categories of conditioning matrices: concrete, bitumen and polymers. Does the panel agree that these are good enough, or is more R&D necessary to come up with better answers for conditioning than the materials we have at this moment?

G. BAUDIN, CEA Fontenay-aux-Roses

I think actually the three matrices you mentioned have been worked on sufficiently. We are not putting those into question. What we are asking you to do, is look into the process of their evolution, rather than thinking of more and new matrices. The knowledge of the evolution of these matrices, in particular under storage conditions, is where we still have a lot to do.

H. KRAUSE, KfK Karlsruhe

I did not want to say that we do not need new processes anymore. I said we must set ourselves clear targets and march firmly towards them.

J. LEWI, CEA Fontenay-aux-Roses

It may seem possible to establish criteria, if you look at the parameters characterising the waste when it is formed or in the very short term. But if you take into account the long-term evolution of the matrices, is it then possible to define criteria?

R. SIMON, CEC Brussels

For long-lived wastes we have to assume, that the waste form is not the only barrier. In a geological repository the natural barriers, taking into account their enormous depth and durability, can be relied upon to remain in place and, essentially, intact as long or longer than the critical nuclides. In defining waste acceptance criteria, our prime concern for the long- term evolution must therefore be to avoid any risk to the geological barriers, which may arise from placing the waste into them. In certain repositories, this also applies to the engineered barriers such as to buffer materials, linings and seals provided to prevent water flow. The waste form criteria to assure long-term safety against release must therefore protect the other barriers against possible damage by heat, gas release, swelling or subsidence, corrosion and, if necessary, radiation.
Leachability is not the critical issue, as long as the other barriers are only affected by natural evolution. It only becomes important if disrup-

tive events like major earthquakes or human intrusion breach those other barriers. The probability of such events must be minimized by repository siting and design criteria.

P. GERONTOPOULOS, AGIP Bologna

Mr. DUNCAN said that taking away the alpha-waste and putting it into glass might not be worth the trouble. From the other side we have the CASTAING report which makes a recommendation in this sense. I would like to find at least a clear indication of whether it is useful or not.

A.G. DUNCAN, Dept. of Env., Romney House, London

This morning, Mr. CECILLE had said that the amount of glass can be reduced by treating the reprocessing concentrate by chemical precipitation and incorporate in the high level waste only the removed residue. That involves a fairly complex precipitation process with undoubtedly the generation of secondary wastes, with more decommissioning wastes and with more operator exposure.
I was suggesting that within the total waste management system, there is a point where it is worth producing more glass in order to avoid the additional complexity, cost and operator exposure, and beyond which it is then worth introducing additional process steps. The only way you can do that is by looking at the system as a whole.
The second point I picked up from Mr. ORLOWSKY's statement yesterday, where he spoke about the work that had been done in the Commission, which shows that the proposal to remove actinides and treat them by transmutation is questionable at least. He said that what we could look at as a practical proposition is the removal of actinides from alpha-contaminated waste for inclusion in high level waste. That may be fine in certain disposal scenarios, but I can think of certain disposal scenarios where what you are doing in effect is just moving the actinides from one drum to another one. You are just changing the waste from your one stream to another, with no benefit. Furthermore, there is a scenario which, in fact, would give you a net disbenefit, where doing that would actually be of detriment to the environment.

G. BAUDIN, CEA Fontenay-aux-Roses

In the report of the CASTAING committee you should not confuse suggestions for research and improvement of processes with the conclusions of the report. The CASTAING committee suggests that alternative processes be examined, but it is not condemning the existing ones, so do not confuse the suggestions for research with the conclusions of the report!

SESSION IV

CHARACTERIZATION OF WASTE FORMS

Mechanisms of leaching and corrosions of vitrified radioactive waste forms

Behaviour of vitrified radioactive waste under simulated repository conditions

Radiation, thermal and mechanical effects in HLW glass

Behaviour of intermediate-level waste forms in an aqueous environment

Radiation, thermal and mechanical effects on low and medium active conditioned waste

The challenge of quality assurance

MECHANISMS OF LEACHING AND CORROSIONS OF
VITRIFIED RADIOACTIVE WASTE FORMS

F.Lanza (CEC), R.Conradt (FhG), A.R.Hall (UKAEA)
G.Malow (HMI), P.Trocellier (CEA), P.Van Iseghem (CEN/SCK)

Summary

The estimation of the risk connected with the storage of radioactive waste in geological formations asks for reliable extrapolation of the data for leaching and corrosion of glasses to very long times. As a consequence the knowledge of the physico-chemical mechanisms which dominate the leaching phenomena can be very useful.

In the corrosion due to aqueous solution three main mechanisms can be identified: ion exchange, matrix dissolution and formation of a surface layer.

The work performed in the different laboratories has allowed to evaluate the relative importance of the various mechanism. The alkali ion exchange does not seems to be predominant in defining the release of the various elements, the matrix dissolution being the most important. The surface composition is important as the compounds present could dominate the matrix dissolution kinetic. Besides the surface layer could form an impervious layer, which, if stable in time, could protect effectively the glass.

1.- INTRODUCTION

The main option of storage of high level wastes (HLW) assumes that the radioactive wastes are conditioned in a matrix composed by borosilicate glasses and stored in a suitable geological formation. In these conditions leaching and corrosion of the glass by the ground water are the only way which could allow a release of the radioactive nuclides towards the biosphere.

The estimation of the risk connected with the repository of radioactive waste in geological formations asks for an evaluation of the resistance to leaching of the vitrified matrix for very long times so that the possibility of the normal experimentation can be exceeded. As a consequence the knowledge of the physico-chemical mechanisms which determines leaching and corrosion phenomena can be very useful for obtaining reliable extrapolations.

In the specific literature there is a certain tendency to abuse the term leaching. In this report it will be used in its more precise meaning, that is selective extraction of one element or of one class of elements. For the general phenomena of attack and alteration of the surface the term corrosion will be used. Moreover to take into account not only that part of the glass which goes into solution but also that which is altered forming the surface layer, the term glass degradation will be used.

In the corrosion due to aqueous solutions three main mechanisms can be identified, namely:

- alteration of a region below the surface of the residual glass by ion exchange;
- dissolution of the glass network;
- formation of a reaction layer on top of the surface of the residual glass.

In order to summarize the work performed in the different laboratories this scheme of presentation will be followed.

2.- ALKALI ION EXCHANGE

One of the most commonly observed mechanism of leaching is the ionic exchange between Na^+ (or other alkali) and H^+ (or H_3O^+) ions which gives rise to a progressive depletion of Na in the subsurface layer and a release of Na to the solution.

At the FhG, investigations have been performed on the glass SM 58 LW 11. Its detailed composition is reported in ref. (1). The leachant has the composition of the invariant point Q in the system $NaCl-KCl-MgCl_2-MgSO_4-H_2O$ at 55°C according to d'Ans. Additional amounts of NaCl are added in order to achieve salt saturation at higher temperatures. The initial pH at room temperature is pH = 5.4 $\pm$ 0.3. The respective final values, even after long corrosion times, are 5.8 $\pm$ 0.4 in this system. The experiments are carried out, in a closed system, at temperatures from 80 to 200°C and at a ratio of the sample surface area SA to the leachant volume V (SA/V) of 0.034 cm^{-1}.

After the leach tests, the samples appear to be covered with a layer of reaction products which is removed by a "rubber policeman". The subsurface zone has been examined by two different surface analytical methods: sodium and calcium profiles (2) have been obtained with the ESCA method coupled with Ar^+ sputtering (fig.1) while hydrogen profiles have been obtained using the N^{15} resonant nuclear reaction technique (fig.2) made available by Rauch at the Frankfurt University.

It can be seen that the surface zone which is altered by ion exchange process is very thin due, probably to the complex composition of the glass. Besides, it appears clearly that the ion exchange depth reaches rapidly, a value which stays constant in time, at each temperature. Typical ion exchange depths range from 0.06 μm at 80°C to 0.2 μm at 200°C, whereas the depths of dissolved glass after 30 d are 0.2 μm at 80°C and 31 μm at 200°C. Figure 2 demonstrates that the ion exchange

zones do not grow with time: the amounts of incorporated protons remain
nearly constant.

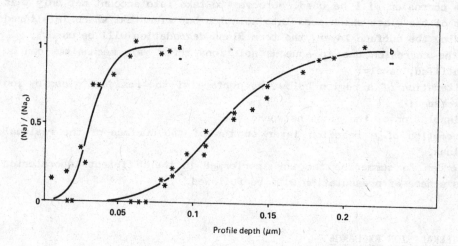

Fig. 1: Ion exchange profiles of Na in the residual glass in samples leached
in quinary salt brine for 8 d at 80°C (curve a) and 200°C (curve b).

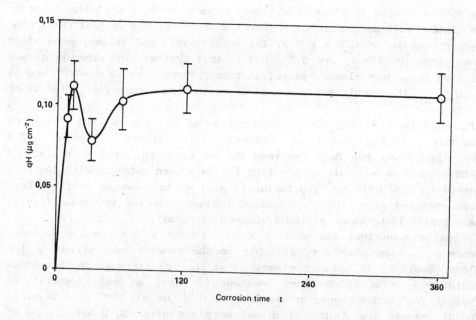

Fig. 2: Amount qH of protons in the residual glass of samples leached in quinary salt brine at 160°C as a function of time (days).

It has to be noted that simultaneously to the $Na^+ - H^+$ interdiffusion is acting matrix dissolution is also occurring. The constant value of the hydrogen which has been observed can then be considered as a value corresponding to an equilibrium between these two different mechanisms. Such equilibrium is obtained so rapidly that only the influence of temperature on penetration is apparent.

Taking into account the very low thickness of the zone in which ion exchange is occurring, and the competition with the matrix dissolution, it can be assumed that the glass degrades congruently.

Tests were performed at Harwell on glasses UK 189 and UK 209 (3) in a closed system using distilled water at 20°C. It was noted that at the beginning of the leaching operation Na and Li were released in proportion to $\sqrt{10^{-pH} \cdot t}$ in the whole pH range which implies a predominantly diffusion-controlled mechanism. For longer times however the alkali are removed proportionally to time due to the influence of matrix dissolution. It is interesting to note that boron leaches diffusively and congruently with the alkali at all values of pH. A similar effect was noted at FhG. An analysis of the boron content of the subsurface layer showed a depth profile similar to that of sodium.

A mathematical model of the leaching of alkali has been proposed by Harwell considering leaching as a three-stage process:

(a) Diffusion of some reactive species, A, from the leachant into the glass;

(b) Reaction of A with the particular element or group containing the element, M, in the glass structure to give a reaction product, P, and an altered glass structure, G;

(c) Diffusion of the reaction product, P, outward into the leachant.

In the simplest case, that is the interdiffusion of uncharged species with constant diffusion coefficients D_A and D_p, the equations describing the process are:

$$\frac{\partial [A]}{\partial t} = D_A \frac{\partial^2 [A]}{\partial x^2} - k[A][M] \qquad \ldots\ldots (1)$$

$$\frac{\partial [P]}{\partial t} = D_p \frac{\partial^2 [P]}{\partial x^2} + k[A][M] \qquad \ldots\ldots (2)$$

where $[A]$, $[M]$, etc. are the concentrations of A, M, etc. with initial values $[A_o]$, $[M_o]$ and k is the reaction constant. Equation 1 was solved numerically for $[M]$, concentration of unreacted species remaining in the glass and for Q, the total quantity of M released. This was found to be

$$Q = 2 \; [M_o] \sqrt{\frac{[A_o] \; D_A \cdot t}{[M_o] \; \pi}} \qquad \ldots \ldots \ldots \ldots (3)$$

which is in agreement with the experimental data for Lithium (3). The shape of the profile for M is governed by the parameter $k[M_o]/D_A$. Figure 1 shows the sodium profile data obtained by Conradt et al (2) together with computed profile curves. The total alkali in this glass, $|M_o|$, was 0.0126 mole cm^{-3} and the leachant pH was ca. 5, so that $[A_o] \sim 10^{-8}$ mole cm^{-3}. The profiles shown in Figure 1 were calculted using the constants shown in Table 1 and assuming (4) that protons are consumed only by alkali ion exchange.

TABLE 1

Constants used in computing the curves in Figure 2

TEST CONDITIONS	D_A $(cm^2 s^{-1})$	k $(cm^3 mole^{-1} s^{-1})$
80°C, 62 days	1.3×10^{-12}	120
200°C, 8 days	2.5×10^{-12}	1800

These calculations are only intended to illustrate how the model can be used to interpret such depth profiles. The actual values obtained are unlikely to be meaningful because in this particular case the original glass surface has receded and the depth of matrix dissolution is not taken into account in this model.

The model described thus far assumes a constant D, and does not take account of effects due to potential gradients generated by the ion exchange process of alkali leaching.

Ion exchange theory is well understood and will not be repeated here. Briefly, the ion fluxes generate a potential gradient across the interdiffusion zone, the effect of which is to accelerate the slower-moving species and retard the faster-moving. A frequently-quoted equation is:

$$\tilde{D} = \frac{D_{H^+} \; D_{M^+} \; ([H^+] + [M^+])}{D_{M^+} \; [H^+] + D_{H^+} \; [M^+]} \qquad \ldots \ldots \ldots (4)$$

If, as is probable, D_{H+} is much larger than D_{M+} then equation (4) reduces to

$$\tilde{D} = D_{M+} \left(1 + \frac{[H^+]}{[M^+]}\right) \quad \ldots\ldots(5)$$

There is also an effect caused by the dependence of D_{M+} on the structural alterations taking place as leaching proceeds. The bulk diffusion coefficient for alkali ions in an unleached glass (UK189), obtained from electrical resistivity measurements, is ca. 10^{-19} cm^2 s^{-1} at 25°C which is ca. 7 orders of magnitude lower than the effective diffusion coefficient for alkali leaching.

One effect of this structural dependence is that the released alkali ions are barred from diffusing inwards into the unaltered glass and are forced to diffuse outwards into the leachate.

3.- MATRIX DISSOLUTION

A second important mechanism is network dissolution. Following this mechanism there is a direct attack of the Si - O - Si bond by the OH presents in the solution.

A large number of laboratories have investigated the kinetic of this dissolution in closed systems, using different glasses. Tests have been performed in distilled water (5, 6), concentrated NaCl solution (7), and quinary brine (2) solution. The corrosion phenomena have been followed by the analysis of the leaching solution and/or by the weight loss measurements of the samples.

The results in general when disposed in a log-log diagram plotted vs. time can be fitted by straight lines. As a consequence the matrix dissolution is discussed in terms of a power relation as

$$W = Kt^n \qquad \ldots\ldots (6)$$

The value of the t exponent can be indicative of the corrosion process. A power factor of 1 is typical of a dissolution process while a power factor of 0.5 can be indicative of corrosion dominated by diffusion, or by a corrosion kinetic dominated by the reprecipitation of various compounds.

In the discussion of matrix dissolution it is necessary to separate the phenomena which occur in pure water, from those which occur in concentrated NaCl solution or in Q brines. In pure water, indeed due to leaching of alkalis, pH variations are possible giving, as a consequence, changes in the phenomenon of matrix dissolution. In the other two solutions, on the contrary, it has been verified that only negligible pH variations were occurring even at high SA/V values.

In distilled water various conditions were explored and dissolution was found under conditions which gave rise to the lowest concentration of SiO_2 in the water. In fact, in the Ispra work (6), which uses a value of SA/V in between 0,22 and 0,5 cm^{-1}, dissolution is verified at 40°C. At 90°C on the contrary, a power factor of 0,5 was obtained.

At SCK/CEN an extensive study (5) was peformed on five different European glasses. Three of them (UK209, SON58 and SON64) were already studied to some extent during the previous joint CEC programme (8). SAN60 and SM58 were introduced during the second Joint CEC programme (5). Glass compositions can be found in the cited references. Most of the tests were performed at 90°C. When a value of SA/V of 0.1 cm^{-1} was used direct dissolution was obtained. At SA/V of 1 cm^{-1} diffusion conditions or intermediate ones were encountered.

Table 2 presents the results obtained on the five selected glasses for the main constituents of the glass for a SA/V ratio of 1 cm^{-1} (5).

TABLE 2

Slopes in the elemental (log $NSWL_i$ vs log t)
plots during corrosion in distilled water at 90°C, SA/V = 1 cm^{-1}

	SAN60	UK209	SM58	SON58	SON64
Si	0.5(2'----4d)	0.5(2'--80d)	0.5(2'--27d)	0.65(10'--1d)	0.65(10'--1d)
Na	0.3(2'--3d)	0.5(2'--1d)	0.5(2'--1d)	0.65(10--27d)	0.65(10'--27d)
		1.0(1d--9d)	1.0(1--27d)		
Li	0.5(2'--240d)	0.5(2'--1d)	0.5(2'--1d)	-----	-----
		1.0(1d--27d)	1.0(1--27d)		
B	0.5(8h--4d)	0.5(8h--1d)	0.5(8h--1d)	0.65(40'--9d)	0.65(40'--9d)
		1.0(1--80d)	1.0(1--27d)		

It can be seen that in most cases a slope of 0.5 is obtained. In order to determine if such a slope is due to diffusion of alkalis in the glass an attempt has been made to calculate the diffusion coefficient of Na using the following formula |10|:

$$D = \frac{\pi}{4} \left(\frac{V_s}{SA}\right)^2 m^2 \qquad \ldots\ldots\ldots(7)$$

where V_s is the volume of the sample and m the slope of the plot of the normalized weight losses against the square root of time. For glasses UK209 and SM58 values of 2 and $5 \cdot 10^{-14}$ cm^2/s respectively were obtained which compare favourably with the value of $5 \cdot 10^{-13}$ cm^2/s of measured at 100°C for high silica glass (11). Alkali diffusion in the bulk glass as the rate controlling process in an alkali ion exchange leching mechanisms, therefore could be responsible for the observed slope.

When tests are conducted for longer periods of time the n power factor appears to decrease or even to have reached a zero value as it corrosion has stopped. However, in one case, in which corrosion was continued for a longer period it appeared to have assumed again a slope of 1 typical of direct dissolution.

The SA/V ratio strongly influences the corrosion rate. However the use of SA/V as an accelerating parameter is hampered by the observation that the change of the SA/V ratio can also induce the change of the pH of the leachate. For instance, at 90°C for large SA/V values (ca. 7 cm^{-1}) the equilibrium pH, in case of SAN60, is about 10, while at a lower SA/V (1 cm^{-1}) a pH value of about 8.75 is reached (5). Therefore the influence of SA/V might not only be to accelerate the corrosion, by accelerating the accumulation of dissolved species in the leachate, but also to alter the corrosion processes.

In the quinary brine, tests were performed at temperatures from 80°C to 200°C at FhG. The network dissolution was followed determining the mass losses of the treated samples after having removed the surface layer. It has to be noted that the mass losses of the residual glass samples agree very well with boron or lithium release data. Only minor pH changes of the brines are observed. The final values are pH = 5.8 + 0.4. The mass losses (per surface area), plotted vs. time, exhibit straight lines in a double logarithmic plot. The correlation is well obeyed during the observation time. The intercepts at t = 1 d and the slopes of the fitted straight lines (calculated under a minimum condition for the standard deviation) are shown in Table 3.

Table 3: Intercepts K at t = 1 d and slopes n of the fit $W = Kt^n$ for the normalized mass loss W from static leach tests and the time t in a double logarithmic scale; α is the standard deviation of the individual values and i the number of data points

T in °C	Time interval	K in $mg \cdot cm^{-2}$	n	α%	i
80	8 to 360 d	0.01	0.48	17	11
120	60 to 360 d	0.03	0.60	9	13
160	6 to 360 d	0.12	0.70	19	25
200	6 to 120 d	0.55	0.79	22	25

The slope increases markedly with temperature. No reduction of n with time towards smaller values is observed.

In concentrated NaCl solution, tests have been performed at HMI with a surface to volume ratio of 0.014 cm^{-1} at 200°C up to 1000 days. Time dependence of leaching was derived from weight loss measurements, after removal of the surface layer, and quantitative ICP analysis of solutions. The mass loss values show that the leaching process slows down with time. Boron, lithium, calcium and strontium follow the same trend of the measured mass loss. Magnesium, silicon and aluminium show rapidly a value which is constant or increases slowly.

The surface was covered with a layer of crystals of various types. Assuming that the largest crystals (analcime) have a protective effect then the surface exposed to the corrosion will decrease in time using these assumption. Experimental mean losses and solution concentrations were fitted by a first order reaction kinetic.

The limit of solubility of SiO_2 in the solution depends strongly on the pH. At neutral pH, or better for pH lower than 9, SiO_2 tends to be dissolved in a non-ionic form H_4SiO_4, the solubility of amorphous silica being higher than that of quartz by a large factor. For higher pH silica also dissolves in a ionic form so that its theoretical solubility limit increases. Solutions of high ionic strength on the contrary tend to decrease this limit.

In general it has been found that when the SiO_2 content in the leachate has reached a pseudo-saturation value, such a value is lower than that corresponding to the solubility limit of the amorphous silica.

Such an effect could be due to either a low kinetic of dissolution of SiO_2 or to the formation of equilibria due to precipitation of complex silicate.

Saturation tests were performed at Harwell using distilled water on three British glasses (189, 209, MW) at 45°C for up to 400 days, 90°C for up to 120 days and 150°C for 30 days. Glass samples were in the form of crushed and sieved granules of a narrow size range. These were leached in Teflon containers with just sufficient water to fill the interstices in the granule bed and cover the sample, giving a SA/V of ca. 200 cm^{-1}. Solids precipitates from the leachate were filtered off and analyzed. Precipitates of hydrous magnesium aluminium silicates formed in the leachates in varying amounts and compositions according to the test conditions and glass composition. X-ray diffraction patterns agreed with sepiolite ($2MgO$, $3SiO_2$, $2H_2O$) or pilolite (Al_2O_3, $4MgO$, $10SiO_2$), these two minerals presenting very similar X-ray diffraction patterns. Precipitation begins at silicon concentrations well below the saturation limit, so that there is no possibility that leaching will cease as the leachate becomes saturated in silicon.

It has to be noted that the level of concentration of pseudo-saturation depends also on the presence in the glass of other elements. For instance, at SCK/CEN it was noted that during the leaching of TRUW silicate glasses (12), when the content of Al_2O_3 in the glass was $\geqslant$ 5 mol%, the pseudo-saturation concentration of Si was lowered by a factor of ten.

If we assume that the level of silicon is controlled by the formation of some new silicate phases then we must accept as a possibility that the dependence of the mass losses from the square root of time is not due to diffusion phenomena but to reprecipitation equilibria.

A mathematical model which describes the consequences of such an assumption has been developed at HMI for NaCl concentrated solution (7).

It is assumed firstly that the disintegration of the borosilicate is a congruent process (if the surface layer is taken as leached material), secondly that the hydrolysis of the Si bonds is the rate determining step of glass dissolution and finally that a certain number of solid phases reprecipitate.

Following Grambow (13) it is considered that the glass degradation can be expressed as

$$\frac{dm}{dt} = Sk^+ \left(1 - \frac{C_{si}}{C_{sat}}\right) \quad \ldots\ldots\ldots(8)$$

where C_{sat} is the equilibrium concentration of silicon in respect to glass and k^+ is the forward reaction rate. Both coefficients depend on the pH of the leaching solution. However, as mentioned previously, in concentrated NaCl solution pH changes are negligible so that these two parameters can be considered constant. Assuming that no reprecipitation of phases occurs the integral of the equation is

$$\xi = \frac{m}{V} = \frac{C_{sat}}{W_{si}}\left(1 - \exp\frac{-W_{si} \, K^+\tau}{C_{sat}}\right) \quad \ldots\ldots\ldots(9)$$

where V is the volume of the leachant, W_{Si} the weight fraction of silicon in the glass and $\tau = (S/V)\cdot t$ is a reduced time scale which takes into account the accelerating effect due to the surface to volume ratio.

If we take into account the precipitation of phases containing silicon, the following relationship between the concentration, C_i of various elements of the precipitating compounds needs to be added to equation (9):

$$\prod_i C_i^{n_{ij}} = K_j \qquad \ldots\ldots(10$$

where n_{ij} are the stochiometric coefficients of the different elements and K_j the equilibrium concentration products.

A numerical calculation has been performed using for K^+ and C_{sat} empirical data obtained from a tests performed at 200°C up to a maximum time of 1000 days (fig. 3).

Three phases have been taken into consideration: Analcime, Phyllosilicate and Willemite. It appears that Analcime already precipitates after about 4 days keeping the silicon and aluminium concentrations almost constant. After about 60 days it is followed by the phyllosilicate causing a small but significant temporary decrease in the silicon concentration, not to be seen in fig. 3 due to the poor resolution of the drawing, and resulting in an increase in the leach rate between 60 and 120 days. Finally the willemite is saturated after about 400 days. A comparison with the experimental data shows that the model is in good agreement with the experiment.

4.- FORMATION AND COMPOSITION OF THE SURFACE LAYER

The borosilicate glasses which have been studied have a very complex composition. As a consequence of the dissolution of the SiO_2 matrix a surface layer is formed containing a large number of elements. As numerous radionuclides will be concentrated on the surface layer, a large number of investigations have been performed concerning its composition. X-ray analysis and surface techniques such as ESCA, direct observation of resonant nuclear reaction, Rutherford backscattering spectrometry and electron or nuclear microprobe have been utilized for these studies. When the corrosion of the glass is performed at high temperatures, crystalline forms can be identified on the surface, the type of crystals being dependent not only on the glass composition but also on the solution composition.

In distilled water in a test conducted at SCK/CEN using the glass SAN 60 on both monolithic ($SA/V = 1\ cm^{-1}$) or powdered samples, extensive surface crystallization occurred when corroding at 150 or 190°C (5). Powdered samples were even completely altered at 190°C. The crystalline phases generated at both temperatures are not identical. At 150°C cubic analcime ($NaAlSi_2O_6.H_2O$) and a tetragonal zeolite ($Na_3Al_3Si_5O_{16}.6H_2O$) crystals were observed; at 190°C three crystal types could be detected: analcime, orthorhombic eucriptite ($LiAlSiO_4.2H_2O$), and a needle-shaped, presently undetermined Ca, Si based crystal.

This surface crystallization was found to have a large impact on the corrosion of the glass. Monolithic glass samples corroded in preconcentrated solutions (resulting from a 28 d interaction of powdered SAN 60 with $SA/V = 42\ cm^{-1}$) at 150 and 190°C, show about 25 x larger weight losses (excluding the surface layer) than samples treated in distilled water.

An extensive study has been performed at the HMI on surface morphology of samples at the glass C-31 - 3 - EC - SPF - Na corroded in concentrated NaCl solutions at 200°C. (7)

On the surface of the samples it was possible to identify cubic crystals of analcime. The morphology of the surface and subsurface layer around these crystals suggests that analcime may form a protective rim. Semi-quantitative electron microprobe analyses on crystal faces indicate that this phase has variable Si:Al ratio. Factors which control the composition of analcime are not yet defined, but high reaction rates appear to correspond to high Si:Al ratios. It also appears that the Si:Al ratio of analcime is a reflection of the Si:Al ratio of the parent glass.

An accelerated test using glass powder was conducted at 200°C for 20 days. In these conditions, beside analcime, it was possible to identify crystals of zinc silicate, barium and calcium molybdate, barium chromate a calcium arsenic phase and a 12 Å phyllosilicate.

At CEN, Saclay two simple glasses containing uranium and thorium respectively were leached in distilled water at 20°C up to 14 days. The uranium shows a release in the solution higher than thorium. However both elements were enriched at the surface (fig. 4). An analysis of the

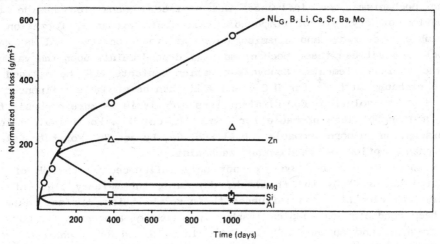

Fig. 3: Measured and calculated normalized elemental and total mass losses of leaching C-31-3 glass in saturated NaCl brine at 200°C.

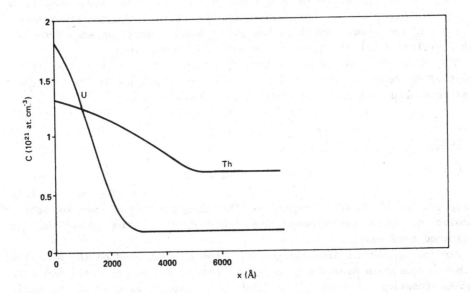

Fig. 4: Depth profile of Th and U concentrations in the surface layer of simple glasses corroded at 20°C for 14 days in distilled water.

silicon content in the surface layer obtained using the resonant nuclear
reaction ^{28}Si revealed an accumulation of Si in the layer. It seems
reasonable then that uranium is fixed in the gel layer in the form of a
uranium silicate, as for instance coffinite $(U(SiO_4)_{1-x}(OH)_{4x})$. An
analysis of Na and Si in a complex borosilicate glass shows that the
surface layer presents a double structure with a peak near the glass,
suggesting that under such conditions a cyclic mechanism of formation of
the surface layer cannot be discarded.

Other mechanisms can influence the surface composition beside the
formation of crystals or compounds: viz. ionic exchange, formation of
insoluble hydroxide and chemisorption of ions on the silica gel.
Competition between these mechanisms will depend mainly upon the valency
of the various elements. Monovalent cation contents will be defined by
ionic exchange with H^+ (or H_3O^+) and will then be strongly influenced by
the pH of the solution. For bivalent ions hydrolysis is more probable for
Zn and Sr but less probable for Mn, Fe, Co, UO_2. For trivalent ions
chemisorption is more probable for Al, Fe, Co, La and Ce. For tetravalent
ions chemisorption is the dominant mechanism.

The composition of the layer is not only influenced by the pH of the
solution but also by the Eh which, determing the valency state of the
multivalent elements, can influence their solubilities. Surface analysis
has been conducted on samples leached in a Soxhlet apparatus at 100°C in
the presence and absence of air (14). The XPS and RBS techniques were
applied to analyse the content of the silicon, iron and uranium while the
nuclear reaction method was utilized to analyse the hydrogen content.

A larger amount of iron and the smaller amount of uranium observed after
leaching in the presence of air confirm what was expected, namely, a
larger solubility of the lower valences of iron and the higher valences
of uranium. The iron present in the glass matrix, as it appears from the
peak position of the XPS spectra, is in the bivalent form.

The formation of less soluble iron hydrates would justify the larger
amount of hydrogen in the samples surface layer leached in the presence
of air compared with those leached in an argon atmosphere.

5.- CONCLUSIONS

5.- <u>CONCLUSIONS</u>

The analysis of the interdiffusion between Na^+ and H^+ (or H_3O^+) shows
that the amount of alkali released by diffusion becomes rapidly
negligible in time with respect to the amount released due to matrix
dissolution. As a consequence the degradation of the glass can be
considered congruent.

The studies on matrix dissolution have been concentrated on the analysis
of the corrosion in closed systems. This phenomenon can be described with
a power function of time. For conditions which give rise to small
concentrations of SiO_2 in the solution a corrosion increasing linearly
with time was observed. In other conditions the corrosion could depend,

either on diffusion, or on the kinetic of reprecipitation of complex silicate. In some cases the corrosion increased linearly even under conditions close to saturation, though at a smaller rate than under conditions far from saturation. In every case it is apparent that no equilibrium conditions are possible and that the glass will continue to degrade.

The analysis of the surface has made it possible to identify various crystalline compounds which could dominate the degradation kinetic. The composition of the layer depends on the glass composition, on the leaching solution composition and particularly on its pH and Eh. The corrosion in concentrated NaCl solution has indicated that analcime crystals could have a protective action on the glass. The question whether there could exist a surface layer stable in time protecting effectively the glass, is still open.

6.- REFERENCES

1. Scholze H., Conradt R., Engelke H., Roggendorf H. - Determination of the corrosion mechanism of high-level waste conteining glass - Mat. Res. Soc. Symp. Proc. - 11, 173-180 (1982).

2. Conradt R., Roggendorf H., Scholze H. - Chemical durability of a multicomponent glass in a simulated carnallite/rock salt environment - Mat. Res. Soc. Symp. Proc. - 26, 9-17 (1983).

3. Hall A.R., Hough A., Marples J.A.C. - Leaching of vitrified high level radioactive waste - Mat. Res. Soc. Symp. Proc. - 11, 83-91 (1982).

4. Burker B.C., Arnold G.W., Beauchamp E.K., Day D.E. - Mechanisms for alkali leaching in mixed Na-K silicate glasses - J. Non Cryst. solids - 56, 295-322 (1983).

5. Van Iseghem P., Timmermans W., De Batist R. - Parametric study of the corrosion behaviour in static distilled water of simulated European reference high level glasses - Scientific Basis of Nuclear Waste Management VIII (1984).

6. Lanza F., Manaktala H., Parnisari E. - Leaching of a borosilicate glass in confined systems - European Appl. Res. Rept. - Vol. 5 N. 3, 363-398 (1983).

7. Haaker R., Malow G., Offerman P. - The effect of phase formation on glass leaching - Scientific Basis of Nuclear Waste Management VIII (1984).

8. Altenheim F.K., Marples J.A.C., Lutze W., Sombret C. - The leaching of solidified high level waste under various condition - I Eur. Conf. on Rad. Waste Management - Luxembourg 1980.

9. Van Iseghem P., De Batist R. - Corrosion mechanism of simulated high level nuclear waste glasses in distillated water - Rivista della St. Sper. Vetro n. 5, 163 (1984).

10. Mendel J.E. - Review of leaching test methods and the leachability of various solid media conteining radioactive wastes - BNWL 1765 (1973).

11. Takata M., Tomorewa M., Watson E.B. - J. Am. Cer. Soc. 65 (2), 91 (1981).

12. Van Iseghem P., Timmermans W., De Batist R. - Corrosion behaviour of TRUW base and reference glasses - Mat. Res. Soc. Symp. Proc. - 26, 527 (1984).

13. Grambow B. - A general rate equation for nuclear waste glass corrosion - Scientific basis of Nuclear Waste Management VIII (1984).

14. Manara A., Lanza F., Ceccone G., Della Mea G., Salvagno G. - Application of XPS and nuclear technique to the study of the gel layers formed under different redox conditions on leached glasses - Scientific Basis of Nuclear Waste Management VIII (1984).

DISCUSSION

C. McCOMBIE, NAGRA Baden

Is it realistic to speak about protective layers considering the period of time of 10^4 to 10^5 years the long-lived radionuclides should be safeguarded?

F. LANZA, JRC Ispra

The protective function of the reaction layer, even on the long-term, is an assumption of predominant importance, for it constitutes the only condition under which the leach rate is reduced. HMI observed the protective property of analcime crystals formed on the surface under certain circumstances, but whether these crystals resist for periods of time as you stated or whether they remain protective on the long-term and under other conditions has still to be demonstrated.

H. GRAMBOW, HMI Berlin

Referring to the power factor (n) of the time (t) dependency of the leach rate, giving an indication of the underlying mechanism, may I ask whether you could recommend an n-factor for long-term predictions?

F. LANZA, JRC Ispra

Since there is no evidence that the kinetics do not resume initial values the exponent of (t) should be assumed to be 1 for long-term extrapolations. A better, even if still conservative approach, is to take the power factor resulting from the evaluation of the kinetics of reprecipitating species.

BEHAVIOUR OF VITRIFIED RADIOACTIVE WASTE
UNDER SIMULATED REPOSITORY CONDITIONS

R. DE BATIST[1], N. JACQUET-FRANCILLON[2], F. LANZA[3], G. MALOW[4],
and J.A.C. MARPLES[5]

1) S.C.K./C.E.N., MOL (Belgium)
2) CEA, MARCOULE (France)
3) JRC, ISPRA (Italy)
4) HMI, BERLIN (Germany)
5) UKAEA, HARWELL (UK)

Abstract

Vitrified radioactive waste products, designed either for the incorporation of high activity fission product wastes from reprocessing
plants or for the conditioning of α contaminated wastes through slagging
incineration, have been tested to investigate their behaviour under
simulated repository conditions. Three types of deep geological formation
have been considered: rocksalt, hard rock and clay. Both inactive and
lightly doped materials were studied. Information has been obtained
about the alteration of the waste forms (crystallization, surface layer
formation) and about the diffusion of the radionuclides in some of the
geological materials. This type of information is needed to validate
modelling codes which have been proposed for describing the long term
behaviour of vitrified waste in repository surroundings and which are
based on geochemical thermodynamic data concerning the system under
consideration. The results obtained so far indicate that the behaviour
of a waste form in a repository will depend on essentially all system
components and may drastically differ under different circumstances.

1. Introduction

One of the main objectives pursued during the 1980–1984 programme of
the European Atomic Energy Community on "Management and Storage of Radioactive Waste" was the investigation of the behaviour of solidified radioactive waste forms in conditions relevant for the envisaged repository
environment. In the present paper the work related to vitrified waste
forms is reviewed. Vitrification is considered worldwide a viable option
for solidification of liquid reprocessing wastes containing the bulk of
the high activity fission products (1); slagging high temperature incineration is another technology producing a vitreous waste form which is a
candidate for the conditioning of α-contaminated wastes (2). The most
likely disposal route presently conceived for these types of waste form is
in a deep geological formation. Within the European Community, rocksalt,
hard rock and clay are the three types of geology which are being studied.
To evaluate the suitability of such geological environments for containing
conditioned high activity and/or long-lived radioactive waste forms, an
extensive date base is required concerning the physico-chemical behaviour
of vitrified waste forms in contact with the geological medium. Since the
most likely pathway for the radioisotopes into the biosphere is via the

water percolating through the geological formation, data about the leaching behaviour in the presence of a geological environment are of prime importance. Consequently, the EC joint research programme on characterization of vitrified waste forms was conceived both to continue the study and modelling of the basic leaching mechanisms in well-controlled conditions (3) and to investigate the effects of the presence of a geological environment on the leaching of the waste forms. The experimental difficulties associated with large scale testing in underground laboratories precluding extensive parametric test programmes being carried out in-situ, the experimental results to be discussed in this paper are obtained from laboratory scale experiments using simulated repository conditions.

2. Waste forms

As mentioned already, essentially two types of vitrified waste are being investigated. The slagging high-temperature incinerator developped at S.C.K./C.E.N. for the treatment of α-contaminated waste yields an almost completely amorphous granulate fairly rich in silica and in ferrous oxide. Based on the average composition obtained for these slags, a reference glass has been determined for laboratory experiments. Its composition is given in Table I as WG124. To further evaluate the influence of iron oxides on the behaviour of this glass, a number of simplified base glasses have also been included in the experimental programme (Table I, WG119, 122, 123). Finally, granulate from actual incinerator runs has been hot compacted into monolithic form to compare with the base and reference glasses. The laboratory glasses have also been prepared with the addition of a small amount of UO_2 and/or PuO_2.

The second type of vitrified waste form considered in this programme is the boro-silicate glass devised for the incorporation of the high activity reprocessing wastes. Based on the experience of the preceding EC five-year programme, a number of reference HL waste forms have been selected for further study in several of the laboratories participating in the joint characterization effort. These reference waste forms are: UK189 and 209 (designed for vitrification of the waste from Magnox fuel reprocessing in the UK), B1/3 and C31/3 (the glass ceramic designed by HMI; also included in the programme is the parent glass from which the glass ceramic is prepared), SON58 and SON64/G3 (two of the glasses prepared in the French waste vitrification programme at Marcoule). Other glass formulations have also been investigated to a lesser extent. Among these one can mention various other French glasses as well as two glasses designed for the vitrification of the LEWC and the HEWC (low, high enriched waste concentrates) stored at the Belgoprocess plant at Dessel (Belgium). These two glass formulations are, respectively, the SM58 and the SAN60. Simplified compositions of the most representative of these glasses are also given in Table I.

3. Geological formations

As mentioned before the three types of geological formation considered as candidates for a disposal site in the European Community are rocksalt, hard rock and clay. Within the frame of the presently discussed five-year programme, studies concerning rocksalt have been performed in the FRG by both the Hahn-Meitner-Institut and the Fraunhofer Institut. These laboratories have investigated the compatibility between various vitrified waste forms and up to ten types of brines; three of the more widely studied solutions are given in Table II (see also (4)).

Hard rock effects have been studied in France (CEA) and at the UKAEA

Table I.   Principal components of waste forms (in wt %).

A. Borosilicate glasses

	UK189	UK209	B1/3	C31/3	SON58	SON64	SM58	SAN60	I117
SiO_2	41.5	50.9	28.0	34.7	43.6	47.2	56.9	43.4	48.0
B_2O_3	21.9	11.1	6.4	4.1	19.0	18.4	12.3	17.0	15.0
Na_2O	7.7	8.3	1.6	1.1	9.4	12.5	4.6	10.7	18.5
Li_2O	3.7	4.0	2.4	1.0			3.7	5.0	
Al_2O_3	5.0	5.1	12.8	10.3	0.1	1.7	1.2	18.1	5.0
CaO			4.0	3.8			3.8	3.5	
BaO			14.8	14.5					0.3
TiO_2			4.6	2.8			4.4		
MgO	6.2	6.3	1.2	1.4			2.1		
Fe_2O_3	2.7	2.7	1.5	1.5	0.6	5.1	1.2	0.3	3.3
FP oxides	9.6	9.7	14.6	15.2	22.7	13.1	6.1	1.2	10.0

B. Incinerator slag simulants

	WG119	WG122	WG123	WG124
SiO_2	56.0	57.2	59.1	60.7
Na_2O	5.3	5.4	5.5	3.6
K_2O	8.0	8.2	8.5	1.3
Al_2O_3			7.6	2.9
CaO	4.7	4.9	5.0	4.1
BaO				4.4
TiO_2				1.1
MgO	3.4	3.5	3.6	2.9
Fe_2O_3	22.6			
FeO		20.8	10.7	12.5

Table II.   Composition of salt solutions (25°C).

Type of brine	Salt content (mole/l)	NaCl	KCl (in mole %)	$MgCl_2$	$MgSO_4$
NaCl	6.2	11.1	–	–	–
Q	5.2	0.68	1.28	6.83	0.53
Z	6.1	0.12	0.18	10.2	0.55

by investigating the leaching behaviour of waste glasses in contact with either several kinds of geological well waters or a water equilibrated with a column of crushed granite. As a matter of fact, other water types were also included in the French experimental programme , namely sea water and synthetic interstitial clay water. The analysis of these waters is presented in Table III.

More detailed experiments related to clay have been performed at JRC, Ispra and at S.C.K./C.E.N., Mol. The interest in clay-waste form interaction processes derives not only from the possibility of using a

Table III. Analysis of the leachants used (ppm).

	Tap water (Vulcain)	Silica water (Mont Dore)	Granite water (Charrier)	Sea water (Toulon)	Clay water (Synthetic)*
pH	8.29	8.58	6.46	8.04	7.31
Dry residue at 110°C	285	1150	34	31850	18500
SiO_2	3	182	4	0	5
Ca	77	55	5	480	380
Mg	8	34	1	2020	1878
Na	10	350	3	8255	4335
K	2	48	0.4	432	532
Fe	0.0	0.1	0.01	0.01	0.1
Al	0.4	0.2	0.01	5	0.3
HCO_3^-	201	915	12	159	213
Cl^-	30	197	5	17200	130
SO_4^{--}	47	88	4	3700	15900
NO_3^-	4	0	4	0	0
B total	0.2	0.1	0.1	5	0

*This interstitial clay water composition corresponds with oxidizing conditions.

deep clay formation as a repository site (5), but also from the fact that clay has a very high retention capacity for radionuclides and hence can be expected to be a useful backfill material. The construction of an underground laboratory in the Boom clay formation beneath the Mol nuclear site has allowed the use in the compatibility experiments of clay excavated during the construction period. Some of the characteristics of this clay are given in Table IV. The presence of pyrite in the clay explains its strongly reducing redox potential ($\cong$ – 200 mV). Exposure to ambient air, however, rapidly results in oxidizing conditions in the clay and also in drastic changes in the composition of the interstitial clay water.

4. Experimental approaches

The evaluation of the behaviour of vitrified radioactive waste in repository conditions is based on the assumption that leaching of radioactivity through contact with an aqueous environment is the critical factor to be examined. In the experimental approaches considered for this research programme, various configurations can be and have been used, going from very simple and well-characterized to extremely complex systems which are hence more difficult to define precisely. As a direct sequel of the parametric studies in pure water, such as described e.g. in the preceding contribution (3), experiments have been performed in which glass specimens are brought in contact with an aqueous environment representative of the geological formation. Effects of temperature, pressure, loading of the water with geological material, pH and E_h, specimen surface area to leachant volume ratio, water flow rate etc. can all be studied in this type of experiment. In another type of laboratory experiment, an attempt is made at investigating the synergistic effects of having different

Table IV. Main properties of Boom clay from Mol site.

Mineralogical composition (%)	illite 25, smectite 20, vermicullite 30, illite-montmorillonite interstratified 15, chlorite + chlorite-vermicullite interstratified 10.
Chemical composition of dry material (wt %)	SiO_2 64; Al_2O_3 14; Fe_2O_3 5.9; K_2O 2.2; Na_2O 1.4; CaO 0.6; TiO_2 0.5; MgO 0.7.
Water content (wt %)	26
Ion exchange capacity	20 meq/100 g
Composition of interstitial clay water prepared under anoxic conditions (g/l)	Na_2SO_4 1.207; $NaCl$ 0.058; K_2SO_4 0.187; $MgSO_4$ 0.021; Na_2CO_3 0.339; $NaHCO_3$ 1.529; $CaCO_3$ saturated ($\cong 7.5$)
Anionic composition of the clay water mixture (100 g/1H$_2$O) (in ppm)	SO_4^{--} 10-100; PO_4^{3-} 70-140; NO_3^{-} 0.3-2

classes of material present in the same environment. Repository conditions are thus simulated by setting up hydrothermal conditions in a system containing not only glass but also solid host rock material, container material with its corrosion products, backfill material, etc. As a final step in this hierarchy of repository simulation tests one may conceive of in-situ experiments performed in an actual geological environment. Variation of the temperature (to simulate the effect of the heat load accompanying introduction of the radioactive waste in the repository) is to be included as a parameter in the experiment.

A further very important aspect in the evaluation of a repository site is the radioactivity of the waste glass to be disposed of. In the case of high-level waste forms, experimentation with the fully active materials poses serious technical problems and has to be restricted to a relatively small number of tests. Therefore, most of the experiments are performed using either simulated, non-radioactive material or materials doped with small concentrations of specific radioisotopes. The use of spiked waste glasses allows one to attempt a complete mass balance of the radioactive material and so to obtain very valuable information about the ultimate fate of the various radioelements incorporated in the glass: they can be retained in the surface layers of the glass matrix, they can remain in solution in the aqueous medium (in ionic or in colloidal form), they can be precipitated as an insoluble solid, they can be immobilized in the geological host rock material, etc.

Clearly, the radiation emanating from the radioisotopes can also be expected to influence the interaction processes governing the behaviour of the waste products in the disposal environment. This aspect is discussed in contribution IV.3. (6).

For the interpretation of the experimental results, data about the overall mass changes of the glass specimens have to be completed with a detailed chemical (and possibly radiochemical) analysis of the leachate, with an analysis of the solid precipitates when present in the leaching vessel or on the specimen surface (both composition and crystal structure are needed) and with an analysis of the surface layers generated on the glass specimen during leaching.

Of course, the ultimate aim of this exercise is the modelling over

very long time periods of the behaviour of the radioactivity which is
introduced in the repository. Applying geochemical codes adapted to the
conditions used in the experiments, it has been possible to model fairly
accurately the observed time dependence during hydrothermal leach tests in
salt brine (7).

5. Rocksalt-waste form interaction experiments

The German final repository concept foresees disposal of the waste in
salt formations. Model calculations of heat release from HLW glass blocks
yield temperatures up to 200°C at the canister/salt interface (8,9).

The composition of the glass C31/3 and the preparation of samples are
described elsewhere (4, see also Table I). Glass beads, chips and powder
were used for static hydrothermal leaching experiments in Teflon lined
autoclaves at a leaching temperature of 200°C. Compositions of the brines
are given in Table II. The pressure was about 15 bar, corresponding to the
equilibrium pressure of the salt solutions. The ratios of sample surface
area (S) to solution volume (V) were between 0.013 cm^{-1} and 10 cm^{-1}.
Samples were leached between 3 and 1000 days (12).

The weight losses of the glass beads and chips were measured and the
solutions were quantitatively analysed by ICP optical spectroscopy. The
surface layers on the glass samples were mechanically removed and used for
quantitative analysis. In parallel experiments the layers were not removed
but prepared for investigation by scanning electron microscopy (SEM),
electron microprobe analysis (EMPA) and X-ray diffraction (XRD). Some
leaching experiments were carried out with glass powder of a grain size
< 60 μm (11,12).

5.1. Specific mass loss

The results of preliminary leaching experiments in various salt
solutions showed no obvious dependence upon chemical composition of the
leachants and values of the specific weight loss are within one order of
magnitude (4). NaCl-, Q- and Z-solutions were selected for a more detailed
investigation of the leaching process (10-14).

The time dependence of leaching in NaCl-brine was derived from weight
measurements and solution analysis. The experiments were carried out with
beads giving a S : V ratio of 0.014 cm^{-1}. The specific mass loss NL_G was
derived from the weight difference of the sample before leaching and after
removing the surface layer. The leachates were analyzed by ICP. Based on
the behaviour of the components "i" the normalized elemental weight losses
NL_i were calculated. The curves are given in (11,12) and show that the
leaching process slows down with time. During the first 30 days NL_G had
the largest increase. For the elements Al, Si, Mg and Zn the curves show a
maximum at about 30 to 90 days. After about 100 days NL_{Si} and NL_{Al} are
constant or increase slightly. After 90 days the normalized specific
weight losses of these elements are considerably smaller than NL_G and the
NL_i values of B, Li, Ca, Sr, Ba and Mo. The leachates were also analyzed
for Zr, Ti, Fe, Nd, Ni, Ce and U. Their concentrations were below the
detection limits. Some semi-quantitative analyses (10,13,14) of the
surface layers revealed enrichments of these elements and other elements
whose NL_i values do not fit into the NL_G curve. Constant or decreasing NL_i
values mean constant or decreasing ion concentrations in the leachates,
due to saturation and oversaturation effects, respectively. In the leaching
process solubility limits were obviously reached for some elements within
10 days. As a consequence, concentrations of those elements were controlled
by solid phases formed upon leaching. It has been shown that leachate

concentrations for various elements can be interpreted with thermodynamic data if one assumes the presence of appropriate solid phases (14,7).

5.2. Sample surface analysis

5.2.1. Relative composition of surface layers (10,12)

Fig. la exhibits an SEM photomicrograph of a sectioned glass bead after 10 d leaching in NaCl. The leached surface consists of two different layers. The inner layer is about 15 µm thick and looks like a typical gel layer with drying cracks, whereas the outer one appears more dense and is relatively thin with a thickness of about 1.5 µm.

Fig. lb to p show EMPA concentration profiles in the form of X-ray line scans obtained at the white line marked in Fig. la. Elements such as Ca, Ba, Cs are nearly completely leached from the layers, whereas Mo and Zn are strongly depleted in the inner layer, but their concentration increases again in the outer layer. Si, Al, Mg are also depleted in the inner layer but again have a higher concentration in the outer layer, whereas U, Ni, Fe are enriched in the outer layer when compared with the pristine glass. The elements Ti, Zr, La, Ce are enriched in both layers.

The results from semiquantitative wavelength dispersive EMPA of the surface layers are given in Table V. The elements Ca, Ba, Cs, Mo and Zn are strongly depleted in the layers leached in salt solutions, whereas Zn is strongly enriched in the outer scale of the water leached surface. Al and Mg are enriched in the Q- and Z-leached surfaces and also in samples leached 30 d in NaCl, whereas the samples leached in water are higher in Al and Mg after 3 d. Ti and Zr concentrations were high in layers formed from all solutions. La, Ce, Nd and Pr are enriched only in the layers leached in NaCl and are strongly depleted in Q- and Z-leached surfaces. It thus appears that more elements are enriched in the H_2O-leached surface and that enrichment is least in Q- and Z-leached surfaces. This means that the tendency toward selective leaching decreases and congruent dissolution increases in the following order: H_2O, NaCl, Q, Z.

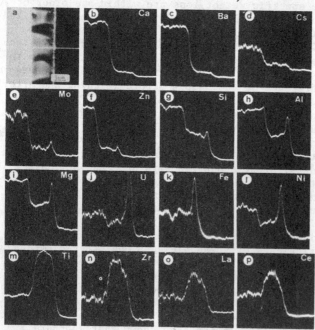

Fig. 1.
SEM photomicrograph and X-ray line scans of a glass bead leached 10 d at 200°C in saturated NaCl-solution.
(a) SE-micrograph with line scan position;
(b)-(p) line scans of elements as indicated (10,13).

Table V. Relative compositions of surface layers on glass, formed during leaching at 200°C in various solutions (10,13).

Element	NaCl-sol./ 10 d.		NaCl-sol./ 30 d.		Q-sol./ 10 d.		Z-sol./ 10 d.		H_2O / 3 d.	
	inner	outer scale	inner	outer scale	inner	outer scale	inner	outer scale	inner	outer scale
Ca	--	--	--	--	--	--	--	--	-	-
Ba	--	--	--	--	--	--	--	--	-	-
Cs	--	--	--	--	--	--	--	--	--	--
Mo	--	--	--	--	--	--	--	--	--	--
Zn	--	-	--	--	--	--	--	--	-	+ +
Si	--	-	--	o	-	--	--	--	--	-
Al	--	-	-	+	+	+ +	o	+ +	-	+ +
Mg	-	o	-	+ +	+, +	+ +	+ +	+ +	-	+ +
U	--	+ +	-	+	--	--	--	--	o	+
Fe	o	+ +	-	+ +	-	-	--	--	o	+ +
Ni	-	+ +	-	+ +	+ +	+ +	-	o	-	+ +
Ti	+ +	+ +	+ +	+ +	+ +	+ +	o	+ +	+ +	+ +
Zr	+ +	+ +	-	+ +	+ +	+ +	+	+ +	-	+ +
La	+ +	+ +	+ +	+ +	--	--	--	--	+	+
Ce	+ +	+ +	-	-	--	--	--	--	-	-
Nd	+ +	+ +	+ +	+ +	--	--	--	--	+ +	+ +
Pr			+ +	+ +	--	--	--	--	+	+
Cr					+ +	+ +	+ +	+ +	o	o
Cl	+	+ +	o	+ +	+ +	+ +	+ +	+ +		
S					+	+ +	+ +	+ +		
K					+	o	+ +	+		

-- strongly depleted: $c_l/c_g < 0.5$
- depleted: $0.5 < c_l/c_g < 1$
o unchanged: $c_l/c_g = 1$
+ enriched: $1.5 > c_l/c_g > 1$
+ + strongly enriched: $c_l/c_g > 1.5$

c_l : concentration of element in the surface layer
c_g : concentration of element in the unleached glass.

In all types of surface layers, independent of the salt brine, Cl was present; also S and K are found in layers leached in Q and Z. When comparing the concentrations of Si and Mg in the layers it was found that the Q leached layers contained 1.6 w/o Si and 13 w/o Mg in contrast to 16 w/o Si and 0.8 w/o Mg in the unleached glass. This means that only 10 % of the original silicon amount is in the layer, but that the Mg-content was increased about a factor of 10. In terms of the glass structure then only every 10th SiO_4-tetrahedron, the main network former, is still present and this is certainly insufficient to preserve the glass structure. Therefore, when leaching in Q, the glass network is nearly completely destroyed and the layer formed on the surface has no longer any structural connection with the original glass structure. This conclusion is further supported by the fact that the surface layers are partly crystalline and do not appear to be homogeneous.

5.2.2. <u>Surface morphology and phase identification</u> (11,14)
In addition to the concentration profiles which were measured on sectioned beads, surface morphology was also investigated by SEM, using

samples as removed from the autoclaves without further preparation. The surface was covered with a layer of various crystal types, which were qualitatively analysed by EDS. Figs. 2 and 3 show the surface after 10 and 30 days. The cubes in Fig. 2 and 3 could be identified as analcime, $Na(AlSi_2O_6).H_2O$. Furthermore two different crystalline phases can be seen: light clusters in the middle of Fig. 2b and light small needles in the foreground of Figs. 2d and 3d. Qualitative EDS analyses yielded the following elements with decreasing intensities:

    clusters   Si, Al, Mg, Fe, Ti, Mn
    needles   U, Ti, Si, Na, Fe, Zn, Ca

Fig. 2c shows the cross section of the surface layer after 10 days. The bottom is pristine glass. Next is the amorphous "gel" layer detached from the glass and cracked as a result of drying. The uppermost layer is mostly crystalline with large analcime crystals on top. The presence of crystals after 5 h leaching shows that solubility limits are reached shortly after the process has begun (14). In the course of the leaching process, the layer grows but the amorphous gel does not grow around the crystals (4), indicating that the amorphous layer is growing only inwards, replacing the glass phase.

The crystal phases in Figs. 2 and 3 are visible already after three days but they are small and continue to grow as leaching continues. X-ray patterns and semiquantitative analyses of the crystalline species were obtained, but it was not possible to identify the crystalline phases except for analcime. Chips and spheres yielded insufficient amounts of surface layer material for the powder X-ray diffraction technique. To get more corroded glass material in short times, glass powder with a grain size < 60 µm was used in addition to chips. In these experiments approximately 10 g glass per liter leachant was decomposed. Surface layer morphologies suggest that analcime may form a protective rim (11,12,3). On the basis of these experimental results a reaction kinetics model has been developed (11,12,3), describing leaching in the presence of solid phases. Phases controlling the concentrations in the leachate of Si, Mg, Al, Mo, Zn, Cr and Fe were identified in the surface layers (3).

5.3. Leaching of actinides

UK209 and C31/3 glasses were doped with americium and leached in saturated NaCl and Q-brine at 200°C for 3 to 47 days. In NaCl the initial leach rate is very high, but it decreases significantly already after 3 days. Total mass losses in Q-brine are smaller than in NaCl and the leach rates of the glass UK209 are always smaller than those measured for the glass C31/3 (13).

The Am-contents of the solutions and of the surface layers were measured by gammaspectroscopy. After leaching in NaCl more than 99 % of the Am corresponding to the total mass loss was detected in the surface layers. After leaching in Q-brine the opposite was the case. Only a few percent of the total Am remained in the surface layer, more than 90 % was in solution for glass C31/3 and more than 98 % for the glass UK209 (12).

Alpha-spectra of the surface layers were also recorded. In layers grown during leaching in Q-brine almost no Am could be detected, in agreement with the observation that most of the americium is in solution.

6. Hard rock – waste form interaction experiments

To examine the likely effect of disposal in hard rock, two series of experiments have been carried out. The first, at Marcoule, examined the effect of different waters at relatively high flow rates. The second, at

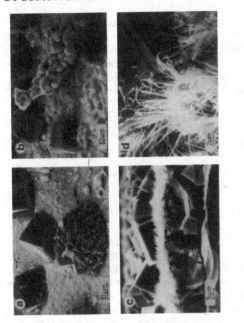

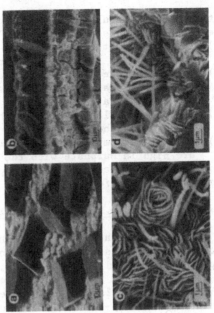

Fig. 2.
SEM-micrographs of C31/3 glass
surface leached 10 days at 200°C
in saturated NaCl solution.

Fig. 3.
SEM-micrographs of C31/3 glass
surface leached 30 days at 200°C
in saturated NaCl solution.

Harwell, endeavoured to mimic a repository, holding glass samples in a
small chamber above a column of granite.

6.1. The effect of the nature of the water

An active glass block of SON622024F3 was leached by the Marcoule
method where water is pumped into the specimen chamber for 30 seconds,
held for 30 seconds and drained out for 30 seconds. The 1,5 minute cycle
is then repeated continuously but the leachate is changed and radio-
chemically analysed every 24 hours. The leach rates for the main radioisotopes and for total β were calculated from the results.

Marcoule tap water was the reference leachant for these experiments,
being used for an initial 5 month period and for 10 day periods between
each of the geological waters (ca. 1 month each). The analyses of the
waters are given in Table III.

During the two year experiment, variations occurred in the leach-rate
both due to changing the leachate and to changes in temperature with the
latter effect being rather larger. However, the effect of the leachate was
separated out and is given below as the ratio between the leach-rate in
each natural water and that in the reference tap-water.

Element	Silica rich water (Mont Dore)	Granite water (Charrier)	Sea water	Clay water
β	0.5	1.6	2	0.5
Cs	0.5	1.3	2	0.4
Sr	0.5	1.7	3	0.6
Ce	< 1	2.2	1.8	–
Sb	< 0.25	1.5	2	0.35

Although these results are somewhat uncertain for the reason noted above, the leach-rate does not seem to be affected by:
(a) The total salt concentration - compare the high leach-rates in granite water (least salts) and sea water (most salts).
(b) The sum of the alkali and alkaline-earth concentrations - these are similar in sea and clay waters but the Cs and Sr leach rates are different by a factor of 5.
(c) The pH - the leach rates of the granite water and the sea water are higher than those of the clay water and the silicious water.
(d) Sulphate - compare the leach rates in sea water and clay water which are high in sulphate with the others.
Additional experiments at various pressures have shown that at room temperature the effect of pressure can be neglected up to 200 bars.

6.2. Leaching in a repository simulation

Samples of glass were made containing a full spectrum of inactive isotopes of the fission products in the correct proportions and were then doped separately with Sr-90, Tc-99, Cs-137, Np-237, Pu-239 and -240 and Am-241. For Cs and Sr the dopant levels were selected arbitrarily, but for Tc, Pu and Am the levels were those expected to occur in the real vitrified waste while for Np-237, in order to achieve a sufficient count-rate in the leachates, the level was ten times this, even assuming that all the Np produced in the fuel will become concentrated in the high level waste stream.

Discs of these glasses with a surface area of about 3.5 cm^2 were positioned in a small chamber (volume 1.5 ml) at the top of a column of crushed granite saturated with water. There was no water flow through the chamber but at about monthly intervals 1 ml samples of water were withdrawn from the chamber, evaporated to dryness and counted to determine their radioactivity content. The flow past the specimen could thus be said to be 1 ml per month: before reaching the specimen chamber the water to replace the sample had been in contact with granite for many months. In this apparatus, there was nothing to prevent the activity leached from the glass diffusing back down the column and indeed, as will be shown below, this did occur.

The leach rate was calculated by dividing the activity in each monthly leachate sample by the specific activity of the specimen, the specimen surface area and the time since the last sample was taken, to give a leach rate expressed in grams (of glass) per cm^2 per day.

Four glass compositions have been tested by this technique: UK189, UK209 and B1/3, both as the parent glass and as a glass ceramic after heat treating at 610°C (3 hours: nucleation) and 800°C (10 hours: crystal growth). As an example, the leach rates for the B1/3 parent glass are shown in Fig. 4 for a period of a year. The "leach rates" obtained in this way when the system had reached equilibrium are given in Table VI. These values are reduced from the "true" ones because of adsorption on the granite and on the rest of the apparatus; a similar phenomenon will also occur in a full scale repository.

After about a year the apparatus was dismantled and the granite columns cut into thick slices. The apparatus was rinsed with acid (HNO$_3$/HF) to take the adsorbed radionuclides into solution and the granite powder in each slice was similarly treated. Aliquots from each solution were then dried and "counted" and divided by the specific activity to give a notional weight of glass adsorbed in each location, this method being used for easier intercomparison between the isotopes. The glass samples were dried and weighed and the overall leach rate calculated both before and after

Table VI. Apparent leach rates at 60°C in granite repository simulation (in g.cm^{-2} day^{-1}).

Isotope	UK189	UK209	B1/3 glass	B1/3 glass ceramic
Sr-90	3×10^{-8}	1.3×10^{-8}	5×10^{-9}	
Tc-99	2.5×10^{-6}	1.7×10^{-7}	4×10^{-8}	
Cs-137	1.5×10^{-8}	2×10^{-8}	7×10^{-9}	2×10^{-7}
Np-237	8×10^{-9}	2.5×10^{-8}	4×10^{-9}	8×10^{-9}
Pu	6×10^{-8}	2×10^{-8}	2×10^{-9}	2×10^{-9}
Am-241	6×10^{-8}	2×10^{-10}	8×10^{-10}	2×10^{-9}
Weight loss	6×10^{-6}∓	✻	$1.6 \pm 0.6 \times 10^{-6}$	$6.0 \pm 2.0 \times 10^{-6}$

∓ 2 specimens gained weight
✻ all specimens gained weight

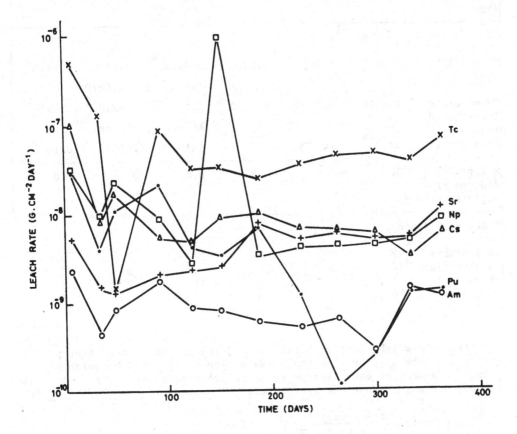

Fig. 4. Leach rate at 60°C of German glass B1/3 in repository
       simulation.

the gel-layer had been removed.

The results are given in Table VII, and the distribution of activity down the columns for each isotope is shown in Fig. 5. The recovery of the activity from the system is not good: the "weight loss" calculated from the activity measurements is smaller than that obtained by direct weighing by about a factor of 4 to 7. The leach rates calculated from weight losses should have been the same for each specimen because the amount of dopant added was not sufficient to affect the leach rate: the differences are in fact not very great.

Table VII. Leach rates and mass balance for German glass B1/3 after a year at 60°C in a simulated granite repository.

Dopant Isotope	Sr-90	Tc-99	Cs-137	Np-237	Pu	Am-241
Weight loss (mg)	1.84	1.52	1.25	3.62	1.35	2.26
Layer weight (mg)	0.25	0.35	0.65	0.45	0.50	0.60
Total loss (mg)	2.09	1.87	1.90	4.07	1.85	2.86

Leach rates ($g.cm^{-2} day^{-1}$):

	Sr-90	Tc-99	Cs-137	Np-237	Pu	Am-241
From weight loss	1.5×10^{-6}	1.2×10^{-6}	1.0×10^{-6}	2.9×10^{-6}	1.1×10^{-6}	1.8×10^{-6}
				Average $1.6 \pm 0.6 \times 10^{-6}$		
From weight loss including layer	1.6×10^{-6}	1.5×10^{-6}	1.6×10^{-6}	3.3×10^{-6}	1.5×10^{-6}	2.3×10^{-6}
				Average $2.0 \pm 0.6 \times 10^{-6}$		
From counting	5×10^{-9}	4×10^{-8}	7×10^{-9}	4×10^{-9}	2×10^{-9}	8×10^{-10}

	Sr-90	Tc-99	Cs-137	Np-237	Pu	Am-241
Weight loss deduced from activity meas. (mg)	0.3	0.5	0.5	0.6	0.4	0.8

% Activity found:

	Sr-90	Tc-99	Cs-137	Np-237	Pu	Am-241
in leachate	2	15	2	18	2	0.1
on container	10	30	17	2	8	2
in water	5	5	8	46	2	5
on granite	76	36	33	1	5	5
in layer	7	13	40	33	83	88

From the weight loss measurements and the amount of each isotope retained in the gel layer on the surface, leach rates can be calculated using the amount of each isotope released from the specimen, i.e. including the amounts adsorbed on the granite etc.:

$$\text{leach rate} = \frac{\text{total weight loss}}{\text{area} \times \text{time}} \times (1 - \varepsilon)$$

(with ε the fraction retained in the layer).
These values are given in Table VIII and are those which should be used in safety assessments where allowance is made for adsorption on the backfill

and other components of the near field.

Table VIII. True leach rates at 60°C in repository simulation (in g.cm^{-2} day^{-1}).

Isotope	UK189	UK209	B1/3 glass	B1/3 glass ceramic
Sr-90	1.8x10^{-6}	1.2x10^{-6}	1.5x10^{-6}	
Tc-99	1.2x10^{-5}	1.0x10^{-6}	1.3x10^{-6}	
Cs-137	5.5x10^{-7}	3.2x10^{-7}	9x10^{-7}	4.4x10^{-6}
Np-237	3.2x10^{-6}	4.6x10^{-7}	2.2x10^{-6}	8.5x10^{-7}
Pu	6x10^{-7}	4.6x10^{-7}	2.5x10^{-7}	3.4x10^{-6}
Am-241	4x10^{-7}	1.1x10^{-7}	2.8x10^{-7}	6.0x10^{-6}

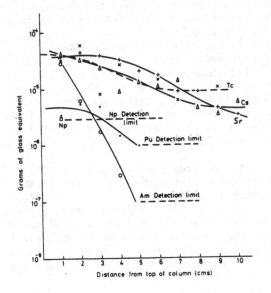

Fig. 5.
German glass (B1/3) in simulated granite repository at 60°C: distribution of isotopes down the granite column.

7. Clay-waste form interaction experiments

For the investigation of the interaction between vitrified waste forms and argillo-aqueous media, static laboratory experiments have been set up under constant conditions of temperature, pressure and specimen surface area to water volume ratio (S:V). For the study of the Boom clay, the main experimental parameters were the characteristics of the clay-water medium and the temperature. In the equipment designed by JRC Ispra (15,16), a number of porous materials were also studied in interaction with a laboratory glass simulant; in addition to information about glass alteration, this experiment also yields data on the diffusion properties of the porous medium, which could be either a backfill material or a geological formation.

7.1. Influence of clay-water characteristics

Using an MCC1 type experimental set up (17), the interaction between some of the waste forms given in Table I and various clay-water media has been studied as a function of the duration of contact at 90°C and for an initial S:V of 1 cm^{-1}. The effect of variations in S:V has been studied in detail with distilled water as leachant (3). Except for a rather limited number of tests with synthetic interstitial clay water (cfr. Table IV), most of the experiments were performed using either solid, wet clay (WC) or clay mixed with water (clay-water mixture, CWM: 100 g clay in 1 l H$_2$O, Table IV; diluted clay-water mixture, DCWM: 10 g l^{-1}; concentrated clay-water mixture, CCWM: 500 g l^{-1}). Both the incinerator-slag reference glasses and the HLW simulants yield the same variation in leach rate with respect to the clay concentration in the leachant. This is illustrated in Fig. 6., which shows the specific weight loss (SWL) gradually increases from leaching in distilled water (DW) over DCWM, CWM to CCWM and WC. The time dependence of SWL also changes with the leaching medium, reflecting changes in the leaching mechanism. The saturation effects which play an important role during leaching in relatively small volumes of DW (3) are gradually suppressed in the presence of clay, leading to essentially constant leach rates. Fig. 7 illustrates furthermore that the relatively large differences observed between the different HLW glass simulants in DW are considerably reduced in argillaceous leachants.

The Boom clay formation in situ is characterized by a strongly reducing redox potential. This negative redox potential can be preserved in the laboratory only under controlled atmosphere. The effect of the redox conditions on the leaching behaviour is shown in Fig. 8., where SWL data are shown for leaching in CWM and in WC under either oxidizing or reducing conditions. Whereas in wet clay there is hardly any effect of the redox potential (+ 460 mV or about - 150 mV with respect to a AgCl/KCl reference electrode), in CWM the SWL's are larger in reducing than in oxidizing conditions.

To complement the information obtained from the over-all SWL data, surface layer observations using infrared reflection spectroscopy (IRRS) (18) as well as SEM coupled with energy dispersive X-ray analysis (EDAX) and (EMPA) techniques were also performed. Except for glass SAN60, surface layers having a fairly uniform thickness between 15 and 70 μm are found on the HLW simulants following eight months of corrosion in WC or one year in CWM. These altered glass layers adhere fairly strongly to the bulk glass, which is usually strongly corroded, and correspond with several times (up

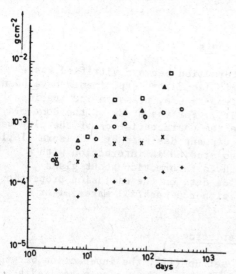

Fig. 6.
Effect of clay-water ratio on SWL for SAN60
(90°C; S:V = 1 cm^{-1}).
+ DW; X DCWM; O CWM; Δ CCWM; □ WC.

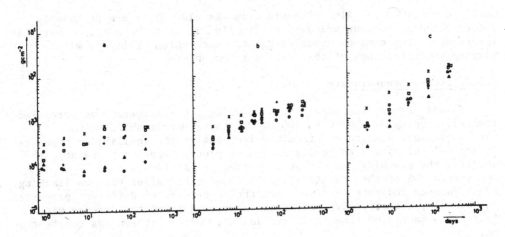

Fig. 7. SWL versus corrosion time in DW (a), CWM (b) and WC (c)
(90°C; S:V = 1 cm^{-1})
□ SM58; X SON58; O SON64; + UK209; Δ SAN60; ● C31.

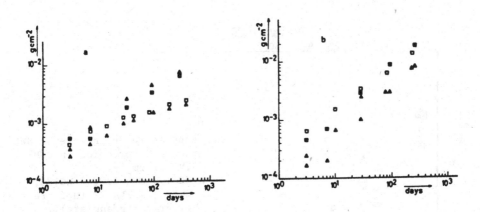

Fig 8. Effect of redox conditions on SWL for SM58 and SAN60 in
CWM (a) and in WC (b) (90°C, S:V = 1 cm^{-1})
Squares: SM58; triangles: SAN60; open symbols: oxidizing
conditions; filled symbols: reducing conditions.

to five times) the measured weight losses. The composition of these
surface layers reveals that alteration of the glass is not a congruent
process. In the HLW simulants, elements such as Cs, Mo, Zn and Mg are
found in smaller concentrations in the surface layers than in the pristine
glass, whereas the surface layers are enriched in elements such as Zr, Al,
Ti, Fe, Nd. In the incinerator-slag simulants, on the other hand, Mg is
frequently also observed in the outer layer, and the total layer thickness

rarely exceeds 40 μm. Also elements from the clay (S,K) are frequently incorporated in the surface layers. Finally, it is to be noticed that the layers are often complex, consisting of a succession of layers with varying concentrations of the different components.

7.2. Effect of temperature

In addition to the experiments at 90°C, other temperatures were also studied, ranging from 40°C to 200°C. For temperatures above 90°C, autoclaves were used to maintain the leachant at the equilibrium pressure. In general, no single thermally activated process appears to be able to describe the leaching, as can be seen from the Arrhenius-type representation of the results (Fig. 9). Data taken after various leaching times further indicate in some cases the occurrence of different processes in sequence; the effective activation energy is found to increase with increasing corrosion time in DW and in CWM, whereas it remains constant in WC.

The analysis of the surface layers formed during exposure at elevated temperatures confirms that it is dangerous to draw very general conclusions irrespective of the glass composition. In terms of glass components, it appears nevertheless that for a specific glass, the elements retained within the surface layer are independent of the leachant (of course, in clay media, elements from the clay are often incorporated in the surface layer). After one month at 200°C, the specific weight losses in WC and CWM are rather similar for the WG type glasses, whereas WC is at least 5 times

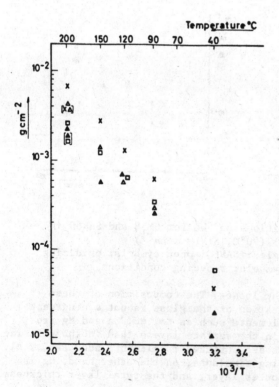

Fig. 9.
Temperature dependence of SWL after 7 d corrosion in CWM
$(S:V = 1 \text{ cm}^{-1})$.
X WG119; O WG122; △ WG123; □ WG124; ▲ FLK 77 (hot pressed incinerator granulate).

as aggressive as CWM for the HLW glasses SM58 and SON60. Yet, layer thicknesses are comparable. Maybe the most remarkable difference between high temperature leaching in either DW or clay media is the very pronounced surface crystallization observed in DW (3) as opposed to its absence (at least for times up to one month) in CWM or in WC.

7.3. Leaching in wet porous media

At JRC Ispra, the leaching behaviour of a HLW glass simulant (I117, see Table I), was investigated in contact with different types of wet porous media: montmorillonite mixed with sand or mixed with sand and Fe_2O_3; sand; clay; sea sediments. Both simple capsule tests yielding SWL vs time data as well as more involved column tests, allowing also the determination of the diffusion coefficient of various elements through the porous medium, have been performed. The time dependence in both geometries is very nearly linear, in general agreement with the results obtained at S.C.K./C.E.N. Mol. The degradation rates determined from these static degradation experiments at 80°C and 75°C are about $(5-10)~10^{-5}~g~cm^{-2}~d^{-1}$. The suppression of leachate saturation effects, which lead to a decrease in leach rates, is ascribed to the adsorption by the porous medium of SiO_2 released from the glass. Only after longer leaching times could it be expected that saturation of the surrounding porous medium might lead to diffusion within this medium being rate determining for the continuation of glass degradation. Since SiO_2 is the main constituent of the waste glasses, one may distinguish between elements which are less soluble in the leachate than SiO_2 and those which are at least as soluble. Clearly, solubility restraints will influence only the low solubility elements (such as the actinides), for which the release from the glass will be governed by the solubility limit of the migrating species and by its diffusion coefficient. This will lead in many cases to retention of such elements in the surface layers of the glass or in the contact layers of the surrounding porous media. Using the column type experiments, diffusion profiles in solid Boom clay have been determined using neutron activation analysis. The diffusion coefficients (at 75°C) derived in this manner for a number of elements are given in Table IX.

Table IX. Diffusion coefficients in clay.

Element	$D~(cm^2/s)$	Element	$D(cm^2/s)$
Cs	$5.2.10^{-7}$	La	$1.8.10^{-9}$
U	$1.2.10^{-8}$	Ce	1.10^{-9}
Co	$4.8.10^{-9}$	Eu	$3.9.10^{-10}$
Fe	2.10^{-9}	Hf	$7.6.10^{-11}$

8. Conclusions

The laboratory scale experiments performed during this 5 year research effort have produced valuable data which will be useful in the safety analysis and feasibility assessment of radioactive waste repositories in the three types of deep geological formations considered in the EC. The cooperative effort of the various contributing laboratories has allowed the scanning of a relative wide scope in terms of waste forms and geological formations as well as in terms of experimental parameters and approaches.

Further modelling of the phenomena will require more information about the thermodynamics of the interaction processes in the complete system, including all components which will conceivably be present in a real repository. A logical extension of the research effort will have to explore these interaction effects by carrying out experiments involving more and more components. Of course, a judicious design of the experiments will be absolutely essential to avoid unnecessary multiplication of experimental conditions and parameters. Clearly, a coordinated effort within the frame of the EC can strongly contribute to the efficiency of such a programme.

References
1. HENCH, L. et al. 1984, Nucl. & Chem. Waste Management 5, 149
2. VAN DE VOORDE, N. et al. 1976. Proc. IAEA-NEA Intern. Symposium on Management of Rad. Waste from the nuclear fuel cycle, Vienna, p. 141
3. LANZA, F. et al. this conference, Paper IV1
4. DE BATIST, R. ed. 1981. Joint Annual Report 1981, EUR 8424 EM
5. BAETSLE, L.H. and MITTEMBERGHER,M. 1980. in Proc. 1° European Conf. on rad. waste management and disposal, Luxemburg
6. LUTZE, W. et al. this conference, Paper IV3
7. STRACHAN, D.M. et al. 1984. Nucl. & Chem. Waste Management, 5, 87
8. ROTHEMEYER, H. 1980. IAEA Vienna 297
9. DELISLE, G. 1980. Z. Deutschen Geol. Ges. 131, 461
10. MALOW, G. 1982. Scientific Basis for Nuclear Waste Management 5, ed. W. Lutze, p. 25
11. HAAKER, R.F. et al. 1984. to be published in Scient. Basis for Nuclear Waste Manag. 8
12. MALOW, G. ed. 1983. Joint Annual Report, EUR
13. ENGELMANN, C. ed. 1982. Joint Annual Report, EUR 9268 EN
14. LUTZE, W. et al. 1983. Scient. Basis for Nuclear Waste Management 6, ed. Brookins, D.C., p. 37
15. LANZA, F. and RONSECCO, C. 1984. Glass leaching in wet porous media, EUR 9215 EN
16. LANZA F. et al. 1984, Study of the release of various elements from HLW glasses in contact with porous media. Presented at Workshop on the Source Term for Radionuclide Migration from HLW or Spent Nuclear Fuel under realistic Repository Conditions, Albuquerque, NM.
17. MENDEL, J.E. Nuclear Waste Materials Handbook – Test methods DOE/TIC 11400
18. VAN ISEGHEM, P. and ROTTI, M. 1983, Proc. European Workshop on Physical Techniques for Studies of Surfaces and Subsurface Layers of Glasses; Lanza, F. and Manara, A., eds., p. 19, JRC Ipsra.

DISCUSSION

R. KOSTER, KfK Karlsruhe

On the transparency (Fig. 23) from Harwell, showing isotope specific leach rates at 60°C of the German glass B1/3 under repository simulation, there was a peak for Np over three orders of magnitude (from 10^{-6} to 10^{-9} g/cm^{2}d). Is there any explanation for this behaviour?

J.A.C. MARPLES, AERE Harwell

The most probable explanation is that a small piece of gel-layer came loose and was taken out with the leachate sample for counting.

C. COOLEY, US/DOE

Could you comment on the solubility limits controlling your leach rates?

R. DE BATIST, SCK/CEN Mol

Solubility limits are not constant and depend strongly on the chemistry of the system. Since in the joint EC-action we investigate different repository environments, the resulting kinetics vary considerably and consequently also the corresponding solubility limits.

M.S.T. PRICE, AEE Winfrith

Have repository barrier materials, such as cement, been taken into account in the characterization programme?

R. DE BATIST, CEN/SCK Mol

Tracing back the joint EC efforts in this field, we started some ten years ago with comparative studies of different glasses in water. During the second five-year programme, emphasis was placed on testing under simulated repository conditions, in particular under the presence of the various candidate host rock materials. The next step will be to concentrate on the engineered barrier materials and first experiments in this field have already been launched. It is, however, too early to draw any conclusions.

RADIATION, THERMAL AND MECHANICAL EFFECTS IN HLW GLASS

W. LUTZE[+], A. MANARA[++], J.A.C. MARPLES[+++], P. OFFERMANN[+],
and P. van ISEGHEM[++++]
[+]Hahn-Meitner-Institut für Kernforschung Berlin GmbH.,
Glienicker Strasse 100, D-1000 Berlin 39
[++]Commission of the European Community - Joint Research Centre Ispra,
I-21020 Ispra (Varese)
[+++]United Kingdom Atomic Energy Authority,
Atomic Energy Research Establishment,
Bd. 220.22, U.K. Harwell, Didcot, OXON OX11 ORA
[++++]Studiecentrum voor Kernenergie, SCK/CEN, Boeretang 200, B-2400 Mol

1 Radiation Effects

1.1 Pu-238 doped sammples

It has been pointed out by various authors that the most likely cause of any radiation effects in the vitrified waste is the α-decay of the incorporated actinides. The α-particles displace some atoms from their positions in the glass network whilst the recoiling actinide nuclei lose virtually all their energy in this way. Assuming a displacement energy of 25 eV each α-particle will displace c 150 atoms and each recoiling nucleus c 1500, the latter being concentrated in a 'spike', somewhat akin to a fission spike.

To test the long-term effect of this on the glass, the most realistic if rather laborious way, is to dope them with a short half-life α-emitter (Cm-244 or Pu-238) so that as many α-decays will occur in the glass in a few years as will occur in the real waste in many millenia.

Accordingly, as part of the first CEC sponsored programme (1), some samples of glass, with compositions suggested by the collaborating laboratories, were doped with 2 1/2 wt.% Pu-238 at AERE, Harwell. Initially the densities, leach rates, stored energies and helium release were investigasted (1) to see if the radiation produced any effects. The stored energies were small and released over a wide temperature range such that no self sustaining temperature rise could occur. Most of the helium from the α-particles was retained in the glass and this also was not seen as a problem. Changes in leach rate and density have continued to be monitored.

After storage at room temperature for almost 6.5 years, the Pu-238 doped samples had reached a dose of about $2.8 \times 10^{+18}$ α-disintegrations per gram. This is approximately equivalent to the following times for the real waste (assuming 0.1% of the Pu and all the Am and Cm are incorporated in the glass): Glasses 189 and 209 (Magnox waste) 1.5M; glass ceramic B1/3 and glass VG 98/3 (LWR waste) 70K; glass SON 58.30.20 (LWR waste) 10K.

The much longer equivalent times for the glasses containing magnox waste are because of the smaller amount of fission products in the glass and the much lower fuel burn-up, leading primarily to a smaller amount of Am in the waste.

1.1.1 Leach rates

The leach rates of samples of all six glasses, stored at room temperature have been measured at intervals using the Soxhlet technique to detect whether any changes had occurred. The results are plotted in Fig. 1 and show increases of about x4 for SON 58, x2 for VG 98/3 and 189 and no effect on the others. It seems to be the softer glasses that are more affected by radiation but it is not known why this should be so.

1.1.2 Densities

The densities of these glass and glass ceramic samples were also measured at intervals and the values fitted to the equation

$$\Delta\rho / \rho_o = A(1-\exp(-\alpha D))$$

where $\Delta\rho$ is the change in density, ρ_o is the original density. A is the saturation value of $\Delta\rho/\rho_o$. D is the dose and α is a constant.

The values of A and α that give the best fits to the data are given in Table 1 and an example in Fig. 2: the data obey the exponential equation very closely. The borosilicate glass SON 58.30.20 and the phosphate glass have continued to contract with increasing dose whilst the other borosilicate glasses and the glass-ceramic expand.

Since the radiation damage to glasses by α-decays is almost entirely in the form of heavily damaged zones round the track of the recoil nuclei, the build up of damage consists essentially of the increase in the number of such zones within the glass.

Let r = rate of damage (α-decays cm^{-3} sec^{-1}), v = volume of damaged zone (cm^3) and F = fraction of sample volume occupied by damaged zones.

In the absence of any recovery:

$$dF/dt = vr(1-F) \tag{1}$$

Integrating we have:

$$F = 1 - \exp(-vrt) \tag{2}$$

If we assume that the density change $\Delta\rho$ is proportional to the volume fraction damaged we have

$$\Delta\rho = \Delta\rho_{sat}(1-\exp(-vrt)) \tag{3}$$

The dose at time t is rt/ρ decays/gram. However, there is the possibility that recovery is occurring simultaneously. For simplicity this is assumed to be first order so that equation (1) becomes:

$$dF/dt = vr(1-F) - kF \tag{4}$$

On integration:

$$F = \left(\frac{vr}{vr+k}\right)(1-\exp(-(vr+k)t)) \tag{5}$$

Thus similar exponentially saturating behaviour is predicted as before but with the saturation value reduced by a factor vr/(vr+k).

Again assuming the recovery is first order, the irradiation induced density change $\Delta\rho$ which is assumed to be proportional to F should recover exponentially with time:

$$\Delta\rho = \Delta\rho_o \exp(-kt) \tag{6}$$

The samples were annealed isochronally after a dose of $2.8 \times 10^{18} \alpha$ -
decays per gram. In this technique, the samples were heated to successively higher temperatures, spaced at about $25^{\circ}C$ intervals for a constant
time (here, 15 hours) and the densities measured (at room temperature)
between each annealing: the results are given in Figure 3.

In radiation damage studies of metals, for example, this technique
sometimes reveals the presence of different types of defect in the structure when they anneal out at different temperatures. When this does occur,
the isochronal annealing curve will have 'steps' in it. However, there is
no sign of this in the present case. The annealing therefore shows either
a single broad annealing stage or, perhaps more probably, the superposi-
tion of a large number of overlapping stages.

At each temperature, k was determined from equation (6) and is shown
as an Arrhenius plot in Figure 4. The lines drawn through the points in
this figure extrapolate to values of k at $20^{\circ}C$ as follows (in units of
sec^{-1}):

189	209	VG98/3	CELSIAN	SON 58.30.20
1.2×10^{-7}	1.2×10^{-7}	5.8×10^{-8}	2.6×10^{-8}	2×10^{-11}

Using ion-bombardment techniques, Dran et al. (2) found a large increase in leach rate in some glasses after a critical ion dose. This has
been attributed to a dose-rate effect (3). Equation (5) shows that if
recovery is fairly rapid, i.e. the recovery constant k > vr then the
fraction, F, damaged will always be small since its maximum value is

$$F_{max} = vr/(vr + k)$$

Table II gives values of this fraction for various values of k and
for values of vr appropriate to the real waste disposal case, to the present Pu-238 doped glass and to the ion-bombardment experiments.

For the values of k found from the isochronal annealing experiments,
tabulated above, in the real situation, damage will only build up to a
small extent in the French glass and hardly at all in the other substances. Partial annealing will occur in the Pu-238 doped samples and little
annealing will occur in the ion bombardment experiments. Thus, on this
simple theory, the ion bombardment experiments are likely to give mis-
leading results and any changes observed in the doped glasses should be
regarded with caution.

There are, however, various observations that suggest that the real
situation is more complicated.

(a) At higher temperatures, the Arrhenius plots of k for some of the
glasses appeared to show a second stage with a much higher activation
energy.

(b) When some of the glass samples were annealed isothermally, i.e.
when these were held at a constant temperature for a longer period, the
densities only followed an exponential decay curve for a short period before becoming constant, i.e. there was a proportion of the radiation induced density change that was stable at each temperature (4).

(c) After annealing, the Pu-238 doped sample of glass VG98/3 has been
held at $130^{\circ}C$. The density changes compared to those found during the
original hold at room temperature are shown in Fig. 2.

The version of equation (5) that applies to density changes is:

$$\Delta\rho = \Delta\rho^{0}_{sat} \ \frac{vr}{vr+k} \ (1 - exp(-(vr+k)t)) \tag{5a}$$

where $\Delta\rho^{o}{}_{sat}$ is the saturation value of $\Delta\rho$ at low temperatures where the recovery constant k is small compared to vr. Using values of k for $130^{\circ}C$ and for $20^{\circ}C$ taken from Fig. 4 suggests that the density changes at $130^{\circ}C$ should be much smaller compared to those at $20^{\circ}C$ than were actually found and also that the density changes at $130^{\circ}C$ should approach saturation much more quickly than at $20^{\circ}C$: again this is not borne out by experiment as may be seen by comparing the observed values at $130^{\circ}C$ with the calculated curve in Fig. 2. Because of the complicated structure of glasses it seems most likely that many types of defects will occur. Each will have its own relaxation constant, with different activation energies and will affect the density to a different extent.

The equations are difficult to deal with, and separating out the various terms may well prove impossible.

1.2 Ion bombarded samples

Experiments with silicate glasses were performed at JRC, Ispra, to investigate the basic damage mechanisms in terms of production, saturation and annealing of atomic defects and to analyse some effects of radiation damage on the structural stability of glasses (i.e. microstructural changes and etching rate variations) and possibly to relate them to atomistic aspects. Main techniques used were optical absorption and electron microscopy. Optical absorption was used as a means of monitoring the formation and annealing of the different types of point defects as a function of dose, mass and energy of damaging particles; electron microscopy to observe microstructural changes during in situ irradiation.

Etching rate measurements of samples irradiated with particles of different mass and energy have also been performed before and after annealing at different temperatures. A correlation between the annealing of excess etching rate and the annealing of some types of defects was attempted.

Samples used were pure silica and borosilicate glasses with different content of sodium. Sources of irradiation were electron, proton, α-particles, neon and nickel ions.

1.2.1 Point defects

Spectrophotometric measurements were performed in the UV up to 10 eV. Colour centers were used (5, 6, 7) to monitor defects which were already present before irradiation as well as new defects created during particle bombardment. It was found that all types of particles introduced atomic defects. Their concentration attained saturation. However, different masses and energies gave rise to different saturation levels. Typical results obtained in amorphous silica irradiated with Ni, Ne, He, H ions and electrons are shown in Fig. 5. From the observation of these data, it appeared that the general dependence of the saturation level on the mass of the impinging particles was very similar for the major absorption bands at 5.05 eV (B_2-centre); 5.80 eV (E_1-centre) and 7.15 eV (not observed with electrons and protons). In particular, one notices the tendency to reach a constant value when the impinging mass is higher than that of neon. An exception is represented by the high efficiency of protons in producing E_1-centres: this has already been explained in terms of H-O interactions, causing the rupture of oxygen bonding and an increase of dangling bonds and silicon sites.

In order to clarify the general pattern of damage produced by different particles, we measured the optical spectra of silica glass bombarded with heavy and light ions, sequentially applied to the sample. The prominent effects are illustrated in Figs. 6 and 7, showing the colouration of samples after single or combined irradiations. In both figures the results concerning B_2 and E' bands were obtained from spectral fitting, details of minor bands have been omitted for the sake of clarity. The optical density values in each case were recalculated to take into account the different ranges related to different particles, so as to illustrate the net effect introduced by the second irradiation. In all cases saturating conditions were realized. One may notice that in the proton re-irradiated sample the B_2 band is drastically reduced, in agreement with what was reported by Arnold (8) after Xenon implantation. However, He^+ ions are also seen to reduce the amplitude of the B_2 band. Therefore we are prone to attribute some importance to radiation annealing mechanisms, in addition to chemical annealing postulated by Arnold for the proton case.

1.2.2 Etching rate

The production of stable, isolated defects after irradiation with sufficient heavy ions was found to correlate with an increase of etchability: The increase of the etching rate is larger when the concentration of point defects at saturation is larger (7,9). The results of etching experiments after proton, neon and nickel irradiation are shown in Fig. 8. In all cases it was found that the etching was linear with time. The etching rate of the irradiated part is higher than that of the unirradiated part. Heavy particles produced a substantially larger effect than light particles: for instance, the enhancement of the etching rate was about 2 with protons, but nearly 4 with Ne or Ni ions (with an almost identical value in the latter two cases). Finally, the etching rate enhancement disappeared when the damaged layer was consumed; this is demonstrated in Fig. 8 by the knee appearing in the etching characteristics of Ne-irradiated samples (for the proton case a similar effect was observed at a depth of 16.5 μm, as expected from range calculations).

A series of experiments was also devoted to the study of the annealing behaviour of the enhanced etching rate and its relation to that of colour centres. For this purpose, irradiated samples were subjected to isochronal annealing at increasing temperatures, the etching rate was measured after each step. Fig. 9 shows the etching rate to decrease with temperature up to 650°C for a Ni irradiated sample. Apparently, the annealing curve does not correlate with thermal bleaching of any of the colour centres. Typical temperatures for the reduction of the enhanced etching rate were between 450°C and 650°C, 200°C above thermal bleaching, but rather close to the region where a major peak of thermal release was observed in calorimetric measurements.

1.2.3 Microstructural changes

Radiation damage to borosilicate glasses of different compositions have been studied using a High Voltage (1.0 MV) Electron Microscope (10, 11,12), to concurrently generate and observe microstructural changes. Electron fluxes during irradiation ranged from 1.10^{21} to 3.10^{24} em^{-2}s^{-1}. Glass discs were cooled to 100K or heated up to 1000K. In pure silica samples, observations failed to reveal any visible clustering effect between RT and 750K, up to a dose of 7×10^{27} e/m^2. In sodium borosilicate glasses containing about 15 at.% of Na, low intensity irradiation generated planar crystalline regions. High intensity irradiation almost immediately nucleated a high density, $\sim 10^{18}$m^{-3}, of spherical pores, each of which grew progressively with increasing dose. The sequence incu-

bation, nucleation and linear swelling was evident for a wide range of irradiation temperature and damage rates, however, fluxes below a critical
value, depending on temperature, failed to nucleate pores. These features
are similar to those observed during void swelling of irradiated metals.
We interpret the observed pores as gas-filled bubbles similar to what
has been established in metals.

Since local alkali depletion under irradiation is a well known
effect in alkali silicate glasses, a more detailed study was performed
to investigate the role of Na in conjunction with the bubble swelling
effect. Glasses containing Na_2O between 3.5 and 25% ($SiO_2/B_2O_3/$
$Na_2O/Al_2O_3/K_2O$ = 80/12,9/3,6/2,1/1,5 and 75/15/10/-/- and 70/15/15/-/-
and 65/15/20/-/- and 53/16,3/25/5,7/-) were irradiated and observed
under the microscope. The results showed that sodium atoms moved during
irradiation from the center of the irradiated volume towards the
periphery, where they probably concentrated in colloidal form. The
concentration of Na in the irradiated region was measured with the help
of an EDAX spectrometer attached to the electron microscope and local
enrichments were detected. Even before bubble nucleation, it was discov-
ered that the concentration of Na decreased in the center of the irradiated volume by a factor of about 3.

The influence of sodium on the critical flux ϕ_c for bubble nucleation, on the swelling rate $dS/d\phi$,and on the maximum swelling is shown in
Fig. 10. On the basis of the experimental results, the following model can
be outlined to combine ionization and displacement damage in these alkali
borosilicate glasses.

A local intense electric field ($\sim 10^5$ V/m) is generated by secondary electrons as a consequence of a gradient in the primary electron beam
density within the irradiated volume. Under the influence of this driving
force, displaced alkali ions and related vacancies are subjected to enhanced diffusion in opposite directions. The high vacancy concentration
present in the irradiated volume is responsible, together with the concomitant oxygen displacements for bubble nucleation and growth, leading to
a swelling behaviour similar to that of a metal. This model explains the
absence of any microstructural changes in irradiated pure silica.

2.   Thermal effects

It is well known that most glasses are crystallized to a small extent
(13,14). Crystallization may be enhanced upon post-conditioning (14,15) or
due to accidental conditions. The waste glass processing itself will not
induce crystallization, since cooling from the melting temperature to temperatures beneath the glass transformation range occurs sufficiently rap-
idly.There is experimental evidence that partial devitrification has no
(16,17) a positive (18), a negative (16) or no influence on the waste
form durability - depending on waste form composition.

The crystallization behaviour of some reference and base borosilicate
HLW- and TRUW-glasses, and its influence on the corrosion stability have
been evaluated at S.C.K./C.E.N., Mol. The HLW borosilicate glasses refer
to those designed to incorporate the HLW stored at the BELGOPROCESS plant
in Belgium (glasses SM58 - actually a precurser of the glass SM513LW11
which will be used in the PAMELA process - and SAN60). The TRUW silicate
glasses (WG119, 123 and 124) refer to the basaltic slag produced by the
prototype FLK incinerator (19).

Information on the glass preparation may be found elsewhere (4,19).
The TRUW silicate glasses were either inactive or contained 2 wt.% UO_2

(denoted UWG). Prior to the thermal annealing experiments, characteristic glass temperatures were determined by thermal analysis. DTA (differential thermal analsis) and DSC (differential scanning calorimetry) were performed with a DuPont 990 Thermal Analyser.

Isothermal heat treatment in the present study was identical to that applied in a previous investigation by AERE Harwell and HMI Berlin (16). All glasses were heated at 500, 600, 700 or 800°C for 2 h, 1, 10 and 100 d.

The corrosion stability of the partially devitrified glasses was studied either under flowing (Soxhlet test: 14 d exposure at 99°C to distilled water with a flow rate of about 1 ml min^{-1}) or under static conditions. The static experiments run over long times (2 years is in progress), in different media: distilled water (DW), a mixture of Boom clay with water (100 g l^{-1}; CWM) or wet clay (WC). The test temperature was either 90 or 200°C. For 90°C, DW, two surface area to solution volume ratios were used (SA.V^{-1} = 1 and 0.1 cm^{-1}). All experiments were conducted in an oxidizing atmosphere.

The thermal analysis data are listed in Table III. They reveal a wide spread in T_g (glass transition temperature, determined by DSC), from 465 (SAN60) to 645°C (UWG124). The M_g (dilatometric softening point) and T_c (crystallization temperature) data, determined by DSC (M_g) or DTA (T_c) also exhibit large differences.

Heat treatment confirmed the different thermal behaviour suggested by thermal analysis. At 500 - 600°C, no crystallization could be measured for both SAN60 - SM58. Strongest crystallization occurred at 700°C (SAN60), 700 - 800°C (SM58; WG119), 800 - 900°C (WG123) and 800 - 1000°C (WG124). The identity of the crystals formed is listed in Table IV, see also (20).

Partial crystallization did influence the corrosion stability of the waste forms. In general, considering all test conditions (different flow rates, different temperatures and different media) the corrosion of WG119 and 124 decreases, of SAN60 and WG123 did not vary and of SM58 increased upon partial crystallization. Table V summarizes results obtained in a Soxhlet experiment (16) and in the presence of clay (similar tests were performed on other reference glasses. The deterioration (SM58) or improvement (UWG119, 124) of the corrosion resistance is within a factor 5. The changes in corrosion resistance can probably be related to changes in glass composition upon crystallization (SM58: substraction of SiO$_2$; UWG glasses: removal of relatively more network modifiers such as Na, Fe, Ca, Mg). The resistance of the residual glass matrix seems to dominate corrosion bevahiour.

3.   Mechanical Effects

At AERE, Harwell, in an investigation of the amount of cracking occurring when glass cylinders are cooled, glass frit was melted in 18/9/1 stainless steel canisters, 0.25 m diameter, 0.75 m long with a wall thickness of 5 mm. The melt was 0.6 m deep and weighed about 80 kg and was allowed to stand at 950°C for at least 24 hours to clear any bubbles. The furnace power was then switched off and the glass cooled following the exponential equation:

$$\Delta T = \Delta T_o \exp (-0.0184 \ t) \tag{7}$$

where ΔT is the excess temperature above ambient (°C) and t is in hours. Above T_g (446°C) the glass is soft and no stress occurs. Below this

temperature, stresses (and hence fracture) only occur due to interaction
with the canister or change in temperature profile.

After cooling, the canisters were cut away from the glass, the fragments were examined and their total surface area estimated by laborious
measurement assisted by computer programme. In the first instance, the
glass adhered extensively to the cylinder and this had apparently caused
much of the cracking. The surface area had increased from 0.57 m^2 for
an intact cylinder to 5.6 m^2, an increase by a factor of 10. For the
second run, the cylinder was lined with a graphite mat and this decreased
the adhesion except near the top of the cylinder where it had oxidised
away. The surface area had decreased to 2.9 m^2, a worthwhile improvement
of about a factor of two.

Unless cooling times of several months are applied, there will remain thermoelastic stresses in the glass. Additional stresses will arise
from the decrease of the heat production rate due to the radioactive decay. Although these stresses are probably not able to initialize spontaneous failure of the glass block, the possibility of stress release by slow
crack propagation has to be considered - a phenomenon well known in glasses, where velocities as little as 10^{-9} m/s were observed. This is a value
still high enough to destroy an initially intact glass cylinder of the dimensions envisaged. Therefore, it is not only necessary to quantify the
remaining stresses in a nuclear waste glass after cooling, but also to
characterize the flaws in the surface of the waste form as they act as
initial cracks. The velocity of the propagation of a crack tip has to
be determined also.

Work performed at HMI, Berlin, will be reported here. At low stresses
the velocity can be described by an empirical law (21)

$$v = \frac{dc}{dt} = A K_I^n \tag{8}$$

the so-called K_I-v-curve. K_I is the stress intensity factor and connected
to the stress field σ by (22)

$$K_I = \sigma c^{1/2} f \tag{9}$$

where c is the length of the crack and f a correction factor depending on
the geometry of the material and the crack itself. A and n in equation (8)
are material constants which may depend on the chemical environment (gas,
liquid). From the combination of equations (8) and (9) a maximum stress
can be calculated which will not increase the crack surface by more than
a factor c_f/c_i (c_f = final crack length, c_i = initial crack length)
within the time t:

$$\sigma_{max} = \frac{1}{f} \left[\frac{2}{t(n-2)A} \right]^{1/n} c_i^{(2-n)/2n} \left[1 - \left(\frac{c_f}{c_i} \right)^{(2-n)/2} \right]^{1/n} \tag{10}$$

The parameters A and n of the K_I-v-curve (equation (8)) were measured
by means of the double-torsion-method using different simulated waste
glasses. The method and the apparatus used are describedin (23). Results of
the complete curve are given in Fig. 11 for the glass SM513 used in the
PAMELA plant/Mol and in Fig. 12 for the German reference glass C31-3 and
the glass ceramic B1-3. There are three regions in the curve easy to distinguish. The first region - probably above a fatigue limit - is at low
values of the stress intensity factor corresponding to the exponential law
of equation (8), the second region where a plateau is reached, and when

the velocity might be determined by transport processes in the neighbour-hood of the crack tip, the third region is the so-called vacuum curve, which is independent of the chemical environment. The vacuum curve is only determined by the binding forces in the material and ends at a velocity leading to spontaneous failure. The corresponding K_I-value is the so-called fracture toughness K_{Ic}. In Table VI measured values of the fracture toughness ($v = 10^{-3}$ m/s) are given together with the parameters of the K_I-v-curve. Generally, the properties of the nuclear waste glasses are between those of commercial glass and ceramic materials. The high values of n - typical of brittle materials - lead to high changes in the velocity for only a small increase of the stress intensity factor. This is the reason why glasses fail almost instantaneous if a crack length reaches a critical value.

To determine a typical initial crack length c_i it is necessary to characterize the surface of the waste form. As 'real' surfaces were not available and experimental methods usually ask for a plane surface, samples were prepared from simulated nuclear waste glasses using differ-ent surface treatments. The initial crack was calculated from equation (8) using the fracture toughness K_{Ic} for the stress intensity factor and the fracture stress σ_c which is necessary to initiate spontaneous propaga-tion of a failure flaw. For the data presented it was determined by means of the Hertzian indentation method. The method and the apparatus used are described in (23). Typical results are shown in Fig. 13 for the glass UK189 and different surface treatments. The data of the fracture stress follow in most cases a Weibull-distribution. 50% values of all materials investigated are given in Fig. 14 for mechanically polished (7μ), fire polished and fractured surfaces. The differences are not very significant, however, mechanically polished surfaces yielded always the smallest, i.e., most unfavorable value. The values are given in Table VII. They were used together with the fracture toughness (Tab. VI) to calculate an initial crack length on the surface (f = 2 was used as an upper limit and corresponds to a scratch on a large surface). The results are given in Table VIII and are typically in the order of magnitude of several microns which was expected due to the polishing procedure.

Finally, equation (10) was used to calculate maximum allowable stresses if an increase of the surface area of a crack by more than a factor of c_f/c_i = 10 is not desired within 100 and 1000 years resp. (Tab. VIII). Due to the high values of n the calculated values for one material are not very different, but they are significantly less than the fracture stress. Stresses in the material of the same order of magni-tude will lead to almost instantaneous failure if a certain crack length is reached. Therefore, it is necessary to estimate a reliable value of the maximum remaining stress in a nuclear waste form after complete cooling taking the decay heat into account.

REFERENCES

1. MALOW, G., MARPLES, J.A.C. and SOMBRET, C. (1980), in Radioactive Waste Management and Disposal; R. Simon and S. Orlowski Eds., Harwood Academic Publishers Luxemburg, p. 341

2. DRAN, J.C. et al., Science 209 (1980), 1518 and Scientific Basis
    for Nuclear Waste Management; Ed. J.G. Moore, Vol. 3 Plenum Press
    (1981), p. 449
3. BURNS, W.G., HUGHES, A.E., MARPLES, J.A.C., NELSON, R.S. and
    STONEHAM, A.M., (1982), J. Nuc. Mat. 107, p. 245
4. Testing and Evaluation of Solidified High Level Waste Forms. R. de
    Bastit Editor; (1983), European Commission Report EUR 8424 En, p.103
5. ANTONINI, M., CAMAGNI, P., GIBSON, P.N. and MANARA, A. (1982),
    Rad. Effects 65, p.281
6. ANTONINI, M., CAMAGNI, P., GIBSON, P.N. and MANARA, A. (1982),
    Rad. Effects 65, p. 289
7. MANARA, A., ANTONINI, M., CAMAGNI, P. and GIBSON, P.N. (1984),
    Nucl. Instr. and Methods, p. 475
8. ARNOLD, G.W., (1973) IEEE Transact. Nucl. Sci. Vol. NF 20,
    p. 220
9. MANARA, A., GIBSON, P.N., ANTONINI, M. (1982), Scientific Basis
    for Nucl. Waste Management, 5, Vol. 11, W. Lutze Ed., p. 349
10. ANTONINI, M., MANARA, A., BUCKLEY, S. (1982), Rad. Effects 65,
    p. 295
11. ANTONINI, M., MANARA, A., BUCKLEY, S.N. and MANTHORPE, S.A. (1985),
    J. Physics Chem. Solid 46
12. ANTONINI, M., CAMAGNI, P., MANARA; A. and SACCHI, "Glass ...Current
    Issues", J. Dupuy Ed., Nato-ASI Series, Martinus Nijhoff Pub.,
    The Hague
13. MENDEL, J.E. et al. (1981), PNL 3802
14. DE BATIST, R. et al., (1980), Scientific Basis for Nuclear Waste
    Management, 2, C.J.M. Northrup Jr. Ed., Plenum Publ., p. 351
15. DE, A.K. et al. (1976), Ceramic Bulletin 55, p. 500
16. MARPLES, J.A.C. et al. (1981), EUR 7138
17. JANTZEN, C.M. et al. (1984), Nuclear Waste Management, Advances in
    Ceramics, Vol. 8, G.G. Wicks and W.A. Ross Eds., Am. Cer. Soc.
    Publ.30
18. FLINN, J.E. et al. (1981), Scientific Basis for Nuclear Waste
    Management, 3, J.G. Moore Ed., Plenum Publ.,p. 201
19. VAN ISEGHEM, P. et al., (1984), EUR 8979, p. 27
20. VAN ISEGHEM, P. et al., SCK report in preparation
21. EVANS, A.G. and WIEDERHORN, S.M. (1974), Int. J. Fracture, 10,p. 379
22. IRWIN, G.R., (1958), Handb. Phys.
23. MALOW, G. et al., (1984) Joint Annual Progress Report 1982,
    EUR 9268 EN
24. RICHTER, H. and OFFERMANN, P., (1982), Characterization of Mechani-
    cal Properties of Nuclear Waste Glass, Scientific Basis for Nuclear
    Waste Management, Vol. 11, W. Lutze Ed., 229
25. OFFERMANN, P. and RICHTER, H., (1984), Slow Crack Growth in Nuclear
    Waste Glasses, Adv. in Ceramics, Vol. 8, G.G. Wicks and W.A. Ross
    Eds., p. 677

Acknowledgements

    The authors acknowledge contributions to this paper and discussion
with K.A. Boult, J.T. Dalton, A.R. Hall, A. Hough, A.E. Hughes and
H. Richter.

Table I: Fitted values of the constants for the density/dose curves

Glass	$(\rho - \rho o)/\rho o$ at saturation	$\alpha \times 10^{17} (g^{-1})$
M5	$-0.0039_7 \pm 0.0001$	0.15 ± 0.01
M22	$-0.0075_5 \pm 0.0004$	0.051 ± 0.005
F SON 58.30.20.U2	$+0.0061_6 \pm 0.0002$	0.061 ± 0.004
G. Celsian B1/3	-0.00484 ± 0.00006	0.118 ± 0.004
VG98/3	-0.0078 ± 0.0002	0.110 ± 0.007
Phosphate	$+0.0058 \pm 0.0006$	0.052 ± 0.011

Table II: Values of vr/(vr+k) for various values of k and vr in sec^{-1}

		vr/(vr+k)				
	vr	$k=10^{-11}$	$k=10^{-10}$	$k=10^{-9}$	$k=10^{-8}$	$k=10^{-7}$
Real waste	10^{-12}	10^{-1}	10^{-2}	10^{-3}	10^{-4}	10^{-5}
Pu-238 doped glasses	3×10^{-8}	0.9997	0.997	0.97	0.75	0.23
Ion bombardment (Dran et al. (2))	10^{-5} to 10^{-1}	1 1	1 1	0.9999 1	0.999 1	0.99 1

Table III: Characteristic temperatures (in $^{\circ}C$)

Glass	T_g	M_g	T_c
SAN602519L$_3$C$_2$	465	495	655
SM58LW11	505	535	-
UWG119	535	570	850
UWG123	575	600	870
UWG124	645	670	775;815

Table IV: Crystalline phases

Glass	T($^{\circ}$C)	Main phase	Additional phases
SAN602519L$_3$C$_2$	700	$0.7NaAlSi_3O_8$, $0.3CaAl_2Si_2O_8$ (plagioclase)	(Ca,Al,Si) rich
SM58LW11	800	SiO_2 (tridymite)	
UWG119	700– 800	$(NaFe^{3+})_x(Ca,Mg)_{2-2x}Si_2O_6$ (pyroxene)	UO_2; K-rich
UWG123	700– 900	$(NaFe^{3+})_x(Ca,Mg)_{2-2x}Si_2O_6$ (pyroxene)	UO_2; K-rich
UWG124	700–1000	$(FeNi)(Fe,Cr)_2O_4$ (spinel)	UO_2 (Si,Fe)-rich (Si,Ba)-rich

Table V: Corrosion data for different glasses showing the effect of heat treatment (based on weight loss) (amorphous = before, part.cryst. = after)

Glass	Physical state	Soxhlet (14d) ($gm^{-2}d^{-1}$)	Static clay-water mixture (182d) (gm^{-2})	Clay (gm^{-2})
SAN60	amorphous	5.7	18.3	76.1 (240d)
	part.cryst.	5.2	28.4	53.9 (182d)
SM58	amorphous	4.2	20.6	130 (240d)
	part.cryst.	12.6	39.6	260 (182d)
UWG119	amorphous	5.0	21.9	251 (365d)
	part.cryst.	1.1	10.5	40.2 (182d)
UWG123	amorphous	0.36	13.1	47.1 (365d)
	part.cryst.	0.45	8.2	24.3 (182d)
UWG124	amorphous	1.4	14.3	118 (365d)
	part.cryst.	0.32	–	12.3 (182d)

Table VI: Fracture toughness K_{I_c} and parameters of the K_I-v-curve of nuclear waste glasses, a soda-lime-silica glass and two ceramic materials

Material	fracture toughness $K_{I_c}/Nmm^{-1/2}$	K_I-v-curve parameters	
		n	- logA [+)]
Soda-lime-silica glass	24	18	21
UK209	22	113	149
C31-3	25	66	87
GP98/12	27	13	16
SM513	31	29	36
B1-3	42	76	112
Eurcryptite	50	129	189

+) for $v/10^{-6}ms^{-1}$

Data from references (23, 24 and 25)

Table VII: Fracture stresses (50%-value) of nuclear waste glasses measured by the Hertzian indentation method on polished surfaces

Material	fracture stress σ_c/Nmm^{-2}
Soda-lime-silica glass	323
SM513	363
UK189	396
UK209	397
SON58	428
C31-3	447
GP98/12	454
B1-3	558

Data from references (23, 24 and 25)

Table VIII: Maximum stresses yielding a tenfold crack surface after 100 resp. 1000 years (f = 2)

Material	initial crack size $c_i/10^{-3}mm$	maximum stress σ_{MAX}/Nmm^{-2}	
		t=100y	t=1000y
Soda-lime-silica glass	3.1	61	54
GP98/12	2.4	50	42
SM513	5.0	93	86
C31-3	2.1	256	247
B1-3	3.8	287	279
UK209	2.1	300	294

Data from reference (25)

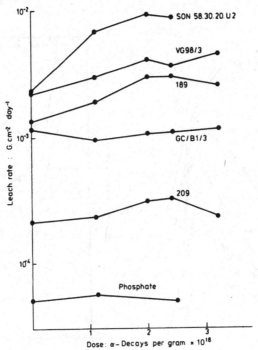

Fig. 1:   Soxhlet leach-rate vs. dose

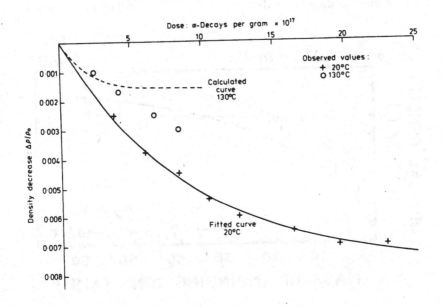

Fig. 2:   Density changes in VG98/3 vs. dose

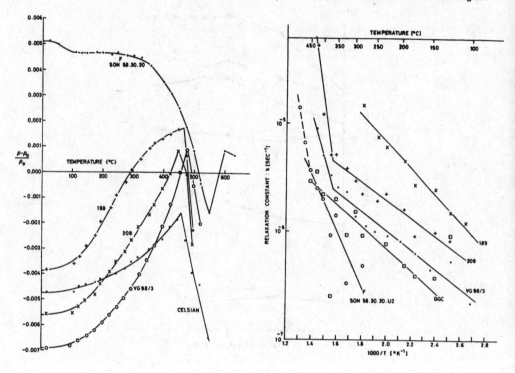

Fig. 3:   Isochronal annealing   Fig. 4:   Arrhenius plot of
           of Pu-238 doped samples   relaxation constant

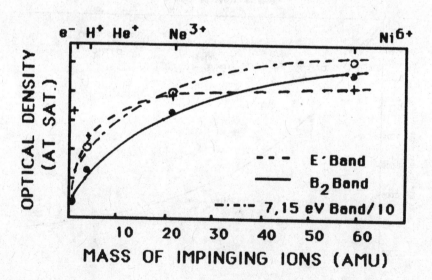

Fig. 5:   Saturation value of various intrinsic colour centre bands in
           irradiated SiO$_2$ as a function of the impinging particle mass

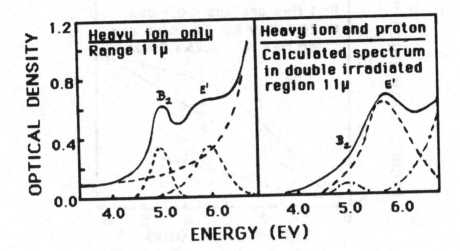

Fig. 6:   Effects of double irradiation on vitreous SiO_2. Left: after
           Ni-ion irradiation. Right: after further proton irradiation.
           The optical density scale is recalculated in each case with
           reference to 11 µm of heavy-ion penetration.

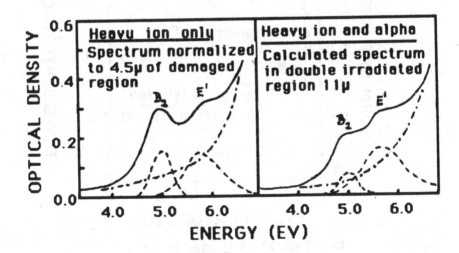

Fig. 7:   Effects of double irradiation on vitreous SiO_2. Left: after
           Ni-ion irradiation. Right: after further alpha-irradiation.
           The otpical density scale is recalculated in each case with
           reference to 4.5 µm of alpha penetration.

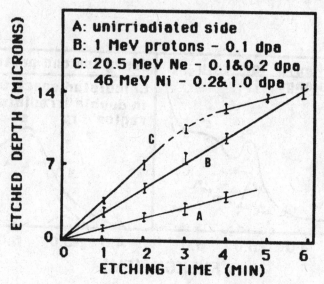

Fig. 8: Effect of various irradiations on the etching/time characteristics of vitreous SiO_2. Experiments refer to the same standard treatment in HF acid solution.

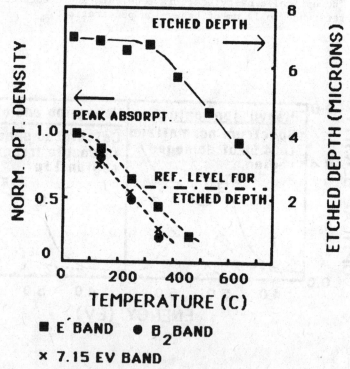

Fig. 9: Isochronal annealing for the excess etching (right scale) and for intrinsic colour centres (left scale) in Ni-irradiated vitreous SiO_2

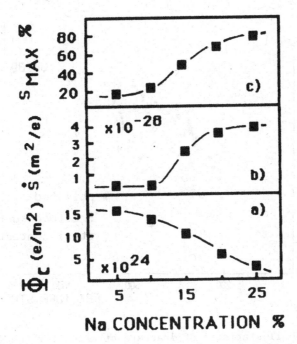

<u>Fig. 10:</u>   Influence of alkali content upon damage production at different
         stages: a) incubation-nucleation; b) bubble growth; c) swelling
         saturation.

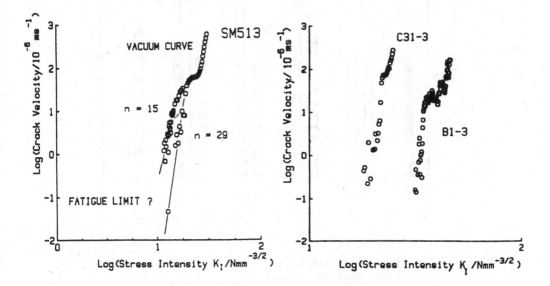

<u>Fig. 11:</u> K_I-v-curves of the glass SM513
         (two runs on the same specimen)

<u>Fig. 12:</u>   Comparison of the
         K_I-v-curves of the
         glass C31-3 and the
         glass ceramic B1-3

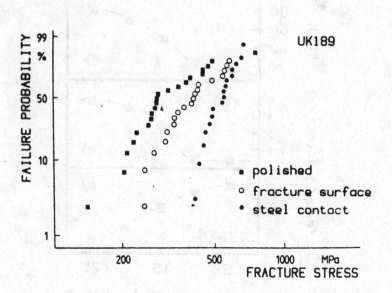

Fig. 13: Weibull-statistic of fracture stresses for the glass UK 189 measured after different surface treatments

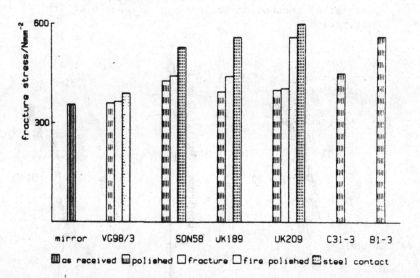

Fig. 14: 50%-values of fracture stresses for different nuclear waste glasses and surface treatments

DISCUSSION

C. McCOMBIE, NAGRA Baden

It seems that neither irradiation, nor thermal or mechanical effects have
a great influence on the stability of the glass. With this in mind, can we
consider research in this field as being concluded or, otherwise, how are
the prospects for future activities?

W. LUTZE, HMI Berlin

With regard to crystallisation it could be presumed that we are now in
position to recommend glass compositions which, in conjunction with the
respective wastes, are not leading to any significant effect on the
release mechanisms. In view of radiation damage, Dran et al. reported a
considerable increase of glass corrosion due to irradiation. This could
later on be explained as a dose-rate effect. However, in this field some
phenomena have still to be clarified, even though there is no evidence of
significant influence on the leach resistance, i.e. not more than an
increase by a factor of 2 to 5.

BEHAVIOUR OF INTERMEDIATE-LEVEL WASTE FORMS IN AN AQUEOUS ENVIRONMENT

S. AMARANTOS (NRC); R. DE BATIST (SCK/CEN); K. BRODERSEN (NRL);
F.P. GLASSER (Aberdeen U); P.E. POTTIER (CEN Cadarache);
R. VEJMELKA (KfK); E. ZAMORANI (JRC)

Summary

Under Action 1 of the Second Community Programme (1980-1984),
study continued of the behaviour of low and medium activity waste
matrices using 10 reference waste forms (RWFs) representative of
the main waste packages produced in the Community. The aim of this
paper is to outline the main results for three types of matrix :
cement and derived forms, organic polymers and bitumens. The
results include data on diffusion coefficients, leach rates and
waste form volume changes and mass losses. They constitute a
considerable advance in knowledge of confinement properties but
bring to light the need for further study of radionuclide release
mechanisms for the purpose of constructing long-term models of
waste form behaviour in the presence of water.

1. INTRODUCTION

From the outset of the Second Community Five-Year Programme, it was
decided that work under Action 1 should deal exclusively with the
characterization of low and medium-level waste forms. Ten types of reference waste form (RWF) were selected, representing the main types of
waste produced in the Member States. The aim of the action was to concentrate the research effort, standardize the materials for study and
allow the exchange of samples. The waste forms are produced by immobilizing PWR and BWR wastes (ion exchange resins, evaporator concentrates,
pond sludges etc.) and fuel-reprocessing-plant wastes (concentrates and
chemical coprecipitation sludges) in matrices such as cement, organic
polymers and bitumens. Some of the results from the studies have already
been published (cf. (1), (2), (3) and (4) in the bibliography). The
characterization of behaviour in hte presence of water, which is almost
always found at disposal sites, is one of the most important safety
criteria. All the behaviour studies include data on waste-form leach
rates and dimensional stability. The division of the studies among Community Laboratories is shown in Table I.

Table I

Division of work to study waste-form behaviour
in the presence of water

Aspect studied	Cement	Polymers	Bitumen
Leaching mechanisms	A.D.F.R.	D.F	F.R
Swelling in the presence of water	-	D.C	M.R
Full-scale leach tests	F.K	F	F
Leaching under disposal conditions	K.C	C	K.F.M.R

A: University of Aberdeen (UK); C: ENA, Casaccia (I);   D: NRC
Demokritos (Gr); F: CEA (F); K: KFK Karlsruhe (GER);
M: SCK/CEN Mol (B); R: RNL Risø (DK).

The most representative or salient results of the studies carried
out in the past five years are summarized below. It has not been possible to show the results of each laboratory separately, but this will be
done in the final report.

2. BEHAVIOUR OF CEMENT-BASED WASTE FORMS

The reference waste forms in question are the following :
RWF 1 : BWR evaporator concentrates. Body responsible : ENEA/I.
RWF 2 : PWR evaporator concentrates. Body responsible : CEA/F.
RWF 6 : Magnox fuel pond sludge.   Body responsible : UKAEA/UK.
RWF 8 : Reprocessing concentrates.   Body responsible : KFK/GER.
The four main fields of study were the following :
- diffusion and leaching mechanisms;
- effect of modifiers on these mechanisms;
- laboratory-scale and full-scale leaching of radionuclides and of the
  inactive constituents of the waste forms;
- post-leaching examination of waste form core samples.

2.1 Study of leaching and diffusion mechanisms

These studies were conducted on small samples (20 to 70 mm in
diameter), at temperatures ranging from ambient to 90°C, primarily with
waste forms RWF 1 and 8. They concentrated on the release of caesium,
strontium, sodium and calcium. Portland cement was used. The samples
were cured in a humid atmosphere prior to the tests (> 30 days).
Leaching conditions were varied as follows : with and without frequent
renewal of the leachant, different thicknesses of water film, stirring
of the leachant, exposure to or protection from atmospheric CO_2 etc.
Cs (and Na) is released at relatively high leach rates which would
appear to be little affected by the frequency with which the leachant is

renewed or agitated, or by the presence or absence of atmospheric CO_2. It has been demonstrated relatively clearly that during the initial phase (10-30 days) leaching of alkaline ions can be represented by a straight-line function of $\sqrt{t}$. This would seem to indicate that release during this phase is governed by Fick's diffusion law. The activation energy for caesium diffusion in the waste form is calculated at 8 to 11 Kcal.mole^{-1}. If one compares this figure with that for diffusion in pure water (4 to 5 Kcal.mole^{-1}), it becomes evident that diffusion is not the only process at work. This would suggest that the microstructure can be affected by two factors : the reduction in the number of pores that remain accessible to the leachant, and the gradual blockage of the other pores as leaching progresses.

As already reported during full-scale leach tests (3), the release of strontium (and calcium) is far more affected by atmospheric CO_2 (see Figs 1 and 2). One of the explanations most frequently put forward is that deposits from the reaction between Ca^{++} and HCO_3^- form in the pores and internal channels of the waste form and obstruct the pores, thereby permitting local coprecipitation of Sr^{++} ions with the Ca^{++} ions. Table II shows the effect of different leaching conditions on the diffusion coefficients.

Table II

Effective diffusion coefficients (De x 10^{-12} cm^2.s^{-1}) for Cs and Sr leached from cement waste forms (RWF 1) Cements used : Portland (OPC) and pozzolan (Pz)

Cement	Isotope	CO_2	Water flow (cm.d^{-1})			
			0,54	0,09	0,009	Zero
OPC	^{134}Cs	Yes	1 500	2 500	2 000	2 100
OPC	^{134}Cs	No	2 100	1 700	1 500	1 400
Pz	^{134}Cs	Yes	-	440	130	-
Pz	^{134}Cs	No	-	400	310	-
Pz	^{85}Sr	Yes	-	0.024	0.002	-
Pz	^{85}Sr	No	-	0.800	0.100	-

Other diffusion coefficient measurements were carried out on simulated waste-form samples with CEN evaporator concentrate waste in nitrate form (90% by weight of the dry extract, together with 1 to 2% of the following sodium salts : chloride, sulphate, phosphate, oxalate and EDTA).

Table III

Diffusion coefficients (De x $cm^2.s^{-1}$)
measured for cement only and for nitrate + various salt waste forms

Constituent	Cement only		Waste form ($\frac{waste}{cement}$ = 0.20)	
	at 70°C	at 90°C	at 70°C	at 90°C
Ca	2.1×10^{-14}	2.2×10^{-11}	1.6×10^{-11}	2.5×10^{-11}
SO_4	-	-	2.7×10^{-9}	7.7×10^{-9}
NO_3	-	-	4.3×10^{-8}	2.8×10^{-8}

2.2 Effect of modifiers on leaching mechanisms

The effect of fume silica (fine particles of SiO_2 of the order of 0.1 μ m) was studied with different tupes of cement, notably Portland and SRPC, which is resistant to sulphates. The effective diffusion coefficient for ceasium, which is 1 to 5 x 10^{-9} cm^2 x s^{-1} for a mixture of Portland cement and sand, was improved to 1.5 x 10^{-10} cm^2 x s^{-1}, with microsilica accounting for some 20% by weight of the mixture. The microstructure is improved by lower porosity, and particularly by a very high proportion of very small diameter sealed pores (closed porosity). One of the effects of this modifier is that the structure obtained attains maximum quality after several months (3 to 6). A similar effect was also obtained in the laboratory using pulverized fly ash (PFA). Improved performance was also achieved by replacing between 40% and 80% of the Portland cement by blast-furnace-slag cement (BFS).

The addition of microsilica and slag cement alters the overall chemistry and mineralogy of the cement and allows the development of mechanisms whereby the caesium is incorporated in the solid phase. Initially, these materials with a high specific surface area absorb the Cs in alkaline solution for a longer period. When these materials are consumed by reaction with $Ca(OH)_2$, the product obtained in tests with optimum mixtures is a calcium silicate hydrogel with a CaO/SiO2 ratio approaching 1. The caesium sorption properties of this silicate are superior to those of Portland cement, which has a CaO/SiO2 ratio approaching 1.7. Table IV shows the effect of modifiers designed to remove Cs from pore fluids.

Table IV

Analysis of pore fluids in blast furnace slag (BFS) and
"ordinary" portland cement (OPC) mixed in various proportions

Composition (% weight)		Residual caesium found in pore fluid (mg/1)
BFS	OPC	
0	100	306
15	85	321
25	75	270
40	60	138
60	40	78
90	10	42

The immobilization of strontium by cement is much more complex. During normal hydration, without CO_2, strontium is not incorporated either into the gel phase or the lime but is incorporated to a more significant degree into the crystallized, hydrated sulpho-aluminate structures (AF_m and AF_t phases).

The duration of these phases during leaching is not known, nor the effect on their existence of variations in pH and the addition of chlorides, sulphates and carbonates. However, with a quinary brine leachant (a concentrated solution of Na, Mg and Ca chlorides and sulphates), the presence of ettringite has never been established.

Clinoptilolite and other zeolites are often used to improve Cs retention in cement. They react with the constituents of the waste form (cement and waste) in various stages : first, there is a rapid ion exchange reaction betwen the Cs and Sr and the K, Na and Ca found in the cement. There then follows a slower, pozzolanic reaction between the $Ca(OH)_2$-rich components of the cement and the zeolite which results in the dissolution of the zeolite (cf. Fig. 3).

2.3 Full-scale and laboratory-scale waste form leach tests

In addition to the tests designed to study leach mechanisms and the improvement of composites, laboratory tests were carried out for the purpose of determining radionuclide release under simulated final disposal conditions.Waste forms RWF 1, 2 and 8 were tested at temperatures of 40 and 50°C with leachants in the form of a concentrated solution of NaCl or quinary brine (Mg, Na, K and Ca chlorides and sulphates, 32% salts, major constituent Mg/Cl_2). The results obtained after one year of leaching with RWF 1 and 2/2 are summarized in Fig. 4. One can see that the annual caesium fraction released exceeds 1%, that the behaviour of composite RWF 2/2 is superior to that of RWF 1, and that leaching is slower in quinary solution.

Tests on laboratory-scale and full-scale samples were carried out on RWF 8 at 40°C, using quinary brine. The caesium releases are compared in Fig. 5. There is a relatively close correlation between the annual fractions released by the laboratory-scale samples with the $\frac{S}{V}$ adjusted to 0.096 cm^{-1}) and the full-scale samples. Here again, Cs release exceeds 1% per year. Another laboratory, using full-scale RWF 2 and RWF 8 samples at 20–23°C and water from a drinking-water supply system, obtained the following results :

Table V

Annual fractions released by full-scale packages leached
in drinking water at 20-23°C

Type of waste form	Isotope	Initial activity (Ao and Ci)	Duration of test (days)	$\frac{\Sigma\ an}{Ao}$ per year
- RWF 2	^{137}Cs	6×10^{-3}	1 215	1.5×10^{-2}
$(V \sim 100\ 1)$	^{90}Sr	6×10^{-3}	1 215	1×10^{-4}
- RWF 8	^{137}Cs	2.04	1 500	1×10^{-2}
$(V \sim 200\ 1)$	^{90}Sr	0.26	1 500	1×10^{-3}
	^{239}Pu	0.16	1 500	1×10^{-6}
	^{241}Am	0.81	1 500	2×10^{-6}

In the case of α emitters, the fraction released includes deposits on the walls of the equipment and at the bottom of the leaching container. The α activity deposited in this way was 5 to 7 times higher than that remaining soluble or in suspension in the leachate. Another series of complementary tests using quinary brine was carried out on RWF 8 at 40, 55 and 90°C, at 1 and 130 bars, using Portland and blast-furnace-slag cements. One of the major conclusions of this study is that corrosion of the waste forms by quinary brine results in the formarion of Friedel salt (C_3A - $CaCl_2$ - $10H_2O$) ($Mg_2(OH)_3Cl$, $4H_2O$) and of several phases of $CaSO_4$ $2H_2O$, which partially obstruct the cement pores, with a concomitant increase in volume which can destroy the samples.

Efforts are currently being made to determine the Pu equilibrium concentration using waste form samples similar to RWF 8. The samples have been prepared with a Pu content of 0.3% by weight, the Pu having first been mixed with the concentrate in the form of Pu (NO_3)$_4$ in an acid environment. The leach tests were conducted on 0.5 to 1 mm granules obtained by crushing after curing, the leachants being distilled water, concentrated NaCl solution and quinary brine at ambient temperature and at 50°C. Table VI summarizes the results to date.

Table VI

Static leaching of 0.5-1 mm RWF 8 granules
(10 % salts, Portland 35 F cement, 0.3 wt % Pu, water/cement
ratio : 0.4 - 4 g granules/40 ml leachant)

Leachant	Concentration Pu mg/l	
	Ambient temperature	50°C
Distilled water	0.1	0.2
NaCl solution	0.4	0.1
Quinary brine	2.6	2.4

2.4 Examination of core samples of full-scale blocks after leaching

In order to measure directly changes in the concentrations of γ - emitting isotopes after leaching, samples were taken of full-scale blocks of RWF 2 and RWF 8. Figure 6 shows the results obtained with a GeLi detector set at 20 and 2 mm and reveals the very great influence exerted by this setting. The depth leached was some 6 mm in the case of 60 Co and 30 mm in the case of 137 Cs. The results also confirmed the good homogeneity of the waste form.

In the case of RWF 8, the sample obtained proved to be very friable in the area close to the surface of the block where it had been corroded by leaching. It was therefore impossible to make gammametric measurements. The reasons for this loss of mechanical strength are the subject of an additional study which includes chemical and radiochemical analysis.

3. BEHAVIOUR OF WASTE FORMS PREPARED WITH POLYMER MATRICES

3.1 Study of release from a cellulose acetate matrix

A study of the leaching mechanisms of soluble (NaCl) and insoluble salts (Ca/SO$_4$ and SrSO$_4$) was undertaken to study elution of cations simulating isotopes incorporated in polymer matrices. Even if the matrix studied, cellulose acetate, is not one of the matrices planned for use in RWFs, the results obtained can certainly be usefully compared with those for other polymers. The tests comprised the preparation of cellulose acetate film ranging in thickness from 0.15 to 0.30 mm containing 10 to 70% by weight of salts. Samples with a surface area of 3 x 3 = 9 cm^2 were mounted on agitators (250 to 300 rpm) and placed in 200 ml of distilled water at 25°C. Releases were measured at different intervals and the whole test lasted between 1 and 2 days.

In the case of the insoluble salts (Ca/SO$_4$ and SrSO$_4$), the total concentration C_S' of the salt, (including the concentration of the pore fluid and the absorbed fraction at equilibrium) is much lower than Co, the initial concentration of the salt in the membrane. One can therefore assume that the release process is governed by Higuchi's equation :

$$\Sigma\, a_n = S\sqrt{2\ DC_0\ C_s'\ t\,},$$

where S is the apparent surface area and D, the diffusion coefficient. In this case, there is a substantial difference between D and the diffusion coefficient D_0 of the salt in the leachant. The relationship $\tau = \dfrac{D_0}{D}$, representing the tortuosity of the pores, was established; this ratio falls very rapidly as C_0 increases. In the case of the soluble salts (NaCl), progress of release shows a clear departure from the time law, probably as a result of the impregnation of the samples by the leachant retained by the waste form salts.

3.2. Measurement of diffusion in polymers by laser probe mass spectrometry (LPMS)

This method of measurement offers the advantage of measuring concentration profiles up to a depth of tens of μm. Sample preparation time is also reduced, since the polymer samples need spend only a short time (several days to several weeks) in the aqueous environment containing the ions being tested. The diffusion tests were conducted with inactive samples, the diffusion measurements, lasting several hours, being made during the transitory stage of element diffusion.
The detection limits were determined using calibration samples with concentrations between 200 and 2 000 ppm. The detection levels depend on the operating conditions of the LPMS spectrometer and the nature of the elements studied, the usual limits being : Co : 150 ppm; Cs : 100 ppm; Sr : 800 ppm.
The vertical concentration profiles are obtained following diffusion of electrolyte solutions in polymer-membrane pellets, with subsequent freeze-drying. An operating method has been developed which allows reduction of the background created by carbonized fragments during each laser shot. The vertical diffusion profiles are established from the calibration curves obtained for each element. The validity of the method was checked using epoxide membranes 150 μm thick. One can obtain up to three profiles (Co, Sr and Cs) at the same time in a single membrane. Figure 7 shows the theoretical diffusion profiles of caesium in an epoxide membrane.
The diffusion coefficient of an element is calculated from a concentration profile obtained experimentally which is then compared in graph form with a theoretical profile. Using the Fick diffusion models, values were obtained which generally correspond to the currently accepted diffusion coefficient (D) values.

D Cs : between 3×10^{-13} and 3×10^{-14} cm$^2 \times$ s^{-1};
D CO : $\sim 4 \times 10^{-16}$ cm$^2 \times$ s^{-1};
D Sr : of the order of 3×10^{-16} cm$^2 \times$ s^{-1}.

However, these measurements remain imprecise and efforts are being made to improve precision by altering calibration and LPMS analysis parameters.

3.3 Leaching studies of full-scale packages and laboratory samples

Leaching tests were carried out on samples (V < 1 l) and full-scale packages (V from 100 to 200 l) of RWF 3 (PWR concentrate in polyester or epoxide matrix) and RWF 5 (ion exchange resin polyester or epoxide matrix). Drinking water (full-scale package) and deminineralized water (laboratory samples) were used as leachants. The results are summarized in table VII.

Table VII

RWF 3 and RWF 5 : annual fractions leached

(values adjusted to $\dfrac{\text{apparent } S}{\text{sample } V} = 0.096 \text{ cm}^{-1}$)

ISOTOPE	LEACHANT	Concentrates (RWF 3) Polyester		Epoxide		Ion exchange resin (RWF 5) Polyester		Epoxide	
		Full-scale	Lab.-scale	Full-scale	Lab.-scale	Full-scale	Lab.-scale	Full-scale	Lab.-scale
^{137}Cs	DeW	1×10^{-2}	-	2×10^{-3}	-	1×10^{-2}	-	1×10^{-2}	-
	DrW	-	1×10^{-4}	-	2×10^{-3}	-	6×10^{-3}	-	3×10^{-3}
^{60}Co	DeW	5×10^{-3}	-	1.5×10^{-2}	-	1.5×10^{-2}	-	3×10^{-3}	-
	DrW	-	4×10^{-3}	-	5×10^{-3}	-	3×10^{-2}	-	2×10^{-4}

(DeW = demineralized water; DrW = drinking water)

The results show a relatively wide spread in the values of the annual fractions released (from 3×10^{-2} to 1×10^{-4}), although the effect of scale ($\frac{S}{V}$ sample) was taken into account. The wide spread and the orders of magnitude of the releases were confirmed in another laboratory. The method of preparation of the samples may have played a part, but the type of leachant (demineralized water or drinking water) would appear to be of little importance. The variations in weight and volume (slight increases) were minor (a few percent). Epoxide composites having recently been improved to allow immobilization of waste containing free water (ion exchange resin suspensions), new leach tests are being carried out, in particular to study the effects of γ irradiation.

4. BEHAVIOUR OF WASTE FORMS USING A BITUMEN MATRIX

4.1 Swelling in the presence of water

When cast, the bitumen waste forms have a very low water content (generally less than 1%, with a maximum of 4% in the case of ion exchange resin waste forms). Water uptake by bitumen waste forms inevitably occurs prior to leaching and can give rise to difficulties as regards the volume stability of the waste form. A major laboratory study was carried out on various simulated RWF 7 and RWF 9 waste forms, and on waste forms containing BWR concentrates and ion exchange resin. A parallel laboratory study was carried out on active waste form samples from the Eurobitume installation at Mol (RWF 7-9).

The simulated wastes were prepared with Na/NO$_3$ crystals of controlled granular size, chemical precipitation sludges and Mexphalte 40/50 bitumen in different proportions.

The real waste forms are from Eurochemic waste in the form of de-cladding and decontamination effluents and various concentrates. They contain approximately (by weight) 25% $NaNO_3$, 15% salts and insoluble compounds and 60% Mexphalte R 85/40. Their average activity is 2×10^{-3} Ci α/kg and 0.13 Ci $\beta\gamma$/kg.

The water uptake speed of a waste form depends on three factors :
- the driving force accounting for the transfer of water in the material;
- the thickness and quality of the bitumen film surrounding the waste particles;
- the diffusion coefficient and the solubility of the water in the bitumen.

The driving force accounting for the transfer of water may be defined as the difference between the water vapour pressure in the waste particles and in the leachant or surrounding air. This suggests that the water uptake mechanism may be hydration (ion exchange resin), the formation of hydrated salts (Na_2SO_4, $Na_2H\,PO_4$) and the formation of saline inclusions ($Na\,NO_3$). This also explains the low rate of water uptake from concentrated saline solutions.

The thickness of the bitumen film decreases as a function of the % of the salts encapsulated and also depends on the size and form of the waste particles.

The water diffusion coefficient in pure bitumen was measured at 3×10^{-8} $cm^2 \times s^{-1}$ in the case of Danish Mexphalte 40/50 and 2×10^{-8} $cm^2 \times s^{-1}$ in the case of French bitumen R 90/40. The test method used combined the measurement of the transfer of tritiated water through a fine bitumen membrane and measurement of weight gain by samples stored under water. During the tests, the solubility of the water in the bitumen was found to be 0.2 to 0.5% by weight.

In theory, knowledge of these three parameters should be sufficient for the preparation of a water uptake model, but the reality is much more complex. The danger of waste particles touching increases the higher the percentage of dry extract and is accompanied by a higher water uptake rate than that accounted for by diffusion alone. Another reason is thought to be that the swelling of the rehydrated particles imposes stress on the bitumen film, causing microfissures which reduce the protective properties of the film. Experimental studies of ion migration have shown that this is a very variable and complex phenomenon. Depending probably on the faults and microfissures in the waste form, ion transfer may be quicker or slower than water diffusion. An example is given in Fig. 8, which compares Cs and Co migration with that of tritiated water.

It has been demonstrated, in the study of the effect of Na NO_3 particle size (from < 0.1 to 1 mm), that fine-grain waste forms have rapid water uptake and a low Na^+ release rate; conversely, large-particle waste forms have a rapid Na^+ release rate and a slower water uptake rate. This may be due to the fact that the structure of the waste form is better adapted to the presence of numerous small inclusions than to a few large vacuoles. Table VIII shows the relationship between diffusion coefficients, Na^+ leach rate and type of waste form.

Table VIII

Effective diffusion coefficients for water uptake and Na leaching
for various waste forms based on Mexphalte 40/50

Waste form (% weight) Sludges : Ba SO_4, Fe Cy Ni Na NO_3 : (a) < 0.1 mm (b) 0.5/1 mm		Effective diffusion coefficients $(10^{-12} cm^2 s^{-1})$		
		Water uptake		Leaching of Na$^+$ Water
		Saturated humid air	Water	
Bitumen	100	0.1	0.5	-
Bitumen Sludges	80 20	0.7	1.4	-
Bitumen Sludges Na NO_3 (a)	60 20 20	210	240	11
Bitumen Sludges Na NO_3 (a)	40 20 40	10,000*	360*	1,500
Bitumen Na NO_3 (a)	80 20	15	60	1.3
Bitumen Na NO_3 (a)	60 40	150*	80	10
Bitumen Na NO_3 (a)	40 60	650*	220	9
Bitumen Na NO_3 (b)	60 40	0.6	3*	0.8

(*) Water uptake rate higher than f $\sqrt{t}$.

Samples of actual Eurobitume waste forms were subjected to static
leach tests in demineralized water (DeW) and clay water (CW) at 49° and
23°C. The clay water was prepared by mixing 100 g of clay from the formation level below the site at Mol (which is currently being assessed as
a potential final repository) with 1 litre of degassed demineralized
water. This operation was conducted in a glove box in a nitrogen atmosphere in order to preserve the reducing conditions that exist in the
clay formation itself. This suspension has a pH of 9.4 and an Eh of −135
mV (ECS).

The results (Fig. 9) show a high degree of swelling, particularly
at 40°C, combined with high releases of Na NO_3 (measured Na NO_3 particle
size was some 40 μ m).

Tests conducted with samples with different $\frac{S}{V}$ ratios showed less swelling after 200 days at 23°C :

$\frac{\Delta V}{V_0}$ = 25% when $\frac{S}{V}$ = 1.4 cm^{-1}, $\frac{\Delta V}{V_0}$ = 135% when $\frac{S}{V}$ = 3.3 cm^{-1} and $\frac{\Delta V}{V_0}$ = 200% when $\frac{S}{V}$ = 6.5 cm^{-1}.

There would appear to be an almost linear relationship between the $\frac{S}{V}$ ratio and the swelling for any given waste form. Extrapolating for a volume of 200 l, equivalent to a full-scale drum, one obtains a swelling of 2 to 5% in one year, when $\frac{S}{V}$ = 0.1 cm^{-1}. In order to try to assess the potential for pressure build up in the waste form as a result of water uptake and swelling, an experimental device was tested on samples of inactive waste forms based on nitrates or ion exchange resin + Mexphalte 40/50. The results are presented in Fig. 10 and show pressures between 20 and 40 bars after the same period of induction as in Fig. 8.

One of the conclusions that can be drawn from results to date from studies of the swelling of bitumen waste forms can be summarized as follows : tests conducted with samples of different sizes have shown that water uptake decreases in direct proportion to the $\frac{S}{V}$ ratio. Assuming an effective diffusion coefficient of 2.4 x 10^{-10} cm^2 x s^{-1}, extrapolation shows that swelling after one year is of the order of 1% for a full-scale package, with $\frac{S}{V}$ = 0.1 cm^{-1}.

4.2 Leaching studies of laboratory-scale and full-scale samples

Laboratory tests were conducted on the above-mentioned Eurobitume samples and on inactive samples comprising 60 wt% bitumen B 15 and 40 wt% Na NO$_3$.

With samples of waste forms of the Eurobitume type, rapid leaching of Na NO$_3$ was observed, with a higher leach rate in the clay mixture (CM) than in demineralized water (cf. Fig. 11). Nitrate ion releases can be described and shown graphically by means of the expression

$$\frac{C}{C_0} = 1 - e^{-\alpha t}.$$

The test results obtained (both at 23°C and 40°C) are : α = 0.006 d^{-1} for demineralized water and α = 0.0115 d^{-1} for clay water. Sulphate ion releases (from Ca SO4, the coprecipitation agent of Sr^{++}) are markedly different : after 90 days at 23°C, approximately 20% of the SO$_4^{2-}$ has been leached out in the demineralized water and 40% in the clay water. The release of α emitters is relatively low at 0.003% at 23°C in distilled water and 0.06% in clay water at 40°C (measured over 360 days). The same is not true of uranium, with 0.025% released at 23°C in demineralized water and approximately 7% released at 40°C in clay water during one year of leaching. No clear time-dependence could be established for the leaching of αemitters. The release of $\beta\gamma$ isotopes is largely proportional to leaching time. The main results are shown in Table IX.

Table IX

Leach rates (R) for $\beta\gamma$ emitters in
Eurobitume samples (g cm^{-2} d^{-1})

Leaching conditions	β Total	Sr	Co	Cs
Clay water				
23°C	2×10^{-4}	**	5×10^{-5}	*
40°C	2×10^{-4}	2×10^{-4}	5×10^{-5}	*
Demineralized water				
23°C	**	**	3×10^{-5}	*
40°C	7×10^{-5}	5×10^{-5}	2×10^{-5}	*

* The R values for caesium, made insoluble by nickel ferrocyanide, are $< 1 \times 10^{-6}$ g $\times$ cm^{-2} d^{-1}.

** The initial release, at 23°C, would appear to follow the law $f(\sqrt{t})$.

The results shown in Table IX were calculated by including in the leached fraction the total activity measured in the leachate, on the walls of the container and on the clay particles in the clay-water suspension.

Tests on inactive waste forms containing 60% bitumen BF15 and 40% Na NO_3 were carried out using the concentrated NaCl solution and quinary brine leachants already mentioned in Chapter 2 at pressures of 1 and 130 bars and a temperature of 40°C. After a year, Cs leaching proved more rapid (1×10^{-3} cm day^{-1}) in quinary brine than in the concentrated NaCl solution ($< 3 \times 10^{-5}$ cm day^{-1}). Pressure seems to have no notable effect on leach rate.

Leach tests on full-scale blocks were carried out with RWF 7 and RWF 9. In addition, numerous water resistance and leaching tests were carried out in the laboratory on active and inactive waste forms and on a full-scale inactive sample of RWF 9 containing > 10 wt% nitrate. Active waste forms RWF 7 and RWF 9, which underwent long-term leach tests (> 3 years), had been manufactured industrially and had had their metal packaging removed prior to leaching. RWF 7 contained both Na NO_3 (35 to 37 wt%) and coprecipitation sludges (2 to 3%). RWF 9, prepared at Marcoule, had a dry extract content of 40/42 wt%, with a Na NO_3 content of 4 to 5%. These percentages relate to the finished waste form. The results of the leach tests are summarized in Table X below.

Table X

Annual fractions released during leaching
of RWF 7 and RWF 9
($20\text{-}23^{\circ}$C - Drinking water)

Type of waste form and volume	Isotope	Initial activity A_o (Ci)	Duration of test (days)	Annual fraction $(\frac{\Sigma \text{ an}}{A_o})$ released
RWF 7	^{239}Pu	0.27	1245	1.6×10^{-4}
(V = 170 1)	^{241}Am	0.20	1245	1.5×10^{-5}
RWF 9	^{239}Pu	0.013	1600	1.7×10^{-5}
(V = 110 1)	$^{241}Am/^{238}Pu$	0.013	1600	2×10^{-5}
	^{137}Cs	0.8	1600	1×10^{-4}
	^{90}Sr	0.026	1600	2×10^{-3}

(The annual fractions released include the deposits and sludges formed). Tests carried out on inactive blocks of RWF 9, containing 10 to 15% nitrates and 20 to 25% dry sludge precipitation extract, showed a very low degree of swelling (waste form $\frac{S}{V}$ ratio : ~ 0.1 cm^{-1}). The bitumen used was type R 90-40. After 360 days immersion in water, swelling varied from 0 to 1.25%.

Other laboratory tests were carried out on RWF 9 consisting of R 90-40 bitumen and 40% dry extract, 1/3 of which was sodium nitrate and 2/3 coprecipitation sludges to which β emitters (90 Sr) and   emitters (137 Cs, 106 Ru etc.) had been added. Specific activity varied from 3 to 30 Ci x kg^{-1}. The samples had a diameter of 50 mm and a height of 50 mm. After 360 days of leaching in drinking water at ambient temperature there was no trace of swelling, with leach rates (g x cm^{-2} x d^{-1}) from 1 to 2 x 10^{-5} in the case of Pu and Am, and 5 x 10^{-5} in the case of uranium.

4.3 Behaviour of bitumen waste forms when exposed to microorganisms

Behaviour of bitumen waste forms was tested in order to try to evaluate the risks of corrosion by microorganisms in the soil. Two main types of test were conducted : simulation of burial conditions and measurement of bacterial metabolism.

RWF 9 samples of approximately 1 litre were placed for a period of almost three years in water-saturated soil in conditions simulating those at a surface disposal centre. The samples were inactive and some of them had been subjected to external γ irradiation of 2.8 x 10^6 rad x h^{-1}, with a cumulative dose of 1 x 10^9 rad. The main conclusions of the study were the following :

-   In the presence of bitumen waste forms, whether in an aerobic or anaerobic condition, there is a marked increase in the biological activity of the soil. Ammonifying and nitrogen-fixing bacteria predominate, and one also finds fungi and actinomycins.
-   Samples examined after over 32 months in water-saturated soil revealed surface deterioration (microfissures, fine cracks etc.)
-   Water uptake by the waste forms (from 1 to 3 wt%) was also found, together with a reduction in sodium nitrate concentration of up to 50% on the periphery of the blocks.

However, these conclusions, which partially confirm the data in 4.1 and 4.2, are not sufficient to allow quantification of the effect of microorganisms on the waste forms. Such quantification is made all the more difficult by the fact that 100 to 200 l blocks of RWF 9 were recovered without apparent damage after being buried for approximately 10 years in topsoil.

Other laboratory tests were conducted in order to attempt to quantify the effect of microorganisms on bitumen and polymer organic matrices. After a number of fruitless tests based on the detection of ATP (adenosine triphosphate) to indicate biological activity, various tests were carried out involving the detection and measurement of gases released (CO_2, N_2O etc.) and the consumption of oxygen by various systems (pure bitumen on an inert medium (bitumen + sand or silica gel), bitumen fractions (maltene) on an inert medium, and waste forms in powder or pieces) in aerobic and anaerobic environments and at soil pH and higher pH levels in order to take into account OH ion release by the cement.

The results to date (the tests are continuing) show that (Fig. 12):
-   Microorganic activity does exist which metabolizes the bitumen hydrocarbons in the form of CO_2 both in an aerobic and in an anaerobic environment (measurements made by respirometry and gas chromatography). In the case of anaerobiosis, the formation of N_2O was detected by gas chromatography.
-   With thermosetting polymers, the constituents of the polymer have varying degrees of inhibiting effect (total in the case of hardeners containing aromatic amines), but the polymer itself loses this inhibiting property. Epoxy resins, on the other hand, show little corrosion.

5. CONCLUSIONS AND PROGRESS OF THE CHARACTERIZATION STUDIES

The study of the behaviour of the different types of reference waste form in the presence of water has yielded a great many results, both positive and negative, that it is impossible to cover fully at this type of conference. The summary report that will be published at a later date will go into the work of each laboratory in more detail. We should also point out that the Commission was the joint organizer of a seminar on the problems caused by the behaviour of coated wastes in the presence of water held at Cadarache in November 1984. This seminar, which was open to experts from outside the EEC (USA, Canada, Sweden, Switzerland and Japan) also provided numerous data which is to be published half way through 1985.

If the ultimate aim is to have models that allow extrapolation of the long-term behaviour of waste forms, the Third Programme (1985-89) must continue the work already started on the effect on waste forms of the action of water and the soluble salts and gases they contain. We already appreciate the full importance of waste-matrix reactions, the effect of $\beta\alpha$ irradiation or internal α irradiation (Dr Philips will discuss this aspect in the following paper), and of the reactions between the waste form and the container, the concrete and the clay. The Third Programme must also seek to achieve data quantification, if possible by the standardization of test methods, since, as we have seen on several occasions, the results are within a broad range, sometimes because of the imprecision of the detection levels (transuranic emitters) and sometimes because of the way in which samples are prepared. Lastly, although we have had neither the time or the opportunity to discuss the subject, we must continue efforts to identify the radionuclides released, since the base data for safety studies depend primarily on identification of the physico-chemical form of these radionuclides.

BIBLIOGRAPHY

1.   R.A.J. SAMBELL, Characterization of low and medium level radioactive waste forms, First annual progress report of the CEC 1980-'84 Programme EUR 8663 EN 1983.

2.   P. VEJMELKA, R.A.J. SAMBELL, Characterization of low and medium level radioactive waste forms, Joint annual progress report CEC 1980-'84 Programme EUR 9423 EN 1984.

3.   W. KRISCHER, R. SIMON, Testing, Evaluation and shallow land burial of low and medium radioactive waste forms, Proceedings of a seminar held at GEEL (CBNM) 28-29 September 1983 Harwood Academic Publishers 1984, CEC Series of Monographs and Tracts Vol 13.

4.   R. de BATIST, R. KÖSTER, P. POTTIER, R.A.J. SAMBELL, R. SIMON, Caractérisation des déchets conditionnés et des colis de faible et moyenne activité dans la communauté européenne en tant que contribution à l'assurance qualité. AIEA. CONF. CN. 43/115. SEATTLE (Wa) 16-20 Mai 1983.

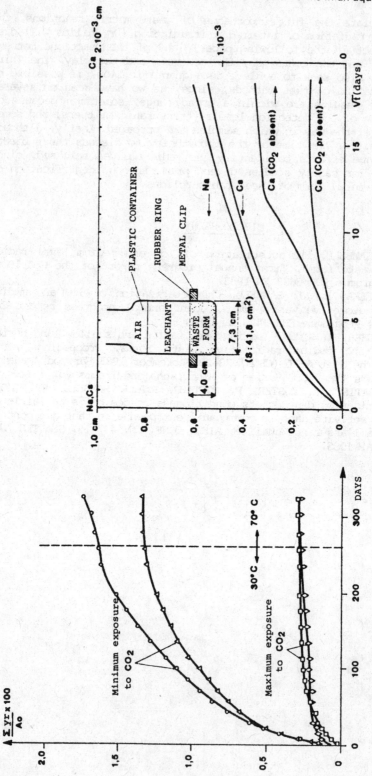

Fig. 2   Equivalent depth leached (Risø Nat. Lab.)
         (cement Na NO₃)

Fig. 1   Fraction of strontium leached
         (NRC Demokritos)
         (cement 6.6wt % Sr SO₄)

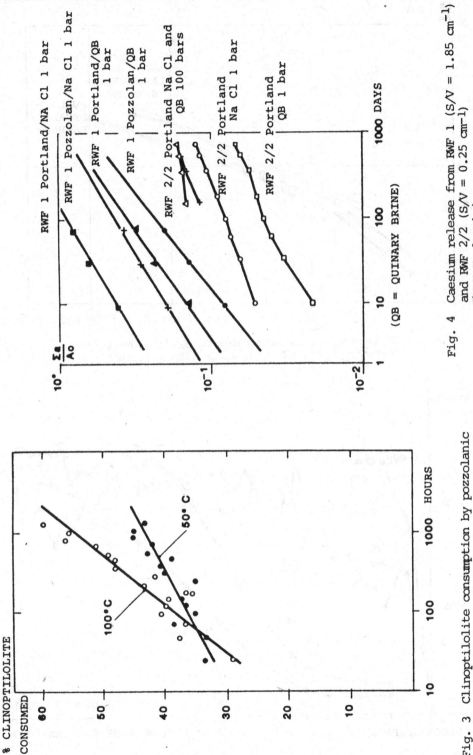

Fig. 4   Caesium release from RWF 1 (S/V = 1.85 cm^{-1}) and RWF 2/2 (S/V = 0.25 cm^{-1}) (KFK Karlsruhe)

Fig. 3   Clinoptilolite consumption by pozzolanic reactions (Univ. of Aberdeen)

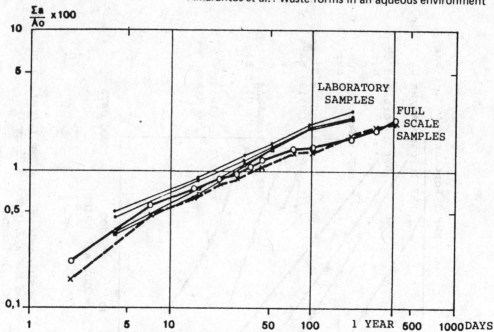

Fig. 5   Comparison of caesium release $S/V : 0.096 \; cm^{-1}$
         (quinary brine RWF 8   8–40°C) (KFK Karlsruhe)

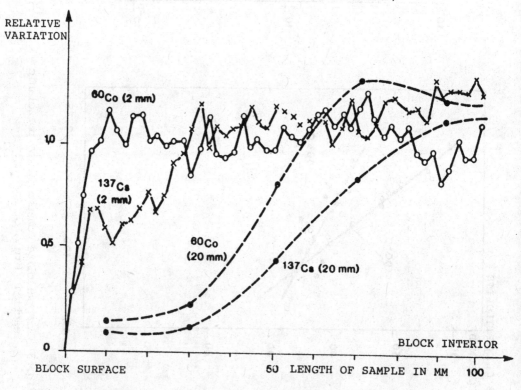

Fig. 6   Gammametric analysis of a RWF 2 sample after leaching
       (collimation 2 and 20 mm)   (CEN Saclay)

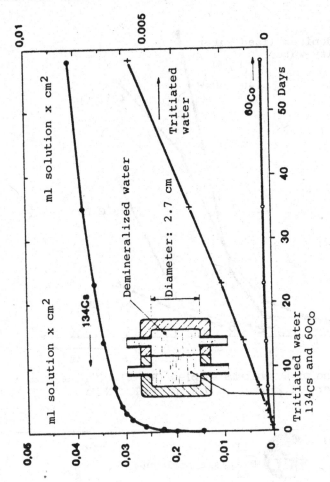

Fig. 8 Diffusion at ambient temperature
R.90.40 bitumen membrane Thickness 0.68 mm
Solution : 0.02 M CsCl2 and CoCl2 + tracers
(Risø Nat. Lab.)

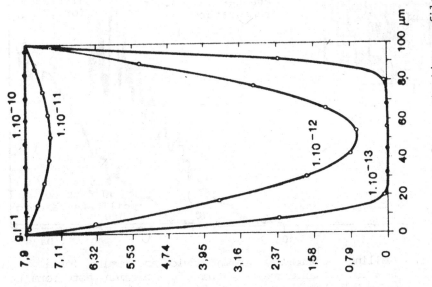

Fig. 7 Theoretical concentration profiles
Epoxide thickness 100 µm
Diffusion over 34 days of a solution
containing 7.9 g x l⁻¹ caesium
(CEN Grenoble)

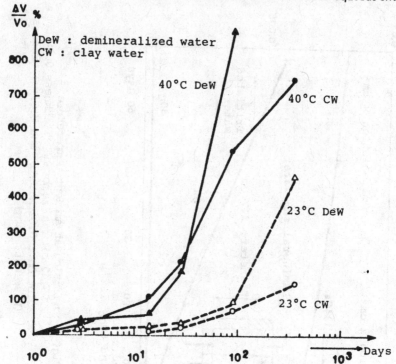

Fig. 9   Swelling of Eurobitume samples (RWF 7)   (Sample size :
diameter ~ 1.25 cm; h ≃ 0.6 cm) (SCK/CEN Mol)

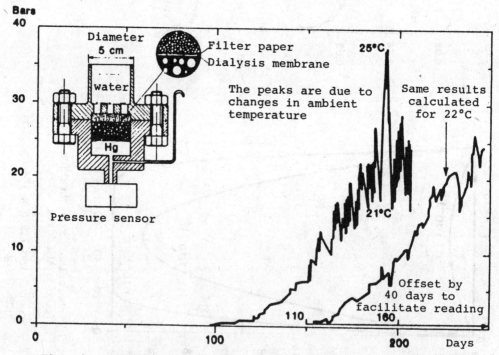

Fig. 10   Swelling pressure measurement device and results for a
50 % ion exchange resin (cation)/50 % bitumen waste form
(Risø Nat. Lab.)

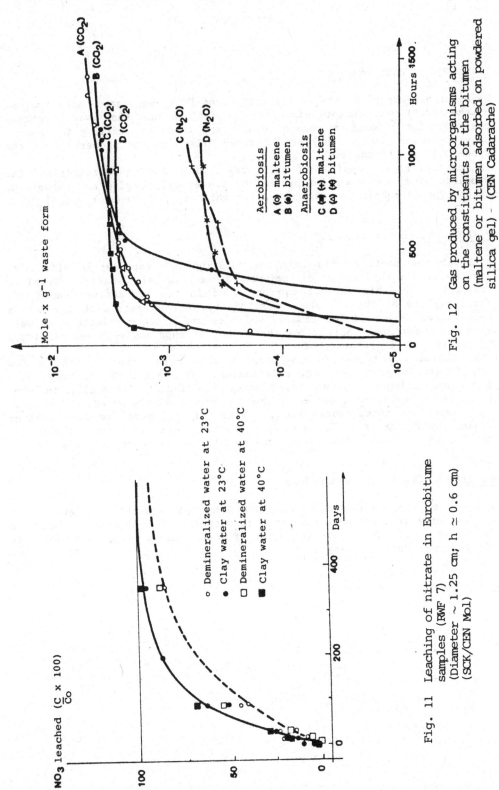

Fig. 12   Gas produced by microorganisms acting on the constituents of the bitumen (maltene or bitumen adsorbed on powdered silica gel) - (CEN Cadarache)

Fig. 11   Leaching of nitrate in Eurobitume samples (RWF 7) (Diameter ~ 1.25 cm; h ≃ 0.6 cm) (SCK/CEN Mol)

DISCUSSION

<u>T. MARZULLO</u>, ENEL Roma

Have the effects of borate, originating from PWR- evaporator concentrates,
on the duration of cement being investigated? We know that borate retards
the hydration process in case of cement and leads to swelling in case of
polymer resins, but whether this has any effect on the properties of the
product is questionable.
In general, is it possible at this stage, after having characterized the
products of the various immobilization processes, to give an indication,
which type of matrix is best suited for which type of reactor waste?

<u>P. POTTIER</u>, CEA Cadarache

Although not mentioned in this presentation the retardation effect of
borate on the hydration of cement was extensively studied. Detailed
information on this subject can be found in the respective annual and
final reports of the EC. The compatibility of the waste with the matrix
material was examined for borate, sulphate and nitrate solutions as well
as for ion exchange resins. For the latter it apparently proved difficult
to obtain a satisfactory product.
What concerns the second question, it should be stressed that a few
immobilization processes e.g. cementation, bituminization and incorporation into polymer resins have reached the industrial scale, in some
countries. For certain wastes, certain matrices are favoured, e.g. cement
for evaporator concentrates and thermosetting polymers for ion exchange
resins. On the other hand, there is certainly a need to improve the cement
matrix, in particular its retention capacity for radionuclides like Cs.

<u>R. ATABEK</u>, CEA Fontenay-aux-Roses

A study to improve cement matrices for the solidification of evaporator
concentrates is under way and first results indicate that the release of
Cs can be decreased by a factor of 10 if pozzolana is being added (40%) to
the ordinary Portland cement.

RADIATION, THERMAL AND MECHANICAL EFFECTS ON LOW AND MEDIUM ACTIVE CONDITIONED WASTE

D.C. PHILLIPS   UKAEA Harwell, United Kingdom
R. KOESTER   KfK, Karlsruhe, Federal Republic of Germany
G. DE ANGELIS   ENEA, Casaccia, Italy

SUMMARY

Irradiation experiments have been carried out on a wide range of simulated, European waste forms based on bitumens, polymers, cements and polymer modified cements. The main effect on bitumenisates is a swelling of up to 10 volume % at 1 MGy. Alpha irradiation is more severe than gamma irradiation. There is no significant effect on leach rates at 10 MGy. The behaviour of polymer waste forms is more complicated and these can either swell or shrink by amounts which depend on the type of polymer and the amount of water present in the waste, but which are typically between 2% swelling and 8% shrinkage at 10 MGy. Cement waste forms are less affected by radiation and show no changes in strength or dimensions which can be attributed to radiation, although one waste form disintegrates under high dose rate radiation. Laboratory techniques for characterising the mechanical and thermal properties of cement and polymer waste forms are described. Full-scale drop and fire tests on 200 litre drums of cemented waste simulate have been carried out and the quantities of activity which would be released from real waste forms inferred.

1   INTRODUCTION

Conditioned, immobilised waste in its container will be subject to aging processes, which may alter its properties. It may also be at risk from accidents, particularly prior to its ultimate disposal to a repository, which again may affect its properties and release radionuclides to the immediate environment.
     Aging processes could include:
         Self-irradiation from the contained radionuclides.
         Waste-matrix chemical interactions.
         Continued chemical reaction of the matrix components and the environment.
         Biodegradation.
     Accidents could include:
         Mechanical damage due to impact.
         Thermal damage as a result of fire.
         Immersion in water.
This paper describes work aimed at obtaining an understanding of the

consequences of self-irradiation, mechanical impact and thermal damage, while the consequences of water immersion and biodegradation are addressed in parallel work described in the paper by Pottier et al (1). The extent of the programmes was such that only a brief summary can be given here.

2   RADIATION EFFECTS

The work, which has been carried out at Harwell, has had two overall objectives. An immediate objective of providing comparative data on the effects of radiation on properties through accelerated radiation testing; and the longer-term objective of improving the confidence with which the effects of long-term, low dose-rate radiation can be predicted from high dose-rate, accelerated laboratory experiments.

Irradiation of the waste forms is inevitable. Table 1 shows the waste forms studied and their approximate levels of activity (2,3). The highest levels imply approximate integrated doses of the order of 7 MGy at 100 years, and ~ 20 MGy at 1000 years (4), with correspondingly lower doses for the less active materials. The total dose and dose rates which will be received by different types of waste forms vary widely depending on the radionuclide inventory, and it cannot be assumed that a low activity waste form will receive a low dose, unless it can be guaranteed that it will be shielded from neighbouring, higher activity, waste forms.

Table 1   Waste Forms Studied in Radiation Characterisation Programme (2,3)

No.	Waste/Matrix	Typical Activity Ci m^{-3}	
		$\beta\gamma$	α
RWF1	BWR evaporator concentrates/cement or pozzolana cement	0.55	–
RWF2	PWR evaporator concentrates/cement or pozzolana cement	0.5-50	$<10^{-3}$
RWF3	PWR evaporator concentrates/polyester or epoxide	0.5-50	$<10^{-3}$
RWF4	Ion exchange resin/polystyrene	$<10^3$	$<5 \times 10^{-6}$
RWF5	Ion exchange resin/polyester or epoxide	5-500	$<10^{-6}$
RWF6	Magnox fuel pond water sludge/cement	35-70	1-2
RWF7	Reprocessing concentrates/bitumen	60	2
RWF8	Reprocessing concentrates/cement	10-40	approx. 1
RWF9	Reprocessing sludges/bitumen	70-850	0.3-0.8
RWF10	Incinerator slags		
WF17	Ion exchange resin/polyester or vinyl ester		

Two types of radiation-damaging processes can occur in encapsulated ILW's: atomic displacements resulting from α particle interactions, and chemical effects resulting from radiolysis of loosely bound molecules. Of these atomic displacement is unlikely to be of importance over timescales of interest because of the low concentration of α emitters present in ILW's. In general, waste forms based on polymers and bitumens are more affected by radiation than those based on cement because of the susceptibility of organic materials to damage by ionising radiations. However, the hydrated inorganic compounds in cement continue to react for a very long time and the addition of waste adds a further set of reactants to an already complex system. Radiation can affect the chemistry of the cement/waste interactions and this needs to be understood to predict the radiation stability of the waste forms.

Research has been carried out on all ten reference waste forms, together with some other waste forms, under the Sheet 1 characterisation programme; and measurements have also been made on the polymer modified cement systems developed under the Sheet 3 processing programme. Only limited results were obtained on RWF10, the incinerator slag, and those data are not reported here. Radiation experiments have also been carried out on the matrix materials alone.

Most of the work has been carried out on inactive or tracered simulates using an external gamma radiation source, but in addition high dose rate alpha damage has been studied in the bitumenisates through the incorporation of ^{241}Am, and some irradiations have been carried out on fully active bitumenisates. The effects of irradiation have been measured on dimensional stability, mechanical properties, gas evolution and leaching.

2.1 Bitumens

Bitumens and bitumenisates can display substantial swelling when subjected to high dose rates of gamma or alpha irradiation. This is due to the nucleation and growth of bubbles of radiolytic gas, primarily hydrogen (5). For any given material the rate and amount of swelling increases as either the dose rate or the sample size increases. Prediction of this swelling under long term, low dose rate conditions, is necessary both to ensure that the container is not at risk and also to enable the laboratory preparation of radiation-aged specimens for radionuclide release measurements.

For gamma irradiation it has been shown that, over a wide range of conditions, similar amounts of swelling will occur in samples of the same bitumenisate of different sizes, in cylindrical containers of depth L, irradiated at different dose rates I, provided IL^2 is constant. Thus by irradiating small samples at high dose rates in the laboratory, similar swelling effects should be obtained as under the large container, low dose rate, storage conditions, provided the IL^2 value for storage conditions is used (5). Figure 1 shows data obtained from RWF7 and its matrix bitumen R85/40 at an IL^2 value corresponding to an activity $\sim 1Ci\ell^{-1}$. The behaviour of all the bitumens and bitumenisates studied was similar, swelling commencing at around 50 kGy and saturating somewhere in the range 0.2-1.0 MGy. The maximum amount of swelling depends on the type of bitumen and the nature and quantity of the waste. The softer grades of bitumen with low quantities of added waste can swell several tens of percent. However the swelling of RWF7 and RWF9 is much less at around 10% for the former and less than 8% for the latter. This swelling can be accommodated easily within a container and does not present a hazard. An anomaly which has been observed is that RWF9 appeared to exhibit two types of swelling behaviour. This may be a reflection of the sensitivity of swelling to the presence of bubble nucleation sites, this in turn depending on the exact processing procedures used in manufacturing the bitumenisate. The swelling which occurs on alpha irradiation is more pronounced than at the same dose of gamma radiation and again is very geometry dependent.

Gamma irradiation experiments to 10 MGy have shown no significant effects of radiation on leach rates for ^{137}Cs. The nature of these experiments, however, was such that the specimens did not swell under irradiation. A more accurate experiment would require samples to be cut from swollen blocks. Leaching experiments are currently under way on alpha-irradiated materials which have swollen and these will provide more insight into the relationship between swelling and leaching. Measurements on fully active materials are broadly in agreement with those on simulates.

2.2 Polymers

Polymer matrix waste forms behave differently to bitumen waste forms because of their different intrinsic properties, and the different nature of the encapsulated wastes (6). In the bitumenisation process most of the water associated with wet wastes is driven off and there is little free water in the final waste form. Most of the polymer matrix waste forms contain substantial quantities of water, and this affects their properties. Waste forms of this type are RWF4, RWF5 and WF17, wet ion exchangers immobilised in epoxide, polyester and vinyl ester resins and in polystyrene.

When exposed to the atmosphere these waste forms slowly lose water at rates which are similar whether irradiated or unirradiated, depend on the sample size, and are lowest for the epoxide resin system. Typical values of water loss from open containers at room temperature over one year are between 1% and 5% although much higher values have been observed.

When irradiated, the epoxide resin waste forms, either containing water or dry, undergo a slight swelling, typically up to 1% at an accumulated dose of around 4 MGy. Preliminary work indicates that this swelling can be described theoretically by an extension of the bubble growth model developed for bitumenisates (9).

The polyester, vinyl ester and polystyrene wet waste forms behave differently, all shrinking. Shrinkage occurs whether irradiated or unirradiated, but is substantially greater under irradiation. The amount of shrinkage varies from material to material and the maximum observed value is around 8 volume % at 10 MGy. It is attributed to a combination of water loss and cross-linking of the polymer during irradiation (6).

Figure 2 shows typical swelling data for some polymer matrix waste forms. These dimensional changes do not present a problem to the container, although it is necessary to understand these effects in order to be able to obtain accurately radiation-aged materials for, for example, leaching measurements. The complicated swelling behaviour is believed due to a combination of mechanisms including bubble growth, cross-linking and water loss, and there has been some success in modelling aspects of these.

Large shrinkage could cause cracking of the waste form but this has been observed only in specimens undergoing the largest shrinkage. In general the mechanical strengths of polyester and vinyl ester waste forms irradiated to 10 MGy are reduced to about 60% of their unirradiated strengths, while the epoxide resin waste forms are less affected.

RWF3, a simulated PWR evaporator concentrate in epoxide or polyester resin, contains dry waste. However the polyester version of this too has

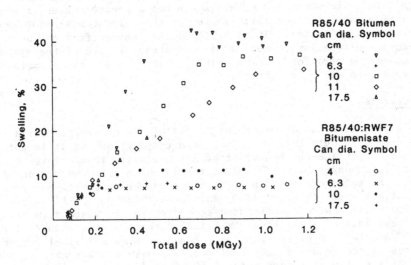

Fig. 1 γ- RADIATION INDUCED SWELLING OF UNFILLED BITUMEN R85/40 AND
BITUMENISATE SIMULATED WASTE FORM RWF7

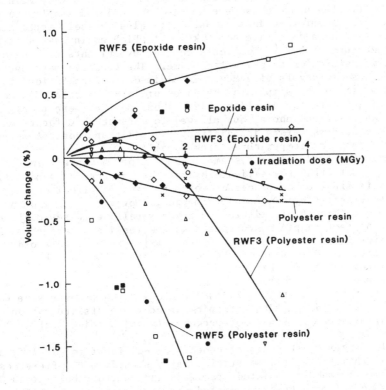

Fig. 2 VOLUME CHANGES FOR WASTE SIMULATES WITH POLYESTER AND EPOXIDE
MATRICES DURING γ IRRADIATION

been observed to lose weight and to shrink whether irradiated or
unirradiated, the shrinkage effect again being greater under irradiation.
In addition some, but not all, samples of the polyester version exuded
waste under irradiation. It is believed that this effect may be a
consequence of the development of a poor state of cure of the system,
indicating the need for quality control. The epoxide resin waste form
displayed none of these effects.

2.3 Cements

The reactions which occur during radiolysis of cement matrix waste
forms depend in detail on the waste form composition and it is not yet
possible to predict the behaviour of any new waste form. However, this,
and other, research have identified trends which allow some general
conclusions to be drawn and a distinction can be made between processes
which affect waste form stability and those which do not (7).

Waste forms were manufactured according to the waste form
specification, using inactive simulates of the wastes and cured for 28 days
at 20°C and 100% RH before testing. Leach samples contained tracer
isotopes of the required radionuclides. Irradiation was carried out by an
external γ source at a dose rate of 10 kGy h^{-1}, in intervals of 3 MGy up to
the required dose. At this dose rate the samples reached a steady state
temperature of 50°C. Batches of samples were divided so that for each set
of samples irradiated a corresponding number of controls were maintained at
20°C and 50°C. Leaching tests were carried out using 2 samples in 150 ml
distilled water, contained in a borosilicate glass vessel, irradiated
during the test at a dose rate of 10 kGy h^{-1} which again gave a steady
state temperature of 50°C. Duplicate control experiments were maintained
at this temperature for an equivalent time. The leachate was analysed and
changed at dose intervals of 1 MGy. In addition to irradiation of
simulated waste forms, samples of hydrated ordinary Portland cement paste
were included to provide baseline data. These samples were prepared in the
same form as described above, but allowed to cure at 20°C for times ranging
from zero to 28 days before irradiation. In addition, control samples were
stored at a range of temperatures from 50-56°C.

Irradiation produces no systematic variation in waste form dimensions.
This may indicate that the silicate hydrate gel structure is not affected
by radiolysis since other processes which do have an effect on it, such as
drying are reflected by dimensional changes. However, expansion of the
material due to growth of existing or new crystalline phases might not be
detectable by macroscopic measurements as the tensile strain to failure in
cement pastes is of the order of 0.02% and expansions of this order are
below the limit of detection of the measurements. Thus cracking would
become evident before the expansion could be measured.

No changes in compression strength which can be attributed to
radiation have been observed for doses up to 9 MGy. However some waste
forms have exhibited cracking, chipping or complete disintegration during
irradiation although this was not preceded by any warning such as loss of
strength.

Small changes in strength of cement are difficult to identify because
of the scatter in data, and no statistically significant differences have
been observed between irradiated specimens and controls held at 50°C.
There is some evidence that irradiated material does show a slightly higher
degree of hydration, as indicated by the calcium hydroxide contents, as
shown in Table 2. However, comparison with controls held at slightly
higher temperatures shows that the small discrepancy could easily be

accounted for by the small temperature fluctuations which occur during
irradiation.

Table 2   Effect of Radiation on the Hydration of Ordinary Portland Cement

Curing Period (days)[a]		Relative Hydration[b]
0		1.01
3		1.12
7		1.06
28		1.16
3	Irradiated	1.12
3	50 °C Control	1.00
3	52 °C Control	1.29
3	56 °C Control	1.11

(a) Curing period at 20°C prior to irradiation (or storage for controls).
(b) Relative hydration = $Ca(OH)_2$ content $\div$ $Ca(OH)_2$ content in 50°C control sample.

An example of a waste form which damages during irradiation is RWF2,
PWR evaporator concentrate borate solutions in cement. Samples of this
waste form disintegrated at doses between 6 and 9 MGy. The effect was
reproducible and no decrease in strength, nor dimensional change was
observed in the dose intervals prior to failure. Microscopic, X-ray and
thermal analyses have not revealed any new phases in the irradiated
material and no satisfactory explanation has yet been found.

During γ irradiation, hydrogen is evolved, and oxygen absorbed for the
majority of waste forms. The quantity of hydrogen evolved appears to be a
linear function of absorbed dose up to the maximum of 9 MGy. $G(H_2)$ values
(molecules/100 eV) for the reference waste forms tend to lie between 0.05
and 0.13, although for RWF8 are as low as 0.01. All of these are less than
the published $G(H_2)$ for pure water of 0.4-0.5 molecules per 100 eV. It is
not easy to quantify the oxygen absorption as only a small amount was
available in the specimen container during each dose interval and this was
completely consumed. Exceptions to this are cemented wastes containing
nitrates, for example RWF8. These tend to have lower $G(H_2)$ values and
noticeably reduced oxygen absorption. Carbon dioxide is also present in
the sample container in significant concentrations after irradiation. The
relative concentrations of oxygen and carbon dioxide vary but adding
together the molecular oxygen and that present in carbon dioxide, shows
that this waste form generates a net increase of oxygen in the higher dose
intervals. Complex radiolytic gas evolution with nitrate bearing wastes in
cement has been reported in the literature (10).

A preliminary experimental survey of the effect of radiation on the
leaching of radionuclides from the reference waste forms is not yet
complete. At present data are available on only a restricted number of
waste forms. Although differences in behaviour between irradiated and
control experiments are not found in all cases, some have been detected.
These differences are generally small but to date they have each shown a
tendency towards lower leach rates during irradiation. This is illustrated

in Figure 3, which shows the equivalent depletion depths for Cs and Sr
leaching from RWF1/2. This waste form exists as RWF1/1 and RWF1/2, the
former using OPC and the latter OPC and a pozzolana. It is interesting to
note that equivalent depletion depths for Cs and Sr are identical in this
material for which an effect of radiation may be evident. This equivalence
is not observed during the leaching of other waste forms.

2.4 Polymer Modified Cements

Polymer modified cements are hybrid materials obtained by
incorporating substantial quantities of polymer additive into a hydraulic
cement at the mixing stage. Their principal advantages are a higher strain
to failure and toughness than unmodified cement, and a somewhat lower
permeability. They are not widely used for the immobilisation of ILW
outside France, where bitumen modified cement has been used for some years,
but have been investigated under a Sheet 3 programme and some of the
results are summarised here.

The polymer additives studied were styrene-butadiene and bitumen
emulsions, and polyurethane resins. A polymer to cement weight ratio of at
least 0.2 is required to give consistent, significant improvements in
strain to failure over unmodified cement. The polyurethane modified cement
system contained a very high volume of polymer. Irradiation of these
materials was carried out at high dose rates ~ 10 kGy h^{-1} in a spent fuel
pond facility and the temperatures rose to 50°C-60°C. At that temperature,
temperature-induced effects as well as radiation effects occur and
therefore unirradiated controls were held at 50°C. Table 3 shows the
results for tensile failure strain in flexure.

Table 3   Effects of Irradiation on Tensile Failure Strain (%) of
Polymer Modified Cements (Standard Deviation in Brackets)

Additive Type	Polymer: Cement Ratio	Water: Cement Ratio	Radiation Dose (MGy)					
			3		10		30	
			Control	IR	Control	IR	Control	IR
OPC	None	0.3	0.048 (0.005)	0.045 (0.005)	0.032 (0.004)	0.032 (0.004)	0.035 (0.010)	0.018 (0.007)
Styrene-Butadiene	0.2	0.2	0.243 (0.027)	0.090 (0.007)	0.141 (0.008)	0.064 (0.015)	0.150 (0.015)	0.051 (0.018)
Bitumen	0.22	0.5	0.052 (0.009)	0.042 (0.004)	0.050 (0.002)	0.044 (0.002)	0.035 (0.004)	0.045 (0.003)
Epoxide	0.2	0.3	0.093 (0.008)	0.077 (0.007)	0.072 (0.006)	0.075 (0.013)	0.081 (0.008)	0.078 (0.008)
Poly-urethane	1.0	0.22	2.511 (0.542)	1.093 (0.141)	2.191 (0.25)	1.181 (0.085)	1.336 (0.205)	0.608 (0.057)

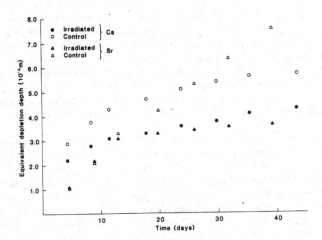

Fig. 3   LEACHING OF C_s AND Sr FROM RWF 1/2

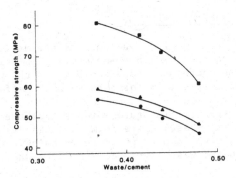

Fig. 4   COMPRESSIVE STRENGTH OF CEMENTED EVAPORATOR CONCENTRATES FROM
BWR (SULPHATES): AFTER 28 DAYS CURING (●); AFTER A PERIOD
(1 MONTH) OF IMMERSION IN WATER (▲); AFTER A FURTHER PERIOD
(7 DAYS) OF HEATING AT 110°C (■)

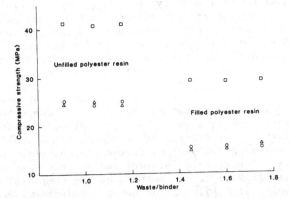

Fig. 5   COMPRESSIVE STRENGTH OF SAMPLES OF FILLED AND UNFILLED POLYESTER
RESIN EMBEDDING BEAD ION - EXCHANGE RESINS; AFTER NORMAL CURING
(○); AFTER A PERIOD (1 MONTH) OF IMMERSION IN WATER (△); AFTER A
FURTHER PERIOD (7 DAYS) OF HEATING AT 110°C (□).

There is much scatter in the data and some uncertainty in the results but in general these, and other, results show that the strain capability of the polymer modified systems remain significantly higher than that of unmodified OPC even after a dose of 30 MGy. The polyurethane resin modified system remains the best system for use with dimensionally unstable wastes as the strain bearing capacity after 30 MGy is still an order of magnitude higher than that of the epoxide resin modified system or the styrene-butadiene modified system, but it presents formidable processing problems.

Compression strengths of polymer modified cements increase as the irradiation dose increases, compression strength of the controls also seems to increase slightly with time at high temperature except for the polyurethane modified samples. Irradiation to 30 MGy has only a small effect on the compression strength of unmodified OPC and bitumen or epoxide resin modified OPC. The compression strengths of the irradiated samples of the styrene-butadiene and polyurethane resin modified OPC systems are substantially higher than for the equivalent controls. This increase in compression strength is probably due, at least in part, to irradiation induced cross-linking of the polymer component.

3   MECHANICAL AND THERMAL EFFECTS

The ultimate objective of mechanical and thermal testing is to predict the consequences of credible accidents. This includes providing information about the consequent handleability of the waste form, and a radionuclide-release source term for assessing the effect on the environment.

Comparative small-scale laboratory tests have been carried out on cement-based and polymer-based waste forms at ENEA Casaccia, and full-scale tests together with laboratory scale tests have been carried out on cement and bitumen based waste forms at KfK Karlsruhe.

3.1   Small-scale Tests

Table 4 lists the laboratory tests used by ENEA, which are typical of tests used by laboratories for characterising and comparing waste forms during the development stage. They have two objectives: to ensure that the waste form is of acceptable quality; and to give an initial assessment of the effects of a hazardous environmental impact by indicating the behaviour of waste forms under degradation conditions ranging from modest to severe (11). Thus, for example, the compressive strength is evaluated not only after the normal curing, but also after a period of immersion in water and a further period of heat exposure, Figures 4 and 5.

Tensile, flexural and compressive tests are performed in order to characterise the waste forms fully. These data are not used in specifications of waste forms but can be useful in comparing the quality of products. The behaviour of a waste form depends on the nature of the incorporating matrix. For example an increasing load applied to cement products causes cracking followed by fragmentation, while polymer products are progressively, irreversibly deformed and as cracks appear the encapsulated water flows out. In both cases the compressive strength depends on the waste loading, and for cement products also on the water/cement ratio and the curing time. These differences in behaviour can make the comparison of waste forms with different matrices difficult.

Table 4   Mechanical and Thermal Tests Employed at ENEA, Casaccia on
Solidified Waste Forms

Test	Standard	Concreted Wastes	Polymerized Products
Compressive strength	UNI 6132-727	+	+
Drop test		+	+
Tensile strength	ASTM C190-77 ASTM D638-80	+	+
Flexural strength	ASTM C348-80 ASTM D790-80	+	+
Impact resistance	ASTM D256-81		+
Abrasion resistance	ASTM C131-81	+	
Heat resistance		+	+
Flame test	ASTM D635-81	+	+
Heat deflection temperature	ASTM D648-72(78)		+
Vicat softening temperature	ASTM D1525-76		+
Brittleness temperature	ASTM D746-79		+
Freeze-thaw cycles		+	+

A range of tests simulate accident conditions. For mechanical impact
there are the drop test and the impact test. The drop test is usually
survived by plastic materials, which rebound after the impact onto a
target while cemented wastes often break. More important is the impact
test, which is carried out by means of a pendulum-type hammer mounted in a
standardised machine. However the impact behaviour is better evaluated on
full-scale samples (12). A flat circular end of a 3.2 cm diameter steel
rod weighing 6 kg is dropped through the distance of 1 m; the sample must
not break, crumble or shatter. Abrasion resistance can be measured (13)
using the Los Angeles Abrasion Machine, which is more commonly used for
measuring the abrasion characteristics of gravel. Normally a standard
specimen is rotated up to 1000 revolutions in the machine with a charge of
six standard steel balls. All cement pieces whose diameter are greater
than one inch after the test are weighed and compared with the original
sample and the results are expressed as weight loss per revolution. The
reproducibility of this test can be assured only for cemented waste forms.
Thermal effects are assessed on the basis of the damage they cause to
the final products, i.e. weight loss, progressive degradation, burning, and
of changes in the mechanical properties. The thermal test is performed by
heating at 800°C in a muffle furnace for 10 minutes; the sample should not
melt, sublime or ignite.

Thermogravimetric analysis (TGA) is helpful for interpreting the phenomena which take place during heating from room temperature to 1000 °C. With cement products the following main steps are recorded, Figure 6:

25-200 °C : loss of sorbed, capillary and crystallisation water.
200-250 °C : water removal from hydrated iron oxide (Fe_2O_3. x H_2O →
Fe_2O_3 + xH_2O).
350-450 °C : decomposition of calcium hydroxide ($Ca(OH)_2$ → CaO + H_2O).
500-800 °C : decomposition of calcium carbonate ($CaCO_3$ → CaO + CO_2).

Water is removed also from polymerised wastes, which are combusted to gaseous products, Figure 6, typically at less than about 550 °C for polyester resins. Further phenomena associated with the presence of a filler can occur such as oxidation with weight increase or decomposition. In Figure 6 the weight increase is due to oxidation of iron filler to iron oxide.

The fire test is more severe for organic compounds. These exhibit a variety of behaviour and can be classified as inflammable, non-flammable, or self-extinguishing. However, serious damage can be suffered by cemented wastes as well. Some cracking with loss of mechanical integrity can occur, further calcium oxide formed during heating can rehydrate to calcium hydroxide with deleterious effects. In some cases reaction between waste and cement can occur and cause conflagration of the waste form even when the matrix is cement. Heat and frost affect the mechanical properties of all materials. Various tests have been designed for evaluating the behaviour of plastics under well standardised conditions. The temperature at which a flat-ended needle of 1 mm^2 circular cross section will penetrate 1 mm inside a sample with a specific load applied on it is called Vicat softening temperature. The heat deflection temperature is the one at which a plastic bar of rectangular cross section deflects 0.25 mm with a load applied at its centre. The brittleness temperature is the temperature below zero at which 50% of (at least) 10 standard specimens exhibit brittle failure under impact.

Also very significant is freeze-thaw cycling, usually between -40 °C and +70 °C with a relative humidity of more than 95%. The moisture which penetrates inside materials increases in volume by freezing, thus causing mechanical stress. Due to their physical nature, being less porous and more elastic, plastics are far more resistant than cements, for whose failure a number of different theories have been formulated (14).

Tests of these types at the laboratory scale are potentially useful because they show that modifying the waste form, for example the addition of fillers, can help in improving some characteristics, but care should be taken in choosing fillers, because of the potential disadvantages of degradation of other properties, an increase of costs, or a complication in the process.

3.2 Full-scale Tests

Normally transport of radioactive waste packages is regulated by the IAEA transport regulations (12) which includes a consideration of impact and fire hazards. The present state of design of an underground repository in the FRG includes the possibility of the waste package being subjected to more severe impact than contemplated by the IAEA transport regulations. Accordingly full-scale tests have been carried out at KfK to assess the effects of impact from large heights on activity release from waste packages (8,15). In addition full-scale fire tests have been carried

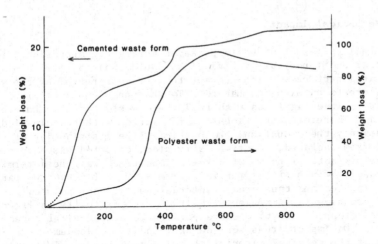

Fig. 6   TGA OF CEMENTED WASTE FORM (SULPHATE WASTE) AND POLYESTER WASTE
FORM (ION EXCHANGE WASTE). NOTE DIFFERENCE IN SCALES·

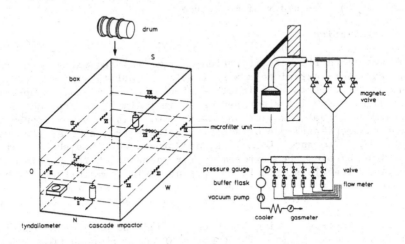

Fig. 7   SCHEME OF THE EXPERIMENTAL SET UP FOR THE COLLECTION OF AIRBORNE
FINES FROM THE WASTE PACKAGE AFTER MECHANICAL IMPACT

out (8,15). The waste form corresponded to RWF8, a cemented waste form containing evaporator concentrates from reprocessing of LWR-fuel elements, main component $NaNO_3$.

3.2.1 Mechanical Impact

Mechanical impact of waste packages during the operational phase of the underground repository in principle could lead to release of small, contaminated particles which are apt to become dispersed by the ventilation and thus create an airborne hazard. To that problem, up till now, no quantitative investigations with full-scale waste packages have been performed. Therefore, experiments were started with the aim of determining quantitatively the amount and the particle size distribution of the dust released from packaged, inactive, simulated, cemented waste form by mechanical impact. A point of special importance was the determination of the release of fines with a particle size $\leqslant 10$ μm because the particles could be inhaled and thus create internal radiation exposure.

The experimental arrangement is shown schematically in Figure 7 (8). Drums were dropped from 65 m corresponding to a mechanical impact energy of 3×10^5 J. On impact drums were damaged and the lids came off. Sieve analyses of the released material showed a good reproducibility. The fraction of the crushed material collected from the target with a particle size < 0.125 mm amounted to 518-608 g, the fraction < 0.063 mm amounted to 82-202 g.

The fraction of the material with particle sizes $\leqslant 10$ μm, carried by air and collected on the installed micro filter units, amounted to $\leqslant 1.0$ g which is equal to 10^{-5} to 10^{-6} of the total inventory. Hence, for a homogeneous product, the fraction of radionuclide released corresponds also to 10^{-5}-10^{-6} of the inventory. Measurements using Tyndallometer equipment resulted in the same order of magnitude.

3.2.2 Fire Testing

To determine the activity release from cemented waste forms containing $NaNO_3$, laboratory experiments with inactive simulated and tracered samples as well as experiments with inactive simulated full- scale samples were performed. The small-scale experiments included Thermogravimetric and Differential Thermal Analyses on inactive materials, and measurements of the temperature dependence of activity release from samples tracered with ^{137}Cs, ^{85}Sr, ^{144}Ce and ^{60}Co. This, and other, work has shown that the way a waste form behaves when exposed to fire depends on the possible waste-matrix reactions as well as the intrinsic flammability of the matrix. By combining the results obtained from different experiments it should be possible to cover a wide range of conditions for which the activity release could be predicted.

Full-scale fire testing has been carried out on bitumen (RWF7) and cement (RWF8) waste forms. The results of the work on bitumen have been reported elsewhere (16) and only the work on cement is reported in detail here. Twohundred litre steel drums containing simulated cemented waste forms were exposed to an oil fire of 30-60 minutes duration at 800°C flame temperature. Several different types of experiments were performed. First, the principal behaviour of the waste packages in an oil fire of about 40 minutes was tested by subjecting closed and open drums to the fire. After the test, the weight loss of the package and the state of the waste forms were determined. This showed that a weight loss of only about 10 kg occured, whether the steel drum was open or closed. This is equal to

10% of the total water content of the package. The lid sealing of the
closed drum leaked due to the internal pressure build-up when two to three
bars were reached. After the test, the waste forms showed only a few
radial cracks.

In a second series of experiments the detailed temperature profile in
the waste form was determined. For this purpose 30 thermocouples were
incorporated into the waste form so that the time dependence of the
temperature profile from the outside to the centre of the drum could be
measured at each of three levels, $\frac{1}{4}$, $\frac{1}{2}$ and $\frac{3}{4}$ from the top surface (8). In
a 40 minute oil fire, at the highest level the maximum temperature achieved
at the drum surface was about 550°C. A large temperature gradient
developed and the temperature in the centre of the waste form reached a
maximum of only 80°C. This maximum occurred 10 hours after the end of the
fire. At the other measurement levels the temperature at the drum surface
reached a maximum of 700°-800°C. These results indicated that water is
released only from the outer parts of the waste form and that in the
temperature range which has to be considered most of the water released is
free pore water.

The next stage in the experimental series was the quantitative
collection of the released water using distillation equipment to determine
the time dependence of water release as well as the cumulative water
release during the test. Inactive simulated waste forms (OPC, W/C = 0.35)
containing 10 weight % $NaNO_3$ and 1 weight % $CsNO_3$ were used for these
tests. The collected fractions of the condensate were analysed for their
Na and Cs content and the pH-value was measured. The results indicated a
very strong connection between water release and Na/Cs release and suggest
that under the experimental conditions the obtained Na/Cs release is caused
only through mechanical carry-over by evaporated water. The total amount
of the condensate amounted to 9.8 l, in good agreement with the weight loss
of the first series of experiments, and the pH-value of all condensate
fractions was uniformly 9.5.

From the analyses of condensate fractions a cumulative Cs release of
37 mg and a cumulative Na-release of 120 mg was calculated. As the total
Cs content in the waste form is 2,727 g and the Na content 9,700 g, the Cs
and Na release amounts to about 0.01 parts per thousand of the
Na/Cs content. The Na/Cs ratio of 3.2 in the condensate corresponds very
well with the Na/Cs ratio of 3.6 in the waste form.

Based on these results and because of the low temperatures arising in
the waste form during the experiments the Na/Cs release is caused only by
the water release. Clearly, if tritiated water is also present this too
would be released in proportion to the fraction of water released.

The results of the experiments give a clear indication that any
activity release from cemented waste forms under thermal impact is very
low.

4   FURTHER WORK

Over the last five years considerable progress has been made through
these, and other European programmes, in understanding the significance of
severe mechanical impacts and fires on full-scale, drummed, waste forms; on
the relevance and usefulness of the various small-scale laboratory tests;
and on the effects of radiation.

Of the three aspects considered in this paper, i.e. radiation, thermal
and mechanical, the radiation damage problem differs from the others in
that it is a very long-term process. Although short-term, high dose rate,
accelerated tests can provide much useful information, there is a real need

FIG. 8.   FLOW DIAGRAM OF IRRADIATION PROGRAMME

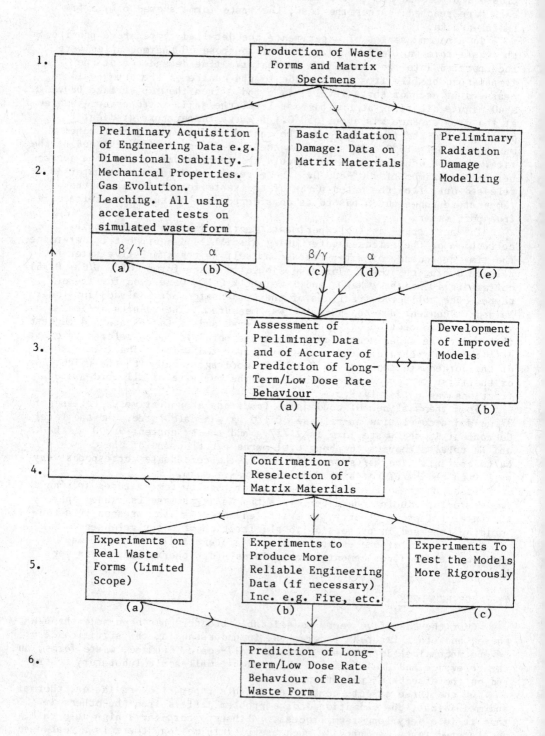

for the development of an understanding, adequate to enable long-term, low dose rate effects to be predicted with confidence. Of necessity the radiation damage programme has to be iterative as shown in Figure 8 which outlines a flow diagram for the programme. Most of the activities in the programme have been accomplished to some extent for bitumenisates but for cements, thermosets and polymer modified cements the programme has only reached level 3 or 4 and there are deficiencies in the theoretical modelling. More specifically the work to date has also shown some anomalies and unexplained phenomena, such as anomalous swelling behaviour of some bitumenisates; disintegration of a cemented borate waste form; and exudation of waste from a polyester waste form. Further work is needed to explain these phenomena and to complete the programme outlined in Figure 8 for important waste forms.

The Karlsruhe full-scale tests have shown that activity release from single drums of cemented waste under very severe drop impact conditions is small. Future plans envisage several drums being contained in more robust containers and therefore that activity release will be even lower in practice. Consequently, KfK regard the full-scale drop testing as being virtually completed. On a more general note, however, the various waste treatment and disposal organisations in Europe have to consider the consequences of accidents involving fire and/or mechanical impact, and full-scale tests which represent such accidents are going to form an important aspect of their programmes. The techniques and data developed by KfK are clearly of help in this respect.

As far as laboratory-scale testing is concerned, new waste forms are being developed as additional waste streams are considered. Continued laboratory scale testing, employing techniques of the type used at Casaccia, is important in choosing the most appropriate matrix and assessing the properties of the waste form prior to embarking on more expensive full-scale tests. There is a need here too for the development of theoretical modelling techniques to relate the laboratory-scale tests to potential full-scale incidents.

ACKNOWLEDGEMENTS

The work summarised here has been carried out through the team-work of many of our colleagues at AERE, KfK and ENEA. It is not possible to list them all here but in particular we mention Mr. R.A.J. Sambell who co-ordinated the Harwell work, Dr. G. McHugh who led the studies of radiation damage of cement and Mr. P. Vejmelka of KfK. Part funding of the UKAEA work was made by the UK Department of the Environment and in that context the results of the work may be used in the formulation of, but at this stage do not necessarily represent, UK Government policy.

REFERENCES

1. POTTIER, P. et al. (1985). Stability of low and medium level waste immobilised in bitumen, cement and polymers. Paper IV-4 of this conference.
2. SAMBELL, R.A.J. et al. (1983). Characterisation of low and medium level radioactive waste forms, 1st joint annual report to the CEC. AERE-R10773 and EUR 8663 EN.
3. VEJMELKA, P. and SAMBELL, R.A.J. (1984). Characterisation of low and medium level radioactive waste forms, 2nd joint annual report to the CEC. AERE-R11349 and EUR 9423 EN.

4.   JOHNSON, D.I. and PHILLIPS, D.C. (1983). The use of organic materials
     for the immobilisation of intermediate level radioactive wastes.
     AERE-R10639.
5.   PHILLIPS, D.C., HITCHON, J.W., JOHNSON, D.I. and MATTHEWS, J.R.
     (1984). The radiation swelling of bitumens and bitumenised wastes.
     J. Nuc. Mater. 125 (1984) 202-218.
6.   PHILLIPS, D.C., JOHNSON, D.I., BURNAY, S.G. and HITCHON, J.W. (1983).
     The effects of radiation on organic matrix waste forms, in Testing,
     evaluation and shallow land burial of low and medium radioactive waste
     forms. Proceedings of a symposium at Geel, Sept. 1983, Ed.
     W. Krischer and R. Simon. EUR 8979.
7.   McHUGH, G., SPINDLER, W.E. and MATTINGLEY, N.J. (1983). Radiation
     effects in hydraulic cement matrix waste forms. IBID.
8.   VEJMELKA, P., JOHNSEN, P. and KOESTER, R. (1983). Activity release
     from waste packages containing cemented ILW under mechanical and
     thermal stresses. IBID.
9.   MURPHY, S.M. (1985). A theroretical model of the swelling of
     γ-irradiated epoxy resin. AERE-R11616. In press.
10. MOECKEL, H. and KOESTER, R. (1982). Gas formation during γ radiolysis
     of cemented low and intermediate level waste products. Nuclear
     Technology 59(1982)494-497.
11. PIHLAJAVAARA, S.E. and AITTOLA, J-P. (1978). Solidification of
     nuclear wastes with Portland cement. 80th Annual meeting of the
     American Ceramic Society, May 1978.
12. Regulations for the safe transport of radioactive materials. IAEA
     Safety Series No. 6, 1973, Revised Edition, IAEA, Vienna.
13. LERCH, R.E. et al. (1979). Treatment and immobilisation of
     intermediate level radioactive wastes. Management of low-level
     radioactive wastes, Pergamon Press, Vol. 1, 513-550.
14. LEA, F.M. (1970). The chemistry of cement and concrete. 3rd Edition,
     Pub. Arnold, London.
15. VEJMELKA, P., JOHNSON, P., KOESTER, R. and BRUNNER, H. (1985).
     Activity release from waste packages containing cemented waste forms
     under mechanical and thermal stresses. Waste Management Symposium,
     Tucson, Arizona, USA, March 1985.
16. KLUGER, W., VEJMELKA, P. and KOESTER, R. (1982). Investigation of
     activity release from bitumenised intermediate level waste forms under
     thermal stresses. IAEA Symposium on the conditioning of radioactive
     wastes for storage and disposal. Utrecht, June 1983. IAEA-SM261-18.

DISCUSSION

T. MARZULLO, ENEL Roma

Is there any technical reason why cylindrical drums are preferred to cubic containers, although the latter would be less space consuming in the repository?

R. KÖSTER, KfK Karlsruhe

The dropping tests at scale 1:1 with 200 l cylindrical drums obtained different results in dust evolution and Cs-release depending on whether the drum fell on its front or its side or on the edge. Based on this, a geometry effect on the activity release can be derived leading to a re-lease increase in case of cubic containers of approximately one order of magnitude.

M. VANDORPE, Belgonucléaire Brussels

Has the test programme also included fire tests with thermosetting resins or polymer modified cements?

D.C. PHILLIPS, AERE Harwell

The fire tests carried out at KFK were restricted to cement and bitumen, but outside this programme fire tests have been conducted also with poly-mer resins$_*$.

B. VERKERK, ECN Petten

There was one cemented waste form which desintegrated under irradiation. Did it contain borate solution and, if so, is it possible that boron gives rise to a   /n-reaction?

D.C. PHILLIPS, AERE Harwell

It was not necessarily the boron which caused desintegration. It should be remembered that as borate retards the cement setting it is necessary to use an accelerator. We experienced similar effects of accelerators with other waste forms, outside this programme. This suggests that we have to investigate accelerator reactions more thoroughly.

---

* See report EUR 7415 FR

THE CHALLENGE OF QUALITY ASSURANCE

R. SIMON, Commission of the European Communities, Brussels
M.S.T. PRICE, Atomic Energy Establishment, Winfrith
W. KRISCHER, Commission of the European Communities, Brussels

Summary

The safe terminal containment of hazardous wastes cannot only rely upon the geographic remoteness and the arid climates of the repository sites. Radioactive and permanently toxic chemical wastes must be prevented from returning to the human environment by natural and engineered barriers. The long-term integrity of these barriers and the safety of waste emplacement operation will be controlled by systematic actions under the common objective of Quality Assurance (QA).
The following paper presents the aims of QA in the design and production of waste packages. It lists the most relevant acceptance criteria and regulatory requirements, investigates the institutional and technical problems of carrying out Quality Assurance and presents suggestions for establishing suitable organisational structures and technical programmes to provide adequate confidence in the safe nature and the performance of waste packages. In view of the technical difficulties of verifying the compliance of industrially produced waste forms with the acceptance criteria, the CEC has laid emphasis on the development of appropriate test methods in its last R&D programme. First results of the work are reviewed in the context of international progress in this field.

1. Introduction
    There are several senses in which the word 'quality' can be employed. The term Quality Assurance (QA) as used in this paper comprises all planned or systematic actions necessary to provide adequate confidence that a structure, system or component will perform satisfactorily in service. Within this context therefore the term is used to denote 'fitness for purpose'.

    Quality Control is a part of Quality Assurance; it refers to those actions related to the physical characterization or the chemical nature of a finished product which provide a means to control its quality to predetermined standards.

2. Quality Assurance in Management and Disposal of Radioactive Waste
    The purpose of disposing radioactive waste by such elaborate means as discussed at this conference is to protect mankind from the effects of radioactivity. The materials and components of waste isolation systems must therefore perform their albeit static task for long periods of time ranging from hundreds to thousands of years, depending upon the radioactive decay characteristics and toxicity of the particular radionuclides. It should also be borne in mind, that geological repositories, after an operational period of construction and filling, will be sealed definitely. Therefore any component failure will be irreparable and all releases of radioactivity to the human environment will be irreversible.

It is necessary therefore not only to provide highly redundant safety features, but also to assure, that they are faithfully implemented in design, construction and operation of the real disposal facility. In the sense of providing a product, which is fit for its purpose, the requirement of quality management for radioactive waste disposal systems is no different from that for any other product such as household appliance, a motor car or a control rod drive for a reactor. In fact the long time scales involved and the irreparable nature of a geological repository for radioactive waste provide a much stronger argument for QA than in the case of consumer goods. Regrettably, however, it is not always recognized, that these same arguments must also be applied to the management of other (non-decaying) toxic or otherwise harmful substances as well as to other human activities which can damage the human environment permanently e.g. mining, regulating rivers or deforestation.

3.   The Multibarrier System

With the possible exception of low level waste, all categories of radioactive waste will eventually be disposed in a solid, stable form which will generally be monolithic i.e. free-standing.

The combination of primary container and the waste form, which can be termed the package, acts as the first series of barriers against radionuclide release.

In addition further engineered barriers, the backfill and the liner of the cavern, as well as, for the case of an underground repository, a formidable geological barrier will constitute successive lines of defence for the protection of the human environment.

The nature and number of engineered barriers designed around the waste package varies in the different concepts. The requirement for Quality Assurance extends to all engineered barriers and, to some extent, to the choice and layout of the geological barrier. Although most repositories are still at the conceptual stage, radioactive waste is accumulating. It has therefore become rather urgent to define specifications for waste packages which will be acceptable for the future disposal facilities.

This paper concentrates on the waste package, which is a main component of the 'near-field' part in the multibarrier disposal system.

4.   Waste Acceptance Criteria

Waste acceptance criteria are those technical requirements which will have to be met by a waste package to qualify for final disposal in a repository. They will be drawn up by the licensing authority or the agency responsible for the disposal facility and constitute the basis for the producers specification of the waste. Compliance with the waste acceptance criteria is the primary objective of QA.

Waste acceptance criteria will be defined for each of the major waste categories. Criteria presently found in the literature are established for particular facilities or repository projects. In this section only a few more or less common criteria are quoted to discuss the aims and means of QA.

The first general requirement for each waste package is that it can be clearly assigned to one of the authorized waste categories. Waste categories are distinguished primarily by the level of radioactivity.

Overall radioactivity cannot however be directly related to hazard and therefore most often separate alpha and beta/gamma limits are defined, or a derived parameter such as radiotoxicity is applied.

In the US, extensive lists of radionuclides figure in the classification criteria.

For comparison of some national criteria, only limits of alpha-activity are listed in Table I.

Table I: Segregation Criteria: Solid Waste Categories
          and Disposal Routes

Country	- Activity limits for conditioned waste (Ci/t)					
	.01	0.1	1	10	100	1000
F [1]	I Shallow land burial	II ?	III deep repository			
UK [2]	LLW Shallow trench burial (Drigg)	ILW Trench or mine	no heat removal	HLW deep repository after		
D [3]	Cat.1 ** Mines: Konrad, Asse		Cat. 2 **	∝ - waste ** HLW ** deep repository: Gorleben		
USA [4,5]	A, B, C intruder Land disposal	TRU deep repository (WIPP for defence w.)		HLW		

*
**   > 0,5 Ci/t : Trench disposal only as exception
     Classification by origin, activity and matrix.

A second major condition for disposal of waste packages is, that all mechanisms leading to a breach of the other isolation barriers must be excluded. Such a general rule has to be made applicable by formulating specific criteria on the composition of the waste. Table II lists the most common restrictions on waste constituents.

Table II: Criteria Protecting the Barrier System against Damaging
            Effects Originating in the Waste

Hazardous Component Effect	a) Required precaution or b) Limitation (example)	Relevant disposal route, phase
Explosives Compressed gas	a) Complete exclusion	all repositories
Pyrophoric substances	a) transformation to inert form	all repositories
Flammable material	b) in cement waste form: < 25 vol% organics (5)	mainly shallow land burial and preclosure phase
Corrosive	a) complete neutralization corrosion protected container (for sludges)	storage + retrievable emplacement
Gas release	a) overall organic content: < 220 kg/m^3 waste (5) or b) gas release: < 10 M/$_\text{m}$3 5 $< 0,3$ cm^3/g (1)	less stringent in porous host rock
Swelling trenches	a) allow void in package for expansion b) in bitumen: < 20 vol% water induced ($_{(I)}$) (RFS) < 10 vol% radiolysis (RFS)	hard rock cavities and concrete
free liquid	a) add twice water volume of absorbent (5) b) interstitial water < 4 vol%	saline host rock

   The third, but very important group of criteria aims at minimizing
the release of radionuclides from the waste package. So far attention and
research effort has largely focused on release in aqueous media, but the
less complex airborne dispersion must not be neglected in defining
acceptance criteria. Examples of criteria related to nuclide release are
given in Table III.
   The criteria and quantitative limits are only presented here to
illustrate the nature and sensitivity of tests required to prove
compliance with the criteria.

Table III: <u>Example of Quantitative Criteria Related to Nuclide Release</u>

Requirement	Example of Limitation
Release limit	France/Cat. I [1] : Leach rate : 5×10^{-6} cm/d : 5×10^{-5} cm/d US/Cat. A+B [4] : Leaching index LI < 6 * US/HLW [6] : Annual release fraction < 10^{-5} **
Absence of chelating agents	US/HLW [5] : Criteria discounted for WIPP US/LLW [4] : < 0.1% chelating agents in waste
Indispersible (monolithic) waste form ***	US/WIPP [5] : < 1 wt % particles below 10 u < 15 wt % particles below 200 u

*   ANS 16.1 leach test

**   10 CFR 60: 1/100000 of each nuclide present 1000 yrs after emplacement

*** for low level solid wastes, this requirement can be accepted, if high integrity containers are used.

    The fourth important set of prescriptions concerns the mechanical strength and resistance of waste packages. These criteria will primarily cover transport and handling requirements, i.e. that a package must retain its integrity after a drop test and that it must support a compressive load when stacked in storage. Whether a high compressive strength is required in a sealed geological repository, depends upon the concept and the characteristics of the host formation. Whereas high level glass packages must resist the lithostatic pressure in all known concepts, low and medium level waste packages must only remain incompressible in hard rock formations and shallow land burial, where subsidence may create water pathways into and out of the waste. In plastic host formations like salt and clay, fracturing of the waste package will not lead to a breach of the geological barrier.
    There are, however, two further reasons for requiring mechanical strength of waste packages. Firstly, to discourage human intrusion and secondly to keep the waste retrievable at least for a few decades.
    Finally, some of the most elementary and well known waste package criteria concern the safety and the practical implementation of transport, handling and storage operations:
    - surface dose rates
    - surface contamination levels
    - exclusion of toxic substances

as well as:
- provision of slinging or lifting accessories
- maximum size and weight
- correct colour-coding and labelling
- comprehensive and meaningful hand-over documents.

This last point, the papers that certify the origin and nature of the waste as well as work tests, the treatment processes and the licenses of the producer must be the main evidence on which waste packages will be admitted to a repository.

The long list of criteria compiled from regulatory or pre-regulatory literature probably contains all the important plus a few trivial requirements for making waste packages safe and compatible with most land disposal routes.

In all but a few cases the question remains, as to how to assure, that these conditions are satisfied. Although compliance with the criteria can, in most cases, be controlled during the management and treatment phase, checking the contents of the waste in a sealed waste package by non-destructive methods will be a very difficult task.

To render quality checking for alpha-waste possible a further series of new criteria is suggested, e.g.:
- waste packages as delivered should be cylindrical and not exceed dimensions permitting alpha assaying and other NDT methods
- wherever practicable, waste packages should allow visual inspection of waste.

5.   Statutory Requirements and Regulatory Guidance

A comprehensive and well known example of regulatory requirements covering QA in nuclear energy is contained in Appendix B of the US Code of Federal Regulations 10 CFR 50. Most of the 18 criteria listed therein apply to the production of radioactive waste packages. In essence the producer should be responsible for establishing and executing a clearly defined QA programme. This programme will, inter alia, cover control of design, procurement, operating procedures, processes, tests, test equipment and records. Although 10 CFR 50 does not specify the role of regulatory authorities it must be anticipated that they will be involved in periodic audits.

In the Federal Republic of Germany and in the UK, QA requirements for the nuclear sector have been established in the KTA-rule 1401 and the BS 5882-1980 respectively. The latter is compatible with the IAEA Code of Practice for Quality Assurance 50-C-QA and is based on a draft standard of the ISO on QA principles (DIS 6215).

The specific application of QA principles to Radioactive Waste Management has only recently become a subject of regulations and official guidelines. A very comprehensive detailed example is the set of guidelines and requirements drafted for WIPP Waste Package Qualification and QA.

The UK Dept. of Environment and the French Ministry of Industry and Research have distributed instructions to waste processes [7,1].

The licensing authority, a government department, will control the application of Quality Assurance in the following ways (Fig. 1): Initially by making the license i.e. STET to operate the waste processing plant conditioned of a satisfactory Quality Assurance Programme and Organisation. The actual examination of the programme will normally be carried out by an official inspectorate like the NII in Britain, the TUV in Germany or the Service Central de Sûreté des Installations Nucléaires in France (SCSIN).

During the operation of the waste conditioning facility, the licensing department will be informed of the occurrence of waste packages not conforming to the criteria as well as the nature of the non-compliance.

Finally the licensing authority or the competent inspectorate will supervise the testing and inspection procedures to be carried out on representative samples of waste packages.

In addition to these checks, which are performed by the waste processors own QA department, verification tests could be made in independent laboratories commissioned by the licensing authority or the agency in charge of waste disposal.

The latter organisation will, in any case, examine not only the documents certifying the quality of the waste packages, but will also check the containers for dose rate, contamination and signs of damage. Random checks of radionuclide content will also be carried out upon reception at the disposal facility.

6.   Important Elements of a Quality Assurance Programme

6.1. The Concept of Total Quality Assurance
     The need for Quality Assurance has many origins and derives from consideration of features such as service conditions, implications of premature failure as well as investigative reasons following abnormal events. With simple manufactured items it is often quite adequate to rely on post-manufacture (or output) controls to assure the quality of the product. But as processes and products become more complex the required range of output controls needs to be increased. The cost and difficulty of carrying them out can then become important considerations. Where quality has to be assured over an extended time-span, the response time for an output control system can be very protracted, unless reliable accelerated tests are available.

     There are obvious difficulties in carrying out post-manufacture controls when radioactive waste has been immobilised in a matrix such as cement. First, any operation involving the preparation of samples by destructive techniques, such as trepanning, will require remotely controlled operations carried out within a shielded cell. Secondly, the waste is at that stage in its final form for disposal and remedial or corrective action, although possible, is difficult. Thirdly, as has been touched upon above, any field trials such as leaching or lysimeter tests will take a significant time to carry out, posing a problem of what to do with the production line during the period it takes to carry out the field trials.

     A complex radioactive waste immobilisation process with the need for comprehensive safeguards necessitates an alternative approach in which the input parameters of the process are controlled and monitored and these results are then complemented with a certain level of output control
(Fig. 2). This is another way of defining the concept of 'Total Quality Assurance' in which a picture of quality is built up from a compilation of data throughout all stages of the manufacturing process from raw materials to the finished product.

     The availability of modern computerised systems to operate and check the performance of processes and accumulate a data base on a particular batch or package is a powerful aid to Total Quality Assurance. The object is then to evaluate whether, as a result of all the data collected, it is reasonable to assume that the product is not significantly different from previous batches/packages of product.

6.2. Documentation and Records

It will be necessary to document all the important phases of a waste manufacturing campaign from initiation of the project, through process design, formulation of manufacturing specifications, materials processing, packaging, quality control and acceptance/rejection procedures.

All Quality Assurance documents need to be controlled by written procedures which should cover aspects such as document approval, document revision and distribution. The overall approach should be embodied in a Quality Assurance Plan which sets out the specific quality practices to be employed.

6.3. Organisation

Within the organisation of the waste producer the Quality Assurance functions are three-fold:

- establishment and execution of an appropriate QA plan,
- identification of quality problems,
- verification that the QA plan is being compiled with correctly.

These duties can run counter to the functions of the production staff who, although responsible for product quality have other constraints imposed by cost, resources and programme requirements, to ensure that the persons carrying out QA have adequate authority and organisational freedom from possible pressures from the production staff.

6.4. Product Evaluation or Qualification

The logic of Total Quality Assurance and the long timescales required to assess the quality of the finished product imply that an extensive product evaluation or qualification programme will have to be carried out before any waste is sent for disposal. For a land burial site the concept will be simplified if a disposal model involving 'near field' and 'far field' effects is used. The objective of the product evaluation programme is then to show by modelling and experiment that the product complies with the near-field requirement.

Results of the EC product evaluation and characterization studies were presented earlier in this session. With wastes containing radionuclides with relative short half-lives (for example up to Cs-137) it is possible to consider relying on a metallic containment barrier. Such an approach orients the qualification programme towards assessment of containment particularly high integrity container corrosion behaviour during both storage and disposal for periods up to 10 half-lives, say 300 years.

The assessment of wastes containing longer half-life materials can also benefit from initial containment in a metallic drum. To model behaviour beyond that time it will be necessary to evaluate the solubility and leaching behaviour of the radioactive waste in its matrix. Land disposal sites are likely to have been chosen to comply with a variety of technical criteria including low water flow rate. Evaluation of radionuclide migration is therefore frequently based on nuclide solubility in a static environment consisting of natural wastes or brines, the waste form and the barrier materials.

It will also be necessary to evaluate the irradiation behaviour of the consolidated waste. Accelerated tests reported in this session have shown that for most of the common low and medium active waste forms, radiation damage will only become a critical item for exceedingly high dose rates. The product evaluation results will need to be used in a radionuclide transport model to demonstrate to the regulatory authority that the fatal risk to human beings at any time in the future is adequately low. These risk levels are likely to be ca. 10^{-6}/year, significantly lower than for the majority of other human activities affecting the safety of human beings.

6.5. Control of Test Equipment

Measurements and tests play a crucial role in the application of Total Quality Assurance. The methods and equipment selected to control the quality must therefore be evacuated carefully when qualifying the process. Their use and characteristics must be identified in the QA plan.

The QA plan will also specify procedures for calibration, adjustment and periodic recalibration.

In the EC research programme, work has so far concentrated on two important items:

Firstly to evaluate the available methods of assaying alpha-activity in solid waste. The present status of an inter-laboratory test campaign on identical reference waste containers has been reported earlier at this conference by Monsieur Bosser.

A second action consistent of developing a versatile leach test – ACL Autoclave Leach Test – for HLW specimens and carrying out a 'Round Robin Test' to determine the accuracy and reproducibility of such a test by using strictly identical specimens, leaching apparatus and test procedures. The tests were carried out by 12 laboratories in nine different countries.

Fig. 3 shows, that reasonably steady results of matrix corrosion under static conditions can already be measured after 28 days, which is quite appropriate for Quality Control.

Fig. 4 compares the results of elemental measurements: whereas the reproducibility limits and inter-lab standard deviation remained in single figure percentages for glassformer (Si, B) and Cs divergence in the analysis of Sr and Nd reached 50%: concentrations of these elements in the leachate were too close to detection limits to allow any high precision.

The new (1985-89) EC research programme has given high priority to QA-related research. In the field of testing, some of the intended actions are:

- application of the autoclave leach test to active specimens,
- development of standard tests for radiation damage and radiolysis,
- improvement of alpha-assaying techniques for conditioned waste,
- evaluation of tomographic inspection methods,
- establishment of model product testing procedures and test protocols,
- demonstration and evaluation of an operative QA programme on LLW and MLW conditioning plants.

6.6. Quality Control

Quality Control is the operational arm of QA. It includes those actions related to physical and chemical characteristics which provide a means to control quality to predetermined standards. It follows from the concept of Total Quality Assurance, stated earlier, that quality controls are applied at all stages of the process. Process controls are instituted as an aid to the satisfactory operation of the process; those which relate to quality become part of the set of quality controls.

For a radioactive waste immobilisation process, such as cementation, the controls fall into three categories:
- purchasing controls
- quality-related controls during processing
- output controls.

Purchasing Controls

The scope of QA must inevitably include control of the purchase of raw materials. In the case of a cemented waste in a high integrity drum this calls for control of the supply of:
- drums and lids
  . material specification(s)
  . manufacturing specification(s)
  . number of drums
  . sampling and testing to assess compliance with specification

It is also necessary to ensure that the raw materials arrive at the processing site in a satisfactory state and that they are correctly received and stored. As an example, if the strategy involves radioactive decay of the waste in a high integrity drum for, say 300 years, special care must be taken to avoid damage to the drum before use.

Quality-Related Controls

Processes for the immobilisation of radioactive waste will be operated as a series of stages at each of which it will be necessary to carry out a range of checks to confirm that the process is proceeding satisfactorily. Some of these checks are related to the ability to operate subsequent stages of the process (e.g. availability of raw materials) and are not related to the achievement of quality. However, other checks are quality-related and information on these becomes part of the data base. Modern computer-assisted control techniques can be used to operate the processsess and check or log the quality-related information.

The quality-related controls can be considered under the following headings:
- (i)   process preliminaries
- (ii) quality-related controls of the process
- (iii) output control checks.

Process Preliminaries

For the cementation process cited earlier the process preliminaries might include checks on:
- drum identification
- cleanliness (radioactive contamination level) of plant
- correct functioning of weighing apparatus
- water temperature
- availability of sufficient raw materials to complete the process
- availability of all equipment needed to complete the process.

Quality-Related Controls of the Process

The most important controls on waste immobilization processes are those on nuclide inventory and chemical composition. Radiochemical and chemical analysis of the raw waste will determine:
- the amount of waste to be immobilized per package

- the pretreatment requirements (setting additives, pH-adjustment,
- the operating conditions (dosing rate, mixing time, temperature) and provide essential certification data such as nuclide specific inventory and the concentrations of organics and chelating agents.

If pretreatment is carried out, a further analysis will control the operating parameters of the immobilisation process itself. In addition to these chemical analyses, controls on homogeneity, viscosity, setting characteristics as well as DTA will be applied on specimens.

The controls available at reasonable cost in a remotely operated process such as cementation will inevitably be very simple and should preferably be non-intrusive or at least non-destructive. There will be considerable reliance on simple techniques, such as weighing, to provide the initial weight of drum (and paddle if a lost paddle mixing system is being used, as well as the individual weights of cementitious ingredients, water, waste and the total weight.

It is important to have early warning of any quality malfunction so that remedial action can be initiated. Work at AEE-Winfrith, during the latter part of the first EC programme has been aimed particularly at the early stages of processing and has concentrated on the measurement of electrical impedance and torque.

The measurement of the torque applied to the paddle in a lost paddle mixer can provide useful evidence that the process is being operated normally. Different paddles show different torque/time behaviour and varying the rate of addition of the cement produces a different torque/time 'signature' (See Figure 5).

Methods of detecting 'set' of a cemented waste are available but difficult to apply. A penetrometer technique could be adopted in some circumstances e.g. if the process permits the lidding operation to be carried out at a later stage. Another possibility is to measure the electrical impedance of the mix using a simple probe at a frequency of 1 KHz. The impedance/time curve exhibits a characteristic form as shown in Figure 3. An advantage of this test is that it can be applied from the commencement of mixing or grouting. A disadvantage in the case of a mixing process is that it will require the insertion of the probe after completion of the mixing stage. An acoustic (pulse velocity) technique could possibly be used to monitor setting but there is difficulty in locating the probes on the outside of the drum.

Output Quality Control Checks

In order to be meaningful, Output Quality Control should be carried out in two phases:
- before closing the primary container to verify setting, absence of free liquid, absence of surface cracks, gas generation and filling level. All this can be done remotely.
- the second part of the output checks will take place after sealing the package. These inspection checks will include:

. drum identification
. surface contamination check
. surface dose rate check
. overall weight
. summation of data covering purchasing controls and quality-
   related controls of the process.

It must be anticipated that from time to time the regulatory authority will wish to take a sample of process output for more detailed
examination at a national QA laboratory or consultant's facility. Verification by the operator of the disposal facility is an alternative. The
destructive and non-destructive tests which will be applied are likely to
include:

- verification of actinide contents and fission product activity
- radiography or tomography (i.e. spatial distribution/homogeneity
  by high energy X-ray or linear accelerator)
- solubility of radionuclides (static equilibrium tests)
- leaching behaviour (although in low flow rate conditions, this
  will mainly relate to solubility)
- mechanical properties
- irradiation behaviour.

Such tests will be expensive and may force producers to set up
similar facilities.

6.7. Quality Audits

In any major Quality Assurance programme an audit procedure is
required. Because the disposal of radioactive waste concerns the liability
of the waste producer to the public and because of the long timescales
involved, the audit function is properly placed under the control of the
Regulatory Authority. Quality Audit is then defined simply as an independent verification of quality. A typical arrangement of regulatory authority, waste producer and disposal organisation is illustrated in Figure 6.
This figure indicates that the audit function of the Regulatory Authority
may be remitted to other organisations such as a national QA laboratory or
to an independent consultant firm. Audits will have to be continued
throughout the operational lifetime of the plant. In practice the type of
audit will range widely from simple scrutiny of process control records
and quality control data in relation to the written specifications referred to earlier, all the way to independent assessment of product
quality e.g. by destructive trepanning of the product.

Public anxieties and political pressures might induce regulatory
authorities to adopt an excessively rigorous approach to quality assurance. On the other side, waste producers and disposal operators may believe, that the quality standards of waste management could be modelled on
the practices applied thirty years ago to chemical wastes and nuclear
defence wastes. We must be aware of both attitudes.

Experience indicates that the best results are obtained when the
audit procedure is seen by the producer as a useful confirmation of
product quality and not as an additional piece of bureaucracy, in other
words that the auditor is concerned that the producer is successful.

6.8. Non-Conformance

Although it is essential that radioactive waste packaging processes
aim to deliver a product which is 'right first time' rather than rely on
post-manufacture controls to sort the good from the unacceptable, it must
be recognised the non-conformance may happen e.g. due to plant malfunction. The QA plan must establish measures to control such materials, parts

or components to prevent their inadvertent use. These measures involve the corrective action to be taken with the actual material, part, component or finished product, the documentation to be prepared and the procedure for notifying the appropriate management level and, where necessary, the regulatory authority.

Technically, there are a number of options to deal with non-conforming waste packages, e.g.:

- Cemented waste which fails to set due to decontamination liquors can be removed from the original drum and mixed to a lower water/-cement ratio.

- Excess porosity in cemented waste due to gas release from metal/alkali reactions can be overpacked.

- Where the activity or the nuclide solubility exceed the requirements for a particular waste category, such packages may be overpacked and/or emplaced with the next higher category of waste, e.g. in a concreted trench instead of a tumulus.

- Packages damaged during transport or by corrosion must be overpacked.

These examples show that satisfactory pragmatic solutions to offset non-conformity can be found. Corrective action may be expensive, but such penalties are ultimately necessary for the enforcement of adequate quality.

7. Conclusions

The results of risk analysis carried out not only in the CEC, but also in Sweden, Switzerland, Canada and the USA agree that disposal of HLW, MLW and LLW will place neither the present nor any future society at risk, if implemented according to the underlying model assumptions. Therefore, although these predictions generally contain comfortable safety margins for individual risk, it is essential that the systems, components and materials will perform as expected.

This will require carefully planned and systematically applied Quality Assurance. We cannot rely only on visual inspection and available non destructive test methods are unable to reveal all possible deviations from the acceptance criteria.

Therefore, confidence in the long-term performance of waste packages can only be based on the concept of 'Total Quality Assurance'. Here 'total' does not imply that every trivial detail on all waste packages will be examined and tested, but rather that the efforts to reliably produce and demonstrate the required properties of the waste begin with the qualification of the conditioning process and cover the production itself as well as the control of input and output. The measurement data from each step of the process and from each test performed on the product together with the analysis of the waste and the properties of the process additives have to be compiled and, like a mosaic, assembled to present a coherent description of the product. The properties derived in this way can be compared with the the requirements or acceptance criteria derived.

The compilation, correlation and comparison of the processing data can best be handled by an appropriate computer system, which will also take care of most documentation requirements. These procedures will all be carried out by the operator of the waste conditioning plant, who will

also certify the conformity of his products with the regulatory requirements. The producers QA procedures must be verified periodically by
independent quality audits and the waste packages themselves should also
be subjected to an inspection by those, who will take responsibility for
the safety of disposal.

This entire procedure will doubtlessly appear very cumbersome and
may, if applied in a bureaucratic manner, turn out to be as counterpro-
ductive as certain licensing procedures for nuclear power plants. We must
therefore remember, that unless proven otherwise by safety analysis, the
long-term safety of a disposal facility does not rely on the performance
of individual waste packages but rather on their average quality and the
sum of their nuclide inventories.

It should also be borne in mind, that not all criteria are of equal
importance. Quality control and quality checking should concentrate on
critical properties taking into account the waste type, the most sensitive
parameters of the process and the particular requirements of the disposal
route as identified by waste analysis. Individual acceptance criteria
should not require waste package performances to be "as high as achievable" but should specify quantitative and verifiable thresholds.

For long-lived wastes some acceptance criteria are so stringent that
available process control techniques are inadequate so that verification
of compliance with present measurement methods is imposssible.

The development of standard tests, process control technology and
non-destructive product testing methods is amongst the top priorities of
the European Communities R&D programme. It should also become a principal
field of international collaboration.

REFERENCES

1.   REGLE FONDAMENTALE DE SURETE No I.2 du 19 juin 1984, relative aux
     objectifs de sûreté et bases de conception pour les centres de
     surfaces destinés au stockage à long terme etc.
     Bulletin Officiel du Ministère de l'Industrie et de la Recherche
     No 3, 1984.

2.   RADIOACTIVE WASTE MANAGEMENT ADVISORY COMMITTEE,
     Fifth Annual Report, HMSO, June 1984.

3.   GESELLSCHAFT FUR STRAHLEN- UND UMWELTFORSCHUNG MBH (GSF) and
     KERNFORSCHUNGSZENTRUM.
     Research Project for the determination of the suitability of the
     'KONRAD' mine as a final repository for radioactive waste products.

4.   U.S NUCLEAR REGULATORY COMMISSION (NRC)
     10 CFR Part 61: Licensing Requirements for Land Disposal of Radio-
     active Waste. (Proposed Rule).
     U.S. Federal Register/Vol. 46. No 142, 1981.

5.   WESTINGHOUSE ELECTRIC CORPORATION (for WIPP)
     TRU Waste Acceptance criteria for the Waste Isolation Pilot Plant
     WIPP-DOE-069, Rev.1, 1981.

6.   U.S. NUCLEAR REGULATORY COMMISSION (NRC)
     10 CFR Part 60: Disposal of High-Level Radioactive Wastes in Geologic
     Repositories,
     U.S. Federal Register/Vol 48, No 120, 1982.

7.   UK DEPT. OF THE ENVIRONMENT
     Quality Assurance in Processing Radioactive Waste for Land Disposal,
     Radioactive Waste Management: Information Note No 2, 1984.

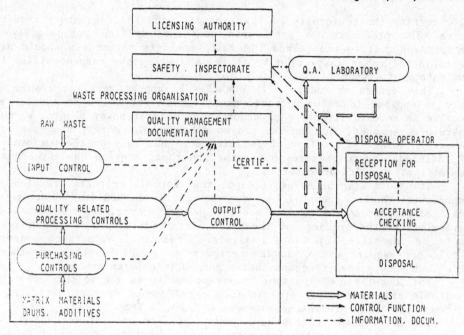

Fig. 1 - Quality assurance scheme for waste packages

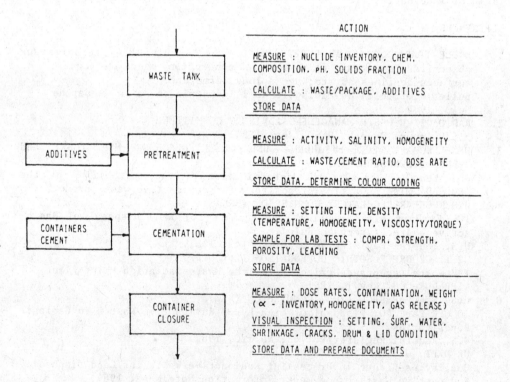

Fig. 2 - Quality - related processing control

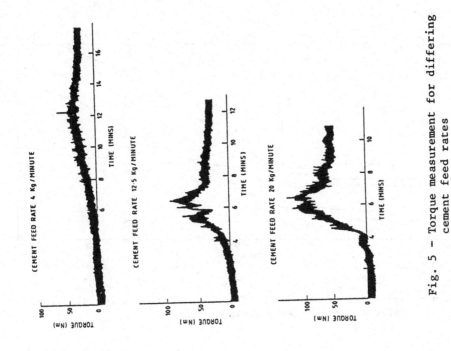

Fig. 5 - Torque measurement for differing
cement feed rates

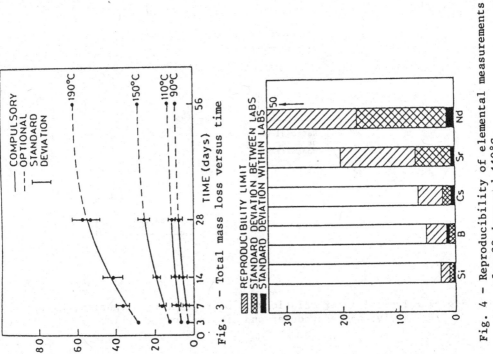

Fig. 3 - Total mass loss versus time

Fig. 4 - Reproducibility of elemental measurements
for 28 days and 110°C

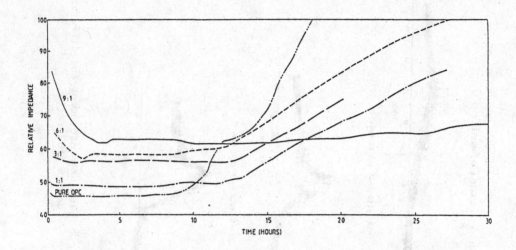

Fig. 6 - Variation of impedance with time, contoured for different
         BFS/OPC ratios   (Water/Cement ratio 0.35)

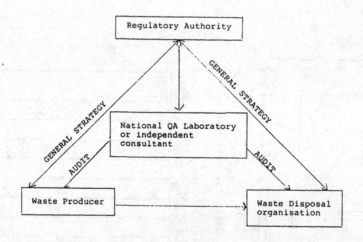

Fig. 7 - Typical inter-relationship of regulatory authority, radio-
        active waste producer and disposal organisation

DISCUSSION

A. DONATO, ENEA Casaccia

You have referred to the US regulatory limits on activity concentrations
and you have recommended, that precise limits should be set. The US
limits have been set on the basis of a risk analysis for specific sites.
How can we establish such limits in Europe, where in many countries the
sites are not yet identified and site conditions are likely to vary
considerably from one site to another?

R. SIMON, CEC Brussels

The American regulations 10 CFR 61 set out limiting criteria for (shallow)
land disposal of non-alpha bearing waste regardless of which licensed
site, Nevada, Henford, Savannah River, etc., will accept them.
For deep geologic disposal, criteria would have to be less generic, but
rather formation specific, without having to be tailored to a single site.
Existing risk analysis shows that gas generation is a much more serious
problem for an impermeable host formation like salt or soft clay than in a
hard rock like granite. In a repository confined by hard, brittle material, swelling of the waste could lead to fracturing and opening water
pathways. For hard rock caverns or concrete silos as described so nicely
in the KBS study, swelling of the waste is a critical hazard.
At least for well defined common waste categories like reactor IX resins
and concentrates it should be possible to formulate generic acceptance
criteria either for shallow land burial or for one of the main geological
formations.

R. KÖSTER, KfK Karlsruhe

You have stressed the importance of developing non-destructive testing
methods in the third EC-programme. May I draw your attention to another
significant issue, the 'pre-production performance tests. Extensive
investigations and testing of waste forms and packages prior to the actual
industrial production will normally reduce the amount of quality-related
processing controls and product testing to some rapid spot-checks. In
particular for cement waste forms such initial performance testing for
each conditioning campaign is decisive for the product quality. An
increased effort in this field will certainly be beneficial.

R. SIMON, CEC Brussels

I absolutely agree and I wish to repeat: the qualification tests are of
vital importance, as we have discussed this morning. Most of the methods
and to a large degree the results of the characterization studies carried
out under the last two EC-programmes are also applicable to the performance testing of industrial products. The establishment of the correct
formulation and the right processing parameters must be done during these
tests to determine qualification standards and set points for the production.

T. MARZULLO, ENEL Roma

I believe the methods for maintaining a reliable production quality should be documented and widely published.

R. SIMON, CEC Brussels

We will certainly do our best to publish all useful information in this field.

SESSION V

MULTI-BARRIER SYSTEMS

HLW container corrosion and design

The MIRAGE project : actinide and fission product
physico-chemical behaviour in geological environment

The MIRAGE project : large scale radionuclide transport
investigations and integral migration experiments

The geoforecasting approach and long-term prediction of
 evolutive nuclide migration

Technology for the improvement of shallow land burial

Shallow land burial analysis

HLW CONTAINER CORROSION AND DESIGN

G.P. MARSH*, G. PINARD-LEGRY[+], E. SMAILOS[#], F. CASTEELS[++]
K. VU QUANG[≠], J. CRIPPS[o] and B. HAIJTINK[**]

* AERE Harwell (UK), [+] CEA Fontenay-aux-Roses (F), [#] KFK Karlsruhe (GFR),
[++] SCK/CEN Mol (B), [≠] CNRS Vitry (F), [o] Ove Arup and Partners (UK)
and [**] CEC Directorate General for Scientific Research and Development

SUMMARY .

Two approaches have been identified for producing long lived
containers for HLW which could act as an engineered barrier in
geological repositories for radioactive waste. The first is
based on the use of corrosion resistant metals, and two candidate
alloys namely Ti-0.2% Pd and Hastelloy C4 are under evaluation.
The second approach envisages the use of carbon steel containers,
which have wall thicknesses sufficient to prevent corrosion
penetration. Present results indicate that Ti-0.2% Pd is
resistant to general and localised corrosion in all three disposal
media, but further work is needed to assess its susceptibility to
hydrogen embrittlement. Hastelloy C4 exhibits good general
corrosion behaviour but is susceptible to localised corrosion in
salt brine and granitic water under strongly oxidising conditions
of the type which could be produced by γ-radiolysis. Consequently
Hastelloy C4 could be used as a container material if measures
were taken to avoid strongly oxidising conditions and to reduce
the γ-radiation dose. The carbon steel corrosion allowance
concept appears feasible for all three rock formations, although
further work is needed to define more precisely the thickness of
metal needed as a corrosion allowance, and to evaluate the long
term risk of hydrogen embrittlement. Design studies for corrosion
resistant and corrosion allowance containers are described, and
recommendations for future studies are given.

1. INTRODUCTION

High level nuclear waste (HLW) will inevitably be encapsulated in
some form of container if only to facilitate handling during storage and
disposal. Such containers, if made of metal, have the attraction of
providing a totally impermeable barrier between the waste and the disposal
environment for as long as they remain intact. Consequently, if it was
feasible to produce containers having an endurance of a few hundred years,
these could form a useful component in the multiple system of barriers
which it is envisaged will provide radiological protection. For example
containers lasting for about 100 years would keep open the option for
retrieval during the period of repository operation, while containers
lasting for the order of 1000 years would permit the waste to pass through
its high thermal period before contacting the disposal medium, and would
also allow the more mobile fission products such as Cs and Sr to decay to
low levels. Additionally, the potential role of containers in winning

public acceptance for disposal should also be considered since their barrier function is probably more easily perceived than, for example, sorption processes. General considerations of the chemical and physical conditions likely to prevail in waste repositories has led to the view that the main threat to container integrity is corrosion induced by contact with any natural waters permeating the host rock. Accordingly several research programmes have been mounted in Europe and N. America to study the long-term behaviour of potential container materials.

Within the Research Programme of the European Communities, container corrosion studies have been undertaken in the Federal Republic of Germany, Belgium, France and the United Kingdom. Work in the FRG, carried out at KFK Karlsruhe, has concentrated on materials for disposal in salt formations, while the Belgian programme at SCK/CEN Mol has been concerned with disposal in the clay formations underlying the Mol site. Work in France and the UK has focussed on disposal in granite formations, but because neither country has identified a specific site, these studies have been more generic in nature. The participating laboratories in the latter two countries have been CNRS and CEA (France) and AERE Harwell (UK). These national programmes have been part financed by the Commission of the European Communities, and since 1980 the Commission has coordinated the projects and promoted information exchange by forming a working group consisting of representatives from the participating laboratories.

This paper describes progress in the research programmes to December 1984, and highlights areas of uncertainty requiring further study. The results from an engineering study aimed at examining the implications of materials selection on container design are also presented.

2.   SELECTION OF CANDIDATE MATERIALS

Up to end-1982 the bulk of the work done in the participating laboratories was concerned with screening a wide range of potential container materials. This work has been thoroughly reported elsewhere[1,2,3,4,5] and will not be described here, except to list the main results.

(a)   Containers possessing the required endurance can in principle be produced either by using metals of high corrosion resistance, or by using metals of lower resistance which can be used equally economically in sufficient thickness to make allowance for corrosion penetration. These alternative approaches are referred to as the Corrosion Resistant and Corrosion Allowance container concepts.

(b)   Despite wide differences in the test environments and techniques used, which reflected differences in the geological formations under consideration, the test programmes all indicated that the Ti-0.2% Pd alloy and the Ni-Cr-Mo alloy Hastelloy C4 were the leading candidates for Corrosion Resistant containers.

(c)   On the basis of cost, ease of fabrication and its corrosion characteristics carbon-steel is the most appropriate choice as a candidate Corrosion Allowance container material.

In accordance with these findings, a more closely coordinated programme was set up early in 1983 aimed at making a detailed evaluation of these candidate alloys[6,7]. Samples of the selected alloys, coming each from one melt, were provided by the Commission of the European Communities, and reference test conditions were defined facilitating a comparison of the results. The following sections describe the results obtained so far from these studies.

3.   EVALUATION OF CANDIDATE MATERIALS

Because of the wide differences in the physical and geochemical conditions pertaining in repositories mined in clay, granite and salt, different test techniques and environments are needed to evaluate materials. For this reason the results are described separately for each rock type.

3.1   Clay

The lead laboratory for this programme was SCK/CEN Mol, although CNRS have also undertaken studies to aid in characterising the protective surface layers formed on the corrosion resistant metals. Design studies for a repository in clay envisage that the waste containers will be placed in holes drilled from the main galleries, and which will need to be lined to prevent creepage of the clay. Under these conditions the most probable environment contacting the containers will be humid air contaminated with volatile organic and inorganic compounds released from the clay. Studies of the pH and chemical composition of the moisture escaping from clay under oxidising atmospheres have shown that the main contaminants are chlorides, sulphates, fluorides, ammonia, CO_2 and hydrocarbons[4]. The release of these compounds increases and the pH of the condensate falls as the temperature of the clay is increased from 25 to 300°C. On the basis of these observations a test method has been developed which involves conditioning air with moisture of a similar composition to that in equilibrium with clay at 300°C (i.e. 4.9 ppm HCl, 15 ppm H_2SO_4, 0.4 ppm HF, 8 ppm NaCl, 0.21 ppm CaF_2, 0.44 ppm KF, 0.1 ppm $MgCl_2$ and pH 3.5). This air is passed at a rate of 14.5 litres/h over specimens heated to 50, 150 and 300°C. Additionally the corrosion of the metals by direct contact with clay has also been investigated with capsule tests at 25, 100, and 150°C.

The possibility that the containers may contact an aqueous environment either due to the release of interstitial water from the clay or due to the intrusion of water from overlying aquifers cannot be dismissed. Consequently immersion tests have also been conducted in synthetic solutions representative of these natural waters (i.e. clay water 13.9 Na_2SO_4, 0.87 K_2SO_4, 19.2 $MgSO_4$ 3.6 $CaSO_4$; intrusive water 0.0398 Na_2SO_4 0.015 $MgSO_4$, 0.0101 NaF, 0.0742 Na_2CO_3 1.344 $NaHCO_3$, 0.394 KCl, 0.0199 NaCl all in gm/litre) at temperatures of 49 and 98°C.

3.1.1   Corrosion Resistant Metals

Tests in the humid clay environment have achieved exposure periods of up to 2000 days, and have shown that the rate of corrosion falls with temperature. This is because of the influence of temperature on the relative humidity of the air contacting the specimens which is the principal factor governing corrosion. At 50°C the average general corrosion rate over a 35,000 test period was of the order of 0.18 µm/yr for both Ti-0.2% Pd and Hastelloy C-4, and at 300°C the corrosion rates of both metals were about 0.01 µm/yr. Surface analysis has indicated the species controlling the corrosion under these conditions are mainly sulphur compounds and to a lesser extent Cl^- ions.

The capsule tests with specimens in direct contact with clay have completed exposure periods of up to 330 days. At the maximum test temperature of 150°C the general corrosion rates of both Ti-0.2% Pd and Hastelloy C4 have been < 1 μm/yr.

The experiments in synthetic groundwater and interstitial clay water have achieved exposure times of between 170 and 1400 days. In both environments, and at 40 and 98°C, Ti-0.2% Pd and Hastelloy C4 have exhibited high resistance to attack with general corrosion rates not exceeding 0.5 μm/yr.

In addition to plain coupon specimens the experiments in the humid clay atmosphere at 50°C, and in the interstitial clay water and groundwater at 98°C, contained self-loading stress corrosion test specimens of several corrosion resistant metals. Metallurgical inspection of these specimens after test periods of up to 1450 days has not detected any indication of cracking in commercial purity Ti or Hastelloy C4. However, Type 316 stainless steel was subject to cracking in the synthetic groundwater while sensitised Type 304 steel cracked in the humid clay atmosphere.

The CEN programme has also included some 'in-situ' tests in a clay formation at Terhagen, Boom. Gravimetric analysis of specimens exposed for up to 3 years at 13 and 50°C have indicated general corrosion rates for Ti-0.2% Pd and Hastelloy C4 of ~ 1 μm/yr, confirming the high corrosion resistance of these alloys.

3.1.2 Carbon Steel

Relevant work to assess the carbon steel corrosion allowance concept has so far been limited to tests at ambient temperature with direct contact with clay for test periods up to 330 days. Under these conditions the steel was subject to general corrosion at a rate of 50 μm/yr, but this should be regarded as a pessimistic result because the clay was oxygenated when it was packed into the capsules.

3.2 Salt

The lead laboratory for this programme was KFK Karlsruhe, but CNRS undertook supporting electrochemical and surface analytical studies of the corrosion resistant candidate materials. A repository mined in salt should be practically dry since the water content of salteriferous formations is typically of the order of only 0.05 ^{W}/o. However, under certain postulated accident conditions water intrusion may occur resulting in the formation of brine solutions. It is this condition that the German work has concentrated on with tests in a quinary brine consisting of 26.8% $MgCl_2$, 4.8% KCl, 1.4% $MgSO_4$, 1.4% NaCl and 65.7% H_2O. The tests have been conducted at temperatures of 90, 170 and 200°C reflecting the high disposal temperatures anticipated in a salt repository.

3.2.1 Corrosion Resistant Metals

Tests of up to 560 days have shown that the rate of general dissolution of Hastelloy C4 increases with temperature, although even at 200°C the rate is comparatively low at 0.3 μm/yr. It is also noteworthy that the rate of attack remains fairly constant with time. The metal proved resistant to pitting and stress corrosion cracking, but did exhibit crevice attack to depths between 20 and 100 μm in tests exceeding 360 days. In tests at 90°C with a γ-radiation dose rate of ~ 10^5 Rad/h, Hastelloy C4 was subject to quite severe pitting and crevice corrosion at pitting rates

up to 0.45 mm/yr (i.e. averaged over 606 days)[8]. These latter
observations are in line with the results of the electrochemical studies of
CNRS which indicated that Hastelloy C4 was susceptible to pitting in brine
at 90°C under oxidising conditions[8].

The corrosion behaviour of the Ti-0.2% Pd alloy was characterised by
an initially low rate of general attack which declined even further with
time to an insignificant level (i.e. < 0.2 μm/yr after 550 days). The rate
of attack was not influenced significantly by test temperatures up to
200°C, and the alloy also proved resistant to pitting, crevice corrosion
and stress corrosion cracking. The alloy also proved resistant to
localised corrosion in tests at 90°C under γ-irradiation, although the
average general corrosion rate was increased slightly to 0.7 μm/yr.
Electrochemical and surface analysis has shown that the high corrosion
resistance of the alloy is attributable to the formation of a protective
TiO_2 surface film. A similar film was also found on the specimens exposed
under irradiation, although in this case a second over-layer consisting of
Mg and O was also detected, and the overall film thickness of 600 nm was a
factor of 10 greater than that formed on the unirradiated specimens[9].
Electron diffraction studies of the film formed on Ti-0.2% Pd in brine and
$MgCl_2$ solution has shown that this has an amorphous structure at 20 and
90°C and a crystalline structure at 170°C[8].

3.2.2 Carbon Steel

Tests with a fine grained 0.17% C structural steel showed that this
was susceptible to general dissolution with rates of the order of 35, 100
and 600 μm/yr at 90, 170 and 200°C. The rate of attack at 90°C was
increased further under irradiation to 465 μm/yr. The rate of corrosion
without γ-radiation is determined by the acidity of the brine which
increases with temperature (e.g. at 90°C pH-3.6). The influence of oxygen
on corrosion over the long test period (i.e. > 70 days) must be small
because the low oxygen content of the brine (i.e. < 0.1 ppm at 90°C) will
be consumed by corrosion in the first few days of the test. Finally, since
carbon does not passivate in brine it has been concluded that it is
extremely unlikely that corrosion allowance containers will be subject to
stress corrosion cracking or deep pitting/crevice attack.

3.3 Granite

The two lead laboratories in this case have been CEA Fontenay-aux-
Roses, which has concentrated on Corrosion Resistant metals, and AERE
Harwell which has worked mainly on the carbon steel Corrosion Allowance
concept. In considering corrosion in granitic repositories it has been
assumed that the repository will be below the water table and will
therefore flood with groundwater transported through fissures in the rock.
To place the work on a good comparative basis it was agreed that both
laboratories would test in the same synthetic groundwater the composition
of which was arbitrarily set at 35 ppm Cl^-, 24 ppm $SO_4^=$, 244 ppm HCO_3^-, 1.9
ppm F^-, 19 ppm $SiO_2^=$, 106 ppm Na^+, 40 ppm Ca^{2+} and 12 ppm Mg^{2+}. It was also
agreed, however, that this composition could be modified if necessary to
evaluate certain corrosion processes. For example additional NaCl and HCl
were added for studies of crevice corrosion and stress corrosion cracking.

3.3.1 Corrosion Resistant Metals

Although the CEA study concentrated mainly on the two reference alloys tests were also undertaken with commercial purity titanium, Zircaloy 4, Types 304L and 316L stainless steel, the high Ni-Cr alloys Inconel 625 and Incoloy 825 and Hastelloy C276. France has been particularly interested in the possibility of using a single vessel for glass casting and disposal, therefore the work with Hastelloy C4 and C276 included tests with specimens given aging treatments to simulate the thermal treatments experienced during casting and welding[3]. The experimental programme has consisted of long term (3000h) tests to evaluate general corrosion, and shorter term mechanical and electrochemical tests to make a comparative assessment of the alloys susceptibility to pitting, crevice corrosion and stress corrosion cracking. Results from these tests, which are conducted at 80 and 170°C are given in Table I.

Main points to note are the excellent general corrosion resistance exhibited by Ti-0.2% Pd and commercial purity titanium, although the possibility that these metals may be subject to hydrogen embrittlement has been raised. Pitting corrosion was evaluated in solutions which were enriched in chloride (i.e. 17,700 ppm Cl^- at 80°C and 1900 ppm Cl^- at 170°C) to take account of possible concentration processes. Even so the tests confirmed the good resistance of the Ti alloys at 80 and 170°C. Hastelloy C4 and C276 were both resistant to pitting at 80°C but were susceptible at 170°C. The Hastelloys also proved susceptible to stress corrosion both at 80 and 170°C, in a solution acidified to pH 1.3 and containing 17,700 ppm Cl^-.

Table I
Summary of Results from Tests on Corosion Resistant
Metals in Granite Solutions

Material	80°C				170°C		
	Gen Corr µm/yr	Pitting	Crevice Corr	SCC	Pitting	Crevice	SCC
Type 304L		VS	VS	S			
Type 316L		VS	VS	C			
Alloy 825		VS	S	R			
Alloy 625	0.4	R	S	S	S	VS	
Zircaloy 4	0.03	LS					
Commercial Purity Ti	0.15	R	R	P	R	R	
Hast C276	0.15	R	R	S	S	VS	S
Hast C276 (aged)			S	S	S	VS	S
Hast C4	0.15	R	R	S	S	VS	S
Hast C4 (aged)			S	S	S	VS	S
Ti-0.2% Pd	0.15	R	R	P	R	R	

S = Susceptible, VS = Very Susceptible, LS = Low Susceptibility, R = Resistant, P = Possible

The Harwell contribution to the work on corrosion resistant metals has been restricted to electrochemical tests to assess the effect of γ-radiation (10^5 Rad/h) on the crevice and pitting corrosion of Hastelloy C4 and Ti-0.2% Pd. These tests, which were conducted in the standard synthetic groundwater and in the same water strengthened with 1900 and 17,700 ppm Cl$^-$ all at 90°C have confirmed the susceptibility of Hastelloy C4 to crevice corrosion, but significantly only at highly oxidising electrode potentials which are unlikely to pertain in practice. Furthermore, it is likely that radiation will enhance the probability of crevice breakdown by causing an increase in the metals rest potential. Ti-0.2% Pd was resistant to localised corrosion both with and without radiation[10].

3.3.2 Carbon Steel

The aim of the work at Harwell has been to estimate the metal thickness required to ensure that carbon steel containers are not penetrated by corrosion over the first 1000 years after disposal[3]. Coupon tests lasting up to 3000h in the synthetic granitic groundwater composition given in the previous section have shown that in the absence of oxygen the rate of general dissolution is < 0.1 μm/yr. Similar tests, but with a γ-radiation dose of 10^5 Rad/h, gave a slightly accelerated corrosion rate of ~ 1 μm/yr. Other tests have been set up which seek more closely to reproduce conditions in a repository by burying coupons in 10 cm beds of either crushed granite or bentonite paste, which are in turn covered by aerated groundwater. These tests are still in progress, but polarisation resistance monitoring indicates that after test periods of the order of 500-600 days the corrosion rates have settled at roughly 30, 13 and 5 μm/yr at 90, 50 and 25°C respectively. The tests have included electron-beam-weld specimens and there is no evidence to suggest these are corroding at a different rate to the parent metal[10,11].

In addition to general corrosion electrochemical studies have shown that carbon steels may also be subject to localised corrosion in certain groundwaters[2]. Accordingly tests have been undertaken to study the rate of pit propagation as a function of time for periods of to 400 days. The maximum pit depths found in these tests have been adjusted using Extreme Value Statistics to take account of the small area of the specimens compared to waste containers. These adjusted values fit an empirical expression of the form[11]

$$P = 8.35 \ t^{0.46}$$

which relates the maximum depth of pitting P (in mm) to time T in years. The expression indicates a maximum pit penetration of 200 mm over 1000 years.

Work has also been undertaken to consider what precautions are needed to ensure the containers are not subject to stress corrosion cracking. The approach adopted here has been to accept that carbon steel could conceivably be subject to cracking in the repository environment, and therefore to estimate the degree of stress relief needed to reduce fabrication stresses etc. below the minimum needed to sustain cracking. Preliminary results from stress corrosion tests in 0.5M Na_2CO_3 + 1M $NaHCO_3$ at 90°C have shown that with normal 0.12% carbon steel stress relief to just below the yield stress should be sufficient to prevent cracking. However, with extra low carbon steels (e.g. ~ 0.04C) a much greater degree of stress relief will be needed[2].

One underlying principle of the UK programme is that extrapolation of experimental results to periods of up to 1000 years can only be justified if it is supported by a good mechanistic understanding of the processes involved. A good method for developing such a theoretical understanding and testing its validity is to formulate mathematical models for the processes and to compare the model's predictions with experimental results. If the model proves correct, it can be used to predict behaviour over longer periods. Such a model has been developed on a one dimensional basis for the general corrosion of carbon steel containers which considers the kinetics of surface reactions and the diffusion of reactants and products through the backfill surrounding the container[12]. The model has shown that when transport occurs only by diffusion the corrosion produced by reaction with oxygen is low (< 4 μm/yr at 90°C) and that most corrosion should occur by direct reaction with water. At 90°C the overall rate of attack is estimated as 59 μm/yr which is very high compared to the experimental dissolution rates measured in deaerated groundwater (i.e. < 0.1 μm/yr). This poor correlation is thought to arise because the model does not take account of stifling of the corrosion reaction by the accumulation of corrosion products and is being modified accordingly.

A second mathematical model is now under development to investigate the kinetics of pit propagation as a function of time and pit depth.

CEA Fontenay has recently begun work on carbon steels, and have indicated that they believe the possibility of hydrogen embrittlment, either in the form of slow crack growth or 'blister' cracking, requires evaluation.

3.4 Generic Studies of Corrosion Resistant Metals

As well as the applied studies described so far, work has also been conducted by CNRS Vitry to gain a more general understanding of the effects of environment and alloy composition on the pitting behaviour of corrosion resistant metals. Tests with Hastelloy C4 and also Type 304 and 316 stainless steels, Incoloy 800 and Inconel 625 have shown that the level of Cl^- needed to cause susceptibility to pitting corrosion is strongly influenced by the presence of other ions (e.g. Figure 2). More specifically it was found that $CO_3^=$, HCO_3^- and $SO_4^=$ had an inhibitive effect such that the Cl^- level required to produce susceptibility increased in the presence of these ions according to relationships of the form[8,12]

$$\text{Log } [Cl^-] = A + B \text{ log [inhibitor]}$$

With Hastelloy C4 in carbonate solution at 170°C the constants A and B were 0.97 and 0.56 respectively. This observation is particularly relevant to granite waters which are usually high in $CO_3^=$ or HCO_3^-. It is also noteworthy, however, that in line with the results in section 3.3.1, Hastelloy C4 only pitted at strongly oxidising electrode potentials.

Analysis of the results from all the metals listed above has highlighted the strong beneficial effect of Mo and Cr alloy additions on resistance to pit breakdown. Linear regression analysis has shown that the effect of the main alloy additions on the pit initiation potential E_{np} can be expressed by the empirical expression

$$E_{np} = A + B[Mo\%] + C[Cr\%] - D[Fe\%] - E[Ni\%]$$

The constants are specific to the test environment, and in deaerated salt brine at 90°C were -165, 20, 15, 7 and 4 respectively.

Electrochemical studies have also been made of the rate of corrosion of the metals in acidic solutions characteristic of those present in active pits and crevices. This has shown that the rate of dissolution is enhanced by chromium but reduced with increasing concentrations of Mo, Ni and Cu. The implication of this observation is that the rate of pit and crevice corrosion penetration should be slowest with Hastelloy C4 which contains more Mo and Ni and less Cr than the other alloys tested.

Studies of Ti-0.2% Pd alloy have shown that it is essentially immune to pitting corrosion in granitic and clay waters and brine. The alloy does exhibit a 'transpassive' type of general breakdown in passivity, but even at 90°C the potential associated with this exceeds +1000 mV (SCE) and therefore is not likely to occur in practice. It is also significant that this breakdown potential increases with increasing chloride concentration[8].

4.   CONTAINER DESIGN ASPECTS

While it is clearly important to identify container materials possessing the required corrosion resistance, it is also important to consider the implications of such materials selections on container design. This aspect of the programme has been conducted by Ove Arup and Partners (UK) who have considered the overall design of waste packages for salt, clay and granitic repositories using both corrosion resistant and corrosion allowance outer containers[14]. In making this study it has been assumed that fabrication and welding techniques which are currently technically feasible (e.g. electron beam welding of thick C-steel sections) will have reached the engineering applications stage when repositories become operational. Furthermore, when considering mechanical requirements the following sources and levels of external pressure have been considered based on reference repository designs.

Table II
Sources and magnitudes of External Pressure

	Clay	Granite	Salt
Ambient Lithostatic Pressure	2.5	9.0	18.0
Peak Thermal Pressure	0	0	10.0
Peak Hydrostatic Pressure	2.5	10.0	0
Total Design Pressure (MPa)	5.0	19.0	28.0

The study has developed four overall waste package designs as shown in figure 3, all of which are based round the COGEMA vitrified waste canister. Type I is made from corrosion resistant metal and is filled with lead or cement grout to withstand the disposal pressures given above. The 6 mm thickness of the outer container is much greater than that needed to prevent corrosion penetration, and is specified for robustness and to allow for fabrication tolerances. Type II is similar to Type I except that a corrosion allowance, arbitrarily set at 75 mm of carbon steel has been used. Types III and IV are designed to be 'self-supporting' against external pressure thus avoiding the need for void filling if so desired. Type III uses a 3 mm layer of corrosion resistant metal supported by a steel shell which will have a maximum thickness of 70 mm in the case of salt. Type IV combines the thicknesses of carbon steel required for corrosion allowance and pressure resistance as a single vessel.

Stress analyses of the above designs have shown that thermally induced stresses could be particularly high if lead is used as a filler. It was also concluded that although the overall levels of stresses were of an acceptable level further studies were required of local areas of potential tensile stress to consider the effects of residual fabrication stresses.

A full costing of these designs was not feasible because it involves considerations of the capital cost of plant for filling and sealing the outer containers as well as materials and fabrication costs. However, it is signifciant that preliminary estimates indicate that there is unlikely to be any significant cost benefit to be gained from using the cheaper steel material.

5   DISCUSSION

The current state of knowledge and the need for further studies of Corrosion Resistant and Corrosion Allowance metals will be discussed separately.

5.1 Corrosion Resistant Metals

One notable feature of the results obtained so far is the high resistance to general and localised corrosion exhibited by the Ti-0.2% Pd alloy in all three environments under consideration. The question remaining to be answered for this alloy are:
(a) Will it be subject to hydrogen embrittlement?
(b) Is there any possibility of passive film breakdown over long
     timescales?
With regard to (a) laboratory tests, in some cases lasting up to 600 days, have not detected any susceptibility. It is possible, however, that hydrogen arising from corrosion and also γ-radiolysis could accumulate in the metal over much longer time scales and ultimately result in embrittlement. It is the view of the authors that this potential process can only be evaluated by the development of a mathematic model to predict the build-up of hydrogen with time in Ti-0.2% Pd, and to relate this to the level of hydrogen needed to cause embrittlement.
With regard to (b) the work at CNRS has shown that if the metal is polarised to extremely oxidising potentials general breakdown of the passive film can occur. This process, which has also been observed with zirconium[15], is generally referred to as 'breakaway corrosion' and is associated with the thickening of the passive film to the point that cracking occurs. The potentiality for this to occur under more realistic oxidising conditions in waste repositories should be addressed by work to measure the rate of film thickening with time. Already work at CNRS and KFK has shown that the film growth is influenced by ions such as fluoride and by irradiation, and this work needs to be extended into a more systematic study.
Turning to Hastelloy C4, this metal has also exhibited excellent resistance to general corrosion, but there is some doubt over its localised corrosion behaviour, particularly in salt brine in which severe pitting occurred under irradiation. Studies at CEA and Harwell have also shown

that the alloy may not be totally resistant to pitting corrosion in granitic environments, although it should be conceded that the tests were conducted under pessimistic conditions, and even then pitting only occurred at potentials which were considerably more oxidising than those likely to prevail in a repository. The alloy has also been shown to be susceptible to stress corrosion cracking, but only in the sort of acid solution that could develop in a pit or crevice. Overall therefore there must be some doubt about the suitability of Hastelloy C4 as a container metal. At this stage, however, it would be wrong to reject the alloy, particularly in view of the present uncertainty over the possible hydrogen embrittlement of Ti-0.2% Pd. It is therefore proposed that future work on Hastelloy should concentrate on evaluating what corrosion engineering measures could be taken to improve its performance. For example since the alloys main problem is localised corrosion induced by high oxidising potentials, it may be possible to eliminate the possibility of such potentials by including getters such as carbon/steel shot in the backfill. Alternatively the high γ-radiation dose rate which could, through water radiolysis, produce high oxidising potentials, could be reduced by adopting a container design such as Type III which would introduce a thick carbon steel wall and hence reduce the radiation dose rate.

5.2 Carbon Steel Corrosion Allowance

The use of carbon steel as a corrosion allowance containment material depends on there being a limited rate supply of reactant to the container. The corrosion studies with carbon steel in brine used a specimen area/corrodent volume ratio of 0.2 cm^2/ml to investigate the case of excess corrodent, and therefore it is not too surprising that high ratios of attack were measured. Even so the rates, at least under inactive conditions, were not so great that they would preclude the use of carbon steel containers in brine. Further work is currently in progress with more limited volumes of brine to study the corrosion due to brine migration to the containers. In this case it would be anticipated that, due to the consumption of acidity in the brine by reaction with the steel, the long term corrosion rate would be much reduced. the other significant result to come from the work in brine was the great acceleration in the rate of corrosion produced by γ-radiation. It should be stressed, however, that this too need not preclude the use of carbon steel containers since the radiation dose rates used were typical of those at the surface of the glass canisters. The high rock pressures operating in salt repositories will require the use of thick-walled containers and this means that the radiation dose rate may well be reduced to a level which has a negligible effect on corrosion.

In the clay and granitic environments tests have shown that up to typical maximum repositories temperatures of 80-90°C rapid corrosion will only occur in aerated solutions. Both experimentation and mathematical modelling have shown that if transport occurs only by diffusion then corrosion due to oxygen should not exceed 5 μm/yr. Additional corrosion will occur due to direct reaction with water, and here the mathematical model predicts a maximum rate of ~ 60 μm/yr, while experiments give a much lower rate of < 1 μm/yr. It is thought that this discrepancy arises because the model does not consider the partial corrosion protection given by accumulations of corrosion products and it is being modified to do this. Overall, however, it would appear that quite modest thicknesses of steel will be needed to make allowance for general corrosion over 1000 years.

The work has also highlighted the possibility that carbon steel may passivate in certain groundwater repositories, thus raising the risk that the containers could be subject to localised corrosion which could penetrate at a much faster rate than the general corrosion discussed above. Experimental work to investigate the rate of pit penetration as a function of time has indicated that maximum pit depths of the order of 200 mm could be produced over 1000 years. Whilst the production and seal welding of carbon steel containers of this thickness is quite feasible, it is not clear at this stage what reliability should be placed on the above estimate since it is based on relatively short term work with shallow (~ 3 mm) pits. Consequently work has been initiated to develop a mathematical model to investigate the growth kinetics of deep pits, and it is proposed to continue this in the future along with experimental studies on deep artificial pits.

Another form of localised corrosion which needs to be considered with carbon steel containers in clay and granite is stress corrosion cracking. The approach currently being pursued is to accept that such a risk exists, and to seek an engineering solution by stress relieving the containers so that they contain insufficient fabrication stress to support the cracking process. However, the design study has shown that the containers may also be subject to thermal stresses, and therefore it is important to examine whether these can be kept to acceptable levels through suitable container design.

Finally the possibility of hydrogen embrittlement either in the form of slow crack growth or 'blister' cracking has also been raised, and this too needs further consideration in the future, both with regard to the amount of hydrogen likely to be introduced into the steel and the possibility of selecting a steel with maximum resistance.

6. SUMMARY

(1) Two concepts have been identified for producing waste containers which could act as an engineered barrier in the geological disposal of nuclear waste. These are based on the use of either Corrosion Resistant metals, or by the provision of a Corrosion Allowance by using thick walled carbon steel containers.

(2) The alloys with greatest potential for corrosion resistant containers are Ti-0.2% Pd and the Ni-Cr-Mo alloy Hastelloy C4.

(3) The Ti-0.2% Pd alloy has exhibited a high resistance to general and localised corrosion in environments representative of all three rock formations (i.e. salt, clay, granite) in tests lasting up to 600 days. It is recommended that future work should concentrate on assessing degradation processes which could occur over prolonged periods, and which therefore are not easily investigated by direct experimentation. Two such processes have been identified, namely hydrogen embrittlement and oxide film breakaway, and it is suggested that these could be evaluated by appropriate mathematical models.

(4) Hastelloy C4 has shown good general corrosion behaviour in all three geologic environments, but has proved susceptible to pitting and crevice corrosion in salt brine, particularly with irradiation. Electrochemical tests have also shown that the alloy may be subject to pitting and crevice corrosion in granitic environments, but only at potentials which are more oxidising than those likely to prevail in repositories. It is suggested that the feasibility of using corrosion engineering techniques to improve the performance of Hastelloy should

be assessed. For example it may be possible to ensure that the environment can never be sufficiently oxidising to induce localised corrosion by placing carbon steel shot in the backfill to act as an oxygen getter. Also the potentially powerful oxidising effect of γ-radiation could be much reduced by adopting a container design which provides shielding.

(5) The study has continued to support the feasibility of using carbon steel corrosion allowance containers in all three rock types. It is recommended that future work should aim to develop and refine the models on general and localised corrosion of carbon steel so that they can be used to support the experimental predictions of the thickness of metal needed to prevent penetration over long periods.

(6) Work on carbon steel should also examine whether the risk of stress corrosion cracking can be avoided by designing the containers and using stress relief heat treatments to ensure that the fabrication and thermal stresses do not reach the levels needed to support cracking. The likelihood of hydrogen embrittlement which could be manifested in the form of slow crack growth or 'blister' cracking also requires further evaluation through experimental and modelling studies.

(7) Designs have been prepared for both corrosion resistant and corrosion allowance containers.

REFERENCES

1. SMAILOS, E., KOSTER, R., and SCHWARZKOFP, W., (1983), Corrosion Studies on Packaging Materials for High Level Wastes, EUR 8657 EN, European Appl. Res. Rept. Vol. 5, No. 2, 175-222.

2. MARSH, G.P., TAYLOR, K.J. and BLAND, I.D., (1983), Corrosion Assessment of Metal Overpacks for Radioactive Waste Disposal, EUR 8658 EN, European Appl. Res. Rept., Vol. 5, No. 2,

3. PLANTE, G. et al., (1984), Etude de corrosion des matériaux de conteneurs pour le stockage des déchets radioactifs en site granitiques, Commission des Communautés Européennes, Sciences et Techniques Nucléaires, EUR 8762.

4. HEREMANS, R. et a., (1984), R&D programme on radioactive waste disposal into a clay formation, Ch. VII: Corrosion of structural and canister materials in clay environments, Commission of the European Communities, Nuclear Science and Technology, EUR 9077.

5. VU QUANG, K. et al., (1984), Etude de films superficiels et du comportement électrochimique de certains alliages à base de Ni et à base de Ti, Commission des Communautés Européennes, Science et Techniques Nucleaires, EUR 9136.

6. ACCARY, A., Corrosion behaviour of container materials for geological disposal of high level radioactive waste. Commission of the European Communities. Nuclear Science and Technology (to be published).

7. HAIJTINK, B. and ACCARY, A., (1984), The Joint European Testing Programme on HLW Container Materials, Proceedings Waste Management 84, March 11-15, Tucson (USA).

8.   Corrosion Behaviour of Container Materials for Geologic Disposal of
     High Level Waste, Annual progress report 1983, Commission of the
     European Communities. Nuclear Science and Technology, EUR 9570
     (1985).

9.   PFENNIG, G., MOERS, H., KLEW-NEBENIUS, H., KIRCH, G. and ACHE, H.J.,
     (1984), Surface Analytical Investigation of Corroded Ti-Pd Proposed as
     Container Material for the Disposal of High Level Wastes, Poster
     Presented at International Conference on Nuclear and Radiochemistry,
     Lindau, FRG, 8-12 October 1984.

10. TAYLOR, K.J., BLAND, I.D., SMITH, S. and MARSH, G.P., Corrosion
     Studies on Containment Materials for Vitrified Heat Generating Nuclear
     Waste, Annual Report for 1984, AERE-G3422 (to be published) (1985).

11. TAYLOR, K.J., BLAND. I.D., NAISH, C.C. and MARSH, G.P., (1985), AERE
     Report-G3217, Corrosion Studies on Containment Materials for Vitrified
     Heat Generating Nuclear Waste, Progress Report to June 1984
     (unpublished).

12. MARSH, G.P., TAYLOR, K.J. and HARKER, A.H., Corrosion of Carbon Steel
     HLW Containers after Geological Disposal, 166th Meeting of the
     Electrochemical Societh, Extended Abstracts, Vol 84-2, pp 403-404,
     1984.

13. VU QUANG, K. and JALLERAT, N., (1984), Proceedings of 9th Inter. Cong.
     on Met. Corr., Toronto, June 3-7.

14. CRIPPS, J. (1984), Geological Disposal of High Level Radioactive
     Waste-Container Design, Commission of the European Communities,
     Nuclear Science and Technology (to be published).

18. WANKLYN, J.N., (1965), Corrosion of Zirconium Alloys, ASTM Special
     Technical Publication 368.

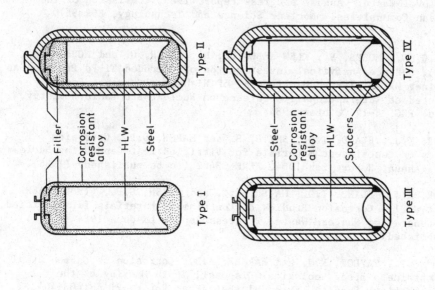

Figure. 3. Schematic diagrams for four alternative waste container designs.

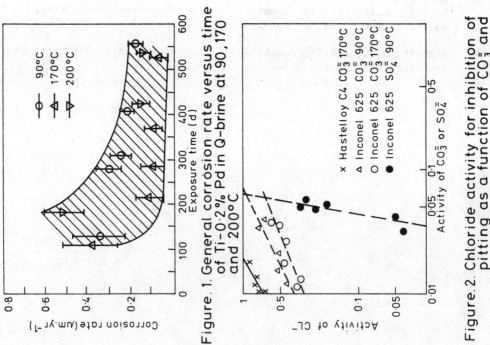

Figure. 1. General corrosion rate versus time of Ti-0·2% Pd in Q-brine at 90,170 and 200°C

Figure. 2. Chloride activity for inhibition of pitting as a function of $CO_3^=$ and $SO_4^=$ activity

DISCUSSION

M.S.T. PRICE, AEE Winfrith

How is the position on corrosion affected by thermal effects induced during welding and also by the presence and properties of weld metal? How does it differ for carbon steel compared with the corrosion resistant alloys?

G.P. MARSH, UKAEA Harwell

No significant differences have been detected in the corrosion behaviour of parent and weld metal samples of either Ti-0,2% Pd or Hastelloy C4. In general carbon steel welds also exhibited corrosion behaviour similar to the parent metal with the exception of Q-brine at 200°C in which there was some preferential attack on the weld. However the main concern with carbon steel is that the weld heat-affected-zones may possess enhanced suscepti- bility to stress corrosion cracking, and this may necessitate post weld heat treatment.

J. HADERMANN, NAGRA Baden

For the validation of predictive models on long term corrosion behaviour, have possibilities on the basis of archeological findings or even natural analogues been investigated?

G.P. MARSH, UKAEA Harwell

This is only possible to a limited extent. The difficulty is that we are dealing with heat emitting waste, therefore the temperature is rather different and the presence of radiation is something rather unique. There might be some advantages in applying this approach, however the problem is that archeologists never can tell us very precisely what the environmental conditions were which the artifacts experienced during their time under the ground, under the sea or wherever they may have originated from.

A. BARTHOUX, ANDRA Paris

You said that detailed costing assessments of containers have not yet been made. Nevertheless I think that cost is an important factor for the orientation of further research. If the cost is low you could apply a large safety factor and discontinue further research, however in the opposite case you have to optimize.

G.P. MARSH, UKAEA Harwell

Costing of containers is very complicated as it is not only a matter of costing of the materials used for the fabrication of the containers, but also of the filling and remote welding techniques used. This exercise was beyond the scope of the design study performed, however I agree with the sense of your comment.

THE MIRAGE PROJECT : ACTINIDE AND FISSION PRODUCT PHYSICO-CHEMICAL BEHAVIOUR IN GEOLOGICAL ENVIRONMENT

A. AVOGADRO
Commission of the European Communities - Joint Research Centre - Ispra
A. BILLON
Commissariat à l'Energie Atomique - Fontenay-aux-Roses
A. CREMERS
Katholieke Universiteit Leuven
P. HENRION
CEN/SCK - Mol
J. I.KIM
Technische Universität München
B. SKYTTE JENSEN
Risø National Laboratory Roskilde
P. VENET
Commission of the European Communities - Brussels
P.J. HOOKER
British Geological Survey, Keyworth

Abstract

This review deals with experimental results obtained by various laboratories working under contract within the CEC actions in Radioactive Waste Management Programme. The objective of this research is to generate the basic knowledge on the geochemical behaviour of actinides and fission products in aqueous systems relevant to real ground waters. The subject comprises, therefore, thermodynamic and kinetic data evaluation, measurement of solubility in real ground waters, speciation of dissolved species, generation of colloids and study of primary sorption mechanisms of radionuclides on geomatrices. The study should ultimately provide an insight into the migration mechanisms and then input data for realistic migration models needed for safety assessment over very long time periods.

1.1 Introduction

This paper summarizes together experimental results obtained in the research area of the Community co-ordinated project MIRAGE (MIgration of RAdionuclides in the GEosphere) dealing with laboratory studies on physico-chemical behaviour of long-lived radionuclides in geological media.

The project itself has been described in previous Community reports (1,2) which can be referred to for further information. MIRAGE concentrates on the study of the transfer of radioactivity from conditioned waste through the different barriers up to the interface between the geosphere and the biosphere.

Assuming that a reasonable background knowledge exists of flow parameters, rock properties and the geological chemical environment, the principal factors required to describe quantitatively the migration potential are : 1) the release source-term, and 2) the retardation processess during the transport. For both phenomena, thermodynamic and kinetic data are needed for understanding and predicting the environmental behaviour of the released nuclides. This research area considers basic chemical studies both in homogeneous (solution) and heterogeneous systems (colloids and mineral adsorbing materials).

The participating laboratories in these studies are :

CEA, Département R&D déchets, CEN de Fontenay-aux-Roses   (F)
CEC, Joint Research Centre, Ispra Establishment   (I)
Risø National Laboratory, Chemistry Department   (DK)
Technische Universität, München, Institut für Radiochemie   (FRG)
CEN/SCK-Mol and Katholieke Universiteit Leuven   (B)
British Geological Survey, Keyworth, Nottingham   (U.K.)

The objective of the research is to generate the basic knowledge on the geochemical behaviour of selected actinides (Np, Pu, Am) and fission products (Tc, Sr), in aquatic systems relevant to real groundwaters. The subject comprises, therefore,

- determination of thermodynamic and kinetic data
- measurements of solubility in real groundwaters
- speciation of dissolved species
- studies on generation and characterization of colloids
- study of primary sorption mechanisms on geomatrices

The results from these studies should ultimately provide an insight into the migration mechanisms as well as the knowledge that facilitates the prediction and modelling of the transport in a given groundwater system.

Important research successfully carried-out on this first phase (1983-84) of the Mirage project includes the complexation, speciation and colloid studies; solubilities in various real aquatic solutions have been measured while looking into dissolution mechanisms. As for the sorption phenomena, ion exchange in a multionic system has been studied for different exchangers; retention of nuclides on halite has also been investigated.

This presentation focuses on progress in common research subjects rather than on activities of each laboratory separately.

2. Work review

2.1 Solubility Studies

The solubility data for actinide oxides, hydroxides and carbonates, in geochemical fluids in the Gorleben site are primary quantities for the assessment of source terms for migration modelling. For this reason, a special effort is being made in this field by the Technical University of München, taking account of important geochemical and physical properties, e.g. Eh, pH, temperature, radiation dose, pressure, and also colloid generation and complexation (3,4).

Experience has shown that the solubility of actinides in real natural fluids cannot be estimated theoretically with a reasonable accuracy. The reasons are the following :

- geochemical characterization of deep groundwaters, being in equilibrium with surrounding geomatrices, is hampered by technical difficulties, e.g. in situ measurements are possible only for a few parameters, pH, Eh etc. and the transport of material to the laboratory disturbs the original equilibrium states.

- thermodynamic quantities, such as solubility products, stability constants of complexes etc. are available for some actinides but are still insufficient or often inaccurate for describing their important geochemical reactions in a given aquifer system

- colloid generation influences the actinide solubilities drastically through formation of either real-colloids and pseudo-colloids. All groundwaters contain natural colloids which interact with soluble actinides and generate pseudo-colloids. Quantification of colloid generation is still difficult to realize and hence its influence on the actinide solubility cannot be assessed easily

- in the near field, the radiation effect on the actinide solubility is considerable, since this provokes many different chemical processes. The radiolysis of fluids, especially saline or carbonate containing solutions, changes the redox potential (Eh) and pH of the system and thus affects the process of dissolution of actinide compounds. The autoradiolytic effect, particularly α-radiation, also produces microcolloids, which remain stable in solution and can be mobile in many aquifer systems.

A typical example for the solubilities of Pu oxide in different solutions as a function of pH is given in figure 1. This figure compares a number of large experimental data sets. From these experiments, it is observed that the solubility is not only controlled by complexations, hydrolysis and redox reactions but also by colloid generation (real-colloids and pseudo-colloids). The radiation effect appears also an important factor for the dissolution process of actinides in all solutions under investigation, by changing the Eh value of solution. The self-discharging α-radiation may generate micro-colloids (< 10 A) which are inseparable by ultrafiltration. Whereas the complexation and hydrolysis reactions involved in a solubility equilibrium can be assessed easily by relevant thermodynamic functions, the quantification of colloid generation and radiation effects in a given system is not easy to realize.

2.2 Complexation reactions

Carbonate and organic complexation is considered of major concern due to the fact that these natural ligands are ubiquitous in geological environments. This subject is being investigated intensively and coordinated among five laboratories : CEN-Fontenay-aux-Roses (5,6) Technical University of München (3,7), Joint Research Centre of Ispra (8,9) CEN/SCK-Mol and Catholic University of Leuven (20,21). The first three institutes are mainly involved in actinide carbonate complexation, the latter two in reactions with organic material typical of the interstitial water of the Boom clay. During the two years of the MIRAGE programme the carbonate complexation, in parallel with hydrolysis reactions, for Np(V), Pu(IV), Pu(VI), Am(III) and Am(V) have been investigated, as shown in Table I and II, with considerable successes. Based on literature data and available values reported on the Tables, it is possible to envisage theoretical speciations of actinide ions in groundwater conditions. Further investigation on this subject is, however, necessary and beneficial for improving the accuracy of the available data.

Carbonate complexation reactions are difficult to study, because the formation of carbonate complexes is accompanied by hydrolysis reactions in neutral and alkaline solutions (7) and also because the functional variables, pH and CO_3^{2-}, are mutually dependent in the system.

Carbonate complexation has been studied at the CEN-Fontenay-aux-Roses and Technical University of München by solubility experiments and at the JRC-Ispra by a solvent extraction procedure. The formation constants from the three laboratories are in good agreement with one another.

Concerning organic complexes: their influence on the observed Kd values is presently well established for the sorption of Eu, Am, Pu, Np and Tc on Boom clay, as a consequences of the interactions of these radionuclides with organic matter. Detailed analysis of the Kd dependence on solid/liquid ratios can give a clue to the relative importance of organic matter. The water suspension of sediment-originating organic matter, appears to be the factor governing the solid-liquid distribution of radionuclides. This is confirmed by optical density measurement at 280 nm on the extracts.

Table III shows absorbance and Kd values as a function of dilution factors for Np. It appears that the total amount of organic matter in the extract is constant, irrespective of L/S ratio, as indicated by the constancy of the products "absorbance x dilution factor". This clearly suggests that the organic matter distribution is an important factor in governing radionuclide behaviour. Such dependence is also clearly illustrated in figure 2 showing neptunium liquid/solid distribution versus absorbance of the extract. Implicit in this line of thought is that

TABLE I : Stability constants of carbonate complexes of Np(V) and Am(III)

Species	log β for $M^{m+} + iOH^- + jCO_3^{2-} \rightleftharpoons M(OH)_i(CO_3)_j^{m-i-j}$			
	TUM	CEN-FAR	JRC-Ispra	Literature
$NpO_2CO_3^-$	-	4.72 (0)	4.13 ·0.03 (0.2)	4.49 (1.0) [10]; 5.9 (0.05) [11]
$NpO_2(CO_3)_2^{3-}$	-	6.65 (0)	7.06 ·0.05 (0.2)	7.11 (1.0) [10];
$NpO_2(CO_3)_3^{5-}$	-	5.86 (0)	-	8.54 (1.0) [10]; 17.4 (0.05) [11]
log Ksp ($NaNpO_2CO_3$)	-	-10.69 (0)	-	
log Ksp ($Na_3NpO_2(CO_3)_2$)	-	-14.19 (0)	-	
$AmCO_3^+$	5.08 ·0.92 (0.1)	5.36 (0)		5.7 (1.0) [12]
$Am(CO_3)_2^-$	9.27 ·2.2 (0.1)	10.50 (0)	11.4 (0.2)	9.4 (1.0) [12]
$Am(CO_3)_3^{3-}$	12.12 ·0.85 (0.1)	12.34 (0)	-	
$Am(OH)CO_3^0$	12.15 ·0.15 (0.1)	-	-	
$Am(OH)(CO_3)_2^{2-}$	16.16 ·0.14 (0.1)	-	15.6 (0.2)	
$Am(OH)_2CO_3^-$	18.29 ·0.17 (0.1)	-	-	
log Ksp ($Am(OH)CO_3$)	-21.03 ·0.11 (0.1)	-	-	
log Ksp ($NaAm(CO_3)_2$)	-	-17.38 (0)	-	

( ): Ionic strength (M); [ ]: Literature

TABLE II : Hydrolysis constants of Np(V), Am(III) and Am(V)

Species	log β for $M^{m+} + iOH^- \rightleftharpoons M(OH)_i^{m-i}$		
	TUM	JRC-Ispra	Literature
NpO_2OH	2.30 ·0.62 (1.0)	4.16 (0.2)	4.93 (0.1) [15]; 3.92 (0.2) [18]
			5.1 (0.02) [19]; 4.68 (1.0) [10]
$NpO_2(OH)_2^-$	4.89 ·0.05 (1.0)	-	
log Ksp (NpO_2OH)	- 8.81 ·0.05 (1.0)	-	- 8.85 (0.1) [15]; - 9.2 (0.1) [16]
			- 9.0 (0.2) [18]; - 9.73 (0.2) [19]
$Pu(OH)^{3+}$	12.20 (1.0)	-	13.1 (0) [14]
$Pu(OH)_2^{2+}$	24.79 (1.0)	-	25.3 (0) [14]
$Pu(OH)_3^-$	35.42 (1.0)	-	36.1 (0) [14]
$Pu(OH)_4^0$	43.42 (1.0)	-	45.7 (0) [14]
log Ksp ($Pu(OH)_4$)	-54.15 (1.0)	-	-53.2 (0) [17]; -61.2 (0) [14]
$Am(OH)^{2+}$	6.44 ·0.35 (0)	-	> 5.96 (0.003) [13];
$Am(OH)_2^+$	13.80 ·0.25 (0)	14.6 (0.2)	10.94 (0.003) [13];
$Am(OH)_3^0$	17.86 ·0.11 (0)	-	>14.54 (0.003) [13]
log Ksp ($Am(OH)_3$)	-28.89 ·0.11 (0)	-	-24.62 (0.003) [13]
$AmO_2OH^•$	1.0 ·0.2 (3.0)	-	-
log Ksp (AmO_2OH)	-9.2 ·0.2 (3.0)	-	-

radionuclides are actually present as organic matter complexes in the extract. In the case of Tc this can be demonstrated directly, using gel permeation chromatography; the association of Eu with organic matter in the extract can also be demonstrated by ultrafiltration combined with absorbance measurements. No speciation evidence has been obtained up to now for Am, Pu, and Np but similar dependence of Kd on L/S ratio and the

TABLE III : Effect of solid-liquid ratio on absorbance ($\lambda = 280$ nm) of extract and Kd(Np) in Boom clay (gallery level) at in situ conditions; extracting solution is $NaHCO_3$: 0.02 M; pH $= 9$-9.2

Liquid/clay ratio	Absorbance	Kd	Absorb. x dilution factor
2	12.23	134	26.9
5	5.09	247	26.5
10	2.65	635	27.0
25	1.18	1052	29.7

linear dependence on organic matter content in the extract may be considered as sufficient evidence that all radionuclides considered are present as organic complexes.

To study the competition between carbonate and organic ligands in the interstitial waters, Europium has been adopted as a test case. Dialysis equilibria between organic extracts and $NaHCO_3$ solutions have been performed and the Eu-organic matter stability constant was found to be in the order of 10^{10}. This allows to make some predictions as to the Eu distribution under in-situ conditions, which indicates that the organic matter fraction will be the sole determining factor and that the Eu fraction present as carbonate complexes can be expected to be in the order of 0.1%.

2.3 Generation and characterization of colloids

Colloids are defined as being of three kinds : real-colloids, groundwater-colloids pseudo-colloids. Real-colloids are produced by the aggregation of hydrolysed metal ions, groundwater-colloid are ubiquitous in all groundwaters and pseudo-colloids are generated by sorption of radionuclides on groundwater-colloids. The size of colloids under discussion is defined as being less than 0.45 μm, which remain suspended in groundwater. Those metal ions which tend to hydrolyse will either generate real-colloids or pseudo-colloids. To this category belong all actinide ions; the tendency is strongest with trivalent and tetravalent ions, followed by hexavalent ions and least with ions of pentavalent state. In near neutral solutions of low salt concentration, the generation of real-colloids of tri-, tetra- and hexavalent actinide ions is prevalent, each to a different extent. Since the solubility of these ions in such solutions is generally very low, the quantification of real-colloids can only be persued by ultrafiltration (22, 23, 24, 25).

Under its own radiation field (especially alpha-radiation) the size of real-colloids becomes smaller with time and can reach dimensions of less than 10A. This type of colloid, called "Micro-colloid", is not even separable by ultrafiltration with the smallest pore size available and remains stable even in saline water of high NaCl concentration. The generation of such micro-colloids increases the apparent Pu solubility to levels higher than those determined from thermodynamic considerations as was shown in figure 1 for the case of ^{238}Pu.

The generation of pseudo-colloids increases actinide solubilities in natural aquatic solutions, while decreasing their retention on geomatrices. The chemistry of actinides in groundwater is largely associated with the chemistry of pseudo-colloids. It is, therefore, of cardinal importance to clarify whether the migration behaviour of actinides in a given aquifer system is typical of pseudo-colloids or not and if so, how to quantify it.

The colloid characterization in natural aquatic solution requires a combination of different analytical methods. Much effort in this direction has been concentrated by the TU-München and JRC-Ispra. The particle spectrum can be obtained by ultrafiltration or ultracentrifugation, or a combination of both. A typical example can be seen in figure 3.

2.4 Development of speciation methods

The migration of radionuclides in natural aquatic system is primarily governed by the physical and chemical states of elements concerned, which determine their interaction processes with surrounding matrices. The speciation of physical and chemical states of dissolved species, e.g. colloids, charge of species, oxidation states etc. is an important subject in the migration study. In the framework of the MIRAGE project three different speciation methods are being developed for application to natural aquatic systems. These methods are free-liquid electromigration (JRC-Ispra), laser-induced thermal lensing spectrometry (joint venture of JRC-Ispra and CEA-Fontenay-aux-Roses) and laser photoacoustic spectrometry (TU-München)

a) Free-liquid electromigration

This method is based on the electrochemical transport system and capable of speciating charge states of dissolved species (26, 27). It can eventually be used to diffentiate colloids based on their surface charge distributions.

Experiments with leachates from a simulated HAW-glass spiked with ^{99}Tc, using a synthetic ground-water relevant to the aquifer overlying the Boom-Clay in Mol, Belgium, has shown that in the leachate the predominant species is the TcO_4^- ion, regardless of whether experiments are carried out under aerobic or anaerobic conditions. Further study with the NpO_2^+ ion demonstrates that this ion tends to form both cationic and anionic species, while the anionic species becomes predominant in carbonate solutions of pH = 8-9. The probable anionic species as reported in Table I are $NpO_2CO_3^-$ and $NpO_2(CO_3)_2^{3-}$ (9).

The applicability of this method is limited to solutions of relatively low conductivity. Whether the method is applicable to real groundwaters containing a substantial amount of colloids is yet to be tested.

b) Laser-induced thermal lensing spectrometry

The principle of this method (28, 29) is as follows : a laser beam focused into the absorbing medium creates a local temperature gradient and hence a gradient refractive index. The spacial distribution of the refractive index makes the absorbing medium behave as a thin optical lens. This can then be observed and measured by the deflection of the probe laser beam or by its defocusing.

The joint experiment of the JRC-Ispra and CEA-Fontenay-aux-Roses (28) has successfully developed a method which is capable of measuring the dissolved uranyl ion in nitric acid with a sensitivity of 4×10^{-6} M/L. Further development is in progress to set up a differential dual-beam thermal lensing spectroscopy system, using either pulsed or continuous excitation (30). The development to a differential operation will improve sensitivity to about 2-3 orders of magnitude better than possible by conventional spectrophotometry.

c) Laser-induced photoacoustic spectrometry

The difference between photoacoustic and thermal lensing spectrometry is in the detection technique. On absorption of photons, the electronic states of actinide ions, or their complexes, are raised to excited energy levels, which decay by radiation emission or non-radiative relaxation. The non-radiative process causes a temperature rise of the solute. In an adiabatic and isobaric expansion, the temperature change creates pressure waves which are, for the photoacoustic techniques, detected with a piezoelectric crystal. The photoacoustic signal thus produced is directly proportional to the concentration of the actinide ion in the solution.

This method developed by the TU-München (31,32) is capable of providing spectral information for a wide range of absorption bands of actinide ions in liquid samples and hence to speciate their oxidation states with the sensitivity an average of 3 orders of magnitude higher than attainable by conventional spectrophotometry.

Further development and improvement of speciation methods are very desirable, especially direct methods which provide high speciation sensitivity; the above mentioned spectrometric techniques, developed in the MIRAGE project, are already capable of speciating not only oxidation states but also complex species and colloid generation in very dilute solutions.

2.5 Sorption phenomena

The study of sorption phenomena amasses all basic geochemical knowledge which has been discussed in the previous chapters. The successful geochemical modelling of migration depends largely upon the knowledge of basic sorption mechanisms. It is impossible to describe sorption mechanisms in a general way which is applicable to a wide variety of chemical systems, but it is not impossible to understand important primary mechanisms that govern interactions of radionuclides with geomatrices less dependent on geochemical surroundings. Such studies are being conducted in the Risø, Fontenay-aux-Roses and Nottingham laboratories.

At Risø National Laboratory the mutual influence of a multication system towards synthetic and natural exchangers has been investigated. Based on the rigorous thermodynamic treatments, taking into consideration of thermodynamic quantities for the hydration of each metal ion and the electric potential of the exchange system, a semiempirical equation has been derived, which facilitates prediction of the behaviour of the Sr^{2+} ion in a given multicomponent solution.

A second Risø study concerns the sorption of radionuclides on halite using column techniques. In the case of pure halite only europium and americium are found to be retained on the solid, whereas other ions (Cs, Sr, Co, Tc) are eluted with very minor, if any, retention. The retention of Eu(III) and Am(III) is interpreted as a surface sorption, which takes place on the surface of any material in neutral solution, not necessarily only on halite. The effect may not be formulated as simple electrostatic attraction, but also as a result of the incorporation of impurities in the natural material.

Exchange isotherms have been determined at the CEA-Fontenay-aux Roses (6) for the argillite, which has the exchange capacity of 36.6 meq/100 g, using two different pairs of ions : Ca^{2+}/K^+ and Ca^{2+}/H^+ with the total concentration of 0.01 and 0.1 N. It is observed that the exchanger contains two exchange sites having different capacities and exchange constants, thus resulting in a bi-functional exchange process. The results are illustrated in figure 4. Exchange capacities and equilibrium constants of the Ca^{2+}/K^+ and Ca^{2+}/H^+ pairs, for the two possible exchange sites evaluated by mathematical treatment, appear to be considerably different. The reason is not elucidated, but it is, however, conceivable that the pH of the Ca^{2+}/K^+ system, being different from the Ca^{2+}/H^+ system, may be attributable to the different exchange capacities observed. This study is intended to continue for NpO_2^+/Ca^{2+} and the NpO_2^+/K^+ pairs but has not yet been completed.

An understanding of the migration behaviour of the actinides and fission products in the geosphere has also been furthered by the British Geological Survey on a multi-disciplinary front, embracing a wide variety of analytical techniques including geomicrobiology (33). Emphasis has been placed on examining the sorption/desorption properties of argillaceous sediment by means of quasi-equilibrium batch sorption experiments with the aim of addressing the mechanistic processes involved.

A programme with the Scottish Universities Research and Reactor Centre (34) has focussed on the investigation of clay-rich lake sediments as a natural analogue of long-term radionuclide migration through argillaceous material. The salient feature of the sediment sequence studied is a marine band deposited 6900-5400 years ago which has been regarded as a source term for the diffusion of elements into the overlying freshwater lacustrine deposits. The most striking result has been the detection of significant enrichments in the marine horizon of U, Ra-226, I, Br, and Sb which correlate positively with the total organic carbon and nitrogen contents; this provides strong evidence for organic fixation as a mechanism for retardation.

It must be noted, however, that long-term predictions of radionuclide transport based on complementary batch sorption/desorption experiments are at variance with these findings. For example iodine-131 and Np-235 both showed small but significant uptake onto the marine material (Kd's=3ml/g) but similar small values are also measured for the lacustrine sediments. Parallel experiments (35) with Lower Oxford Clay containing 4,5% organic carbon and 0.13% organic sulfur are being performed in order to control the potentially reactive effect of this material towards redox sensitive elements such as Tc and Np.

The study of natural analogues is considered very valuable and needs to be promoted further. It is expected that such research will produce important information concerning transport mechanisms on a geological time scale.

3. Conclusions and perspectives

Based on discussions made with the scientific representatives from the participating laboratories and also based on the present state of knowledge assessed for the subject from the literature, the following suggestions are summarized for future studies. These are considered to be an indispensable complement to the work described for reaching a better understanding of migration mechanisms of actinides and fission products.

- Colloid generation
  . generation mechanism of pseudocolloids in macro/micro sizes
  . characterization (e.g. composition, stability etc.)
  . quantification (as a RN transport medium)
  . interpretation for modelling

- Speciation (development of methods)
  . direct methods, e.g. advanced spectroscopic detections
  . indirect methods, e.g. new chemical procedures

- Complexation with natural inorganic and organic ligands
  . organics : humic/fulvic acids, citric acid etc.
  . inorganics : hydroxide, carbonate ions etc.
  . effects on sorption of radionuclides
  . effects on colloid generation

- Sorption phenomena (primary mechanisms)
  . physical or chemical sorption
  . multicomponent competition
  . filtration in different aquifer systems

- Translation of basic knowledge for the geochemical modelling
  . how to correlate individual experimental results with geological
    variables
  . how to formulate sorption coefficients as functions of geochemical
    variables
  . how to couple the results to modelling

    The intention of the first three subject is to clarify our
understanding of the chemical behaviour of actinides and fission products
in natural aquatic systems. This will thus help improve our knowledge of
sorption phenomena. The ultimate goal of the basic chemical studies is to
directly support the optimization of geochemical modelling.

4. <u>REFERENCES</u>

1. The CEC Project MIRAGE : Programme document EUR 9403 (1984)
2. MIRAGE project : first summary report (Jan-Dec. 1983); EUR 9543 (1985)
3. J.I. Kim et al. Progress report of the contract CEC 359-83-7 WASD, RCM-01484, July 1984
4. J.I. Kim and M. Bernkopf, RCM-02884, Oct. 1984
5. P. Vitorge IAEA Seminar, Sofia (6-10 Feb. 1984) paper IAEA-SR-104
6. A. Billon : annual report 1984 of the contract CEC 361-83-7-WASF
7. J.I. Kim, M. Bernkopf, Ch. Lierse and F. Koppold, ACS Symp. Ser. 246, 115 (1984)
8. G. Bidoglio, Radiochem. Radioanal. Lett. 53, 45 (1982), 19 M.S.
9. G. Bidoglio, G. Tanet and A. Chatt, accepted for publication in Radiochemica Acta (1985)
10. L. Maya, Inorg. Chem., 22, 2093 (1983)
11. H. Nitsche and N.M. Edelstein, the paper presented in "12-Journées des Actinides", Orsay (1982)
12. R. Lundqvist, Acta Chim. Scand., 36, 741 (1982)
13 D. Rai, R. Strickert and D. Moore, PNL-SA-10635 (1982)
14. C.F. Baes, Jr. and R.E. Mesmer, "Hydrolysis of Cations", John Wiley + Sons (1976), P 188
15. K.A. Kraus and F. Nelson, AECD-1864 (1948)
16. K.A. Kraus, Proc. Int. Conf. Peaceful Use of Atomic Energy, Geneva, A/Conf. 8 (1956), P 73

17. J.A. Perez-Bustamante, Energ. Nucl. (Madrid), 13, 23 (1969)

18. A.I. Moskvin, Radiokhimiya, 13, 681 (1971)

19. E.P. Sevostyanova and G. V. Khalturin, Radiokhimiya, 18, 870 (1976)

20. P.N. Henrion, M. Monsecour, A. Fonteyne : Characterization of the Boom Clay submitted to "Radioactive Waste Management and Nuclear Fuel Cycle"

21 A. Maes and A. Cremers : Speciation - 85 Seminar : Speciation of fission and activation products in the environment (CEN and NRPB) Christchnick Oxford (U.K.) (1985) in press

22. A. Avogadro, G. De Marsily : Mat. Res. Soc. Symp. Proc. Vol. 26 (1984) Elsevier Science Publishing Co., Inc.

23. G. Bidoglio, A. Avogadro, A. De Plano and G.P. Lazzari, Radioactive Waste Management and Fuel Cycle, 5, 311 (1984)

24. J. I. Kim, G. Buckau, F. Baumgärtner and H. C. Moon, "Scientific Basis forNuclear Waste Management" Vol. VII, Ed. G.L. McVay, North-Holland (1984) p 31

25. J.I. Kim, W. Treiber, Ch. Lierse and P. Offermann, 1984-MRS-Meeting, Nov.26-30, 1984, Boston, Paper No. N 3-6, in Print

26. G. Bidoglio, A. Chatt, A. De Plano and F. Zorn, J. Radioanal. Chem., 79, 153 (1983)

27. A. Chatt, G. Bidoglio and A. De Plano, Anal. Chim. Acta 151, 203 (1983)

28. T. Berthold, P. Manchien, N. Omenetto and G. Rossi, Anal. Chim. Acta, 153,   265 (1983)

29. J.P. Gordon,, R.C.C. Leite, R.S. Moore, S.P. Porto and J.R. Whinnery, J. Appl. Phys., 36, 3 (1965)

30. Th. Berthaud, N. Delorme and P. Manchien, submitted for publication

31. W. Screpp, R. Stumpe, J.I. Kim and H. Walther, Appl. Phys., B 32 207 (1983)

32. R. Stumpe, J.I. Kim, W. Schrepp and H. Walther, Appl. Phys., B, in print

33. J.M. West et al. "Radioactive Waste Management and the Nuclear Fuel Cycle" 6(1), 1-17 (1985)

34. A.B. Mc Kenzie et al. Proc. 1984 Cicago, Natural Analogue Workshop (US-DOE/SKBF) KBS Tech. rept. in press

35. I.G. McKinley and J.M. West : Radioactive Waste Management and the Nuclear Fuel Cycle 4 (4), 379-399 (1984).

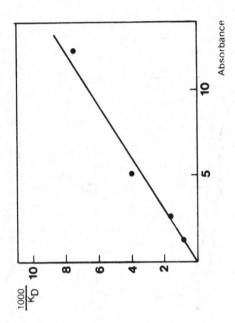

Fig. 2.: Reciprocal of $K_D(Np)$ in Boom clay versus absorbance (280 nm) of the extract

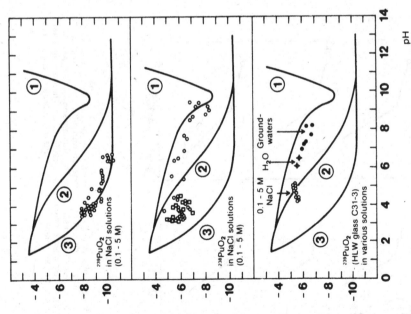

Fig. 1.: Pu solubilities as a function of pH. The lines are solubility curves for comparison: $^{239}PuO_2$ in HCO_3^-/CO_3^{2-} solution at $I = 1.0$ M (1); freshly prepared $^{239}Pu(OH)_4$ in 1 M $NaClO_4$(2) and $^{239}PuO_2$ in 0.1 M $NaClO_4$(3). The phase separation is made by ultrafiltration at 1 nm for all samples (TUM)

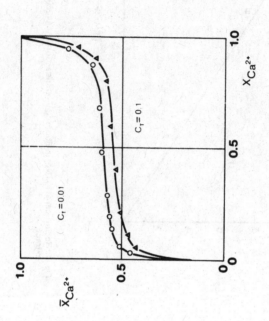

Fig. 4.: Isothermal ion, exchange of the Ca^{2+}ion on argillite. X and $\bar{X}$: mole fraction in solution and in argillite, respectively (CEA-FAR).

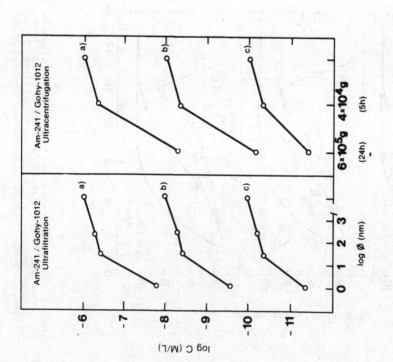

Fig. 3.: Pseudocolloids of $^{241}Am(III)$ in Gorleben groundwater fractionated by ultrafiltration and ultracentrifugation with different pore sizes and centrifugation forces respectively. The ordinate shows the ^{241}Am concentration remained in solutions after fractionation. Three different Am concentration are initially introduced to the groundwater a, b, c (TUM)

DISCUSSION

M. PRINS, RIVM 'Bilthoven

How do you explain the large dependence of the K_D value on the liquid/-solid ratio in the batch experiments?

P. HENRION, CEN/SCK Mol

The sorption on clay has been studied with synthetic groundwaters which did not contain humic acids. Humic acids present came exclusively from the clay and were diluted by the synthetic added ground water. Since there seems to be a quantitative complexation of the radionuclides with the humic acids, you simply get a volume effect.

THE MIRAGE PROJECT : LARGE SCALE RADIONUCLIDE TRANSPORT INVESTIGATIONS AND INTEGRAL MIGRATION EXPERIMENTS

B. CÔME
Commission of the European Communities - Brussels

G. BIDOGLIO
Commission of the European Communities - Joint Research Centre - Ispra

N. CHAPMAN
British Geological Survey - Keyworth

J. DELMAS
Commissariat à l'Energie Atomique - Cadarache

F. EWART
United Kingdom Atomic Energy Authority - Harwell

M. FRISSEL, M. PRINS
Rijksintituut voor Volksgezondheid en Milieuhygiëne - Bilthoven

M. JEBRAK
Bureau de Recherches Géologiques et Minières - Orléans

A. FLAMBARD, G. MARX
Freie Universität - Berlin

Summary
Predictions of radionuclide migration through the geosphere must be supported by large-scale, long-term investigations. Several research areas of the MIRAGE Project are devoted to acquiring reliable data for developing and validating models. Apart from man-made migration experiments in boreholes and/or underground galleries, attention is paid to natural geological migration systems which have been active for very long time spans. The potential rôle of microbial activity, either resident or introduced into the host media, is also considered. In order to clarify basic mechanisms, smaller scale "integral" migration experiments under fully controlled laboratory conditions are also carried out using real waste forms and representative geological media. All these results will be used to improve the reliability of predictive radionuclide migration modelling.

1. Introduction

Transport by mobile groundwaters is the most likely process by which disposed radionuclides can return to man's environment. On their way through the geosphere these radionuclides will interact with geological media and be retarded; an effective disposal system will ensure that they have decayed to innocuous levels before they enter the biosphere.

Any reliable prediction of these transport and retardation processes must be supported by large-scale and long-term investigations; some mechanisms must also be clarified by laboratory experiments in fully controlled conditions. The Community MIRAGE Project on Migration of Radio-nuclides in the Geosphere encompasses a wide variety of such investigations (1). The seven research areas identified in this Project are outlined in Table I. Those areas concerned with hydrogeology (n°4), natural migration systems (n°5) and micro-organisms (n°6) are mainly devoted to in-situ investigations, whereas supporting laboratory experiments are the subject of the so-called "integral" simulations (n°2). The complete results can be found in Project Progress Reports (2); salient achievements are presented in this summary paper.

Table I
Research Areas of the MIRAGE Project

1. Basic actinide and fission product chemistry
2. Integral simulation experiments
3. Laboratory studies related to specific sites
4. Hydrogeology
5. Natural geological migration systems
6. Micro-organisms
7. Modelling studies

2. Full-scale radionuclide transport investigations

A reasonable control of experimental conditions may be obtained by tests in which radionuclide solutes or tracers are pumped and recovered via boreholes drilled into geological formations. However, more reliable information can be gained from natural occurences of geological migration processes ("natural analogues"). The very long time spans involved are comparable with the estimated durations of the processes controlling the effectiveness of a disposal system.

2.1. Experiments in boreholes and underground laboratories

These tests are being performed for the specific sites identified for the MIRAGE Project : Fanay-Augères (F), and Troon (Cornwall, UK) for granite. Mention will be made of the in-situ experiment in a sedimentary sequence at Drigg (UK). As these investigations are the subject of other papers in this conference, only a summary of their most important features is given here.

In an existing haulageway of the French uranium mine at Fanay-Augères near Limoges, the Commissariat à l'Energie Atomique (CEA, Fontenay-aux-Roses) is carrying out a programme of large-scale permeability and tracer dispersion measurements in the fractured granite surrounding the gallery (3). This research complements the investigations at Troon, Cornwall, and at Chalk River, Canada, where single fracture migration experiments using tracers are being carried out by the United Kingdom Atomic Energy Authority (UKAEA, Harwell) (4). Similarly, the British Geological Survey (BGS) operates a tracer test in shallow glacial deposits at Drigg, UK (5). In these latter tests, tracers are injected in one borehole and recovered in one or several boreholes at various distances. The groundwater flows and retention properties of the geological media can be derived from these tests, although the limited time span (maximum some months) may be insufficient for some mechanisms to be completely quantified.

2.2. Geological evidences of transport processes

In order to gain reliable information as to the possible distances and durations involved in radionuclide migration, the processes involved must be investigated by studying their natural analogues which have operated at comparable scales of time and distance. From this point of view, natural mineral deposits exhibit several properties common to those of disposal sites, such as size, geochemical characteristics and duration of emplacement and later evolution. No completely analogous model exists however. It is therefore necessary to distinguish the various stages in the evolution of a disposal site and to study each migration process separately.

It is generally recognized that elements such as rare earths, and also uranium, can be considered as useful analogues for the radionuclides of interest in the MIRAGE Project (6).

2.2.1. Hard rock analogues

The first set of investigations concentrated on granitic analogues of disposal sites, and was carried out by the Bureau de Recherches Géologiques et Minières (F). The study undertaken revealed two essential stages : (a) a hydrothermal stage resulting from the reaction of the radioactive waste with its surroundings at "high" temperatures (100° - 200°C); and (b) a supergenic stage, related to the lowering of the erosion level and associated weathering.

The behaviour of radioactive elements under conditions of hydrothermal alteration is still poorly understood. Therefore, it was decided to investigate analogous elements such as the lanthanides, whose geochemical properties are similar to those of the radioactive elements. Detailed studies have been made at two sites.

The first site is situated in the Vosges (F) on the eastern edge of the Ballons granite in the Langenberg. This is a monzogranite of the Visean age, enriched in U, Th and lanthanides, and which has been affected by two hydrothermal episodes : a late magmatic one at about 250° - 300°C, and another at about 150°C during the late Permian. It has been possible to characterize this latter system in detail : it resulted from a thermal anomaly related to the opening of Permian grabens, and was marked by the formation of quartz-fluorite veins at a depth of about 500 to 1,500 m. The fluids responsible were aqueous and slightly saline.

The late magmatic hydrothermal system provoked pervasive propylitic alteration and is evidenced, according to the abundance of carbonates, by chlorite-epidote-illite or calcite-albite-illite-leucoxene parageneses. A fracture system containing calcite and/or corrensite (regularily interlayered) is also associated. Despite the considerable mineral transformations, the geochemical budget remains constant. The lanthanide spectra show no significant variations, although there is a slight decrease in the total rare earth content. There has therefore been no significant migration of the lanthanides under these conditions.

The hydrothermal system limited to the fluorite-bearing veins resulted in clay mineral - carbonate alteration, in places with metavermiculites, in other areas with siderite-kaolinite-beidellite-potassic illite, and carbonate veins. The alteration was restricted to the proximity of the hydrothermal conduits and has affected only 0.5 -1 m either side of the mineralization. Geochemically, a lowering of the rare earth fractionation was observed, implying differential mobility of the elements under low-temperature conditions.

The second site is situated in Brittany, in Cambrian acid volcanic rocks enclosed in the syncline of La Telhaie. This has also been affected by two hydrothermal episodes : (a) an early one at about 220°C, in which fluids of about 9 % NaCl equivt. wt. were active, causing propylitic alteration, and (b) a late alteration at between 80°C and 150°C that resulted in argillic alteration and Pb–Ag bearing quartz-carbonate deposits. The propylitic alteration is diffuse and accompanied by no significant trace element variations. The argillic alteration however (kaolinite, beidellite-illite-carbonate), caused considerable leaching of the lanthanide content over several metres on either side of the mineralized zone.

These two "analogues" thus show increased mobility of these elements under low temperature conditions. The higher mobility at La Telhaie could have resulted from the more carbonic nature of the hydrothermal fluids there, an observation which has been confirmed in various environments at higher temperatures (7).

Apart from their predictive interest, these studies are also useful for the choice of criteria upon which the selection of disposal sites should be based : pre-existing alteration would in effect allow selective absorption of the elements and reduce permeability by sealing fractures.

Another part of this work, on supergene alteration of analogous mineral deposits, was limited to a bibliographic study directed towards selection of the best possible "analogues". Apart from uranium deposits, two important families of ore deposits can be considered : those associated with highly evolved acid magmatism, enriched in Li, Nb, Ta and the lanthanides, such as lithium-bearing pegmatites, and ore bodies associated with alkaline magmatism enriched in the lanthanides, Cs, Nb, Zr and Hf. Investigation of their alteration is difficult however because of the autometasomatism generally affecting them. Nevertheless they do show that the mobility of the elements is strongly dependent on the mineralogical characteristics of their host-minerals.

2.2.2. Sedimentary analogues

A well characterized series of clay-rich (> 80 %) sediments from Loch Lomond (Scotland) and adjacent freshwater lochs have been studied by the British Geological Survey (BGS) and the Scottish Universities Research and Reactor Centre (8). These sediments were laid down under oscillating freshwater and marine conditions and consequently the pore-water chemistry and certain diagenetic reactions were expected to vary from marine to freshwater bands. A detailed geochemical profile would reveal which elements were mobile in the pore waters since sedimentation occured, or which processes may have been responsible for fixing them. The time-scale is very well fixed by palynological, palaeomagnetic and radiocarbon dating. The marine band of interest is about 1 m thick, lies at 3 m depth, and was deposited between 6900 – 5400 years B.P. Core from a neighbouring loch which was not subject to marine transgression but which received sediments from the same catchment area was used to check whether changes in sedimentation during the marine period might confuse the analysis of post-depositional element movement.

Deposition of the Loch Lomond sediments extends from the present day back to about 8000 years. In the marine band the original and evolved pore water composition was different to that in the under and overlying freshwater sediments; as sedimentation progressed redistribution of elements would occur by diffusion in the pore water and upwards pore water advection caused by progressive compaction of the sediments. The latter process was shown to result in movement about twenty times slower than diffusion. The system is analogous to a concentration gradient around a leaking waste container surrounded by porous saturated sediments (e.g. a clay formation) where diffusion profiles (modified of course by sorption/retardation) which have been established over several thousand years can be compared with anticipated profiles for waste elements derived from laboratory models. The model takes into account the rate of sedimentation which is slower than diffusion and constantly changes the distance to the free-surface.

Concentration profiles in the solid-phase were determined for 27 elements including rare earth elements and natural series radionuclides. Laboratory sorption studies provided data on the retention properties of the core for Sr, Cs, Co, Ce, I, Tc and Np. A simple diffusion and sorption model then assessed how elements may have moved from one band (the marine unit in particular) to another.

The following principal conclusions were drawn :

(i) Sorption processes on the core were commonly irreversible. The model frequently used to assess migration through clays assumes diffusion plus instantaneous, reversible sorption. This is clearly not the case here, throwing doubt on its ability to predict transport rates in sediment sequences, and emphasising the need to study sorption kinetics in detail.

(ii) Evidence for post-depositional migration of elements out of the marine band was not found (with the possible exception of ^{234}U). This is very re-assuring with regard to nuclide migration, especially as the time-scale involved is several thousand years.

(iii) Certain elements, some of which (e.g. iodine) were expected to migrate quickly, were retained in the thin marine band. There is strong evidence for organic fixation of I, Ra and U on biogenic particulates in the clay, emphasising the need for a better understanding of the rôle of sediment organics in retardation.

The second analogue study of clays has been focussed on the sedimentary sequence at Orte (I) and operated by the Comitato nationale per la ricerca e per lo sviluppo dell'energia nucleare e delle energie alternative (ENEA). It involves the study of migration processes which have been active in the upper levels of a thick clay sequence, where redox conditions have dominated the bulk geochemistry. The clay sequence is overlain by a sand unit, in turn covered by a layer of volcanics with a high natural series radionuclide content. Groundwaters percolating down through these volcanics are expected to have leached rare earths, U, Th, and Ra, and transported them down through the sands to the upper part of the clay. The whole sequence is less than 10 million years old and the migration process can be considered to have been operational since the deposition of the volcanic tuffs between 60 and 100 thousand years ago. These tuffs and the sands are extensively oxidized, as is the uppermost part of the clay. Since oxidizing conditions are known to favour the migration of the actinides in particular, the intention is to profile the distribution of rare-earth elements and the natural series radionuclides to see how far they have moved through the sand, permeated into oxidized fractures in the clay and then diffused, and presumably have been more strongly retarded, in the reducing zone of the main clay body.

The study could thus be considered as a direct analogue of diffusion processes in the far-field (reducing conditions) and in the near-field where the possibility exists of a radiolysis-induced oxidation zone surrounding leaking waste containers.

The work has consisted of carrying out a detailed sedimentological analysis of the site (a clay pit, la Fornace, at Orte) to determine the interrelationship of the various clay and sand units, and to assess the present day pattern of groundwater movement. A full palaeo-environmental study has been performed and a stratigraphic scheme set up. This has allowed reconnaissance sampling of the site, and eight specimens of clay and sand have been analyzed for their major and trace element chemistry and Fe II/III and Mn II/IV oxidation ratios. A system for pore-water extraction is now available, so that in future the solid phase chemistry can be compared to the fluid phase.

2.2.3. The uranium series disequilibrium method

The study of groundwater circulation patterns, flow rate and residence time in a given geological setting is one of the central themes in assessing the suitability of a disposal site. Should mixing occur, the mixing patterns could be described and quantified by characterizing the isotopic signatures of different water masses in the region of interest. Nuclides of the uranium series provide one of the most sensitive isotopic signatures in terms of their specific activities (disintegrations per unit mass of water). In particular, uranium content and activity ratios of various soluble and insoluble nuclides in the uranium decay series have been found very useful in groundwater studies (9). Uranium isotope disequilibrium (departure of the $^{234}U/^{238}U$ activity ratio from unity) has been observed to occur commonly in circulating groundwaters and associated rocks, and is known to be related generically to the geological environments through which the waters flow.

This method has been used to characterize the groundwater flow patterns in the sedimentary layers at the Mol site (B), by the UKAEA-Harwell. Part of the Belgian programme is a study of the regional hydrogeology in north-eastern Belgium. To this end observation wells have been drilled to various depths at more than 12 sites, two of which are at Mol. The current programme has been enlarged to include a comprehensive study of regional hydrochemistry incorporating major and minor ions, some trace elements, stable isotopes of hydrogen, carbon and oxygen, and ^{14}C. The measurements of hydrogeological parameters such as piezometric heads, porosity, transmissivity and storage coefficent are serving to provide the data for a hydrological model designed to realistically describe the groundwater flow on a regional scale particularly with respect to the Boom clay formation. To supplement the modelling effort, hydrochemical and isotope data are expected to shed further light on the validity of the model findings. The uranium series disequilibrium study was designed to supplement the above programme by delineating bodies of water in the region in terms of uranium, thorium and radium isotopic content. Specifically, the current programme is expected to yield information on uranium series disequilibria in the groundwaters deriving from sand horizons above and below the Boom clay formation and establish unambiguously whether there is a groundwater movement through Boom clay and at what rate. Ultimately, by using this method, the Mol site could be considered as an analogue for itself, which is the preferred situation.

Samples of groundwaters derived from shallow and deep sand layers above and below the Boom clay formation at five locations have been collected and analysed for uranium, thorium and radium isotopes. The younger surface waters in the sand layers above the clay are characterized by low $^{234}U/^{238}U$ activity ratios (1 to 2) and variable uranium content which is dependent on the amount of leachable uranium locally available on the surface of sand particles in the system. An older component obtained mainly at greater depth is characterized by higher $^{234}U/^{238}U$ activity ratios in the range of 2 to 4 and again variable amounts of uranium in solution. The old end-member waters have not been sampled and analysed to date making the considerations of mixing patterns rather difficult. However, recoil/leaching considerations in conjunction with some simple aging models have led to the conclusion that the residence times of the sampled groundwaters do not exceed greatly 10×10^3 years.

The thorium content of these waters is very low reflecting a combination of low thorium mobility and the low-thorium lithology in this environment. No discernable trend with depth exists in the radium isotopic data obtained so far.

The radiometric data obtained from the analysis of two sand samples at Wuustwezel (one above and the other below the Boom clay layer) and a Boom clay sample at Mol indicate uranium precipitation in the sand layers at both horizons but no mobility of thorium and radium. In contrast, the analysis of bulk clay sample indicates no mobility of any of the uranium, thorium and radium isotopes in that material, confirming its potential usefulness as a natural barrier to radionuclide migration. However, it has been suggested that a detailed analysis of different mineralogical phases in the clay may reveal the exchange mechanisms at work between the slowly moving water through the clay layer (in either direction) and some minority minerals such as iron oxyhydroxides which may not be detectable in the bulk analysis data.

A more comprehensive sampling and analysis programme both on a regional and localised scale would facilitate numerical modelling of the disequilibria data and thus, supplement hydrological modelling. The present pilot study however has confirmed the value of the uranium series disequilibrium methods in validation programmes of potential geological formations for radioactive waste disposal.

2.3. Rôle of micro-organisms in radionuclide migration

The potential for micro-biological activity to modify near-field geochemistry (thus affecting rates of corrosion and leaching, nuclide speciation, and possibly even migration mechanisms), has been under study for about four years in the context of deep repositories. Preliminary studies by the British Geological Survey (BGS) / Napier College Edinburgh, and the Commissariat à l'Energie Atomique (CEA) at Cadarache have not indicated any conclusive evidence for the presence of active micro-organisms in deep granitic groundwaters (10). The main thrust of the work being carried out in the United Kingdom, France and Italy concentrates on characterising populations of organisms which will be introduced into an active repository, and defining the energy and nutrient requirements and potential effects. A preliminary study of microbial populations of disused mines in the United Kingdom, and of underground research laboratories and potential disposal facilities in the EC countries and Sweden has been made by BGS/Napier College and has demonstrated viable populations in a variety of environments from granitic rocks to salt mines (11). A parallel study of microbial activity in montmorillonite has shown the potential for contamination of the canister surface by the backfill, with possibly enhanced corrosion (12). A joint project is underway between BGS/Napier College and UKAEA to quantify actual rates of corrosion of steel by microbial activity. Modelling studies are now attempting to put the role of micro-organisms on a quantitative footing by assessing repository system energetics and attempting mass balance calculations (13).

In Italy, ENEA is studying the progressive invasion of a new excavation in Plio-Pleistocene clays by micro-organisms as part of its underground laboratory programme at Pasquasia. In addition studies are being made of latent micro-organisms (spores) in the same clays in quarries in

Lazio and Toscana.

Having located and characterised populations in the first phase of the studies, future work will concentrate on defining long-term viability of populations and their geochemical effects. To the first end preliminary studies of pressure, temperature and radiation tolerance of selected microbes are currently underway at the British Geolocial Survey (BGS).

3. Control of processes by integral simulation experiments at smaller scale

These experiments can be considered as being small-scale "laboratory analogues" of a repository; groundwater is passed over a waste form and through columns which contain material typical of the near-field and the geosphere in order to obtain direct measurements of the retardation of radionuclides and to reveal any mobile species.

This part of the MIRAGE Project has made use of common samples being made available to the laboratories involved : (a) for high level reprocessing waste, the French doped borosilicate glass R7T7 (CEA-Marcoule); (b) for medium level waste, two actual waste forms, mainly a reprocessing concentrate in cement (WAK-Karlsruhe) and a reprocessing sludge incorporated in bitumen (CEA-Marcoule). The behaviour of technetium (a long-lived fission product) and plutonium, americium and neptunium (representative actinides) has been investigated in contact with samples from the specific geological media considered in the other areas of the Project. Attention was paid to the creation of test conditions similar to those which could be found in deep geological formations. Flow-through experiments were carried out at ambient temperature.

3.1. Vitrified high-level waste in sediments and granite

Experiments being carried out by the CEA at Cadarache and Marcoule are divided into two sections, the first dealing with the mobilization of nuclides from the source-term followed by speciation of the resulting leachates, and the second concerning the percolations of these leachates through columns filled with geological material in order to simulate radionuclide migration.

The source-term consists of a glass spiked with ^{237}Np, ^{238}Pu, ^{239}Pu and ^{241}Am. Solutions to be used for the leaching process are prepared by allowing groundwater to come to equilibrium with the geological media (granite and glauconitic sand) over a one month period. The leachates are subsequently prepared by contacting these solutions with the glass for 28 days at a temperature of 90°C and under a pressure of 10 MPa. Speciation tests consist primarily of the determination of the colloidal-size distributions.

A first series of results was obtained for the static leaching of ^{238}Pu with groundwater from the Mol sands. The leach rate of ^{238}Pu was 1.5 10^{-6} g/cm^2 over 28 days, whereas the bulk leach rate was 3.18 10^{-3} g/cm^2, again over 28 days. The total activity found in the vessel amounted to 3.15 10^4 Bq. After leaching, 56 % of the activity was found on the vessel itself and the geologic material, and 44 % in the leachate. 20 cm^3 of this leachate were sampled for further analysis. The activity could be categorized as follows : 4 % fixed on the sampling bottle, 84 % fixed on particles greater than 0.8 µm, 2 % on particles smaller than 0.8 µm and greater than 25 nm, and 10 % in solution. After filtration of the fraction greater than 0.8 µm, the remaining fluid was analysed by alpha-spectrometry. Results are given in table 2.

Table 2

Analysis of a Mol groundwater leachate (prior to column experiment)

Radionuclides	Activity (Bq/m^3)
total alpha	$2.74.10^7$
$238_{Pu} + 241_{Am}$	$2.55.10^7$
$234_U + 237_{Np}$	$1.59.10^5$
239_{Pu}	$4.45.10^5$
238_U	$3.4 .10^2$

The leachates are then passed through columns of length 20 cm filled with site-specific geological material. Samples of granite from Fanay-Augères (F) and Cornwall (UK), weathered granite from Auriat (F) and glauconitic sand from Mol (B) are employed, together with pure quartz as reference. Both oxidizing and reducing conditions are being used for these experiments. In addition, parallel experiments involving reference tracer solutions are also being carried out.

In order to compare migration behaviour of statically and dynamically obtained leachates, a joint experiment is performed with the JRC Ispra. Flow-through leaching of glass spiked with 241 Am is studied in oxidizing and reducing conditions.

Such columns are also considered as components of a prototype probe "FORALAB" now under development, to be used later for in-situ migration tests in boreholes in which actual Eh and pH conditions will be found. Therefore, the effective porosity of the columns is determined by measuring the breakthrough curve of tritiated water. Experiments with calibrated inactive molecules (DEXTRAN) are also performed in order to define the column porosities and the separation limits in the absence of sorption. Differential refractometry is used for detection in this case.

With tritiated water, the column effective porosities were determined as follows : 40 % for pure quartz, 41 % for Mol sand, 36 % for granite from Fanay-Augères, 29 % (average) for Cornwall granite, 25 % for Auriat granite (all granites crushed).

3.2. Vitrified high-level waste in clay and salt environments

These experiments have been carried out at the JRC Ispra. Conditions associated with the Boom clay formation in Belgium and the Gorleben salt dome in Northern Germany have been simulated by employing site-specific natural geological material. Results of percolation tests and speciation experiments of Am, Np, Pu and Tc in the glass leachates, have been used to describe the major mechanisms affecting the distribution and behaviour of radionuclides in the selected systems.

Borosilicate glasses doped with the radionuclide under examination were used as the source term. The soil material used in the experiments was uniformly packed in glass columns of 20 cm length and 2.5 cm diameter and equilibrated with groundwater. The glass leachate was then pumped continuously through the column assuring an upward direction of the flow. Samples of the effluent were collected and the radionuclide concentration measured either directly or after a coprecipitation step with cerium hydroxyde. In the case of experiments with saturated brine a small pre-column consisting of salt granules were inserted immediately before the inlet of the soil column. At the end of the percolation tests the distribution profile of the radioisotope were measured either by cutting the column into thin sections or by using a gamma scanning device.

In groundwater of low ionic strength (representing the clay condition), most of the Am(III) activity leached from the glass matrix was determined by ultrafiltration to be associated with colloidal material (14). After percolation through a glauconitic sand column (from the Mol site in Belgium), the column effluent was found to contain only soluble anionic species of americium; evidence indicating these to be carbonate complexes. The retention of americium was subsequently modelled assuming both a filtration effect of the polydispersed colloïds released from the glass and a sorption saturation mechanism establishing a constant concentration of americium in the final section of the column (15). Percolation tests of ^{237}Np through glauconitic sand columns were performed under oxic conditions, simulating the worst situation which could take place in a geological repository. Despite reports of a high mobility of Np(V)(the main valence state occuring at oxidizing Eh values), all the activity was found to be retained by the soil column. Experiments are currently under way in order to account for this behaviour which appears to be a chromatographic movement of the cationic neptunyl (V) species (16).

Americium shows a different behaviour when the leaching solution is a saturated brine (from a salt dome). In contrast to freshwater systems, colloidal fractions were determined to be negligible, because of the very high ionic strength of the brine solution (5.4 M NaCl). Owing to the mass action of alkali and alkaline earth metal cations, a saturation of the sorption capacity of the soil (collected from the overlying strata of the Gorleben salt dome) was also observed. Good agreement was found between distribution coefficients determined from batch and column experiments suggesting that sorption reactions were at equilibrium in the column. The average retention factor R_f of americium was found to be 168. The predominating species are soluble cationic chloro-complexes of americium whose transport seems to be governed by a simple Langmuir isotherm (17). Similar experiments are being carried out with ^{238}Pu and ^{237}Np (18).

Further work is needed in order to investigate the effect of reducing environments on the geochemical behaviour of radionuclides (19). First results on the influence of the redox conditions were obtained with technetium, which was seen to exhibit a higher retention in glauconitic sands (clay condition) when the experiments were operated in the absence of oxygen. These tests are being carried out in anaerobic chambers simulating conditions representative for the repository. To prevent a pH increase, a constant partial pressure of CO_2 is being kept in the nitrogen atmosphere.

The main mechanisms characterizing the migration of radionuclide in the geosphere have been identified as follows :
(i) Tri-and tetra-valent actinides in fresh groundwaters exhibit a strong tendency to form colloids which are filtered by the soil column.
(ii) Colloidal fractions are negligible in groundwaters of high ionic strength, only soluble species existing.
(iii) Complexation with natural ligands is a process competing with adsorption sites for dissolved radionuclides. Carbonates are the main inorganic complexing agents in fresh groundwaters.
(iv) Reducing conditions, such as those encountered in geological repositories, lead to higher retardation effects.

3.3. Intermediate level waste in granite and clay environments
These series of tests have been carried out by UKAEA Harwell.
The geological materials for these were Carnmenellis granite (UK), Fanay-Augères granite (F), and Anversian glauconitic sand supplied by CEN/SCK, Mol. Groundwater for the experiment was obtained from Fanay-

Augères, and from the same aquifer as the Anversian sand.

Concrete, which is used in these experiments to represent a typical backfill, was obtained from a 20 to 30 year-old weathered slab from the Harwell site. The container for the waste was represented by stainless steel (to minimize ferric hydroxide formation) and its corrosion products by ferric oxide. The waste forms used were intermediate level concentrates from WAK, Karlsruhe, immobilized in cement and intermediate level waste from Marcoule, immobilized in bitumen.

The experiment consists of a source of water, which flowed through a redox potential control cell to a chamber where it contacted the waste form again with control of the redox potential. The flow was then divided between a number of columns (four to each cell, of length 75 cm) which contained sequential layers of canister or corrosion products, backfill and geological material. The outlet from the column flowed through a magnetic valve into a sample bottle.

The columns themselves consisted of either PTFE (for conditions of high Eh) or of acryl (for condition of low Eh). The different construction materials were used because PTFE was found to be permeable to oxygen. The columns were constructed for upward water flow in order to minimize the blocking effects of any fine particulate matter.

The high redox potential columns were packed with a coarse 250 μm polypropylene mesh, 25g of Fe_2O_3, 40g of concrete and filled with granite. The final free volume in the column was 41 % of the total volume for the Fanay-Augères granite and 48 % for the Carnmenellis granite. The low Eh columns were packed with a 550 μm stainless steel mesh, 20g of concrete and filled with either sand or granite. The final free volume in the columns was 40 % for the Fanay-Augères granite, 43 % for the Carnmenellis granite and 35 % for the Anversian sand.

The header tanks and the redox control cells were filled with the groundwater as supplied and the waste form was loaded into the contactor cell. After the water in the header cell had attained the correct redox potential (a period of a few days), the columns were then filled, bypassing the contactor cell. After some five days the bypass was closed and the water from the contactor cell was admitted to the columns.

The flow through the columns was not continuous but in the form of a sample taken with the flow control system; these samples were equivalent to approximately 0.01, 0.05 and 0.2 free column volumes. Each sample was counted by gross gamma counting and stored for actinide and technetium analysis. The sampling was carried out on five days in each week, the columns being closed off at weekends. The flow to some of the columns was interrupted for longer periods (40 days in an extreme case) when difficulties had arisen in the redox control cells.

Samples were taken of the water leaving the contactor cells. These samples were counted for gross gamma activity and were reserved for actinide and technetium analysis.

The column feed and effluent solutions were analysed for actinides by alpha spectrometry. The analyses of the feed solutions showed that there was an initial release of activity which lasted for some ten days in which the plutonium levels reached 0.5 $Bq.ml^{-1}$ (10^{-9} M) and americium 0.2 $Bq.ml^{-1}$ (6.10^{-12} M). Neptunium was not detected in any of the feed solutions at concentrations above 2.10^{-4} $Bq.ml^{-1}$ (3.10^{-11} M). These levels were largely independent of the waste form or water chemistry and were attributed to surface contamination of the wastes. After this initial release the actinide levels dropped rapidly to < 2.10^{-4} $Bq.ml^{-1}$ (4.10^{-13} M) and $\pm 10^{-3}$ $Bq.ml^{-1}$ (3.10^{-14} M) for plutonium and americium respectively from both the cement and the bitumen waste. The analytical procedures adopted to detect these concentrations have prevented separation of the colloidal fractions. No plutonium has been found at concentrations greater than 2.10^{-4} Bq.ml^{-1} in the effluent solutions even after 150 days flow.

Americium was found in the column outlet solutions at concentrations of about 10^{-3} Bq.ml^{-1} (3.10^{-14} M), i.e. substantially the same concentration as the inlet solution.

The caesium, in contrast, has been rapidly, and in the case of the cement waste, almost totally released from the waste forms. It has passed through the geological columns with a velocity almost equal to the water front. Such behaviour is not unexpected for a highly soluble and poorly sorbed nuclide.

3.4. Intermediate-level waste in salt

Experiments at the Freie Universität, Berlin, have included investigations into model actinide reference systems' behaviour as well as those of conditioned waste forms (waste simulates containing inactive fission products). Four saturated salt solutions (pure NaCl solution, Na3 γ salt solution, Na3 γ in a Gorleben groundwater, and a quinary solution derived from oceanic salt deposits at its equilibrium point Q, the "Q-solution"), have been prepared and characterized.

The equilibrium solubilities of standard americium and plutonium hydroxides in these saturated salt solutions have been determined together with the tendency towards colloid formation. Results have indicated solubility limits for americium to lie between 10^{-7} and 10^{-3} M.dm^{-3} (2,20). Investigations have also shown that with the exception of a freshly-prepared plutonium hydroxide dissolving in strongly acidic solution, little evidence of colloid formation (< 100 nm) could be observed. Experiments into determining the influence of the Fe^{2+}/Fe^{3+} couple upon these solubilities (in order to simulate the presence of corroded canister material), have proven extremely difficult as a result of strong pH changes brought about by the hydrolysis of Fe^{3+}.

Waste leaching experiments have been carried out with a cemented simulated waste form. Only very small amounts of activity could be leached into rock salt solution (plutonium concentration : 10^{-9} M.dm^{-3}) although for the Q solution the amount of activity which could be leached was found to be ca. three orders of magnitude greater. As with inactive cement matrix (21), strong changes in pH were observed during the course of these experiments. For the rock salt solutions a pH of 12 was attained after a period of only hours, whereas the final pH of the quinary solution remained at ca. 6.8 as a result of the system's ability to buffer. This important difference is held largely responsible for the somewhat extreme variation in waste leaching levels. As before, the question as to the extent of the influence of the presence of canister corrosion products remains open due to problems associated with the hydrolysis and precipitation of iron at these pH values. Significantly, the ultrafiltration of all leachates have indicated practically the complete absence of colloidal material within the size-range 800 to 1.3 nm (nominal filter porosities).

For the migration experiments, the column themselves were commercially available chromatographic columns consisting of precision borosilicate glass tubing of diameter 2.5 cm. The column packing consisted of either pure sodium chloride, or geological material specific to the Gorleben site: Na3 γ salt and sand from the overlying strata. This material was effectively held between two adaptors allowing columns of various lengths to be easily facilited, although the standard length adopted was 10 cm.

Column porosities were determining by measuring breakthrough curves for
saturated salt solutions containing tritiated water as tracer.

Column migration experiments with the model systems have shown that
whereas for americium no significant differences between the retention
properties of pure sodium chloride or rock salt could be observed, there
were some indications as to this effect in the case of plutonium (22).
This may be a result of the somewhat more complex chemistry of plutonium
in solution (23). Further experimentation has indicated that for both
americium and plutonium, the extent of actinide recovery in the column
effluents to be strongly dependent upon the nature of the solution placed
onto the columns. Thus with actinide nitrate solutions high recovery
levels could be observed (in some cases close to 100 %), although for
solutions derived from the hydroxides very much lower levels were found.
In all cases however, the K_D values remained similar (generally below 0.05
$cm^3 . g^{-1}$). As well as the actinide complexation, solution pH is also
believed to play an important role in determining sorptive behaviour.

The results for column migration experiments involving leachates from
the simulated cement waste form are given in Table 3. These were first
filtered through membranes of pore size 800 nm in order to remove coarse
cement particles. Thus, whereas similar K_D values to those determined for
the model systems were observed, the recovery levels were very much higher
than those previously seen for the hydroxides. At a pH of 12 therefore, a
large fraction of the leached activity was able to pass through the
columns. Ultrafiltration of the column effluents indicated no significant
amounts of particulate matter of size greater than 1.3 nm. For experiments
involving sand columns, it may be seen that very little activity could be
found in the column effluents indicating extensive retention behaviour.
Furthermore the K_D values are increased when compared to those for salt.

Finally, the effect of competitive sorption on the mobility characteristics of actinides through salt has been investigated, whereby solutions
have been doped with lanthanides (caesium and samarium) before being
passed through the columns. Remarkable increases in actinide recoveries
have been observed (24), which may assist in the explanation of the
findings for waste leachates passing through salt. The importance of such
a result when considering models for radionuclide migration in geological
systems need not be stressed, especially after consideration of the
complexities of the waste inventories.

Table 3

Results of integral migration experiments with leachates derived from
simulated ILW in cement

Column material	Effective Porosity (%)	Actinide Recovery (%)	R_F	K_D ($cm^3 g^{-1}$) .
NaCl	36.6	33	1.1	0.02
NaCl	36.6	68	1.2	0.04
RS	38.7	88	1.2	0.04
RS	38.7	72	1.2	0.06
Sand*	40.0	0.3	1.4	0.12
Sand*	40.0	0.5	1.5	0.12

RS : rock salt Na3γ
* : Leachate in RS solution

3.5. Supporting investigations

In the Rijksinstituut voor Volksgezondheid en Milieuhygiëne (NL), the influence of the salt concentration on the migration behaviour in samples of sandy deposits containing glauconite was investigated. For Am and Pu the adsorption may decrease with factors up to 20 when the salt concentration increases from "fresh water" concentrations to 1 molar. For Np such an influence is absent. It takes more than 3 weeks before the adsorption equilibrium is established. Additionnally, the impact of micro-organisms on the solubility of Pu was investigated in closed vessels which allowed control of pH, pCO_2 and growth of bacteria. The pH and E_H were measured continuously. The CO_2 pressure was controlled by the composition of a gas mixture supplied to the system. The growth of micro-organisms was controlled by pulse application of glucose. The results suggest that the main impact of micro-organisms results from their ability to lower the redox potential of the sytem. Probably they are able to reduce PuO_2 to Pu^+ or $PuCO_3^+$.

4. Conclusions

A large body of useful experimental data has been collected for the areas of the MIRAGE Project considered here, both in the field and in the laboratory. Two main qualitative conclusions can be drawn :

(i) The examples of natural geological migration systems considered in the Project indicate that radionuclide analogues migrate only short distances over very long time-spans.

(ii) The laboratory migration experiments have shown that the amount of radioactivity left in groundwater after the leaching of waste forms and subsequent percolation through columns of geological materials is extremely low.

Generally speaking, these data indicate that the geological media considered for radioactive waste disposal do act as efficient barriers vis-à-vis radionuclide migration.

Furthermore, quantitative estimates of transport and retardation processes are now becoming possible. This improvement is accompanied by the parallel development of more adequate calculation tools (25). All these results can now be used to improve the reliability of predictive radionuclide migration modelling.

REFERENCES

1. Commission of the European Communities (1984)
   The CEC Project MIRAGE. Programme document.
   CEC Report n° EUR 9403, Luxembourg.

2. Project MIRAGE : first summary progress report (1985).
   CEC Report n° EUR 9543, Luxembourg.
   (Second Summary progress report under publication as EUR report)

3. BARBREAU, A. et al (1985)
   Determination of the characteristics of deep granites.
   Paper n° VII-1, ibid.

4. BOURKE, P. et al (1985)
   Associated research and development : measurements of characteristics determining radionuclide movement through, and mechanical behaviour of fractured rock.

Paper n° VII-2, ibid.

5. HEMMING, C.R. et al (1985)
    Shallow land burial analysis.
    Paper n° V-6, ibid.

6. CHAPMAN, N.A. and SARGENT, F.P., editors (1984)
    The geochemistry of high-level waste disposal in granitic rocks.
    Proceedings of an AECL/CEC Workshop, Minster Lovell, September 1983.
    CEC Report n° EUR 9162/AECL Report 8361, Luxembourg.

7. Rare Earth Symposium, SFMC, Paris, December 1983.

8. MacKENZIE, A.B., RIDGWAY, I.M., SCOTT, R.D., HOOKER, P.J., WEST, J.M.
    and McKINLEY, I.G. (1984)
    A natural analogue study of long-term ($10^3 - 10^4$ a) radionuclide
    migration in saturated sediments from Loch Lomond, Scotland.
    (attribution to follow - in press).

9. IVANOVICH, M. and HARMON, R.S. (1982).
    Uranium series disequilibrium : applications to problems in Earth
    Sciences.
    Oxford University Press.

10. CHRISTOFI, N., WEST, J.M., ROBBINS, J.E. and McKINLEY, I.G. (1983)
    The geomicrobiology of the Harwell and Altnabreac boreholes.
    Rep. Inst. Geol. Sci. FLPU 83-4.

11. CHRISTOFI, N., WEST, J.M., PHILP, J.C. and ROBBINS, J.E. (1984)
    The geomicrobiology of used and disues mines in Britain.
    Rep. Brit. Geol. Surv. FLPU 84-5.

12. PHILP, J., CHRISTOFI, N. and WEST, J.M.
    The geomicrobiology of calcium montmorillonite (Fuller's Earth).
    Rep. Brit. Geol. Surv. FLPU 84-4.

13. McKINLEY, I.G., VAN DORP, F. and WEST, J.M. (1984)
    Modelling microbial contamination of a deep geological repository for
    HLW. In Proceedings of Workshop on "The role of micro-organisms on the
    behaviour of radionuclides in aquatic and terrestrial systems and
    their transfer to man", Brussels (in press).

14. AVOGADRO, A., MURRAY, C.N., DE PLANO, A. and BIDOGLIO, G. (1981)
    Proc. Int. Symp. on Migration in the terrestrial environment of
    long-lived radionuclides from the nuclear fuel cycle.
    IAEA, Knoxville.

15. SALTELLI, A., AVOGADRO, A. and BIDOGLIO, G. (1984)
    Nuclear Technology, 67, 245.

16. BIDOGLIO, G., DE PLANO, A., AVOGADRO, A. and MURRAY, C.N. (1984)
    Inorganica Chimica Acta, 95, 1.

17. BIDOGLIO, G., AVOGADRO, A., DE PLANO, A. and LAZZARI, G.P. (1984)
    Radioactive Waste Management Nuclear Fuel Cycle.

5, 145.

18. AVOGADRO, A., BIDOGLIO, G. and SALTELLI, A. (1984)
    Workshop on the source term for radionuclide migration under realistic
    repository conditions.
    US-DOE/OECD-NEA, Albuquerque.

19. BIDOGLIO, G., CHATT, A., DE PLANO, A. and ZORN, F. (1983)
    J. Radioanal. Chem.
    79, 153.

20. MARX, G., ESSER, V., FLAMBARD, A.R. and KEILING, Ch., in Fachband Nr.
    14 zum PSE-Abschlussbericht, 3-1, (1985)

21. FLAMBARD, A.R. and FUSBAN, H.U. (1984)
    The influence of cement degradation on the density and pH of concen-
    trated salt solutions, FU Berlin.

22. FLAMBARD, A.R., FUSBAN, H.U. and MARX, C. (1984)
    Sorption von Plutonium und Americium im Versatz eines Grubengebäudes
    eines Endlagers in einem Salzstock, Abschlussbericht, FU Berlin.

23. Plutonium Chemistry, Eds. Carnall, W.T. and Choppin, G.R., ACS
    Symposium Series 216 (1983).

24. MARX, G., FLAMBARD, A.R., FUSBAN, H.U. and KEILING, Ch., in Fachband
    Nr. 14 zum PSE-Abschlussbericht, 4-15 (1985).

25. de MARSILY, G. et al (1985).
    Modelling of radionuclide release and migration.
    Paper n° VIII-3, ibid.

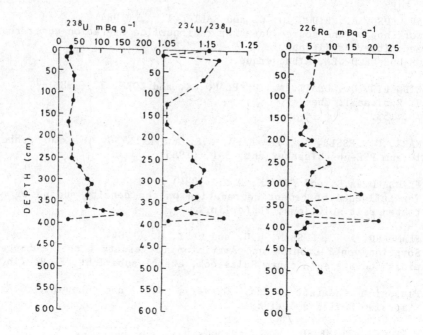

FIGURE 1 : Elemental profiles in Loch Lomond Sediments

FIGURE 2 : Column set-up at JRC ISPRA for integral migration experiments

DISCUSSION

J. LEWI, CEA-IPSN-SAED, France

Do you think that natural analogues can be used for validation of models?

B. COME, CEC Brussels

Studies on natural analogues may permit a better understanding of processes and consequently improve predictive modelling. However they cannot give all the answers, as the environmental conditions are not the same as those which will be present in waste repositories. Moreover the radionuclides are not the same, as in nature one does not find the radioelements of main interest in radioactive waste disposal studies. For example the study of Loch Lomond sediments has shown that contrary to what one could expect the Iodine has remained fixed in the marine band in which it was originally deposited. It is also possible to establish a relation between organic matter present in that layer and the retention of Iodine. So we are now forced to examine the reasons of the observed phenomena.

R. KOBAYASHI, JGC Corporation, Tokyo

You mentioned that in granite the mobility of rare earths decreases with temperature increase. How do you explain this?

P. OUSTRIERE, BRGM Orléans

We have observed this behaviour of Lanthanides in the laboratory as well as at a site at Langenberg. We cannot yet exactly explain the mechanisms which make the Lanthanides less mobile at higher temperature, but it has to do with the different alteration processes which is of the argillic type at low temperatures of about 100°C and of the propylitic type at higher temperatures around 200-250°C.

R. KOBAYASHI, JGC Corporation, Tokyo

You used crushed granite in your column experiments, but in actual circumstances you will have granite with some fractures. Do you have some relationship between crushed granite and fractured granite?

B. COME, CEC Brussels

The column experiments performed with crushed granite are of a preliminary nature and are not entirely representative for the real conditions. They permit however a better understanding of migration processes, but for the moment it is not possible to give a simple relationship between crushed granite and fractured granite. An experiment is planned to be performed with core samples presenting a longitudinal fracture.

J. DELMAS, CEA Cadarache

At Cadarache a research project has been started in which the crushed granite used in column experiments will be replaced by real overcored fractured granite samples. This project is part of a programme using the FORALAB probe developed by CEA. This probe permits borehole experiments by maintaining the water-gas-substrate equilibrium. The programme aims to reproduce as realistically as possible a saturated fractured environment by replacing the presently used filling materials by clays from the site and replacing the materials by real overcored fractured granite.

K. KUHN, GSF Braunschweig

Concerning the experiments on sorption of Americium by salt samples from the Gorleben site, you said that this was related to impurities in the salt. Could you say what kind of impurities were involved and what was the ratio between sorption by salt and by impurities.

B. COME, CEC Brussels

Experiments performed by the University of Berlin on pure crushed salt samples and on crushed salt samples from the Gorleben site have shown better retention properties of the latter. The explanation given is that the impurities, e.g. clay particles, play a role in better fixing the Americium. This effect is even stronger with plutonium.

A. AVOGADRO, JRC Ispra

The same observation has been made at experiments performed at Risø (DK). Natural salt samples have better sorption properties with regard to trivalent radionuclides than pure salt.

THE GEOFORECASTING APPROACH AND LONG—TERM PREDICTION OF EVOLUTIVE NUCLIDE MIGRATION

P. PEAUDECERF (B.R.G.M.)
M. D'ALESSANDRO (C.C.E.)
M. CANCEILL (B.R.G.M.)
J. FOURNIQUET (B.R.G.M.)
P. GODEFROY (B.R.G.M.)

Abstract

Safety analysis of long-lived radioactive waste repositories is dependent on natural site evolution scenarios. This is a complex problem given that there may be a large number of interdependent factors whose occurence probability is usually very difficult to quantify. A major methodological effort has therefore been undertaken in this field under the terms of the Community's R & D programme. Research teams from the JRC and the BRGM tackled the problems by way of two complementary approaches; A critical analysis of existing methods was also undertaken at the request of the Commission. This paper sets out to summarize the work and, as far as possible, make a methodological evaluation.

Clearly, any **safety analysis** of high level and long-lived radioactive waste burial sites must take into account what happens to the various barriers over the full period of nuclide activity. The evolutive behaviour of the geological barrier is particularly important over the very long term.

It is therefore essential to :
a) define site-specific evolution **scenarios** and,
b) evaluate the **consequences** of site evolution for possible nuclide migration conditions.

The geological medium may be subject either to natural events (i.e. gradual phenomena or sudden events) or to the repercussions of human activities. This paper considers only natural factors (the effects of human activities have not yet been studied to the same degree). A variety of approaches are used to tackle the above problems; they may roughly be divided into three main types (13) :

1. Determinist studies based on specific geological containment failure hypotheses carried out in terms of what are felt to be the most plausible scenarios; "worst case analysis" studies are a special case.

2. "Probabilist" studies based on fault tree analysis (FTA) or event tree analysis (ETA) based on the hypothesis that the repository/geological medium system can be broken down into a series of independent elementary components behaving according to a binary logic (i.e. fault or no fault) quantifiable in terms of occurence probability; the aim here is to set out in graphic terms all the possible links in the chain

through which geological barrier failure may occur, the fault itself being the top event working down to a series of primary events via a cascade of intermediate events; it is thus possible to calculate the probability of a top event by reference to primary event occurence probabilities; in other words, FTA is a deductive process. ETA, on the other hand, is an inductive technique starting from primary events and progressing logically to their immediate effects and ultimately to system failure; studies along these lines have been extensively developed, in particular by research teams from the Ispra JRC, where this method has been gradually improved to good effect (4, 16).

3. "Mixed" studies whereby, in parallel with a probabilist approach, some stages in the safety analysis process are tackled in determinist terms and an attempt is made to quantify the disposal risk by weighting estimated damage with an occurence probability factor; these studies are often based on stochastic simulation techniques for formulating fault scenarios (e.g. Monte Carlo, Markov Chain, Latin hypercube, adjoint state method, etc.).

All these approaches have their advantages and disadvantages (or insufficiencies) given the complexity of nature and our relative state of knowledge (15). To bring out the complementarity of the various approaches, we shall describe here the methodological studies pursued under the Community research programme by reference to a probabilist approach adopted by the Ispra JRC and geoprospective modelling of the effects of scenarios based on past events formulated by the BRGM.

EVALUATION OF THE PROBABILITY OF GEOLOGICAL EVENTS

To enable a quantitative evaluation to be made of the risk involved in the disposal of radioactive waste in deep geological formations, we need to develop models capable of predicting the consequences of possible leak mechanisms. As these consequences are liable to vary widely depending on the various release modes, it is essential to find out which of the possible disposal site evolution scenarios have the highest occurence probability. In other words, we have to select, from the many possible scenarios, a reasonable number with a relatively high occurence probability.

It therefore follows that all the possible future site geology configurations must be classified in terms of occurence probability, an essential exercise given that probability evaluation is a major element in the geological appraisal process. The current state of the art is such that the predictive dimension of geology as a science is still limited (it is more concerned with describing the past history of the earth than with its evolutive future). Another point to be borne in mind is that, since most of the events have only very low occurence probability ratings, it is rarely possible to deduce probability values from a statistically representative set of data. The result is that, for a large number of processes looked at in terms of risk analysis, occurence probability evaluation is largely dependent on expert's subjective judgement, based in turn on a long process of quantitative and qualitative evaluation drawing on the current state of knowledge. In other words, the probability of geological events amounts to a quantitative measurement of the expert's "degree of conviction" based on the sum total of his own knowledge. As experts' knowledge of the subject

is liable to change in the light of new experimental findings or theoretical innovations, their opinions are liable to change accordingly,
with the result that a more sophisticated probability evaluation can be
adopted. It is important to note that, according to the "subjectivist
school of thought", probabilities depending essentially on relative
frequency should not be given any more weight than those based on subjective appraisal. There is no such thing as a "true" probability; all
probabilities are in fact "quantitative measures of opinions" (12).

Generally speaking, the range of events liable to trigger off a
fault in the geological containment is fairly wide; from sudden random
events (whose probability can be statistically evaluated) to more
complex events for which occurence probability evaluation will have to
be based on certain basic hypotheses and subject to clearly defined
limits. Tectonic events capable of rupturing the geological barrier can
normally be classed as random events, belonging to the group of phenomena for which a statistical data set is available (or at least obtainable).

Fault formation frequency within a selected zone may be regarded
as constant over a tectonic phase, and occurence may roughly be regarded
as instantaneous on a geological time scale.

By contrast, certain geomorphological processes capable of setting
in motion a direct radionuclide release up to the surface are often due
to the combined action of a number of events; the gravity of their
effects normally depends on the occurence of a particular sequence of
phenomena.

For instance, a glaciation cannot readily be regarded as a random
event; nonetheless, intuitively, its effects in a zone which is already
affected by pleistocene glaciations may appear substantial. The probability of this kind of event must therefore be carefully evaluated even
if - as in this case - statistical data are usually lacking. A new
glacial phase at a disposal site is a slow and continuing process which
may occur with varying degrees of severity (like any other erosion
process), with the result that it may be difficult to evaluate the
probability factor in the light of the event frequency alone.

Where the scenario is due to the combined action of a number of
events and/or conditions, application of the fault tree analysis method
should have certain advantages. For one thing, it is easier to deal with
the most complex phenomena by reducing them to a set of elementary
events which readily lend themselves to study. The slow and continuing
processes may be analysed by classifying them according to a speed
scale, each being identified by its probability value (4).

But however much special techniques may be useful in rationalizing
the approach to be adopted, estimation of the probability of geological
processes will always rest (in part at least) on the expert's subjective
judgement. This does not mean to say, however, that such limitations
render the results of probability estimation meaningless : the expert's
professional experience, coupled with the data available, is a highly
valuable tool in carrying out the evaluation exercise.

The probability of geological processes must therefore be estimated first of all on the basis of an analysis of the relevant documentation, followed by an in-depth study of the geological history of the
region, with special reference to the Quaternary, so as to identify the
geomorphological processes and the events which are likely to be important in terms of the site.

Finally, the average or external conditions to be expected over the period of time considered must be evaluated. A probability rating can therefore be postulated on the basis of all these elements.

Once the possible scenarios have been classified, the risk analysis procedure may be continued by studying the consequences. Each form of release will require a special model with a view to evaluating consequences in terms of population doses. At this stage, the most plausible scenario(s) will have to be refined so that the consequence simulation stage can follow on correctly. To this end, most of the data collected for the probability of evaluation exercises will be reused. Ultimately, the two objectives, i.e. release occurence probability and the relative consequences in terms of population dose, will constitute a relative damage yardstick for a given scenario.

GEOFORECASTING APPROACH

The 'prospective' approach sets out to project past natural trends into the future and thus to compare situations in the context of consistent and plausible scenarios, where the projection horizons may be at any stage in the future.

Generally speaking, this approach does not set out to predict future events, but simply to examine possible future configurations with a view to defining, from a practical point of view, the best ways of tackling the range of conceivable hazardous situations. It is thus essentially an operational (i.e. practical) approach.

Application of this approach to containment performance analysis for radioactive waste deposited in deep geological formations was first suggested at the first Community conference on radioactive waste in 1980 (3).

Prospective scenarios are initially drawn up on the basis of historical analysis of the geosphere and biosphere systems with a view to defining the major evolutive trends, the major cyclical phenomena and random events to yield characteristic traits to be projected into the future in the context of consistent and plausible scenarios.

At the same time, the prospective approach prompts study of the natural and artificial environments with a view to ascertaining their interactions in a purely deterministic context (e.g. interaction between waste, repository, surrounding rock, receiving massif, etc.).

The scenarios are therefore drawn up in terms of increasing complexity involving both natural evolution and the impact of the installation on the geological medium; there are four basic scenario types to facilitate comparison (13) :

- The stable environment scenario (strict determinism), which presupposes that the current natural conditions are immutable and that the medium only evolves under the impact of the repository. This hypothesis is plausible to a horizon of several centuries.
- The retrospective scenario (historical determinism), which presupposes the creation of a waste repository 10 000, 100 000 and a million years ago, and evaluates the impact of the repository on the geosphere bearing in mind the natural evolution of the paleosite revealed to us by a study of the Quaternary (e.g. climate, geology and biology).
- The tendential scenario (subjective probability = plausibility), which takes account of natural tendency-based evolution phenomena determined by historical analysis of the environment at various levels of complexity (convergence of tendential events).

- The disaster scenario (strict probability), which complicates the above scenarios by taking into account sudden catastrophic events wherever the defined evolution context permits (e.g. faulting or reactivation of an existing fault in the repository, volcanic eruption, major earthquake, accidental human intrusion, etc.); purely catastrophic scenarios may be formulated for the operational phase of construction of the repository and waste disposal : e.g. accidental flooding, collapse of a gallery, etc.

The research work carried out at the BRGM in this field in the context of the Commission's indirect action programme have been restricted so far to the prospective approach to the natural geological evolution of a disposal site.

In this initial phase, the following points have been covered :

1 - Inventory and analysis of the external and internal geodynamic factors dictating the natural evolution of a site, characterized by scalar parameters or spatial or spatio-temporal distributions,

2 - definition of the interrelations between these factors,

3 - selection and ranking of the most significant factors, bearing in mind the location of the sites considered,

4 - construction of (initially relatively simplified) scenarios.

The study was carried out over as wide as spectrum as possible, concentrating however on the most plausible locations for French or north-west European sites. By way of illustration, the instrument developed for setting up scenarios has been applied to actual geological locations with real site potential.

INVENTORY AND QUANTITATIVE ANALYSIS OF FACTORS

Initially, a detailed analysis was carried out for each type of geological phenomenon with potential for affecting the natural evolution of the site, concentrating on the following points :

1 - the nature and appearance or triggering conditions of the phenomenon (dependence vis-à-vis other factors),

2 - the quantitative parameters likely to affect the evolution process and the variability thereof in spatial and temporal terms,

3 - the effects of the said phenomena on the state of the site.

A separate report was drawn up for each study (5, 6, 7, 8, 9, 10, 11, 12). Detailed analysis of these external or internal geodynamic factors enabled researchers to select and classify the most significant factors for the various types of geological medium and to specify how they should be integrated into the site evolution scenarios.

Fig. 1 shows how the various factors interact. It is thus possible to make a choice of simple parameters to show temporal evolution :

- of the site ("state variables"; e.g. altitude)
- of the characteristic waste containment characteristics : essentially the waste-to-outlet transfer time.

SETTING UP THE SCENARIOS

A special programme entitled CASTOR (Construction automatique de scénarios d'évolution d'un site de stockage de déchets radioactifs) has been set up to guarantee the following four criteria : (14)

- Generalization : to take into account the maximum number of phenomena and situations
- Flexibility : to make techniques simpler or more sophisticated depending on requirements and to make changes to the process by a system of coupled modular programmes

- Reproducibility : to repeat the scenarios with variations in certain parameters (important in terms of sensitivity analysis)
- "Neutrality" :   to enable the user to decide whether or not to apply certain mechanisms which are not systematically imposed important in terms of controversial mechanisms).

The site is characterized by situation variables listed in Table I. The CASTOR programme controls a time iteration loop t enabling the evolution of the variables to be defined according to time steps fixed by the user.

The interactive dialogue facility with the user or with the automatic control facility are designed to enable two distinct types of site evolution mechanism to be processed :
- "Identified" mechanisms, corresponding to ineluctable processes which can be readily simulated (e.g. erosion).
- "Controlled" mechanisms, where the triggering and subsequent evolution conditions are known to be contentious, so that the mechanisms have instead been left under the user's control; foolproof devices have simply been introduced to verify the consistency of the information introduced by the user; This will be the case, for instance, in introducing a glaciation factor.

In most cases, very simple calculation rules have been applied, initially at least, to simulate evolution mechanisms :
- the evolution of a phenomenon is modelled in terms of a simple time function (e.g. uniform speed)
- its consequences are restricted to those phenomena considered essential in terms of the site and the repository.

The importance of the automatic programme is therefore to be found more in the management of these various modules (e.g. linkage, consistency, etc.), and the facility for multiplying the tests. The only relatively sophisticated module calculates variations in hydraulic parameters, especially permeability, as a function of the state of the site. This particular module was adapted from a calculation programme written at the BRGM for modelling the coupling of hydraulic, thermal and mechanical phenomena in a fractured medium.

The programme was applied experimentally at an actual site chosen arbitrarily in a granite massif in the west of France (Massif Armoricain), in an area of old baserock, at low altitude and within 30 km of the present coastline.

The hydraulic configuration is very simple (Fig. 2), enabling the system to be represented by a two-dimensional vertical grid system with triangular finite elements. The programme has been adapted to calculate three-dimensional permeability to cope with the eventuality of sub-vertical fractures (14).

The current geomorphological state of the site is set out in Fig. 3. To illustrate how the method is used, a series of diagrams (Figs 3-6) shows the evolution of the site through one of the scenarios under study, viz. the effects of the establishment and melting of an ice cap in 100 000 years. Although the occurence of a new global glacial epoch is more or less a matter of certainty within the stated period, the development of an ice cap on this particular site would count as a much less likely event. Nonetheless, it is of methodological interest; progressive glaciation (in 7 000 years) is, according to this scenario, accompanied by fluctuations in sea-level, isostatic compensation, glacial and fluvatile erosion and changes in the permeability of the fractured massif. All these phenomena are relevant to the time scale of radionuclide migration from the repository to the point of release. It therefore follows that the containment parameters vary, and the simulation values at different time steps are set out in table 2.

Automatic construction of natural site evolution scenarios, according to the principles adopted here, is only at a preliminary stage. Only a few simple tests have so far been carried out at an actual site chosen in an actual geological medium.

However, the few tests which have been carried out have underlined the importance of this kind of approach in identifying the interactions between the various evolution mechanisms and taking account of the progressivity factor. The groundwork has now been done for a new approach to supplement the conventional event-related analyses.

At this stage, a large number of improvements are still envisaged, and must be incorporated as the research programme develops :
- at the scenario construction stage : by improving the sophistication of certain specialized modules (e.g. incorporating a hysteresis phenomenon in vertical movements) or of the scenarios themselves, by incorporating the occurence of "catastrophic" events;
- at the result analysis stage : by introducing value bands and quantification of site variable-related uncertainties into the various parameters involved in the specialized modules, using sensitivity analysis to highlight the critical parameters and mechanisms.

These results will only be available in conjunction with detailed site studies to enable the sophistication of modelling techniques to be brought into line with the degree of detail of the processed data.

We are restricted at present to estimating the natural site response to natural events. To be able to attain our proposed operational objectives, we shall have to evaluate the way the repository responds to evolution and vice versa, and ultimately express the global response (i.e. site and disposal) in terms of risk.

REFERENCES

1. DE FINETTI, B. (1974). Theory of probability John Wiley, N.Y.
2. APOSTOLAKIS, G. (1978). Nuclear safety, 19, 305
3. MASURE, P. et VENET, P. (1980). Some considerations on radioactive waste disposal in continental geological formations. Proceedings of the first European Community Conference on radioactive wastes, May 20-23, 1980, Luxembourg
4. D'ALESSANDRO, M. and BONNE, A. (1981). Radioactive waste disposal into a plastic clay formation : a site specific Exercice of probabilistic assessment of geological containment, Radioactive Waste Management, Vol. 2, Harwood Acad. publ., Paris
5. FOURNIGUET, J. (1982). Etude géoprospective d'un site de stockage ; les mouvements verticaux : causes et quantification, rapport B.R.G.M. n° 82 SGN 1012 GEO, Orléans, France, septembre 1982
6. GADALIA, A. et VARET, J. (1982). Etude géoprospective d'un site de stockage ; l'activité volcanique, rapport B.R.G.M. n° 83 SGN 010 STO, Orléans, France, décembre 1982
7. BILLAUX, D. (1983). Etude géoprospective d'un site de stockage ; étude des variations de la perméabilité avec le temps, rapport B.R.G.M., n° 83 SGN 189 GEG, Orléans, France, mars 1983
8. GODEFROY, P. (1983). Etude géoprospective d'un site de stockage ; la prise en compte de l'activité sismique, rapport B.R.G.M., n° 83 SGN 301 GEG, Orléans, France, avril 1983
9. GROS, Y. (1983). Etude géoprospective d'un site de stockage ; tectonique prospective : durée des phases compressives et distensives récentes, évolution du champ de contraintes dans les 100 000 ans à venir, rapport B.R.G.M. n° 83 SGN 210 GEO, Orléans, France, juin 1983
10. COURBOULEIX, S. (1983). Etude géoprospective d'un site de stockage ; climatologie : évolution du climat et glaciations, rapport B.R.G.M. n° 83 SGN 143 GEO, Orléans, France
11. LAVILLE, P. et LAJOINIE, J.P. (1983). Etude géoprospective d'un site de stockage ; mécanismes d'altération et d'érosion, rapport B.R.G.M., n° 83 SGN 541 GEO, Orléans, France, août 1983
12. BILLAUX, D. et ROBELIN, C. (1983). Etude géoprospective d'un site de stockage ; dômes de sel : étude bibliographique sur les conditions de leur formation, rapport B.R.G.M., n° 83 SGN 657 GEG, Orléans, France, septembre 1983
13. MASURE, P., GODEFROY, P. et IMAUVEN, C. (1983). Les analyses de sûreté du confinement géologique des déchets radioactifs à vie longue ; revue critique des méthodes existantes et proposition d'approche prospective, rapport EUR 8587 EN/FR, Office des publications officielles des communautés européennes, série : Sciences et techniques nucléaires, Luxembourg.
14. CANCEILL,M. et al (1984). Etude géoprospective d'un site de stockage ; simulation de l'évolution d'un site à l'aide du programme "CASTOR", rapport B.R.G.M., n° 84 SGN 229 STO, Orléans, France, juin 1984
15. BILLAUX, D. et al, (1984). Evolution naturelle d'un site de stockage de déchets radioactifs à vie longue, essai de modélisation dans le cadre d'une approche géoprospective, 27ème congrès géologique international, Moscou, août 1984
16. BERTOZZI, G., D'ALESSANDRO, M., SALTELLI, A. Waste disposal into a plastic clay formation : risk analysis for probabilistic scenarios. Report EUR in press

TABLE 1
SITE VARIABLES, CONTAINMENT AND EVOLUTION PARAMETERS

SITE-SPECIFIC VARIABLES	1. Altitude (m) 2. Depth of repository (m) 3. Morphology: 0 under water; 1 plain of hillside; 2 mountain 4. Thickness of ice (m) 5. Alterite thickness (m) 6. Average density of overlying rock 7. Young's module (bar) 8. Poisson's coefficient 9. Isostatic load (bar) 10. Vertical stress (bar) 11. Maximum horizontal stress σ_{Hmax} (bar) 12. Minimum horizontal stress σ_{Hmin} (bar) 13. Stress conditions 14. Azimuth σ_{Hmax} (bar) 15. Number of fracture families in each permeability zone 16. Average direction ⎤ 17. Gradient for each 18. Average fracturation density ⎟ family 19. Average fracture opening ⎦ 20. Groundwater above (m) 21. Groundwater below (m) 22. Salt dome depth
SITE-SPECIFIC EVOLUTION PARAMETERS	23. Erosion rate (m/year) 24. Sedimentation rate (m/year) 25. Isostatic movement rate (m/year) 26. Rate of epirogenesis (m/year) 27. Rate of ice cap formation (m/year) 28. Melting rate (m/year) 29. Rate of sub-glacial erosion (m/year) 30. Rate of salt dome rise
REGIONAL SITE VARIABLES	31. Sea-level vis-à-vis t_o = 0 (m)
REGIONAL EVOLUTION PARAMETERS	32. Sea-level variation rate (m/year)
CONTAINMENT PARAMETERS	33. Permeability tensor (first order) 34. Permeability tensor (second order) 35. Porosity (%) 36. Lentgh of upstream flow line (m) 37. Length of downstream flow line (m) 38. Transfer rate (m/year) 39. Transfer time (years)

TABLE 2

EXAMPLE OF EVOLUTION OF CONTAINMENT PARAMETERS

IN A GLACIATION SITUATION

TIME AFTER to (years)		PERMEABILITY (zone 1) (m/s)	PERMEABILITY (zone 2) (m/s)	TRANSMISSION VELOCITY (m/year)	TRANSFER TIME (years)
	10 000	$0.65 \ 10^{-10}$	$0.52 \ 10^{-10}$	1.15	858
	20 000	$0.65 \ 10^{-10}$	$0.52 \ 10^{-10}$	1.15	857
	30 000	$0.65 \ 10^{-10}$	$0.52 \ 10^{-10}$	1.15	857
	31 000	$0.37 \ 10^{-10}$	$0.30 \ 10^{-10}$	0.65	1 500
Initiation of glaciation	32 000	$0.31 \ 10^{-10}$	$0.17 \ 10^{-10}$	0.32	2 617
	33 000	$0.12 \ 10^{-10}$	$0.10 \ 10^{-10}$	0.22	4 550
	34 000	$0.7 \ 10^{-11}$	$0.6 \ 10^{-11}$	0.12	7 894
	35 000	$0.4 \ 10^{-11}$	$0.3 \ 10^{-11}$	0.07	13 676
	36 000	$0.2 \ 10^{-11}$	$0.2 \ 10^{-11}$	0.04	23 676
	37 000	$0.1 \ 10^{-11}$	$0.1 \ 10^{-11}$	0.024	40 992
	40 000	$0.1 \ 10^{-11}$	$0.1 \ 10^{-11}$	0.024	40 934
	60 000	$0.1 \ 10^{-11}$	$0.1 \ 10^{-11}$	0.024	40 550
Ice melting	61 000	$0.9 \ 10^{-11}$	$0.8 \ 10^{-11}$	0.16	5 917
	62 000	$0.66 \ 10^{-10}$	$0.53 \ 10^{-10}$	1.16	843
	63 000	$0.66 \ 10^{-10}$	$0.53 \ 10^{-10}$	1.16	843
	64 000	$0.66 \ 10^{-10}$	$0.53 \ 10^{-10}$	1.17	843
	65 000	$0.66 \ 10^{-10}$	$0.53 \ 10^{-10}$	1.17	842
	66 000	$0.67 \ 10^{-10}$	$0.53 \ 10^{-10}$	1.15	854
	70 000	$0.67 \ 10^{-10}$	$0.53 \ 10^{-10}$	1.16	853
	100 000	$0.67 \ 10^{-10}$	$0.54 \ 10^{-10}$	1.16	851

N.B. The fine degree of accuracy has no absolute significance; it simply serves to bring out any minor variations in the values.

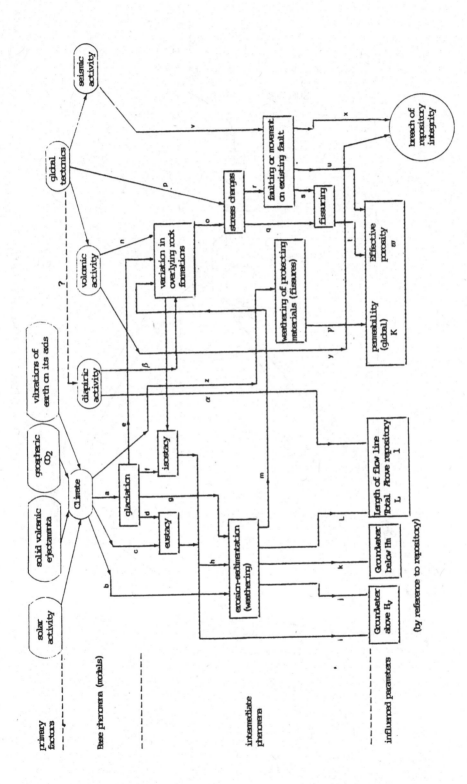

FIG. 1 – INITIAL INVENTORY OF GEODYNAMIC FACTORS AND POSSIBLE EVOLUTIONS

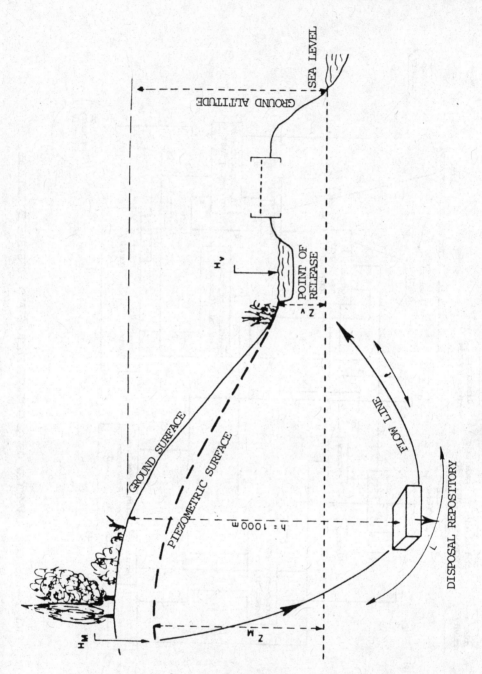

FIG. 2 – HYDRAULIC CONFIGURATION

country rock of brioverian (upper proterozoic) schists

intrusive cadomian (lower paleozoic) granite

main fractured rock gaps

15 m alterite giving uniform zone
cover (simplified ideal case)

hornfels

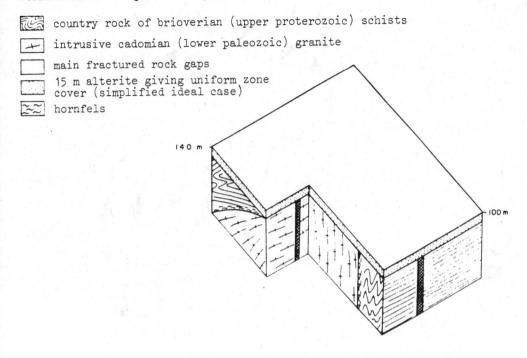

FIG. 3 – EVOLUTION OF MODEL SITE STAGE t_O

surface alterite

local collapse basin with sand-clay fill

undifferentiated granodiorite

hornfels

brioverian schists

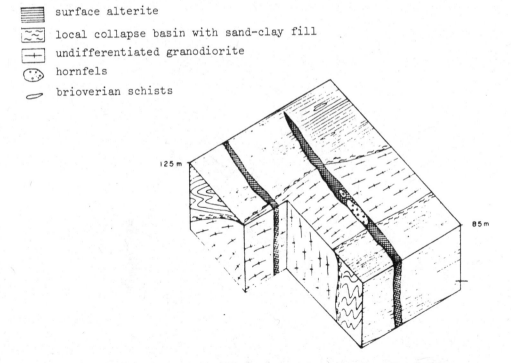

FIG. 4 – EVOLUTION OF MODEL SITE STAGE (t_O + 300) YEARS

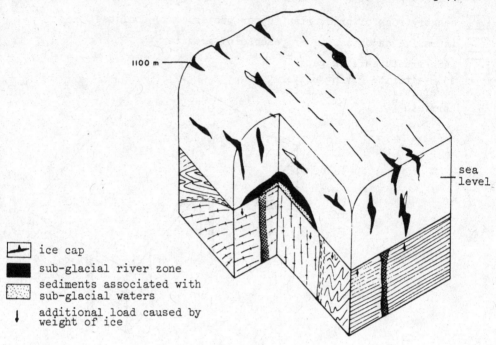

ice cap
sub-glacial river zone
sediments associated with
sub-glacial waters
additional load caused by
weight of ice

1100 m

sea
level

FIG. 5 – EVOLUTION OF MODEL SITE STAGE (t_0 + 67 000) YEARS

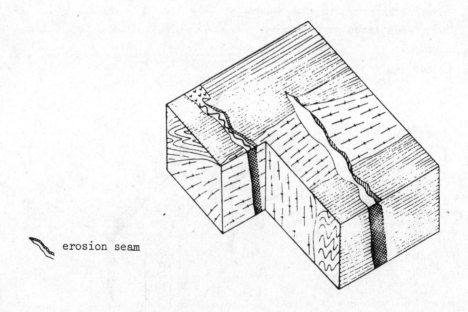

erosion seam

FIG. 6 – EVOLUTION OF MODEL SITE STAGE (t_0 + 80 000) YEARS

DISCUSSION

R. KOBAYASHI, JGC Corporation, Tokyo

The code CASTOR that you mentioned, is it 3-dimensional in treating hydrological problems?

P. PEAUDECERF, BRGM Orléans

The code CASTOR is very flexible as one can introduce modules in function of the case studied. In the case I presented, the hydrological model was rather simple and two-dimensional.

R. KOBAYASHI, JGC Corporation, Tokyo

What was the significant parameter resulting from your sensitivity analysis?

P. PEAUDECERF, BRGM Orléans

We have not performed a systematic parameter sensibility study, so the results are only partial. However the significance of parameters will be site specific.

F. GIRARDI, JRC Ispra

I agree with you on the interest to test your method on specific sites. Do you have any projects on particular sites?

P. PEAUDECERF, BRGM Orléans

We are at present testing the method for the evolution of clay formations on a fictive site in the Parisian "bassin" . As no drillings have been carried out we don't have real data. Furthermore we are testing the method on some granitic massifs, but we are open for other proposals.

TECHNOLOGY FOR THE IMPROVEMENT OF SHALLOW LAND BURIAL

K. BRODERSEN and R. ANDRE JEHAN
NRL Risø   CEA ANDRA
Denmark   France

Summary

Low cost and easy operability are the main arguments for disposal of low level radioactive waste by shallow land burial. The risk of activity release is partly determined by waste properties and site characteristics, but it is also influenced by the general construction of the burial facility and by the properties of materials used in engineered barriers. From none to many such barriers may be present in a burial system. There is a continuous transition from real surface disposal to systems situated at considerable depth. To evaluate the balance between cost and risk residue for various alternatives, series of documented burial concepts are needed. Three variants of a fairly advanced generic system developed at Risø are described. Tentative cost figures are given. Results from material studies of various types of concrete and of bitumen proposed as barrier materials are presented. The importance of the development of methods for the study of interaction processes and long-term behaviour of barrier materials is underlined. As an example of the documentation and possible improvement of barriers at an operating site, French experiments with water penetration into the soil layer on top of the tumuli at the Centre de la Manche are described.

1. INTRODUCTION

Enormous volumes of ordinary waste: refuse, sewage sludges, fly ash, industrial wastes and mining wastes are disposed of every year often by some type of shallow land burial. Compared with these volumes the amounts of low- and medium-level radioactive wastes generated by the nuclear fuel cycle are small. Based on the CEC analysis from 1982 (1), the present annual volume is about 60,000 m^3. It is about 0.2 to 0.5 kg per year and person in the Community, much less than the about 500 kg conventional waste per year and person typical for industrialized countries. (These figures do not include conventional mining waste, nor uranium mill tailings, the only really bulky waste product from the nuclear fuel cycle.).

Although the volumes of radioactive waste are relatively small, they still have to be handled and disposed of. The increase in nuclear power production and the reduced possibilities for sea dumping makes this requirement more acute. All the major European countries are therefore working on projects for the disposal of low- and medium-level radioactive waste by shallow land burial (France, UK) or by disposal in mines (W. Germany) or rock caverns (Sweden). Some European shallow land burial sites have been operated for many years, e.g. the Centre de la Manche in France and Drigg in UK, but the capacity of these sites are limited.

In Denmark the problem is not acute since there are no nuclear power reactors and only an accumulative amount of about 600 m^3 research waste in storage. However, also this waste has to be disposed of, so the interest taken in development of disposal systems is not entirely academic, although there are no specific plans for implementing the proposals made in the following.

Table I. Radioactive waste types which are or might be disposed of by shallow land burial.

	Long-term activity concentration	Source	Volumes
$\beta\gamma$ low α + $\beta\gamma$ low α low α	Zero or low	Reactor operation Decommissioning Research, etc. U-fuel fabrication U-mine tailings	Medium Medium Small Medium Large
$\alpha\beta\gamma$	Medium	Accidents Reprocessing	Small (hopefully) Medium
α	Medium	Pu-fuel fabrication	Medium

Table I gives an overview over the types of low- and medium-level radioactive waste which are or could be disposed of in shallow land burial if sufficient safety can be demonstrated. For waste containing significant amounts of longlived α-activity some type of geological disposal may be preferred, but from an optimalization point of view, and maybe also for emergency purposes, there is an incentive to investigate the limits of safe operation of the various disposal options. The presence of longlived activity will require documentation of the long-term behaviour of the systems.

Shallow land burial is often the only economically feasible option for the disposal of bulky waste, e.g. uranium-mill tailings. In the case of other radioactive wastes such constraints may not appear, but ideally the selection of disposal systems should always be based on a balance between cost and risk, so that unnecessary use of resources is avoided.

Fig. 1 shows in a very schematic way the risk/cost correlation as dependent on some of the major parameters which can be influenced by waste strategies and by site selection and design of the disposal system. The cost will increase with the volume to be disposed of, but a waste strategy leading to lower volumes with higher activity concentrations may give rise to a higher risk to personnel handling the materials. Use of engineered barriers around the waste can be rather costly and the risk estimates will depend on the possibilities for documentation of a sufficiently low degradation rate. In the same way the disposal at increased depth is normally

more costly than disposal at or near the surface. The risk of activity release by accidental intrusion decreases with depth, while the general risk
estimate also will depend on the possibilities for reliable modelling of the
migration in the environment.

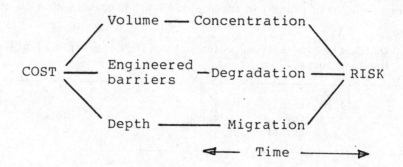

Fig. 1. Some parameters influencing the cost/risk correlation.

From none to many engineered barriers may be present in a burial
system, and there is a continuous transition from real surface disposal to
systems situated at a considerable depth. To evaluate the balance between
cost and risk residue for various alternatives, series of documented burial
concepts are needed.
Work on improvements of engineered barriers in shallow land burial
systems and on the risk of degradation of such barriers has been made in
the last 5 years within the frame of the CEC coordinated projects. The
work includes:
- Danish generic studies and material investigations which may be used –
  some time in the future – in the construction of new disposal facilities
  (2,3,4).
- French work on the documentation and/or improvement of the properties of
  the cover material used in already operating or planned shallow land
  burial systems (5).

2.   DANISH CONCEPTUAL DESIGN

A concept for shallow land burial of reactor waste, based on the French
system with burial in concrete cells, as done for the more active waste at
the Centre de la Manche, was discussed in Denmark in 1977. It was later
used as one of three reference cases in an Inter-Scandinavian system- and
risk-analysis study of storage, transport and disposal of reactor waste
(6). Under contract with the CEC further work has been made since 1981 on
evaluation and improvement of such systems.

2.1 Construction of a hexagonal overpack container

The handling and emplacement operations are much simplified if the
waste, as far as possible, is delivered to the disposal site in form of
standardized units. A regular hexagon, see Fig. 2, was selected as the outer
shape of the standard container since such units can be placed together with

very little interspace between them. Full-size containers, which could be
used directly for conditioned waste or as overpack for standard 210 l drums,
were cast from ordinary concrete (based on sulphate-resistant Portland
cement, SRPC) or from Densit*, a high strength, low permeability concrete
containing silica-fume particles as additive. The concrete was reinforced
with steel net and by addition of 12 and 25 mm long steel fibers. After
positioning of the drum the interspace between the drum and the container
was filled with concrete. The lid was cast in the same operation. The
material costs were in 1982 about 500 Danish Crowns per unit.

   Empty full-size units have been exposed to 1.2 and 5.5 m drop tests
without significant damage. Some units were tested against internal pressurization and were found to crack at 3 to 5 bars water-pressure. A slight
change in design, which is under investigation, may improve the resistance
against overpressure. The units were not air-tight probably due to small
cracks caused by slight contraction during hardening of the concrete.

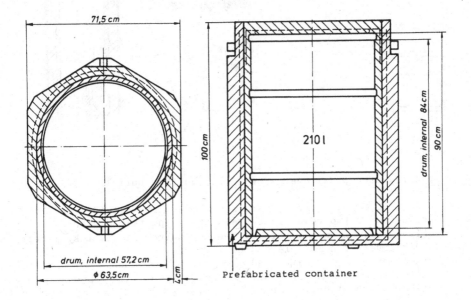

Fig. 2. Hexagonal concrete container used as overpack on standard 210 l drum.

2.2 Emplacement in 3 different types of near-surface disposal facilities

   The hexagonal standard units can be disposed of in many different
types of repositories. It has been pointed out that accidental intrusion in
an abandoned near-surface disposal facility must be taken into account when
evaluating the long-term risk of disposal. Three conceptual designs have
therefore been made where the soil cover is 2, 10 or 20 m thick, respectively. See Fig. 3.

_____

* Protected trade name. The product was developed by Aalborg Portland
   Cement Factories. It is now marketed by Densit A/S, Sølystvej, DK-9220
   Aalborg Øst, Denmark.

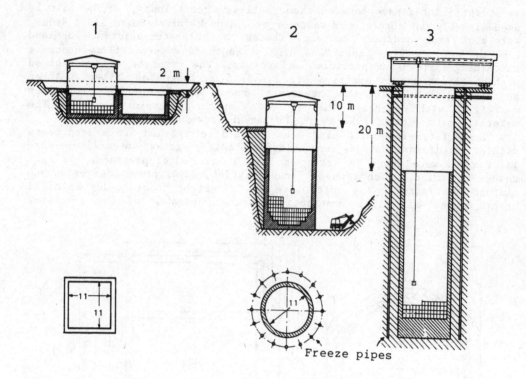

Fig. 3. Three conceptual designs for shallow land burial.

In concept No. 1 square concrete bunkers with wall thickness about 1 m are constructed in two rows beside each other and filled with units stacked in five layers on top of each other. The emplacement is made by a travelling crane under cover of a movable shelter.

In concept No. 2 cylindrical (and therefore much stronger) concrete constructions, e.g. with internal diameter 11 m, are placed in rows in a ~ 30 m deep excavation. The units are stacked in 21 layers. One cylinder may contain about 3300 standard units corresponding to the volume of reactor waste from the production of 1 to 2 GW·year.

In concept No. 3 the cylindrical constructions are supposed to be made using the freezing technique employed for the construction of mine shafts in unconsolidated formations. The method is well established and was used e.g. for the sinking of the shaft to the underground laboratory at Mol, Belgium. In principle this system permits the positioning of the repository at any desired depth, but the cost will increase rapidly with depth. The diameter and the height of the construction usable for disposal can be optimized against the thickness of the desired soil cover, but this is a question for a project at a more advanced stage.

All three types of structures can be made using conventional construction technology. A welded steel plate used as inner mould is supposed to be left in place as part of the structures and as an additional barrier against penetration of water into the interior.

The cost of the construction work necessary for disposal of one standard unit has been estimated to 1400, 1800 and 2900 Danish Crowns for concept No. 1, 2 and 3, respectively (1983 prices).

As a special factor it is proposed to use molten bitumen for the seal-ing of the interspace between the individual hexagonal units. The required amount of bitumen is reasonably small (~ 200-300 Danish Crowns/unit) and the use of bitumen gives some major advantages compared with concrete. The bitumen can be expected to fill all crevises completely, it has a low permeability for water, and it is a soft material which will not crack due to relative movements of the units caused e.g. by uneven settling of the ground under the repository. There is, of course, a fire risk involved in handling of large amounts of molten bitumen, but this can be circumvented by the use of a CO_2 cover gas.

The feasibility of bitumen sealing has been shown on a limited scale by a demonstration burial of 21 empty standard units, see Fig. 4. The structure is covered by ~ 0.3 m soil. The bitumen layers on the outside of the outer units are in direct contact with the surrounding water-saturated soil. The development in electrical resistivity in the buried concrete con-tainers and over the bitumen membranes are followed and the humidity in the air in some of the empty drums determined occasionally. The trends are in agreement with what can be expected from material investigation, see Fig. 8. So far the drums have been dry.

γ-doses to personnel could be a serious problem during emplacement of a large number of radioactive waste units in a repository of one of the types shown in Fig. 3. A model for calculation of dose rates as a function of distance from single standard units and groups of units with known contents of γ-emitters was developed. The results for a single unit were verified experimentally, and it was found that the model overestimates the

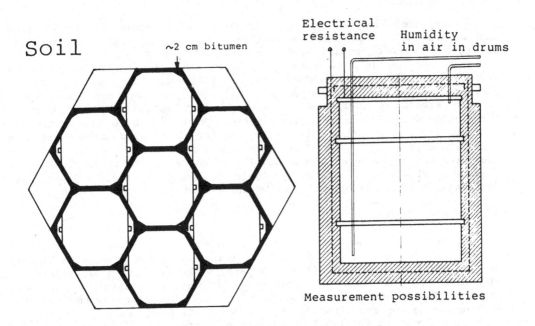

Fig. 4. Configuration of demonstration burial. 21 empty units placed in 3 layers with 7 in each. The interspace between the units is sealed with bitumen.

dose rates with about a factor 2, especially at larger distances. The conclusion of this part of the investigation was that facilities for remote emplacement of some or all of the units will be necessary, at least for reasonably fresh reactor waste.

The long-term risk associated with a repository for radioactive waste will depend on 1) the isotope inventory, 2) the properties and long-term stability of the engineered barrier system and 3) the hydrology and retention properties, etc. of the surroundings as determined by the siting of the system. Risk evaluation is the topic of the following paper to this conference, but some main point must be mentioned here as background for the investigation of material properties.

3. MATERIAL STUDIES

No releases will occur if it is possible to keep water out of contact with the waste for sufficiently long time, and only very slow releases can be expected if the transport out of the repository is exclusively by diffusion after water saturation has taken place. It follows that the problem is to document that major defects in barriers will be slow to develop, and that significant water flow driven by hydraulic gradients, solution density differences or gas pressures will be improbable. This is a complex task requiring much knowledge about the long-term development of the material properties as influenced by the interactions with the other materials in the repository and the surroundings.

3.1 Waste materials

As reported in Session IV of this conference, extensive work has been made within the CEC projects on leaching, volume- and radiation stability of various types of conditioned low- and medium-level radioactive waste. Two phenomena may lead to interaction with outer barriers: The first is that the internal water chemistry can be influenced very much by leached macro-components from the waste, e.g. Na^+, SO_4^{--}, organic components, etc. The second is that swelling due to water uptake in some types of bituminized materials may generate forces large enough to crack external barriers around the waste. This is the case for bituminized dry ion-exchange resin where pressure up to 20-30 bars has been measured, and much larger pressures can be foreseen theoretically. It may also occur, due to osmotic phenomena, with bituminized (and probably also polymer solidified) materials containing soluble salts, but experimental evidence is lacking in this area.

3.2 Concrete barriers

Portland cement concrete is a long known and much used material. In comparison, the variant called Densit is quite new with a production history of less than 10 years. This is a draw-back, since evaluation of the properties must rely on relatively short-term laboratory experiments. However, the important point is that possible deleterious reactions in the specific environment of the repository must be understood. Such knowledge will often be of quite new origin also for ordinary concrete.

Densit paste is produced by mixing e.g. sulphate resistant Portland cement with silica fume*, an organic superplasticizer** and water in ratios about 100:20:2:25. Sand and coarser materials are added to form a concrete easily workable even at this low water/cement ratio. The reason is that the superplasticizer permits a much closer packing of the particles in a still fluid system, see Fig. 5. It is only by use of such additives that Densit type products can be produced, where the small silica particles are packed closely in between the much larger cement particles.

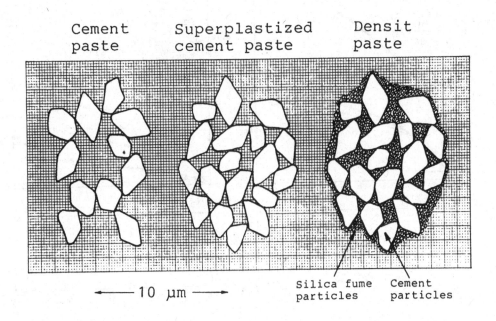

Fig. 5. Effects of superplasticizer and silica-fume additives on fresh cement paste, after (7).

Experimental comparisons between various properties of Densit and ordinary concrete have been made.

Measurements of the hydraulic conductivity of ~ 1 cm thick samples gave low and very variable results in the same range of values (10^{-9} to 10^{-11} cm/sec) for both Densit and superplasticized concrete, indicating that the transmission is mainly through micro-cracks and not through a uniform pore system. The permeability decreases with time of exposure to the hydraulic gradient. This is not the case for ordinary concrete without superplasticizer, where the permeability is stable with values depending on water/cement ratio, e.g. 10^{-8} cm/sec at w/c = 0.34.

The diffusive properties of the two materials have also been investigated. Figure 6 shows examples of the results from leaching experiments using a system developed for the characterization of cemented waste materials. For one-dimensional diffusion-controlled leaching from thick samples the effective diffusion coefficient is given by $D_e = \pi \cdot \alpha^2/4$, where α is the slope of the leach curves plotted against $\sqrt{t}$, as done in Fig. 6.

* A by-product from ferro-silicium production. It consists of nearly pure SiO_2 in form of spherical particles with diameters about 0.1-0.2 μm.
** Na-salt of a naphtalene-sulphonic acid formaldehyde condensate.

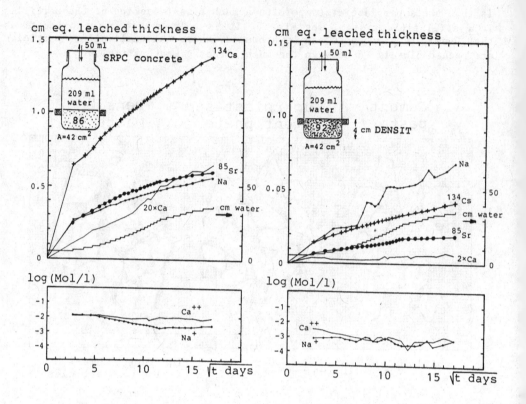

<u>Fig. 6.</u> Leach curves for samples of the two types of concrete.

The conditions of the experiment are indicated on the figures. The step-curves show the amounts of water which have been passed through the systems. This parameter is of minor importance for Cs- and Na-leaching, while the Sr- and Ca-leach curves show indications of solubility limitations.

Effective diffusion coefficients can also be obtained from experiments with diffusion through slabs of the material as shown on Fig. 7. Break-through curves, normalized to ml of the strong solution on one side transmitted per cm^2 surface area to the weak solution on the other side, are shown as a function of time. The effective diffusion coefficient is in this case obtained from $D_e = x^2/6\lambda$, where λ is the so-called time lag and x the sample thickness. It is seen that the transmission of Cs is much more irregular than the transmission of tritiated water and in some cases takes place at a higher rate. To the right in Fig. 7 the [134]Cs concentration profile is shown for the two slabs after the experiments. The scale is logarithmic which means that the concentration profile should have followed the stippled curve, if this was a case of simple Fick's diffusion.

Table II gives some typical values for effective diffusion coefficients obtained with the two methods. It is seen that especially the values for [134]Cs are different for the two materials and the two methods. This and other features of the experimental results, e.g. the concentration profiles in the samples, can be explained by reasonable assumption about the pore structure in the materials as demonstrated by model calculations in (4).

Table II. Typical effective diffusion coefficients.

$10^{-12} \dfrac{cm^2}{sec}$	w/c	Diffusion cell		Leaching
		TOH	^{134}Cs	^{134}Cs
SRPC concrete	0.34	30 000	60 000	20 000
Densit	0.25	20 000	10 000	100

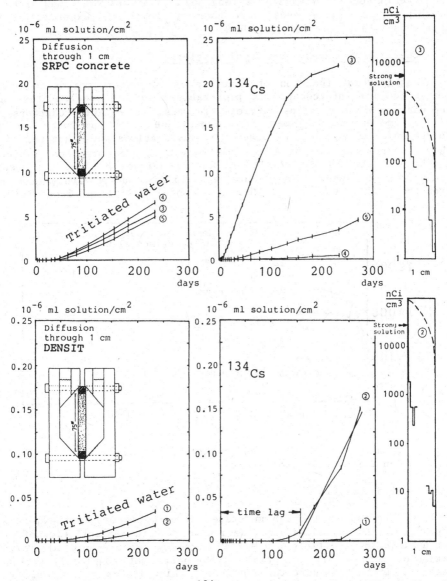

Fig. 7. Break through curves for ^{134}Cs and tritiated water for 3 identical samples of SRPC concrete and 2 of Densit.

The break-through curve for [134]Cs for sample No. 3 in Fig. 7 shows, after a period of rapid transport, a tendency to decline. This behaviour can only be explained by closing of a micro-crack by some precipitation reaction. Similar effects have been noticed in experiments with [36]Cl[-] diffusion through ordinary concrete as well as Densit. Other types of experiments do also indicate that there is a tendency to precipitation of materials, which under suitable conditions may lead to the self-healing of cracks in concrete. Further documentation of such effects will be of much value for the safety evaluation of concrete barriers.

A question which also must be evaluated is the long-term stability of concrete in contact with e.g. strong salt solutions which may arise due to leaching of waste in a repository. However, preliminary experiments indicate that this is not a serious problem.

3.3 Steel reinforcement and steel barriers

An important function of concrete is to act as corrosion protection for embedded reinforcement and preferably also for steel plates in drums and casings which serve as barriers. In Densit pozzolanic reactions between silica fume and $Ca(OH)_2$ may lead to decrease in pH and increased risk of corrosion of embedded iron. However, investigations show that this is probably of minor importance. A contributing factor to corrosion protection is the high electrical resistivity of Densit. Figure 8 shows the results of resistivity measurements on laboratory samples compared with values obtained from the full-size units in the experimental burial mentioned previously.

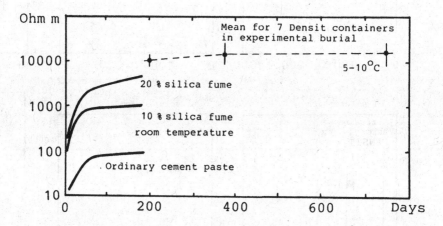

Fig. 8. Development of electrical resistivity in laboratory samples and in Densit in full-size units.

3.4 Bitumen barriers

Water transport through membranes of pure bitumen has been studied in connection with waste characterization work. Water solubilities of 0.2-0.5% and a diffusion coefficient $D = 2 \cdot 10^{-8}$ cm[2]/sec were determined by a combination of methods. The material is therefore not completely water impermeable, but the transport through 1-2 cm thick layers will be slow. It has been shown that high pH values, as typical for water conditioned by contact with concrete, will retard if not prevent microbial degradation of bitumen.

4.1   FRENCH INVESTIGATIONS OF PROTECTIVE COVERS

Adequate safety of shallow land burial of low- and medium-level waste may be attained without the use of a complex system of surrounding barriers. However, the facilities should preferably be protected against percolating rain water by some type of water-tight covering. ANDRA, the organisation responsible for disposal in France, has therefore with support from the CEC initiated a program for development and testing of such coverings.

In parallel with theoretical work, experimental equipment has been installed to study the hydrological and physico-chemical conditions in cover materials under the environmental conditions at the Centre de la Manche burial site. The experimental set-up is of similar type as the lysimeters used in agricultural research. The purpose of the investigation is to show that the system and the installed measuring equipment can be used to document the quality of coverings on the so-called tumuli, i.e. stacks of drums and other waste units placed above ground and then covered by soil, one of the burial principles employed at the la Manche site.

The main feature of the system (see Fig. 9) is a 6 ⤬ 4 m rectangular collector plate of stainless steel with edges rised to a height of 0.5 to 1 m. The plate is inclined 10% so that percolated water is collected at one side from where it is led to the measurement chamber situated below. The collector is nearly filled with high permeability materials: in the bottom a layer of pebbles, then gravel and finally a layer of fine sand. The cover material is spread in a thickness of 3-4 m above the sand layer in the collector. The edges of the collector extend some way up into the cover material which also is placed over a considerable area around the collector.

The installation is built into the side of an ordinary tumulus. From about 300 m^2 slightly inclined surface area on top of the system, run-off water is collected and led to the measurement chamber.

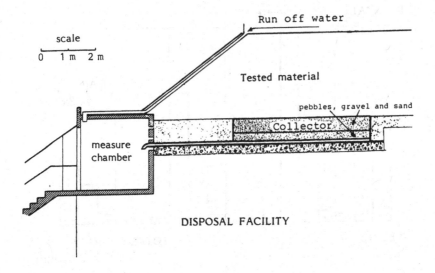

Fig. 9. Section of the French test facility.

The amounts of percolated and run-off water are measured continuously by a decantation system. The following physico-chemical measurements are made: Conductivity, pH, Eh, temperature, turbidity, dissolved oxygen and major an- and cations. Water collected in the recording rain-gauge situated on the site has also been analysed occationally.

In addition to the water-collecting systems some further installations permit measurements in the cover material: 5 vertical 2.5 m long aluminum tubes are used to measure water content by help of a neutron source and suitable detectors. 6 groups with each 5 tensiometers have also been installed at depth between 0.3 and 2 m. The water tension, i.e. the pressure balancing the water suction out of a water-filled porous porcelain tube is a measure for the degree of unsaturation of the cover material.

The system has only been in operation for about one year. The results for the first material investigated were obtained in the 6 months from February to the end of July 1984.

4.2 Infiltration rates

The recorded relationship between rainfall and collected water is shown in Fig. 10. There is indications that two types of infiltration occur: - A more or less continuous seepage of water shows that the cover material has a not negligible hydraulic conductivity, determined by the pore system. The percolation rate declined from 0.9-0.5 mm/day in February to 0.1 mm/day at the end of July, but was never zero. - An increase in the amount of percolated water followed 6 to 24 hours after rainfalls, probably due to percolation via macro-pores or channels in the material. In addition, after periods with heavy rain, a general increase in percolation rate could be seen with a delay of 12-15 days. This is ascribed to a higher degree of saturation of the deeper layers due to water infiltration via macro-pores.

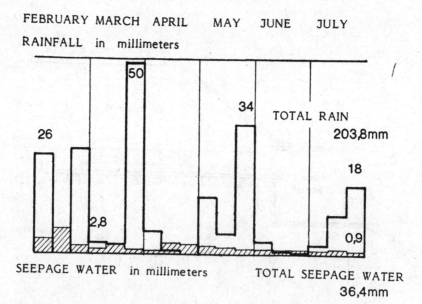

Fig. 10. Correlation between rainfall and percolated water for the first tested material.

The over-all water balance for the period was: 204 mm rain distributed as 23 mm run-off (11%), 36 mm percolate (18%) and evapotranspiration estimated as about 70%.

4.3 Water chemistry

The continuous analyses of the seepage water show that the pH fluctuates between 7.9-8.3 for the whole period. No correlation with water flow could be seen since the soil acts as a buffer material. The conductivity increases from 400 to 600 µS/cm during the experiment and varied slightly with rainfall. These changes in conductivity are related to the residence time for the water in the soil and the higher degree of saturation of major ions attained when the percolation rate is low. The Eh fluctuates from 550 to 450 mV and even 400 mV in July, i.e. there is a tendency to less oxidizing conditions when the water flow is low.

The chemical composition of the seepage water was relatively constant. Values are given in Table III.

Table III. Water chemistry of seepage water.

meq/1			
Cations		Anions	
Ca^{++}	0.9-0.5	HCO_3^-	2.8-3.3
Mg^{++}	0.9-0.6	Cl^-	1.5-2.0
Na^+	4.5-5.3	SO_4^{--}	0.9-1.5
K^+	~ 0.45	NO_3	0.1-0.15
NH_4^+	0 -0.03	SiO_2	0.32

The Na^+ concentration tended to increase, while Ca^{++} and Mg^{++} decreased during the period. There was no significant change in the ionic strength of the solution.

4.4 Result and perspective

The experimental facility has made it possible to demonstrate the applicability of various measurement systems for the study of cover materials. The test has been made in a scale large enough to permit extrapolation to larger areas. The installation gives data concerning the hydrodynamics of a cover layer, the geochemistry of the percolated water, and the run-off characteristics for surfaces on the site.

On background of the extended French program on disposal of low- and medium-level waste it was thought to be of interest also to use the test facility for a cover material useable for real water-tight covers. The new cover investigated in the winter 84/85 consisted of a 3 m layer of compacted clay on top of which was placed a layer of sand and a layer of soil to support revegetation. Although the rainfall during the winter was about 590 mm, only ~ 0.3 mm seepage water was collected in the first month of operation, declining to zero in the following 4 months. This demonstrates that water-tight covers can be made if proper materials are selected.

5.   GENERAL CONCLUSIONS

It is probable that engineered barriers can contribute significantly
to retardation of the release of radionuclides from shallow land burial
facilities. Something can often be attained simply by use of more suitable
materials as exemplified by the French studies. Additional safety may also
be ascribed to the use of systems with more or less redundant barriers as
in the Danish concepts. However, it may be difficult to prove the redundancy
due to the complex long-term interactions between the materials in such
multi-barrier systems. The problem can be alleviated by further fundamental
material studies, but some uncertainties will probably always remain. This
is a draw-back for multi-barrier systems since it is difficult to motivate
investments in engineered barriers which are not given any specific value
in safety assessments due to doubts about long-term behaviour.

6.   REFERENCES

1.   Analysis of the present situation and prospects in the field of radio-
      active waste management in the Community. Community plan of action in
      the field of radioactive waste. CEC 1983.
2.   BRODERSEN, K., LARSEN, V.V. Development of waste unit for use in shal-
      low land burial. Final report 1981-1983 for CEC contract No. 195-81-6-
      WASDK (G).
3.   Ibid. Progress report 1983 for contract No. 331-83-63-WASDK (G).
4.   Ibid. Progress report 1984 for contract No. 331-83-63-WASDK (G).
5.   CEA-ANDRA, CEA-ISPN, Etude des données hydrologiques et géochimiques
      des couvertures des sites de stockage subsurface. Progress report 1983
      for contract WAS 239-83-6-F(RS).
6.   Nordic study on reactor waste. NKA/AO(81)5, 1981.
7.   BACHE, H.H. Densified cement/ultra-fine particle-based materials,
      2.int.conf. on superplasticizers in concrete. Ottawa 1981.

DISCUSSION

D.J. SWALE, BNFL Sellafield

What kind of clay was used in the experiments for the determination of a suitable top layer on the tumuli at the La Manche Centre which more or less stopped seepage?

R. ANDRE-JEHAN, ANDRA Paris

It concerned a multiple layer of tertiary clays which was deposited in such a way to obtain maximum density.

SHALLOW LAND BURIAL ANALYSIS

C R HEMMING
Formerly of National Radiological Protection Board, UK
Now of Department of the Environment, UK

V VOLDBRO LARSEN
Risø National Laboratory, Denmark

M LOXHAM
Laboratorium Voor Grondmechanica, Deflt, NL

G M WILLIAMS
British Geological Survey, UK

Summary

This paper describes work on the analysis of shallow land burial of radioactive waste carried out under sheet 6 of the Radioactive Waste Management Programme. The work described encompasses a wide range of topics aimed at improving understanding of the influence of the barriers in the shallow land burial system, which are intended to control the release of radioactive material from the waste to man's environment. A qualitative description of the near-field barrier system is given and the phenomena and mechanisms which could influence the integrity of the system are identified. A detailed consideration is then given of the geologic barrier. The results of in-situ radio-nuclide migration experiments, which test the validity of the use of laboratory sorption measurements and migration models to predict radionuclide migration, are presented, and a theoretical study of the influence of ground heterogeneity on radionuclide migration is des-cribed. It is, however, the barrier system as a whole which deter-mines the impact of shallow land burial of waste on man. Some exam-ples are given of the application of a methodology which has been developed to evaluate the radiological impact of shallow land burial of wastes, taking account of the whole barrier system.

1. Introduction

The four studies described briefly in this paper cover a diversity of aspects of the analysis of shallow land burial of radioactive waste, but have as a common objective the search for an improved understanding of the influence of the barriers in the shallow land burial system, which are intended to control the release of radioactive material from the waste to man's environment. The first study concerns a qualitative description of the near-field barrier system and the identification of the phenomena and

mechanisms which may affect its integrity. The following two studies focus
on the geologic barrier and the prediction of radionuclide migration
through the geosphere. Firstly, consideration is given, through the use of
in-situ migration experiments, to the validity of predicting radionuclide
migration using laboratory sorption measurements and migration models.
Secondly, the influence of heterogeneities in the geosphere are investigated by means of a theoretical model. Finally, some examples are given of
the application of a methodology for evaluating the radiological impact of
shallow land burial of waste. This methodology considers the whole barrier
system, and can be used to help identify those parts of the system, which
have the greatest influence on the radiological impact, and hence guide
further research.

2. Near-field phenomena in a multi-barrier system

A hexagonal waste unit has been developed at Risø, DK, for use in
shallow land burial of low- and intermediate-level waste. An example of a
conceptual burial facility for the units is shown in Fig 1. The corresponding barrier system in disposal facilities of this type is also illustrated
in Fig 1.

In this figure, the type of fault which most probably would occur and
give rise to an imperfection in the barrier is indicated at the bottom.
Under normal conditions activity can only be released from the waste inside
the barriers to the surroundings, if water is able to penetrate from the
outside to the interior of the system. This is likely to take a considerable time with the barrier system proposed.

If water penetrates, and there is no hydraulic gradient over the system, no flow of water is possible. the only possible release of activity
is then by diffusion of water-soluble radioactive species through water-
filled cracks or pores.

If a hydraulic gradient is present, and there are at least two interconnected defects through all the barriers, the result is transport of
soluble species by water flowing through the barriers and an increased risk
of significant release of activity to the surrounding formation. The presence of a hydraulic gradient will create very complicated conditions, especially when the repository is placed above ground water level, or in a
zone which is changing between occasionally saturated and unsaturated conditions. While the questions about the hydraulic system are mainly site-
specific in nature, the questions about the development with time of the
properties of an ageing barrier system are mainly complex functions of the
material properties, material compatabilities and construction features.

In the following, some phenomena and mechanisms which could influence
the integrity of the barrier system are described. The presentation is
based on the barrier system shown in Fig.1, but could be easily modified to
be applicable to other combined barrier systems.

a) The first barrier consists of the waste material itself. Its properties depend on the type of waste, conditioning etc. These will determine
the release due to leaching from the material, but the behaviour of the
waste may also have a profound influence on the other barriers; swelling
due to water uptake can, for example, generate large internal pressures
with some types of bituminized waste.

b) The steel drum is the next barrier. It will probably remain in
good condition for a long period due to the surrounding conditions in this
system, ie, almost dry environment and high pH.

c) Concrete is the main barrier of the waste unit itself. It can be
made of either ordinary concrete or the specially developed concrete
Densit, which has lower diffusion rates for various dissolved materials and

radioisotopes than ordinary concrete. The risk of crack formation and the general long-term stability of this new material is uncertain, but this is in some ways also the case with ordinary concrete.

d) The next barrier is the bitumen sealing between the individual units. Bitumen is a good sealing or back-fill material for various reasons:

- Molten bitumen is easy to cast in the relatively narrow crevices between the units, although care must be taken to avoid inclusion of large air bubbles, and to cast the material at a sufficient rate that premature cooling is avoided. The system should be dry when the casting is made so that steam bubbles are not formed.

- The bitumen layers act as nearly water impermeable membranes around each unit. Some slight water migration by diffusion may take place. The permeability coefficient is about 10^{-10} cm^2/sec. This may result in the movement of up to 1 g water per year through 1 m^2 of a 1 cm thick bitumen layer.

- Bitumen is soft and able to compensate for some relative movements of the units. This is an advantage compared with the use of concrete as back-fill material, since concrete would crack under such circumstances. Relative movements may be caused by mechanical failure of the outer barriers.

The main draw-backs associated with the use of bitumen are the risk of fire under the sealing operation and the possibility of formation of organic degradation materials due to bacterial attack on the bitumen which might promote the migration of some radioactive species.

e) The outer construction is the next barrier. It consists of a ca. 1 m thick concrete wall which is the main constructive feature of the engineered repository. It is likely to be built from ordinary reinforced concrete and will therefore be slightly permeable to water and diffusing species. The strength of this concrete construction determines its integrity when mechanical forces from the surrounding formation are acting on the system.

f) The final barrier is the surrounding formation which consists of a layer of disturbed materials just outside the engineered system followed by the undisturbed formation. It will show all the retention phenomena known from general migration theory.

A qualitative description of a barrier system as given above is the first step in the formulation of a safety analysis for the system. It provides a framework for selection of relevant degradation mechanisms and for collection of the necessary data.

3. In-situ radionuclide migration studies

3.1. Introduction

Prediction of radionuclide migration from shallow land burial sites is often based on laboratory measurements of the equilibrium distribution of a radionuclide between rock and aqueous phases. The distribution coefficient, or Kd, thus obtained can be used to determine the retardation experienced by a radionuclide species compared to the average pore water velocity of the advecting water, assuming that the Kd value is constant over the range of radionuclide concentrations expected, that the sorption mechanism is reversible, and that equilibrium is attained instantaneously. No other chemical reactions, eg precipitation, are assumed to occur.

In order to validate the use of laboratory batch sorption measurements in predicting retardation in the field, a small scale tracer test has been undertaken in which sorbed and non-sorbed radioactive tracers (^{85}Sr and ^{131}I respectively) have been introduced into a radially divergent flow field around a recharge well in a confined unconsolidated glacial sand aquifer. Comparison of the breakthrough times of the tracers in a number of monitoring points gives a direct measurement of retardation which can then be used to determine the effective Kd of the aquifer.

Results for distribution coefficients determined by laboratory and field methods can therefore be compared. In order to predict the ground-water flow field around the recharge well, a hydraulic model has been developed in collaboration with the Delft Soil Mechanics Laboratory. This has assumed the aquifer to be homogenous and is based on transport para-meters determined from aquifer tests on the whole aquifer thickness.

3.2 Results

The aquifer comprises of about 80% quartz, with up to 10% alkali and plagioclase feldspar. Minor amounts of calcite (1%) and organic carbon (0.6%) are present while the remaining clay component consists of mica, kaolinite and chlorite with trace amounts of smectite. Surface area meas-urements (16-40m^2/g) suggest a relatively low capacity for sorption.

Results of batch equilibrium sorption experiments provide a linear isotherm for ^{85}Sr with no hysteresis.(Fig 2) The average of Ksorp and Kdesorp values for ^{85}Sr are shown for 15cm core intervals in Table I, for one concentration only; iodine was not significantly sorbed.

Breakthrough curves for Sr and I are shown in Fig.3 with breakthrough times and retardations in Table I. Using averaged physical properties of 15% for porosity and 2.25 gm/cc for bulk density the Kd values derived from the tracer experiments are also given in Table I.

3.3 Discussion

Laboratory values for ^{85}Sr sorption are consistently higher than those derived-from the field tracer test. However, the greater the residence time of the tracer in the aquifer, then the nearer is the convergence bet-ween laboratory and field values. The reason for this may be a kinetic effect on sorption which is more pronounced in the high groundwater flow conditions near the tracer release well. This effect is being studied further by undertaking field tracer tests at varying groundwater flow rates and by laboratory experiments to determine sorption kinetics at representa-tive groundwater temperatures.

The breakthrough times for ^{131}I predicted by the hydraulic model,which assumes a homogeneous sand aquifer, are widely different from those observed, and result from heterogeneity within the sediment. A more sophisticated model is necessary, based on a layered aquifer system in which different values for permeability, porosity and dispersivity can be assigned to the layers. Interchange of solute between individual layers may also need to be included in the model to fully represent the observed field results.

3.4 Conclusions

The following conclusions can be drawn from the above results:

i) Field and laboratory results for sorption of ^{85}Sr based on the Kd concept are in relatively good agreement.

ii) Kinetic controls on sorption may be responsible for the lower sorption observed in the field experiment in the zones of high groundwater flow.

iii) Flow models which assume an aquifer to be homogenous may be grossly inaccurate when applied to heterogeneous systems.

Future work will therefore attempt to reconcile field and laboratory results taking into account time dependant sorption. A flow model based on more detailed characterisation of the hydraulic properties is considered necessary to allow for aquifer heterogeneity. Sorption mechanisms are also to be studied by desorption experiments with Sr contaminated aquifer material.

4. <u>Influence of ground heterogeneity on radionuclide migration</u>.

Almost all existing and most proposed shallow burial sites are situated in hydrogeological environments that are heterogeneous in nature. That is to say, the soils found on the path between the site and potential targets are physically structured on a length scale that is significant when compared to the path length itself. The impact of this soil structure on the migration patterns of the components leaching out of a generic site has been investigated.

It is well established that the sorption of radionuclides on soil, and the associated apparent retardation of the breakthrough front, is one of the most important factors that has to be taken into account in the overall safety assessment of a shallow burial site.

This retardation is caused by sorption on clays and organic materials. The sorption capacity is in turn a function of the surface area and availability of these materials to the sorbing species in question. In real soils, and especially those of sedimentary origin, the adsorbant components are to be found as discrete lenses of low permeability rather than as an evenly distributed phase, as is essentially assumed in the laboratory methodologies for determining retardation factors. In these lenses penetration to the bulk of the sorption sites is restricted to diffusion and small residual convection fluxes.

The important quantity is then the effective mobilisation of the laboratory measured retardation capacity of a given soil system, on the time scales associated with those of the convective transport through the more permeable and less adsorbing strata. This is the subject of this investigation.

The central objective of the study is to demonstrate the importance of heterogeneity in the migration analysis. In order to do this, a generic site was assumed, with parameters that correspond to many existing sites. The site was assumed to be situated some 25 m above the groundwater table and 500 m from an abstraction point defined as the target. (see Fig 4).

Leachate from the site migrates downwards through the unsaturated zone at a small (eg 5 mm/y) rate, depending on the assumed infiltration through the site. It then penetrates the aquifer and is convected along to the target. The water velocities in the aquifer were assumed to be of the order of 25 m/y. The travel times in the two zones were of the same order.

The adsorbing component was assumed to constitute 25% of the system volume and was distributed either evenly (in the homogeneous base case) or as discrete lenses with chosen (constant) thicknesses. A permeability was attributed to the lenses, ranging from values the same as the rest of the aquifer to a factor 1000 smaller.

In the absence of field data for similar systems, the breakthrough curves at the target were determined by calculation. For the saturated aquifer a numerical solution of the conventional convection-dispersion equations was used. In order to model the sorption fluxes to the lens surfaces properly, a very fine space discretization had to be used, leading to very long computing times.

For the unsaturated zone, such a procedure was not practically feasible and analytical solutions using Greens functions were sought. It was assumed that the leachate flowed down through the more permeable zones without significant dispersion and then horizontally along the lens surface, cascading in this way from lens to lens through the unsaturated zone. Sorption then took place from the film on the lens into the bulk adsorbant. In this way sequences of lenses and permeable zones could be modelled. Whether or not this model is physically realistic awaits experimental confirmation.

Typical results from this study are shown in Fig 5. Here the influ-

ence of the lens thickness for a clay and sand system on the time for 1%
breakthrough at the target is shown. The time axis has been normalised by
the retardation for the same species in the homogeneous case, giving the
effective retardation factor ratio. In this way the effective mobilisation
of the sorption barrier is shown.

As can be seen, the effective retardation factor ratio for the saturated case falls from unity in the case of an infinitely thin lens to only
0.3% for thicker lenses. Similar behaviour is observed in the unsaturated
zone. In the numerical case considered, this result implies breakthrough
after 30 rather than 10000 years!

This and many similar results lead to the conclusion that in this sort
of system, with typical shallow burial site parameter values, it is the soil
heterogeneity, if present, that dominates the breakthrough behaviour. Equally, the laboratory data for retardation factors can only be applied to
field scale calculations with great care, and there is a real danger of
wildly overestimating the retardation barrier function of a heterogeneous
system.

The practical consequences of this conclusion can be far reaching, both
in the recognition of the detail required in the site investigation stage
of a siting exercise, and in the formulation of site selection criteria in
general.

5.   Radiological assessments of shallow land burial

The radiological impact of the disposal of radioactive wastes is one
of the factors influencing decisions on the management of these wastes.
While detailed studies of particular aspects of a waste disposal system,
such as those described above, are necessary for an improved understanding
of the processes involved and improved modelling, it is also important to
consider these aspects in the context of their influence on the overall
radiological impact of the disposal system. The study described in the
following concerns the development of a methodology for assessing the radiological impact of the shallow land burial of wastes and its use in preliminary illustrative calculations. In particular, the results of calculations, which give an indication of the influence of incorporating various
engineered features into burial facilities, and of the retardation capacity
of the geologic media surrounding burial sites, are described.

Results are presented here for three types of waste in two different
types of facility. The first type of waste consists of general low level
waste, most of which is in the form of biodegradable materials. The mean
specific activity of these wastes was assumed to be about 10^9 Bq m^{-3}, the
majority being due to ^{137}Cs and ^{90}Sr. These wastes were assumed to be
disposed of in simple trenches excavated in reasonably permeable weathered
material above the water table and covered with a layer of soil. The
other two waste types are reactor operating wastes, namely, the sludges
and resins arising during the operation of Magnox reactors and ion exchange
materials, filters, concentrates and sludges from LWR reactors. The mean
specific activity of these wastes is about 10^{12} Bq m^{-3} with ^{137}Cs and ^{60}Co
the most abundant radionuclides. These wastes were assumed to be incorporated in cement and packaged in 200ℓ steel drums prior to disposal in an
engineered facility. A concrete lined trench placed between 18 m and 11 m
deep on a generic clay site was considered.

A comprehensive assessment of the radiological impact of the disposal
in a particular facility of specific radioactive wastes should consider all
the potential radionuclide release and transport mechanisms that contribute
to the overall radiological risk. It should also include an estimate of the
uncertainty in the results. In the preliminary generic assessment presented

here only the release and transport mechanisms likely to be the main cont-
ributors to the radiological impacts were included. These are release by
trench fire (low level wastes only), contact by water followed by radionuclide migration, and disturbance in the future for building and farming
purposes.

The results of the calculations are summarised in Table II. The
nuclides which make the most significant contributions to maximum individual doses and risks, for each combination of waste type and disposal
facility design, are listed in Table II as a function of exposure route
and time. The results obtained show a number of features. Firstly, the
important exposure route at times up to about 250 y after closure of the
site is intrusion into the facility for building purposes, and the most
important radionuclides are ^{137}Cs and $\propto$-emitting radionuclides such as
^{239}Pu and ^{241}Am. At later times, from about a few hundred years onwards,
release of radionuclides into water and subsequent migration to man's
environment becomes more important, and the radionuclides giving rise to
the highest doses are generally those that are long-lived and assumed
to be poorly sorbed on geologic media, for example ^{99}Tc, ^{129}I and ^{237}Np.

A number of conclusions can be drawn from these results. Firstly,
it is clear that engineered barriers which prevent contact of wastes by
groundwater would only be effective in reducing long-term doses and risks
if they retained their integrity for periods of the same order as the
half-lives of the radionuclides which contribute most to doses via water
and farming pathways. Since these periods are very long (for example, the
half life of ^{99}Tc is about 210^5y), it would be difficult to guarantee that
a barrier would retain its integrity for this length of time, and thus
difficult to show that the barrier would reduce long-term risks. However,
sensitivity analyses carried out as part of the study have shown that the
use of barriers, which reduce the rate of release of radionuclides into
groundwater, could reduce the potential doses and risks from radionuclides
that are poorly sorbed on the geologic media in which the burial facility
is located. In cases where it is not, for some reason, possible to locate
a burial facility in a medium with a high sorption capacity, such barriers
would have considerable advantages in reducing doses via water and farming
pathways. In situations where the rates of migration of most radionuclides
are very low, such barriers could reduce the doses and risks to individuals
from those few radionuclides which are poorly sorbed. A further more
detailed study of the influence of various engineered barriers, such as
impermeable liners and trench caps and high integrity infilling material,
has recently been completed.

Other sensitivity analyses have shown that the retardation assumed for
the radionuclides making the most significant contribution to the doses
from release into groundwater has a considerable influence on these doses.
It is therefore important that the sorption interactions between the
radionuclides of interest and the geologic strata at and around burial
sites are well understood. The simple equilibrium approach adopted in
most radiological assessments may not be adequate for comprehensive assessments. More work is required on models incorporating more complex mechanisms and, in particular, on research to determine appropriate parameter
values.

6. Conclusions

This paper has considered four different approaches to the study of
various aspects of the shallow land burial of radioactive wastes. Detailed studies have been carried out of some of the processes affecting the

performance of the engineered barriers of a typical disposal facility and the retardation experienced by radionuclides migrating through the geosphere, and the results have been briefly summarised here. Studies such as these are important in obtaining a better understanding of these processes and the factors which influence them. It is, however, also important that these processes are considered from the point of view of their influence on the overall radiological impact of disposal. A methodology for evaluating the radiological impact of shallow land burial has been developed and some examples have been presented of its use in guiding further research into engineered barriers and retardation mechanisms.

Table I

Results of in-situ migration study

Borehole 7 (x = 0.53m, y = Om)

Sample depth (m,bgl)	I_{peak}(hrs)	Sr_{peak}(hrs)	Sr/I	K_d Field	K_d (lab)
6.68	3.75	53	14.1	1.03	
6.83	2.5	25	10.0	0.71	
6.98	2.5	27	10.8	0.77	
7.13	2.75	24	8.7	0.06	
7.28	4.5	72	16.0	1.18	
7.43	11.5	281	24.4	1.84	

Borehole 8 (x = 1.25m, y = 0.4m)

6.42	17	394	23	1.73	2.18
6.57	14	265	18.9	1.40	1.94
6.72	12	239	19.9	1.48	2.63
6.89	12.75	239	18.7	1.39	2.09
7.02	11.25	289	25	1.88	2.63
7.17	13.25	257	19	1.41	2.37
7.32	20	517	25.6	1.93	2.05

Borehole 13 (x = 1.75m, y = Om)

6.37	27	577	21	1.57	
6.52	26	643	24.5	1.84	
6.77	23	450	19.5	1.45	

K_d in ml/g

Table II

Nuclides making the most significant contributions to individual doses
received via the various exposure routes considered

Exposure route	Important nuclides
a) Minimum engineered facility, Inventory 1	
Trench fire	^{234}U, ^{238}U
Use of site stream	
- 50y after site closure	^{99}Tc, ^{129}I
- 5 10^3 y after site closure	^{237}Np
- 10^5 y after site closure	^{210}Pb, ^{226}Ra
Building on the site	
- less than 250 y after site closure	^{137}Cs
- beyond 250 y after site closure	^{234}U, ^{238}U
b) Fully engineered facility, Inventory 2	
Use of site streams	
- 5 10^4 y after site closure	^{237}Np
- 10^6 y after site closure	^{210}Pb, ^{226}Ra, ^{234}U, ^{238}U
Farming on the site	
- 5 10^4 y after site closure	^{237}Np
- 10^6 y after site closure	^{210}Pb, ^{226}Ra, ^{234}U, ^{238}U
Building on the site	
- less than 250 y after site closure	^{137}Cs
- beyond 250 y after site closure	^{239}Pu, ^{240}Pu, ^{241}Am
c) Fully engineered facility, Inventory 3	
Use of site streams	
- 5 10^2 y after site closure	^{14}C, ^{99}Tc, ^{129}I
- 5 10^4 y after site closure	^{59}Ni, ^{237}Np
- 10^6 y after site closure	^{210}Pb, ^{226}Ra
Farming on the site	
- 5 10^2 y after site closure	^{14}C, ^{99}Tc, ^{129}I
- 5 10^4 y after site closure	^{59}Ni, ^{237}Np
- 10^6 y after site closure	^{210}Pb, ^{226}Ra, ^{231}Pa, ^{234}U, ^{235}U, ^{238}U
Building on the site	
- less than 250 y after site closure	^{137}Cs
- beyond 250 y after site closure	^{137}Cs, ^{239}Pu, ^{241}Am

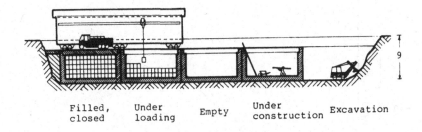

(a) Conceptual burial facility

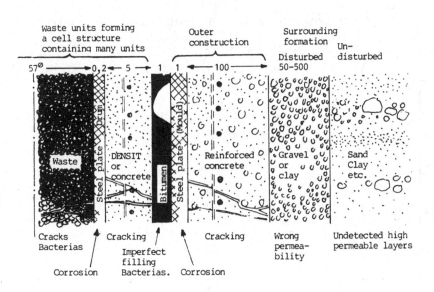

(b) Near-field barriers

Fig.1 The near-field barrier system

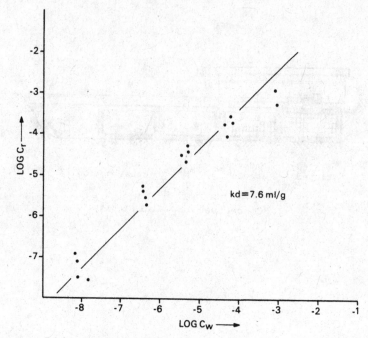

FIG 2.   LINEAR ISOTHERM FOR STRONTIUM Cr (g/g); Cw (g/ml)

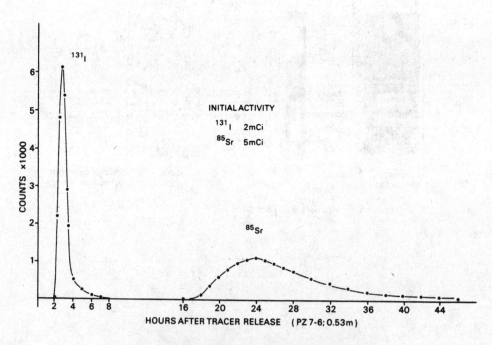

FIG 3. EXAMPLE OF BREAKTHROUGH IN SAMPLING POINT
(BOREHOLE 7, PIEZOMETRE 6; 0.53m FROM RELEASE WELL)

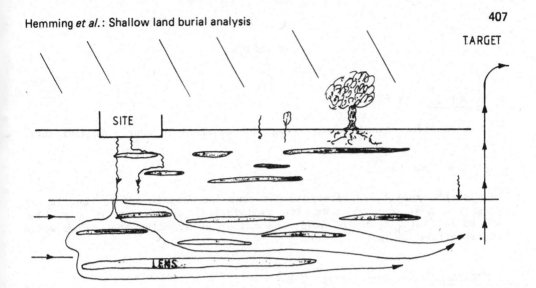

Fig.4   Characteristics of heterogeneous site considered

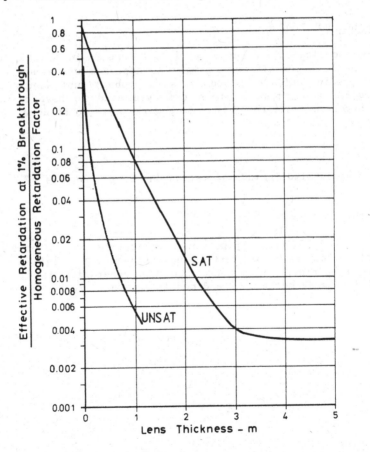

Fig.5   Influence of lens thickness on effective retardation

DISCUSSION

T. MARZULLO, ENEL Roma

Did you consider seismic events (e.g. earthquakes) in your risk analysis?

C. HEMMING, DOE London

The risk analysis I described did not consider seismic events, but it was recognized that any comprehensive radiological assessment of a shallow land burial facility would need to consider events like that which would have an effect on barriers.

R. KOBAYASHI, JGC Corporation, Tokyo

How did you treat precipitation which might occur during release?

C. HEMMING, DOE London

In the radiological assessment I mentioned, in the calculation of migration of radionuclides a simple K_D approach was used, which assumed that no precipitation occurred. This is probably not adequate, but for a radiological assessment of this type you need to have a fairly simple model. Nevertheless I think that more work needs to be done.

MR. X, Belgium

Did your study also include migration due to gas formation?

C. HEMMING, DOE London

The study I described did not include release of radionuclides by gas generation. At present at NRPB a study is underway looking specifically at engineered barriers and I know that gas generation is recognized as a potential release mechanisms and models for this are being developed.

SESSION VI

DISPOSAL IN SALT AND CLAY FORMATIONS

In situ-investigations in salt formations

Specific investigations related to salt rock behaviour

Modelling of the thermomechanical behaviour of salt
rock

Research programmes in underground experimental
facilities : Konrad - Pasquasia - Mol

Characterization and behaviour of argillaceous rocks

Natural evolution of clay formations

IN SITU-INVESTIGATIONS IN SALT FORMATIONS

K. KüHN and B. VERKERK

Gesellschaft für Strahlen- und Umweltforschung mbH (GSF)
Institut für Tieflagerung, Braunschweig
Bundesrepublik Deutschland

and

Stichting Energieonderzoek Centrum Nederland (ECN)
Petten, Nederland

Summary

The Federal Republic of Germany and the Netherlands continued their in situ-investigations for radioactive waste disposal in salt formations during the second CEC program in the Asse salt mine. The central test of the German program, Temperature Test Field 5, showed that crystal water from the hydrated salt mineral polyhalite is liberated only at 235 °C. The quantities of water liberated from rock salt and from polyhalite are negligible from a safety point of view. In the same test it was proved that rock salt creeps drastically at elevated temperatures and that no micro-cracks are formed which could increase the permeability of rock salt. Additional geophysical measurements confirm these results. Valuable data could be achieved by in situ-investigations for the scenario "water or brine intrusion" within safety analyses. The Dutch in situ-test program concentrated on convergence and pressure measurements at normal and elevated temperatures in a 300 m deep borehole which was drilled in the Asse salt mine. Construction and testing of a large diameter roller-bit body was started which will allow the dry drilling of boreholes up to 1 m in diameter. Both countries decided to continue their successful cooperation during the forthcoming third R&D-program of the CEC.

1. Introduction

It was already mentioned by the same authors in their paper at the First European Community Conference on Radioactive Waste Management and Disposal in 1980 [1], that the two countries Federal Republic of Germany and the Netherlands started their R&D-programs for the disposal of all categories of radioactive wastes in salt formations relatively early. During the second CEC-program which was performed between 1980 and 1984, both countries put main emphasis on in situ-investigations, because most of the scientific and technical data and parameters necessary for the design, construction, and operation of a repository can only be gained in an adequate underground laboratory.

Cooperation and work-sharing between the two countries already started during the first CEC-program. This cooperation was very successful so that it was intensified during the second CEC-program. Consequently, the Dutch in situ-investigations were again performed in the German

Asse salt mine. It will be shown in this paper that the respective experiments of the two countries harmonized and that the achieved results complement each other. So, it is a mere logic consequence that the Federal Republic of Germany and the Netherlands are again preparing a close cooperation during the forthcoming third CEC research and development program on radioactive waste management and disposal. The main objective of this cooperation is the performance of a test disposal of vitrified high-level radioactive glass blocks in the Asse salt mine which will actually start in early 1987.

2. German in situ-investigations

Main emphasis of the German in situ-investigations during the second CEC-program lay on problems connected with the disposal of heat generating high-level wastes. The most important experiment in this respect was the operation and evaluation of Temperature Test Field 5. Geophysical monitoring methods were applied during this test, too. In addition, they were and are further developed in order to control a repository with regard to certain aspects of mine stability. Valuable data were also produced by in situ-investigations for the scenario "water or brine intrusion" which is considered within safety analyses for repositories in salt formations.

2.1 Temperature Test Field 5

2.1.1 Objectives

The disposal of solidified high-level radioactive waste into rock salt formations causes heating of the host rock due to the decay heat and irradiation of it due to γ-radiation of the waste. One question to be solved in this connection is the thermally induced liberation and migration of volatile components. It is known that rock salt may contain minor amounts of water in three different forms [2, 3]:
- crystal water of hydrated salt minerals, like polyhalite
  ($K_2MgCa_2(SO_4)_4 \cdot 2H_2O$) or kieserite ($MgSO_4 \cdot H_2O$),
- water adsorbed on grain boundaries of rock salt,
- liquid inclusions.

JOCKWER has shown that the average value of water content of Older Halite (Na2β) from the Asse salt mine is 0.04 weight-% [2]. The three different forms of water show different behaviour of liberation mechanism at elevated temperatures. Liquid inclusions which mainly occur in coarsely crystalline salt, migrate along a temperature gradient towards the heat source and are only liberated at the borehole wall [4]. Crystal water of hydrated salt minerals is liberated at distinct temperatures into the intergranular pore space and then migrates due to gradients of pressure, concentration, and temperature [2]. The migration of adsorbed water depends on its occurrence in liquid or in vapour phase. In the liquid phase it may migrate due to the temperature gradient like liquid inclusions. In the vapour phase - which is also the main migration phase form for the liberated crystal water of the named hydrated salt minerals - a high vapour pressure exists which decreases towards the borehole wall, i.e. the "drier zone". Due to this pressure gradient the water vapour starts migrating towards the heat source and is finally liberated into the borehole.

The liberated water or gas components may influence the corrosion behaviour of the HLW canisters or may cause a pressure build up in the sealed borehole. Therefore, an exact knowledge of the quantity and rate of water is necessary. Because of the specific strain/stress-conditions in a repository this knowledge can only be gained by in situ-experiments.

2.1.2 Layout

The question of thermally induced water liberation from rock salt containing polyhalite under in situ-conditions was investigated in Temperature Test Field 5 (TTF 5). For the performance of TTF 5, an area above the 775 m-level in the Asse salt mine was chosen which is located in the so-called "Polyhalitbänkchensalz (Na2P)". This part of the Staßfurt halite (Na2) is especially rich in polyhalite with an average of 3.65 weight-% and a maximum value of 13.2 weight-%. For better later accessibility three electrical heaters were installed in a horizontal borehole of 7.0 m length and 0.28 m diameter. The heaters had a total length of 3.04 m and a diameter of 0.205 m so that an annulus of 3.75 cm width existed in the beginning.

From laboratory investigations it was known that the liberation of hydrated water occurs at distinct temperatures [2]. For this reason the heaters were not operated at constant power. The power was rather varied in that way that constant temperatures of 100 °C, 150 °C, 200 °C, 230 °C, and 270 °C were achieved at the borehole wall for a period of 60 days in each case.

The quantity of liberated water was measured in a cold trap. This technique was developed in the Institut für Tieflagerung and was described by ROTHFUCHS [5]. The gas-tight sealed heater borehole (Fig. 1) was connected by pipes and valves with the cold trap. A membrane pump ventilated the hot and humid air through the cold trap where the water was condensated and measured. The dried air was then pumped back into the borehole.

An important parameter for the migration of water or gas through rock salt is its permeability. Therefore, four boreholes with a diameter of 40 mm (named P1 through P4) were drilled parallel to the central heater borehole at a distance of 0.43 m and 0.99 m, respectively. These boreholes were also sealed gas-tight by packers and valves. Whereas boreholes P2 and P3 were left under atmospheric pressure, boreholes P1 at a distance of 0.43 m and P4 at a distance of 0.99 m were filled with neon at a pressure of 5 bar. At the same time the heater borehole was filled with helium at a pressure of 1 bar. By taking periodic gas samples from all five boreholes it was investigated if helium was migrating out of the heater borehole or neon into it. These measurements should also prove if micro-cracking would occur in the near-field around the heaters.

Further measurements covered temperature and displacement. Temperature in the surrounding rock salt was measured at 32 different places and at three places each on the surface of the electrical heaters, in the annulus, and on the wall of the heater borehole. Displacement was measured with three rod extensometers of 24 m total length each. Four of them were installed parallel to the electrical heaters and one perpendicular.

All generated data were hourly recorded on magnetic tape for further processing in a computer.

2.1.3 Results

The electrical heaters were switched on on April 19, 1982, and operated without any major failure for the foreseen period. They were switched off again on February 18, 1983, after a total heating period of 305 days.

Three examples for temperature measurements are given in Fig. 2. It can be seen that the above mentioned borehole-wall target temperatures (A1) could be reached in all five periods after some days. Measured and calculated temperatures - calculated after the test with the actual power input - are in excellent agreement for A2 and A3. This is not true for the temperature values on the borehole wall (A1). This difference may be caused by difficulties during calculation – taking into account an annulus or not – and/or by uncertainties of the exact contact of the thermocouple on the borehole wall.

The time- and temperature-dependent liberation of water into the heater borehole is shown in Fig. 3. The amount of 100 g liberated in the very first days of the test is caused by drying off adsorbed water from the borehole wall which was open and in contact with the mine's atmosphere for about 18 months before it was sealed. In heating phases No. 1 through No. 4 the water liberation rate achieved a nearly constant value of 0.5 to 0.7 g/day after some 20 days. This is not true for heating phase No. 5 at 270 °C. Here, the rate was nearly constant at 0.9 to 1.0 g/day after 30 days, but was still decreasing until the end of the test. In Fig. 3 a second curve is shown for comparison where the above mentioned 100 g are subtracted from all measured values. This curve shows that a total amount of 300 g water was liberated during the test and 80 g thereof during heating phase No. 5 at 270 °C.

These results confirm JOCKWER's earlier result from laboratory tests that liberation of crystal water from polyhalite is only starting at a temperature of 235 °C [2].

It can also be seen from Fig. 3 that using the Knudsen model for water vapour migration does fit the measured values better than using the Darcy model. The Knudsen model is:

$$v = -c_K \frac{\sqrt{T}}{p} \quad \text{grad } p$$

v = velocity of flow (cm/sec)

c_K = Knudsen factor (cm^2/$\sqrt{K}$·sec)

K = permeability of rock salt (cm^2)

T = absolute temperature (°K)

p = partial pressure of water vapour (bar)

One example of the permeability measurements is shown in Fig. 4. It shows the gas pressure in borehole P1 which was located at a distance of 0.43 m from the heater borehole and which was originally filled with neon at a pressure of 5 bar. Gas samples were taken at the end of each heating phase with simultaneous equalizing the pressure drop (exemption: heating phase No.1). It can clearly be seen from the curve that no pressure drop occurred but that gas pressure increased markedly with increasing temperature. This proves that no microcracks were generated during the whole test. After the end of the test, however, when the electric heaters were switched off after 305 days a complete and sudden pressure loss was observed caused by cracking of the suddenly cooling salt. This condition, however, will never occur in a repository for radioactive wastes.

Analyses of the taken gas samples show that the permeability of rock salt at elevated temperatures is not zero in the sense of physics. With values of about 1.4 x 10^{-23} m^2 or 1.4 x 10^{-11} Darcy at 200 °C, however, they are only of interest for the calculation of water vapour migration.

The results of the displacement measurements also confirm that no cracks were created in the salt by heating because the displacements are evenly distributed over the total test period (Fig. 5). The five heating phases, however, can be clearly recognized in the displacement curves. At the very end of the test, i.e. after switching off the heaters at day 305, sudden and strong displacements can be seen which were caused by cracking.

Very important results were achieved when the three electrical heaters were cut free after the end of the test. It could be shown by this action that the originally present annulus of 3.75 cm width had complete-

ly disappeared over the total length of the heaters and was "replaced" by
rock salt which crept onto the heaters (Fig. 6). The pressure which was
thereby executed lead to a strong deformation of the two former concen-
tric tubes of the heaters. And even these deformations of the heaters
were filled with salt. In spite of this extreme strain of the salt no micro-cracks could be detected which would have increased the permeability
of the salt.

Another very important result is given by comparing the temperatures
on the borehole wall with those on the heater surface at heater midplane.
It can be seen from Fig. 7 that after 180 days both temperatures show the
same value of 220 °C. At this time at the latest the creeping salt has
contacted the heater.

2.2   Geophysical monitoring

2.2.1 Objectives

The local heating of a rock mass by an underground temperature test
leads to changes in stress distribution and may cause fracturing of or
microcracks in the rock salt. Therefore, a microseismic monitoring system
was installed in the Asse salt mine in order to detect and to locate
small local stress releases. Seismic transmission measurements are also
used to monitor selected areas in the mine for eventually occurring microcracks.

2.2.2 Microseismic monitoring

A network of seven geophones was installed in 1979 on several levels
in the Asse salt mine (Fig. 8). In 1983, two horizontal transducers each
were added at locations No. 1, 2, 3 and 7, so that three-dimensional registration is now possible there. Examples of different microseismic
events which can be determined with this monitoring system are shown in
Fig. 9. The most frequently registered event during recent years – about
five events per month – is roof fall in old empty rooms of the mine. A
typical roof fall is announced by a preliminary signal which arrives
1.7 s before the main signal. This 1.7 s represents the height of 15 m of
old rooms in the mine. By sophisticated processing of registered roof
fall data it is meanwhile possible to locate these events with an accuracy of about 50 m within the mine.

Another result of microseismic monitoring is shown in Fig. 10. In
this figure one can see the microseismic events projected into a profile
through the Asse salt anticline with roof fall events being deleted. The
seismic activity shown in this figure is assisting the results from rock-
mechanical monitoring of the mine. These results tell that a large scale
equalizing flow of salt does occur from the deep core of the anticline
towards the system of rooms and pillars existing in the southern flank of
the anticline, cf. Fig. 11 [6].

2.2.3 Seismic transmission measurements

Active seismic transmission measurements are performed in situ with
the hammer blow method as a non-destructive method for the investigation
of selected parts of the mine. One example is given in Fig. 12. The in
situ seismic velocities of rock salt (v_p = compression wave; v_s =
shear wave) shown here for the immediate vicinity of Temperature Test
Field 5 were measured five times before, during, and after performance of
the test. It can be seen that even after switching off the electric heaters of TTF 5 only very limited variations of the seismic velocities occurred. The very slight decreases at distances 1 – 6 and 3 – 6 are very
close to the measuring uncertainty. In consequence, also these results
confirm that no cracks or fissures were produced during TTF 5.

The installation of further developed instruments for microseismic observations and transmission measurements at another in situ heater test with very high power input in the Asse salt mine is completed and will deliver further valuable results.

2.3 Water or brine intrusion into a repository

Safety analyses for radioactive waste repositories in salt formations take into account a scenario of water or brine intrusion into the sealed repository. Therefore, different laboratory and field experiments are performed by the Institut für Tieflagerung since many years in order to know the physical, chemical, and mineralogical processes which could take place in such a scenario. Relevant in situ-investigations are performed at different locations:
- in flooded shafts of former salt or potash mines,
- in the Brine Test Field set up on the 800 m-level of the Asse salt mine,
- in the former potash mine Hope north of Hannover which is presently filled with NaCl-solution.

By taking continuous density-, pressure-, Eh-, pH-, conductivity-, sound velocity-, and flow logs in 13 water and brine filled shafts of former potash and salt mines in Northern Germany down to a depth of 1000 m and by analyzing the simultaneously taken samples, a good knowledge was accumulated during the last ten years with regard to physical and chemical equilibrium within these flooded shafts. The most important results are that in all shafts very distinct segments of different chemical composition exist which depend on the mineralogical composition of the relevant rocks [1], that very distinct boundaries exist between these segments which can only be penetrated by migrating ions by very slow diffusion processes, and that a salt plug is slowly crystallizing at the density boundary of $MgCl_2$/NaCl-solutions which is forming an additional barrier. These results could be confirmed by laboratory investigations and by a large-scale tracer test performed in the field.

The Brine Test Field in the Asse salt mine was put into operation in November 1984. Objectives of this in situ-test are to investigate the performance of a HLW-borehole seal at elevated temperatures and completely covered with brine and to control the heat dissipation front within the brine itself and in the adjacent rock salt. First results show that the measured temperature values in the brine as well as in the rock salt are lower than predicted by model calculations.

The former potash mine Hope, about 30 km north of Hannover, is presently filled with NaCl-solution which is generated by solution mining processes of gas storage cavities in another salt dome. It was and is possible by this project to perform an extensive R&D-program with regard to events which take place before, during, and after flooding this mine. These events are:
- physico-chemical dissolution processes between NaCl-solution and different salt rocks
- mode and time necessary to achieve physico-chemical equilibrium
- rock-mechanical behaviour of drifts and rooms during and after flooding
- generation of spallings, fissures, or microcracks caused by flooding
- investigation of the effectiveness of a specifically designed seal bulkhead which was constructed at a special site in the mine before flooding.

A further objective of this R&D-program was and is the development, testing, implementation and long-term behaviour of tooling, probes, and electronic measurement equipment in highly concentrated salt solutions under hydrostatic pressures between 3 and 8 MPa.

Flooding of the Hope mine started on March 12, 1984 and has meanwhile reached the former main level which is located at -500 m. First results show that great difficulties occur with the electronic measurement
equipment especially at the seal bulkhead. A report with detailed description of the total project was recently published [7].

3. The Dutch in situ-test program

In the development of a Dutch concept for a high-level waste repository in a salt dome, the use of long boreholes, spaced on a rather wide
grid, was considered for reasons of limiting maximum salt temperatures
and the more efficient use of the vertical dimension of the formation
[8]. As a proof of the validity of this concept, the feasibility of drilling long holes, up to 300 m length, with a dry drilling technique, had
to be shown.

In close cooperation with GSF and under contract with the CEC, ECN
could have made this 0.3 m diameter hole by a German firm, from a specified location at the 750 m-level of the Asse salt mine.

The mining inspectorate required a geologic reconnaissance drilling
in close proximity and in this way the experiment served three purposes:
- additional information about the deeper regions of the formation;
- demonstration of the dry-drilling technique;
- the realization of a facility to measure thermomechanical properties of
  salt in situ under rather undisturbed conditions.

The hole was used, during the Second Program 1980 - 1984, for a series of experiments:
- a long lasting free convergence measurement near the bottom of the
  hole, starting immediately after completing it, in order to follow the
  first stages of creep at ambient temperature;
- a series of five free convergence tests at different depths in the hole
  and at elevated temperatures, using an electrically heated, reusable
  probe;
- two pressure tests, at depths of 262 and 80 m and elevated temperatures, by allowing the salt to creep up to the probe wall.

The set up of these experiments and the test results will be described here. The theoretical thermomechanical development of the tests
will be presented in the paper of ALBERS and others [9].

3.1   The experimental equipment

The equipment used for the first convergence rate determination was
already described in the paper by KÜHN and VERKERK during the first Luxemburg conference [1]. In addition, the results of this experiment also
have been reported [10] so that in this paper only the subsequent heated
convergence and pressure tests are described.

3.1.1 Data collection and transmission

The probes used for the heated tests were coupled to an instrumentation box (CIB), which was kept cool by providing good thermal contact
with the salt wall using phosphor-bronze spring clips. Further the CIB
was kept away about 4 meters from the heated probes that also had non-
heated parts at both ends. In this way the electronics in the CIB could
be kept at about 50 °C during tests (salt temperature ∿42 °C). The instrument box contained the following systems:
- power supply for the strain gauges and transducers,
- amplifier for the strain gauge and transducer signals;
- scanners;
- analog digital transformers.

To improve the reliability of the electrical system, the equipment was split up into four independent signal groups and corresponding cables. A very important feature of this part of the total experimental set-up was the coupling-decoupling device that allowed to free the expensive instrument box from the first high pressure probe (HPP-I), stuck in the borehole after the test, and to use it again. The digitalised data from the CIB were transmitted by electrical cables to the data collection and processing system, located at the 750 m-level in a small cabin, and consisting of a computer and a disc unit, placed in a cooled cupboard. The computer controls the heat input to the probe, alarms in case of malfunction, presents the measuring data on a monitor, prints and stores the information on a disc unit. The experiment is watched from ECN Petten by a telephone line.

If a power failure occurs, and this happened a number of times during the tests, the power for the heaters is automatically reduced to zero. After return to normal power, the heat input is slowly brought back to the original level.

3.1.2 The Heated Free Convergence Probe (HFCP)

This probe consists of a straight stainless steel pipe, diameter 0.18 m, to give a sufficient radial space between probe and borehole wall to allow creep to proceed over about 50 mm. The heating system, applied over a length of 3 m, consists of a thermo-coax wire placed in a groove (double pitch), machined in the outer surface of the cylindrical wall. To guard the wire against mechanical damage and possible chemical attack (brine), the wire is covered with a lead string, which is forced into the groove. At both ends an additional 1.5 m length is not heated. The maximum power that can be applied is 6000 W, corresponding to about 1710 W per meter borehole length.

Since in this type of experiment rather large salt displacements occur in a relatively short time, a simple swing arm, coupled to a transducer, was used as a measuring device, in this case applied as 30 single units over the probe length. Thermocouples measured the temperatures of the probe itself (max. 360 °C) and at the ends of the swing arms touching the salt. Due to heat radiation from the probe, the latter couples did not indicate the real salt temperature and corrections for this effect had to be applied.

3.1.3 The Heated Pressure Probes (HPP)

The first of these probes (HPP-I), was used in a test at 262 m depth in the hole, the second, somewhat improved one, HPP-II in a test at 80 m. The HPP's consist of a straight tube, sufficiently strong to withstand the pressure build up of the salt due to the induced heat flow. The tube is divided in 5 parts, three in the middle to be heated, the outer two thermally insulated from the heated area. The application of these outer parts was necessary in order to avoid the deformation of the salt over the end of the heated area, which would lead to a lower value of the maximum pressure and to difficulties in the interpretation of results. The heating system is constructed as described for the HFCP's. For the purpose of calibration in an oil-pressure jacket, the ends of the heater wire are welded in the groove over 8 cm to prevent a leak path. In order to obtain a high level of reliability of the pressure measurements at the elevated temperature of about 200 °C two different systems, strain gauges and transducers, are applied to measure the displacement ΔD of the cylinder of the probe. They are mounted in the Displacement Measuring Device (DMD) as shown in Fig. 13. The strain gauges are glued on the outside of the cylindrical ring which is compressed during the test due to the inwards directed displacement of the cylinder of the HPP. A full

bridge strain gauge arrangment is applied to avoid temperature compensation problems and to obtain a relatively high input signal for the data
transmission system. The displacement transducer is of the induction
type, the application of some ferritic material is then essential.

The temperatures are measured with thermocouples of the chromel-alu-
mel type. The thermocouples are located as follows:
- The centre part of each of the 23 DMD's is equipped with one thermocouple for temperature control of the gauges and displacement transducers.
- 13 internal thermocouples are placed between the wall and a number of
  DMD's.
- 12 external couples are placed at the outside of the probe in such a
  way that a direct contact with the salt surface is obtained. These
  couples are situated as closely as possible to the corresponding internal couples in order to get an indication when contact between the
  probe surface and the salt formation has been established.

The DMD's were calibrated with a prescribed displacement system at
temperature levels of ∿20, ∿100, ∿150, and ∿200 °C. The strain gauges and
the transducers showed minor temperature dependency in the applied temperature range.

The transducer circuit in the CIB, however, showed a considerable
temperature dependency. To be able to compensate for this effect non-ac-
tive transducers are mounted in the CIB. During the calibration pressure
test of the HPP-II by means of an hydraulic jacket the DMD's did not show
hysteresis effects. This proved that the design change in the HPP-II in
comparison with HPP-I which showed a considerable hysteresis was successful.

After transport to the Asse salt mine the whole assembly was tested
and calibrated again. These calibration results are used as zero values
in the final transformation of electrical signal to deformation results.

3.2 Some results

In the following some results are presented of the various tests
performed. The results in the first place served to verify the thermomechanical models and calculations. This work is presented in [9].
Fig. 14 shows the complete series of tests with some details about power,
location and time periods.

3.2.1 The free convergence at elevated temperature

In Fig. 15 the measured displacements (dots) are shown for the HFCP
test at 231 m depth in the hole. The maximum temperature at the salt wall
at the end of this test was 190 °C. For two days the electronics were
tested without heat input. Then heat was applied, up to 6000 W and the
test continued for 20 days, when power supply was stopped to prevent the
probe from sticking in the hole. The ultimate radial salt displacement at
the central part was about 40 mm. With the measuring television vehicle,
as shown in Fig. 16, a device constructed to check the overall shape of
the salt wall and to view the wall surface, the local diameter as shown
on the continuous line in Fig. 15 was measured. Both types of measurement
agree reasonably well. The fast convergence at 190 °C means that high-level waste containers placed in a borehole will be completely embedded in
the salt within a period of some weeks.

3.2.2 The pressure tests

When the HPP-I test started there was a radial space of some millimeters between the probe (diameter 296 mm) and the wall. The power was
set at 4715 W. During the first 3 days of heating the strain gauges and

transducers showed an unclear picture until pressure build-up started.
During the following 9 days the deformation of the probe and the temperature increase until a stabilization is reached. The pressure build-up is
shown in Fig. 17, where a maximum pressure of about 350 bar remained
stable over about 40 days.

Concerning the test results with HPP-II the following can be said.
In general the construction of the HPP-II is much alike compared with the
HPP-I layout. The diameter of the HPP-II is somewhat larger because of
the larger diameter of the hole at the 80 m-level, compared with the
260 m-level where the HPP-I test was executed.

The arrangement of the measuring system was revised in view of the
results obtained from the HPP-I test. During the latter experiment it was
observed that cylindrical cross-sections of the probe were deformed into
an oval-shaped cross-section. It can be shown that at least 3 "diameters"
under 120° in one cross-section must be measured to establish the real
pressure on the probe.

The complete set of measuring data is not available at this time. In
Table I a relevant selection of the values of the experiment with the
HPP-II are given. Underneath Table I a qualitative presentation of the
pressure build-up versus time is given.

From these preliminary results the following conclusions can be
drawn:
1. P_{max} was not reached at the several heat input levels.
2. P_{min} was not reached at the 0-power levels.
3. Thermal expansion of the probe during the test runs has a great in-
    fluence on the results.
4. Lithostatic pressure of the salt at the depth where the test was done
    is <100 bar.

After "no. 8" a short cut in the power system caused a definitive
shut-down of the experiment. Measurements to get an indication of the lithostatic pressure are still going on. Later it was theoretically shown
that the unexpectedly low lithostatic pressure was due to the large number of big excavated rooms that exist in the Asse salt mine.

3.3   Work on a larger diameter hole
In view of the fact that reprocessing plants in Europe envisage the
use of HLW-containers with 0.43 m outer diameter, to which for purposes
of strength and corrosion an overpack should be added, boreholes
of ∿ 0.6 m diameter will be needed. These large diameter holes cannot be
drilled with the dry-drilling technique applied for the 0.3 m hole.
Therefore, at ECN a different method is developed at present. It is the
intention to drill, again in the Asse salt mine, a 600 m deep hole, with
a diameter of 1 m in the upper part and 0.6 m in the lower part.

3.3.1 Test facility
A new test facility was designed to study and measure the flow phenomena in the entrance area, under the roller-bit body, and the exhaust
system.

Table I: Tentative results of HPP-II experiments

No.	Time in days	Power total (W)	Mean Temp. (°C)	Pressure in section no. (bar)		
				II	III	IV
1	12	3000	91	105	122	79
2	25	3000	94	145	154	114
3	37	0	40	109	107	81
4	83	0	36	91	89	68
5	95	3000	93	156	157	130
6	110	4000	118	178	182	151
7	112	5000*	136	186	189	157
8	119		128	169	182	125
9	135	0	40	135	115	97
10	160	0	36	120	102	86

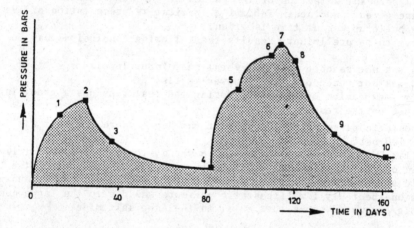

The application of "loose" sand as a substitute for salt has certain disadvantages to study the pneumatic transport. In order to create a more realistic situation large sand bodies were fabricated. The salt was "glued" together by means of waterglass to get a solid body. The sand bodies, with a diameter of 1 meter and a height of the same size were inserted in a cylinder of about the same diameter ("A" in Fig. 18). At four places in the circumference of the cylinder windows are made to study the drilling operation and sand dust transport. In the right hand side of Fig. 18 the roller-bit configuration can be seen [B]. In Fig. 19 a more detailed picture of the roller-bit is presented. In order to get a controlled airflow under the roller-bit configuration with sufficient velocity, a guide-plate is mounted. In this picture the exhaust pipe, where

*) After 2 days at 5000 W a short circuit occurred.

the sand-air mixture is sucked, can also be seen. Some difficulties were encountered to fabricate sand-bodies which were not too hard, in order to minimize the "drilling-energy". After some fabrication test-runs this problem was solved.

3.3.2 Tentative results

As stated earlier the purpose of this test is in the first place to demonstrate the feasibility of dry drilling of large diameter holes, in particular focused on the removal of the cuttings. The tests so far have shown that the cuttings can actually be removed by the air flow.

In preliminary tests it proved possible to apply this air driven removal process. The rotational speed of the drill in this test was 20 rev/min, whereas the air flow amounted to 600 Nm^3/min.

As the results obtained so far indicate that the vertical speed is limited by this factor, further tests are aimed at optimizing the parameters governing the horizontal transport mechanism. These are, for example, the air flow, the guide plate geometry, and the distance of the exhaust tube to the sand surface.

REFERENCES

[1] KÜHN, K. and VERKERK, B. (1980). Disposal in Salt Formations. Proc. of 1st Europ. Community Conference on Radioactive Waste Management and Disposal, Luxembourg, May 20-23, pp. 385 - 419

[2] JOCKWER, N. (1981). Untersuchungen zu Art und Menge des im Steinsalz des Zechsteins enthaltenen Wassers sowie dessen Freisetzung und Migration im Temperaturfeld endgelagerter radioaktiver Abfälle. GSF-Bericht T 119

[3] BRADSHAW, R. L. and McCLAIN, W. C. (1971). "Project Salt Vault: A Demonstration of the Disposal of High Activity Solidified Wastes in Underground Salt Mines." ORNL-4555

[4] JENKS, G. J. (1979). Effects of Temperature, Temperature Gradients, Stress and Irradiation on Migration of Brine Intrusions in a Salt Repository. Oak Ridge National Laboratory, ORNL-5526

[5] ROTHFUCHS, T. et al. (1983). Simulationsversuch im Älteren Steinsalz Na2β im Salzbergwerk Asse - Temperaturversuchsfeld 4. GSF mbH München, Institut für Tieflagerung. Abschlußbericht im Rahmen der Verträge WAS 015-76-1 und 058-78-1 D des Indirekten Aktionsprogramms der Europäischen Gemeinschaften

[6] DÜRR, K. (1984). Markscheiderische Überwachung der Standsicherheit eines Versuchsbergwerks für die Lagerung radioaktiver Abfälle. Das Markscheidewesen, 91, Nr. 2, pp. 388 - 392

[7] HERBERT, H.-J. et al. (1985). Vorgänge vor, während und nach der Flutung des Salzbergwerkes Hope. GSF-Bericht T 12/85

[8] HAMSTRA, J. and VERKERK, B. (1981). The Dutch Geologic Radioactive Waste Disposal Project. Final Report, EUR 7151 EN

[9] ALBERS, G. et al. (1985). Modelling of Thermomechanical Behaviour of Salt Rocks. Conference Paper VI-3, 2nd Europ. Community Conf. on Radioactive Waste Management and Disposal, April 22-26

[10] DOEVEN, I. et al. (1983). Convergence Measurements in the Dry-Drilled 300 m Borehole in the ASSE-II Salt Mine. Europ. Appl. Res. Dep. Nucl. Sci. Technol., vol. 5, no. 2, pp. 267 - 324

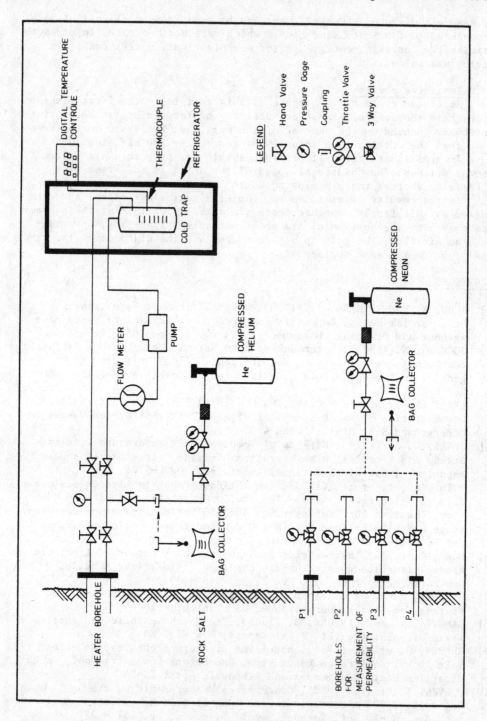

Fig. 1 — TTF 5 — Cold trap and permeability measurement

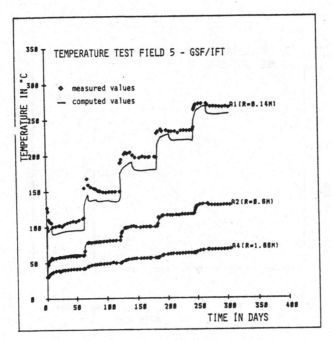

Fig. 2 - TTF 5 - Rock temperatures with time at different distances
(A 1 ≙ borehole wall)

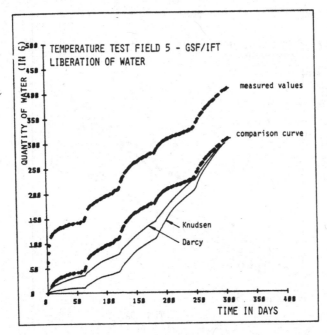

Fig. 3 - TTF 5 - Water liberation with time - comparison of measured
values with calculated values using the Knudsen- and Darcy-
model

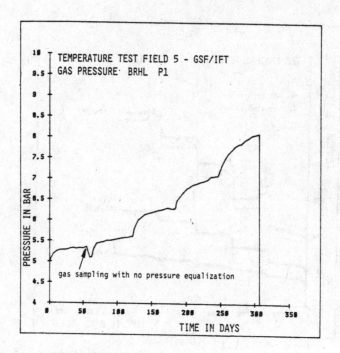

Fig. 4 - TTF 5 - Gas-pressure increase in borehole P1

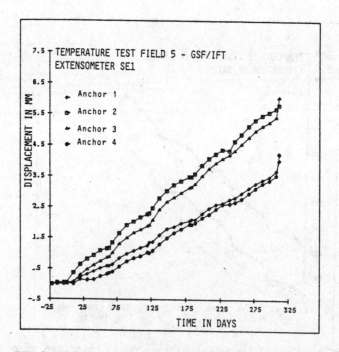

Fig. 5 - TTF 5 - Displacement/time diagram of extensometer SE1

Fig. 6 – TTF 5 – Electrical heaters cut free after the test

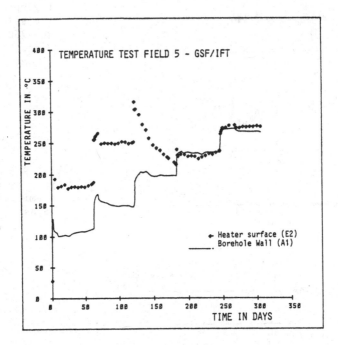

Fig. 7 – TTF 5 – Temperatures on the borehole wall and on the
heater surface at heater midplane with time

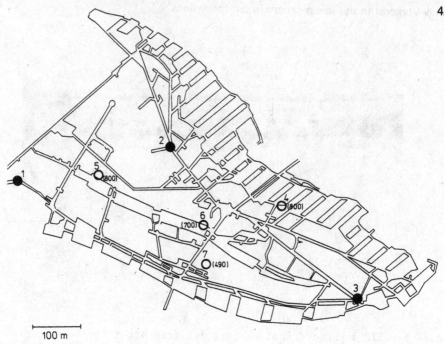

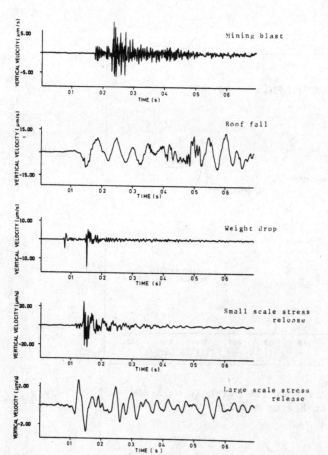

↑ Fig. 8 – Locations
of geophones in the
Asse salt mine
● = 750 m-level
○ = other levels
at ( ) m

←Fig. 9 – Different
types of micro-
seismic events in
the Asse salt mine

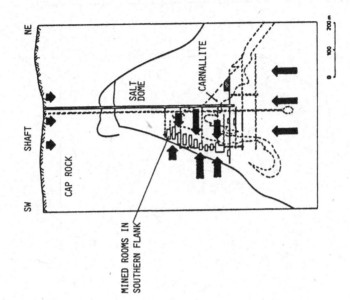

Fig. 11 – Directions of large scale flow of rock salt masses in the Asse salt mine

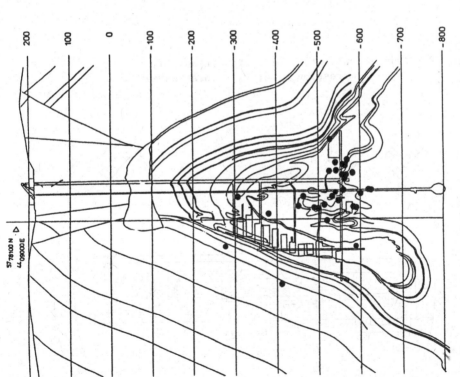

Fig. 10 – Locations of microseismic events in the Asse salt mine (projected into a profile with roof fall events being deleted)

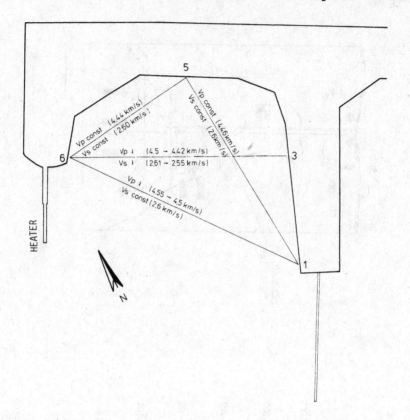

Fig. 12 – Seismic velocities of rock salt near TTF 5
(v_p = compression wave; v_s = shear wave)

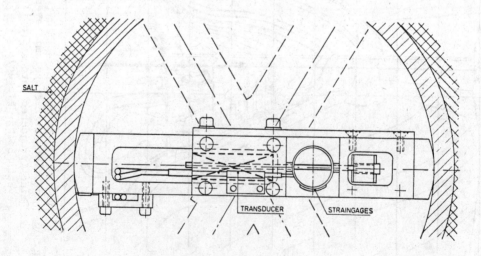

Fig. 13 – Displacement measuring device (DMD)

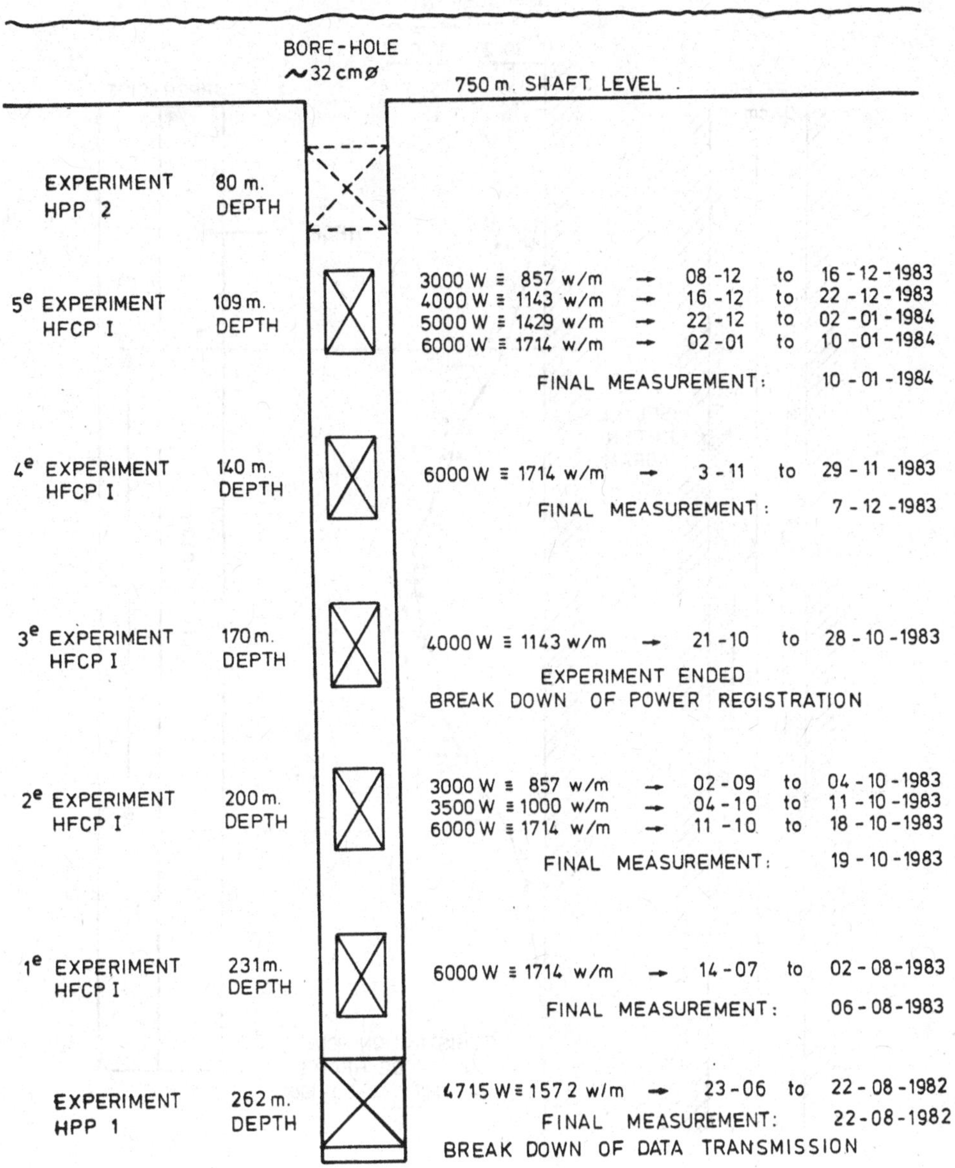

BORE-HOLE
~32 cmØ

750 m. SHAFT LEVEL

EXPERIMENT   80 m.
HPP 2   DEPTH

5^e EXPERIMENT   109 m.
HFCP I   DEPTH

3000 W ≡ 857 w/m	→	08 -12	to	16 - 12 -1983
4000 W ≡ 1143 w/m	→	16 -12	to	22 - 12 -1983
5000 W ≡ 1429 w/m	→	22 -12	to	02 - 01 -1984
6000 W ≡ 1714 w/m	→	02 -01	to	10 - 01 -1984

FINAL MEASUREMENT: 10 - 01 -1984

4^e EXPERIMENT   140 m.
HFCP I   DEPTH

6000 W ≡ 1714 w/m   →   3 - 11   to   29 - 11 -1983

FINAL MEASUREMENT :   7 - 12 -1983

3^e EXPERIMENT   170 m.
HFCP I   DEPTH

4000 W ≡ 1143 w/m   →   21 -10   to   28 - 10 -1983
EXPERIMENT ENDED
BREAK DOWN OF POWER REGISTRATION

2^e EXPERIMENT   200 m.
HFCP I   DEPTH

3000 W ≡ 857 w/m	→	02 -09	to	04 - 10 -1983
3500 W ≡ 1000 w/m	→	04 -10	to	11 - 10 -1983
6000 W ≡ 1714 w/m	→	11 -10	to	18 - 10 -1983

FINAL MEASUREMENT: 19 - 10 -1983

1^e EXPERIMENT   231m.
HFCP I   DEPTH

6000 W ≡ 1714 w/m   →   14 -07   to   02 - 08-1983

FINAL MEASUREMENT: 06 - 08-1983

EXPERIMENT   262 m.
HPP 1   DEPTH

4715 W ≡ 1572 w/m   →   23 -06   to   22 - 08 -1982
FINAL MEASUREMENT: 22 - 08-1982
BREAK DOWN OF DATA TRANSMISSION

Fig. 14 - Experiments with HFCP 1 and HPP in the Asse salt mine

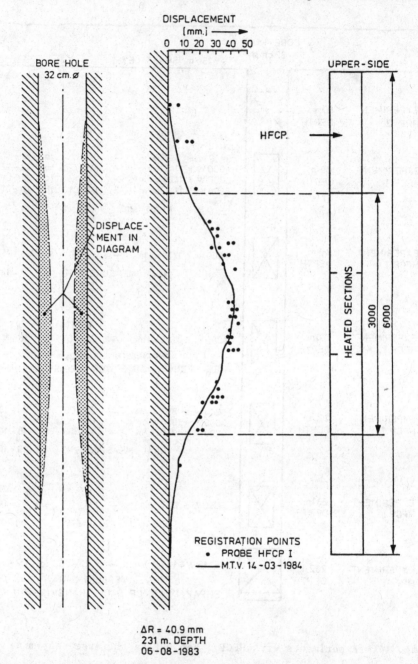

Fig. 15 – Comparison between HFCP values versus the measuring
television vehicle measurements

Fig. 16 - Measuring television vehicle

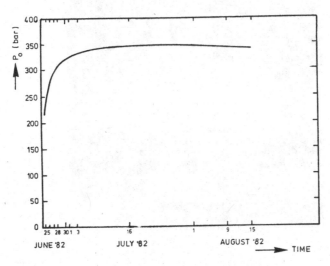

Fig. 17 - Pressure build-up on probe HPP I

Fig. 18 - The test facility, partly dismantled

Fig. 19 - Detailed picture of roller-bit configuration

DISCUSSION

R. KOSTER, KfK Karlsruhe

The pressure build-up is of significant importance, in particular to the HLW-disposal experiments in ASSE salt. Could you give additional information about the accuracy of your pressure measurements? You mentioned the result of 350 bars and in our experiments we had an accuracy of 15%.

L.H. VONS, ECN Petten

The accuracy was of the same order - say $\pm$ 10%.

R. KOBAYASHI, JGC Corporation Tokyo

In your presentation you mentioned, that after some time the pressure dropped suddenly. Did you simulate the actual cooling period?

K. KUHN

We controlled all parameters during the heating phase and it was confirmed, that there was no micro-cracking during that phase. But at the end of the test, on day 305, we abruptly switched off the heaters. This sudden drop of power provoked micro-cracks by contraction of the salt. Of course such a temperature transient cannot occur in a repository.

R. KOBAYASHI, JGC Corporation Tokyo

Do you plan to simulate the cooling down period?

K. KUHN

Yes, in the next test, the gradual heating will be followed by slow cooling.

SPECIFIC INVESTIGATIONS RELATED TO SALT ROCK BEHAVIOUR

L.H. Vons
Netherlands Energy Research Foundation (ECN)
(editor)

SUMMARY

In this paper results are given of work in various countries in rather unrelated areas of research. Nevertheless, since the studies have been undertaken to better understand salt behaviour, both from mechanical and chemical points of view, some connection between the studies can be found. In the French contribution (A) the geological conditions have been investigated that might promote or prevent the formation of salt domes from layers in view of possible use of the latter type of formation.

This was done theoretically by the finite element method, and a start was made with centrifuge tests. The density of a number of samples from salt and overburden from the Bresse basin was measured and it was shown that a favourable condition exists in this region for waste disposal.

In the German contribution (B) various subjects are touched upon, one being the effect of water on the mobility in the early stages of salt dome formation. Evidence was found for an anisotropy in salt.

One Dutch contribution (C) describes results of studies on the effect of small amounts of water on the rheology of salt. The results imply that flow laws obtained for salt at rapid strain rates and/or low confining pressure cannot be reliably extrapolated to predict the long term behaviour of wet or even very dry material under natural conditions.

Preliminary results on the effect of water upon ion-mobility indicate a certain pseudo-absorptive capacity of salt e.g. for Sr.

Another Dutch contribution (D) presents results of laboratory and in-situ measurements at ambient temperatures in the Asse-mine, with an acoustic method to detect the extension of cracks in stress-relieved locations, such as gallery walls etc.

A somewhat related type of work is presented in the Danish contribution (E) where the absorptive capacity of powdered salt for a range of radio-nuclides of different valency was measured both for pure salt and salt with addition of hematite and anhydrite impurities.

A. STUDY OF SALT DOMES

PROF. A. ZELIKSON and L. CHARO
Laboratoire de Mecanique des Solides, Ecole Polytechnique

Conditions for salt dome formation are studied in LMS, Ecole Polytechnique, with respect to the safety analysis of a nuclear waste repository in a salt layer.

The development of domes is possible when the density of a salt layer is less than its overburden, as it is shown hereby:

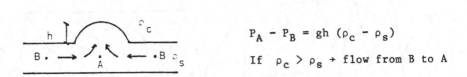

$$P_A - P_B = gh \, (\rho_c - \rho_s)$$

$$\text{If } \rho_c > \rho_s \rightarrow \text{flow from B to A}$$

Nevertheless, the salt flow does not occur if the density contrast and the dimensions of the irregularity of interface are not high enough to exceed the strength of overburden and that of salt itself. This is the reason why it is necessary to do a parametrical study in which the main parameters are the density, the viscosity and the yield point of the materials, as well as the dimensions of the irregularity at the salt-overburden interface. So, it has to be determined in which cases a salt dome can develop and if so, what growing rates are likely.

To do such a study we use a finite element elastoviscoplastic code using the following constitutive law for the materials involved:

$$\dot{\varepsilon} = \dot{\varepsilon}^{el} + \dot{\varepsilon}^{vp} \quad \text{and} \quad \dot{\varepsilon}^{vp} = \frac{\langle F-S \rangle^n}{\eta} \, \frac{\delta Q}{\delta \sigma}$$

where F, Q and S are respectively a viscoplastic criterion, a viscoplastic potential and a yield point.

As an example of the results obtained, Figure A-1 shows how an initial irregularity of 100 m high and 600 m diameter develops during a 6000 years period. The values of mechanical parameters were taken as:
- densities   : ρ_c =   2,35 ρ_s =   2,15 g.cm^{-3};
- yield points: S_c =   3 MPa S_s =   0,5 MPa
- viscosities : η_c = 31.700 MPa.an^{-1} η_s = 317 MPa.an^{-1};
- viscoplastic criterion and potential: Von Mises.

In order to validate the numerical calculations, an experimental study with a centrifuge is also done: the centrifugal force simulates the gravity, main factor in dome development. For this purpose, an experimental model has been designed.

To complete this study, it is important to determine the properties of the materials existing in situ, and especially, in the sites that may become waste repositories in the future. Figure A-2 shows a summary of laboratory measurements taken on samples cored from the Bresse Bassin. A very small difference is observed between the average rocksalt density and that of its overburden: 2,18 vs 2,19. From the safety point of view, this fact should be regarded as a favourable factor restricting the possibility of dome formation at this site.

Laboratory tests are also done in order to determine the rheology of the rock salts from the potentially favorable sites. A part of these tests is done under a program supported by the European Communities. An important difficulty remains in extrapoling results of tests lasting a limited time to periods of several hundreds of centuries. The results of our creep tests under different loads and temperatures appear to be very similar to those obtained by other teams (eg. Langer, 1979). See figure A-3.

REFERENCES
LANGER, M. (1979). Rheological Behaviour of Rock Masses. General Report, 4th International Congress on Rock Mechanics, Montreux. Vol. 3, p. 29, Balkema Editor.

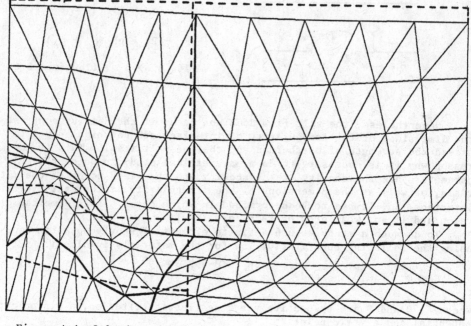

Figure A-1: Salt dome development   ---- initital grid (t = 0)
           Numerical results   —— final grid (t = 6000 years)
           Scale: 1 : 10000

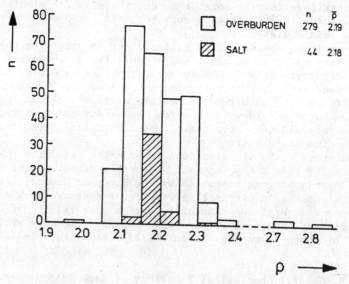

Figure A-2: Density of salt and its overburden
            Bresse Bassin

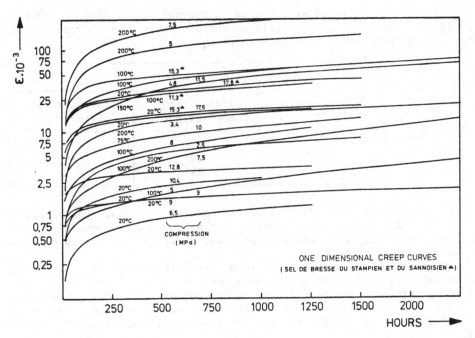

Figure A-3

B. ASSOCIATED STUDIES IN ROCK SALT FORMATIONS

Prof. H. GIES
GSF, Institut für Tieflagerung, FRG

1. Some notes on water in salt minerals

The uplift of salt diapirs and their particular mobility is most probably enforced by the presence of water and possibly by the presence of
gases. The amount of water (0.04 wt%) found in the Stassfurt Halite may
thus be considered as the final result of a late stage in a purifying process, since the water content of younger but already diagenetic salt layers
is considerably higher (1-10%). It must further be borne in mind that an
essential part of those 0.04 wt% is hydrated water which is released only
in parts of salt rock which are stress relaxed (Kern 1980).
In Northern Germany the Stassfurt-Leine-Halites, which are separated by
the Stassfurt seam, represent the main potential disposal horizons. It was
possible to study the distribution of water in a profile through the upper
parts of Na_2 by means of a 300 m deep borehole drilled by ECN and GSF on
the 750 m level of the Asse mine. This showed that the Stassfurt seam is
underlain by a transition zone where the Stassfurt halite has higher contents of sulfate minerals, particularly polyhalite, thus rendering higher
contents of crystalline water (appr. 0.25 wt%) which will be thermally
released if stress relaxation takes place in the salt. The deeper parts of
the halite - the so-called "Main Salt" - once again shows the average content of 0.04 wt% of water (Figure B-1).
It could also be shown that the chemical analysis of drilling fines, including those of dry drillings, provide a good possibility of showing the
average composition of a geological profile reflecting stratigraphical
differences.
To ensure that the maximum temperature in the Stassfurt carnallite

seam will not be exceeded at any time, a thermal safety distance must be upheld. Depending on the maximally acceptable temperature in carnallite the following distances have to be expected.

	CARNALLITE : $K Mg Cl_3 \cdot 6H_2O$		
Table B-1	Rock pressure (bar)	Water release temperature (°C)	Thermal safety distance (m)
	1	~ 85	~ 60
	40	139	~ 40
	100	145	

In Temperature Test Field 5 of the Asse mine it could now be proved for the first time, in situ as well, that by a stepwise increase of temperature the hydration water from polyhalite will not be released at temperatures below ~ 250 °C.

The thermal release of water from rock salt and hydrated minerals is accompanied by release of gases. Here both the naturally occurring and thermally produced gases will come under the influence of radiation in a HLW borehole. Consequently, irradiation experiments were performed which showed that the production of gases in some cases clearly depends on the dose.

Table B-2   Gas release as a function of the absorbed dose for a Rock salt specimen with: 85% Halite, 13% Polyhalite and 2% other components

	H_2	O_2	CO_2	HCl	CH_4
not irradiated		$2,5.10^3$ppm $9,2$ Nl/m³ Rocksalt	50 ppm $0,2$ Nl/m³ Rocksalt	240 ppm $0,9$ Nl/m³ Rocksalt	5 ppm $0,02$ Nl/m³ Rocksalt
irradiated upto 10^8 rad	$3,2.10^3$ppm $11,7$ Nl/m³ Rocksalt	$2,4.10^3$ppm $8,8$ Nl/m³ Rocksalt	$1,1.10^3$ppm $4,2$ Nl/m³ Rocksalt	458 ppm $1,7$ Nl/m³ Rocksalt	124 ppm $0,44$ Nl/m³ Rocksalt
irradiated upto 10^9 rad	$24,7.10^3$ppm 92 Nl/m³ Rocksalt	$4,1.10^3$ppm $15,0$ Nl/m³ Rocksalt	$12,6.10^3$ppm 46 Nl/m³ Rocksalt	554 ppm $2,0$ Nl/m³ Rocksalt	84 ppm $0,3$ Nl/m³ Rocksalt

A current in situ test in the Asse mine using heaters combined with Co-60-sources had the following borehole atmosphere in such an actively loaded borehole:

Table B-3   Composition of the atmosphere in actively loaded borehole (US/FRG-Brine Migration Test)

Components	H_2	O_2	N_2	CO	CO_2	CH_4	C_2H_6	C_3H_8
Concentration (Vol.%)	0,57	0,3	91,3	0,35	6,45	0,23	0,046	0,03

Comparisons with nonactive boreholes lead to the conclusion that the produced hydrogen in the borehole is primarily formed by corrosion reactions of the lining materials.

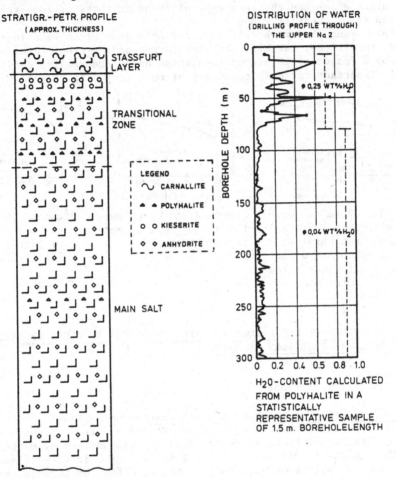

Figure B-1: Stratigraphic profile and distribution of water content in 300 m bore hole

2. Mechanical anisotropy

The petromechanical behaviour of salt and its uniaxial strength depend on the direction in which it is loaded. It could be shown by uniaxial pressure tests (Gessler 1983) that a correlation exists between the angle of stratification (here: angle between layers and the direction of pressure) and the measured values of compressive strength. The results vary between ~35% for this strength and ~80% for strain.

The same is valid for the compressive strength of rock salt showing an orientation of the longer axis of its grains in such a way that the uniaxial strength parallel to the orientation of the grains was, on the average, 20% smaller than normal to it (~28,5 N/mm^2 to 35,5 N/mm^2). Salt texture exhibits the same clearly anisotropic behaviour as extruded synthetic salt with a pronounced fibre texture in <100> direction when investigating its creep behaviour.

The results show that the specimens compressed at room temperature

parallel to the fibre axis exhibit significantly higher strains than those
normal to it (Figure B-2).

Firstly performed uniaxial long term tensile tests revealed extension
time curves which had the same typical shape as those of long term compression tests (Gessler 1983).

This means that rock salt does not only show creep behaviour under
tensile conditions, but also that the corresponding creep rates are at
least by a factor of 10 higher than those under compressive conditions.
This is important for the fracturing of rock salt (Figure B-3).

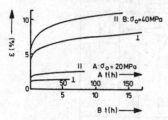 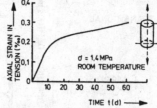 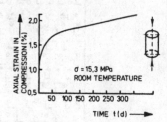

Figure B-2: Room tempera-
ture creep curves for salt
specimens cut parallel
and perpendicular to the
extrusion axis (σ_0 = stress
at the beginning of
deformation

Figure B-3a: Creep curve
for a long term direct
tension test

Figure B-3b: Creep
curve for a long
term test in uni-
axial compression

C. THE INFLUENCE OF FLUID-ROCK INTERACTION ON THE RHEOLOGY OF SALT ROCK AND ON IONIC TRANSPORT IN SALT

C.J. Spiers
Rijksuniversiteit Utrecht

1. Effect of brine of rheology and deformation mechanisms

It is well established that polycrystalline materials containing
small quantities of fluid at grain boundaries can deform by matter transport through the liquid phase (1,2). This mechanism involves dissolution
of material at points of high normal stress, diffusion through the grain
boundary fluid, and precipitation at points of low normal stress. Constitutive equations derived theoretically for this type of creep predict that
the mechanism should become important in halite rocks (containing as little as 0.1 - 0.01 wt% H_2O at low strain rates, and should lead to a substantial weakening effect (see Figure C-1). Solution-precipitation reactions occurring across grain boundary brine films are also predicted to
lead to low temperature dynamic recrystallisation by fluid-assisted grain
boundary migration (3).

Although it has long been known that increased atmospheric humidity
causes accelerated creep in salt (4), water weakening effects have received little attention. Recent experimental work at Utrecht has confirmed
that significant water weakening effects, similar to those predicted, can
indeed occur in salt (5,3). In particular, stress relaxation tests performed in the range T = 20-200 °C, P = 2.5-20.0 MPa (using salt rock from
the Asse Mine) have shown that at strain rates below those normally investigated in the laboratory (i.e. < 10^{-7} to $10^{-8}s^{-1}$), wet material (i.e.
deformed in the presence of brine at 1-10 Mpa pressure) undergoes a marked
weakening effect. This is seen as a change in constitutive behaviour from

$$\dot{\varepsilon} = A \cdot \frac{VCDF}{kTd^2} \cdot \sigma \qquad \text{(see Stocker and Ashby [4], also Raj [3])} \qquad (1)$$

where $\dot{\varepsilon}$ = strain rate, σ = applied stress, d = grain size
$\quad$ k = Boltzman's constant, T = absolute temperature
$\quad$ V = molecular volume of diffusing species
$\quad$ C = solubility of diffusing species
$\quad$ D = liquid diffusivity of diffusing species
$\quad$ F = volume fraction of liquid in system
$\quad$ A = structure parameter, value 10-100, taken as 20 here.

Assumptions: Diffusion is rate controlling and fluid has bulk liquid
$\qquad\qquad\quad$ properties.

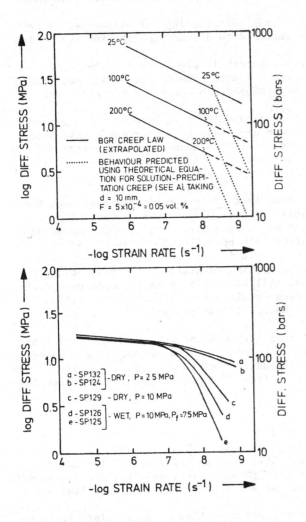

Figure C-1 A) Theoretically derived constitutive equation for steady state deformation by solution-precipitation creep [2,1]. B) Creep diagram showing the BGR creep law for salt, plus the constitutive behaviour predicted using equation 1 (C-1A) applied for salt rock with a grain size (d) of 10 mm and a brine content (F) of 0.05 vol.% (typical for "dry" salt rock). Note that solution-precipitation creep is predicted to become important at strain rates of 10^{-8} to $10^{-9} s^{-1}$, the accompanying change in constitutive behaviour leading to a significant weakening effect.

Figure C-2 Typical rheological data (after Spiers et al. (5)) for Speisesalz salt rock tested in stress relaxation mode at 150 °C under the following conditions:
a,b – "Dry" (i.e. containing inherent brine only),
$\qquad$ P = 2.5 MPa (dilatant conditions)
c – "Dry" (i.e. containing inherent brine only),
$\qquad$ P = 10 MPa (non-dilatant conditions)
d,e – Wet (i.e. brine added),
$\qquad$ P = 10 MPa,
$\qquad$ P_f = 7.5 MPa, (non dilatant conditions).
N.B. P = confining pressure, P_f = pressure of added brine (pore fluid).

the usual power law type behaviour (stress exponent n ≥ 6-8) above $10^{-7} s^{-1}$, to power law creep with n = 1-2 at lower rates (c.f. Figure C-1). The transition has been shown to be associated with the onset of fluid film assisted migrational recrystallisation, this being almost certainly accompanied by solution-precipitation creep (note n = 1 for this mechanism, see Figure C-1). At confining pressures sufficiently high to suppress dilatancy (i.e. ~ 10 MPa), the weakening effect also occurs in "dry"

material through the activity of inherent brine present in fluid inclusions and films at grain boundaries (this totals only 0.02-0.05 wt%). The
effect does not occur in material deformed dry (i.e. without added brine)
under unconfined or dilatant conditions. Typical data are shown in Figure
C-2. The results imply that flow laws obtained for salt at rapid strain
rates and/or low confining pressure (i.e. much of the pre-existing rheological data) cannot be reliably extrapolated to predict the long term behaviour of wet or even very dry (0.1-0.01 wt% H_2O) material under natural
(non-dilatant or elevated pressure) conditions. Clearly, far more data are
needed before extrapolation to long term behaviour can be confidently
made. However, if taken as valid for the temperature range 20-200 °C, the
Utrecht data have a number of implications for radwaste disposal in salt.
These include the following:

a) Formation Water Content - The long term rheological properties of any
   given salt formation can be expected to depend upon the inherent brine
   content.
b) The long term residual stresses generally predicted for the region
   around a waste repository in salt can be expected to decay away considerably faster than previously thought, to lower and therefore even
   more favourable levels.
c) Thermally induced bulging above a repository in layered salt may occur
   over a wider area but with lower amplitude than previously calculated.
d) Creep closure of galleries and backfilled tunnels can be expected to be
   to some extent accelerated by flooding.

2. Ion Transport

Recent work at Utrecht has shown that truly intact Asse salt rock has a
vanishingly small (gas) permeability of $< 10^{-20} m^2$ even at confining pressures of only 2 MPa. Other experiments have shown that creep-induced dilatancy and permeability are strongly suppressed when brine is added to
the system (5). This suggests that even if brine does gain access to a
repository area, permeable pathways (e.g. backfilled galleries, dilated
wall rock, dilatant shear zones) will be unable to remain open except when
the fluid pressure is more or less equal to the lithostatic confining
pressure. Thus it seems likely that fluid can probably only escape by
means of compaction expulsion.

Finally, it should be noted that although not yet conclusive, preliminary ion transport experiments (in brine-saturated, deformed, and permeable salt) indicate that recrystallisation-related ion exchange effects may
occur, such that permeable salt will exhibit pseudo-absorptive properties
for certain elements (e.g. Sr).

REFERENCES
1. RAJ, R. (1982). Creep in polycrystalline aggregates by matter transport
   through a liquid phase. J. Geophys. Res., 87, 4731-4939.
2. STOCKER, R.L. and ASHBY, M.F. (1973). On the rheology of the upper
   mantle. Review of Geophysics and Space Physics, Vol. 11, no. 2, 391-426.
3. URAI, J.L. (1983). Deformation of wet salt rocks. Thesis, University of
   Utrecht, Netherlands, 221.
4. HORSEMAN, S.T. (1985). Moisture content - A major uncertainty in storage cavity closure prediction. Proceedings of the Second Conference on
   the Mechanical Behaviour of Salt, Hannover, September 1984 (in press).
   Abstract available in conference abstracts volume.
5. SPIERS, C.J., URAI, J.L., LISTER, G.S., BOLAND, J.N. and ZWART, H.J.
   (1983). The influence of fluid-rock interaction on the rheology of salt
   rock and on ionic transport in the salt. Periodic Report, contract no.
   WAS-153-80-7-N(N), January - June 1983.

D. ACOUSTIC P-WAVE VELOCITY MEASUREMENTS IN ROCKSALT AND THEIR INTERPRETA-
TION IN TERMS OF STRUCTURAL EFFECTS

J.P.A. ROEST and J. GRAMBERG
Laboratory for Rock Mechanics
Delft University of Technology, The Netherlands

1. Introduction

Acoustic measuring tubes proof against 150 °C have been developed for
the in situ research on structural changes in rock salt. The test involved
so called cross hole measurements in two parallel boreholes, containing
the measuring tubes. One tube is provided with transmitters, the other one
with receivers. The occurence of microcataclasis in the side wall of a
relatively recent gallery was shown. It was also demonstrated that macro-
cataclasis, i.e. large cracks and fissures did not occur in the gallery
wall. This test also showed the necessity of very careful drilling the
holes and of accurately determining a divergence from parallellity. As
future measurements are part of a project with a running time of several
years, the measuring devices have been automated. The apparatus is adapted
to, for example, the long term monitoring of a pillar or to the long term
observation of the structural changes due to a heating test. With respect
to the latter subject, such heating tests have already been carried out in
the laboratory on a small scale. The structural changes in the rock salt
could be followed clearly by means of acoustic traveltime measurements.
During the test decompaction and also recompaction of the rock salt were
detectable; the forming of macro cracks was clearly observed, when the
heater was suddenly switched off. A situation not representative for a
repository.

2. Acoustic velocity measurements
2.1. Location

In the northern wall of room 8 on the 725 m level of the Asse II
mine, two horizontal 66 mm ϕ measuring holes were drilled parallel to each
other to a depth of about 13 m. The room had been worked about 60 years
ago, and was refilled up to half the height. The measuring holes were
drilled very carefully; they were so straight that the bottom was almost
completely visible; by means of theodolite survey a divergence of 0,37 degree was determined.
The first 11 m consisted of Halite ("Alteres Steinsalz"), whereas from
about 9 m a large number of thin Polyhalite bands occurred. After a depth
of about 11 m Carnallite was expected. However this could not be confirmed
by endoscope, because this apparatus did not reach so far.

2.2. The measurements

The measuring device in each hole consisted of two connected rubber
tubes, together 4 m long. By means of steel pipe the rubber tubes can be
pushed as far into the measuring hole as desired, in this case up to 12 m.
This depth of the measuring holes permitted three successive measuring
series, A, B and C. The 4 m tube lengths were provided with 11 transmitters, respectively 11 receivers. The first measurements were carried out
by relais switching, whereas the transient recorder was operated manually.
Seismograms were displayed by X-Y recorder. In the second round the transient recorder and relais were controlled by a microcomputer. The transient recorder registered the average of 256 pulse arrivals, pertaining to
one measurement. By means of a "fast handshake" the data were transferred
to the computer. Data reduction was achieved by storing only a limited
length of 50 μsec of the wave train. The sample frequency was 1 sample/-

μsec. The computer made also a provisional estimate of the first arrivals and the first amplitudes. After a series of 121 measurements the collected data were transferred to the tape of a built-in cassette recorder. One tape may contain 25 to 30 series of 121 measurements.

2.3. The readings
    Three types of readings are available.
A.: Special software has been developed in order to obtain more accurate readings of the first arrivals. To that purpose the 50 μsec section of the wave train, containing a length of dead time, the first amplitude together with the zero intersection, was stored on a tape. In the laboratory this 50 μsec section is displayed on a screen, blown up to a length of 21.5 cm. Now the determination of the first arrival point can be carried out very accurately, providing the definitive figures.
B.: It is mentioned before that the computer, while measuring proceeds, already makes provisional estimates of the first arrival points. This is done instantaneously, but without optimal accuracy. The estimates of the amplitudes on the contrary have at once a good precision.
C.: Manual operation and display on X-Y recorder shows the complete seismogram, which might be needed to get insight in the quality of the signal. The reading of the first arrival however will be less accurate, for a section of 50 μsec will be displayed in only one cm.
    The acoustic velocities, calculated from the more accurate data, obtained by readings from the screen, are shown in Figure D-1. Series A proves that open fissures can be marked excellently by this measuring method. The acoustic signal will not be transferred across an open fissure of several mm's wide. However, within the block, bounded by open fissures, the signal response is quite normal. From transmitter 4, in series A, a fan of measurements shows a decrease of the velocity values, the more so as the angle between the measuring direction and the room side wall increases. This phenomenon points to the occurence of microcataclasis with a tendency of parallellity to the side wall. Two brown coloured bands of about 3% Polyhallite seem to have no influence on the velocities. Comparing the fans of measurements in the series A and B, it seems that the effect of the microcataclasis will decrease according to the increase of the distance from the side wall.
Series C shows lower velocities, probably because of the presence of a great number of Polyhalite bands, and over 11 m depth the possibly presence of Carnallite.

3. Relationship between microcataclasis and physical properties
    Research is carried out on the relationship between microcataclasis, Young's modulus, Poisson's ratio, cataclastic planar pore porosity and acoustic velocity. From laboratory measurements in 1983 on a cube of rock salt, with edges of 302 mm, loaded under triaxial "hydraulic" conditions, a series of the above mentioned data were determined. In the time however the measuring devices did not yet function optimally. Therefore we will call the results a preliminary first approach.
For the calculation of the linear part of the cataclastic planar pore space (εl_{pp}) it was postulated that the planar pores were closed during the highest applied triaxial pressure of 30.7 MPa (relative condition cracks closed, cc). As a result the values E=44.0 GPa and μ=0,329 at that pressure are supposed to be relevant for "massive" rock salt. Further the material of the block was supposed to be isotropic, i.e. with random microcataclasis. At every triaxial pressure (see tabel D-1) the value of εl is calculated from the relevant E and μ. This value will consist of a

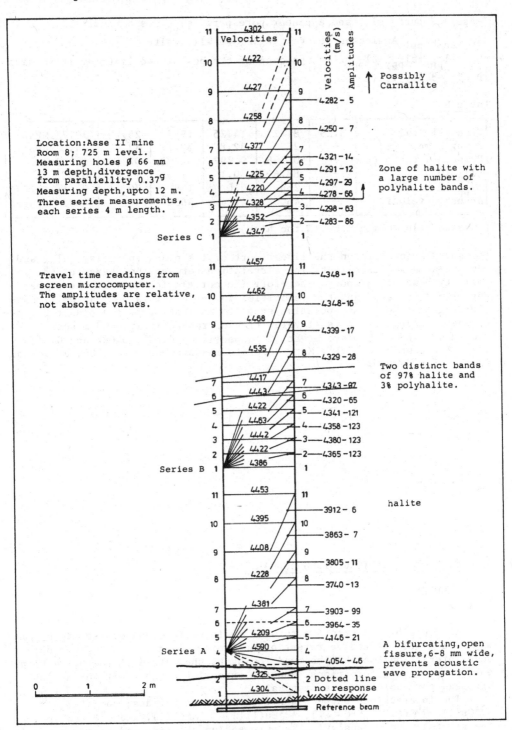

Figure D-1: Acoustic crosshole measurements

"massive" part, εl_{cc} and a planar pore part, εl_{pp}. As a result $\varepsilon l - \varepsilon l_{cc} = \varepsilon l_{pp}$. Analogous to $\varepsilon V = \varepsilon_x + \varepsilon_y + \varepsilon_z$ we will write $\varepsilon V_{pp} = \varepsilon l_{ppx} + \varepsilon l_{ppy} + \varepsilon l_{ppz}$. In the case of the postulated isotropy it results in $\varepsilon V_{pp} = 3 \cdot \varepsilon l_{pp}$.

Table D-1

Triaxial load	3.29	8.77	14.25	19.7	25.22	30.7 MPa
Young's mod. (E)	–	26.2	32.8	38.1	42.1	44.0 GPa
Poisson's ratio (μ)	–	0.268	0.292	0.302	0.318	0.329
εl_{pp} 10^{-6}	–	88.6	70.1	49.5	23.2	0
εV_{pp} 10^{-6}	–	266	210	148	70	0 (isotropy)
Acoustic velocity	–	4600	4624	4650	4627	4632 m/sec

(average values, might be 1% too high)

The correlation between the linear part of the planar porosity, εl_{pp} and the pressure was excellent. The correlation between the pressure and the velocity however was poor. Therefore the correlation between the planar porosity data, εV_{pp} and the velocities in Table D-1 is poor as well. However a rather bold extrapolation seems to indicate that the planar porosity due to microcataclasis at velocities of respectively 4500 m/sec, 4400 m/sec and 4300 m/sec might be respectively 0.07%, 0.08% and 0.09% i.e. of the order of 0.1%. More accurate measurements might be carried out in the near future.

E. RETENTION OF RADIONUCLIDES IN HALITE

LARS CARLSEN
Risø National Laboratory

The possible interaction between radionuclides and halite (rock salt) may be a highly effective retention mechanism, even in cases where only weak interactions can be observed, owing to the nature of the brine transport in rock salt formations, and the extension of the salt formations proposed for nuclear waste disposal.

The interactions between a series of radionuclides, comprising ^{134}Cs(I), ^{85}Sr(II), ^{60}Co(II), ^{154}Eu(III), ^{241}Am(III), and ^{99}Tc (as pertechnetate), and a) analytically pure sodium chloride and b) analytically pure sodium chloride, with purposely added impurities as hematite (Fe_2O_3) or anhydrite ($CaSO_4$) in low amounts, in order to simulate naturally occurring rock salt, have been studied by a colomn technique. Typically 20 cm columns (ϕ : 1 cm) of crushed material were eluted with saturated sodium chloride solutions (250 ml) containing the appropriate radionuclide. Studies using ^{22}Na spike were carried out as reference, demonstrating the applicability of the system.

In the case of pure halite only europium and americium were found to be retained by the solid material, whereas all other ions under investigation were eluted with very minor, if any, retention.

Parallel brine-migration-simulation studies revealed that no incorporation of other cations than sodium itself into the NaCl-lattice could be observed. Hence, the observed retention of Eu(III) and Am(III) is rationalized in terms of surface effects only.

However, it appears that the effect cannot be formulated as simple electrostatic attraction, since desorption experiments disclosed that only very minor amounts of the sorbed europium/americium could be leached with pure sodium chloride brine (Figure E-1). In Table E-1 the results are summarized.

Table E-1   Sorption characteristics for Eu(III) and Am(III) on a halite column (flow rate 1.5 mL/min)

| | SORPTION[1] | | SORPTION/DESORPTION[2] | |
	eluted (%)	sorbed (%)	eluted (%)	sorbed (%)
Eu(III)	51	22	58	21
Am(III)	45	25	54	20

[1]   250 ml Eu/AM containing brine (10^{-7} M)
[2]   250 ml Eu/Am containing brine (10^{-7} M) + 250 ml pure brine

Based on available stability constants (1,2) it can be calculated (3) that europium and americium will, in saturated sodium chloride solution, predominantely exist as the chloride complexed species MCl_2^+ (M=Eu,Am). Thus, it is suggested that the retention of Eu(III) and Am(III) is a result of interaction between these complexes and "free" chloride atoms, which will be negatively charged, at the halite surface, i.e. formation of solid MCl_3 (M=Eu,Am) crystals, A very slow dissolution/precipitation process in the brine/halite system is responsible for the apparent lack of desorption.

The apparant lack of sorption of Co(II), which also forms chloride complexes, is most probably to be explained by the formation of "higher" chlorides, which are neutral or negatively charged species (4), and, hence, will not be retained.

Incorporation of "impurities" in the halite caused drastic changes in the sorption behaviour. In Figure E-2 the europium concentration per gram moist column material (expressed relative to the applied concentration) throughout a trisected column consisting of pure halite/halite with 1% Fe_2O_3/pure halite (10/5/10), respectively, is depicted as a function of total moist column material withdrawn from the column after passage of 250 ml of a 10^{-7} M Eu solution. The very high sorbing efficiency of the iron oxide is noted. In this type of columns also Co(II) was found to be extensively retained, as ca. 75% of the applied cobalt was refound on the column material. A similar, but less pronounced effect is seen in cases, where anhydrite is used as "impurity".

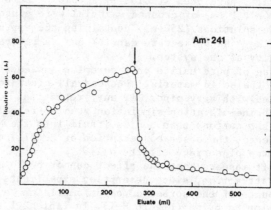

FIGURE E-1

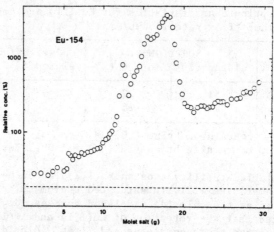

FIGURE E-2

Conclusion

It can be concluded that Eu(III) and Am(III), and probably actinide (III) elements in general are strongly sorbed to halite and desorbed only with difficulty. The interaction between the pure solid sodium chloride and the actinide elements involves chloro complexes of the type MCl_2^+. Impurities as hematite and anhydrite apparantly strongly improve the efficiency of rock salt as a natural barrier towards actinide migration.

REFERENCES
1. SMITH, R.M. and MARTELL, A.E. (1976). Critical Stability Constants. Vol. 4 (Inorganic complexes) Plenum, New York, 1976.
2. PHILIPS, S.L. (1982). Hydrolysis and Formation Constants at 25 °C. DOE report, LBL-14313, Lawrence Berkeley Laboratory. Berkeley CA, 1982.
3. SKYTTE JENSEN, B. and JENSEN, H. Complex Formation of Selected Radio-nuclides with Ligands Commonly Found in Ground Waters: Low Molecular Organic Acids. European Appl. Res. Rept. Nucl. Sci. Technol., in press.
4. NILSSON, K., SKYTTE JENSEN, B. and CARLSEN, L. The Migration Chemistry of Cobalt. European Appl. Res. Rept. —Nucl. Sci. Technol., in press.

DISCUSSION

J. HAMSTRA, AVORA B.V. Bergen

It has always been an aim to look for as pure as possible halite formations for disposal. Can we conclude from your results, that we should now
prefer less pure salts because of their better retention properties?

L.H. VONS, ECN Petten

One does get this impression, but there are, in my opinion many good other
arguments to stay in the main salt formation.

Chairman

Were the sorbtion characteristics that much better?

L. CARLSEN, RNL Risø

That is correct, but the tests were carried out by passing 250 ml of
solution containing a 10^{-7} molar concentration of radionuclides (Eu or Am)
through a column of 25 g of crushed salt. In this experiment 25% of the
radioactivity will remain sorbed to the salt surface. Taking into account
the mechanism of brine migration, salt will be a very effective natural
barrier, even though only this relatively small fraction is sorbed. In a
second type of experiments, where the nuclides pass through a salt column
as a pulse, like in normal column chromatography, quantitative sorbtion of
Eu and Am will be observed, which will not be eluted.

MODELLING OF THE THERMOMECHANICAL BEHAVIOUR OF SALT ROCK

G. ALBERS, RWTH; V. GRAEFE, GSF; E. KORTHAUS, A. PUDEWILLS, KfK;
J. PRIJ, ECN; M. WALLNER, BGR

Summary

The aim of thermal and mechanical calculations describing the
load-bearing behaviour of salt rock and the integrity and geological stability of a salt formation is to provide a realistic picture of the strains and stresses caused by thermal loads, a hitherto largely unknown aspect of rock mechanics.
Suitable calculation methods and programmes are now available, as
are consistent material laws for the mechanical behaviour of salt
rock. Since predictions as to the long-term isolation of radioactive waste from the biosphere require experimentally reproducible
short- term processes to be extrapolated over long periods of time
by means of appropriate calculations, the verification of the
model used assumes crucial importance.
Some examples of calculations carried out in parallel with tests
show what progress has been achieved with the calculation programmes. In addition, some results of modelling calculations for a
repository are presented by way of illustration.

1. INTRODUCTION

The German and Dutch concept for the disposal of radioactive waste
from nuclear installations calls for its final disposal in a salt formation to afford full protection for the biosphere.

Vitrified, highly radioactive waste in particular releases heat due
to radionuclide decay, raising temperatures locally for a limited period
of time in the salt formation. The result is thermally induced strains
and stresses.

To guarantee the safe functioning of the repository and the isolation of waste from the biosphere, safety and other analyses need to be
carried out to demonstrate that these thermomechanical loads do not
represent a long-term danger to the integrity of the barrier system
(comprising natural and engineered barriers) as a whole (11).

2. AIM OF THERMOMECHANICAL MODELLING

The main aspects covered by a stability study for waste disposal in
salt are the load-bearing behaviour of the strata, the long-term protection provided by the plastic salt rock and the geological stability of
the salt formation. The strains and stresses caused by thermal loads,
together with the resulting stability questions, are still a largely
unexplored area in current mining practice.

Modelling thermomechanical behaviour hence has the following aims :
- by means of a mathematical analysis of thermomechanical processes, to evaluate their effects correctly and to add to current mining knowledge;
- to derive rock-mecanical criteria for a stable repository design from the above analysis;
- to adjust the repository design, where necessary, to prevailing geological conditions on the basis of the model calculations or the derived rock-mechanical criteria;
- finally, to provide long-term predictions as to the integrity of the salt formations, which can only be done by means of modelling.

3. MECHANICAL BEHAVIOUR OF SALT ROCK

In recent years, the Bundesanstalt für Geowissenschaften und Rohstoffe (BGR) has carried out extensive research into the mechanical behaviour of salt rock with the aim of evolving a consistent material model (15), (16). Some typical results tests to distinguish between non-fracturing deformations and fractures are shown in Fig. 1.

Samples 4 and 5 show shear fractures. Samples 1 and 2 were tested under the same surface pressures, though at a rate of axial strain lower by a factor of a hundred. In spite of considerable deformations, the latter samples do not exhibit any signs of fracture.

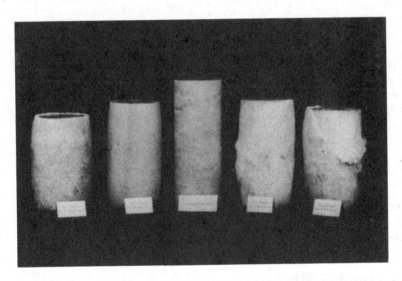

Fig. 1 : Rock salt samples subjected to triaxial tests, Asse Liniensalz Na$_3$

Developed on the basis of exhaustive laboratory experiments by the BGR, a material model covering only the universally valid aspects of strain behaviour is given by the material law proposed in Fig. 2. The overall rate of strain $\dot{\varepsilon}$ t is calculated from the following :
- rate of elastic strain $\dot{\varepsilon}$ el
- rate of thermal strain $\dot{\varepsilon}$ th
- rate of creep $\dot{\varepsilon}$ cr (time-dependent, irreversible deformation)

- rate of fracture strain $\dot{\varepsilon}^f$ (irreversible deformation, independent of velocity, with formation of discrete fracture surfaces and dilatation).

For the rate of creep $\dot{\varepsilon}^{cr}$, only steady-state creep has been considered as the major factor in long-term creep. The mathematical formula for creep is based on a model developed by Munson and Dawson (12), and is the sum of three distinct creep mechanisms :
- dislocation climb at low stresses and high temperatures
- transient dislocation mechanism at repository temperatures and stresses
- dislocation gliding at high-stresses

For the case where the long-term strength criterion is not met, the fracture deformation rate is incorporated in the form of an associated flow rule and an extended drucker/Prager flow condition.

For transient creep behaviour immediately following stress changes, only incomplete measurements are available to date, with the results described phenomenologically in terms of time- or strain-hardening experiments. These cannot be safely generalized to stress conditions other than those tested. For the long-term forecasts required for a repository, however, transient creep should be virtually insignificant.

The material behaviour of crushed salt is currently the subject of intensive investigation. Consistent material laws should be available in the not too distant future.

$$\dot{\varepsilon}^t_{ij} = \dot{\varepsilon}^{el}_{ij} + \dot{\varepsilon}^{th}_{ij} + \dot{\varepsilon}^{cr}_{ij} + \dot{\varepsilon}^f_{ij}$$

$$\dot{\varepsilon}^{el}_{ij} = -\frac{v}{E}\,\dot{\sigma}_{kk}\,\delta_{ij} + \frac{1+v}{E}\,\dot{\sigma}_{ij} \qquad\qquad \dot{\varepsilon}^{th}_{ij} = \alpha_t\,\dot{t}\,\delta_{ij}$$

$$\dot{\varepsilon}^{cr}_{ij} = \frac{3}{2}\,\frac{\dot{\varepsilon}^{cr}_{eff}}{\sigma_{eff}}\,s_{ij} \qquad {}^1\dot{\varepsilon}^{cr}_{eff} = A_1\,e^{\frac{-Q_1}{RT}}\,(\frac{\sigma_{eff}}{\sigma^*})^{n_1} \qquad {}^2\dot{\varepsilon}^{cr}_{eff} = A_2\,e^{\frac{-Q_2}{RT}}\,(\frac{\sigma_{eff}}{\sigma^*})^{n_2}$$

$$ {}^3\dot{\varepsilon}^{cr}_{eff} = 2\,(B_1 e^{\frac{-Q_1}{RT}} + B_2 e^{\frac{-Q_2}{RT}})\,\sinh(D \,{}_<\,\frac{\sigma_{eff} - \sigma^o_{eff}}{\sigma^*}\,{}_>)$$

$$\dot{\varepsilon}^f_{ij} = \frac{1}{\eta}\,<F>\,\frac{\partial F}{\partial \sigma_{ij}} \qquad\qquad F = \text{sign}\,(I_0)\,\alpha\,|I_0|^m + \sqrt{II_s} - k$$

Fig. 2 : Material model for rock salt

4. CALCULATION PROGRAMMES

4.1 Programmes available

In order to calculate the thermal and mechanical impact of a repository in a salt formation, suitable programmes are available for analysing non-stationary temperature fields and thermally induced stress and strain conditions. Table I gives an overview of the programmes currently used by the authors, including some developed as part of the CEC project.

The development of programmes to solve the heat conduction problem can be regarded as largely completed. Progress to date with the pro-

grammes for mechanical analysis is essentially as set out in Table I.
Details of programmes supported by the CEC (GOLIA, FAST, MAUS, TEFELD)
are contained in (1) and (2).

TABLE I - CALCULATION PROGRAMMES

PROGRAMME	ADINA	ANSALT	GOLIA	MARC	MAUS
Authors	Bathe	Wallner/Wulf	Donea/ Prij	Marcal	Albers/ Biurrun
Used at	Kfk/GSF	BGR/GSF	ECN	ECN	RWTH/KfK
Funding	Kommerz.	BMFT	CEC	Kommerz.	CEC
Associated temperature prog	ADINA	ANTEMP	analyt. solution	MARCHEAT	TEFELD FAST
Geometries	1-, 2-, 3-D	1-, 2-, 3-D	2-D	1-, 2-, 3-D	2-D
Order of element	linear + quadrat.	linear + quadrat.	linear	linear + quadrat.	linear + quadrat.
Material models					
Thermoelasticity	X	X	X	X	X
Steady-state creep	X	X	X	X	X
Transient creep		X	X	X	X
Destrengthening Tensile failure Backfill model	X	X X X		X	X X X
Excavation Filling	X X	X X			X X
Selection of time steps		semi- autom.	fully autom.		fully autom.

4.2 Model verification

Any mathematical stability study requires reality to be mapped onto
a model. In doing so, it is of crucial importance to represent the
following factors as realistically as possible :
- material law describing the thermorheological behaviour of the strata
  (time-, temperature- and stress-related deformation and resistance
  behaviour)
- geological structure of the strata
- initial status of strata (stresses, temperature)

- geometry (determination of load-bearing behaviour by means of spatial, plane or rotationally symmetric representation models)
- stress states and changes through time.

The aim of verification is to prove that a mathematical model for a repository in salt rock is realistic by means of a programme comprising several stages with the following main components :
1. Proof of the correctness of the mathematical formulation of the programme by checking analytical examples and comparing with other computer codes (benchmarks)
2. Checking of the formulation of the material law by comparing calculated results with the results of laboratory measurements under clearly defined initial and boundary conditions.
3. Comparison of thermomechanical calculations with the results of thermomechanical in-situ tests.

As predictions as to the long-term isolation of radioactive waste from the biosphere require the extrapolation of experimentally reproducible short-term processes over the long term, the verification of the thermomechanical material law for salt rock assumes crucial importance, since the laws on which the material model is based have to be shown to be consistent over long periods of time as well.

Since November 1984, the European Community has been supporting a "CEC inter-comparison exercise on rock mechanical computer codes for salt", in which an approach is being made towards model verification on the basis of two benchmark examples.

5. CALCULATED RESULTS

5.1 Calculations accompanying tests

Over the last few years, the Asse II mine has been used as an experimental installation for conducting R&D work on the disposal of radioactive waste.

The research work in the mine comprises :
- long-term monitoring of empty mined cavities and cavities filled with radioactive waste
- experiments to test disposal techniques and investigate the occurence of disturbances.

In addition to radiation measurements, long-term monitoring concentrates on :
- seismic location of partings and cracks
- micro-acoustic measurements of creep processes
- deformation and displacement measurements.

Fig. 3 shows the 750 m floor in the mine by way of illustration. The vectors depicted represent the displacements measured from the measuring points of the rock observation programme. The graph was plotted by the newly developed postprocessor program GRAPLOT following digital storage of the mine geometry and the measurement results.

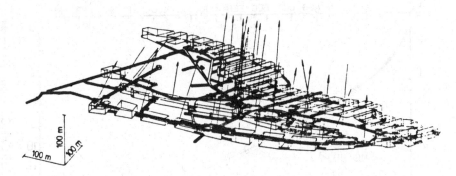

trimetric parallel
perspective

Fig. 3 – Asse II, mined cavities on the 750 m floor, displacements
measured (magnified 2 000 times)

This diagram depicts the displacements from control points measured by
the rock observations programme (K. Dürr) over the period March 1981 to
March 1984. These represent the large-scale convergence of the mine
workings (displacements magnified by a factor of 2000 compared with mine
plan)

Experiments to test disposal technology comprise :
- heater tests to simulate the effect of heat from radioactive waste
  on the salt rock
- brine migration test to examine the movement of small brine inclusions in the non-homogenous temperature field,
- ECN's 300 m dry-drilled borehole as a model for the emplacement of
  high-level waste,
- and (in preparation) experimental emplacement of highly radioactive
  glass blocks in eight cased boreholes.

This range of in-situ experiments offers good prospects of obtaining measurement data for comparison with the numerical calculations. To this end, a new data bank system consisting of the components FILEBUILD and GEOMECH-DBS has been developed to store the data
acquired and enable users to handle data comfortably (9).

In a large dry-drilled 300 m deep into the 750 m floor, ECN
measured convergences at 1 050 m depth and carried out comparative calculations to derive the constants of a law for steady state creep (7),
(13). For a measuring period of 200 days, the following equation gave
the best mathematical fit with the measured results :

$$\dot{\varepsilon}^{cr} = 8.8 \times 10^{-11} \quad 5.5$$

The mathematical prediction thus obtained for convergence up to 800
days agrees closely with the measured values (Fig. 4).

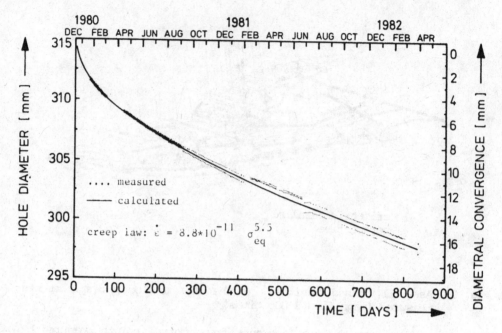

Fig. 4 - Convergence of ECN borehole

Using the same borehole, a heater experiment was carried out accompanied by parallel calculations. In this test, a tube 3 m in length
was heated at 5 kW for 60 days at a depth of 1 010 m. The maximum temperature reached was 460 K (187°C). The calculated and measured curves
for the pressure on the sample cylinder, reaching a maximum of around
350 bar, show a good fit (Fig. 5).

However, the calculations also show that the pressure build-up is
only weakly dependent on the activation energy in the Arrhenius function
and thus cannot be used to derive this factor (2).

The convergence behaviour of individual boreholes provides an opportunity for a systematic experimental and mathematical investigation
to examine the available calculation methods and thermomechanical material laws for rock salt. Accordingly, the aim of Temperature Test
Field 3 (TTF 3) was to measure temperatures and borehole convergence.

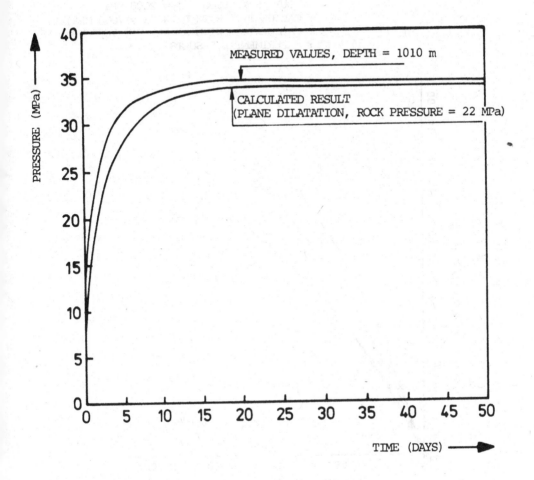

Fig. 5 – Pressure measurements in the ECN borehole

The thermomechanical processes were also calculated with the FE programmes ADINA and MAUS (5). The aim was to check the numerical reliability of their results and the predictive power of the numerical models.

Using realistic mechanical boundary conditions and material values to describe the thermomechanical behaviour of rock salt, results were achieved showing a close fit with the measured values. The borehole convergences calculated with ADINA and MAUS are virtually identical, which indicates that the solutions are numerically correct. In addition, in order to underpin these results further calculations were carried out to determine the sensitivity of borehole convergence to variations in a number of material law parameters. Fig. 6 shows the measured values and the results obtained with the two programmes for borehole convergence at the heat source's mid-plane. Some calculated results for several of the parameter combinations used are also shown.

◇ : A = 2 x A (IN CREEP LAW)
X : YOUNG'S MODULUS (E) = 2000 MPa
+ : CALCULATION WITH MAUS (E = 7000 MPa)
▲ : CALCULATION WITH ADINA (E = 7000 MPa)
☉ : EXPERIMENTAL RESULTS

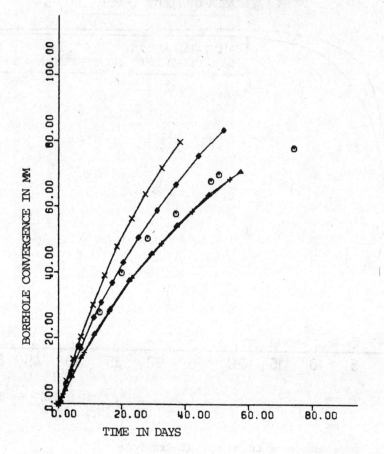

Fig. 6 – Graph showing borehole convergence for various material parameters

However, the agreement observed between calculated and measured
results does not give any precise indication as to the correctness of
the selected material data and boundary conditions. It would also be
possible to achieve just as good a fit using other combinations of
material parameters and boundary conditions. Therefore, when carrying
out similar in-situ experiments, the test site should be selected such
that the initial stress conditions in the surroundings can be estimated
with sufficient accuracy. The remaining uncertainties in the material
law could then be estimated by varying the relevant material parameters.

As an example of a more complex experiment, Temperature Test Field
5 was cross-checked with MAUS (3), (4). The output of a cylindrical
heater placed in a horizontal borehole was increased in stages. The
sketch in Fig. 7 depicts the model. Calculations of the radial displacement of the borehole wall at successive points in time show the gradual

creep of the salt onto the heater. Although the contact process was not measured, there are a number of indications that the calculated contact was both qualitatively and quantitatively a good simulation.

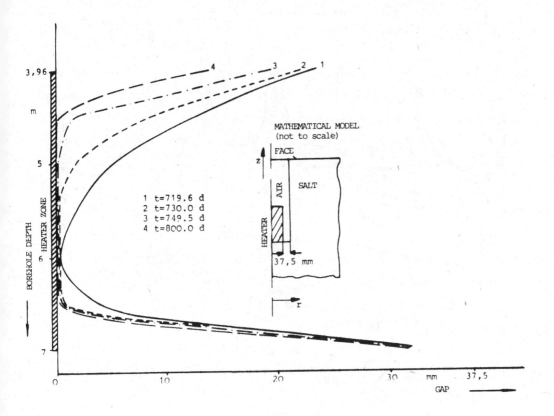

Fig. 7 – Creep of rock onto heater in Temperature Test Field 5

In Temperature Test Field 6 – a hexagonal configuration of 6 heaters around a central heater as a model for a repository – the calculation programmes have already been used to interpret the results. The same is currently being done with the HLW test, an initial experiment with the disposal of highly radioactive glass blocks. Two parallel galleries each contain four boreholes, three of which are filled with glass blocks of differing levels of activity while one contains an electric heater. Fig. 8 shows the results of the three-dimensional temperature calculations.

After five years there is already a clear interaction between the boreholes. However, as this interaction is initially very weak, a rotationally symmetric calculation can be applied to the individual boreholes for this period, yielding a maximum radial loading of around 48 MPa after one week for the backfilled borehole type. It is assumed – conservatively – that the initial stress is given by the weight of the overburden. The borehole casing is to be dimensioned accordingly. Due to the decline in radioactivity, the maximum temperature on the borehole

wall is likely to be reached after one to one and a half years depending
on the borehole filling.

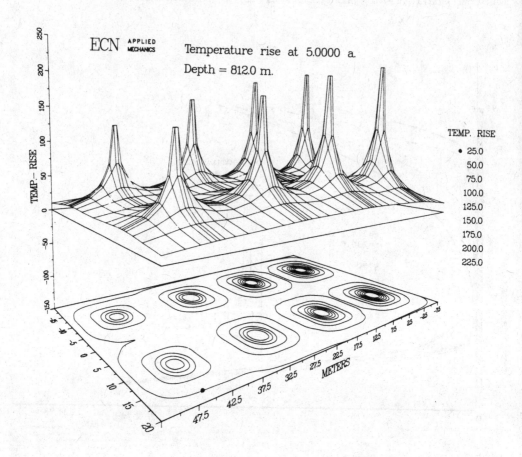

Fig. 8 – Temperature distribution in the planned HLW test after 5 years

5.2 Modelling of the radioactive waste repository

As part of the R&D programmes supported by the Commission of the
European Communities for the disposal of radioactive waste, ECN, KFK and
RWTH Aachen have also carried out some site-independent modelling calculations for a repository (1), (2). Three results are discussed below
as examples.

ECN's calculations are based on a simplified model assuming a cylindrical, homogeneous salt dome with a diameter of 1 500 m. The base if
the dome with a 300 m thick salt cushion is 3 000 m deep, while the top
is at 300 m. The total initial thermal loading as a result of the nuclide inventory is 27.72 MW. The decay curve for heat production was
derived from a INFCE study.

For the calculations, a rotationally symmetric section with a
radius of 4 000 m was considered. The selected finite element network
and the kinematic boundary conditions are shown in Fig. 9a. The creep
behaviour of salt rock was assumed to be in accordance with the creep
parameters determined in the ECN borehole (see above).

The temperature field calculations in the form of a semi-analytical solution yielded maximum temperatures of less than 150°C. The influence of temperature was investigated in a subsequent stress analysis. Figs. 9 b, c and d show characteristic results for the distributions of horizontal, vertical and shear stresses. These results represent the maximum values of thermally induced stresses approximately 50 years after disposal. It will be seen that in this case the stresses in the salt dome are mainly calculated as compressive stresses with small deviatonic components. In the cap and surrounding rock small stress releases result. The overall stresses, i.e. taking into account the initial stresses, remain in the compressive range, however.

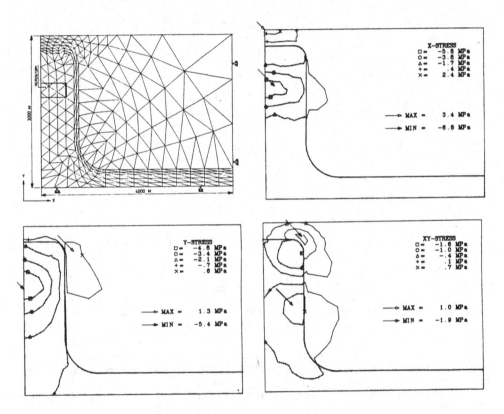

Fig. 9 - Repository modelling, thermally induced stress

KFK has carried out calculations for the repository concept in the Federal Republic of Germany, in which HLW waste is disposed in 300 m-deep boreholes arranged in a regular, square configuration with 50 m between boreholes at depths between 800 and 1 100 m. The heat production per borehole is assumed to be 216 kW at time of disposal.

The thermomechanical processes in the near field of a HLW borehole were calculated with the ADINA programme.

Assuming a lithostatic initial stress condition for the model and deriving material values to describe the thermomechanical behaviour of rock salt from the cross-checking of the in-situ experiments (TTF 3 (5), standard probe (14), borehole convergence was measured up to the closure of the air gap between the salt and waste packages, as were the stress fields in the surrounding rock salt.

Fig. 10 shows the vertical curve for radial displacement at 0.21 m, 1.5 m and 10 m from the borehole axis 150 days after disposal.

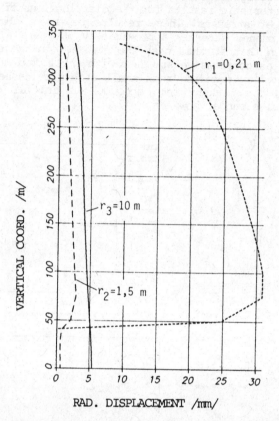

Fig. 10 – Radial displacement at 0.21 m, 1.5 m and 10 m from the bore-hole axis after 150 days (HLW laboratory borehole)

The results of the calculations show that at 40 m above the bottom of the borehole the 5 cm air gap closes after about six months, and at 240 m only after a year.

These studies of thermomechanical processes in the near field of disposed HLW waste are still at the development stage. In order to refine the model, the borehole seal and the pressure build-up on the waste blocks (contact effect) will also have to be taken into account.

To study the convergence of backfilled disposal cavities and the accompanying backfill compaction, a 1-dimensional finite difference programme has been developed for spherical or cylindrical geometry (1). It contains a material law for the compaction of crushed salt, which closely approximates the compactions and compaction rates observed in laboratory experiments at constant and varying pressures.

In order to describe long-term compaction, which cannot be determined directly by experiment, this material model also contains an analytical relation based on a considerably simplified model assumption for already strongly compacted material (Butcher's spherical void model) (6).

In parameter studies, the programme was used to investigate the influence of the model geometry and the relative backfill volume (assuming that the waste stack itself is incompressible) as well as of certain parameters of the crushed-salt material law.

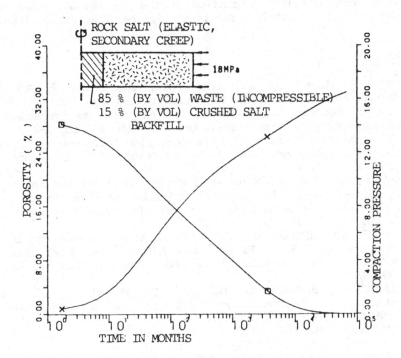

Fig. 11 – Pressure build-up and backfill compaction in disposal cavity with 15% (by vol.) crushed-salt backfill.

By way of an example, Fig. 11 shows the result for a cylindrical cavity at a depth of about 850 m with 15% (by vol.) backfill; The compaction behaviour assumed for the backfill is in line with the results of laboratory experiments carried out by BGR.

In order to obtain more reliable information on the long-term behaviour of crushed salt, further extensive laboratory experiments are currently being carried out covering periods of time as long as possible.

6. CONCLUSIONS AND OUTLOOK

All in all, the calculation methods and programmes currently available appear to be suitable tools for scientific and technical purposes, capable of providing a realistic analysis of the specific thermomechanical effects of the disposal of radioactive waste in salt formation.

As an evaluation of the stability of the repository and the integrity of the salt dome barrier requires long-term mathematical predictions, the verification of the model used to calculate such predictions assumes particular importance. There is a need for further studies to prove that the model is realistic and in particular to demonstrate the consistency of the material law for rock salt. This applies particularly to the behaviour of crushed salt as a backfill material, which is currently being studied in detail.

For large-scale modelling of the repository, it will be necessary to acquire more detailed information on the precise structure of the rock mass in order to obtain a realistic picture of actual conditions.

7. REFERENCES

1. CEC - Kernforschung- und technologie, Entwicklung von Rechenverfahren und Durchführung von Modellrechnungen zur thermomechanischen Wechselwirkung des Salzes mit der Bohrlochauskleidung bzw. mit eingelagerten Abfallblöcken. - EUR 8812 DE, 1984.

2. CEC - Nuclear Science and technology, The thermomechanical behaviour of a salt dome with a heat-generating waste repository. - EUR 9205 EN, 1984.

3. ALBERS, G. : MAUS - A Computer Code for Modelling thermomechanical Stresses in Rocksalt, Computer Modelling of Stresses in Rock. - Proceeding of a Technical Session Held in Brussels (December 6 and 7, 1983), EUR 9355 EN.

4. ALBERS, G., BIURRUN, E., SCHLICH,M. : Kontrollrechnungen zum Temperaturversuch 5. - Bericht zum Auftrag der Gesellschaft für Strahlen- und Umweltforschung mbH, München, Auftrags-Nr. 31/139239/84.

5. ALBERS, G., BIURRUN, E., KORTHAUS, E., PUDEWILLS, A. : Zur Abschätzung der thermomechanischen Auswirkungen der hochradioaktiven Abfälle. - ATW, Nr. 11 (1984).

6. BUTCHER, B.-M. : Creep Consolidation of Nuclear Depository Backfill Materials. - SAND 79-2212, October 1980.

7. DOEVEN, I., SOULLIÉ, P.-P., VONS, L.-H. : Convergence measurements in the dry drilled 300 m borehole in the ASSE-II Saltmine. - European Appl. Res. Rept. Nucl. Sci. Technol., Vol. 5, No. 2, (1983), pp. 267-324.

8. DÜRR, K. : Temperaturversuchsfeld 3. - Versuchsbericht, Clausthal-Zellerfeld, GSF-T 73.

9. GRAEFE, V., FRIEDRICHS, W. : Entwicklung von gebirgsmechanischen Rechenprogrammen zum Festigkeitsverhalten des nicht aufgeheizten Gebirges. - 2. Halbjahresbericht (1983) zum Vertrag WAS 336-83-7 D des indirekten Aktionsprogramms der EG, Institut für Tieflagerung, GSF Braunschweig.

10. KORTHAUS, E. : Effect of Backfill Material on Cavity Closure. - Second Conference on the Mechanical Behaviour of Salt, Hannover, September 1984.

11. LANGER, M. and VENZLAFF, H. : Sicherheitsnachweis und Störfallanalyse für ein Endlagerbergwerk im Salzgebirge. - Geol. Jb. A 75, Hannover, 1984, S. 627-633.

12. MUNSON, D.-E. and DAWSON, P.-R. (1979) : Constitutive model for the low temperature creep of salt (with application to WIPP). - SAND-79-1853, Sandia National Laboratories, Albuquerque NM.

13. PRIJ, J., MENGELERS, J.-H.-J. : On the derivation of a creep law from isothermal convergence. - ECN-89 (1981).

14. PUDEWILLS, A., MÜLLER, R., KORTHAUS, E., KÖSTER, R. : Thermo-mechanical in situ-Experiments and Finite Element Computations. - Proc. 5th Int. Symp. on the Sci. Basis for Rad. Waste Management, 7-10 Juni, Berlin 1982, S. 477-486.

15. STAUPENDAHL, G., GESSLER, K. and WALLNER, M. : Zusammenfassende Darstellung von Versuchsergebnissen des spannungs- und temperatur-abhängigen Festigkeits- und Verformungsverhalten von Steinsalzen. - Proc. ISRM Symp. Rock Mechanics : Cavern and Pressure Shafts, Aachen 1982, Vol. 3, pp. 1115 - 1119.

16. WALLNER, M. : Stability Calculations Concerning a Room and Pillar Design in Rock Salt. - Proc. 5th Int. Congr. on Rock Mechanics, Melbourne, Australia, 1983, D 9 - D 15.

DISCUSSION

UNIDENTIFIED SPEAKER

Is the creep rate or the convergence affected by radiolysis or the presence of brine?

K. KUHN, GSF Braunschweig

I believe the creep rate has a tendency to reduce the brine migration. By
creeping, the salt will, in a few months, closely and completely surround
the waste and then only very small amounts of brine or gas can come into
contact with the waste container.

R. DE BATIST, CEN/SCK Mol

What is the sensitivity of your model calculations to the accuracy of the
material property data; in particular, have your caluculations taken into
account possible variations in these values due to radiation damage?

M. WALLNER, BGR Hannover

This is a very important question. I could not show results of parametric
studies, although these are necessary: temperature, stress and humidity
have an influence on the creep role. Radiation however has not such a
marked effect, as far as we know. It only has an influence for a very
short time.

R. KOSTER, KfK Karlsruhe

You have quite rightly stressed, that the results of the in-situ tests can
be calculated from the FEM models, when boundary conditions and material
properties are correctly introduced. You have also said that a calculated
verification of these experiments can be obtained for a variety of Young's
moduli and initial stress patterns. The calculating model therefore is not
yet completely confirmed. Could you tell us how you will proceed from
here, to identify initial stress conditions with more accuracy?

M. WALLNER, BGR Hannover

This is not only a question of stress conditions. For long-term predictions we must use established physical laws, mathematical methods and
tools which have to be verified entirely. This means we have to check the
numerical part of the calculation and eliminate inconsistencies in our
material property laws. The verification aims at producing sound long-
term forecasts and not only the recalculation of short-term experiments.

It has been going on for some time and since November 1984 such validations are also part of the EC work. Measurement of stresses are a sideline, they are probbaly less important for the long-term evolution in a repository than for the interpretation of short-term experiments.

R.M. KORTHOF, EZ Den Haag

Did you use natural analogs to check the long-term validity of your models?

M. WALLNER, BGR Hannover

No; have you any suggestions?

R.M. KORTHOF, EZ Den Haag

One could use the models to recalculate the natural evolution of a salt dome.

M. WALLNER, BRG Hannover

This has been done. But when you calculate diapirism, other aspects intervene. Rock mechanics of mined cavities depend largely on elastic and viscous behaviour. The evolution of a salt dome is only dependent on viscous movement; it can be modelled but has little relevance to our problems of repository stability.

RESEARCH PROGRAMMES IN UNDERGROUND EXPERIMENTAL FACILITIES

Konrad - Pasquasia - Mol

P. Manfroy (CEN/SCK - Mol), F. Benvegnu (ENEA - La Casaccia),
R. Heremans (ONDRAF/NIRAS - Bruxelles), W. Brewitz (GSF - Braunschweig)

Summary

Three underground laboratories are described. Two of them are implanted in argillaceous formations in Italy and Belgium, the third in an iron ore body in Federal Republic of Germany.

The Italian and German laboratories take advantage of the existence of disused mining plants, the Belgian one has been created only for experimental purposes under the Nuclear Research Center of Mol.

The geology of the sites as well as the different kinds of tests foreseen or already carried out are described putting the focus on the geotechnical aspect of the research.

INTRODUCTION

A very important stage in a R&D-programme on the geological disposal of conditioned radioactive waste is the one which consists of making "in situ" experiments in the selected formation for that purpose.

Several countries have had the opportunity of reaching such a stage in installing underground laboratories in different types of igneous, saliferous or sedimentary rocks.

Some countries have taken advantage of the presence in their subsoil of disused mines which were excavated in a formation close to or identical to the one selected for the final disposal. Other countries, having no mining facilities in adequate formations, were obliged to create experimental facilities out of nothing in the potential host rock, committing very important investments and thus demonstrating the high degree of confidence assigned to the qualities of the selected rock medium.

The purpose of the present paper is to render an account of three important projects in sedimentary rocks having received or receiving financial support of the Commission of the European Communities, in the framework of its indirect action in the field of radioactive waste management:

- the experimental programme led in the disused iron mine of Konrad (Federal Republic of Germany) ;
- the experimental programme foreseen in a test gallery whose construction is going to commence in the potash mine of Pasquasia in the island of Sicily (Italy) and
- the experimental programme in progress in the underground laboratory dug beneath the Nuclear Study Centre in Mol (Belgium).

In the last two cases the geological formations dealt with in the programmes are clay layers, whilst the one investigated in the first case consists of a sedimentary iron ore isolated from the surface by thick claystone layers, so that, this peculiar host medium is much closer to argillaceous formations than igneous or saliferous ones with respect to the nature of the undertaken experiments.

1. RESEARCH PROGRAMME IN THE DISUSED KONRAD IRON ORE MINE

1.1. General

The Gesellschaft für Strahlen- und Umweltforschung mbH (GSF) in Collaboration with Karlsruhe Nuclear Research Center (KfK) has on behalf of the Federal Minister for Research and Technology (BMFT) examined the possibilities and potentials of the disused Konrad iron ore mine for disposal of low-level waste and decommissioning waste from nuclear power stations. From the initial investigations which had commenced in 1975, one year before ore production terminated, no facts could be established which might have indicated the mine to be unsuitable for disposal of radioactive waste in one way or another. After, in 1976, an agreement was reached with the owners on the use of the mine for feasibility investigations, BMFT supported an extensive R&D programme carried out by GSF's Institut für Tieflagerung (IfT). This programme was extended until 1982 and was additionally supported by the CEC within the scope of two contracts between 1978 and 1981. Up to the end of 1981, about DM 60 million of public funds were spent on execution of the research project, including the running costs for keeping the mine open and for large-scale mining tests.

Already before the termination of the R&D programme the Federal Government had declared that in the case of a positive result of the work a licensing application would follow for the Konrad Mine as a waste repository. This procedure was started on August 31st, 1982 and might last up to spring 1987, whereafter construction of the repository will take another two years, the commencement of the disposal operation taking place early 1989.

1.2. Features of the geological structures of the Konrad site

In view of the intended use of the mine as waste repository the following features of the geological structure are of particular importance :

- the rock formations form a simple geological structure and lie horizontally or have a slight slope ;
- the slightly synclinal ore deposit (Upper Jurassic Oolithic Limonite) is completely covered by overlaying cretaceous formations
- no contact exists therefore between the ore deposit and the surface or the groundwater bearing formations close to the surface ;
- fault zones involving major displacement do not occur in the overlaying cretaceous formations ;
- intensive tectonic faulting of the ore deposit is restricted to just few sub-areas in the mine and pronounced micro-fissuration of the ore deposit is only developed to a minor extend ;
- the covering rock strata consists of a thick sequence of homogeneous rocks units in which aquifers are only involved near the surface and are separated from the ore deposit by an extremely low permeability claystone formation extending over a large area and presenting thicknesses of 600 to 1000 m ;
- the pronounced expandable character of the clay minerals of the cover cause self healings of cracks and fissures. In addition a high degree of sorption provides a direct radionuclide retention ;
- the iron ore itself can be classified as impermeable and presents

high sorption capacity. Its mechanical properties permit driving
stable chambers without permanent support of lining.
Fig. 1 gives the geographical situation of the Konrad mine close
   to the city of Braunschweig.
Fig. 2 shows a geological West-East section through the Konrad
   Mine.
Fig. 3 gives the stratigraphic profile of Shaft Konrad 2.

1.3. Aims of the research and development programme

The R&D programme has covered all important aspects for assessment of
the suitability of an underground waste repository
  - Geoscientific investigations
  - Mining investigations, and
  - Waste and waste handling investigations.

By dealing with these aspects it was the intention to establish
whether on account of its geological and mining conditions the disused
iron ore mine is suitable for disposal of conditioned low level waste from
the operation of nuclear power stations, major research centres, waste
storage depots from the federal States, and contaminated or activated
large components from the decommissioning of nuclear power stations,
whether it provides operational and long term radiological safety and
whether the mine can be converted for disposal with reasonable
expenditure.

1.3.1. Geoscientific investigation

To determine the geological suitability, the following subjects were
examined :
  - geological structure of the ore deposit including the petrographic
    and tectonic conditions of the rock strata
  - hydrogeological conditions of the covering rock formation and of
    the mine
  - properties of the geological formations as diffusion barriers for
    radionuclides
  - rock mechanics in the vicinity of mine chambers and galleries and
    of other mine workings
  - seismic safety of the repository during operation and after shut
    down.

1.3.2. Mining investigations (Figures 4 and 5)

Assessment of possible use of the existing mining installation was
carried out on the basis of the following analyses and planning concept:
  - handling capacities in the main shaft and travel ways ;
  - layout of waste repository chambers/galleries and driving
    capacities ;
  - disposal technology for the types of waste containers involved ;
  - disposal capacity in developed and explored mine sections ;
  - disposal performance, subject to disposal technology, conveyance
    and driving capacity.

In view of the technical safety of the disposal operation, the
following important operational stages were investigated :
  - driving of stable disposal chambers ;
  - backfilling and sealing of disposal chambers and mine shaft ;
  - concepts for simultaneous chamber driving and disposal operations
    and for the drainage of the mine ;
  - ventilation in disposal galleries and adjacent mine sections.

1.3.3. Nuclear and waste handling investigations

The points of view relating to radioactive waste described within the scope of the Konrad Research programme concern :
- conditioned low-level waste and decommissioning waste from nuclear power stations including the waste package ;
- quantities of the waste concerned in the Federal Republic of Germany ;
- transport concept for the waste to be disposed.

The investigation carried out from the aspect of radiation protection and repository safety includes both development and assessment of :
- radiation protection in case of operational failure events in the loading station at surface and in the underground repository ;
- safety conception for the postoperational phase.

2. RESEARCH PROGRAMME IN THE FUTURE UNDERGROUND LABORATORY OF THE "PASQUASIA" MINE IN SICILY (Italy)

For diverse reasons the argillaceous formations have been considered as suitable for the disposal of conditioned radioactive waste in Italy.

The first research works were concentrated on the "La Trisaia" site during the late sixties. Then the characterization of samples coming from other argillaceous formations was undertaken. However a broad consensus is quickly appeared on the necessity to have an underground facility allowing the setting up of "in situ" investigations techniques as well as carrying out of experiments and measurements able to bring valuable informations for the future development of a disposal concept of conditioned high level waste in comparable argillaceous formations.

Considering the practical possibilities of the time being, the site of the PASQUASIA Potash Mine in Sicily was selected by the ENEA (Comitato Nazionale Energia Nucleare ed Energie Alternative).

2.1. Geology of the site and construction of the laboratory

The clay formation is of pliocenic age. Its local thickness of about 90 to 100 m ; it is delimited by marly limestones at the base and by clayey sands at the top (Fig.6).

The clay is slightly marly and contains a weak percentage of silty fraction. The clay minerals (illite, kaolinite) represent 50 % of the bulk composition, the carbonate content varies between 25 and 30 %. The clay is quite stiff and presents an undrained shear strength of 1 MPa, a cohesion of 0.1 to 0.15 MPa and an internal friction angle comprised between 30° and 34°. At the level of the future experimental gallery the lithostatic pressure is about 3.6 MPa. The preconsolidation is unknown but had to be very important. The formation is intersected by diverse joint systems which present slickenside structures.

The construction of the gallery will begin from the access ramp at the level -160 m. The 25 m long experimental zone will be separated from the access ramp by a 36 m long gallery. A 6 m long room aimed to gather later all the control instrumentation will be interposed between the buffer gallery and the laboratory gallery. The direction of the gallery is northward and perpendicular to the direction of the clay layers. The location of the underground laboratory in the geological formation, its access way and its dimensions are given at the Figures 7 and 8.

The foreseen gallery lining will be composed of typical NP16 frames with a 0.6 m spacing and a metal wire netting covered with a 10 cm shotcreted layer and with probably a cement furbishing layer (Fig. 9).

2.2. Experimental programme and instrumentation design

2.2.1. Experimental programme

For the time being and because of a lack of characterization data about the Pasquasia Clay formation, it is not possible to elaborate a detailed programme of the experiences to carry on in the underground laboratory.

However, the parameters which have to be analysed are the following :
- mechanical behaviour of the clay,
- heat effect on the behaviour of the clay body,
- heat effect on the physico-chemical properties of the clay,
- heat effect on the interstitial fluids,
- long-lived radionuclides mobility in the clay,
- radiation effects on the diverse components of the system.

Additionally, it might be of great importance to undertake experiments leading to the engineering aspects of a waste disposal as, for example :
- sealing and backfilling techniques for the shaft and cavities,
- special boring and excavation techniques,
- geophysical techniques for the "in situ" study of the clay formation.

2.2.2. Instrumentation design

One of the main purposes of the Pasquasia research programme is the measurement of the characteristics of the host rock and the measurement of the variation of these characteristics during the whole period of experiments. The data acquisition system is foreseen to collect automatically the variations of the signals coming from the sensors installed in the formation before recording and data processing. These sensors are subdivided into different categories :
- thermoresistances for the temperature measurements
- pressure transducers for the interstitial water pressure measurements
- multiple extensometers for the control of the longitudinal deformations
- total pressure cells in contact with the clay body and with the cavity lining.

All these sensors will be connected to a data logger which proceeds automatically and continuously to the auscultation and to the recording of the data on magnetic tapes.

2.2.3. Main specifications

These specifications are based upon the results of the field experiments which have already been carried out in Italy as well as upon economical considerations.

2.2.3.1. Temperature measurements

A variable field temperature measurements with 25 Platinium thermo-resistances, ranging from room temperature up to 200° C has been foreseen with a precision of one tenth of a degree. The distance between the temperature sensors and the heating device is variable between 0.5 m and 2.5 m. The thermo-resistances will be installed in small diameter boreholes up to a depth of 10 m beneath the floor of the cavity.

The duration of the temperature measurements with the heating device is one year.

2.2.3.2. Interstitial pressure measurements

The measuring field will be defined on the base of the field studies which are carried on at the moment at the "Monte Rotondo" site as well as on the base of the peculiar characteristics of the initial underground characteristics. Small electrical piezometers will be used in order to reduce the time lag. The distance between the sensors and the heating device will range between 0.5 m to 2 m.

So far the technical possibilities permit, the piezometers will be placed in the same boreholes and at the same depths as the temperature sensors.

2.2.3.3. Deformation measurements

The deformation field will be previously defined, by the aid of a mathematical model, in the vicinity of the heating device and along the cavity wall as well. The control of the deformations induced on the clay body by the heating will be done by three multiple extensometers in radial position with regard to the heating device.

Three other extensometers will measure the deformations of boreholes drilled from the walls of the cavity.

The spacing between each measuring element of the extensometer is one meter.

2.2.3.4. Total pressure measurements

The total pressure in the clay body will be measured by the aid of four hydraulic cells equipped with compensation circuity in order to assure a good contact with the rock mass. The use of the total pressure cells depends upon the nature of the cavity lining. If it appears that such measurements are not significant or reliable the instrumentation of the gallery will be limited to the extensometers.

2.2.3.5. Measuring and data acquisition system

All the measures will be gathered in a 50 channels data logger extendable to 100 channels. The average measuring frequency will be of one measure each 6 hours. A maximum measuring frequency of one measure per second can also be obtained.

3. REALISATION OF AN UNDERGROUND LABORATORY IN THE BOOM CLAY (CEN/SCK-Mol)

In Belgium, the decision to build an underground laboratory was taken in the late seventies, at the end of the first R&D five-year plan on the waste disposal possibilities in a tertiary clay layer located in the North-East of the country. The difference between that project and the others was that this laboratory should be built on a site which could later be taken in consideration for the construction of a final disposal facility. The site is the one of the Nuclear Research Centre of Mol (CEN/SCK) and no problem occurred for the obtainment of the legal authorizations for the construction and for the exploitation.

3.1. Site Geology and construction of the underground laboratory

The tertiary (oligocene) Boom clay formation, selected in the national catalogue of the geological formations suitable for the disposal of conditioned high level and alpha-bearing waste, stretches at depth in the North-East of the country with a thickness of about 100 meters. It is overlied by water, by miocene, pliocene and pleistocene water-bearing sands and underlied by sand layers from early oligocene.

A simplified stratigraphic cross-section is given at the Figure 10.

The Boom clay is homogeneous and compact and about 50 % of its grain size distribution is constituted by a fraction inferior to 50 μm.

Its main clay minerals are illite, smectite, vermiculite and chlorite-type interstratified minerals. The carbonate and organic materials contents are respectively 1.5 % and 3.5 %. The undrained shear strength measured on core samples varies between 0.33 to 0.86 MPa (triaxial tests on undisturbed samples taken at depth in the laboratory are in progress). The average cohesion and the internal friction angle are respectively 0.12 MPa and 19°. Accurate determinations of a broad number of geomechanical characteristics of samples taken during reconnaissance drillings or during the construction have been done and compared to those obtained on the same clay at the outcropping zone near the city of Antwerp. Prior to the beginning of the construction the zone was seismically prospected in order to make sure that the clay formation was not affected by tectonic features like faults of flexures. The first preparation works on the field began early in the year 1980.

Taking into account the presence of water bearing sands above the clay layer, the ground freezing technique was chosen for the sinking of the access shaft. Because of a lack of experience in the excavation of clay at such a depth and for security reasons, the ground freezing technique was also used for the construction of the horizontal gallery .

The excavation and lining works were ended in July 1983. A schematic view of the whole facility is given at the Figure 11. Details about the different construction phases and the constructional particularities of the facility are given in some recent papers noted in the references 11 and 12.

3.2. Experimental programme

The lack of knowledge on the rheological behaviour of non indurated clay at depth led the CEN/SCK to elaborate, in collaboration with the contractor (Foraky, Brussels), with the industrial architect (Tractionel, Brussels) and with the Laboratoire du Génie Civil of the Louvain-la-Neuve University (Louvain-la-Neuve), an extensive programme of "in situ" geotechnical measurements. Priority was given to that programme because of the paramount importance of its results on the feasability of industrial disposal facilities in the clay with technologically and economically acceptable means.

During the coming months all the other phases of the experimental programme such as corrosion, hydrology, backfilling, migration etc. will be progressively started.

3.2.1. Geotechnical research programme

The principle of the geotechnical programme in the experimental facility was the systematic auscultation of the clay during and after the excavation and lining works. That involved stresses, deformations and temperature measurements.

The measurements of the stresses concerned the natural stresses (lithostatic and hydrostatic pressures) and the stresses in the lining.

The measurements of the deformations concerned the study of the rheological behaviour of the clay body and the deformations undergone by the linings.

In addition, a series of coupled stress-strain measurements have been carried out with a Menard pressiometer both in frozen and non frozen clay. Lastly, temperature measurements, intensive during the freezing phase and less frequent afterwards, constituted an essential complement to the stress-strain measurements for the showing off of the influence of the frozen state on geotechnical parameters.

3.2.1.1. Instrumentation and measurements in the access shaft (Fig. 12)

The instrumentation of the access shaft took place during two different phases : during the digging works and after the completion of the shaft.

- During the digging works 51 total pressure hydraulic (Glötzl) cells were only placed in the clay massive prior to the pouring of the external lining of the shaft.

These cells were emplaced in small recesses dug in the clay in diverse positions with regard to the shaft lining in order to measure the total pressure in all the directions and at different levels around the shaft.

- After the end of the digging works a certain number of measuring devices were installed on the wall of the internal lining for the measurement of the deformations at different levels :
  - 16 vibrating wire extensometers (telemac type)
  - 8 inductance variation extensometers (LVDT type)
  - a large number of reference points, plugged in the shaft wall, for the measurement of the peripheral and diametral deformations by the aid of mechanical deformeters and with inductive sensor rods.

During the pouring of the shaft linings different watertight sleeves had been placed at diverse levels. After the completion of the shaft, some additional devices were inserted through these sleeves in the clay body :

- thermistance canes for the monitoring of the temperatures at different radial distances from the lining,
- hydraulic piezometric Glötzl cells for the measurement of the interstitial pressure in the clay,
- splitted tubes for further insertion of Menard pressiometers both in frozen and non frozen clay.

3.2.1.2. Instrumentation and measurements in the experimental gallery

During the digging and lining operations, before the placement of cast-iron segments, 25 total pressure Glötzl cells coupled with thermistances were installed in recesses dug in the clay in various directions in order to monitor the tangential, radial and longitudinal lithostatic pressures at different positions around the lining. A large number of strain gauges have also been placed on the cast-iron segments in order to control the stress state of the lining in function of the build up of the external pressures.

Piezometric Glötzl cells and thermistances have also been placed at different distances in the clay body for the follow up of the evolution of the interstitial water pressure during the thawing.

3.2.1.3. Additional works in the non-frozen clay

Owing to the fact that until the end of the construction of the gallery, little valuable informations were gained on the rheological behaviour of the non-frozen clay, because of the use of the freezing technique, it was decided to dug a little vertical reconnaissance shaft and a small horizontal gallery, both lined with concrete segments with an excavated diameter of 2 m and an useful diameter of 1.4 m. Before the sinking of the little vertical shaft several measuring devices have been

placed from the main gallery in order to monitor the deformations of the
clay body during the excavation works (Fig. 13):

- 2 vertical and 2 oblique inclinometer tubes for the measurement
  of the radial deformation by the aid of a inclinometric torpedo
  compelled to follow grooves made in the tubes ;
- 3 rod extensometers, anchored at different depths in the clay in
  the axis of the little shaft before its excavation and
  systematically measured with regard to a reference point in the
  main gallery, for the measurement of the longitudinal deformation
  of the clay as the digging proceeded.

During the digging of the little shaft some Glötzl total pressure
cells and load cells were respectively placed in the clay outside the
shaft and between the concrete blocks in order to follow up the evolution
of the pressures in the clay and in the lining in function of the time.

Before the excavation of the little 7 meter long horizontal gallery
at the bottom of the reconnaissance shaft several devices have also been
installed in collaboration with the "Agence Nationale des Déchets Radio-
Actifs" (ANDRA) from Paris and with the technical help of the "Bureau de
Recherches Géologiques et Minières" (BRGM) from Orléans.

- a multiple Telemac Distofor extensometer, placed in a slightly
  oblique borehole drilled from the main gallery, allowing the
  follow up of the radial deformations of the clay massive during
  digging and after lining (Fig. 14),
- a rod extensomter bored in the axis of the gallery for the
  measurement of the longitudinal displacements.

As the digging of the little gallery proceeded, diverse Glötzl total
pressure and load cells were also placed respectively in the clay and
between the concrete blocs for the same measurements as for the reconnaissance shaft. Additionally, after the end of the digging works, the
radial convergences of the lining was systematically measured by the aid
of distomatic (Telemac) device with regard to reference points plugged in
the concrete blocs in diverse directions.

3.2.2. Miscellaneous research programme

Beside the intensive geotechnical programme described above, a series
of very important experiments will take place in the coming months in the
experimental laboratory.

- Corrosion experiments with the purpose of evaluating the
  corrosion resistance of diverse candidate canister materials for
  high and medium level waste both in direct contact with clay at
  depth and in contact with the different gazeous components coming
  from the clay. These experiments are going to last several
  years in order to get a knowledge over the resistances
  capabilities of the tested materials ;
- leachability experiments with the aim to evaluate the leaching
  rate of glass samples in contact with clay ;
- heat transfer tests allowing to monitor all the changes undergone
  by the clay both on a geomechanical and a physico-chemical point
  of view, in function of a temperature rise ;
- backfilling tests aiming to study the confinement performances
  and the physico-chemical changes of plugging materials in
  function of the time and severe pressure and temperature
  conditions ;
- hydrological tests for trapping and analysing of clay conate
  waters ;
- migration and sorption tests with different tracer species in the
  clay body surrounding the experimental facility.

CONCLUSIONS

The interest presented by the availability of an experimental zone at depth in the very type of rock envisaged for the final disposal of radioactive waste is unquestionable and has been already demonstrated by the results obtained in experimental facilities such as Stripa in Sweden, Asse in the Federal Republic of Germany, Hanford in United States etc. Taking into account the lack of knowledge about the properties at depth of some types of sedimentary layers like clays or some types of ore bodies convenient for that purpose, that interest changes into a necessity.

The Commission of the European Communities has well understood this necessity in supporting the research programmes described in the present paper. Two of them, Konrad and Mol have already delivered a considerable sum of knowledge and data. The impending commencement of the tests in the "Pasquasia" mine will allow Italy to confirm its waste disposal options.

REFERENCES

1   TASSONI, E. "An Experiment on the heat transmission in a clay rock" Proceedings of the NEA-workshop on the "Use of argillaceous materials for the isolation of radioactive waste". Paris 10-12 Sept. 1979. p. 23, ISBN 92664-02050-0

2   TASSONI, E. "In situ and laboratory heating experiments in clay". Oral communication : Technical Committee Meeting on "Effects of Heat from High-level Waste on Performance of deep Geological Repository components" - IAEA, Stockholm, 28/08-2/09-1983

3   HEREMANS, R., BUYENS, M. MANFROY, P. "Le comportement de l'argile vis-à-vis de la chaleur". Proceedings of the NEA-workshop on "In situ heating experiments in geological formations". Ludvika/Stripa, 13-15 Sept. 1978

4   "Relazione sugli Studi effectuati sulla zona di Pantanello (Golfo di Taranto"). CNEN/EURAOM contratto 004665-3 WASI - Parte IV. vol. 1

5   BRUZZI, D., GERA, F. (ISMES-Italy) "Geotechnical Instrumentation for Field Measurements in Deep Clays". Proceedings of the CEC/NEA-workshop on "Design and Instrumentation of in situ Experiments in Underground Laboratories for Radioactive Waste Disposal". Brussels 15-17 May 1984

6   NEERDAEL, B. MANFROY, P., HEREMANS, R. "Caractérisation géomécanique de l'argile de Boom". Proceedings of a technical session on rock mechanics "Advance in Laboratory sample testing" CCE - Brussels 27.04.1983. A paraître

7   MANFROY, P. "Rejet des déchets radioactifs en formation argileuse profonde - Construction d'un laboratoire souterrain pour un programme expérimental approfondi". Proceedings of the International Conference "Radioactive Waste Management" paper IAEA 43/54 vol. 3 (Seattle/USA 16-20 May 1983) - IAEA-Vienna 1984

8   BREWITZ, W. "Disused Konrad Iron Ore Mine - a Future Low-Level Waste Repository in the FRG". Nuclear Europe 9/1983

9   BREWITZ, W. "Evaluation of disposal possibilities and potential in the Konrad iron-ore mine based on experiments for the handling and isolation of radioactive wastes. Final Report EUR 9094 EN

10   BENVEGNU, F. "Selezione di un sito adatto alla costruzione di un laboratorio sotteraneo in formazioni argillose. Rapporto finale Eur 9279 IT

11   FUNKEN, R., GONZE, P. VRANKEN, P., MANFROY, P., NEERDAEL, B. "Construction of an experimental laboratory in a deep clay formation" Eurotunnel 83, Conference, paper 9, Basle (Switzerland), June 22-24 1983

12   MANFROY, P. NEERDAEL, R. "Expérience acquise à l'occasion de la
     réalisation d'une campagne géotechnique dans une argile profonde".
     Proceedings of the CEC/NEA- workshop on "Design and Instrumentation
     of in-situ Experiments in Underground Laboratories for Radioactive
     Waste Disposal", Brussels 15-17 May 1984
13   CASTEELS, F., DE BATIST, R. KELCHTERMANS, J. DRESSELAERS, J.,
     TIMMERMANS, W. "In situ Testing and Corrosion Monitoring in
     Geological Clay Formation". Proceedings of the CEC/NEA- workshop on
     "Design and Instrumentation of in-situ Experiments in Underground
     Laboratories for Radioactive Waste Disposal". Brussels 15-17 May
     1984
14   HEREMANS, R., "Expériences en place en terrains argileux" CEC/NEA-
     workshop on design and instrumentation of in-situ experiments in
     underground laboratories associated with geological disposal of
     radioactive waste. Brussels, 15-17 May 1984.

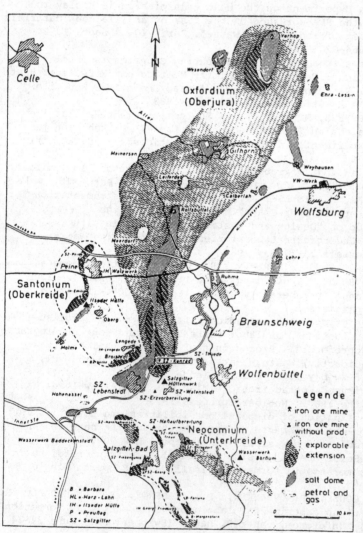

Fig. 1 - shows the
various Jurassic and
Cretaceous iron ore
deposits in the
Salzgitter Peine area.
The Konrad Mine is
located in the Southern
part of the Jurassic
iron ore deposit
(extensive hatched
area between Salzgitter
steelworks in the South
and Vorhop in the
North).

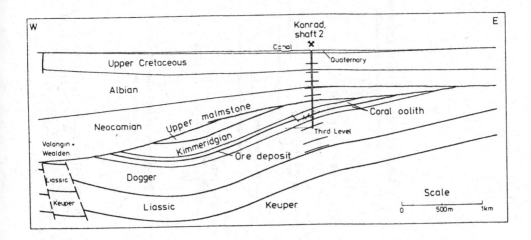

Fig. 2 - shows a geological West-East section (corresponding to section
         line B in Fig. 1) through the Konrad iron ore deposit. The
         synclinal Jurassic rock strata (Liassic-Dogger-Malmstone) are
         covered by overlying Cretaceous strata (Neocomian-Albian-Upper
         Cretaceous). The ore deposit has no outcrop. It is divided by
         large fault zones (Bleckenstedter Fault, Konrad Fault and
         Sauinger Graben) into various sub-areas. These faults vanish
         in the claystones of the Neocomian and do not therefore continue
         to the surface. They are no water pathways.

Shaft profile Konrad 2

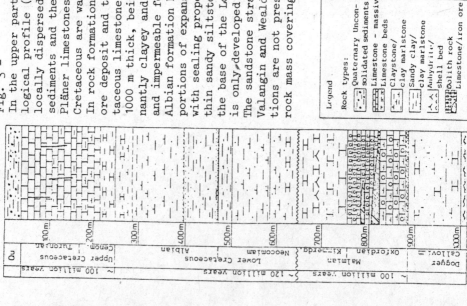

Fig. 3 -
In the upper part of the geological profile (Fig. 3) the
locally dispersed Quaternary
sediments and the fissured
Pläner limestones of the Upper
Cretaceous are water bearing.
In rock formations between the
ore deposit and the Upper Cretaceous limestones are 600 to
1000 m thick, being predominantly clayey and clayeymarly
and impermeable for water. The
Albian formation has high proportions of expansive clays
with sealing properties. A
thin sandy siltstone bed at
the base of the Lower Albian
is only developed in shaft 2.
The sandstone strata of the
Valangin and Wealden formations are not present in the
rock mass covering the mine.

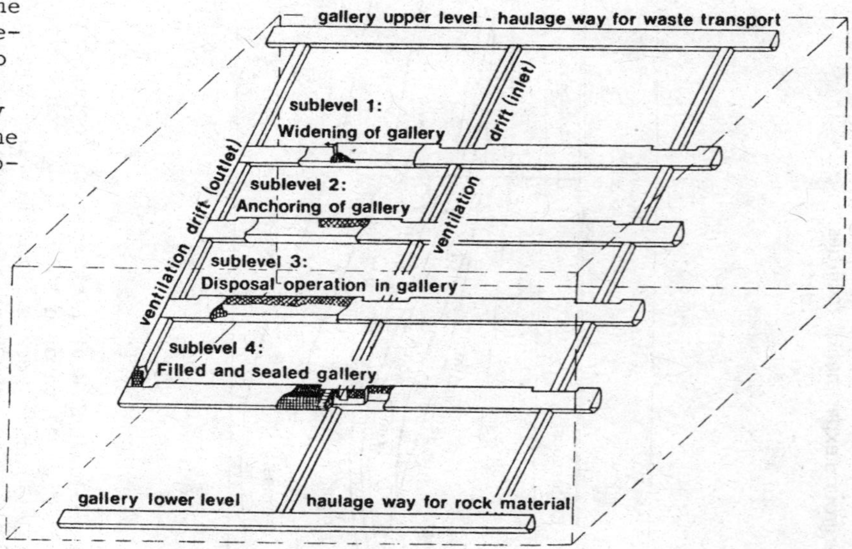

Fig. 4 - Concept of a repository section with disposal
galleries of 40 m2 cross section which are separated
from one another by adequate rock pillars.

gallery upper level - haulage way for waste transport

drift (inlet)

ventilation

sublevel 1:
Widening of gallery

sublevel 2:
Anchoring of gallery

sublevel 3:
Disposal-operation in gallery

sublevel 4:
Filled and sealed gallery

ventilation drift (outlet)

gallery lower level

haulage way for rock material

Legend

Rock types:

Quaternary uncon-
solidated sediments
Limestone - massiv
Limestone beds
Claystone
clay marlstone
Sandy clay
clay marlstone
Anhydrite
shell bed
oolith rock
Limestone/iron ore

Shaft profile column labels:
ou
Cenom. Turon. Upper Cretaceous ~ 100 million years
Albian Lower Cretaceous ~ 120 million years
Neocomian
Kimmeridg. Malm Oxfordian ~ 100 million years
Callovian II Dogger

100m 200m 300m 400m 500m 600m 700m 800m 900m 1000m

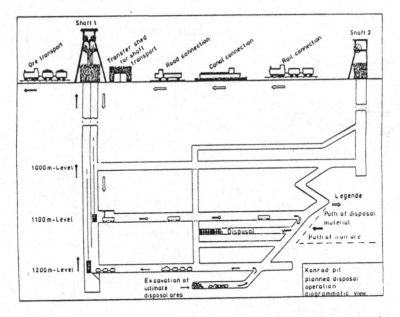

Fig. 5 - Konrad Mine planned disposal operations, diagrammatic view

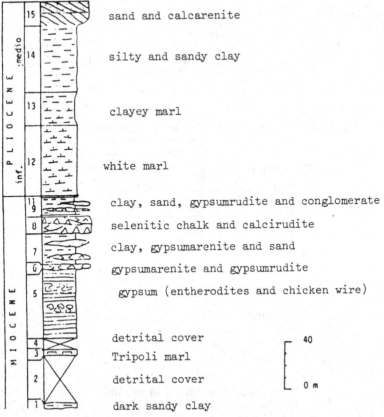

Fig. 6 - Schematic stratigraphical column in the mio-pliocenic
formations of "De Pasquasia"

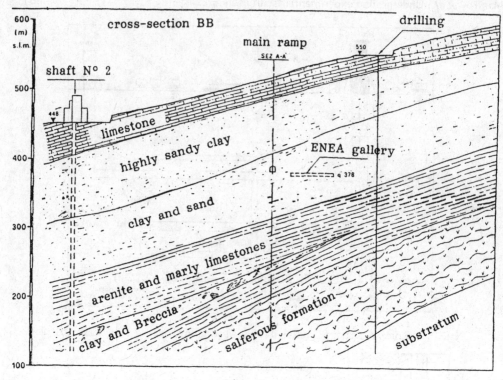

cross-section BB

drilling

main ramp

SEZ A-A'

550

shaft N° 2

600
(m)
s.l.m.

500

448

limestone

highly sandy clay

ENEA gallery

400

q 378

clay and sand

300

arenite and marly limestones

200

clay and Breccia

salferous formation

substratum

100

Fig. 7

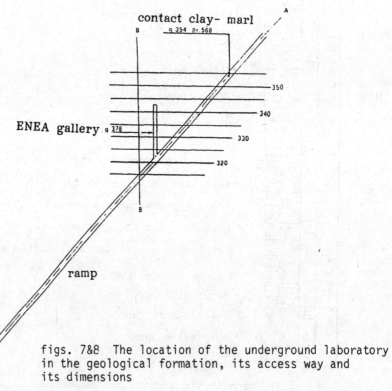

contact clay- marl

B   q 354 pr. 568

A

350

340

ENEA gallery q 378

330

320

B

ramp

A

figs. 7&8 The location of the underground laboratory
in the geological formation, its access way and
its dimensions

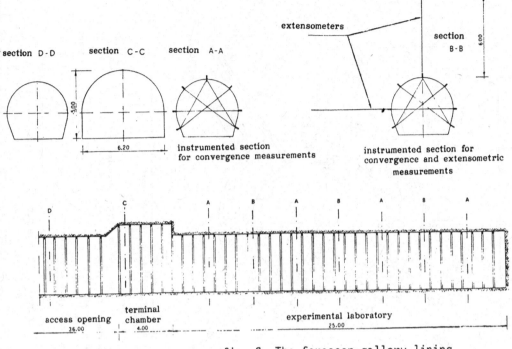

section D-D   section C-C   section A-A

extensometers   section B-B   600

5.00   6.20   instrumented section for convergence measurements   instrumented section for convergence and extensometric measurements

access opening   terminal chamber   experimental laboratory
36.00   4.00   25.00

fig. 9 The foreseen gallery lining

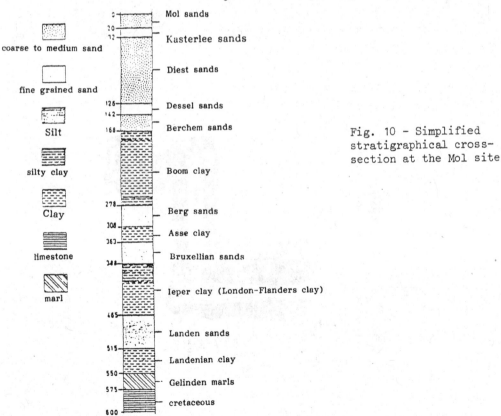

coarse to medium sand

fine grained sand

Silt

silty clay

Clay

limestone

marl

Mol sands
Kasterlee sands
Diest sands
Dessel sands
Berchem sands
Boom clay
Berg sands
Asse clay
Bruxellian sands
Ieper clay (London-Flanders clay)
Landen sands
Landenian clay
Gelinden marls
cretaceous

Fig. 10 - Simplified stratigraphical cross-section at the Mol site

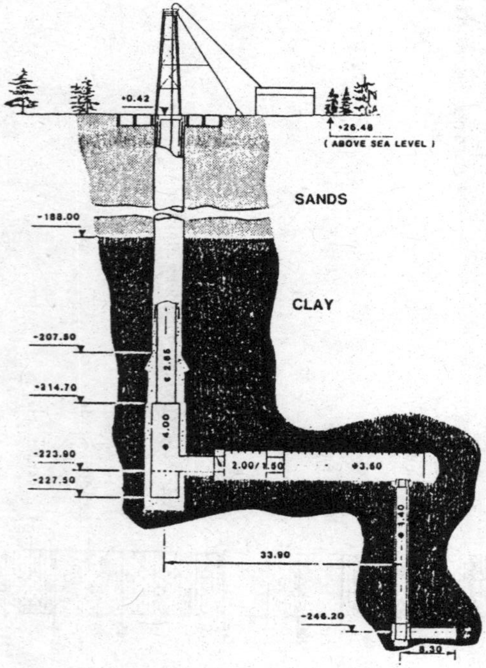

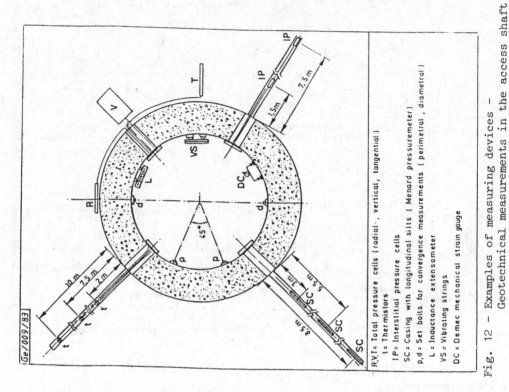

Ge/009/83

R,V = Total pressure cells (radial-, vertical, tangential)
  t = Thermistors
IP = Interstitial pressure cells
SC = Casing with longitudinal slits (Menard pressuremeter)
p,d = Set bolts for convergence measurements (perimetral , diametral)
  L = Inductance extensometer
VS = Vibrating strings
DC = Demec mechanical strain gauge

Fig. 12 — Examples of measuring devices —
           Geotechnical measurements in the access shaft

Fig. 11 — Scheme of the as built
           underground experimental facility

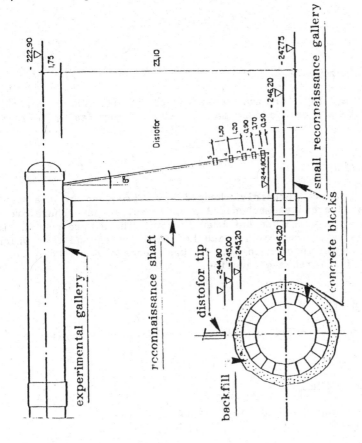

Fig. 14 – Position of the Distofor

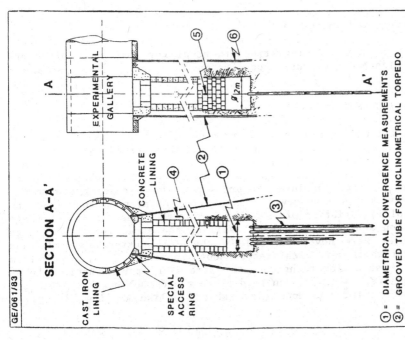

Fig. 13 – Configuration of constraint and deformation measurements

① = DIAMETRICAL CONVERGENCE MEASUREMENTS
② = GROOVED TUBE FOR INCLINOMETRICAL TORPEDO
③ = TENSILE RODS
④ = HYDRAULIC CELLS FOR TOTAL PRESSURE MEASUREMENTS
⑤ = FLAT JACK
⑥ = CASED THERMOPROBE BOREHOLE

DISCUSSION

H. GIES, GSF Braunschweig

Has freezing the clay for the construction of the shaft changed its properties? Will the data measurement in the gallery really be representative?

P. MANFROY, CEN/SCK Mol

Freezing the clay has certainly changed the structure of the clay and hence its resistance and geomechanical properties. We have therefore built a well into the area which had not been frozen. There were no immediately obvious differences between clay which had been frozen and that which was pristine. But more interestingly, it was necessary at that depth to freeze the clay for further excavation.

UNIDENTIFIED SPEAKER

That clay was probably overconsolidated. Could you measure its consolidation factor?

B. NEERDAEL, CEN/SCK Mol

The overconsolidation was low, just above a factor of one, and K_o, the factor of anisotropy, was between 0,6 and 0,5.

G. DE MARSILY, Ecole de Mines Paris

Were the stresses on the lining structure equal for non-frozen areas and for sections which had previously been frozen?

P. MANFROY, CEN/SCK Mol

It is rather difficult to make exact comparisons, as the reinforced lining was not installed quite the same way in the earlier frozen section as in the other parts, in particular filling in voids between the formation and the structures was done differently. But, in those parts, where the clay had been frozen, a rapid build-up of pressure, even exceeding the normal lithostatic pressure, was observed. Due to the excavation technique and the positive volume variation of the clay during the freezing stage, there was a contraction against the external wall of the structure. These high partial pressures were never reached in the small extension shaft and gallery, both of which were excavated without freezing. Thus, a part from the different construction technique, the stress and the mechanical behaviour of the non-frozen clay is quite different and much better. This can be stated at least for the Boom clay, which, unlike kaoline, is rather stiff.

CHARACTERIZATION AND BEHAVIOUR OF ARGILLACEOUS ROCKS

a joint paper by : A. Bonne[1], J. Black[2], F. Gera[3],
P. Gonze[4], E. Tassoni[5] and J.F. Thimus[6]

(1) SCK/CEN - Mol, Belgium ; (2) BGS - Nottingham, U.K. ; (3) ISMES -
Bergamo/Rome, Italy ; (4) FORAKY - Brussels, Belgium ; (5) ENEA - Rome,
Italy ; (6) UCL - Louvain-la-Neuve, Belgium

SUMMARY

In the present paper the main activities concerning characterization and behaviour of argillaceous rocks and their environment are
presented.

In Italy large argillaceous basins have been characterized on the
basis of sedimentological, petrological, mineralogical and geochemical
data. Site specific approaches in Belgium and the United Kingdom on
argillaceous formations underlying nuclear sites, have yielded hydrogeological and geological characteristics. This was enabled by deep
reconnaissance and sampling at these sites (e.g. underground experimental facility at Mol).

In all these countries fundamental and methodological issues related to the characterization of argillaceous media have been investigated: e.g. heat transfer, bore-hole sealing, γ-irradiation, pore water
chemistry, geomechanical characteristics, rheology of clay.

1. INTRODUCTION

Research and development on disposal of radioactive waste in
argillaceous media have been continued during the last five years with
intensified effort on the assessment of two main issues : the technico-
economical feasibility and long-term safety of the option.

Various papers presented at this conference undoubtedly illustrate
the progress that has been made in these assessments based on continued
and in some cases increased support for investigation programmes on
argillaceous formations in Belgium, Italy and the United Kingdom.

The approaches and aims of each of these countries are not identical. They are guided by national priorities and policies which differ
in several aspects. However they all address items related to the two
main issues mentioned above.

2. GENERAL CHARACTERIZATION STUDIES OF LARGE ARGILLACEOUS BASINS

In Italy the structure, sedimentology, and mineralogy of outcropping tertiary argillaceous formations (essentially pliocene clays) have
been studied. The extent of the area surveyed did not allow a dense
sampling, but the representativity of the samples and of the characteristics determined were evaluated by comparing the results of the
large-scale sampling with those obtained from a dense sampling in a few
particular basins such as the Val d'Era basin, the Crotone basin and
the Vasto basin (Fig. 1). Apart from interesting conclusions concerning

the natural evolution of these clay formations (1) it should be mentioned that all the argillaceous formations studied could be classed in
several groups (or "provinces") (2) according to their mineralogical
composition, because compositional differences are linked with distinct
geographical areas. Within each "province" dominant characteristics may
be found. A direct relationship is observed between the clay mineral
content of the outcropping argillaceous rocks and the original source or
parent rocks. Kaolinite bearing formations seem to be derived from
granites and older crystalline rocks, illitic argillaceous rocks are
shown to be derived from crystalline, arenaceous and calcareous rocks.
Smectite bearing rocks are largely related to carbonate rich marls, and
ophiolitic marls are shown to contribute an important chlorite content
to their argillaceous derivates. The variability within each "province"
of outcropping argillaceous formations is linked to the variation in its
source and tectonic environment. The more tectonically active an area is
the more various rock types in that area are potentially exposed to
alteration and erosion.

It is expected that the observed relationship between compositional
characteristics of argillaceous formations and parent rocks may be a
guide for finding favourable formations and areas for hosting nuclear
waste repositories.

In Table 1 an overview is given of some compositional characteristics of Italian clays and of clay from particular areas (Val d'Era,
Crotone and Vasto).

3. SITE SPECIFIC STUDIES

In Belgium and the United Kingdom the investigation programmes have
followed a site specific approach which allows a more comprehensive
characterization of the particular argillaceous environment investigated
and a more direct investigation of the main assessment issues. In the
United Kingdom argillaceous rocks have been considered as potential host
rocks for the disposal of low and intermediate level waste. To date only
one area has been investigated in the United Kingdom, the Harwell Site
(South Oxfordshire), underlain by three argillaceous formations within
the top 400 metres. In Belgium the only final option studied for the
disposal of high level and long-lived waste is the clay option, which is
intensively investigated on the Mol site, where the so-called Boom Clay
is present at depth.

3.1. The Harwell Site

The aims of the investigations at the Harwell site are to assess
the likely migration rates of dissolved species across the low permeability formations and to identify the potential pathway for release of
radionuclides back to the biosphere. Whilst the first aim only
encompasses studies of the clay and mudrock formations, the second has
the wider aim of defining the regional flow system. For these reasons
the programme of investigations includes studies of the wider Harwell
region as well as the detailed work centred on a group of four specially
drilled and completed bore-holes.

The rocks beneath Harwell comprise an alternating sequence of high

Table 1. Compositional characteristics of argillaceous formations studied in Italy. Nation wide survey and detailed surveys in the Val d'Era, Crotone and Vasto area

	Nationwide	Detailed		
		Val d'Era	Crotone basin	Vasto basin
max. number of samples analysed	125	20	47	35
grain size (in %)				
- sand (2-0.063 mm)	0.1-61.7	0.5-13.6	0.8-33.0	0.52- 2.4
- silt (63 μm-2 μm)	26.9-92.3	37.6-95.2	15.7-85.1	48.7 -55.7
- clay (<2 μm)	1.2-61.2	1.7-54.1	1.5-82.3	43.2 -50.7
mineral composition				
- quartz	3 - 30	10 - 18	10 - 18	5 - 20
- K.feldspar	0 - 15	<5 - 7	0 - 10	0 - 5
- Na.feldspar	0 - 15	<1	0 - 5	0 - 5
- calcite	0 - 80	10 - 25	12 - 25	20 - 40
- dolomite	0 - 7	-	-	0 - 5
clayc omposition fraction				
- smectite	+ - >++++	+ - ++	+ ->++++	+ - >++++
- chlorite	+ - +++	+ - ++	+	+ - ++
- interstratified	+ - +++	+ - ++	+ - ++	+ - ++
- illite	+ - >++++	+++ - ++++	++ - ++++	++ - ++++
- kaolinite	+ - >++++	++ - >++++	+++ - >++++	++ - >++++

Classes : + = 0-10 %, ++ = 11-20 %, +++ = 21-30 %, ++++ = 31-40 %

and low-permeability rocks (Table 2) with the thickest unit the Oxford/Corallian Clay reaching 100 m.

In the argillaceous strata, porosity is dominated by intergranular pores, and these represent the major pathways for ground water percolation through the bulk of the rocks. Here the pores are dominantly less than 0.1 μm wide occurring as laminar "films" and oriented preferentially in a sub-horizontal manner by compaction.

The Oxford Clay and Kimmeridge Clay are dominated by an assemblage of illite and show minor amounts of kaolinite. The clays of the Corallian beds are rich in smectite or mixed-layer clay minerals. Opal-CT, zeolite and pyrite are abundant in many parts of the sequence.

Of the four boreholes at the Harwell site, the three deepest are in a line 5 m apart. The screened intervals in these boreholes were positioned in the Corallian limestones, the Oxford Clay and the Great oolite

group (of limestones) respectively. The fourth shallower borehole was
left open-hole apart from some surface casing. Having completed three of
the four boreholes the rock was tested from inside the boreholes using
the BGS developed wire-line, straddle-packer system. The testing
provided reliable values of hydraulic head, hydraulic conductivity and
estimates of specific storage (Fig. 2). The trend of decreasing hydraulic conductivity with depth is clearly seen (3). Of the two zones which
have both field and laboratory measurements of hydraulic conductivity,
the field results are slightly higher. This is wholly in line with the
concept of fissuring adding to the matrix conductivity which was measured in the laboratory. Considering the differing zones of influence
of the field tests, this implies that the rock close to the borehole has
not been significantly altered by the drilling process.

Table 2.   Summary of strata beneath Harwell

Unit Name	Lithology	Depth (m bgl)	Thickness (m)
CRETACEOUS			
Lower chalk	clayey limestone	0 – 62	62
Upper greensand	clayey sandstone	62 – 87	25
Gault	clay	87 – 155	68
Lower greensand	sandy mudrock	155 – 161	6
JURASSIC			
Kimmeridge Clay	mudrock	161 – 187	26
Corallian limestones	limestone/sandstone	187 – 222	35
Corallian beds	mudrocks	222 – 265	43
Oxford Clay	mudrocks	265 – 328	63
Kellaways beds	sandy mudrocks	328 – 335	7
Great oolite group	muddy limestones	335 – 384	49
Inferior oolite group	muddy limestones	384 – 399	15
Lower Lias	mudrocks	399 – 446	47

Measuring the hydrogeological properties of the Oxford Clay in situ
was a more difficult task since it could not be pumped in the normal
sense. After completion the borehole was left full of water. This high
water level was seen to decline by only 2 m in 4 months. A short-term
pulse test (by withdrawal) was performed when the water level was close
to the equilibrium level and a hydraulic conductivity of $4E-12$ m.s^{-1}
was obtained. A consolidation test in the laboratory at the in-situ
stress yielded a value of $5E-12$ m.s^{-1}. This corroboration of the field
measurement of hydraulic conductivity lends credibility to the measure-

ment of head particularly since the pulse test method requires an accurate initial head. The head measurements (see Fig.2) show a distinct pattern of falling heads from the surface towards what seems to be a sink in the Corallian limestones. Below these limestones, the head appears to rise again inferring upward water flow from below. A vertical section of the region has been modelled (4) using a two-dimensional finite element code called NAMMU. The choice of section and boundary conditions was difficult owing to the lack of a well defined southern boundary and the complexity of the system in terms of the number of identifiable units. In the event the section was of limited extent running from the River Thames in the north to just south of the Harwell site. It was aligned approximately parallel to the direction of ground-water flow within the Corallian and forms a sort of blunted triangle (Figure 3). The top surface is the water table within the chalk which is considered to be a constant head boundary. The lower boundary (at the base of the Corallian limestones) is taken as a zero flow boundary owing to the low permeability of the Oxford Clay beneath.

The first simulations with this model were carried out using the material properties measured in the Harwell boreholes since these were considered to be the most reliable in the region. A closer matching of the model results with measured reality was obtained by varying the hydraulic conductivity of the mudrocks according to the thickness of overburden. This achieved the result shown in Fig. 3 and matches all measured head data along the line of section.

In an attempt to check the validity of the chemistry data measured in the programme the flows of Fig. 3 were converted into times for migration across the Gault Clay between the chalk and the Corallian aquifers. The results of this calculation show that water takes a wide range of times to cross the Gault Clay and must subsequently mix within the Corallian. The southern half of the Corallian is modelled with transit times ranging between 2E5 and 2E7 years whilst "helium ages" in that area yield values in the range from 4E6 to 2E7 years.

3.2. The Mol site

The objectives of the investigations on the Mol site are (5) to acquire on a site-specific basis the necessary data to assess the technico-economical feasibility and long-term safety of a disposal system in clay. This investigation programme is focused on the Boom Clay, the uppermost argillaceous formation underlying the Mol site. The Boom Clay is studied in its regional context, from its outcropping area in the west to its "in situ" conditions in the underground experimental facility (6) at Mol.

A hydrogeological investigation in the region around Mol, covering an area of about 2,500 km², is focussed on getting as precise a picture as possible of the groundwater flow system and of the geological structure. By numerous bore-holes in the area the hypothesis of a very regularly gently dipping layered sequence of alternating impervious formations and sandy aquifers has been confirmed. The typical sequence of sedimentary strata as they were encountered at Mol is given in Table 3. Geophysical logging within the bore-holes, before their completion as hydrological observation wells allowed to show in fair detail the lateral continuity of minute lithological variations in the Boom Clay (7).

For the purpose of laboratory determination of the characteristics of the Boom Clay various forms of field sampling techniques were adopted. These includes piston coring, rotary coring, cutting of blocks of clay up to 30 by 30 centimetres side and ordinary sampling of Boom Clay. The main physical characteristics of the Boom Clay are given in Table 4.

Table 3. Summary of the strata beneath the Mol site (location 15)

Unit name	Lithology	Depth (m bgl)			Thickness m
NEOGENE					
Mol sands	quartz sands	0.5	–	19	19
Kasterlee sands	quartz sands	19	–	26	7
Formation of Diest	glauconitic sands	26	–	128	102
Formation of Berchem	glauconitic sands	128	–	148	20
OLIGOCENE					
Voort sand	fine sands (glauc.)	148	–	150	12
Boom Clay	compact clay	160	–	269	109
Berg sands	fine sands	269	–	284	15
Transistion zone	fine clayey sands	284	–	301	15
EOCENE					
Asse Clay	glauconitic clay	301	–	312	11
Wemmel sands	calc. sands	312	–	316	4
Lede sands	calc. sands	316	–	336	20
Brussels sands	sands/sandstone	336	–	346	10
Formation of Ypres	fine sands/silty clay	346	–	443	97
PALEOCENE					
Formation of Landen	sand/siltstone/ claystone	443	–	555	112
Formation of Heers	marls	555	–	570	15
CRETACEOUS					
Maastrichtian	calcarenite	570	->	577	>7

Besides the characteristics of the Boom Clay itself some hydrogeological parameters of the over- and underlying aquifers have been determined because of their importance for the identification of the general ground water flow regime in the layered sequence beneath the Mol site. For the aquifer of the Berg sands, underlying the Boom Clay, permeability values ranging from 3E-6 $m.s^{-1}$ to 3E-8 $m.s^{-1}$ have been obtained, depending upon the type of test and the location. For the neogene aquifer, overlying the Boom Clay, the overall transmissivity has been calculated by a one-dimensional inverse model that has been used to interpret the natural fluctuations of the water levels. The transmissivity values (4E-3 $m^2.s^{-1}$) obtained from this approach match very well with those calculated from classical well tests in this aquifer. A straightforward estimation of the macro-permeability of the Boom clay, from the water infiltration rate into a piezometer installed in the underground experimental facility yields a value which is in fair agreement with those obtained on samples in the laboratory : about $3E-11 \text{ m.s}^{-1}$.

The regional hydrological model around the Mol site assumes a multi-layered system of aquifers and confining layers with seepage through the latter. Using a numerical model (NEWSAM-version of Ecole des Mines, Paris) the observed hydraulic heads in the aquifers above and below the Boom Clay are fairly well predicted. For this prediction the model used the permeability values of the Boom Clay obtained from laboratory experiments. The good match between model and observation leads to the conclusion that Boom Clay permeability on the mesoscale (permeater tests) is identical with its permeability on a regional scale. The ground water flow pattern obtained indicates a westward flow in all aquifers beneath the Mol site. In the two uppermost confining strata a downward seepage is deduced in the eastern part of the Mol area and an upward seepage is deduced west of the site (see Fig. 4).

Table 4. Main physical characteristics of the Boom Clay

Grain size (1), (%)	
<2μm	56.0
2-60μm	39.0
60-200μm	4.0
>200μm	1.0
(<20μm)	78.0
Bulk density (2), (kN.m³)	19.7
Dry density (2), (kN.m³)	16.2
Water content (2), (%)	22.2
Porosity (2), (%)	38.4
Saturation (2), (%)	94
Atterberg limits (2)	
plasticity limit	26.6
liquidity limit	69.9
plasticity index	43.3
Permeability (3), (m.s^{-1})	E-10 - 5E-12

(1) average of samples between 185-200 m bgl. at HADES-location
(2) average of 5 samples between 243-248 m bgl. at HADES-location
(3) geotechnical boring.

This picture of the regional groundwater flow system reflects of course the present day situation. The hydrological model has also been applied as a tool for evaluating the impact of possible future natural events such as climatic change, erosion, faulting, denudation, etc.. The outcome of these possible changes is, that for the Mol site, taking into account reasonable intensities for the natural phenomena, the ground water flow regime will be most influenced by a climatic change. In that particular event the seepage velocity through the Boom Clay could change by an order of magnitude.

4. FUNDAMENTAL AND METHODOLOGICAL RESEARCHES

Besides the generic studies in Italy, and the site specific approaches in Belgium and the United Kingdom, fundamental and methodological investigations have also been carried out in these three countries. The aim of these investigations is to increase the confidence and reliability of the data and derived characteristics needed for the assessments.

Specific questions concerning the interaction between argillaceous host rocks and various repository components have been investigated in Italy and Belgium.

The dissipation of heat, generated essentially by the radioactive decay of fission products embedded in the conditioned waste matrix was a important issue of the research. An experimental research programme, both in the laboratory and in situ has been carried out by ENEA, in order to know the temperature field around a heat source and the effects caused by heat. A field experiment was carried out in an open clay quarry at Monterotondo near Rome (8). The clay studied at that location presents the following composition : clay minerals 55 %, quartz 10 %, calcite 25 %, dolomite 5 %, feldspars 5 %. The experiment comprised an electric heat source buried in the clay at 6.4 metres depth with measurement of the temperature evolution up to 2 metres distant (Fig. 5). After a running time of 1,200 hours the electrical heater power was increased from 250 to 500 watt (3,300 hours).

A similar in situ heating test has been performed by SCK/CEN at Terhaegen with a heat source simulating a vitrified high level waste canister (stainless steel, 300 mm diameter, 1,500 mm height, 6 mm thick wall, powered electrically by inner heating wires). The source was buried about 6.5 m deep in a 500 mm wide hole backfilled with quartz sand. Temperature probes were emplaced around the heat source up to 4.15 metres distance from the source (Fig. 6). The power of the source was increased stepwise up to 1,500 kw. The total test lasted about 1.5 year.

The experiments at Monterotondo and Terhaegen reached quite similar conclusions.

The numerical code MPGST and/or variants of it (developed, at SCK/CEN (9)) was applied in both cases to calculate the temperature increase around a heat source and it demonstrated to be a reliable forecasting tool.

The thermal conductivity deduced by 'curve fitting' from the experimental data in the plio-pleistocene clay at Monterotondo ranges from 1.5 to 1.7 $W.m^{-1}.°C^{-1}$. For the Boom clay at Terhaegen a thermal conductivity of 1.69 $W.m^{-1}.°C^{-1}$ has been derived by the same calculation procedure. These thermal conductivity values, determined by large scale in situ near surface experiments, are in good agreement with laboratory experiments and values cited in literature for clayey materials.

The temperature variation due to forced water convection in the clay itself is virtually negligible. A future experiment to prove this last statement is planned by ENEA.

ENEA has also developed a laboratory-automated method to study, in more detail, the variability of the thermal conductivity and diffusivity in clay samples from various Italian Clay formations. The preliminary conclusions are :
- thermal conductivity is anisotropic (along bedding planes > across the bedding plane) ;
- thermal conductivity decreases with increasing water content ;
- thermal conductivity seems to increase with decreasing clay mineral fraction. On clay samples from the Vasto area no conclusive results could be obtained concerning the temperature dependence of the thermal conductivity (temperature range studied 20 - 60°C) ;
- thermal diffusivity determined on a clay sample from Monterotondo was reckoned to be 17 $m^2.y^{-1}$. This value is near to the value of the thermal diffusivity of the Boom Clay, calculated from the in situ experiment at Terhagen (19 $m^2.y^{-1}$).

Studies by SCK/CEN on the radiolytic effects on Boom clay were undertaken but essentially focused on γ-radiolysis. Moisture content, H_2 and CO_2 pressure build-up and dose were examined in detail on 16 irradiated clay samples (some of these being pre-treated, e.g. drying, moisturing,...).

The most conclusive results obtained may be summarized as follows :
- The hydrogen production increases with increasing γ-dose. CO_2 is produced by radiolysis of organic matter.
- The ratio H_2/CO_2 is about 5 (a few exceptions were found).
- When 97 % of the free water content has been removed from the clay a sharp decrease in radiolytic hydrogen is observed.

Still many questions remain unsolved and will be tackled in the future such as the observation that an excess of moisture content does not correspond with a higher radiolytic H_2-production.

One of the most interesting aspects of these studies are the natural analogues of the modifications in clays caused by heating. This aspect has been studied in Tuscany where small subvolcanic intrusive bodies are encountered in pliocene clays. At Orciatico, in the Val d'Era area, the thermal metamorphism of a clay in contact with a alcaline trachyte laccolith has been investigated. A very detailed geological field survey and mineralogical, petrographical and geochemical studies on drill core led to the following conclusions.
- In the periphery of the intrusive body dehydration effects were observed to distance of 6 metres. Important modifications in the mineralogical composition of the clay and thus of its physical properties were restricted to two metres distance from the contact. The paragenesis of K- and Na/Ca-feldspars indicates a temperature field of about 400°C in the immediate surrounds of the intrusion.
- Clay on the top of the intrusive shows thermal influence and chemical mobilization over a total thickness of 12 - 14 metres. However, mineralogical modifications don't occur more than 4 metres from the subvolcanite. The pyroxene-feldspars paragenesis in the contact zone indicates a temperature rise of the clay in the order of 400 to

600°C. Due to the formation of smectite alongside the feldspars above this zone the metamorphosed clay displays higher sorption capacities than the intact clay.

In the site investigation programme at Harwell attention was also paid to a number of fundamental issues related to field investigation in argillaceous media. The following conclusions can be drawn from the tests and observations.

- Darcyan flow seems to be occurring in the mudrocks, as a response to hydraulic potential differences. This has been deduced from hydrodynamical and geochemical observations.
- Background work in the form of interpreting the borehole geophysical logs revealed the lack of suitable calibration constants and no directly useful insights into mudrock characteristics were obtained (10).
- It was found that conventional cement bond logs were ineffective in defining "goodness of bond" in argillaceous rocks. Laboratory studies indicated that the bond was likely to be poor since there was no chemical bond and in the bore-holes the physical bond was influenced by the presence of drilling fluid. Some mineral constituents such as Opal CT, zeolite and pyrite are believed to react with cementitious materials and degrade the stability and integrity of the cement rock bonds (11).
- Thriving colonies of allochtoneous micro-organisms were identified within the bore-hole which was completed in the Oxford clay (12).

An aspect of utmost importance in the characterization of an argillaceous rock or formation, especially for understanding its geochemical, rheological, corrosive and thermal behaviour, is the knowledge of its mineral composition and the physico-chemistry of the clay pore-water. Researches have been carried out by ENEA on various types of Italian clays for the characterization of the pore-water. A squeezing system has been applied and variation in the composition of the pore water during the successive squeezing steps was less than differences observed between specimens of the same rock. The salinity of the pore-water range from 4 g.l^{-1} to 24 g.l^{-1} and the water chemistry is highly variable ranging from calcium-magnesium sulphate to sodium chloride or to calcium magnesium-sulphate with sodium chloride (see Fig. 7). No significant relation was found between clay mineral species and pore-water composition in the Italian Clays.

As already mentioned above, Boom Clay pore-water can be sampled in the underground experimental facility from holes equipped with piezometric screens beyond the zone of frozen clay. These samples are assumed to be fairly representative of "in situ" pore water. The chemistry of one of the samples is given in Table 5 and it is very similar to the one obtained previously by extrapolation from washing techniques. The Boom Clay pore water is clearly dominated by sodium and carbonate is mildly alkaline and very reducing. The salinity of the pore water in the Boom Clay is about 1.3 g.l^{-1}. In addition a significant proportion of organic substances (up to 340 mg.l^{-1}) is observed and appears to be an important factor in the trapping of radionuclides (13).

Table 5. Composition of Boom Clay pore water (mg.l^{-1})
          (HADES-location)

	sample - 258 m bgl	sample 251 m bgl
Na	405	391
K	11.1	10.9
Mg	2.61	2.34
Ca	5.53	3.25
Fe	0.74	1.8
Si	4.26	(5.13)*
SO_4	14.3	14.5
Cl	36.7	34
F	(3.5)*	3.7
C tot. carb.	927	826
pH	(8.68)*	8.38
Eh (mV)	- 200 to - 260	- 260 to - 280
DOC	206	/

(*) determined on a separate batch

At the Harwell site hydrochemical and mineralogical studies are rather focused on establishing profiles through the alternating sequence of argillaceous and permeable lithological units. As already mentioned above a pattern of falling heads from the surface to deeper aquifer appears to present a sink in the Corallian limestones. To test this hypothesis the chemistry of the clay pore-waters (obtained by squeezing clay samples) has been compared with the waters pumped from aquifers in the sequence. Trends in this system are evidenced by two indicators (see Fig. 8). They are based on averaged samples. It can be concluded that : 1) Water salinity increases with depth and with a maximum of concentration of Na and Ca in the Corallian Beds and 2) The muds display a higher Na and Ca content then the adjacent permeable formations.

The migration and behaviour of radionuclides in argillaceous rocks is discussed in a separate paper (13) and the reader is referred to that contribution for more details about interactions between radionuclides and the clay medium. However it is worthwhile to mention here the study of ENEA concerning the difference in some migration parameters among samples originating from various argillaceous basins in Italy. Results obtained show a clear correlation between grain size of the samples studied and the distribution coefficient for Cs (Fig. 9). Also some correlation was observed between the distribution coefficient for Cs and lithological characteristics (as shown in Table 6).

For the various argillaceous rocks beneath the Harwell site it was observed that their geotechnical characteristics reflect the petrology and the burial history. There is a clear relationship between sample depth and void ratio regardless of the difference in mineralogy as shown in Fig. 10. Other parameters like liquidity index, also show a depth relationship. At depth uniaxial compressive strength is greater than for the same mudrocks nearer the surface though the conditions for shear failure probably only exist in the upper part of the sequence.

Table 6.   Lithological characteristics and some migration parameters
            for Italian clays

basin	dominant clay species	% clay minus % sand	Kd (Cs) $(cm^3 \cdot g^{-1})$	D $(cm^2 \cdot s^{-1})$
Ausonia	illite	0	1335	9.8 E-9
Val d'Era	koalinite/illite	43	1924	8.1 E-9
Crotone	koalinite/smectite	16	2338	7.4 E-9
Monterotondo	smectite	26	3237	4.2 E-9
Vasto	smectite	50	4154	2.4 E-9

Several formations of stiff overconsolidated clays in Italy, are known to be intersected near the surface by a number of fractures. Such observations have led to questions on the persistence of such fractures at depth, their potential impact on the permeability of the formation and fault propagation in clays.

ISMES has approached this study of fractures in clays by three different but related viewpoints : mathematical modelling, experimental research, field observations.

Fracture generation and propagation in a clay medium has been modelled by assuming a loss of continuity due to instability in the stress field. In order to simulate numerically actual stresses, a constitutive model of the "Cam clay" family has been modified for application under high lithostatic pressure.

The experimental work has been carried out in order to provide the required support for theoretical development. Fracturing has been induced in clay samples under three different experimental set-ups. In a high pressure triaxial cell fractures have been generated under undrained conditions up to 9 MPa, that is the maximum confining stress allowed by the system. In addition, fracturing tests have been performed in plane strain cells, designed and built for this purpose. The cells have a mobile section geared to induce a differential displacement on a surface of the clay sample and a transparent side to allow visual observation of the specimen. Different kinds of discontinuities have been observed in these tests, depending on boundary conditions.

As far as field observations are concerned the emphasis has been placed on deep tunnels excavated in Italian argillaceous formations. The fracturing of deep clays seems not to be systematic at depth and rather limited to overconsolidated clays, such as those of Pliocene and Pleistocene. Argillaceous materials of other ages seem to be relatively free from permeable fracture zones.

The geomechanical characterization and behaviour of argillaceous formations has been addressed in various research programmes, but most progress seems to be made in the Belgian programme. This is obviously linked to the fact that at Mol the construction of the underground experimental facility raised particular questions about the "in situ" behaviour of clay. Although ground freezing of the clay was selected for the construction of the underground experimental facility, this technique is not considered as a conditioning technique for the construction of a final repository. However some efforts were undertaken to study the geomechanical and rheological behaviour of frozen clay. Compressive tests (uniaxial) and creep tests (uniaxial and triaxial) run on frozen clay samples at the Foraky-LGC laboratory (Lab. de Génie Civil of the Université Catholique de Louvain) allowed to get values for the rheological parametres of frozen clay. For stress levels having not led to failure the observed linear strain in frozen clay samples remains in the primary state of creep and the experimental curves of deformation are well approximated by Vyalov's law (Fig. 11). At higher stress levels, having led to failure Fish's law matches fairly well the observed strain in Boom Clay samples because the latter law brings into account the tertiary creep and is based upon the deformation rate of samples with time.

Laboratory tests on non frozen clay and parameters derived from these tests get their full significance if they are applied in the modelling of the rheological behaviour as it is now undertaken jointly by SCK/CEN and Foraky, in an exercise of modelling the behaviour of intact Boom clay under the excavation or tunneling conditions at the Mol site. For the time being Vyalov's law $\varepsilon = \varepsilon_0 + A \sigma^B t^C$ (ε = strain ; ε_0 = instantaneous strain ; σ = stress ; t = time ; A, B, C = experimental values) is tested for this purpose. The parameters A, B and C were selected statistically out of the results of triaxial creep tests performed at LGC. Using MPa and hours as units for respectively stress and time the following experimental parameter values were obtained : A = 3E-3 ; B = 0.65 ; C = 0.27. First preliminary results and comparisons between the experimental measurements (closure of a vertical borehole) and the simulated numerical behaviour (by finite elements) in function if time show that matching of the observations and calculational results may be obtained if variances of the A, B and C parameters are taken into account.

CONCLUSIONS

The main conclusions that can be drawn from the researches on the characterization and behaviour of argillaceous rocks are the following :
- Generic and site specific studies have shown that most of the data necessary for the assessments may be obtained by various approaches.
- Generic studies have also shown that among argillaceous rocks important variances may be found (in their composition and thus also in their properties).
- The difference observed between different determination approaches for the same characteristics (e.g. between laboratory and "in situ" conditions) may be understood and explained for most of the characteristics.
- Site specific studies demonstrate our ability to depict and understand the local conditions in a regional system.
- Still future research and tests are needed for covering the behaviour of argillaceous rocks with regard to the repository impact (e.g. in situ tests for α and γ-radiation, heat effects, rheological properties, etc.).

REFERENCES CITED

1. BRONDI, A., et al. (1985). Natural evolution argillaceous formations, this conference.
2. BRONDI, A. (1984). Deduzione dell caratteristiche dei bacini
   argillosi sepolti attraverso indagini superficiali di basso costo.
   Applicazione alle argille italiene. EUR - 9361 IT.,
3. ALEXANDER, J. and HOLMES, D.C. (1983). The local groundwater regime
   at the Harwell research site, FLPU 83 - 1.
4. BRIGHTMAN, M. A. and NOY, D.J. (in press). Finite element modelling
   of the groundwater flow around Harwell. FLPU 84 - 1.
5. BONNE, A. (1985). Clay : Evaluation of Geological Disposal of
   Radwaste in Belgium, Nucl. Europe, n° 2, 29 - 30, 1985
6. MANFROY, P., et al. (1985). Research Programmes in Underground
   Experimental Facilities (Konrad, Pasquasia, Mol), this conference.
7. NEERDAEL, B. (1981). Utilisation des diagraphies dans la détermination des caractéristiques lithologiques des argiles de Boom et leur
   intérêt pour les corrélations dans l'argile. In : Siting of Radioactive Waste Repositories in Geological Formations, NEA/OCDE,
   p. 133 - 149, (1981).
8. TASSONI, E. (1983). Smaltimento Geologico dei rifinti radioacttivi :
   Esperieuze in situ sulla dissipazione del calore in fromazioni
   argillose. ENEA - RT/PROT (83) 17.
9. BUYENS, M. and PUT, M. (1985). Heat Transfer experiments in Boom
   clay. Proc. 9th Eur. Conf. Thermophysical Properties, Sept. 84,
   Manchester.
10. BRIGHTMAN, M. A. (in press). Geophysical logging of the Harwell
    boreholes. FLPU 83 - 11.
11. ROBBINS, N. S. and MILODOWSKI, A. E. (1982). Design and evaluation
    of the Harwell borehole cement systems. ENPU 82 - 9.
12. CHRISTOFI, N. et al. (1983). The geomicrobiology of the Harwell and
    Altnabreac boreholes, ENPU - 83 - 4.
13. AVOGADRO, A. et al. (1985). The MIRAGE-project : physico-chemical
    behaviour of actinides and fission products in the geological
    environment. This conference.
14. HORSEMAN, S. et al. (1982). Basic geotechnical properties of core
    from the Harwell boreholes. FLPU 82 - 7.

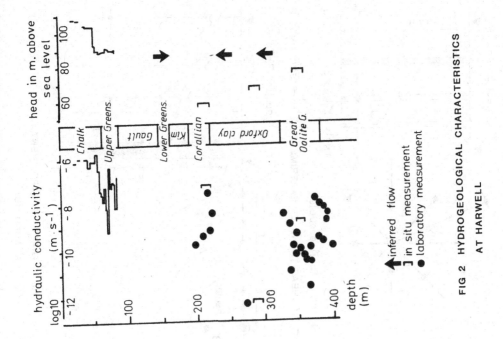

FIG 2 HYDROGEOLOGICAL CHARACTERISTICS
AT HARWELL

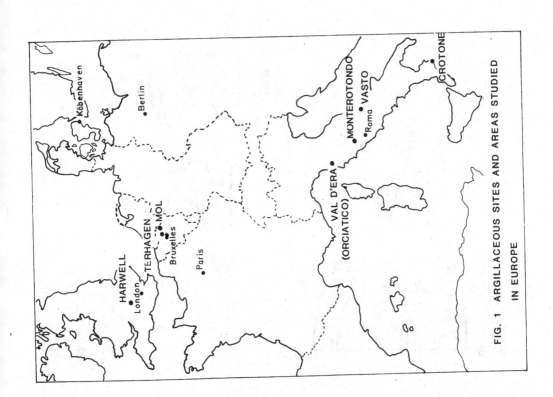

FIG. 1 ARGILLACEOUS SITES AND AREAS STUDIED
IN EUROPE

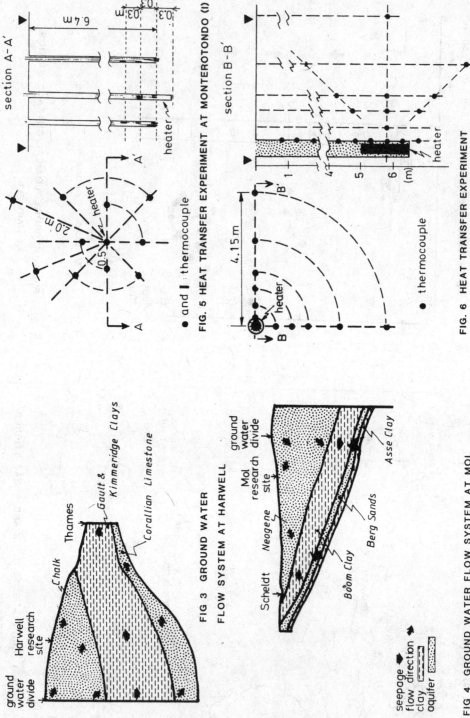

● and ▮ : thermocouple

FIG. 5 HEAT TRANSFER EXPERIMENT AT MONTEROTONDO (I)

● thermocouple

FIG. 6 HEAT TRANSFER EXPERIMENT AT TERHAGEN (B)

FIG 3 GROUND WATER FLOW SYSTEM AT HARWELL

seepage ▶
flow direction
clay
aquifer

FIG 4 GROUND WATER FLOW SYSTEM AT MOL

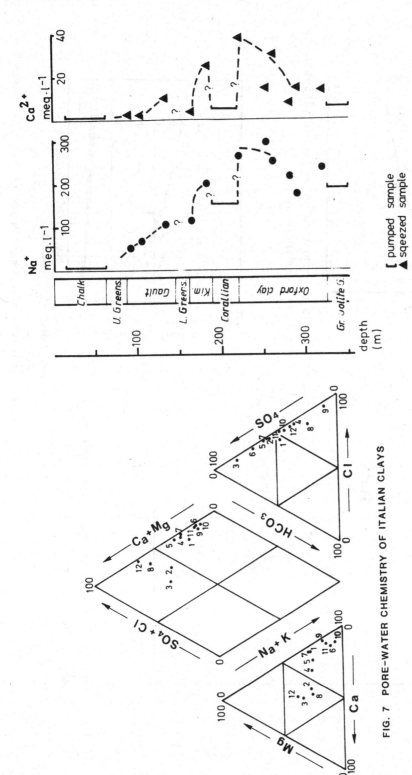

FIG 8 MUDROCK PORE-WATER CHEMISTRY PROFILE

AT HARWELL

FIG. 7 PORE-WATER CHEMISTRY OF ITALIAN CLAYS

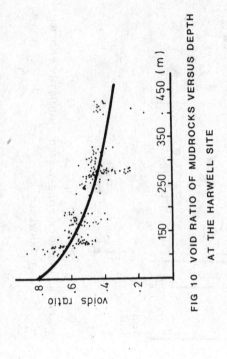

FIG 10 VOID RATIO OF MUDROCKS VERSUS DEPTH
        AT THE HARWELL SITE

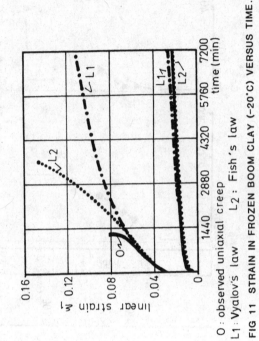

O : observed uniaxial creep
L1 : Vyalov's law   L2 : Fish's law
FIG 11 STRAIN IN FROZEN BOOM CLAY (-20°C) VERSUS TIME.

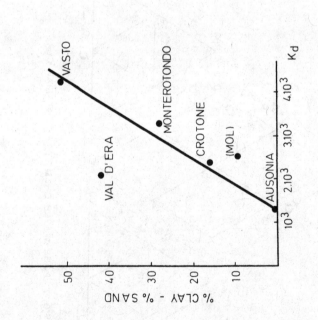

FIG 9 CORRELATION BETWEEN CESIUM RETENTION
        AND GRANULOMETRY

DISCUSSION

R. KOBAYASHI, JGC Corporation Tokyo

In the long term, will the chemical composition of the ground water vary?

A. BONNE, CEN/SCK Mol

There is some evidence, that such natural changes may occur in the long term, but the most immediate changes will be induced by man. Human activities constantly interfere with the ground water and this affects clay strata, which are often lying between aquifers, more than other formations.

W. BREWITZ, GSF Braunschweig

The age, origin and chemistry of the ground water are important issues in safety assessments. Which hydrochemical investigations have been carried out in view of such assessments, in particular in Harwell? I refer here to experiments analysing trace elements and isotopes in deep aquifers.

A. BONNE, CEN/SCK Mol

Such investigations are carried out in many programmes to confirm the hydrodynamical models by chemical measurements. The first requirement for such an approach is a good tracer or indicator. We have tried to confirm our models in this way in collaboration with UKAEA Harwell with uranium and although only few results are available at this time, no contradicting element has been found so far. Carbon-14 can also be used as tracer for dating, when you attempt to establish a ground water flow system. In Mol, C-14 could be present with an age of up to 50000 yrs. In the aquifer above the Boom clay, water is aged about 20000 yrs, in the one below, more than 30000 yrs. Water samples from the clay itself contain very important information. If the water is drained downward, as we assume, we should find samples of intermediate ages in the clay. We have indeed been able to date such samples from the clay layer between 20000 and 30000 years. But these preliminary results should be used with the necessary caution.

NATURAL EVOLUTION OF CLAY FORMATIONS

A. BRONDI, C. POLIZZANO, ENEA, Italy;
A. BONNE, CEN/SCK, Belgium; M. D'ALESSANDRO, JRC Ispra

Summary

Clay basins of EC territory fall into two main types:
1) basins developed in northern Europe around peneplanated rather stable basement (Mol, Harwell) displaying a very extended, regular series of homogeneous sand and clay layers;
2) basins of the mediterranean area (Italy), corresponding to an active orogenic belt, and containing sand-clay series of a thickness and homogeneity variable. Clays generally inherit the geochemical characters of the depositional and early diagenetic phase. They in fact maintain negative values of Eh, which cause trace elements cations, including radionuclides, to be fixed into stable forms. The penetrating oxigenated water may determine the Eh to turn to positive values. As demonstrated from natural evidence in Italy this perturbation may only affect a very thin thickness of clay at the surface and in short superficial fractures. Risk analysis conducted on Boom clay (Belgium) shows that tectonic evolution of the site and development of a new glacial phase may rule the stability of the waste deposit.

Clays are one of the geological formations which are being considered in the research and development activities coordinated by the European Community as a possible host rock for the disposal of radioactive waste. The countries which are actually engaged in the study of these formations are Belgium and Italy, although other Member States are researching specific aspects of clays. The morphology and structure, age and genesis of clay basins differ significantly from one part of European territory to another. Clay sedimentary formations in Northern Europe comprise basins resting on peneplane basements. They usually have the following characteristics:
1) geometric and lithological lateral continuity;
2) a high degree of geopetal homogeneity within the individual formation;
3) little tectonic and seismic activity to disturb their regularity;
4) interbedding in composite series of alternate clay and sand strata.

The Rupelian (Middle Oligocene) Boom clay in Belgium (Fig. 1) under the Mol site is a fairly good example of a depositional environment of this kind. The Boom clay was deposited in a very extensive sedimentary clay province covering Belgium, the Netherlands, Denmark and Northern Germany (13, 17). The lithological composition, structure and palaentological features of the Boom clay unquestionably point to a detrital marine environment with deposits at a depth of approximately 50 m. Studies of the rhythmicity, conducted throughout the whole of the formation, have shown that there is a high degree of continuity over large areas exceeding even 1 000 km^2. The rhythmicity (1) (2) was probably caused by pulsations, possibly of climatic origin, in the depositing of the sediment. The marked prevalence of clay minerals in the formation is unquestionably characteristic. Wherever they occur, beds in carbonates also proved to be very extensive in a horizontal direction.

The Harwell site (Fig. 2) in the United Kingdom is at the top of a sedimentary series including several clay formations. The depositional environment of the argillaceous rocks is in some respects cross between the marine coastal environment of Mol and the intracontinental environments found in Italy. They are nonetheless of the marine type. The Triassic, Jurassic and Cretaceous sedimentary series rests on a palaeo zoic basement, the London Platform, which is covered with carboniferous strata. The clay formation of this series changes progressively in a horizontal direction. The sedimentary series to which the formation belongs was deposited in a basin covering much of Southern England. Within this basin sedimentation was governed locally by a structural relief and by depressions (3) (4). The tectonic structures were active during the sedimentary phase which resulted in lateral and vertical lithological changes, accompanied by brief, intermittent phases of erosion. One mineralogical feature of the sedimentary series of the Harwell site is the existence of significant quantities of carbonates, which are a constituent characteristic of the era of deposition.

Italian clay basins are located in active orogenic areas. The clay series studied so far in terms of their suitability as nuclear waste repositories belong to the Plio-Pleistocenic epoch (1). The clay formations found in the western sector of the peninsula are located in parallel multiple series of deep intracontinental trenches of the graben type (Fig. 3), lying in the direction of the Appenines and having steep slopes. They comprise individual basins of a length of several dozen kilometres and a width of approximately ten. The sedimentary series contained in them vary in thickness from a few hundred to 2 000 metres and comprise intercalations of sand and clay. The individual layers of clay may be a few hundred metres in depth. Clays predominate and are fairly homogenous along the axial parts of the basins and in the centre of the series. The sand deposits become progressively more substantial at the sides of the basins and in the upper parts of the series. Transverse incisions in the opposite direction to the Appennines were the only routes of communication with the open sea. In the more inland parts of the territory the marine basins gradually gave way to lacustrine basins of similar size and shape. The great thickness of the sedimentary deposits in relation to the sizes of the basins was caused by persistent subsidence.

On the eastern slope of the Italian peninsula clay basins are found in the area of the present-day foretrench. They are bounded on the western side by the Appennine chain and are for the most part open to the sea on the eastern and southern sides. They are therefore asymmetrical in shape (Fig. 4) and extend for several hundred kilometres in the direction of the Appennines and for several dozen kilometres in a normal direction. As a result of continued subsidence, the Plio-Pleistocenic sedimentary series of the foreland are extremely thick, in some instances many thousands of metres. The banks of clay may be several hundred metres thick. The uplifting of the Appennine chain and the simultaneous lowering of the bed of the basins, consistently offset by the accumulation of sediment, caused sediment to be deposited in two ways (Fig. 5). The normal deposition of material from the continental mountain range resulted in the formation of sediments comprising granules gradually decreasing in size in relation to their distance from the coast. Because of the rapid accumulation caused by the rapid erosion of the Appennine chain as it rose, the deposits of sediment under the sea were frequently in a state of unstable equilibrium, with the result that there were landslides which caused turbidity currents which distributed the sand at great distances from the coast. Both deposition processes might occur at the same point under the sea, hence sand sediment might be interbedded with pelagic sediment.

The considerable lithological variability of the Appennine chain has resulted in the production of detritus which varies significantly in both mineralogical composition and granule size. Basins of the same shape therefore contain series which differ in respect of both the reciprocal frequency of banks of sand and clay, and their mineralogical composition (1).

The majority of the clay minerals commonly found in deposits of sediment are inherited from the various types of parent rock. The formation of new clay minerals is possible however, especially during the post-depositional phases of compacting and diagenesis. Significant processes which occur during the diagenetic phase in clay deposits are the, at least partial, dehydration of the intergranular spaces, the adsorption of potassium and magnesium from the interstitial solutions, the redistribution of ions within the crystal lattices and the massive build-up of illitic and chloritic minerals (10). All these phenomena gradually increase with their depth within the sediment.

Sudden variations in the degree of compacting of the clay along the vertical series indicate the chronological separation of sedimentation into several phases. An example is the deep clay in the subsoil of the Trisaia Nuclear Research Centre (3) which appears to be excessively consolidated in relation to the load of the overlying sedimentary series. This is because the clay in question originally had a thicker sedimentary series overlying it. The deep clay was presumably uplifted into an erosion area and subsequently lowered and covered by a thinner series.

The chemical composition of the clay medium is one of the factors controlling the behaviour of the trace elements, including radionuclides, which may be present in it. The seawater trapped in the clay at the time of deposition is probably affected by the palaeogeographical conditions existing at the time of sedimentation and, possibly to a greater extent, by the diagenetic processes which occur after the formation of the sediments and the subsequent geological history. In this paper we will merely state that the salinity of the fluids extracted from Italian clays (11) ranges from 0,45 to 24,5 g/l, while the chemical composition ranges from alkaline chlorides to sulphates of alkaline-earth metals.

The oxidation-reduction potential has a significant influence on the chemico-physical conditions controlling the mobility of the radionuclides. Clay, and sand not permeated by oxygenated water have very low or decidedly even negative Eh values, which, as is well known, ensures the immobility of all the radionuclides in cation form. Depending on the chemico-physical condition of the clays, theoretically at least, only the I 129 and the potential anionic complexes of the other radionuclides are mobile. This property is therefore one of the major advantages of clay formations as a repository for the long-term disposal of radioactive waste.

The negative chemico-physical condition is created in the first diagenesis of the sediment. The boundary between the water and sediment restricts or prevents exchanges of fluids and gives the two environments a high degree of independence. Depending on local conditions, the surface with Eh values=0 is above, level with or below the water-sediment boundary (9) (Fig. 6). Usually the Eh=0 surface is slightly below the boundary. The Eh values tend to decline lower down whereas the pH values tend to increase (Fig. 7). These variations in deposits of sediment are caused by colonies of bacteria which decompose organic substances. At the time of the deposition of the sediment, the organic substance is subjected to a process of biodegradation, the Kinetics of which vary significantly, depending on its origin and the availability of oxidizing agents, and thus, in the final analysis, on the amounts involved. As a rule, large organic polymers of limited use to bacteria are eventually (as the accumulation of sediment continues) trapped in the sediment at depths at which the bio-oxidization process becomes less efficient, due to the limited supply of oxidizing agents. This means that the activity of the bacteria is stratified in both quantitative and qualitative terms. However, it seems likely that the bacterial activity does not cease altogether but is merely very considerably retarded. The discovery of live sulphate-reducing bacteria both in deep sea sediments and in fossil Plio-Pleistocenic clays in the Tiber Valley (4) lends weight to this theory. How the bacteria use the organic substances has not been established however; as a rule they use small organic molecules.

To illustrate the important part played by diagenetic processes in determining the geochemical properties of clays, as outlined above, we will indicate some of the specific properties detected in the Mol and Harwell clays. The geochemical properties of the Boom clays at Mol (13, 17), which were deposited gradually in a normally oxidizing marine environment, gradually shifted towards reduction. This is proved by the existence of iron pyrites, which tends to be associated with the presence of organic materials. The carbonate trapped during the sedimentation process is mobilized and redistributed in the form of carbonatic nodules. The pore water must have been changed into the typical interstitial water of clays very early in the diagenetic process. Geomechanical and physico-chemical findings suggest that the Boom clay expelled considerable quantities of water during the compacting and diagenetic processes. As a result of these processes the Boom clay is on the borderline between plasticity and compactness.

Research on the Harwell series (6, 15) concentrated on the Oxford clay, which soon and swiftly moved gives way to strata which are more rich in carbonate and are also interbedded with more permeable strata. In this instance too, the burial and compacting of clay sediments caused the particles to become more closely packed. The formation of pyrites was likewise one of the first diagenetic processes. Geochemical and palaeontological surveys conducted on the whole series showed that orogenic activity played an important part in the diagenetic evolution, particularly in the more permeable strata; this is proved by the bacterial reduction of the sulphates and the precipitation of dolomite and silica.

A possible explanation of the calcite found to be dissolved in the more permeable strata is the percolation of fresh water following the uplifting of the series in the continental area.

The process which could destroy the reducing condition of the clay so that it cannot act as a geochemical barrier to the potential migration of radionuclides is oxidation, which could result from exposure to the atmosphere and penetration by oxygenated water. In practice, however, the limited permeability and porosity of the clay to some extent restrict the likelihood of this happening. Illustrations of some macroscopic examples are given below. These are based on in situ observations and research carried out in Italy, showing how in fact the oxidizing effects of the external environment on clay are somewhat limited even in cases where there is a high degree of exposure.

In situ observations of the state of oxidation of rocks are made easier by the fact that they are grey when in a state of reduction and yellow when in a state of oxidization. The uplifting of the clay-sand series in a continental environment and the consequent erosion processes cause oxygenated meteoric waters to penetrate the permeable layers. The layers which are most liable to oxidization are of course the sand layers which form the top of the sedimentary series. However, as a rule, only the thickness of sand affected by the seasonal and climatic fluctuations of the water table is oxidized to any considerable degree. Below the lowest point to which the water table falls, the sand retains the original grey colour which it acquired after deposition and which indicates the continuation of a reduced condition after the series was uplifted in a continental environment. The natural section visible in Volterra (Fig. 8) illustrates a fossil situation of this kind. In the frequent instances in which the fluctuation of the water table surface reaches the underlying clay, a sharp change in colour from yellow to grey indicates the boundary between the oxidized permeable sand sediments and the reduced clay layers (Fig. 9). At the point where the change in colour occurs there are frequently ochre-coloured crusts which are the product of the precipitation of the sand ions mobilized by oxidation at the point of contact with the sediment which can still have a negative Eh.

The upper part of the Pliocenic series in Orte in the Tiber Valley (4, 5) (Fig. 10) is currently being subjected to geochemical analysis, with a view to the interpretation and assessment of the effects on clay of penetration by oxygenated water. Above the massive layer of cultivated clay, the series comprises layers of sand interbedded with clay (Fig. 11). The sand is completely oxidized, whereas the clay has remained grey, indicating that it is still in a reduced condition. This contrast is also apparent in cases where there is a layer of clay within a totally oxidized bank of sand, which, despite its limited size, appears totally unaffected by oxidation.

As in the case outlined above, at the point of contact between the sand and clay there are almost always crusts of precipitates. The geochemical analyses which have been carried out (5) (Fig. 10) do not indicate significant differences between the content of trace elements of the main self-protected clay body and that of the thin layers immersed in the oxidized sand. The same can be said of some organic substances (2). Within the individual clay layers immersed in the sand the content of trace elements in the clay which is in contact with the oxidized sand does not differ from that in the central parts.

The Orte series has normal faults running through it (Fig. 10). The total absence of traces of oxidation at the sides of the fault plane noted in the cases observed to date (Fig. 12) is evidence of its impermeability to water and therefore its capacity for self-sealing. In contrast, the many simple fractures in the main clay mass have oxidized edges 1 cm thick. This phenomenon is very commonly found in clays which are exposed to erosion (Fig. 12). As a rule these fractures involve no displacement merely run downwards for some dozen metres before coming to an end (Fig. 13). Networks of fractures, always recognizable by the presence of oxidized edges, intersect the aforementioned vertical fractures at right-angles. The most likely explanation of the genesis of these fractures is that the clay mass increased in volume as a result of decompression, brought about by the disappearance, by erosion, of the superimposed and lateral masses. In this instance the fractures would be open, thus causing secondary permeability and permitting the penetration of the mass by surface water.

The reducing nature of clay as illustrated above, is essentially due to the presence of organic substance in rocks of this kind and to the capacity of these rocks to retain it, largely because of their impermeability. This correlation is indirectly confirmed by an example of the reverse situation observed in a quarry in Umbria. Open fractures running through a layer of oxidized sand have grey, reduced edges (Fig. 14). This phenomenon is caused by the presence of organic substances in the form of dead plant roots within the fractures. The progressive destruction of these substances by oxidation resulting from bacterial activity is responsible for the reduction of the surrounding environment.

Orogenic uplifting and the consequent erosion processes have resulted in the clay deposits on Italian territory being broken up. The examples quoted above show that the penetration of the clay system by oxygenated water is controlled by these factors. Striking examples of disturbance (Fig. 15) and erosion (Fig. 16) provide evidence of the speed of erosion processes in Italian basins. It is therefore obvious that the optimal conditions for ensuring the stability of formations in which nuclear waste is to be deposited are likely to be found in stable areas or areas of subsidence. Reliable calculations of the speed of uplifting and erosion should be made if it is decided to opt for areas which are subject to upward movements.

In this instance, account could be taken of the differing degrees of erosibility of clay formations associated with their granulometric and mineralogical composition (Fig. 17).

The JRC in Ispra and the CEN/SCK in Mol have cooperated closely on a "geoforecasting" experiment designed to determine the reliability of a hypothetical waste deposit in the Boom clay, in terms of the likelihood of its being released.

The experiment in question (7, 8, 17) entails the application of Fault Tree Analysis and proves that the most likely cause of environmental pollution, particularly over rather lengthy periods of time (100 000 – 250 000 years) would be the contamination of subterranean waters followed by the migration of radionuclides through the column of sediment. In the shorter term, the most likely cause would be the actual unearthing of the waste as a result of a new ice age.

With regard to the contamination of the acquifers, the most likely causes of release would be tectonic phenomena resulting in the creation of faults. More specifically, instantaneous displacements of 5-10 m appear to play a major role in determining whether the geological barrier is likely to give way. Faults of this size were considered to be capable of fracturing the clay mass, although it seems likely that the lithostatic load, together with the plasticity of the medium, would in a short space of time restore the physical integrity of the formation.

It is impossible to predict the extent of any possible percolation through the fault, since it would depend on many factors, such as the force of the stresses and the relation between the lithostatic and hydrostatic loads.

For 2 000 years following the closure of the deposit, the probability of release due to a fault would account for 50% of the overall probability of release; the remaining 50% would involve possible human activities (boring of wells, etc). The relative importance of the incidence of faults increases rapidly with the passage of time, so that a displacement through the deposit (probability of 10-7 - 10-8 per year), causing the acquifers above and below to come into contact with one another, followed by a flow which would leach the waste, would be the most likely cause of release. The subsequent conveyance of the radio-nuclides through the acquifer, the encroachment of the pollution into the catchment areas of wells and the use of water from the wells for domestic or irrigation purposes would be the stages whereby the radio-activity could affect human life.

Fault Tree Analysis highlighted a second possible cause of environ-mental pollution. This is the formation of an ice cap (12) (Fig. 18) over the site of the deposit and the geomorphological phenomena associ-ated with it. Some of these appear to be capable of destroying the deposit and causing the dispersal of the waste into the environment. This cause was studied on the basis of evidence of Saalian glaciation the effects of which can be seen in part of the Netherlands, which was affected by periglacial phenomena along the south-western edge of the Saalian segment. This possible cause of environmental pollution warrants consideration for the following reasons:

- The occurrence of a phenomenon of this kind is the most likely cause of the release of the waste directly in the biosphere; furthermore although the probability of this is slight, it cannot be overlooked.

- Although the site of the potential deposit is outside the area affected by the quaternary glacial ages, the effects of periglacial phenomena mere 100 km to the north of the potential site are very evident.

- The impact of glacial action appears to be capable of not merely causing the geological barrier to give way but also of dispersing the radioactive matter into the environment.
This possibility has been analysed and discussed at length in a separate report (8).

There are also other circumstances, associated with processes of various kinds, which are thought to be capable of causing pulverized matter to be released into the atmosphere. Although this could cause serious consequences in terms of doses inhaled. Such phenomena are highly unlikely (in the region of 10-11 - 10-12 per year) and therefore no attempt has been made to analyse them. There is one further possible cause of release which should be mentioned and this is land subsidence in the Netherlands, followed by the erosion of the covering sediment and the deposit by tidal gulleys. This possibility has not been considered in the model used to determine the probability of release, since the periods of time considered in the study in question are not long enough to allow the occurrence of such drastic process. For such a phenomenon to have this effect, it would have to be preceeded or followed by other erosion processes and to occur in particular circumstances, which render it highly improbable. As to the consequences in terms of the doses resulting from this method of release, the level of pollution of the marine environment would be very low because it would be diluted. This is yet another reason why this phenomenon has not been considered.

In conclusion, all the studies which have been made of the evolu-tion of clay formations point to the fact that such formations would in themselves, be suitable for the long term isolation of radioactive waste from the biosphere and that, should the physical and physico-chemical barriers give way, this would be due to circumstances affecting the geo-logical environments of which they are part. The selection of sites which would be suitable for the construction of waste deposits must therefore be based first and foremost on a thorough examination of external circumstances. Risk analysis is one of the means of studying the evolution processes to which the geological structures of the individual sites considered are subject. Assessments of the general and local circumstances are in any event a vital means of determining the accuracy and suitability of risk analyses associated with the sites considered.

BIBLIOGRAPHY

1. BRONDI, A. (1984). Deduzione delle caratteristiche dei bacini argillosi sepolti attraverso indagini superficiali di basso costo. Applicazione alle argille italiane. CCE. Scienze e tecniche nucleari. EUR Report 9301 . Brussels

2. CALDERONI, G. (1984). Istituto di Geochimica dell'Università di Roma. (lecture - currently being printed)

3. CCE-CNEN (1980). Contratto 020-77-1 WASI. Final report Rome.

4. CCE-ENEA (1984). Ricerche sulle barriere geochimiche delle formazioni argillose. Studio di un fronte di cava ad Orte (Valle del Tevere). II° Rapporto semestrale 1983. Contratto 375-83-7 WASI. Rome.

5. CCE-ENEA (1984). Ricerche sulle barriere geochimiche delle formazioni argillose. Studio di un fronte di cava ad Orte (Valle del Tevere). I° Rapporto semstrale 1984. Contratto 375-83-7 WASI. Rome.

6. COPE, J.C.W. et al. (1980). A correlation of Jurassic rocks in the British Isles. Geol. Soc. London. Spec. Report, n° 14, 73 pp.

7. D'ALESSANDRO, M., BONNE, A. 1981. Radioactive waste disposal into a plastic clay formation: a site specific exercise of probabilistic assesment of geological containment. Radioactive Waste Management, Vol.2, Harwood Acad. Publ. Paris.

8. D'ALESSANDRO, M., BONNE, A. (1983). Report EUR-8709 EN. Brussels.

9. D'ARGENIO, B., IPPOLITO, F., PESCATORE, S., T. (1979). Elementi di Petrologia. Liguori Ed.. Naples.

10. DUNOYER DE SEGONZAC, G. (1970). The transformation of clay minerals during diagenesis and low grade metamorfism: a review. Sedimentology 15.

11. GRAGNANI, R. et al. (1985). Determinazione delle caratteristiche geochimiche delle acque interstiziali di argille plio-pleistoceniche. Contratto CCE-ENEA 151-81-7 WASI. ENEA. Rome.

12. HEREMANS, R. et al. (1984) R and D programme on radioactive waste disposal into a clay formation. Contract N° 144-80-7 WASB. CCE Brussels.

13. NEERDAEL, B. et al. (1981). Utilisation des diagraphes dans la determination des caracteristiques lithologiques des argiles de Boom et leur interet pour les correlations dans l'argile. Document présenté lors de la réunion de travail sur le choix des sites des dépôts des déchets radioactifs dans des formations geologiques. OECD/AEN, Paris.

14. SALTELLI, A., ANTONIOLI, F. (1985). Radioactive Waste Disposal in Clay Formations: A systematic approach to the problem of fractures and faults permeability; Radioactive Waste Management and the Nuclear Fuel Cycle. Richaland Wash. USA.

15. TALBOT, M. R. (1980). Major sedimentary cycles in the Corallian Beds (Oxfordian) of Southern England. Palaeogeog., Palaeoclimat., Palaecr..

16. VANDENBERGHE, N. (1978). Sedimentology of the Boom clay (Rupelian) in Belgium. Verhandl. Kon. Acad. Wetenschappen, Letteren and Schone Kunsten van Belgie, Klasse Wetenschappen, Jaargang XL, n° 147.
17. VANDENBERGHE, N. et al. (1983). Proc. of NEA Workshop on Radio-nuclide. Release Scenarios for Geologic Repositories, 8-12 Sept. Paris.

ACKNOWLEDGEMENTS

We would like to thank Dr Fiorenzo Benvegnù and Dr Giovanni Giannotti for their assistance with the preparation of this paper and the appended drawings.

Legend

	sand
	clay
	lignite
	carbonatic rock
	basement
	fracture
	fault trace
	oxidation band
	precipitation crust
G	grey colour
Y	yellow colour

Different types of clay basins in some EC countries

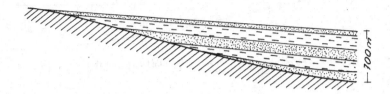

Fig. 1 - Mol, Belgium. Clay basin around and on a peneplanated stable basa-
ment. The sedimentary levels are quite regular in shape, thickness and composition.

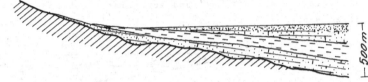

Fig. 2 - Harwell, Great Britain. Sediments deposited on a tectonically active at the time and moderately irregular basemement. Lateral and vertical lithological variations are frequent.

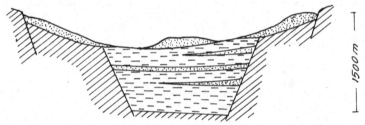

Fig. 3 - Western Italy. Horst-graben basins. Sediments deposited in deep
trenches. Sand levels are frequently intercalated to clay layers. Disjun-
ctive tectonic activity is going on.

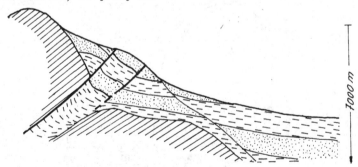

Fig. 4 - Eastern Italy. Sedimentary basins located in a foreland zone. Lateral variations may be important with regard to both thickness and composition.
Compressive tectonic activity is going on.

Fig. 5 - Combined effect of normal and turbiditic deposition in a clay basin.

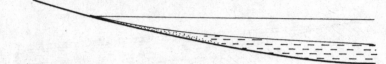

Fig. 5.1. - Normal deposition: sand accumulation inshore; clay accumulation offshore.

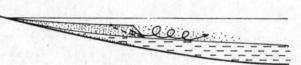

Fig. 5.2. - Turbiditic deposition: sand accumulates offshore because of slumping deplacement.

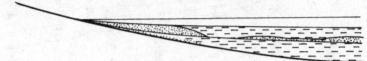

Fig. 5.3. - As a consequence of turbiditic deposition sand and clay are interbedded offshore.

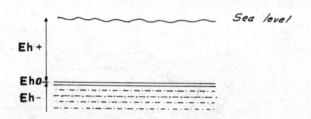

Fig. 6 - The sea floor usually separates positive Eh values in the water from negative ones within the sediments.

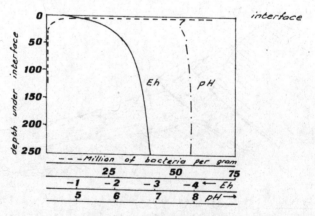

Fig. 7 - Variations of Eh, pH and bacteria count below the seafloor.

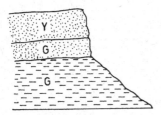

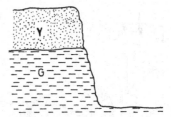

Fig. 8 - Volterra. The lower part of sandy deposits still shows the original grey colour inherited by reducing environment.

Fig. 9 - The lithological boundary usually sharply separates oxidized upper sands from underlying reduced clay.

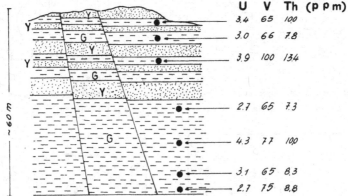

U	V	Th (p p m)
3.4	65	100
3.0	66	7.8
3.9	100	13.4
2.7	65	7.3
4.3	7.7	100
3.1	65	8.3
2.7	7.5	8.8

Fig. 10 - Orte. No significant variation in the content of some trace elements between the massive lower clay deposit and the thin upper clay levels interbedded with oxidized sandy banks.

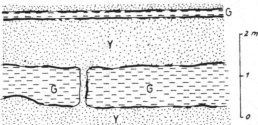

Fig. 11 - Orte. Clay levels interbedded with oxidized sands maintain the original grey colour. Elements leached from oxidized sands precipitate in form of crust at the contact, with clay where reducing conditions occur.

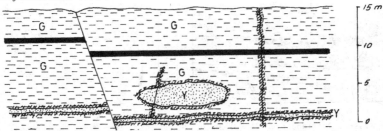

Fig. 12 - Narni. Oxidation bands in clay along fractures and boundaries of sand lenses. No oxidation exists along the true fault.

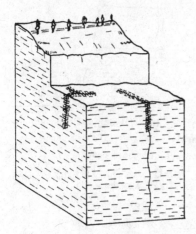

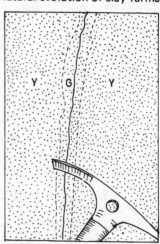

Fig. 13 - S. Quirico quarry. The oxidation band along fractures stops some tens of meters from the surface.

Fig. 14 - Narni. An opposite case: organic matter in the fracture, supplied by tree roots, causes the oxidized environment in the sand to be turned to a reduced condition.

Fig. 15 - Abruzzi foreland. Uplifting of the Appennines has continuosly perturbed the sedimentary series since deposition.

Fig. 16 - Atri. Degradation effects of accelerated erosion caused by the orogenic uplift.

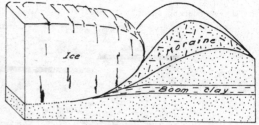

Fig. 17 - Vasto. The landscape is rather flat and scarcely eroded because of the smectitic content of the clay.

Fig. 18 - Mol. Hypothetical disruptive scenario of the clay deposit as a consequence of a possible new glacial age.

DISCUSSION

UNIDENTIFIED SPEAKER

About ten years ago a fairly detailed study of plio-quarternary clays under the Trisaïa establishment in the South of Italy had been carried out. This formation is about 800-900 m thick according to the test drillings. The series is extremely homogeneous. This subsidence is well known to European and Italian geologists. Has this study been continued or why has it been stopped, as the results of the test series showed a very promising formation for a repository?

A. BRONDI, ENEA Casaccia

These studies were not really abandoned, but as clay formations are present practically in all regions of Italy, it was found appropriate to start a general screening of all these layers. In this way, our future choice would be less influenced by the availability of the infrastructure and the location of the waste producing facilities. The series under Trisaïa is very homogeneous, continuous and of great thickness. Once our systematic screening of sites is completed, in about 5-10 yrs, this area will of course also be considered.

SESSION VII

DISPOSAL IN GRANITE AND INTO THE SEABED

Introduction by the session chairman

A study of the characteristics of very deep granite :
test in underground laboratories and using boreholes

Associated R & D : radionuclide movement through and
mechanical properties of fractured rock

Backfilling and sealing of repositories and access
shafts and galleries in clay, granite and salt formations

The feasibility of heat generating waste disposal into
deep ocean sedimentary formations

Costs and ways of financing of the geological disposal
of radioactive waste

INTRODUCTION BY THE SESSION CHAIRMAN

This session covers disposal into granite formations and seabed but when looking at the papers, it is apparent that there is a much wider scope than that and those who are concerned with clay and salt will also be interested. The work which is to be discussed about backfilling and sealing is relevant to all types of geological formations. The work which is undertaken with respect to seabed is quite pertinent to disposal of waste in argilous formations, and the final section dealing with maybe the most critical factor of all, how to finance disposal facilities, should be of interest to everybody. The first papers will indeed address the question of disposal in granite. There is a problem of water ingress, but the effects of hydrostatic pressure and heating could assist in closing fissures.

The work has been progressing for a number of years now; in the EEC it has been somewhat inhibited by the lack of underground laboratories or test disposal facilities in granite when compared with salt or clay. However, outside the EEC programme there has been an extensive amount of work involving more detailed studies. In particular, the American work in Nevada must be mentioned, which is the only instance where actual fuel has been placed in a deep underground system, and has remained in place for several years. It has given one of the best demonstrations available of the effectiveness of deep disposal. In addition, the European Community is now working closely with the Canadians who are developing their own underground laboratory in granite, with the Swiss in the Grimsel mine. In Sweden many European countries are collaborating in experiments within the Stripa mine. So, only part of the worldwide programme is presented here.

The papers on granite cover a wide range of topics. The seabed work which will then follow, has a wider implication than just deep seabed area itself. It could, for instance, be relevant to coastal repositories or to clay systems. Finally, we will consider crucial matters relating to institutional control and financing.

A STUDY OF THE CHARACTERISTICS OF VERY DEEP GRANITE:

Test in underground laboratories and using boreholes

A. BARBREAU; N.R. BRERETON; P. PEAUDECERF
CEA-IPSN/DPT; British Geological Survey; BRGM

Summary

A research programme has been carried out in France and in the United Kingdom concerning the characteristics of deep granite. In France, two boreholes have been drilled in the Auriat Massif (1 000 m and 500 m) for the investigation of petrographical, structural, mechanical and hydrogeological properties. If the granite is highly fractured, permeability is very low (10^{-11} – 10^{-12} m/s) at depth because of backfilling of the cracks. A comparative study of the evolution of fracturing from the surface downwards has been performed in tunnels and in two granitic massifs in order to set guidelines for extrapolating shallow data to depth. This study has shown that it is possible to estimate deep fracturing of a granite massif with the help of surface data. A study of the influence of the scale effect on the measured values of the permeability and dispersion coefficients in underway in a underground laboratory in a uranium mine at Fanay-Augères. This experiment comprises different permeability tests in radial boreholes and the injection of tracers at increasing distances. In the United Kingdom, the Strath Halladale granite in the region around Altnabreac has been studied with a view to establishing a clear understanding of the mechanisms which control the migration of aqueous solutions. There is a statistically significant exponential decrease in hydraulic conductivity with depth. Groundwater is very largely meteoric in origin.

INTRODUCTION

Granitic formations have been studied for a number of years in France and the UK to assess their suitability for future disposal of radioactive waste. For the most part at the research and development stage, these studies have concentrated on the geological, hydrogeological and mechanical properties of deep granite in order to assemble sufficient information for, in particular, an evaluation of the rardioactivity containment capacity of this kind of rock.

1. RESEARCH CONDUCTED IN FRANCE

1.1 Borehole studies (1)

After carrying out an inventory, and surface examination, of a number of granite batholiths in various parts of France, the Institut de Protection et de Sûreté Nucléaire (IPSN) selected a granitic massif in the Auriat region of the Massif Central and commissioned the Bureau de Recherches Géologiques et Minières (BRGM) to drill two deep boreholes, 10 m apart, to 1 003 and 500 m respectively. Using these boreholes, it has been possible to study the structural, mechanical, petrographical and hydrogeological characteristics of the granite together with the geochemical nature of the water at these depths. The following in-situ measurements were made : petrographical analysis of oriented cores, well logging, slug tests, injection tests between packers and tests of fil-tration speeds. Laboratory work comprised chemical and mineralogical analysis of the granite, structural analyses of cores and geotechnical measurements of the extent of fracturing after heating and cooling (the temperature ranging from 100 to 250°C). In addition, potential altera-tion of the granite was studied and chemical and isotopic analyses of deep water carried out.

    - The principal results were as follows :

- **Petrography** : two types of granite were identified, one a pseudopor-phyrytic monzonitic granite, highly granular and with frequent in-clusions of large crystals, and the other a fine-grained monzonitic granite occurring in both pseudoporphyrytic and non-pseudoporphyrytic forms.

- **Fracturing** : of the many fractures, the major ones were studied to determine their form, direction and age. Seven families of fractures were identified, showing a constant orientation over the whole length of the borehole. No significant fall in fracture frequency with depth was observed (Fig. 1). Quartz and calcite infill was usually found in the major faults and three types of hydrothermal backfilling were noted in the horizontal and smaller fractures : clays (montmorillonite and illite), calcite and quartz. There was none of the kaolinite usually typical of minerals produced by meteoric alteration. Frac-tures without infilling are generally closed as is confirmed by high values for the speed of P waves and for the resistivity of the granite.

- **Hydrogeology** : slug tests and tests using injections between packers have proved the existence of the following decline in average per-meability values between the surface and a depth of 1 000 m :
    from 0   to 50 m $K = 6 . 10^{-9}$ m/s
    from 60   to 250 m $K = 7 . 10^{-11}$ m/s
    from 250 to 540 m, K is between $3.4 . 10^{-11}$ m/s (slug test) and $2.7.10^{-10}$ m/s (injection)
    from 540 to 1 003 $K < 1 . 10^{-12}$ m/s
Greater permeability was however observed in two fractures :
between 298 and 308 m $K = 3 . 10^{-9}$ m/s
between 365 and 375 m $K = 9.5 . 10^{-10}$ m/s.

    No hydraulic conductivity was observed between the two boreholes (500 and 1 003 m) (between 75 and 500 m in depth) even though they were only 10 m apart. A tracing experiment between the two boreholes (using rhodamine Wt and uranine) at a 400 % gradient yielded no results what-ever over a period of approximately three months. Total porosity is lower than 1 % at distances of greater than 500 m. The very low per-meability of the Auriat granite is due to filling of the fissures.

- **Thermal behaviour** : three samples of deep granite heated to 100, 150 and 200°C showed a marked increase in pore spaces above a temperature of 150°C.

- **Alterability of the granite** : laboratory tests of relative solution speeds ranked the minerals as follows : albite > chlorite > orthosite. The following minerals are likely to precipitate out : hydroxides of iron, calcite and dolomite, montmorillonite, talc and illite.

- **Geotechnical and thermal characteristics** : there is little variation with depth in the properties of the two facies of granite; specific gravity 2 630 kg/m^3, compression resistance 160 Mpa, traction resistance 160 MPa, Young's modulus 72 800 mPa, Poisson's ratio 0.31, thermal conductivity 3 to 3.5 W/m°C, linear expansion ratio calculated at 6.7 x 10^{-6}/°C. In other words, this is high quality granite. The temperature at 1 000 m is 39°C, giving an average thermal gradient of 29.16°C per km.

- **Geochemistry** : geochemical tests on deep waters were carried out on samples collected using a high-pressure sampling system while the pH was measured in-situ using a special probe. The upper levels of the borehole (down to about 700 m) showed pH values of between 6.5 and 7.5 while in the lower part this rose to between 8 and 8.5. In general, the sampled water seemed to have come from fairly close to the surface.

1.2 A comparative study of deep and surface fracturing in granite (2, 3, 4, 5, 6)

Research in two successive phases (contracts No 148-80-7 and 416-83-7) has focused on the study of deep fracturing in granite rock massifs and has been carried out using resources committed jointly by the French Commissariat à l'Energie Atomique and by the Bureau de Recherches Géologiques et Minières. This work aimed at studying changes in fracturing in granite between the surface and areas of great depth in order to develop guidelines for extrapolating surface data to greater depth so that they could then be used in studies of storage sites. The research work was carried out in the following stages :

- a literature search of available information on deep fracturing in granite. In particular, an analysis was made of structural data from two major alpine tunnels (those of Mont-Blanc and Arc-Isère), from the 1 000 m borehole at Auriat and from the granite massif at Bassiès in the Pyrenees;

- a structural analysis carried out in the granite mines at Saint Sylvestre (in the western part of the Massif Central) where galleries at different depths make it possible to study the vertical evolution of fracturing;

- a statistical and geostatistical study of structural data previously obtained in the Fanay-Augères mine.

Together, this work has led to the conclusions that granite massifs, as is the case with surrounding rocks, are traversed by major faults, several kilometres in length, accompanied by small fractures.

Although not absolutely constant, the orientation and density of these small fractures is not significantly influenced by depth. In particular, the geostatistical study, which concentrated on the "density of small fractures" variable, produced the following results :

- The spatial distribution of fracture densities is not random but presents instead a certain level of spatial correlation known as "structure".

- The geologist's approach is an established one, namely that of studying fracture densities by differentiating those directional families or fracture systems that appeared during given tectonic episodes.

- The study of "structure" (variographic analyses) demonstrates ranges (the maximum distance at which a correlation still exists with fracture density values) of approximately 100 to 250 m depending on the tectonic episode involved together with the likelihood that fractures are grouped in "structures" of varying and "nested" scales involving "structures" tens, hundreds and without doubt also thousands of metres in extent.

- It proved possible to draw up fracture density charts (with isolines): data-smoothing charts which allow estimates to be made (standard deviation charts) and simulation maps giving a detailed picture of variations in fracture density.

- The prerequisites for the analyses of fracturation, that are essential if representative geostatistical results are to be obtained, were determined.

By contrast, the results of structural analyses of the massifs studied suggest that the small fractures (persisting for metres or tens of metres) and mesoscale fractures (tens to hundreds of metres) appeared before the major faults. The result has been a completely new model for fracturing which , is valid both in the vertical and the horizontal plane : major faults of well-defined thickness varying from a metre to several tens of metres cross most of the zones with a high density of small and medium sized fractures, but they also sometimes cross zones with a low fracture density. Some zones with high densities of small and medium sized fractures are, however, to be found in areas not crossed by major faults.

In terms of studies on deep storage sites in granite for radioactive waste it can be concluded that it is possible to assess the level of fracturing at a depth of 1 000 m using the results of fracture studies carried out on the surface over some tens of square kilometres and also that it is necessary to locate not only the major faults but also to study the spatial distribution of the orientation and density of small and medium-sized fractures, these being studied directional family by directional family and system by system.

Both at the surface and at depth, interpolation of the results obtained with virtually ramdom or linear surveys of minor fracturing can be carried out using statistical and geostatistical techniques to produce maps and sections for estimating and simulating fracture orientation and density. The fracture distribution model produced can provide guidelines for data sampling surveys and simplify the interpolations which have to be made between the surface and locations at depth.

1.3 Study of the scale-effect in the underground laboratory at Fanay-Augères

The IPSN's Département de Protection Technique has been responsible for carrying out a number of methodological and fundamental studies on the properties of fissured rock, especially granites. A site with suitable conditions was therefore sought with the final choice being the mine at Fanay-Augères, 20 km from Limoges, in the Crouzille mining region, which is run by the Cogema Company of CEA. This mine is in the Saint Sylvestre (see Fig. 1) granite massif which is a medium to large-grained two-mica leucogranite in a 20 x 30 km mass dating from the Namurian age approximately 320 million years ago. This massif is intrusive in metamorphic formation and crossed by major NW/SE faults which contain uranium deposits. The well documented nature of this mine is due not only to mining-linked research but also to the structural and hydrogeological work carried out by the BRGM and the Paris School of Mines for the DGRST (General Delegation for Scientific and Technical Research) and, in the field of fracture analysis, for CEA-IPSN.

The experimental zone finally selected is a stretch , approximately 100 m in length, of a gallery at the so-called 320 level, which is actually about 170 m below the ground, the topography in this region being very irregular (cf figures 2, 3 and 4).

1.3.1 Evaluatory study of the site (7) : It was then decided to commission the BRGM and the Paris School of Mines to carry out a number of studies to provide a definitive assessment of the site, comprising :

- A detailed structural analysis :

This involved a systematic survey of all discontinuities on the eastern wall of the gallery over a height of 2 m, these then being studied in great detail. In this way, 14 fracture families were identified. Most of these were not persistent (only 6 % exceeded 3.5 m) and the majority were isolated; two thirds had no visible opening, 44 % appeared to be wet and 5 of the 14 families seemed likely to be involved in hydraulic conductivity.

- Borehole tests :

Six cored bores varying in length from 8 to 16.70 m were drilled in the rock surrounding the gallery in order to determine the position of major tectonic features nearby together with the characteristics of natural fracturing and the hydraulic state of the rock mass. All six bores were fitted with piezometers to measure the massif's hydraulic pressure near the end of the bore holes.

Extremely consistent pressure measurements over time, varying from a few decibars to approximately 2 bars, show that the massif around the gallery is thoroughly saturated although pressures are low as a result of disruption from mining work. Pressure distribution is very even.

- Discharge measurement in the gallery :

Suitable equipment was installed in the gallery to measure water discharge flowing from the experimental section. The measured discharge rate (5.75 l. perr minute), together with borehole observations, makes it possible to estimate the permeability of the granitic massife around the gallery as being in the order of 10^{-7} to 10^{-8} m per second.

1.3.2. The research programme on "scale effects" : This work being completed, it was decided that the site was well enough evaluated for the real research phase to begin. The first stage, in which the BRGM will be working with the Laboratoire d'Informatique Géologique of the Paris School of Mines, involves the study of scale-effects on the permeability and the dispersion coefficient of the fractured rock.

This approach was adopted because many of the measuring techniques used in hydrogeology do not yield a good understanding of the behaviour of heterogeneous and almost impermeable areas in a large scale massif. It is therefore necessary to determine what influence scale-effects have on the measured values.

The study is based on measuring discharges and pressures at increasing distances from the gallery by drilling out from it radial boreholes fitted with packers in series that isolate a number of sections of each borehole to form independent chambers linked by tuves to the gallery. These measurements, carried out at a large number of points, should make it possible to determine the permeability tensor.

It is intended to use the same facilities to study hydraulic transmission in the rock by injecting appropriate tracers into the borehole at ever greater distances and then recovering them in the gallery. Moreover, it has been possible to correlate the characteristics of the fracture network, the state of stress of the massif and local permeability as established by water tests in these same boreholes. The research programme comprises the following operations :

- **Permeability studies** (phase 2A of the programme)
  Drilling 10 cored boreholes, each 50 m long, radiating out from gallery 320 :
  These bores are split into three groups, the central one having four boreholes and the two end groups being formed of three boreholes (8) (Fig. 2).

Structural survey of all discontinuities apparent in the oriented cores taken during drilling. Identification of zones of water inflow in the boreholes by placing a single packer in the borehole to allow overall measurements of water inflow in that part of the borehole between the end of the hole and the packer, approximately 2 m stretches being measured at any one time. By subtraction, it is possible to arrive at partial inflow figures for each section of the borehole.

Permeability within the massif is tested by injecting the water into a section of the boreholes sealed off by two packers. Tests were carried out first in approwimately 10 m stretches of the borehole and then in stretches of about 2 m.

The installed arrangement has seven packes in each borehole forming seven separate measuring chambers with the exception of the bores F8 and F9 which have only six and rather different lengths of measuring chamber. Each chamber is linked with the gallery by two 6/4 mm rilsan pipes which allow waterr to be drained off, pressure to be measured and tracers to be injected. Once the multiple-packer system has been installed in each borehole, the 70 measuring chambers are brought up to pressure and it is then possible to measure discharges and stabilized pressures.

The permeability values obtained during the various tests ranged from 1.10^{-6} to 3.10^{-10} m per second (the lowest sensitivity level for the equipement used) at 10 m, and from 5.10^{-5} to a value lower than 6.10^{-10} m per second at 2 m and 2.5 m. Figure 3, in the form of a permeability log shows the results obtained in borehole F 10. The considerable variation in permeability within the massif makes it likely that there are localized flows.

From a structural point of view, there was no direct link between the fracture density within the massif and measured permeability. In particular, the most fractured zones do not turn out to be the most permeable ones. Permeability is essentially a function of the openness of certain fractures and the degree to which they interconnect with others.

Piezometric measuring equipment has been installed in the gallery since the beginning of February 1985. Similarly, an impermeable floor was laid to allow measurement of the discharges into each of the three subsections of the experimental area. Since the beginning of February, piezometric measurements have been carried out at regular intervals in the 68 chambers now created. These measurements will be continued until stable values are recorded.

Initial assessments were based on data collected during the pilot phase and used in an analytical representation of flow that assumes a homogenous and anisotropic medium. This gives an average permeability value of 2.10^{-8} m per second, the anisotropic ratio being 0.3, which the highest permeability being at a orientation of 40° from the horizontal.

A second analysis was made using the METIS bidimensional finite element model which represents a vertical cross section of the massif at right angles to the axis of the gallery and also takes into account other neighbouring mining galleries. The anisotropic ratio remains at 0.3 but matching observed discharges required average permeability to be reduced to 10^{-8} m per second.

- **Research into the effect of scale on dispersion** (phase 2B of the programme).

This ongoing experiment involves a study, using tracers, of the nature of hydraulic conductivity in the rock around the gallery. The tracers are injected into the individual chambers constructed during phase 2A with the following operations then being carried out :

- Tracer tests by injecting tracers into chambers located further and further away from the gallery (scale effects). Different tracers will be used simultaneously to shorten the time-span of the experiment. The tracers will be collected in the gallery itself, with no attempt being made to locate or sample directly from individual fractures because it is precisely the overall effect that the experiment is designed to study.

- Studies of the state of stress in the massif (by hydraulic fracturing in the boreholes drilled during the gallery assessment phase or by overcoring).

- Interpretation of tracing measurements in terms of a continuous three-dimensional model representing conductivity around the gallery and the calculation of variation in the longitudinal dispersion coefficient as a function of distance from the point of tracer injection.

2. STUDIES CARRIED OUT AT ALTNABREAC IN THE UNITED KINGDOM

The study of the Strath Halladale Granite in the region around Altnabreac in Caithness, Northern Scotland, was conceived as being part of a wider research programme into the feasibility of the disposal of high-level radioactive waste into geological formations (9). A need was identified for comprehensive site specific field data from a range of potentially appropriate geological environments, directed principally to the modelling of groundwater flow and the geochemistry of waste mobilisation and migration. In late 1981, the UK Government changed its policy on research into the disposal of high levelradioactive waste and accordingly the drilling programme was abandoned. By that time borehole drilling had taken place only at Altnabreac and so the need for comprehensive data remains (10).

The overall approach adopted during the investigations at Altnabreac was to develop an understanding of the relationships between the petrology and fabric of the rocks, and the hydraulics and geochemistry of the groundwaters. Because the site was being investigated for research purposes, rather than for repository development purposes, and also because Altnabreac was intended to be the first of many sites where a deep borehole investigation philosophy would be developed, the research programme was not designed to answer site specific questions but rather to give an initial feel for the problem.

The Altnabreac area consists of monotonous gently rolling to flat country between 100 and 200 metres above mean sea level which largely owes its geomorphological characteristics to denudation by glacial processes (11). The terrain is mostly covered by peat, which is often underlain by glacial and fluvio-glacial moraine, resulting in poor surface exposure of the underlying bedrocks. The geology consists of a large easterly dipping granite sheet, the Strath Halladale Granite which was emplaced within Moine metasediments during the final stages of deformation (13, 14) (Fig. 4).

Following a preliminary reconnaissance geophysical survey during the drilling phase (15), follow up ground magnetic, very low frequency (VLF) electromagnetic, and refraction seismic surveys were carried out (16). The VLF results were able to define a number of conductive zones which could be related to the presence of fractures. Regional gravity and aeromagnetic data were interpreted in terms of the Strath Halladale Granite being a relatively low density and moderately magnetic body extending with depth to between 3 km and 4 km, and also, at a shallow depth, well to the east of the mapped contact between the granite and the country rocks.

Three fully cored boreholes were drilled in the area, between 2 km and 4 km apart, to depths of about 300 m (12). A further twenty four shallow boreholes were drilled to depths of about 40 m over a much wider area to provide additional regional geological information.

A weathering profile was seen to exist in the borehole cores from surface down to about 40 m (14). In thin section the major effect of the weathering was identified as hydration of micas and oxidation of iron oxides, most probably as a result of midly acidic solutions recharging from the surface. A detailed geochemical analysis of the rocks and of the fracture infilling materials was carried out by Storey and Lintern (28) (31). They concluded that during and since the intrusion of the granite, continuing reactions between residual magmatic fluids and introduced meteoric waters acting on the granite rocks, produced the minerals forming the fracture infills, alteration phases and secondary minerals. The calcites, which are ubiquitous as a fracture infill, were shown not to be precipitated from present day groundwaters but from the hydrothermal solutions associated with late stage magmatic activity.

A study of the fractures and faults in surface outcrops and intersected by boreholes (18, 19) has shown that the granites have five district orientations of fracturing, the majority of which are steeply dipping. The Moinian rocks have different major fracture orientations which are all sub-vertical. McEwen and Lintern (18) note the presence of 50 m wide zones of high fracture density, which may be continuous over distances of at least 1 km and to depths of over 200 m. Mather et al. (10) estimate the frequency of hydraulically significant fractures to be every 10 m to 50 m. Furthermore, the core analysis shows that there is no apparent decrease in fracture density with depth to at least 300 m.

A comprehensive geophysical borehole logging programme was carried out (22) with the objective of assessing the geophysical data relative to corresponding parameters measured on the core and in the boreholes. Many of the logs clearly responded to open fractures, where there is an increase in effective porosity. Fracture indices were calculated from the sonie, neutron and focused resistivity data and a series of multi colour cross plots were used to separately identify fractured and weathered zones. A valuable sonde for fracture identification and orientation is the acoustic televiewer (21). Non of the methods of fracture identification were able to evaluate the groundwater flow characteristics of individual fractures.

Since the majority of groundwater flow through granite will take place through fractures, considerable effort has been placed upon understanding the fracture system. However, the transport of solutes through the rock will be attenuated by diffusion of the fracture fluids into the rock matrix. The degree of attenuation will be a function of the rock porosity. A detailed investigation of the granite (27, 20) indicates correlations between porosity, density, state of alteration, rock fabric and mechanical strength. It was clear that because of the low values being measured, significant differences in porosity can result from the use of different laboratory measurement techniques. This was also true of the different geophysical logging techniques for measuring porosity (22). Even so, Storey and Lintern (28) reported that the major variations in porosity are related to weathering and alteration. There was a marked decrease in porosity between the surface weathered granites (about 2%), the altered granites (about 1%) and the unaltered granites (about 0.5%). The relatively high porosity of the surface weathered granite was ascribed to the expansion of the biotite lattice and the resultant formation of radial fractures.

Laboratory measurements of the stress-strain response of both granite and Moine core samples (20), are characteristic of strong crystalline rocks containing microrocks. Significant features are the initial non-linearity caused by crack closure, the relatively linear elastic region, and the microcrack initation, growth and associated dilatancy. The onset of dilatancy is also accompanied by changes in the porosity, permeability and thermo-mechanical properties of the rock.

A regional hydrogeological assessment of the area identified the most important surface water and groundwater catchments. This was followed up by an airborne thermal infra-red survey which was able to identify a large number of groundwater discharges (springs) for subsequent geochemical sampling. Borehole water level measurements were analysed in terms of flow nets from which it was possible to define the regions most likely to reflect high hydraulic conductivity. Of the estimated 100 mm groundwater infiltration rate about 70% is considered to provide the base flow to streams and rivers while the remaining 30% discharges as distinct springs. In general the surface water and groundwater systems were thought to be in hydraulic continuity and it was evident that the water table is often very close to the topographic surface, and in many places coincident with it.

Straddle packer tests carried out in the three deep boreholes (25, 26) revealed hydraulic conductivities over a range of 1×10^{-12} m/sec to 2×10^{-6} m/sec. The average bulk rock value was 7×10^{-10} m/sec but fracture zones were shown to exceed 1×10^{-6} m/sec. A non parametric statistical analysis of hydraulic conductivity (K) against depth (D) revealed the relationship :

$$K = 1.07 \ D^{-3.95}$$

This general decrease of hydraulic conductivity with depth seems to apply to both fracture zones and bulk rock irrespective of wether the dominant rock type is granite or metamorphosed sediments (32). Hydraulic pressure differences between the borehole fracture zones were usually about 0.5 m or less, but vertical pressure gradients of up to several metres indicated the potential for vertical flow both down to and up from depths of at least 300 m. The actual flow regime seemed to be related to the topographical position of the measurement borehole.

Groundwater and surface water samples were taken from the deep and shallow boreholes and from the streams, rivers and spring discharges (29). Detailed chemical, isotopic and inert gas analyses were consistent with a near surface groundwater flow system in which all waters are of meteoric origin. Most groundwaters sampled from within the 40 m weathered granite zone are dominated by recharge within the last 30 years. However, water sampled from about 300 m in one borehole was dated to be between 1×10^4 years and 2×10^5 years (30). It was further concluded that the dominant control on the groundwater chemistry is the fracture infilling material and that there is little evidence of reaction between the groundwaters and the minerals of the rock matrix.

The results from the hydrogeological and hydrogeochemical studies coupled with the results from the groundwater discharge mapping and preliminary flow modelling suggest a near surface groundwater flow system related to topography. Recharge and downward flow velocities are greatest beneath high ground with steep topographic gradients. Many of the flat plateau-like areas, on the other hand, represent regions where the groundwater surface is coincident with the ground level, which results in permanent saturation. At the breaks of slope down gradient from these areas, groundwater discharges often give rise to spring derived surface water streams and tributaries. These groundwater discharges not only form the headwaters to rivers but also provide the natural drainage from the plateau-like areas, whose groundwater levels are in turn maintained by local topography (24).

The major part of the groundwater circulation consists of recently infiltrated groundwaters being recharged into the higher ground and following predominantly sub-horizontal flow lines within the near surface superficial deposits and weathered granite, to be discharged as base flow in the river valleys. Mature groundwaters upwelling from deep within the granites seems unlikely, except on a localised basis when associated with major faults and fracture zones.

REFERENCES :

1. "Etude des possibilités de stockage des déchets radioactifs dans les
    terrains cristallins. Investigation par forages profonds du granite
    d'Auriat.", 1982, Rapport CCE - EUR 8178 FR.

2. BLANCHIN R., "Etude statistique et géostatistique de la fracturation
    de la mine de Fanay-Augères (Haute-Vienne)", 1984, Rapport
    CCE/CEA/BRGM n° 84 RDM 074 IM.

3. BLES JL., BONIJOLY D., "Fracturation profonde des massifs rocheux
    granitiques : étude documentaire", 1983a, Rapport CCE/CEA/BRGM n° 83
    SGN 858 GEO; Rapport CCE - EUR 9381 FR ; documents BRGM à paraître.

4. BLES JL., DUTARTRE P., FEYBESSE JL., GROS Y., MARTIN P.,"Etude
    structurale de la fracturation du granite de Saint-Sylvestre : mines
    de Fanay-Augères et de Margnac (Haute-Vienne)", 1983b, Rapport
    CCE/CEA/BRGM n° 83 SGN 426 GEO ; Rapport CCE - EUR 8922 FR (1984) ;
    documents BRGM à paraître.

5. BLES JL., "Fracturation profonde des massifs rocheux granitiques (2ème
    phase) - rapport de synthèse : études documentaire, géostructurale et
    géostatistique", 1984, Rapport CCE/CEA/BRGM n° 84 SGN 323 GEO.

6. BLES JL., "Distribution de la fracturation en surface et en profondeur
    dans les granites", Journées sur le granite, Orléans, juin 1984,

7. DURAND E., "Etude de l'effet d'échelle en milieu fissuré - Phase
    pilote : certification du site expérimental de Fanay", 1984, Rapport
    final CCE/CEA/BRGM n° 84 SGN 237 STO.

8. DURAND E., "Etude de l'effet d'échelle en milieu fissuré - Phase 2A ;
    étude des écoulements", 1984, Rapport semestriel 1er janvier au 30
    juin 1984 CCE/ENSMP/BRGM n° 84 SGN 382 GEG.

9. MATHER J.D., GRAY D.A., GREENWOOD P.B., "Burying radioactive waste ;
    the geological areas under investigation", 1979, Nature, Vol. 281,
    N° 5730, pp332-334.

10. MATHER J.D., CHAPMAN N.A., BLACK J.H., LINTERN B.C., "The geological
    disposal of high level radioactive waste - a review of the Institute
    of Geological Sciences research programme", 1982, Nucl. Energy, Vol.
    21, N° 3, pp 167-173.

11. PEACOCK J.D., "Geomorphological predictions with respect to Altnabreac
    area, Caithness", 1980, Report of Br. Geol. Surv., ENPU 80-10.

12. LINTERN B.C., RAINES M.G., "Borehole drilling details and survey data
    for Altnabreac site", 1980, Report of Br. Geol. Surv., ENPU 80-2.

13. McCOURT W.J., "The geology of the Strath Halladale - Altnabreac
    district", 1980, Report of Br. Geol. Surv., ENPU 80-1.

14. LINTERN B.C., STOREY B.C., "Geology of the Altnabreac research site,
    Caithness.", 1980, Report of Br. Geol. Surv., ENPU 80-14.

15. LEE M.K., RICHARDS M.L., "Reconnaissance geophysical surveys at Altnabreac, Caithness.", 1980, Report of Br. Geol. Surv., ENPU 80-3.

16. LEE M.K., SAVAGE G.M., CARRUTHERS R.M., HOUGHTON M.T., BAXTER M.G., "Magnetic, VLF and seismic surveys at Altnabreac.", 1980, Report of Br. Geol. Surv., ENPU 80-16.

17. LEE M.K., "Regional setting of the Strath Halladale granite from gravity and aeromagnetic data.", 1980, Report of Br. Geol. Surv., ENPU 80-15.

18. McEWEN T.J., LINTERN B.C., "Fracture analysis of the rocks of the Altnabreac area.", 1980, Report of Br. Geol. Surv., ENPU 80-8.

19. McEWEN T.J., "Fracture analysis of crystalline rocks : field measurements and fiels geomechanical techniques.", 1980, Report of Br. Geol. Surv., ENPU 80-11.

20. McEWEN T.J., HORSEMAN S.T., LAI S.F., "Geomechanical properties of rocks from the Altnabreac area.", 1980, Report of Br. Geol. Surv., ENPU 80-13.

21. McEWEN T.J., "An analysis of in-situ stress measurements by hydrofracturing and the use of the borehole televiewer at Altnabreac.", 1982, Report of Br. Geol. Surv., ENPU 82-10.

22. McCANN D., BARTON K.J., HEARN K., "Geophysical borehole logging with special reference to Altnabreac, Caithness.", 1981, Report of Br. Geol. Surv., ENPU 81-11.

23. GLENDINING S.J., "A preliminary account of the hydrogeology of Altnabreac, Caithness.", 1980, Report of Br. Geol. Surv., ENPU 80-6.

24. BRERETON N.R., HALL D.J., "Ground water discharge mapping at Altnabreac by thermal infrared linescan surveying.", 1983, Report of Br. Geol. Surv., FLPU 83-7.

25. HOLMES D.C., "Hydraulic testing of deep boreholes at Altnabreac : development of the testing system and initial results.", 1981, Report of Br. Geol. Surv., ENPU 81-4.

26. Soil Mechanics, 1980, "Preliminary results of hydraulic conductivity testing at Altnabreac borehole Al A.", Report of Br. Geol. Surv., ENPU 80-4.

27. ALEXANDER J., HALL D.H., STOREY B.C., "Porosity measurements of crystalline rocks by laboratory and geophysical methods.", 1981, Report of Br. Geol. Surv., ENPU 81-10.

28. STOREY B.C., LINTERN B.C., "Alteration, fracture infills and weathering of the Strath Halladale granite.", 1981b, Report of Br. Geol. Surv., ENPU 81-13.

29. KAY R.L.F., BATH A.H., "Groundwater geochemical studies at the Altnabreac research site.", 1982, Report of Br. Geol. Surv., ENPU 82-12.

30. KAY R.L.F., ANDREWS J.N., BATH A.H., IVANOVICH M., "Groundwater flow profile and residence times in crystalline rocks at Altnabreac, Caithness, UK.", 1984, Isotope Hydrology 1983, IAEA, Vienna, pp 231-248.

31. STOREY B.C., LINTERN B.C., "Geochemistry of the rocks of the Strath Halladale Altnabreac district.", 1981a, Report of Br. Geol. Surv., ENPU 81-12.

32. MATHER J.D., "Geological assessment of crystalline rock formations with a view to radioactive waste disposal.", 1984, CEC Report EUR-8828-EN, Luxembourg.

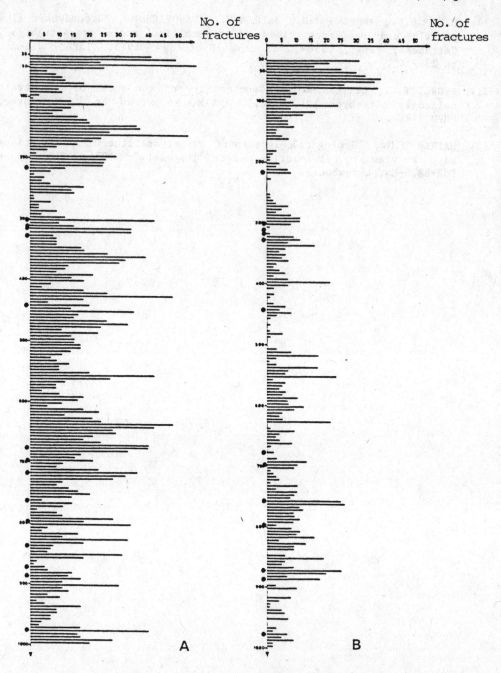

FIGURE 1 – The Auriat borehole : graphs of the number of fractures
within each 5 m stretch (after Gros, 1982).
A = steeply dipping fractures (dip ⩾ 60°)
B = gently dipping fractures   (dip ⩽ 30°)
● = major faults

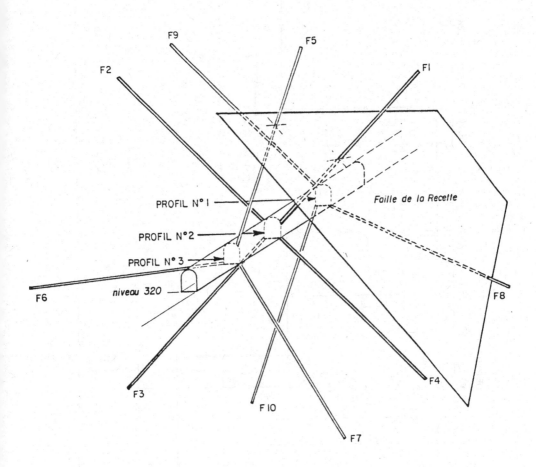

FIGURE 2 – Perspective drawing of the experimental zone in the gallery
at Fanay-Augères and of the three radial borehole-profiles.

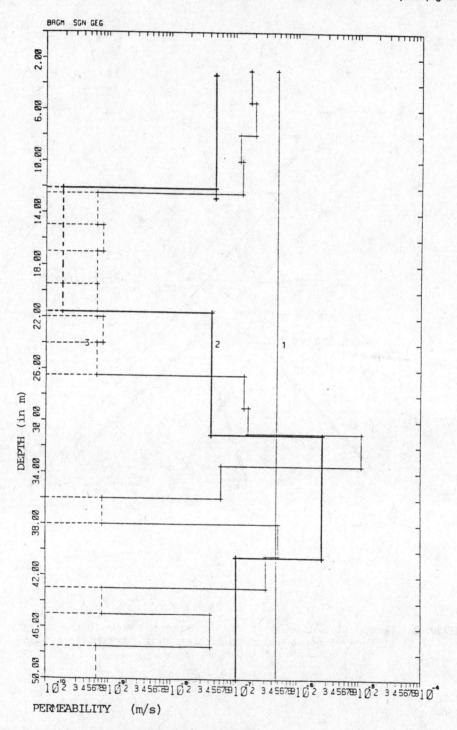

FIGURE 3 - Permeability log from borehole F 10
Graphs 1, 2 and 3 correspond to measured borehole sections
50 m. 10 m and 2 - 2.5 m in length respectively.

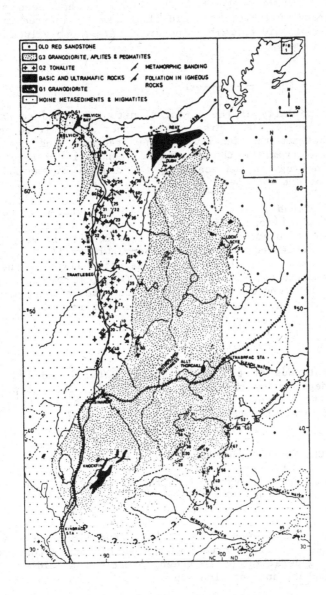

FIGURE 4 – Geological sketch map of the Strath Halladale–Altnabreac
           district

DISCUSSION

J. HADERMANN, NAGRA Baden

In one of the slides, the fracture frequency has been shown. To
perform a safety analysis, it would of course be interesting to know
the percentage of void carrying fractures. Are those values known, and
how can they be measured?

A. BARBREAU, CEA Fontenay-aux-Roses

The cinematic porosity of the Auriat granite was found to be significantly lower than 1%. Of course, it is rather difficult to establish a
correlation between the density of the fractures and the fractures
playing a role from the thermohydraulic conductivity point of view.
Studies are in progress and measuring the density of the fractures is
one of the problems to solve.

It is highly probable that in reality, in a factured formation, the
percentage of water conducting fractures is very small. In the Auriat
formation in particular, conducting fracture density is very low; some
specific zones excepted, the measured permeability is in the order of
10^{-11} to 10^{-12} m/sec, which is at the detection limit of the existing
measurement techniques, like pulse-test, slug-test and injection-test.
In this formation, where a big amount of fractures exists, most seem to
be filled by materials like calcite, clay and silicates.

S. ORLOWSKI, C.E.C. Brussels

What are the trends of the evolution of porosity and fracture as a
function of depth? For the Altnabreac formation for example, it has
been stated that the porosity diminishes with depth. Is this a general
statement, and over what range of depth is it valid?

A. BARBREAU, CEA Fontenay-aux-Roses

It has been observed, going to a depth of 1 000 m in Auriat and to
somewhat less in the case of Altnabreac, that the hydraulic conductivity diminishes over the first hundreds of meters from 10^{-6} or 10^{-8} to
10^{-10} m/sec. Going deeper than 400 or 500 m, hydraulic conductivities
of 10^{-10} to 10^{-12} are reached, 10^{-12} being the limit for the measurement techniques. This is true for the Auriat and Altnabreac formations, and no generalization to other formations should be made. UK
researchers suggest an approximation :

$$K = 1.7 \times D^{-3.95}$$

K being the hydraulic conductivity and D the depth below gound level.
In Sweden, correlations have also been published, giving the permeability as a function of depth. Despite the fact that the trends are
known, it must be stressed that additional measurements are necessary.

Effectively, the value of permeability is important when evaluating the transfer conditions for the geological barrier. There is no particular difficulty in performing such measurements, as the techniques required are available.

ASSOCIATED R&D
RADIONUCLIDE MOVEMENT THROUGH AND MECHANICAL PROPERTIES OF FRACTURED ROCK

P.J. Bourke, Atomic Energy Research Establishment, England
S. Derlich, Commissariat à l'Energie Atomique, France
P. Goblet, Ecole des Mines de Paris, France
D. le Masne, Bureau de Recherches Geologiques et Minières, France

ABSTRACT

This paper reviews work in France and England on four scientific topics associated with the radioactive waste management research programme for the Commission of European Communities. All are concerned with burial of waste in granites.

Radionuclide movement

Convection of radionuclides with flow of water through fractured rock will be both retarded and dispersed by several processes. These are discussed and the data needed to quantify both the flow and convection are specified. The question of how best to model these phenomena mathematically is considered and partly answered.

Field measuring techniques

Established and new geological methods for detecting fracturation, involving measurements from boreholes have been proved. A piezofor method of measuring hydraulic pressure fields without perturbation of these fields by the boreholes used, has been developed.

Mechanical and thermal properties

The main physical and mechanical properties - density, porosity, strengths, elastic limit, moduli, Poissons ratio - have been determined both at ambient and elevated temperatures up to 200°C.

Dissolution of vitrified waste

Data are given for the products of dissolution of borosilicate glass with granite in de-ionised water at 50 MPa pressure and temperatures up to 350°C.

1.   RADIONUCLIDE MOVEMENT THROUGH FRACTURED GRANITE

Convection in any water flow is likely to be the means by which most of the radionuclides leached from waste deeply buried in the earth may return to the surface. Quantification of these returns in terms of the totals, concentrations, times and places of radioactivity arriving at the surface is necessary to assess their biological consequences and much experimental work and mathematical modelling is being done to try to predict these quantities.

This section of the paper reviews experimental work being done in the MIRAGE programme to study water flow and radionuclide movement in granites. It describes the phenomena which have been investigated and suggested to the mathematicians to be important in formulating models and it summarises the data determined in field and laboratory experiments and needed to quantify these phenomena. It also considers the different mathematical approaches which may be taken to modelling the movement and it tentatively suggests the optimum approach.

1.1 IMPORTANT PHENOMENA AFFECTING THE MOVEMENT AND DATA NEEDED
1.1.1 Nature of the water flow

Because the radionuclides are convected with the water flow, it is obviously necessary to have a thorough understanding of this flow. The fracturing of granites generally is such that the movement of water through them is expected to occur mainly as flow in discrete fractures or fractured zones. This leads to the question of how much residual flow occurs as permeation through the intact rock between fractures, which flow will of course depend on the permeability of this rock.

Values of this permeability have been measured for various granites in laboratory tests and, hence for any hydrostatic pressure gradient through the rock, the permeation can be calculated. Comparison of this with the fracture flow cannot however be readily made. The reason for this is that the fractures vary widely in size as measured by hydraulic transmissivity or effective aperture. For example, the distribution of apertures in Cornish granite (Figure 1) shows more than a tenfold variation above the experimental limit of detection of individual fractures. To make a distinction between permeation and flow through the large number of small fractures is therefore experimentally difficult because of the problem of identifying and measuring the flow in all the small fractures.

For quantification of the water flow per se in terms of volume flow rate, this distinction is unimportant and, provided both the flow characteristics of the identified fractures and the residual equivalent permeability of the unidentified fractures and the rock are measured, the total volume flow can be calculated. It is nevertheless necessary to make this distinction or at least to know how most of the flow occurs, if the convection of radionuclides is to be realistically quantified. This is because the important phenomena of retardation and dispersion of this convection are affected differently by fracture flow and permeation, as described below.

Accordingly, techniques have been developed for identifying and locating progressively smaller fractures until data have been obtained for those fractures through which most of the flow occurs. Essentially this is done both by searching for the appearances in one drill hole of radioactive tracers pumped into fractures in other holes in tracer tests (Bourke, 1981) and by reducing the axial pressurised length between packers until two conditions are met in injection tests (Bourke, 1982; Heath). The first condition is that the flow measurements and the logging of the hole and examination of its drilling core should show there to be only one or no water-bearing fracture in most test lengths. The second is that the total flows into the rock between these fractures is small compared with the flows into the fractures. The residual permeation may then be neglected with only small error in modelling of the flow as if it were wholly through discrete fractures.

1.1.2 Data needed to quantify flow

The individual water-bearing fractures are taken to be discrete, planar or at least two-dimensional flow paths. Data for granites show that the fractures have a tendency to occur in sets or groups with orientations in particular directions, but do also occur with other orientations. They therefore intersect each other frequently to form a more or less random three dimensional pattern of two dimensional flow paths.

To quantify the flow through such patterns of fractures, it is necessary to have data for four of their characteristics:

-   , occurrence or separation through the rock
-     orientations in space
-     intersections between fractures
-     hydraulic transmissivities or equivalent apertures.

Before any modelling of the flow, it is however necessary to consider two further points.

Firstly, the fracturing in several granites in which hydrogeological investigations have been made is such that there are both a few major fractures or wide (metre) zones of locally high permeability and large numbers of small fractures with narrow (millimetre) apertures. The former are likely to have rapid flows and the main barrier to return to the surface may be the rock between the buried waste and these major fractures. At any site envisaged for burial, they must be individually located and characterised, and the waste should be put as far as possible from them. The latter are generally so numerous that they have continuous distributions of occurrence (Figure 2), orientation (Figures 3 and 4) and aperture (Figure 1). There will be too many for all to be individually mapped and they will have to be treated statistically in modelling the flow.

Secondly, because of the compressive stress through the rock due to its weight, all fractures will have extensive areas of contact between their faces. These will form islands of no flow around which the water will flow in channels. For example, flow between two holes drilled to lie in the plane of one fracture in Cornish granite was such that most (80+%) of the flow into the receiver hole entered it along only a small (20-%) part of its length (Figure 5) suggesting that channels of flow occupy little (10-%) of the fracture area. For calculating the total volume flow through fractures, hydraulic apertures of equivalent parallel walled slots are adequate. These apertures are, however, insufficient for calculation of the radionuclide retardation and dispersion for which distributions of the dimensions and areas of channels are necessary.

A further important aspect of this second point touches the interpretation of the fracture occurrence data. This is that if the average dimensions of islands of no flow are large compared with the diameter of the hydraulic test holes, then these holes will intersect some water-bearing fractures in the no flow areas and these fractures will not be counted as water-bearing. Given adequate statistics for the dimensions of the no flow areas, this effect can be taken into account in modelling.

1.1.3 Retardation and dispersion of the radionuclide convection

Water contacting the buried waste will leach radionuclides out of the waste matrix material. These will slowly diffuse through the engineered barriers to the surrounding rock to be convected away in any flow through the fractures. This phenomenon will therefore produce a slow rise and later fall in concentrations of radionuclides in the flow until all that have not decayed are eventually released. These radionuclides may be present either in solution or as colloids.

This concentration pulse will tend to move with the flow but will be subject to retardation and dispersion.

Retardation will occur due to diffusion from the flow into stagnant water in the porosity of the lowly permeable rock between fractures (Lever) and by any sorption on this rock followed by desorption and diffusion back to the flow as the concentration pulse moves onward. Colloids may also be retarded by filtering on infilling material in the fractures. These processes are wholly beneficial because in retarding the convection they increase the time for decay and delay and reduce the total amount of radioactivity returning to the surface.

Dispersion will occur due to several processes; diffusion in all directions in the water, hydraulic mixing of the flows both within fractures and at fracture intersections and by slow diffusion into and back out of the rock. Such dispersion is beneficial in that it reduces the maximum concentration in the pulse. The first two processes however also accelerate the convection of some radionuclides to the surface and are therefore partly detrimental.

The extent to which retardation will occur will depend on the kinetics of diffusion and sorption in the rock, on its thickness between fractures and the time available for these processes. This time will be the time it takes the concentration pulse to pass points along the leakage path. At the start of this path it will be the time for release of radionuclides through the engineered barriers. Although this time will be long, the time needed for slow diffusion and sorption throughout large thicknesses of rock may be even longer and the maximum partition of radionuclides between the flow and the rock and, hence, the maximum retardation may not be obtained initially. Dispersion along the leakage path will, however, progressively increase the time of passage of the pulse and hence there is interaction between dispersion and retardation and the latter will increase asymptotically towards the maximum possible.

It may be noted that the above processes are markedly different from those producing retardation and dispersion in highly permeable rocks in which the flows spread throughout their volumes. Hence, for rocks in which most of the flow is through fractures, equivalent permeability and porosity of the fractures are not appropriate for quantifying the radionuclide movement.

1.1.4 Data for Retardation and Dispersion

In addition to data about the geometry of the water-bearing fractures, to calculate retardation and dispersion, data are also required for diffusion, porosity and sorption in the rock.

Techniques have been developed for studying diffusion of non-sorbed radionuclides through intact rocks (Bradbury, 1982; Hemingway). Measurements of the initially transient and latter steady diffusion rates are made with constant concentrations in water at the opposite faces of specimens of various thicknesses. Analysis of the results provides data both about the total porosity of the rock, and its division into interconnected pores which extend long distances and short, cul-de-sac pores and about effective coefficients of diffusion. Values of porosity range from 10^{-2} to 10^{-4} of up to a half may be in cul-de-sac pores. Diffusion coefficients, reduced by the presence of the rock and tortuosity of the pores to much less ($\frac{1}{10\ 000}$) than the corresponding values in water alone were obtained for several granites. These techniques and data are generally considered adequate.

Data for sorption are very numerous but may not be wholly satisfactory. The main reason for this is that they have been mostly obtained by the batch technique. In this, the rock is crushed to small (millimetre) sized pieces and immersed in water with radionuclides in solution. Their concentrations in the water are monitored until they have fallen to a steady values when equilibrium sorption is assumed and its coefficients are calculated. Uncertainties with this technique are that the new surfaces exposed in crushing may increase the sorptive capacity and that slow changes in concentration may make attainment of equilibrium too long for convenient experiments. Further, even the reproducibility of data in batch measurements is poor - in a trial with nominally identical conditions in six laboratories disagreement by factors of 100 could not be explained. So the accuracy of such data must be seriously in doubt.

For these reasons other techniques for measuring sorption are being developed (Bradbury, 1985). These basically involve using large specimens to avoid measurements with large new surface:volume ratios and pumping the water through the specimens with rapid convection rather than slow diffusion to bring the sorbates to the sorption sites and hasten equilibrium. Preliminary results for the sorption coefficients of strontium and caesium now being measured are a hundred-fold smaller than corresponding results obtained using the batch technique.

1.2 TWO APPROACHES TO MODELLING MOVEMENT

For predicting flow and movement in fractured media, two different approaches are being followed:
- the equivalent continuous porous medium (ECM)
- the discrete fracture characterisation (DFD).

These approaches, the incentives to use one or the other, or to combine them, their disadvantages and their different requirements for experimental data are described here.

1.2.1 The equivalent continuous medium

This is the traditional approach to describe fractured media. It is assumed that fractured media can be treated as a continua, when considering volumes greater than some minimum, the Representative Elementary Volume, (REV) which encompasses statistically representative ranges of the variables of the relevant properties.

Incentives
  a)   Numerical and analytical models of porous media are well developed and can be used for prediction in one, two and three dimensions.
  b)   Characterisation of the system only needs the equivalent properties of the medium.
  c)   No complex description of the medium, which is very difficult to collect, is needed.

Disadvantages
  a)   From recent work on the discrete fracture approach, it is clear that the ECM can be applied only if the occurrence and intersection of the fractures are such that a percolation threshold is reached. Below that threshold, their will never be an equivalent medium, and, whatever the size of the REV, the

medium cannot be described by this approach because the degree of connection of the fractures is too low to ensure equivalent porous flow (see Robinson, 1984; Goblet, 1984, 1; Charlaix, 1984; Marsily, 1985).

Further, even when the percolation threshold has been passed, the size of the REV at which the ECM is valid can be shown to be a function of the occurrence and intersection. Close to the threshold, this size can be very large. Also there is the possibility of local clusters of fractures not connected to the main pattern of interconnected fractures which would adversely affect the accuracy of the ECM approach (see also implication on data acquisition for a further discussion).

b)   Even if the ECM is applicable above the percolation threshold, the size of the REV may be large compared with some of the volumes for which predictions are needed. For example, the average flow through the total volume of a repository may be predicted using the ECM but this will not give any information about the probably large variability from the average with position within the repository volume.

Such information may be needed to predict the different corrosion rates of canisters, leachings of the waste limited by leach rate or solubility and convection of radionuclides away from individual packages, all of which depend on the local flow rather than the REV average. Concerning this last point, it must be said however that the variability of the velocity in the medium is represented, in the ECM, by the equivalent hydrodynamic dispersion coefficient. Provided that this can be estimated (see data acquisition), and provided that the distance between the repository and the surface is large enough, compared to the REV, the ECM approach can perhaps still be used. This remains to be proved. Experiments in France on the scale effect in fractured media are intended to do so (Dieulin, 1984; Durand et al., 1984).

c)   Describing radionuclide movement using the ECM is questionable because it does not explicitly take into account one important retardation mechanism; matrix diffusion. Further colloid transport may be poorly described in the ECM but so little is known so far about movement of colloids in fractures that this is not yet a major objection.

Implications for Data Acquisition using the ECM

a)   Field measurement of hydraulic properties of the ECM must be made over volumes comparable with the REV. The Cornish data for deep granite shows the average separation between water-bearing fractures to be about 10m and, taking the REV to have dimensions of ten times this length, suggests that the minimum volume for measurement will have dimensions of about 100 m.

It is highly questionable if local hydraulic tests in boreholes, or even cross-hole test between boreholes (e.g. sinosoidal testing) can determine the average permeability to use in a ECM. It may however be possible to calibrate digital ECM models of sites using pressure measurements in various boreholes and flowrates in the underground drifts to estimate the average permeability. The comparison of the measured and model predicted heads and flow rates in the URL at Whiteshell in the AECL programme will be highly interesting.

b)   Tracer tests, to determine the dispersion coefficient
      representing the variability of the velocity, will have to be
      made at large scale, probably larger than the REV. This may
      result in their requiring inconveniently or impractically long
      times of years, or perhaps tens of hundreds of years to complete.
      Naturally occurring tracers might be used but this is not yet
      established.

c)   Independently of the variability of the water velocity, the local
      volume flow rate will also vary with the fracture variables. As
      described above, its local values are needed to predict the
      behaviour of individual waste packages. It may be possible to
      determine these values from measurements of flows into holes for
      packages when the repository has been built.

d)   Because of the existence of the percolation threshold, it will be
      necessary even using the ECM to determine the fracture occurrence
      and intersection for assessment of whether or not the ECM can be
      justifiably applied. Furthermore, these fracture variables will
      also be needed to predict matrix diffusion and sorption.

1.2.2 Discrete fracture description

In this approach, to determine the volumes, velocities and mixing of
the flow, probability distributions of separations, orientations, effective
hydraulic apertures, intersections and channelling of the water-bearing
fractures are required. These distributions are randomly sampled to obtain
a notional representation of the real fracture pattern and the flow through
this is then calculated. This process is repeated many times and the
calculated volume flows, velocities and mixings are averaged to obtain
estimates of their means and variations, i.e. the Monte Carlo method.

In present developments of this approach, the flow is generally
calculated in each fracture using the cubic laminar flow law, assuming no
head loss at each fracture intersection. Channelling in fractures is not
yet included. The radionuclide movement is assumed to be mixed inside
fractures or at the intersections, or, as a limiting case not to be mixed
at all (Endo, 1984).

The models using this approach are still in an early stage of
development, at least in three dimensions. The various published
two-dimensional models are only conceptual tools, and are unrealistic for
repository predictions.   (Robinson, 1984; Long, 1983; Rouleau, 1984;
Schwartz, Smith, 1981.)

Incentives

a)   Data on fracture variables are being collected; either by
      observation on outcrops, in drifts, and in boreholes, and by
      hydraulic and tracer testing in boreholes (e.g. Bourke, 1984).
      Such data can be be used to obtain a better understanding and
      description of flow in fractured media.

b)   The description of the variability of flux and velocity in the
      medium is easily obtained from the fracture variables, by the use
      of the Monte Carlo method.

c)   Realistic description of matrix diffusion, of sorption on
      fracture walls, and perhaps on colloid transport requires to use
      the DFD.

d)   The DFD is the only possible approach if the network is below or
      close to the percolation threshold. However, it is not clear at
      present if actual fracture patterns are close to the percolation
      threshold: due to their tectonic origin, it may very well be
      that _all_ fracture systems are initially systematically _above_

percolation threshold. In this case, the below-percolation behaviour which has been reported (Marsily, 1985) may be due to partial sealing of the fracture planes by precipitation. See also disadvantages.

Disadvantages

a)   In the DFD, it is in principle necessary to describe the complete fracture pattern to make predictions and this must be done in three dimensions to be realistic. Hence data for the fracture variables over the whole of volume of interest is needed. Fracture variables, for example occurrence (Beucher, 1983; Beucher, 1984), vary in space and it is probably not valid to assume statistical homogeneity.

b)   The channelling inside a fracture plane is still conceptually difficult to introduce in the models (see data acquisition).

c)   The mathematical modelling of flow in three dimensions is not yet adequately developed. Even if feasible, some kind of simplification will probably be necessary in order to make the approach tractable.

Implications for Data Acquisition using the DFD

a)   It is important to obtain required properties of water-bearing fractures singly and this can probably be done only by hydraulic and tracer tests of representative numbers of individual fractures.

b)   It is essential to determine the extent to which channelling occurs in the fracture plane. This will necessitate a large number of single fracture tests rather than one large-scale test (although some assessment of channelling might perhaps be obtained from the interpretation of a large-scale tracer test using a DFD model).

1.2.3 Combined approaches

In view of the disadvantages of both approaches, it is clearly desirable to try to find improvements. Possibilities which can be suggested and seem promising include the following.

Double ECM approach

It may be possible to improve the ECM approach for calculations for large volume by a double medium approach. In this, two ECMs are superimposed, one to represent the fractures and one to represent the matrix, with exchange of radionuclides between the two. Note that it is still essential to obtain field data for separations between water-bearing fractures in order to determine the dimensions of the matrix in which diffusion occurs.

A formulation was recently proposed (Neretnieks, 1984) to introduce matrix diffusion in an ECM with varying block sizes using the concept of a pseudobody inside an REV. The diffusion from the fractures inside blocks is assumed to depend of the distance between the fracture and each point considered in the block All the inner surfaces of the blocks situated at a same distance from the closest fracture are thus agglomerated in an equivalent surface for a single fictitious block. The same procedure for all the possible distances produces an equivalent block, the pseudobody, which has approximately the same diffusive behaviour as the set of blocks it replaces, since a relationship between distance from the fracture and surface offered to diffusion is globally conserved. The diffusion equation inside this pseudobody is then solved by integrated finite differences. If sorption on the fracture walls is also to be included in the ECM, the density of fractures is also required.

Discrete major fractures in an ECM model

Where a small number of major faults or fracture zones are observed in a fractured medium, they can be introduced explicitly in an ECM model. This was done in the Finite Element code METIS (P. Goblet, 1984, 2) by using in two dimension special linear elements between the usual quadrilateral elements.

Use of a DFD for estimating the average properties of an ECM

Where field observation shows the medium to be fractured and data for the fracture variables has been obtained, a DFD may be used to determine both the minimum values over which an ECM is justified and its equivalent properties.

Use of DFD at small-scale, and of an ECM at large-scale

One could very well conceive of a three-dimensional DFD embedded in a porous equivalent medium imposing the boundary conditions at the limit of the DFD. The former would be used to predict, e.g., small-scale repository behaviour, where the ECM is not valid, and the later the large-scale flow (and movement) in the medium away from the repository. This ECM could have some discrete fractures, in the second combination above.

1.3 CONCLUSIONS

(i)   Because of the phenomena occurring, it is essential for any realistic prediction of radionuclide movement through fractured rock to obtain statistically representative data for:
  - occurrence or separation
  - orientation in space
  - intersection
  - apertures
  - channelling
  of the main, individual, water-bearing fracture and for:
  - diffusion
  - porosity
  - sorption of the worrisome radionuclides
  in the rock between fractures.

(ii) Neither the equivalent continuous medium nor discrete fracture approaches to modelling movement is wholly satisfactory because both only approximate reality and ideally require more data than are and, perhaps, can be obtained. A combination of the two approaches with field work to obtain carefully and clearly specified data may be the best - but not necessarily a good - compromise for practical predictions.

2. FIELD MEASURING TECHNIQUES
2.1 GEOPHYSICAL METHODS OF DETECTION OF FRACTURATION

Because of the importance of fractures for radionuclide movement, the Bureau de Recherches Geologiques et Minières has for four years been working on the detection of fracturation by geophysical methods from boreholes.

During the first half of this period, classical logging, together with new electrical dipole-dipole probes, has been used in boreholes to characterise the fracturation in their vicinity (10 cm to some metres). In the second half, a wider (decametric to hectometric) range of investigation has been looked for, using basically electrical methods, with cross-hole or hole-to-surface arrays.

Borehole logging

Field work has been emphasised in the first period, essentially on the two 500 m and 1000 m boreholes of Auriat (Creuse). Electrical (resistivity, SP and IP), nuclear (gamma-ray, gamma-gamma, neutron probe), and sonic logging has been used along with of these diamond drilled boreholes. We could then compare these field results with the direct observation of fracturations (diamond drilling), in terms of fracture frequency (number of open or closed fractures per metre).

Four physical parameters are adequate to characterise the fracturation: neutron-porosity, gamma-density, normal resistivity and propagation speed (sonic). Cross plots between each of these parameters and the observed open fracture frequency have shown good correlation. This result proves the possibility of estimating the fracture frequency in crystalline rocks from geophysical logging.

Correlation between these physical parameters and the closed fracture frequency is not so good, which implies a good discrimination between open and closed fractures.

DIDIER probe

To increase the necessarily small (decimetric) range of investigation from the boreholes of these methods, we adapted from classical surface surveys, the electrical dipole-dipole method, and created a new electrical probe, named DIDIER (see array on Figure 1). Its theoretical response on various conductive thin features has been calculated by one- and three-dimensional modelling. Its main interest is to estimate the lateral extension of a conductive fracture which has been evidenced by classical logging. The lateral range of investigation from the borehole is plurimetric to decametric for this DIDIER probe (depending on the transmitter-receiver length).

Electrical extension methods

Electrical properties of rocks, only, have been used in our second 2-year research programme: we intended to increase the range of investigation of the geophysical methods by using two different locations for transmitter and receiver, either two boreholes (cross-hole technique or a borehole and the surrounding surface (hole-to-surface methods).

In spite of the emphasis of the theoretical part (modelling) of this research, a field work has yet been undertaken in order to compare theoretical results to actual field data.

Two different programs have been created or adapted for electrical one- or three-dimensional (ELEC3D program) modelling. Two general types of models have been taken into account:
- finite, or infinite, thin conductive bodies embedded in a resistive homogeneous (or layered) halfspace, supposed to account for a single fracture;
- finite massive conductive bodies accounting for zones of maximal fracturation in a crystalline matrix.

Cross-hole electrical method

The electrode array for the cross-hole (MIMAFO) technique can be seen on Figure 6. The potential V due to every current source A (one injection at a time) is recorded continuously along the receiving borehole. The knowledge of the complete distribution of potential for numerous sources regularly spaced in the injection borehole leads to an electrical tomography of the boreholes' environment.

We studied (with 3D modelling) the theoretical influence of the thickness, the lateral position, the orientation, dip and resistivity contrast of the conductive heterogeneity, and of the distance between boreholes. Combination of various criteria (potential, vertical electrical field, apparent resistivity) enables us to assess (or not) the continuity of conductive features between the boreholes, or to detect any anomalous zone of higher conductivity in the vicinity of the boreholes. In this respect, Figure 2 gives an interpretation of the conductive features connecting three couples of boreholes on the granitic site of Le-Mayet-de-Montagne (Allier). Another experiment at Beauvain (Orne) led us to confirm the presence of a massive high-conductivity zone detected by a hole-to-surface method on this granitic site.

Hole-to-surface methods

The BIDIFO method array (a transmitting electrical bipole on the surface, a receiving electrode or dipole in the borehole) is described on Figure 1. Three-dimensional (3D) modelling has been applied to this particular array to characterise the influence of many parameters (size, resistivity contrast, dip and depth of the heterogeneity and its distance to the borehole) on the 3D response. Using three orientations (every 60°) of the transmitting bipole allows us to detect conductive bodies lying away from the borehole; we know furthermore the orientation of the anomalies, from the borehole, and their approximate depth. Application of this method has been undertaken on both sites of Beauvain and Le-Mayet-de-Montagne.

The ELECENT method array differs from the former one on one point (Figure 1): the injection is buried in the borehole, and the measurements of the potential and of the horizontal electrical field are done on the surface. Using various sources (one at a time) at different depths in the borehole allows a better precision in the detection and location of conductive bodies. Modelling with conductive cubes or parallelepipeds gave more precise the location (azimuth, depth, distance from the borehole) of conductive anomalies. An example of interpretation is given for Beauvain on Figure 7.

All of these extensive electrical methods experimental during this research (DIDIER, MIMAFO, BIDIFO, ELECENT) could only be effective in the field through the use of a three-dimensional modelling technique. This programme will still further be improved in order to take into account more than one parallelepipedic heterogeneity at a time.

Our aim, to increase progressively the range of investigation of the borehole geophysical methods, has been reached, but these last electrical methods cannot give as accurate an idea of the fracturation as the classical logging. They are better suited to the detection of big fractures and to the location of zones of maximum fracturation in crystalline rocks. Traditional logging and extensive methods are, in fact, complementary.

2.2 MEASUREMENT OF HYDRAULIC PRESSURE FIELD - PIEZOFOR METHOD

In addition to considering the effects of fracture on radionuclide convection with water flow, it is of course essential to know the hydraulic pressure field throughout the rock and to determine its gradients which cause flow. These are primarily dependent on the topography and rainfall but are also to some extent dependent on the permeability and porosity of the ground. Measurements must therefore be made over the ranges of depths and distances of interest.

In practice, the sinking of a borehole for measuring the hydraulic pressure in an aquifer (the virgin pressure, as the oil industry calls it) always involves a risk of modifying the initial conditions.

The principle of the Piezofor method is to sink a borehole and line its walls with a membrane which is leak-tight but flexible over its whole length. When the inside of the borehole is kept at sufficient pressure the different levels containing fluids are isolated. If we insert in the borehole a device consisting of two inflatable diaphragms with which a temporary depression can be created in the section thus isolated, the membrane is deformed and the establishment of equilibrium in the isolated section can be monitored by means of sensors. We then know the hydraulic pressure at this level. In this way the whole length of the borehole can be explored.

Given several boreholes close to one another, one can use this method to determine the pressure field in the volume between them.

Installing a flexible membrane at depths of several hundred metres is a tricky operation and requires an apparatus consisting of steel tubes which are perforated and lined with a flexible membrane.

The walls are isolated by cementing the annulus between the ground and the steel tubes.

Hydraulic pressure coupling between the tubes and the ground is achieved by porous plugs stuck to the tubes at regular intervals. The space between the plugs is filled by the cement.

The instrumentation has now been completed. During the first half of 1985 the special tubing will be installed in a borehole 250 m deep. Hydraulic pressure measurements will be made so that the vertical hydraulic gradient of the borehole can be determined.

3.   MECHANICAL AND THERMAL PROPERTIES
3.1   TEST PROGRAMME

Studies of rock mechanics in the laboratory offer a fairly easy way of obtaining a great deal of information. They are carried out for all public works, mining, oil field and hydrocarbon storage projects. The great difference between these areas and radioactive waste disposal is the introduction of an artificial heat source dissipating a large amount of thermal energy on a timescale of geological order (several hundred thousand years).

When exploratory drilling in various granite massifs showed them to be suitable for burial of waste samples of the granite were taken and measurements were made of their physical and thermal characteristics; the temperature-dependence of their properties was also studied. The results have been published, and this note provides a summary of them.

The following measurements were made with test specimens of various sizes.

(a)   on granite in its initial stage - normal temperature and pressure
- density
- velocity of longitudinal waves
- Young's modulus
- Poisson coefficient
- simple tensile testing
- Brazilian tensile testing
- porosity
- Mohr fracture envelope

(b) on granite in its initial state – measurements made at 18 to
      200°C
         – coefficients of longitudinal and transverse expansion
(c) on granite after heating and cooling for periods of 8 days to
      1 month – measurements at temperatures of 100, 150 and 200°C
         – porosity
         – longitudinal wave velocity
         – simple compressive strength
         – compressibility
(d) tests at temperature
         – simple compressive strength
         – Young's modulus
         – Poisson coefficient
         – intrinsic curve.

3.2 GENERAL RESULTS

–   Natural state
     Physical characteristics
        Longitudinal wave velocity V_L = 4900 ms^{-1}
        Density γ – 2.62
        Porosity 1.5 < n < 2%
        Coefficient of volumetric expansion α_V = 20 x 10^{-6} °C^{-1}
     Mechanical characteristics
        Simple compressive strength σ_c = 160 MPa
        Elastic limit in simple compression σ_E = 125 MPa
        Tensile strength (Brazilian test) σ_b = 11 MPa
        Young's modulus E = 68 GPa
        Poisson coefficient v = 0.27
     Compressibility
        In a spherical stress field ($\sigma_1 = \sigma_2 = \sigma_3$ = P) up to P = 100
        MPa we find K = 50 GPa (modulus of compressibility). We find
        a pressure porosity of 0.55 x 10^{-3} for a sealing pressure of
        50 MPa.
–   After heating
     Damage is considerable after 8 days at 100°C.
     There is an increase in microfissuration and the porosity
     increases by 0.6 vol %.   The simple compressive strength and
     Young's modulus decrease.
–   During heating
     The effect of heat is detectable only at temperatures of more
     than 100°C.
     At 200°C the simple compressive strength and Young's modulus
     decrease by 20%. The porosity changes from 1.5% in the initial
     stage to 2% after 1 month at 100°C.
     The Mohr fracture envelopes curves for samples in the initial
     state and samples heated to 200°C show a loss of strength from a
     containing pressure of 100 MPa upwards (tests carried out at up
     to 250 MPa).

3.3 COMMENTS

This programme of measurements of characteristics and of their
temperature dependence makes it possible to determine the influence of heat
on samples representing the massif.
     Quantitatively these effects appear as relative variations of several
per cent.
     When calculations are unable to determine of the various phenomena
needed in order to understand and quantify the dispersion of radionuclides

escaping from a repository, parametric studies make it possible to determine the importance of each parameter.

When any characteristics are measured in the laboratory, however, one has to assess the extent to which they are representative from two points of view:

1. the accuracy of the measurements;
2. how the values measured on the scale of the matrix (the sample) relate to those likely to occur on the scale of the repository.

On the laboratory scale the scatter of the experiments is always large. Determination of the various characteristics always requires several series of measurements in experimental conditions differing as to:

- dimensions of samples (effect of change of scale),
- experimental procedure (e.g. rate of loading),
- experimental equipment (e.g. stiffness of presses),
- measurement principle (e.g. standard versus Brazilian tensile testing).

The many systematic studies made in order to explain and quantify the influence of all these factors have not led to any final solution. There are too many factors involved.

For any particular case, however, it is possible to improve our understanding of the phenomenon at relatively low cost by supplementary experiments.

The problem of relating characteristics measured on samples to those measured in situ is a harder one to solve. We know that laboratory measurements on samples do not allow for alterations in the matrix (e.g. fractures). One is tempted to carry out much more representative measurements in situ. But at present this also has its disadvantages:

- The installation of instruments always modifies the initial natural conditions.
- Setting up the operation and carrying out the measurements is very onerous and limits the number of experiments which can be made.

Geophysical measurements in boreholes or underground workings make it possible to compare in-situ and laboratory values. They always require the performance or use of mining or drilling operations.

4.   DISSOLUTION OF VITRIFIED WASTE

Four intractions experiments involving a simulated borosilicate waste glass, granite and deionised water have been carried out at 100°, 150°, 200° and 350°C at a total pressure of 50 MPa to simulate the near-field geochemistry of a high level waste repository in granite. Experiments were conducted in gold-titanium cell, direct sampling autoclaves for run durations of 200 days (100°, 150° and 200°C) and 30 days (350°C), during which time solution samples were extracted for the analysis of 25 chemical species. Solid phases retrieved at the end of the experiments were examined using X-ray diffraction and scanning electron microscopy. The high temperature speciation characteristics and degrees of mineral saturation of the fluids were investigated using the geochemical software packages, EQ3 and SOLMNEQ.

Generated fluids were all alkaline (final pHs measured at 25°C) with calculated in situ being: 10.0/9.6 (100°C); 9.6/8.9 (150°C); 9.2/8.4 (200°C); 8.3/7.3 (350°C). Hydrogen ion activity continued to increase with time suggesting that pH was limited by the rate of precipitation of hydroxyl-consuming mineral phases, principally smectite. The oxygen activity of experimental fluid phases were interpreted to be broadly 'reducing' at 200° and 350°C (dominated by ferrous iron) and broadly 'oxidising' at 100° and 150°C (dominated by ferric iron). The oxygen activity of the fluid phase is directly related to the temperature dependent rate of oxygen consumption by ferrous iron dissolved from the granite mineral phases.

Steady-state fluid phase concentrations were achieved for most chemical components during the course of the experiments, the exceptions to this showing close approach to steady-state. Consideration of the XRD, SEM and modelling data suggest the importance of the following solids in controlling the aqueous phase concentrations of various chemical components (in parenthesis): smectite (H^+, Al, Mg); amorphous silica (SiO_2); albite (Na, Al); alkaline-earth carbonates (Ca, Sr, Ba, CO_2); zektzerite-emeleusite (Zr, Fe); lithium-sodium borosilicate hydrate (Li, B); ameghinite (B); willemite/smithsonite (Zn); pollucite (Cs); stillwellite (La, Ce, Nd). The bulk of these phases precipitate in the 200°C and 350°C experiments rather than at 100° or 150°C but an assessment of the thermal stability data of these phases available in the literature suggests that most will be stable at 100°C or less. The precipitation of these phases at 100° and 150°C was therefore prevented/hindered by sluggish kinetics rather than by thermodynamic instability. The granitic mineralogy persisted throughout the experiments, although corrosion of phases was pronounced at 350°C. Only chloride was destroyed during reaction (at all temperatures). Glass persisted at 100°, 150° and 200°C, but was totally consumed at 350°C. Natural analogue assemblages of the alteration mineralogy occur as late stage veins or pegmatites in granitic or alkaline igneous rocks.

Results of the experiments confirm the feasibility of the 'fate of nuclides' approach to source term modelling. Radionuclide release from a high level waste repository is most realistically modelled as a function of solubility and groundwater flux. There is a general coherence of geochemical processes operating across the temperature range investigated, suggesting that high temperature experiments (T>200°C) are a convenient means of accelerating low temperature processes. The observed differences between results at different temperatures are attributable to differences in precipitation kinetics rather than fundamental differences in reaction mechanisms. The effects of increased temperature are to slightly increase solubilities or, in some instances, to decrease solubilities of chemical components.

ACKNOWLEDGEMENTS

The authors wish gratefully to acknowledge contributions to this paper by Professor G. de Marsily of the Ecole des Mines de Paris and Dr. D. Davage of the British Geological Survey. They are also indebted to Dr. D.P. Hodgkinson of the Theoretical Physics Division at the Atomic Energy Research Establishment, Harwell.

The French and United Kingdom governments contributed to the costs of the work reported here. This paper however contains no implication about the policies of these bodies.

REFERENCES

ANDERSSON, J., SHAPIRO, A.M., BEAR, J. (1984). A stochastic model of a fractured rock conditioned by measured information. Water Resources Res., 20(1), p.79-88.

BEUCHER, H. (1983). Contribution de la géostatistique à l'étude d'un milieu fracturé et mise au point d'un modèle hydraulique sur le site de la mine de Fanay-Augères. Ecole des Mines, LHM/RD/83/10, Report for DGRST.

BEUCHER, H., BLANCHIN, R., FEUGA, B. (1984). Application des méthodes géostatistiques à la prévision de la fracturation d'un massif granitique. Paper presented at the "Jourlées sur le Granite", Orléans, 26.06.84. Document BRGM No.84.

BEUCHER, H., MARSILY, G. de (1984). Approche statistique de la détermination des perméabilités d'un massif fracturé. Ecole des Mines, LHM/RD/84/6, Report for CEA and CEC.

BOURKE, P.J., BROMLEY, A.V., RAE, J. and SINCOCK, K. (1981). Technfiques for investigating flow through fractured rock. Proc. of NEA Workshop: Siting Radioactive Waste Depositories, PFaris, Mfay 1981, p.173-190.

BOURKE, P., EVANS, G.V., HODGKINSON, D.P. and IVANOVITCH, M.I. (1982). Water flow and radionuclide transport through fractured rock. Proc. of NEA Workshop: Geological Disposal of Radioactive Waste, Ottawa, September 1982, p.189-199.

BOURKE, P.J., DURRANCE, E.M., HODGKINSON, D.P., HEATH, M.J. (1985). Fracture hydrology relevant to radionuclide transport. Report AERE-R 11414.

BRADBURY, M.H., LEVER, D.A. and KINSY, D.V. (1982). Aqueous phase diffusion in crystalline rock. Scientific Basis for Radioactive Waste Management - V, 1982 Elsevier.

BRADBURY, M.H. (1985). AERE Harwell. Priv. Comm.

CHARLAIX, E., GUYON, E., RIVIER, N. (1984). A criterium for percolation threshold in a random array of plates. Solid state communications, vol.20, No.11, p.999-1002.

DIEULIN, A. (1984). L'effect d'échelle en milieu fissuré. Ecole des Mines, LHM/RD/84/17, Report for CEA and CEC.

DORMUTH, K.W. (1984). Hydrogeological modelling. Paper presented at the fourth Euratom AECL Meeting, Ispra, 12-14 September 1984.

DURANT, E., LEDOUX, E., MARSILY, G. de, PEAUDERCERF, P. (1984). Edute de l'effet d'échelle en milieu fissuré. Phase pilote: certification du site de Fanay-Augères. Report BRGM 84 SGN 237 STO, for CEA and CEC.

ENDO, H.K., LONG, J.C.S., WILSON, C.R., WITHERSPOON, P.A. (1984). A model for investigating mechanical transport in fracture networks. Water Resour. Res., 20(10), p.1390-1400.

GOBLET, P. (1984). General review of methodologies and approaches in mathematical models for interpretation of tracer data in hydrological systems. Ecole des Mines, LHM/RD/84/64, paper presented at an IAEA workshop, Vienna, September 1984.

GOBLET, P. (1984). Extensions du code METIS. Rapport d'avancement au 15.12.84. Ecole des Mines, LHM/RD/85/20.

HEATH, M.J. and DURRANCE, E.M. (1984). Radionuclide movement through fractured rock. AERE-R 11402.

HEMINGWAY, S.J., BRADBURY, M.H. and LEVER, D.A. Effect of dead-cut porosity in rock matrix diffusion. AERE-R 10691, 1983.

LEVER, D.A. and BRADBURY, M.H. Rock matrix diffusion and its implication for radionuclide migration. Proc. of Mineralogical Society Symp.: Mineralogical Aspects of Disposal of Toxic Wastes. London November 1983.

LONG, J.C.S. (1983). Investigation of equivalent porous medium permeability in network of discontinuous fractures. Ph.D. Thesis, Lawrence Berkeley Laboratory, Univ. of California.

MARSILY, G. de. Flow and transport in fractured rocks: conectivity and scale effect. Paper presented at the IAH Symposium on the Hydrogeology of Rocks of Low Permeaility, Tucson, Arizona, 7-12 January 1985.

NERETNIEKS, I., RASMUSSON, A. (1984). An approach to modelling radionuclide migration in a medium with strongly varying velocity and block sizes along the flow path.

ROBINSON, P.C. (1984). Connectivity flow and transport in network models of fracture media. Ph.D. Thesis, Oxford Univ. AERE Harwell TP 1072.

ROULEAU, A. (1984). Statistical characterisation and numerical simulation
    of a fracture system. Application to a groundwater flow in the Stripa
    granite. Ph.D. Thesis, Univ. of Waterloo.
SCHWARTZ, F.W., SMITH, L., CROWE, A.S. Stochastic analysis of groundwater
    flow and contaminant transport in a fractured rock system, paper
    presented at the Symposium on the Scientific Bases of Nuclear Waste
    Management, Mater Res. Soc., Boston 1981.

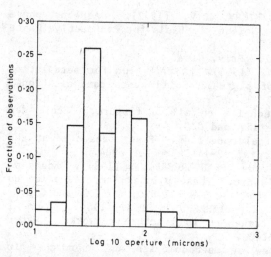

FIG.1. HISTOGRAM OF FRACTURE APERTURES.

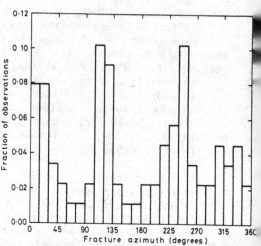

FIG.3. HISTOGRAM OF FRACTURE AZIMUTHS.

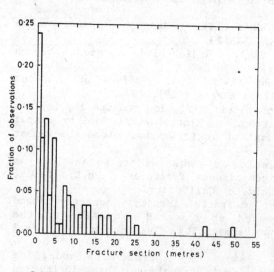

FIG.2. HISTOGRAM OF FRACTURE SEPARATIONS.

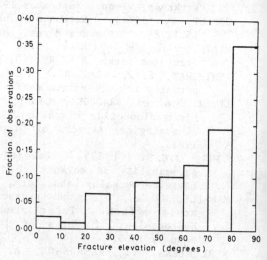

FIG.4. HISTOGRAM OF FRACTURE ELEVATIONS.

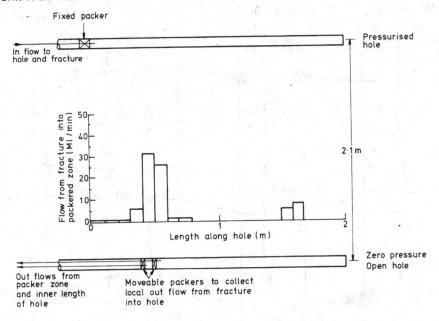

FIG.5. CHANNELLING OF FLOW IN A SINGLE FRACTURE

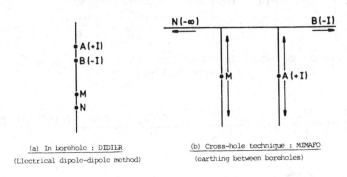

(a) In borehole : DIDIER
(Electrical dipole-dipole method)

(b) Cross-hole technique : MIMAFO
(earthing between boreholes)

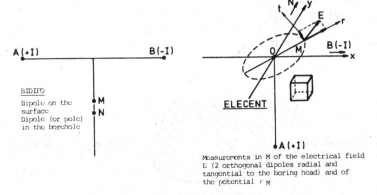

BIDIFO

Bipole on the
surface
Dipole (or pole)
in the borehole

ELECENT

Measurements in M of the electrical field
E (2 orthogonal dipoles radial and
tangential to the boring head) and of
the potential r_M

A and B : Transmitting electrodes
M and N : Electrodes for measurement of potential

(C) Hole - surface

ELECTRICAL EQUIPMENT USED AND MODELLED

Fig. 6

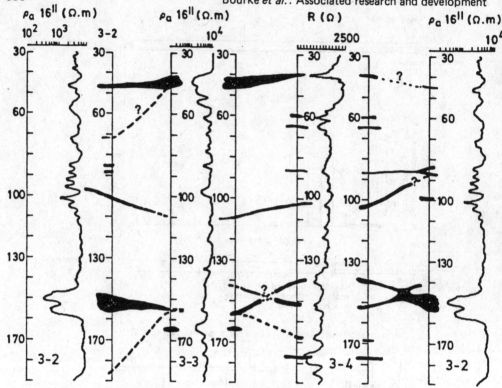

(a) Le Mayet-de-Montagne (electrical connections between boreholes are shown in black)

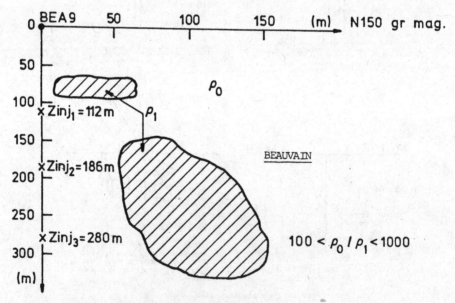

(b) Beauvain : conductive bodies indicated by hatching

INTERPRETATION OF BOREHOLE GEOPHYSICAL SURVEYS

Fig. 7

DISCUSSION

R. ROMETSCH, NAGRA Baden

There is a small number of fractures which must be avoided. In the
preceding presentation, it has been stated that the average conductivity was around 10^{-11} to 10^{-12} m/sec. Is it possible to be more specific? What does a small number of large fractures mean? What is their
dimension and what kind of permeability should be avoided in realizing
a repository?

P.J. BOURKE, AERE Harwell

In Cornwall, a large number of small fractures were found, which show
continuous distributions of the variables mentioned in the presentation. But one very large fracture was detected at about 700 m depth,
which cannot be included in the statistics. Similar observations were
made at other sites. At Whiteshell, going down to nearly a kilometer
depth, three major rubblized zones have been encountered. In Switzerland, the observations available show that there might be a few major
ones, I understood less than 10. The flow in these zones is almost
certainly rapid and may well be directed back to the surface. The
repository should be located as far as possible from them, preferably
half way between two of them. From the limited amount of data available for the sites, the distance between them may be hundreds of
meters. These hundreds of meters of rock with only small fractures may
be the main barrier and it is to be hoped an adequate one, if there is
rapid flow to the surface through the major fractures.

R. DE BATIST, SCK/CEN Mol

In the abstract, some results on dissolution of vitrified waste are
indicated, which have not been mentioned in your presentation. What
were the main results?

P.J. BOURKE, AERE Harwell

The general results of this work, which was done by the British Geological Society, is that dissolution of waste incorporated into glass
showed low dissolution rates which, not surprisingly, increased with
temperature.

<u>BACKFILLING AND SEALING OF REPOSITORIES</u>
<u>AND ACCESS SHAFTS AND GALLERIES IN CLAY,</u>
<u>GRANITE AND SALT FORMATIONS</u>

L.M. LAKE   (Mott, Hay & Anderson, Croydon)
I.L. DAVIES   (Taylor Woodrow Construction, Southall)
F. GERA   (ISMES, Bergama)
M. JORDA   (CEA, Fontenay-aux-Roses)
T. McEWEN   (British Geological Survey, Keyworth)
B. NEERDAEL   (CEN/SCK, Mol)
M.W. SCHMIDT   (Gesellschaft für Strahlen-und-
                 Umweltforsthung, Braunschweig)

<u>Summary</u>

The paper summarises the work carried out under ten Commission
contracts in the field of backfilling and sealing radioactive waste
repositories. It covers theoretical, laboratory and field trials and
experiments involving three potential host types, namely clay, salt
and hard rock. It concludes that maximum opportunity should be taken
over the next 15 to 25 years with a view to obtaining first hand
experience in real ground with real wastes.

1.   <u>Introduction</u>

1.1 Investigatory boreholes and shafts, as well as the shafts, galleries
and chambers used in development and operation of a repository may provide
potential preferential pathways for radionuclide migration to the
biosphere. Attention must be paid initially to the manner in which these
works are carried out to minimise the number and magnitude of potential
pathways and subsequently, in the backfilling and sealing of those
openings essential to repository development. (Ref.1).

1.2 The Community research programme has considered disposal in granite,
clay and salt formations and three important areas have been addressed :-

(i)   Buffer materials in the near field (close to the waste
containers);
(ii)   Backfilling and sealing materials and processes to be used in
galleries and shafts;
(iii) Sealing materials and processes to be used in boreholes drilled
from the surface or from within the repository.

1.3 The subject of the research work to be presented in this paper, fall
under five headings :-

(i)   Identification of the requirements and the selection of likely
backfill materials based upon existing knowledge, literature surveys and
consultation with manufacturers.
(ii)   Laboratory trials with suitable materials including :-
       - host clay mixtures for backfilling in clay;
       - processed clays and pre-placed aggregate concrete for
          backfilling in granite;

        - crushed salt for backfilling in salt.
(iii)   Study of placement techniques and development of prototype scale
test models to investigate practical processes and verify predicted
behaviour.
(iv)   Study of the conditions that arise at the interfaces between the
buffer and backfilling materials, the ground support, and the host rock,
all of which could have an adverse influence on radionuclide containment.
(v)   Investigations into the development of non destructive tests and
remote sensing devices to control and monitor the efficacy of backfill and
sealing materials.

In order that the subject areas presented can be easily identified and
pursued, the research work reported in this paper is presented
sequentially, in alphabetical order by country.

2. Belgium - Clay Host   Ref.2

2.1 The Mol test facility constructed within the Boom clay formation, is
at an advanced stage of development. Between 1980-84 different clay based
mixtures, using Boom clay as the principal component, have been studied as
potential buffer and general backfill for boreholes, freeze pipes and
behind the gallery lining. Initially, testing concentrated upon
formulating and selecting various mixtures and establishing their basic
mechanical properties. Currently, testing is devoted to investigation of
the permeability at the interfaces between selected mixtures and the clay
host.

2.2 The experimental mixtures investigated comprised Boom clay, bentonite,
calcium carbonate and Portland cement. Simultaneously, mixtures of
thermohardening resins (polyurethane or epoxy) with Boom clay were also
investigated. The properties determined included tensile and compressive
strength, setting time, shrinkage, density, porosity and permeability.
Attention was also paid to practical matters relating to placement
conditions, particularly placement water content and emplacement and
injection technology, using commercially available grout pumps. The
Institut National des Industries Extractives at Liege made valuable
contributions to this work.

2.3 The most promising mixtures and formulations derived from the testing
programme comprised :-

		Parts %		
		(i)	(ii)	(iii)
remoulded Boom clay	:	92	87	45
bentonite CLARSOL FTPI	:	3	3	-
limestone	:	5	5	-
cement P50	:	-	5	-
3 component epoxy resin	:	-	-	55
water content (by volume)	:	60%	65%	-

Another mixture, consisting of 50 cc drilling bentonite with one litre
PPz pozzolanic cement with 0.75 litre of water to each litre of cement was
used to seal existing vertical freeze pipes and laboratory tests were also

carried out on this material.

2.4 Currently, research is focussed upon a study of the permeability in the interface zone between the selected mixture and the host clay. To derive the coefficients of bulk permeability of the clay host and selected mixtures, standard soil mechanics consolidometers (oedometers) were used. To determine interface and mass permeability directly, a 'macro permeameter' was developed by the Laboratoire de Genie Civil of the University of Louvain-la-Neuve. The horizontal and vertical permeability can be measured, using cylindrical specimens 168 mm diameter by 150 mm thickness.

2.5 In effecting the test, great care is taken in the preparation of the test specimens and in the manufacture of the watertight cell (to prevent water losses other than through the test specimen). All specimens are thoroughly saturated and the pressure gradient through the specimen is carefully regulated. An important innovation is the fabrication of a 'porous stone' annulus to surround the specimen, (when measuring horizontal permeability), using a sand/resin mixture.

2.6 Preliminary testing confirmed that boundary conditions were controlled (using a special glue to ensure watertightness) and that the measurements made were reliable. Data from tests on the clay/bentonite/limestone mixture showed that no significant hardening occurred and that the vertical mass permeability ranged from 10^{-10} to 10^{-12} m/s, agreeing closely with data obtained from standard oedometric tests. Currently, tests are determining the mass horizontal permeability of the first mixture and experiments of the interface zones are to commence soon. The three remaining mixtures will be subject to similar test programmes.

2.7 In the field, experiments have been carried out to extract a number of the vertical freeze pipes used to sink the test facility shaft. Ten of the thirty two freeze pipes were extracted using hot brine, 100 ton hydraulic jacks and a standard drilling machine. Conventional drilling muds were necessarily used to seal these freeze pipe holes. In conjunction with ISMES attempts are to be made to develop wireless instrumentation to verify the quality of insitu plugs.

3.   France - Synthetic and natural clay materials Ref.3

3.1 It is recognised that there is a need for stress-strain and thermal compatability between waste units, buffer backfill and host ground under varying and complex thermally induced field conditions. Discontinuities forming either within the buffer or host ground or at the interfaces must be avoided both to prevent the ingress of corrosive elements and the egress of radionuclides. Further, the buffer material should possess good regulating physico-chemical attributes to minimise corrosion processes and absorb and/or delay radionuclide migration.

3.2 Two basic groups of materials are being studied by CEA for use as buffer material, a magnesium rich synthetic material ('Georoc') and natural clay. The mechanical properties and physico-chemical properties are being investigated. Studies are also being undertaken to investigate the thermo-mechanical behaviour of the backfilling in relation to the vitrified waste and the host rock (granite).

3.3 The synthetic (Georoc) material under investigation is subject to a
French Patent. It has known expansive properties, low permeability, good
aggregate binding and high strength and modulus characteristics, but very
low thermal conductivity. It also has a low solubility and would create
an alkaline environment, desirable in the near field situation. Studies
are being undertaken into mixtures combining natural clays with this
manufactured complex magnesium oxide and chloride substance in order to
improve overall performance, but there is concern over the manufacturing
capability and hence the supply situation on a major industrial scale.

3.4 Research is also continuing into the substitution of naturally
occurring clays for bentonite clays, in order to achieve the optimum
performance at an optimum cost. Sodium bentonite from Wyoming (Volclay
MX80) is being used as the reference material, because its properties are
well documented. Six clays have been studied, four calcium Smectites,
one sodium Smectite (artificially transformed from the calcium form) and
one illite. In all the naturally occurring clays, the clay sized
particles (argillaceous phase) accounts for at least 75% of the material
and in most between 95 and 100%.

3.5 It has been demonstrated for clays compacted at low moisture
contents (0.5% for the smectitic and 4.5% for the illitic clays)that :

(i)   With low compactive effort, the nature and granulometric
composition of the clay governs achievable density.
(ii)   With high compactive effort (100 MPa) achievable density is
dictated by clay type; illite is more resistant than smectite.
(iii)   The free swelling potential of heavily compacted clay in
unconfined conditions depends upon the clay type (smectites swell more
than, but more slowly than illites).
(iv)   The swelling pressures developed by clays at uniform density when
hydrated in confined conditions depend upon the clay type (6 MPa for Na
smectite, 5 MPa for Ca smectite and 4 MPa for illite).

       Future testing will examine the mechanical and compaction responses
of the clays at different water contents.

3.6   The third element in the research programme has been concerned with
modelling the thermomechanical behaviour of the buffer materials and the
effects upon, and the constraints imposed by, the waste units and the
host ground. Initially, only thermal effects have been considered,
assuming the ambient ground loading is developed in the buffer backfilling
and neglecting the swelling potential of the clay backfill. This approach
was adopted in the absence of any knowledge on the likely availability of
free water in the system or on the coupling of hydration and thermal
effects.

3.7 The repository model used provided 30 m deep emplacement shafts at
35 m centres at a depth of 500 m. Each shaft was assumed to accommodate
15 No. containers, each 0.43 m diameter and 1.33 m high, with a temperature
of 120°C in the waste unit and 115°C in the buffer material continuing for
20 years; the temperature falls thereafter. It was concluded that under
the assumed conditions :

(i)   The thermal effects produce a compressive state of stress in the
backfill which is less than the available compressive strength of the host

ground would not be adversely affected.
(ii)   The stresses induced in the vitrified waste arise from differences
in the coefficient of thermal expansion and induce only a small stress
increase in the backfill.

4.   Germany - Salt   Ref.4

4.1   Research and development in backfilling and sealing studies in
Germany is centred upon the abandoned Asse salt mine although tests are
being made elsewhere. Use is being made of an existing salt mine located
some 750 m below surface to carry out in-situ observations and tests as
well as backfilling and sealing trials. The programme includes :

(i)   Investigations into backfilling disposal chambers, the voids
between waste containers and the galleries and drifts.
(ii)   Investigations into sealing the chambers, galleries and boreholes.
(iii)   Backfilling and sealing the mine openings and shaft.

    The main objectives are to prevent the distribution/release of radio-
nuclides via brine flows, reduce the number of possible brine flow paths
and fill all voids within the repository environment.

4.2 The current backfilling programme falls into three categories :

(i)   Geotechnical in-situ experiments.
(ii)   Complementary laboratory investigations.
(iii)   Handling and placement trials.

4.3 Insitu experiments have included observations on the comparative rates
of convergence of open and backfilled (60 years age) drifts and measuring
the convergence rate and density of the old backfill. Supporting
laboratory testing has measured permeability, compressive and tensile
strengths and creep velocities.

    Observations have also been made on the behaviour of gravity stowed
salt and to determine the influence of the backfill upon the deformation
of stalls and pillars. Extensometers, inclinometers, flat jacks and
settlement cells have been used for these purposes and associated testing
has included drop penetration tests and density measurements.

4.4 The complementary laboratory testing has involved : confined
compressive tests on crushed salt and other process residues from the
factory plant, to determine load deformation behaviour; unconfined and
triaxial compression testing; permeability measurements (using oxygen) on
compacted salt-grit specimens; permeability testing using saturated Na Cl
brine on crushed salt specimens, at varying densities, porosity, grain
size distributions and size ranges.

4.5 In the general field of sealing, work has been carried out into preselection and pre-testing of suitable materials, testing of seals in
chambers, drifts and boreholes, measurements during the construction of
seals and controls of the construction process. The current programme
has involved the construction and monitoring of two chambers and one drift
seal at Asse salt mine and the construction and monitoring of a drift seal
in the flooded Hope salt mine (near Hanover); a constant brine pressure
of 60 Bars applied at the latter. The observations accompanying these

works include measurement of pressure, temperature, permeability,
displacements, humidity in the seal and rock deformation.

5. Italy - Borehole Sealing Ref.5

5.1 Research in Italy has been concentrated in two areas :

(i)   Laboratory testing and evaluation of materials to seal boreholes
and particularly the permeability and bonding properties of the materials
with potential host clays.
(ii)   Development of instrumentation and procedures for the insitu
testing of sealed boreholes.

5.2 Laboratory tests have been conducted on two scales, in cells taking
specimens between 3.8 and 7.0 cm in diameter and in a larger cell taking
a 24 cm diameter specimen. The latter cell can be loaded to 11 MPa and is
fitted with a drill to enable model boreholes to be made and enable plugs
to be installed. The plugged borehole zone can be loaded to the same
intensity or differentially from the rest of the test specimen.

5.3 The results obtained to date show :

(i)   The mass permeability of a cement plug is lower than that of the
clay.
(ii)   When cured under different ambient conditions, but without
external loading, no bond appeared to develop between the clay and the
cement grout or mortar forming the plug. The same observation was made
whether or not an expansive agent was used.
(iii)   The flow of water through a test specimen comprising a clay core
surrounded by remoulded clay is greater than that through undisturbed
clay. This observation is based on tests at a low confining pressure,
maximum 500 kPa.
(iv)   When cured under external loading, good bonding developed between
the clay and cement mortar.
(v)   No significant difference in the water flow occurred in
undisturbed clay and clay with a cement plug.

5.4 Ideally, a borehole plug should not effect any disimprovement in the
vertical permeability of a clay host. However, verifying the effectiveness of plugs in very low permeability materials is extremely difficult
and, conventionally utilises hydraulic systems with connecting tubes or
data transmission cables. To avoid these difficulties, investigations
are being conducted into a wireless communication apparatus. Tests are
scheduled to commence in Spring 1985 and the apparatus will incorporate a
water injection unit, a pressure transducer and an acoustic data
transmission system. The signal detector may be located on the top of the
plug under test or in a second nearby borehole. For transmission over
greater distances it is thought that the acoustic emitter could be
substituted by a very low frequency electro-magnetic wave emitter.

The overall borehole permeability will first be determined using a
conventional double packer water injection system and then repeated
with a plug in position using the wireless apparatus.

6.   United Kingdom - Borehole Sealing Ref.6

6.1 Use has been made of deep boreholes located on the Harwell site to investigate the efficacy of conventional oil industry practices in the cement sealing of boreholes. The principal objectives of the research programme were :-

(i)   To determine the reliability of existing geophysical borehole logging techniques to test seal integrity.
(ii)   To examine the chemical compatibility and bonding properties of conventional oil-field borehole cement grouts in argillaceous host rocks.
(iii)   To study the hydraulic properties of an existing cement seal in semi-plastic clays and relate these to potential shaft and borehole leakages from a deep repository.

6.2 The conventional Cement Bond Logging (CBL) and Variable Density Logging (VDL) were found unsatisfactory for seal evaluation. For this purpose, full acoustic wave train analysis was considered to be potentially more useful and worthy of investigation. Only one in ten acoustic transmission from the sonde is plotted, giving the equivalent of one wave train per 30 cm length of borehole. By direct observation and without further analysis, valuable information about the state of the cement-casing-rock bond could be deduced, far better than the CBL. The wave train signals were used to locate the poor bond sections which were then perforated to enable hydraulic conductivity tests to be carried out. In general the bonding between the cement and argillaceous strata appeared poor.

6.3 More detailed analysis of the geophysical logs has been carried out, producing crossplots of both amplitude and frequency versus transit time in which 100 μ sec windows are used on the original wave train. Also, amplitude and frequency versus transit time plots displayed as individual depth registered functions have been generated. Neither approach has proved particularly useful in indicating the nature of the cement bond, although perhaps more sophisticated wave train analysis may be more successful.

6.4 Cored rock samples from the test borehole had plugs of cement paste inserted and were cured for more than one year. They were then systematically examined with the scanning electron microscope (SEM) for evidence of reaction and bonding. Reaction was common between the cement and the opaline silica (in the Gault and Upper Greensand), resulting in a gel formation which generates expansive forces and tensile fracturing; these gels can be lost by solution in migrating ground waters. Another reaction observed occurred between the dolomitic carbonate in the host rock and alkalis from the cement when in the presence of illite. This reaction is not understood but may be dedolomitisation which also generates an expansive force.

6.5 As the borehole walls were coated with drilling mud before the cement grout was introduced, this may have prevented boundary reaction and the development of bond between the cement and host rock. Certainly, the electron microscope studies are consistent with the borehole geophysics in showing poor contact across the interface in the argillaceous zones.

7. United Kingdom - Design solutions to interface problems Ref.7

7.1 It is envisaged that alpha-bearing radioactive waste will be
disposed of in repositories in geological formations whose properties
are such that isolation of the waste can be achieved. Although there
is every likelihood that suitable formations can be identified, there is
a possibility that the engineering works necessary for construction and
operation of the repository may reduce the efficacy of this isolation.
This study examined the backfilled tunnels of a sealed waste repository
as potential weaknesses in the geological barrier.

7.2 A variety of host materials and of tunnel forms were investigated.
For analytical purposes the host formations were considered as either
fissured or as porous media. The significant zones for groundwater flow
along a backfilled tunnel were identified. For study purposes four
distinct zones were considered - the host rock, the "disturbed zone", the
backfill proper and the interface between the "disturbed zone" and the
backfill proper. Depending on conditions, the "disturbed zone" or the
interface itself were identified as the significant flow paths.

7.3 The processes involved in the creation of a disturbed zone were
examined and it was found that the disturbed zone was a function not only
of the excavation method, but also of the stress field around the
repository and the mechanical behaviour of the host material. The
development of a disturbed zone is inevitable, although techniques can be
applied to minimise the extent of this zone.

7.4 The study showed that the disturbed zone is only a preferential flow
path in dilatant materials, and techniques for controlling flow in these
materials have been considered. It was found that control of water flow
was extremely difficult and so, techniques for the interception of
nuclides and the control of their migration by sorptive processes have
been developed. The relationship between the geometry of disturbed zone
and sorptive collar and the efficiency of retention was investigated, and
theoretical radionuclide breakthrough curves for selected radionuclides
have been generated. The principal radionuclide studied was 237 Np, which
is particularly difficult to retain. Comparative studies with 239 Pu were
made for certain flow conditions, and it was found that techniques
developed to retain 237 Np readily retain 239 Pu.

7.5 Consideration was also given to flow in an interface fissure between
the backfill and the host material. Consideration of the performance of
cement-bound and earthen fill materials led to calculation of a conceptual
width for this separation plane. A sensitivity study on the variation of
system performance over a wide range of fissure widths was performed, and
it was shown that the system would continue to work satisfactorily for
separations substantially greater than the reference case aperture of
100 μm.

7.6 The importance of designing the shape of collars with respect to the
ground stresses and the backfilling process was identified and the
placing techniques for a range of impermeable backfilling materials
discussed in detail. The problems involved in creating a porous but
unfissured medium in the sorptive collar have been examined.

7.7 Detailed study was concentrated upon those parts of the repository

remote from the emplaced waste canisters and although no detailed
consideration has been given to thermal effects or to radial migration
of radionuclides, a limited qualitative discussion of near-field
phenomena has been presented.

8.   United Kingdom - Simulated repository facility   Ref.8

8.1 The objectives of this study were twofold :

(i) To establish the necessary testing programme to select suitable near-
field buffer materials and assess buffer placement techniques.
(ii) To assess the viability of the construction of a large scale model
to investigate the interaction between waste package, buffer material and
host material under realistic repository conditions.

8.2 The study showed that the extent of the coupled interactions within
the buffer/host system is such that extensive decoupled testing of the
various parameters of behaviour would be required. Many of the coupling
phenomena only become evident in long term tests and to obtain results in
a reasonable time would require strict control to eliminate coupled
effects. To this end, a number of subsidiary facilities to investigate
individual parameters or limited degrees of coupling have been proposed.

8.3 The principal parameters of interest in repository design are the
mechanical, thermal and hydraulic behaviour of the buffer material and
the host rock. In the case of particulate media, such as buffer
materials, the bulk properties depend crucially on the nature of the pore
fluid and, at the time of buffer placing, this fluid would be partly air
and partly water. The host material, due to drainage into the repository
during the operational life of the repository, will also be only
partially saturated. This is important since the mechanical, thermal and
hydraulic properties vary strongly with saturation, and saturation in the
repository proper will vary with time.

8.4 One of the principle objectives in creating a large scale block
model was to create a "host rock" of known, homogeneous, isotropic
material. This material would contain embedded instrumentation for the
measurement of material response to waste emplacement. The study failed
to identify a material that is capable of modelling the mechanical
behaviour of the various potential host materials and that could be made
to an adequate degree of homogeneity in the sizes required. The study
also found that within the vicinity of the emplacement hole there would
be instrumentation difficulties.

8.5 A range of modelling difficulties arise with a concrete block, the
mechanical, thermal and hydraulic properties of which vary with pressure,
temperature and moisture content - but not in the same manner as those
of any of the host materials. As an example, the properties of granite
are insensitive to moisture, whereas those of concrete are sensitive; at
least, irrigation experiments are impossible. The properties of salt
are markedly time dependent, those of concrete are not. The properties
of mudrocks are markedly anisotropic, those of concrete are not. More
indirectly, the degrees of coupling differ and where similitude has been
achieved at one temperature or stress in one part of the nodel it may
not be achieved elsewhere.

8.6 The study revealed further problems with the modelling material
arising from the need to hold the block together. Substantial stresses
would be needed to prevent the opening of 'macro-cracks' at the joints
in the model and these stresses would be sufficient to initiate micro-
cracking within the concrete and change its properties.

8.7 In view of the difficulties inherent in modelling, the benefits of
using a substitute host material have been questioned and the viability of
using actual host materials re-examined. Further consideration has been
given to the mechanical and thermal boundary conditions at the outside of
the block, and it is shown that the assumption regarding the boundary
conditions will be a major factor in the parameters measured. It is
pointed out that there would be a real physical 'boundary value problem'.
The possibility of in-situ experimentation has been re-examined.

9. United Kingdom - Role of concrete in backfilling and sealing Ref.9

9.1 Concrete is a widely used engineered material, for which the placement method, strength, durability and permeability can be pre-selected
and assured. Three research approaches have been effected into the
potential use of concrete :

(i)   Mathematical modelling.
(ii)   Laboratory trials with Preplaced Aggregate Concrete.
(iii)   Archaelogical research into concrete.

9.2 It was judged necessary to carry out parametric analysis of the
multiple interactions relevant to repository performance sensitivity and
damage assessment. These included mechanical, thermal, creep and moisture
effects. Finite element analyses were carried out using computer codes
ADINA (stress/strain), ADINAT (thermal) and TEMPOR (moisture front/drying
study). The models took account of thermal, excavation, rock creep,
lithostatic stresses in the backfill and differential expansion and
contraction, due to cement heat of hydration. The postulated drying front
condition generated by heat emitting waste was also studied.

9.3 The analysis indicated that creep in a granite host rock would be
small and should have no significant adverse mechanical affect upon
concrete inclusions. With full lithostatic loading, having a marked
difference in the vertical/horizontal stress field and with waste unit
inclusions, no significant zones of concrete damage were predicted.
However, tensile stresses induced by heat of hydration with large
concrete pours in a 6 m diameter tunnel were at the limits of acceptability
and, therefore of concern. (Pre-cooling the mix water or the use of
additives may improve this situation).

     The investigation of the drying front postulation suggested that at
the temperature levels and gradients likely to be permitted, no front
would form.

9.4 Pre-placed Aggregate Concrete (PAC), compared with conventional
Portland cement concrete, is claimed to have reduced drying shrinkage,
creep, heat of hydration and overall permeability and improved mechanical
properties (density, strength etc). Laboratory trials have been carried
out to develop a mix design for a grout that did not contain organic

additives (flow aids or retarders) but that could still be injected into
pre-placed aggregate to form PAC. A high water content grout is
required to achieve flow and penetration, but this is subject to 'bleed'
(water loss) consequently, a grout incorporating bentonite was developed
to overcome this difficulty.

9.5 A 1.0 m cube of PAC was formed using the developed grout mix and
cores have been taken for testing; these include :

(i)   Compressive and tensile strengths and elastic modulus.
(ii)   Creep and shrinkage.
(iii)   Thermal expansion and conductivity.
(iv)   Permeability.

The tests are continuing, but initial results suggest that the
expected improvements have not been obtained with the concrete produced.
This is attributed to the high cement content, high water : cement ratio
and low sand content which are necessary because the use of additives was
prohibited.

9.6 Pozzolanic and similar cements have been used for thousands of years
and the collection and study of archaelogical concretes is being undertaken, with the objectives of gathering information on durability,
chemistry and "time modification". It is anticipated that the information
will be invaluable to engineers attempting to assess various cements for
solidification of wastes and the durability of engineered barriers. The
oldest material collected to date is some 7000 years old, but information
and/or samples from other engineers and scientists would be most welcome
by the study group.

10. Concluding statement

10.1 The researches presented in this summary paper embrace theoretical
and practical study subject areas and laboratory and field programmes.
They have highlighted the difficulties and limitations in working with
scaled down models and prototypes and the difficulties that have to be
overcome in determining and verifying the performance and efficacy of
components in the engineered barriers. The level of accuracy required in
the measurement of performance and the extrapolation of these observations
to the time scale required for long term repository security are outside
existing engineering experience. Perhaps this points the way to
focussing attention upon the desirability of moving towards controlled
field trials with real wastes, in real ground with realistic engineering
construction and conditions. Such an experiment must of necessity, have
a built in recovery option. In most European countries, large scale
disposal of HLW can reasonably be delayed for say 15 to 25 years and this
available time should be employed gainfully and not simply confirming
research to theoretical studies and experiments carried out in splendid
isolation.

Acknowledgements

The authors of this paper wish to register the major contribution made by
all their fellow workers in the research work being undertaken with the
support of the Commission and their respective Governments or Agencies.

They also wish to acknowledge the special contribution made by Bernard Côme whose enthusiasm and encouragement has been an example to us all.

REFERENCES

1. CEC Contract 204-81-7 WASUK, MHA. Backfilling and sealing of radioactive waste repositories. EUR 9115 EN 1984.
2. CEC Contract 334-83-7 WASB, CEN/SCK. Backfilling studies.
3. CEC Contract 347-83-7 WASF, CEA. Backfilling in hard rocks.
4. CEC Contract 336-83-7 WASD, GSF. Flooding of repositories in salt; testing old backfill.
5. CEC Contracts 199-81-7 WASI, ENEA/ISMES and 401-83-7 WASI, ISMES. Study of clay materials and Wireless data transmission.
6. CEC Contract 346-83-7 WASUK, BGS. Sealing of boreholes.
7. CEC Contract 348-83-7 WASUK, MH&A. Study of interfaces.
8. CEC Contract 349-83-7 WASUK, MH&A. Simulated repository facility.
9. CEC Contracts 350-83-7 WASUK, TWC and 432-84-7 WASUK, TWC. Concrete materials and Historical concretes.

DISCUSSION

L. BAETSLE, SCK/CEN Mol

One possibility, which is to use pre-compressed bentonite as a means of sealing off heat sources and radioactive sources, was not dealt with; obviously, the reason is that no research on this sort of material has been performed in the Community programme. Is this system, which is proposed in Sweden, a reasonable and practical means of backfilling?

L. LAKE, MHA Croydon

There is no doubt that bentonite blocks can be made; they are fabricated in very high compression presses. But putting together blocks, either on an open site or inside a contained chamber, will lead to a lot of voids between blocks. In order to get swelling of the blocks in a controlled uniform fashion, controlled irrigation may be needed. Such a system will probably form one of the barriers, but all problems are not yet solved.

A. BRONDI, Casaccia

The main subject of the presentation was to describe support systems for cavities which form the repositories. It would also be interesting to look at artificial geochemical barriers, because it is known that many elements, and particularly the long-lived ones, the transuranic elements for example, could be blocked by the barriers. Adding organic elements could help the further creation of barriers so as to block the radionuclides, especially of transuranic type. Such barriers have allowed the conservation of metals, for example bronze of old greek art works, which would have been destroyed by exposure to oxidized water or air. Could a geochemical barrier act as a means of backfilling?

L. LAKE, MHA Croydon

Geochemistry is certainly important, but it was adequately covered elsewhere. Of course, it is a major component in protection and sealing.

Archeological approaches have been mentioned, but the main problem is the relatively short life of the archeological remains. In the South of England, a 400-year old boat was raised recently; survival relied upon rapid initial soil burial affording a complete seal. Wooden pipes laid by the Romans, still exist in London, again buried in saturated clay. It is encouraging to have these surviving artefacts, but the time gap is still great.

THE FEASIBILITY OF HEAT GENERATING WASTE DISPOSAL
INTO DEEP OCEAN SEDIMENTARY FORMATIONS

C.N. Murray
Commission of the European Communities
Joint Research Centre - Ispra Establishment
Radiochemistry Division
21020 Ispra (Va) - Italy

A. Barbreau
Commissariat à l'Energie Atomique
Centre d'Etudes Nucléaires de Fontenay
IPSN-DPT, B.P. 6, F - 92260
Fontenay-aux-Roses

J.R. Burdett
Building Research Establishment
Department of the Environment
Garston, Watford, GB - Hertz. WD2 7JR

ABSTRACT

Research is being undertaken by a number of countries of the OECD, including some member countries of the European Community, as a part of the NEA International Seabed Working Group 5-year Coordinated Programme on the feasibility and safety of the disposal of vitrified heat generating waste into deep ocean sedimentary formations.

The main objectives of the research are to assess the long-term safety of the option, to identify characteristic study zones in the North Atlantic and/or Pacific Oceans, to demonstrate the necessary engineering emplacement capability within oceanic geological formations.

Active participation by the Commission of the European Communities in the international investigations of the sub-seabed option began with the Joint Research Centre - Ispra Establishment, multi-annual R&D programme 1980-1983, and the cost-shared programmes DG XII, Brussels, 1980-1984.

The present paper briefly reviews the work undertaken to date by the Commission "Sub-Seabed Program" in collaboration with national programmes of member countries. Special emphasis has been placed on the studies of the characteristics of deep ocean sediments to act as a barrier to the dispersion of radionuclides and the technical investigations carried out to demonstrate engineering feasibility of the option.

1. INTRODUCTION

Research is being undertaken by a number of countries of the OECD, including some member countries of the European Community, as a part of the NEA International Seabed Working Group (SWG) 5-year coordinated programme on the feasibility and safety of the disposal of vitrified heat generating waste and unreprocessed fuel elements into deep ocean sedimentary formations. The main objectives of the research are to assess the long-term safety of the option,   to identify characteristic study zones in the North Atlantic and Picific oceans , to demonstrate the necessary engineering emplacement capability within oceanic formations. At present, two types of emplacement technologies are being investigated: free fall penetrator and drilling.

Three research and development programs of the Commission of the European Communities (CEC) are producing results related to seabed disposal of radioactive wastes:
- the cost-sharing action program "Radiation Protection";
- the cost-sharing action program "Management and Storage of Radioactive Waste";
- the direct-action program "Safety of Nuclear Materials", carried out at the Joint Research Centre (JRC), Ispra.

The indirect-action programs are carried out through contracts with national laboratories in the European Community. The CEC partially finances these laboratories. The staff of the Directorate-General XII for Research, Science and Education (Brussels) devises and supervises the indirect-action programs.

The Seabed Working Group (SWG) is composed of the following countries: Canada, Federal Republic of Germany, France, Japan, the Netherlands, Switzerland, the United Kingdom, the United States of America and of the Commission of the European Communities. Belgium and Italy have observer status. It performs its functions by promoting the exchange of information so that the programs of the member countries may be more efficient by avoiding unnecessary overlap. The studies of the SWG are conducted by a number of Task Groups as shown in Fig. 1. These cover the areas identified at present as being necessary for the assessment of the sub-seabed option.

The study zones in the Atlantic which are at present being extensively investigated especially in Europe, are Great Meteor East in the Madera Abyssal Plain, and the Nares Abyssal Plain.

In order to undertake assessment, identification of the main environmental compartments must be made which will act as barriers to the movement of radionuclides. In Fig. 2 is schematically shown the different pathways which are under investigation. To these natural barriers can be added man-made or engineered barriers which will ensure a further reduction in the ability of radionuclides to migrate; these are:
i) very highly insoluble waste forms which will dissolve (only with great difficulty) in surrounding pore water;
ii) canister material which will isolate the solidified waste from the environment for long periods of time, either through corrosion resistance or by its thickness at known corrosion rates;
iii) backfilling material in the drilled emplacement system which will reduce the ability of the surrounding media to interact with the emplacement system.

Once the natural and man-made barriers have been identified, it is necessary to develop mathematical descriptions of them which correctly model the mechanisms controlling dispersion of the radionuclides contained in the waste. This requires data to be obtained from both laboratory expe-

riments and in-situ investigations /1/ so that input parameters to the models can be defined and eventually the models themselves verified under realistic conditions. Sensitivity analysis can now be carried out to identify
not only the major radionuclides producing a potential exposure to man but
also the main mechanisms and pathways which are important in the control
and dispersion of the nuclides in the environment /2/.

By giving input parameters a weighted range of values, it is possible
to estimate the probability that a given nuclide will produce doses to man
at different periods of time.

A further aspect of the assessment is to investigate different potential accidents which can be identified and to estimate their impact on the
environment and on man. Here, the need is to obtain statistical data on
fish catches, road, rail and ship transportation, population distributions,
etc., and to apply them to models which describe the overall functioning
of the disposal option, from the point of view of production in nuclear
power plants to their final transport and emplacement in a deep ocean sedimentary geological formation.

The present paper will concentrate on two aspects of the assessment:
the sediment barrier and the engineering technology for emplacement.

2. THE SEDIMENT BARRIER

In undertaking the assessment of the sub-seabed option, a major area
of attention has been the deep ocean sedimentary formations and the identification and investigation of the characteristics of these zones. These
areas are considered to represent one of the major barriers to the dispersion of emplaced waste materials and it is thus necessary to develop detailed understanding of the ability of the sediment column to retard radionuclide migration.

Two distinct areas in the sediment column have been considered for
both the penetrator and the drilled option. These are: the near-field and
far-field regions.

2.1 Near field

The near field can be considered to be that region which includes the
reactions and interactions of the canister, the waste form, the sediments
and pore waters under the influence of heat and radiation fields produced
by the initially hot waste. Although this region is physically small, it is
particularly important as it is the source term for the radionuclides moving into and through the far-field sediments. A number of phenomena within the near-field zone are being studied in order to gain a wider understanding of the mechanisms controlling dispersion.

2.1.1 Thermal fields and induced pore water movement

The thermal conductivity of deep ocean clay sediments are such that
the emplacement of heat generating waste will lead to the development of a
zone of significantly higher temperature around the canister compared to
the natural conditions of about 2^oC. Depending upon the age of the emplaced
waste temperatures of up to several hundreds of degrees may occur, leading
to a number of changes in the properties of the surrounding solid media and
associated pore waters. In order to assess the possibility of induced pore
water movement leading to accelerated dispersion through the sediment column, thermal transport models are being developed.

Attempts to characterise such effects have been carried out as part of

the drilled option feasibility study /3/, and in experimental studies in a
centrifuge /4/. Although more investigations of the effects of heat on
deep ocean sediments are required, two significant points resulting from
these studies are worth mentioning:

- the theoretical studies show that the temperature of the near field for
  HGW emplaced after cooling for 5 years is high: 500°C above ambient;
- under certain conditions, in centrifuge studies, the material (kaolin)
  in the vicinity of a heater canister exhibited horizontal cracking.
  Whether the cracks are a result of the experimental procedures or are indicative of a mechanism which could occur in situ in deep ocean sediments, is uncertain at this time;
- predictions suggest that for a 10 kWatt canister of 10 year old waste
  induced water movement in Pacific red clays would be of the order of 1 m
  in one thousand years.

2.1.2 Canister corrosion

In order to emplace and subsequently isolate waste material from the
surrounding media during high thermal output periods, two concepts for
containment are foreseen: the first uses relatively inexpensive mild steel
(or other material) for the canister construction, having very thick walls
making allowance for corrosion losses; the second uses a very corrosion-
resistant material such as titanium in a rather thin-walled canister.

In Europe the study of the thick-walled corrosion allowance option
assumes that wastes will be kept in intermediate storage for 50 years or
more and on final storage the maximum temperature within the sediment column will not be greater than 100°C. In reducing sediments as occurs at
some of the study sites, corrosion rates were some 3 orders of magnitude
lower than in oxygenated pore waters /5,6/.

The US assumption, at present, is that there will only be short interim storage of around 10 years, the temperature expected in the sediments
will thus be considerably higher, in the range of 200-250°C. The potential
canister material being investigated is the alloy Ticode 12 (Ti;0.8% NI;
0.3% Mo). Information to date shows that this exhibits very low corrosion
rates of about 1 µm per year. Investigations on hydrogen embrittlement (to
which Ti is susceptible) in conjunction with stress corrosion have shown
that hydrogen concentrations in excess of 1000 ppm are needed before significant breakdown of the mechanical properties of the alloy occurs.

2.1.3 Waste form

Borosilicate glass is the most favoured form for the solidification
and stabilization of high-level wastes in Europe with a number of plants
either already constructed, such as in France, or under construction. The
US is also considering SYNROC, a titanium oxide-based, ceramic waste form.

Most of the glasses proposed have densities of around 2.5 g.cm^{-3},
however, if they contain large quantities of ZnO or BaO, this can increase
to 3.2. The density of SYNROC is about 4.2.

It seems likely that almost 100% of the non-volatile fission products
will eventually be incorporated in the glass. Possible exceptions to this
are the fission product insolubles that remain as dissolver sludges during
the reprocessing of oxide fuel. The gaseous fission products (Kr and Xe)
and almost all the iodine escape from the dissolver are at present discharged from reprocessing plants. The actinide content of HLW is somewhat
difficult to define. Reprocessing attempts to remove all plutonium and
uranium for re-use whilst leaving the other actinides, are under way.

The rate of leaching is a measure of the speed at which radionuclides
incorporated in the waste motion pass into the pore water of the sediment

matrix. This rate has usually been derived by measuring the samples' weight
loss and dividing by the surface area of the sample, and the time taken
for the experiment. Data obtained up to now have usually been derived from
experiments done under high flow rates or at large leachant volumes, both
conditions which seem unlikely in the case of disposal into seabed sediments. However, a limited number of experiments have been carried out at
much lower flow rates /7/.

2.1.4 Chemical speciation

In order to understand the processes which govern the migration of materials within deep ocean sediments, it is necessary to know the speciation
of the elements in solution and suspension taking into account the physico-chemical properties of pore water around the canister and the alterations caused by the thermal and radiation fields on the near-field sediment mineralogy.

From investigations carried out by US workers, it has been shown that
the thermal loading from 10 year old reprocessed waste caused changes in
pore water characteristics inducing a sharp increase in pH as acid conditions were produced by the precipitation of a magnesium hydroxysulphate
phase and smectite. The formation of acid was then slowly neutralized by
phases such as feldspar.

The redox conditions within the sediment column are thought to be
controlled by the relative amounts of organic carbon and reduced iron-
bearing phases (detrital in origin), and oxidized iron and manganese minerals (authogenic phases).

A phenomenon which is of particular interest is the Soret effect. Investigations have shown that thermally driven diffusion can be very important in the near field. One of the main results is that major cations in
the pore water are able to diffuse up a concentration gradient. Under steady state conditions the concentration of these ions is approximately 40-
50% in pore waters less (in the hot area around the canister) than would
normally be the case /8/.

In order to correctly model specific ion activities of aqueous species, a number of models have been developed. Input data necessary for these
calculations include hydrolysis constants for minerals, thermodynamic equilibrium constants, solubility products and formation constants. It is clear
that at present there is a lack of such data and major effort is being made
to produce this information.

2.2 Far-field radionuclide transport

The far-field region of the sediment barrier can be defined as that
region which is unaware of the existence of buried waste from the point of
view of a significant temperature and radiation field, and beyond the zone
in which reprecipitation occurs.

The movement of radionuclides through the sediment can occur in a
number of different ways; diffusion through the pore water, advection of
the pore water in which the nuclides are dissolved or suspended, physical
displacement of the bulk sediment through thermal or erosional processes.

Erosional processes are known to occur in some parts of the deep ocean,
however, at the sites chosen for the sub-seabed option it can be shown
that the sediments have been both stable and predictable for many millions
of years.

Advection of pore water can occur through natural processes or possibly be induced by the hot canisters during the first few hundred years of
emplacement. It has been estimated that induced pore water movement will

be less than 1 m during the existence of a thermal field (500-1000 years).
Natural pore water advection, on the other hand, must be assessed for each
site /9/. Some recent data, however, has shown the occurrence of acoustic
discontinuities at both the NAP and GME sites. More detailed studies are
being carried out to assess their possible importance /9/. In general, the
values obtained are so low that they are indistinguishable from zero movement of the sites of interest to the seabed option. Movement of radionuclides by natural diffusional processes is extremely slow, even in the case
of soluble mobile ions having very low adsorption coefficients /10/. The
effect of the high ambient pressure which exists at these ocean depths,
is being studied by the JRC /11/.

To assess the efficiency of the sediment as a long-term barrier, it
is thus necessary to study the natural geochemistry of the sediment-pore
water system and to develop mathematically realistic models in order to
understand and predict the effect of the major controlling mechanisms in
the emplaced wastes.

2.2.1 Natural geochemistry

Deep ocean sediments consist mainly of very fine clay minerals with
additional amounts of biogenic carbonate and silicate material in biolo-
gically productive areas. The pore water content of deep ocean sediments
can vary greatly (specially in layered areas) between 20 and 80% (wet
weight) in the upper layers, and is often enriched in dissolved ions as a
result of redox controlled diogenic processes. Pore water movement in the
clay systems is generally extremely slow in the areas of interest to the
sub-seabed disposal of nuclear wastes, in the order of millimeters to cen-
timeters per 1000 years. Thus the redox related processes are the most important control of ion mobility. The supply or organic carbon and the rate
that the organic material is buried controls the distribution oxidized and
reduced sediments. It is these conditions that can affect the mobility of
the emplaced radionuclides, either directly by altering their chemical
form and reactivity, or indirectly by changing the form of naturally occurring ions which are instrumental in controlling the mobility of the isotopes /12/.

Redox ranges are difficult to estimate due to limited data. At some
depths (cm to meters) below the zone under the influence of dissolved
oxygen diffusion (from overlying seawater), organic carbon oxidation uses
nitrate as the electron acceptor and the nitrate is slowly consumed. As
this occurs, solid-phase manganese and iron start to play the role of oxidizing agents. In strongly reduced sediments, dissolved sulphate is reduced to sulphide and bisulphide. By studying in the laboratory the distributions of dissolved nitrate, iron, manganese and sulphate, the thickness
of redox zones of different intensities can be identified. The discovery
of anomalous reaction zones within the sediment column suggests that chemical short circuits may be available which could affect waste leach rates
and ion adsorption capacity. For example, oxidation of reduced Fe and Mn
causes co-precipitation of many dissolved ions. Under certain conditions,
however, precipitation of iron phosphate and manganese carbonate can occur
significantly reducing the concentration of these elements in the pore water. Also, iron silicate mineral formation appears to be an active process
in some deep sea sediments reducing the concentration of soluble silicon
which, in turn, would increase the rate of the vitrified waste.

In order, thus, to assess the properties of the far-field sediments
which act as a major barrier to the mobility of nuclides, it is clearly important to study both the solid and aqueous phases of the system in detail
in order to identify diogenitic, diffusional and physical disturbance processes /10/.

2.2.2 Nuclide distribution coefficients

Studies have been undertaken by CEA in collaboration with IFREMER during an oceanographic research cruise to two sites in the Cap Verdi Abyssal Plain (CV1, CV2) during which cores were taken by a Kullenberg corer to depths of 15 m. The sediment samples obtained were investigated for their capacity of retention of a selected number of actinide elements (Pu, Np, Am), and the fission radionuclide Cs. Among the physico-chemical parameters which may affect retention temperature was chosen in view of the fact that natural values of around $4^{\circ}C$ may be increased due to the thermal output of canisters to around $100^{\circ}C$. Studies on continental material have shown the importance of saturation effects. The retention was evaluated in terms of the distribution coefficient, K_d.

The adsorption of Neptunium was practically invariable between 4 and $15^{\circ}C$ (K_d = 300-500 ml.g^{-1} depending upon the chemical species present), however, between 30 and $50^{\circ}C$ it increased significantly to between 1300 and 2700 ml.g^{-1}. It appeared that these absorption reactions are reversible.

The adsorption of Plutonium increased linearly with temperature, the K_d increasing from 1000 to 3000 ml.g^{-1} between 4 and $80^{\circ}C$. However, the reversibility of the adsorption is inversely proportional to the temperature; at $4^{\circ}C$ it was practically impossible to extract measurable quantities of Pu adsorbed on the sediment originally at $4^{\circ}C$, while at $50^{\circ}C$ there was more or less equilibrium between the sorption and desorption reactions.

The adsorption of Americium was so strong by the sediments that the analytical methods employed were not able to detect the remaining very low levels of concentration in the solution, and thus were unable to evaluate the influence of temperature. Six different types of sediments were obtained from the Cap Verdi site and investigated for their ability to adsorbe americium. The results for the K_d values were, in every case, very high (between $4 \cdot 10^5$ and $8 \cdot 10^5$ ml.g^{-1}, and appeared to be independent from the sediment type (e.g. insensitive to the influence of carbonate).

On the basis of experiments undertaken with sediment samples from other sites of Cap Verde and measurements made in other countries /13-16/, it is possible to conclude that the K_d values of americium and marine sediments lie within the range of 10^5 - 10^6 ml.g^{-1}.

In the range 4-30$^{\circ}C$, the adsorption of Caesium in the absence of a scavenger is very little influenced by temperature, however, between 50 and $80^{\circ}C$, K_d values increase by a factor of 20-40 depending upon the sediment sample. Although phenomena were confirmed experimentally with pure illite in seawater, no explanation is yet available why this occurs. For a sediment sample CV1 it appears that the reaction is reversible for temperatures lower than $30^{\circ}C$, but irreversible at $50^{\circ}C$ in all the samples studied. The adsorption of Cs is sensitive to the mass of the element present; for an equal radioactivity, the adsorption of Caesium-135 is lower than that of Caesium-137. The ratio of the masses of Cs-135/Cs-137 is about 10^5. The addition of a certain quantity of stable isotope, even as low as 10^{-6} g.l^{-1} causes a significant decrease of the K_d of Cs-137.

From the above studies on the different CV sediments it may be concluded that the effect of temperature must be taken into account in any studies on the migration of Np, Pu and Cs in the zone close to emplaced canisters. The sorption of Am, on the other hand, appears extremely strong in all the cases studied, with K_d values ranging between 10^5 and 10^6 ml.g^{-1}. Furthermore, in the case of Cs investigations where the migration of Cs-135 is simulated using Cs-137, the effect of the mass of the isotope must be taken into account.

3. ENGINEERING FEASIBILITY STUDIES

In order to assess the engineering requirements of sub-seabed disposal, the CEC has undertaken a number of studies in collaboration with the UK and France. Taylor Woodrow Construction and Ove Arup and Partners, two UK companies, have conducted studies /17,18/ to assess in broad terms the feasibility of two methods of ocean disposal of HGW identified by the Engineering Studies Task Group as worthy of further study. These methods are the free-fall penetrator and the drilled emplacement options.

3.1 Drilled option

The drilled option study gave special consideration to environmental and geotechnical constraints; possible methods of transfer of radioactive material from the supply ship to the HGW disposal vessel (which may be either a semi-submersible platform or a ship); the problems of forming suitable holes in the seabed; methods of drilling and emplacement of waste canisters; backfilling and final grouting of the drilled holes. In addition, optional logistics, problems of station keeping and procedures which would be adopted under fault conditions were studied.

Although experience of the drilling of holes into the seabed at deep ocean depths is at present limited, the study concluded that most of the operations for emplacing the waste are within the bounds of existing or foreseeable technological development.

The conceptual design of a drilled borehole vault which could be constructed at suitable locations in the deep oceans, was devised to satisfy inter alia the following criteria: to provide a sealed enclosure for 300 or more canisters containing HGW (the canisters having dimensions of 1.3 m length and 430 mm diameter), to reliably support canister deployment operations, to be capable of structurally surviving exposure to deep ocean sediments for periods of the order of $10^2 - 10^3$ years.

The outline design proposed by the study was that the borehole vault would consist of a removable re-entry cone with an opening at the top, measuring 5.5 m in diameter, a conductor casing which would penetrate approximately 25 m into the sediment, and a lower steel-lined vault, 475 m in length and 610 mm in diameter. The vault would be accessed through a removable aluminium or polyethylene casing segment, which can be reamed out and removed prior to the final closure of the vault. Closure of the vault is effected by pumping a special grout and/or reconstituted sediment mixture into the top section. A tentative sketch of the proposed drilled vault design is given as Fig. 3.

It is proposed that the canisters are lowered into the vault on a drill string. The first pipe of the string would be a flexible lead pipe instrumented to enable it to be mated with the re-entry cone to the vault. This would be followed by a number of pipes containing grouted in waste canisters (the load). When the load has entered the hole it is then lowered until it reaches the bottom of the hole and then disconnected from the lowering string. Several options exist for the type of disconnection joint which could be employed. Those which could utilise direct pressure through the lowering pipes or rotation and "jarring" actions are considered to be the most suitable for this purpose.

The study proposes that the space between the pipe string containing the load be filled with delayed setting grout or non-setting material. This material is placed into the hole before emplacing the load. This procedure obviates the need for a separate grouting operation to fill the annulers between the load and the casing. The grout selected must have certain design requirements, namely sufficient delayed setting time, pump

ability, adequate final strength and integrity, density and other properties to enable the load to be emplaced and to provide a barrier to barrier
to the movement of radionuclides. At the present time, the preferred material for grouting the hole would be a cement-based material, but the use of
a granular material in a gel matrix could, with further research and development, be an alternative method. Additional work to examine the range
of possible materials suitable for backfilling and sealing a drilled vault
is planned.

3.2 Free-fall penetrator option

The penetrator option is possibly the simplest of those which have
been considered for the ocean disposal of radioactive waste /19/. In outline the method consists of the loading of one or more canisters of HGW
into a penetrator - a large free-falling torpedo-shaped shell. The penetrators are transported to a sub-seabed repository area in the deep ocean,
released and come to rest within the sediment at depths dependent on the
physical properties of the sediments and on the weights and geometries of
the penetrators.

The primary objectives of the DOE (UK) and CEC joint study /18/ were
to establish in sufficient detail the technical and operational feasibility of the method to enable comparisons to be made with the alternative
options. Disposal using free-fall penetrators is considered to be technically feasible from an engineering standpoint, providing the following
criteria can be met: the requisite embedment depth can be attained, the
hole behind an embedded penetrator is closed, that the state of the disturbed zone (the near field) can be quantified, and repository design criteria can be satisfied.

The study examined and considered several conceptual designs for penetrators covering a range of containment options for the waste canisters.
The canisters themselves were not considered as forming part of the engineered barrier because of their inherently low corrosion resistance.

For both the drilled option and the free-fall penetrator feasibility
studies, the reference designs of canisters chosen were those proposed by
the CEC and the UK. The physical form of these canisters is similar but
the dimensions and weights differ (Fig. 4).

Preliminary radiological assessment /2,20/ of deep ocean disposal has
given indicative evidence that there is little benefit in extending containment life beyond 500 years unless it is extended to several thousands
of years. Consequently, the study took the upper bound on containment design life to be 500 years.

The range of options studied for the engineered containment are classified as: pressure resistant, corrosion resistant, corrosion allowance
pressure resistant and transport overpack (penetrator body) designs. These
designs were selected on the basis that the containment must resist and
remain unbreached by hydrostatic pressure and exhibit adequate corrosion
behaviour throughout the design lifetime. If the containment resists, independently of support from its contents, the full hydrostatic load over
its design lifetime (disregarding any material provided for corrosion allowance), this is called a pressure resistant design.

Corrosion resistant materials examined /21/ were Hastelloy C, titanium and titanium-malladium; corrosion allowance is taken to infer that
the material chosen for the penetrator body corrodes at a predictable rate;
soft iron is the reference material selected for examination of this option. For the corrosion resistant and corrosion allowance options the penetrator body is designed to deform under the hydrostatic loads.

The transport overpack concept is similar to the pressure resistant containment option except that the thickness of the wall is sufficient to provide a degree of radiation shielding which would enable HGW containers to be transported within a safety normal condition.

The reference design selected on the basis of the comparative advantages and disadvantages of these options is the pressure resisting design. Fig. 5 shows the reference design selected, capable of containing five canisters. This reference design satisfies inter alia the following operational and technical criteria: the penetrator must be able to withstand the impact with the seabed to ensure satisfactory emplacement; the design is selected to achieve the greatest practicable embedment depth without incurring unreasonable costs; the penetrator has sufficient structural integrity to be handled during: fabrication, assembly and emplacement operations; the containment is designed to isolate the waste from the surrounding environment for a period of at least 500 years. It is proposed that the void between the canisters and the penetrator body be filled with a material which satisfies the requirements of mechanical strength, and which has good flow properties to aid positioning of the canisters during penetrator assembly. In addition, the filler should not react deletereously with the canister or the penetrator body.

3.2.1 Prediction of penetration depth

A major effort has been made by the Commission of the European Communities, UK and France to develop models which will allow a realistic prediction of the depth of penetration in different clay sediments of known properties. Several methods of approaching this problem theoretically have been attempted. The one used in the UK and JRC studies is based upon simple Newtonian dynamics. During the embedment phase of the motion, the deceleration of the penetrator is assumed to obey the following equation:

$$M z = M_g - F_B - F_D - F_s$$

where: z = depth below seabed; M = mass of penetrator; F_B = buoyant force; F_D = inertial resistance of sediment; F_s = resistance due to soil stresses.

Simplifying assumptions on the mechanisms giving rise to the inertial and stress resistance of the soil have been made. F_s is taken from elementary pile theory to be given by

$$F_s = (N_{ed} S_o + \alpha_d S_b) C_U$$

with: S_o = cross-sectional area; N_{ed} = dynamic and heaving capacity factor; S_b = surface area of penetrator; α_d = dynamic adhesion term; C_U = soil shear strength. The inertial resistance of the sediment is taken as being equivalent to an effective hydrodynamic drag.

This model can be used to estimate the penetration depth of penetrators of known sizes and weights (Fig. 6). One of the objectives of the penetrator field trials has been to validate the predictive utility of this model. The main objectives of the French work were to develop a theoretical model based upon plasticity theory to predict the depth of penetration of a penetrator in a clay sediment (of known properties) and to develop instrumented penetrators as a means of undertaking geotechnical investigations of potential future storage sites. The work /22/ was carried out along the following lines:
- physical analysis of the phenomena of penetration by a torpedo-shaped body into a deep-sea clay to permit the development of a mathematical mo-

del of penetration. The model used is based upon rigid plastic medium and calculation of the energy expanded during penetration;
- the model approach corresponds to a code already operational. Its validation has been confirmed by comparisons with the data obtained from the present programme, as well as data obtained from in-situ experiments carried out by other national programmes;
- centrifuge experiments undertaken with instrumental projectiles in reconstituted clay. These tests, under acceleration of 50 gravities, allowed investigations simulating penetration, of very small instrumental projectiles into a clay medium scaled to a gradient of consolidation equivalent to that assumed to be present in-situ;
- the small instrumental penetrator tests were undertaken in a basin filled with kaolinite and allowed the following:
  . testing of a rapid electronic data acquisition system for eventual use in large penetrators in deep water;
  . comparison of results obtained under presently known conditions (shallow water, under consolidated sediment, etc.) with model predictions.

The mathematical modelling was found to be reasonably simple and could be programmed on a desk-top minicomputer. In this context the "upper bounds method" seems particularly interesting; even though it theoretically only gives indications on the energy dissipated by deformation, it often allows a close approximateion to reality due to the possibility offered by the velocity field parameters to represent flow.

During the penetration of a projectile in a marine clay, four phases can be distinguished:
1) impact, penetration by the head of the penetrator;
2) penetration by the body of the penetrator;
3) penetrator burial, continuation of penetration (with or without hole closure);
4) velocity sufficiently slow that the effects of viscosity are important.

Each of the above phases is governed by a specific behaviour. In phase 1) the clay has no limitation to flow. In phase 2) the behaviour of the clay can be modelled by application of a rigid-plastic law. The criterion adopted, therefore, is that of Von Mises with the limitation of flow as a function of depth. In phase 3) the phenomena of hole closure (or its absence) cannot be studied using a rigid-plastic model, however, for simplification this phenomenon is not taken into account and the model used in phase 2) is used.

Modelling phases 2) and 3) of the penetration is carried out using the following simplifications:
- rigid perfectly plastic medium;
- medium which obeys the criterion of plasticity of Von Mises;
- flow as a function of depth;
- energy dissipated during the moment of impact not taken into account

In order to investigate solutions not directly open to this method of plasticity, an approach using a method of "upper bounds" has been applied which is to say that calculating the excess energy dissipated by plastic deformation of the medium. To undertake this, it is necessary to imagine the velocity field for each point in the deformed medium, which is likely to undergo the physical processes. This velocity field must be
- kinetically realistic in that it must rest within the imposed conditions of the velocity and incompressibility of the medium;
- plastically acceptable, that is to say that at each point it is possible to associate a constraining tensor which is governed by a plastic behaviour law.

By applying the theory of extremes to the velocity (this is deduced

from the theory of maximum work), it is possible to show that the total
energy dissipated by the simulated field is greater than that dissipated
by the penetrator, and thus leading to an underestimate of the depth of penetration. The energy dissipated by the field can be separated into three
terms:
- energy dissipated in the bulk volume;
- energy dissipated by friction;
- energy dissipated by internal interactions.

A comparison has been made using the model with the velocity curves
as a function of depth and deceleration with time for the sea trials undertaken in March 1983 by the Building Research Establishment and the Joint
Research Centre - Ispra, at the Great Meteor East site in the North Atlantic /23/. From this comparison two aspects may be distinguished:
- based on the global measurements of the penetration depth and the rate
  of penetration, the predictions given by the model are 10% less than the
  results reported. The missing 10% can, with certainty, be attributed to
  the lack of realism in the proposed velocity field;
- concerning the precise shape of the velocity-depth curves, it appears
  that the model, which predicts very closely the initial three-quarters of
  the penetration (however, with a loss of velocity very slightly less initially), undergoes a much more rapid deceleration in the latter stages
  of penetration.

One of the weaknesses of this modelling procedure is that it only
takes into account rigid-plastic behaviour and cannot, thus, describe the
eventual closure of the sediment column behind the penetrator. As a consequence, it is in general unable to correctly model the behaviour of the
sediment once the projectile has entered well into the medium.

3.2.2 Physical modelling in a centrifuge

The object of this study was to simulate the penetration of a projectile containing radioactive waste in deep ocean sediments. The simulation
is undertaken by firing small instrumented model penetrators, at realistic
environmental velocities into a reconstituted sediment, in a centrifuge.
This series of experiments was rather delicate to undertake and had the
following aims:
- to re-constitute at a scale of 1/50 depth, a clay sediment normally consolidated, having a realistic cohesion gradient;
- to fire, at real velocities, model penetrators and measure:
  . the degree and duration of the deceleration during penetration;
  . the degree and duration of the applied force on the projectile during
    penetration;
  . to verify, using X-ray radiography, the final position of the projec-
    tile in the clay after the centrifuge has come to a standstill;
  . to study, if possible and if it exists, the form of the track left in
    the clay after the passage of the projectile (using X-ray radiography
    after the centrifuge has come to a standstill);
  . to determine, if possible, the influence of the nose geometry on the
    main parameters of penetration.

The centrifuge experiments were undertaken in three stages:
- investigation of the total system and static tests;
- initial tests in flight of experimental prototype;
- final phase of tests, diring eight instrumented penetrators.

The depth of penetration of the projectiles is measured after the
flight using X-ray radiography. Analysis of the deceleration curves obtained during flight showed that the duration is of the order of 22 milli-
seconds. The axial deceleration at the beginning of penetration is about

630 msec^{-1}.

While the deceleration curves showed a constant signal pattern, it appears that for the force curves a variation of the results occurs around a mean value. This type of behaviour may be explained by a change in the flow patterns around the walls of the projectile. However, the distribution of forces of the soil on the projectile during penetration is not at all well known. The recorded forces in the range of 100-200 N are compatible with the decelerations measured.

On preliminary consideration, not taking into account the effect of Coriolis forces and taking the measured exit velocities from the canon and the depth of penetration measured by X-ray radiography, the distribution of the 8 firings undertaken in the final phase of the experiment, gave results which appear in good agreement with those measured in-situ.

Concerning the closure of the hole behind the projectiles, it was never possible to detect an indication of the presence of a track in the sediment column. The track, if it existed at depth, must have been under the detection limit of the X-ray method.

The curves of acceleration have been integrated two times in order to obtain the velocity and depth of penetration at each time period. The model has been run using the data obtained from the different trials. The density of the saturated kaolinite has been taken as equal to 1.6 based on the results of experimental measurements. The value of the coefficient for the gradient of the critical state has been taken as 0.79; this value corresponds to that found classically.

The penetration curves for the shots agree, in general, to within 10% with the experimental depths measured on the basis of integration of the accelerator sensors and those given by the model (Fig. 7). The decelerations calculated by the programme agree well with the data obtained from the penetrators up to the point of hole closure (supposed). The programme, at present, does not take into account this phenomenon however.

The results of the tests show that "hole closure" causes a change in the slope of the deceleration curve which becomes less during the later phase of penetration.

3.2.3 In-situ penetrator trials

Trials of model penetrators have been conducted at two of the study areas selected by the NEA Seabed Working Group: Great Meteor East and the Nares Abyssal Plain. These trials have been collaborative exercises involving the Building Research Station (Department of the Environment, UK), the Joint Research Centre, Ispra (CEC) in conjunction with the Institute of Oceanographic Sciences (UK), Rijksgeologische Dienst (NL) and Sandia National Laboratories (USA). They form part of the 5-year plan of the Engineering Studies Task Group /24/. The primary objectives of these deep ocean penetrator experiments have been to demonstrate in outline the feasibility of the free-fall penetrator concept of disposal for HGW and to test the theoretically based predictive models for calculating the embedment depth of penetrators in deep ocean sediment. Additional objectives of the continuing penetrator trials programme are to measure the motions of penetrators through the water column and sediment, and to obtain geotechnical, geophysical and geochemical properties of the sediments of significance to both the drilled and penetrator options.

Results from the first set of experiments have recently been reported /23,25/. Four penetrators to a design resulting from a study by Aermacchi /26/ were dropped at two locations in the Great Meteor East region of the North Atlantic. These penetrators, constructed of steel and of dimensions: 3.25 m in length and 0.325 m diameter, were equipped with 12 kHz constant frequency acoustic transmitters. The velocities of the penetrators as they free-fall through the water column and impacted with the sediment, were deduced by measuring the Doppler shift of the received acoustic signal. Data of a remarkably high quality was gained from the first two penetrator drops (in the final two drops the signal to noise ratio was not high enough to enable the velocity of the penetrators to be determined once the transmitters had entered the sediment).

A comparison of the experimental data with predictions from the elementary force balance method of describing the penetration event was made. These showed that the measured penetration depths (calculated by integrating the velocity time profiles during the motion through the sediment) agreed with the predictions to within a few percent.

The second series of trials in March 1984 at Nares Abyssal Plain, were conducted with the intention of extending the database of results. Penetrators of differing geometries and weights/densities were deployed so that the predictive methods used in the analysis of the first trials, could be subjected to verification and test across a wide range of penetrator parameters. A total of 17 penetrators to 8 differing designs were launched. The designs were selected to give a wide range of variation of density, weight, length to diameter ratio and nose sharpness. Two different systems of instrumentation were employed - ADSS (Acoustic Doppler Shift System) using constant frequency 12 kHz Doppler units and EATS (Explosive Acoustic Telemetry System) developed by the Sandia National Laboratories. A schematic figure of the range of penetrators deployed is given in Fig. 8.

Although the overall success of the second series of penetrator trials was somewhat marred by the failure of some of the Doppler units, sufficiently high quality data was obtained to enable tentative conclusions to be drawn:

- The force balance method of analysis again predicted the embedment depth of the penetrators with a degree of "engineering" accuracy;
- Local variations in the lithology of the sediments do not seem to have a marked effect on the penetration behaviour;
- The nose shape of penetrators is not a critical factor in determining the overall resistive force experienced by a penetrator during embedment;
- The inclusion of a velocity dependent fluid drag like term as a resisting force during penetration improves the correspondance of the strength profiles of the sediment derived from data from each penetrator test;
- This derived strength profile may be considered as a characteristic of the sediment - differing from the conventional shear strength obtained by laboratory tests. Once a given site or location has been characterized with a derived strength profile - a penetrator resistance profile - obtained from penetrator trials, the embedment behaviour of penetrators of a wide range of sizes and shapes at that location can be predicted.

The full results of the second set of trials will be published in due course /27/.

The conclusions drawn from the penetrator trials rest on untested assumptions: that the penetrator remains vertical during the embedment event, and that a unique value can be given to the velocity of sound in the acoustically complex region in the sediment behind a moving penetrator. Further series of trials to investigate these as yet untested assumptions on the dynamics of penetrators during embedment, as well as the question of hole closure, are planned. It is expected that the results from these trials will remove some of the uncertainty about the dynamics of the penetrator during embedment. In addition, the decelerations of the penetrators are to be measured directly - using accelerometers - to remove the uncertainty which may be felt to exist in deriving this fundamental quantity from acoustic measurements alone.

4. CONCLUSIONS

1. Two study zones have been identified and are being investigated in the North Atlantic, Great Meteor East and Nares Abyssal Plain, which appear to have characteristics of interest to the sub-seabed option.

2. The results of geochemical investigations carried out to date show that deep ocean sediments potentially form a very effective barrier to the dispersion of radionuclides.

3. Studies of two emplacement options using drilled or penetrator techniques have been made. No insurmountable problems have so far been identified for either technology, although more detailed engineering analysis will be required to determine final technical solutions.

4. The problems of site discontinuities, hole closure and thermal field development are under detailed investigation to assess their possible effects on the long-term stability of the sediment column.

5. Preliminary estimates on the potential cost of disposal using the drilled or penetrator option depend strongly on the number of canisters to be disposed of annually. Unit costs are similar for the two emplacement technologies and appear comparable to land based alternatives.

6. Investigations of other important environmental compartments such as biology and physical oceanography, are being actively studied. A feasibility report on the sub-seabed option is expected to be published in 1988 which will review all available data on the sub-seabed option.

REFERENCES

1. P.A. Gurbett, "Langrangian current measurements and large-scale long-
    term dispersion rates (SOFAR float experiment)". Nuclear Science and
    Technology, EUR 9401, 1984.
2. G. de Marsily, M.D. Hill, S.F. Mobbs, G.A.M. Webb, C.N. Murray, D.M.
    Talbert and F. van Dorp, "Application of systems analysis to the
    disposal of high-level waste in deep ocean sediments". Radioactive
    Waste Management 3 (2), 199-213, 1982.
3. E.K. Duursma, L.A. van Geldermalsen and J.W. Wegereef, "Migration
    processes in marine sediments caused by heat sources: simulation
    experiments related to deep sea disposal of high-level radioactive
    wastes". European Applied Research Reports, 5, 3, 451-512, 1983.
4. C. Sarridon, "Effects of a heat source in saturated clay". Report to
    the Dept. Environment, UK (awaiting approval).
5. G.P. Marsh, "Influence of temperature and pressure on the behaviour
    of HLW and canister materials under marine disposal conditions.
    Part 2 - Corrosion of waste canisters". Report to the Joint Research
    Centre, Ispra Establishment, 21, 1985.
6. F. Lanza, C. Ronsecco, "Leaching of a borosilicate glass in sea sedi-
    ments". In preparation.
7. J.A.C. Marples, "Influence of temperature and pressure on the beha-
    viour of HLW and canister materials under marine disposal conditions.
    Part 1 - Factors affecting the leachability of vitrified waste".
    Report to the Joint Research Centre, Ispra Establishment, 17, 1985.
8. W.E. Seyfried, Jr. and E.C. Thornton, "Seawater-sediment interaction
    at elevated temperatures and pressures and in response to a thermal
    gradient: implications for the near field of a sub-seabed repository
    for high-level radioactive waste". Report to the Joint Research Centre,
    Ispra Establishment, 45, 1985.
9. R.T.E. Schüttenhelm, A. Kuypers and E.J. Th. Duin, "The geology of
    some Atlantic abyssal plains and engineering implications". In "Off-
    shore Site Investigations '85", Graham & Trotman, London (in press).
10. P.J. Schultheiss and J. Thomson, "Disposal in sea-bed geological for-
    mations: properties of ocean sediments in relation to the disposal
    of radioactive waste". Nuclear Science and Technology, EUR 8952, 1984.
11. C.N. Murray, D. Stanners and A. Avogadro, "A laboratory approach to
    the study of radionuclide mobility in deep ocean sediments". Int.
    Symp. "The Behaviour of Long-lived Radionuclides in the Marine Envi-
    ronment", EUR 9214, 119-129, 1984.
12. R. Chester, R.G. Messiha-Hama, "Trace element partition patterns in
    North Atlantic deep-sea sediments". Geochimica et Cosmochimica Acta,
    34, 1121-1128, 1970.
13. D. Rancon, "L'aspect géochimique pour les évaluations des transferts
    de radioélément dans les milieux poreux souterrains". CEA R.4937, 1978.
14. D. Rancon and J. Rochon, "Recherches en laboratoire sur la rétention
    et le transfert de produits de fission et de transuraniens dans les
    milieux poreux". Underground disposal of radioactive wastes, IAEA,
    Vienna, 1980.
15. D.A. Stanners and S.R. Aston, "Factors controlling the interactions
    of ^{137}Cs with suspended sediments in estuarine environments". Int.
    Symp. on "The Impact of Radionuclides Released into the Marine Envi-
    ronment", IAEA, Vienna, 1980.
16. D. Rancon and P. Guegueniat, "Etude de la sorption des éléments dans
    les sédiments marins des grands fonds de l'Atlantique Nord". Nuclear
    Science and Technoloty, EUR 9571, 1984.

17. "A feasibility study of the off-shore disposal of radioactive waste by drilled emplacement". Taylor Woodrow Construction Ltd. Work performed under contract 256-81-7-WAS-UK to the CEC and part-funded by the UK Department of the Environment. DOE report No.RW/83.108; RW/83.114; RW/84.120.

18. "Penetrator engineering study - preliminary feasibility". Ove Arup and Partners. Work performed under contract 394-83-7-WAS-UK to the CEC and part-funded by the UK Department of the Environment. DOE report No.RW/83.094.

19. "Concepts for the disposal of high-level radioactive waste: the deep ocean bed". Atkins Planning. DOE report No.RW/82.015.

20. "Radiological consequences of operational failures and low probability post-disposal events". M.D. Hill and C.E. Delow: Appendix D of 8th Int. NEA/Seabed Working Group meeting, Anderson D.R. (Ed.), Sandia report SAND 83-2122.

21. "Corrosion assessment of metal overpacks for radioactive waste disposal". G.P. Marsh, DOE report No.RW/83.021.

22. J.Y. Boisson, "Etudes des conditions d'enfouissement de pénétreurs dans les sédiments marins". Nuclear Science and Technology, EUR 9667, 1985.

23. T.J. Freeman, C.N. Murray, T.J.G. Francis, S.D. McPhail and P.J. Schultheiss, "Modelling radioactive waste disposal by penetrator experiments in the abyssal Atlantic ocean". Nature 310 (5973) 130-133, 1984.

24. Report on the 3rd Interim Meeting of the Engineering Studies Task Group (1982). Talbert D.M. SAND 82-2701.

25. T.J. Freeman, S.G. Carlyle, T.J.G. Francis and C.N. Murray, "The use of large penetrators for the measurement of deep-ocean sediment properties". Oceanology International, paper 1.8, 8, 1984.

26. "Hydrodynamic analysis and design of high-level radioactive waste disposal model penetrator". Aermacchi SpA (Rep. 419/STE/82, CEC - Joint Research Centre, Ispra, 1982).

27. T.J. Freeman, C.N. Murray and D.M. Talbert, "Penetrator experiments in the Nares Abyssal Plain of the Atlantic Ocean". In preparation.

TASK GROUP STRUCTURE

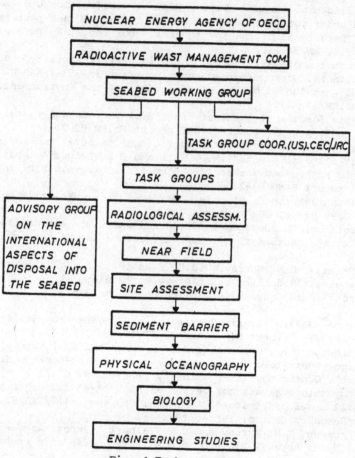

Fig. 1 Task groups

RADIONUCLIDE PATHWAYS

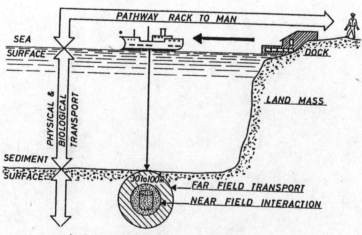

Fig. 2 Pathways under investigation

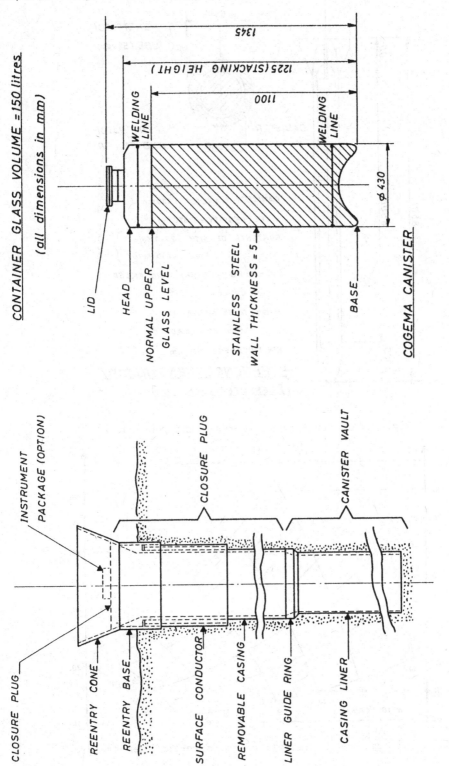

CONTAINER GLASS VOLUME =150 litres

(all dimensions in mm)

1345

1225 (STACKING HEIGHT)

1100

WELDING LINE

WELDING LINE

⌀430

LID

HEAD

NORMAL UPPER GLASS LEVEL

STAINLESS STEEL WALL THICKNESS = 5

BASE

COGEMA CANISTER

Fig. 4 Canister design

INSTRUMENT PACKAGE (OPTION)

CLOSURE PLUG

CANISTER VAULT

CLOSURE PLUG

REENTRY CONE

REENTRY BASE

SURFACE CONDUCTOR

REMOVABLE CASING

LINER GUIDE RING

CASING LINER

Fig. 3 Drilled option

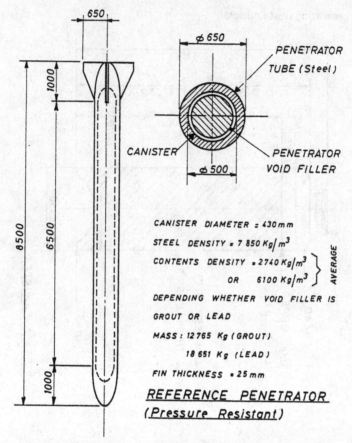

650

Ø 650

PENETRATOR
TUBE (Steel)

1000

CANISTER

PENETRATOR
VOID FILLER

Ø 500

8500

6500

CANISTER DIAMETER = 430 mm

STEEL DENSITY = 7 850 Kg/m³

CONTENTS DENSITY = 2740 Kg/m³

OR   6100 Kg/m³ } AVERAGE

DEPENDING WHETHER VOID FILLER IS

GROUT OR LEAD

MASS : 12 765 Kg (GROUT)

18 651 Kg (LEAD)

FIN THICKNESS = 25 mm

1000

REFERENCE PENETRATOR
(Pressure Resistant)

Fig. 5 Full-scale pressure-resistant penetrator

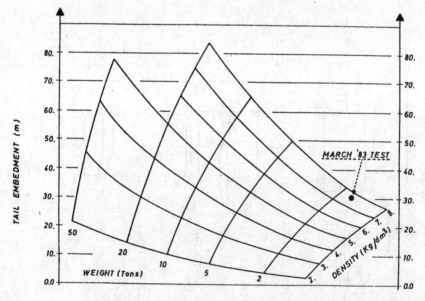

Fig. 6 Tail embedment predictions for free-fall penetrators

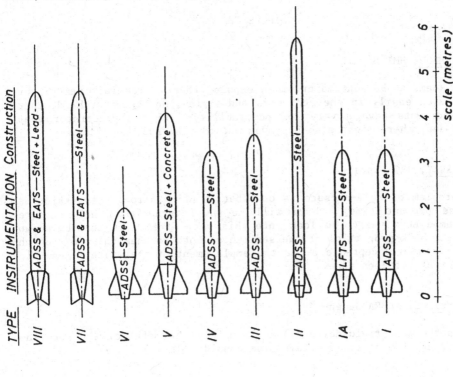

Fig. 8 Range of penetrators deployed

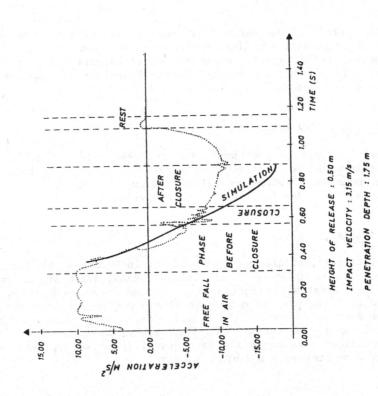

Fig. 7 Comparison between test and model results

DISCUSSION

D. GRAMBOW, HMI Berlin

There seems to be contradiction in showing that a free-fall penetrator
penetrates easily in the sediments and saying, on the other hand, that
the sediments have a very low permeability; if the penetrator goes
very deep, there might also be a way out for the radionuclides.

N. MURRAY, JRC Ispra

The problem of hole-closure is one that is being strongly investigated
at the present time; there will be an international hole-closure
programme next year. To look into this very effect, it must be shown
that the bulk properties of the sediments, both geochemical and geotechnical, remain acceptable after the emplacement of the waste. And this
has not as yet been done.

R. ROMETSCH, NAGRA Baden

How is the near-field defined? Is it mainly the metal of the penetrator, or are the other materials also considered?

N. MURRAY, JRC Ispra

The near-field is defined in the sub-seabed programme as that area
which is under the influence of corrosion products, radiation field and
thermal field, and therefore can be differentiated from the far-field
which is not aware basically of the behaviour of the canister within
the sediment column.

W. NAGEL, KEMA Arnhem

Assuming this project would be realized, a lot of different sorts of
metals would be emplaced the one quite close to the other. There might
be enhanced electro-chemical corrosion leading to dissolution of the
containers. Has this problem been envisaged?

N. MURRAY, JRC Ispra

Effectively, studies were carried out in Europe as well as in America,
using the same approach as for continental disposal, taking the thick-
walled constant corrosion rate approach for rather low temperature
waste, being perhaps old waste that was stored for intermediate times.
The Americans consider the titanium type alloy which is more corrosion
resistant; of course, all the studies were carried out under the
conditions present in the deep ocean, that is oxygenated sediments or
reducing sediments and high pressure.

COSTS AND WAYS OF FINANCING OF
THE GEOLOGICAL DISPOSAL OF RADIOACTIVE WASTE

P. VENET
Commission of the European Communities (CEC), Brussels

L.H. BAETSLE
Centre d'Etude de l'Energie Nucléaire (CEN/SCK), Mol

A. BARTHOUX
Agence Nationale pour la Gestion des Déchets Radioactifs (ANDRA), Paris

H.J. ENGELMANN
Deutsche Gesellschaft zum Bau und Betrieb von Endlagern für
Abfallstoffe (DBE), Peine

Summary

A global approach to the management of radioactive waste must take
into account not only the technological or safety aspects but also
economic and financial considerations. In this study, the costs of
geological disposal of radioactive waste are initially evaluated for
a certain number of representative cases of present tendencies in
the European Community. These expenses comprise research, development and site validation costs, transport and interim storage costs
and finally expenditure relating to various investment and exploitation phases of the disposal site as well as its closure. The possible ways of financing are subsequently reviewed and the financial
charges which resulted are calculated for each considered scenario.

The study is based on the most recent technical knowledge. It has
been carried out by national organisations involved in the management of radioactive waste : ANDRA in France, CEN/SCK and ONDRAF/
NIRAS in Belgium and DBE in F.R. of Germany on behalf of the Commission of the European Communities.

1. Basic hypotheses

1.1. The waste considered is that arising from reprocessing of spent fuel
elements and which requires long-term final storage in geological formations. The fuel elements have been discharged from light-water reactors
and have a burn-up of 33 GWd/ton. The waste is chiefly high-level waste
(HLW) as well as intermediate-level waste (ILW) i.e. hulls, reprocessing
sludges and technological alpha waste. Table 1 gives the characteristics
of this waste and of its conditioning. The vitrified HLW, which contains
11% fission product oxide content, is assumed to be disposed of in geological formation without overpack.

The use of alternative or advanced methods for conditioning ILW can
lead to an appreciable reduction in its volume. The influence of these
methods on the cost of disposal has therefore been taken into consideration : the results will be subsequently published.

Table 1 : Characteristics of conditioned waste

Waste	HLW	Hulls	Medium active waste	Technological Alpha waste
Conditioning	glass	concrete	bitumen	concrete
Conditioned waste volume in m^3/GWe/year	3	19,5	18	40
No. of containers per GWe/year	20	15	100	100

Waste coming from mixed oxide fuel fabrication plants and dismantling of nuclear facilities, such as nuclear reactors and reprocessing plants, as well as geological disposal of spent fuel are not covered in the present study.

1.2. Three continental geological formations are considered : clay, granite and salt domes. In order to study realistic cases, three sets of light-water nuclear reactor generating capacities of 10, 25 and 60 GWe respectively, all in operation for 30 years, are being considered. Three cases are representative of the existing or planned nuclear capacities in the Member States.

Taking the origin of time when the first fuel elements are discharged from the reactor, various scenarios were selected on the basis of the disposal dates of the two major waste categories, respectively 10, 30 and 50 years after reactor unloading. In all the scenarios studied, it is assumed that the rates of waste production, loading and unloading at the interim storage site, waste transport and its emplacement into the repository, is identical. In this publication, it is assumed that waste transport is only carried out between two sites : one site for the reprocessing and conditioning; the other site, 800 km away, for interim storage and geological disposal. The reprocessing of spent fuel takes place five years after discharging from the reactor, and the HLW waste is vitrified ten years after. When the ILW waste is directly dispatched to the interim storage site after conditioning, the HLW waste, after vitrification, is stored for ten years on the reprocessing site before being sent to the interim storage site.

Figure 1 shows, by way of example, the sequence of operations in the case where the geological disposal of the two waste categories is carried out simultaneously thirty years after unloading of reactor fuel elements. The waste disposal is carried out from between 30 and 60 years after the end of the electronuclear power production and the completion of the disposal cycle takes place 60 to 70 years afterwards.

2. Cost of geological storage

2.1. Taking into account the disposal scenarios, the types of geological formations and the nuclear reactor programmes considered, the annual

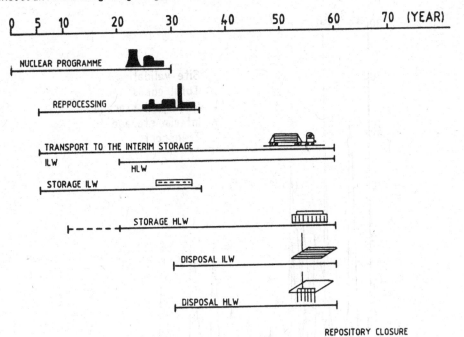

FIGURE 1 : PLANNING OF WASTE DISPOSAL (SCENARIO : 25 GWE FOR 30 YEARS, SIMULTANEOUS DISPOSAL OF ILW AND HLW 30 YEARS AFTER UNLOADING FROM REACTOR)

financial commitment was calculated in ECU (European Currency Unit). This expenditure covers research, development and site validation costs, transport and interim storage costs and finally expenditure relating to various investment and exploitation phases of the disposal site as well as its closure. The costs of waste conditioning have not been taken into account in the study.

    The schedule of expenditure, of extremely variable levels in absolute value, will range from several decades to a century. Figure 2 is an example of breakdown of annual expenses as a function of time. The abrupt fluctuation of the curves is characteristic of the results obtained.

2.2. The interpretation of such curves is difficult and it is easier to analyse the components of the annual disposal cost when these curves have been smoothed out. Figures 3 to 5 show the results obtained in the case of a reference scenario : the quantity of waste corresponds to an installed nuclear capacity of 25 GWe for 30 years, the ILW and HLW being disposed of simultaneously 30 years after unloading of the fuel elements from the reactors.

    The annual breakdown of the cost of interim storage and transport is shown in Figure 3. The costs relate to the value they would actually have had in 1984. The two peaks correspond to the construction of the interim storage facilities of ILW and then HLW. The tabular form of the transport

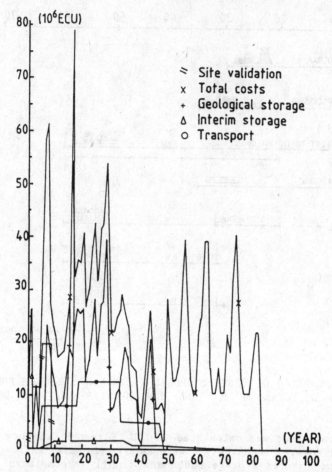

FIGURE 2 : TYPICAL DIAGRAM SHOWING THE DISTRIBUTION OF COST COMPONENTS OF GEOLOGICAL DISPOSAL (DISPOSAL IN GRANITE, 25 GWE FOR 30 YEARS, ILW 10 - HLW 50)

cost curve is to be observed : the first plateau corresponds to the transport of ILW, the upper plateau to the simultaneous transport of ILW and HLW and the last plateau to the transport of HLW waste. This distribution is valid no matter what the geological formation considered.

Figure 4 shows the annual distribution of the costs of site validation (cross-hatching on the figure) and of geological disposal for the three geological formations . Site validation comprises the inventory of sites and their preselection, geological reconnaissance and deep borehole drilling and finally the construction and then the operation of underground laboratories or any underground facilities intended to validate the site. It can be seen that the validation costs for salt is high : in fact, at the repository level, a salt dome must be examined as a whole before the site validation is possible; the cost of constructing the repository is consequently reduced, as the gallery excavation will have already been carried out during the validation phase. The validation costs of the argillaceous and granitic sites are lower in view of the

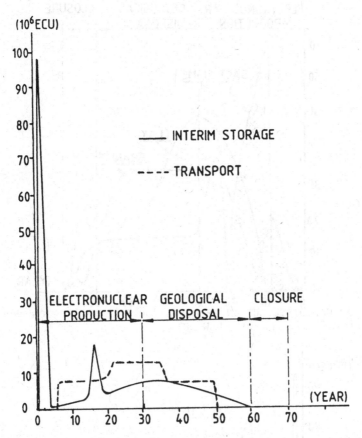

FIGURE 3 : ANNUAL DISTRIBUTION OF THE TRANSPORT AND INTERIM STORAGE COSTS
(25 GWE, ILW 30 - HLW 30)

nature of these formations; their storage cost is, on the other hand, higher and spread out over a longer period of time than in the case of salt.

Figure 5 summarizes the annual distribution of the total costs for the three geological formations. The total cumulative cost of geological disposal has been estimated for each of the three geological formations at about 2 milliard ECU for a capacity of waste disposal corresponding to a nuclear programme of 25 GWe over 30 years.

2.3. Figure 6 visualised the relative levels of the various cumulative costs for each component of geological disposal. If the average cost of site validation is taken as a reference, calculated here at 125.000.000 ECU, the cost of interim storage is twice this average cost, that of transport three times and that of final storage ten times the average cost. In other words, one could say that the transport cost is a third of the disposal costs and that of interim storage is a fifth. The full extent of the transport costs which often tend to be neglected is clearly seen here.

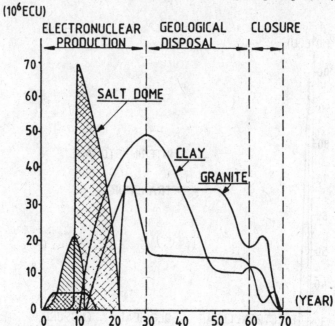

FIGURE 4 : ANNUAL DISTRIBUTION IF COSTS FOR SITE VALIDATION AND GEOLOGICAL DISPOSAL (25 GWE, ILW 30 - HLW 30)

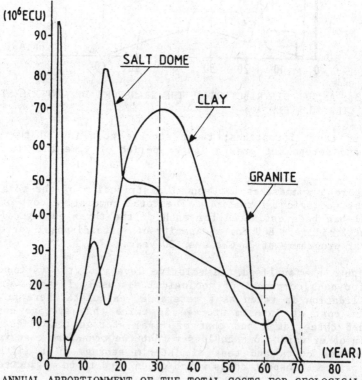

FIGURE 5 : ANNUAL APPORTIONMENT OF THE TOTAL COSTS FOR GEOLOGICAL DISPOSAL (25 GWE, ILW 30 - HLW 30)

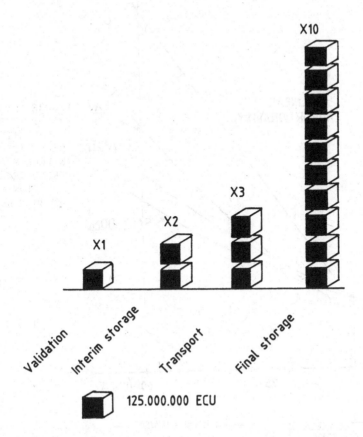

FIGURE 6 : COMPARISON OF CUMULATED COSTS, IN ECU 1984, FOR EACH COMPONENT OF WASTE DISPOSAL (25 GWE, ILW 30 - HLW 30)

Figure 7 shows the scale effect of the nuclear capacity on the total cost of the disposal in the various environments. The straight transverse line represents the theoretical cost if the proportianality had existed. This effect is more marked for the disposal in a salt dome than in an argillaceous formation.

Figure 8 shows the influence of the time at which the waste is disposed of, on the distribution of the total annual costs. The selected example is that of disposal in a salt dome. As it might be expected, the later the burial date, the more the costs are carried forward in time.

Finally Figure 9 indicates the percentage of the total cumulative costs at the end of the period of electronuclear power production, or after thirty years in the example. At this date, it can be seen that if half the expenditure is made in the case of the disposal scenario after ten years for the ILW and fifty years for HLW, this percentage goes up to 60% for the simultaneous disposal scenario after thirty years and is no more than a quarter of the total expenditure for the simultaneous disposal scenario of fifty years.

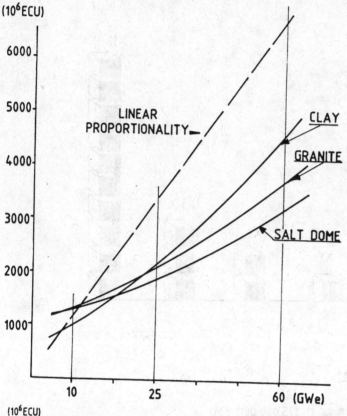

FIGURE 7 : SCALE EF-
FECT OF THE NUCLE-
AR CAPACITY ON THE
TOTAL DISPOSAL COSTS,
IN ECU 1984, FOR DIF-
FERENT GEOLOGICAL
FORMATIONS (
ILW 30 – HLW 30)

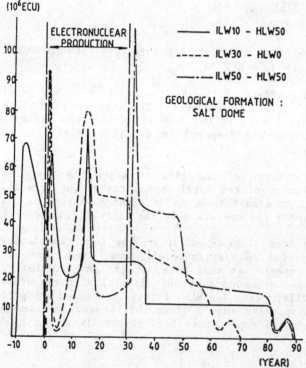

FIGURE 8 : ANNUAL DISTRI-
BUTION OF THE TOTAL DIS-
POSAL COSTS (ECU 1984) IN
FUNCTION OF THE DISPOSAL
TIME (DISPOSAL IN SALT
DOME, 25 GWE FOR 30 YEARS)

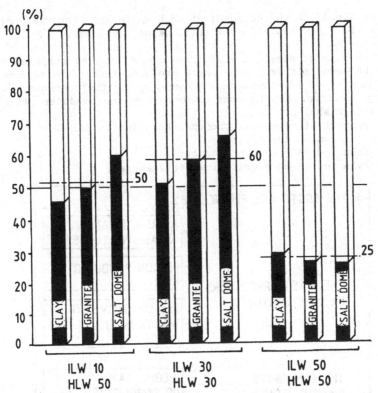

FIGURE 9 : FRACTION IN % OF THE TOTAL COSTS ARISING DURING THE LIFETIME OF THE NUCLEAR POWER PLANTS (YEARS 0 TO 30) IN FUNCTION OF DISPOSAL TIME AND FORMATION TYPES (25 GWE FOR 30 YEARS)

3. Ways of financing

3.1. The geological storage of radioactive waste is not a common operation in today's industrial world. Not so much because of the amounts at stake but in consequence of the exceptional dimension which the time factor assumes.

It has already been seen that it is necessary to count from 70 to 100 years to dispose of radioactive waste for a given nuclear programme. If one supposed that nuclear energy has already existed since the beginning of the 1900's, it is only now that we would have finished burying the final waste produced by our ancestors. It is hardly likely they would have thought of the future economic upheaval which would subsequently occur and its consequence on inflation, interest rates and the parity of currencies. When all is said and done, who knows whether the waste producers of that time wouldn't have disappeared in this turmoil ?

Considerations of this kind should give us cause to use sufficient discretion when interpreting the results such as those arising from the study in view of long-term economic and monetary uncertainties.

3.2. It must be stated that there is not at present a uniform approach by the organisations charged with the management of radioactive waste, on the ways of financing the disposal of such waste and that the application procedures are often imprecise or still being developed.

For the sake of simplicity (Figure 10), it may be supposed that the costs can be borne either by the waste producer (under the circumstances, the manufacturer of electronuclear power), or by the community (i.e. the State), by taxation. If the cost is borne by the producer, three cases can be distinguished :
- prorata payment of expenses, the charges being immediately financed the moment they arise;
- payment by a fee on each kWh product;
- payment by a fee on each unit of waste disposed.

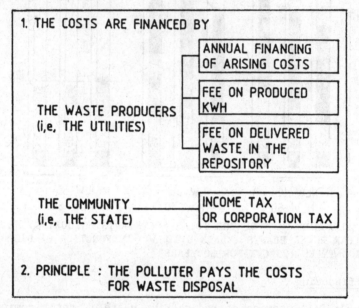

FIGURE 10 : MODES OF FINANCING OF WASTE DISPOSAL

There is general acceptance of the polluter-pays principle, whereby the conditioning and disposal of radioactive waste cost must, in the same way as other industrial waste, be borne primarly by the facilities which produce it. Application of this principle therefore rules out financing by the community which benefits from nuclear energy since equitable distribution by means of taxes is almost impossible.

3.3. The first case considered is the simplest one : the waste producer pays the cost of geological disposal as and when the waste arises. Payment can be made annually or in accordance with multiannual financing plans of geological disposal.

Another way of financing is the payment of a fee on each nuclear kWh produced (Figure 11). This is based on the fact that each kWh produced corresponds to a certain quantity of waste. In this case, a fund would be constituted which could be managed by a specialised body, under State control for example. Since the waste is stored for a period longer than that over which the electricity will be produced by nuclear means, the interest of capital constituted in this way will contribute to covering this future expenditure.

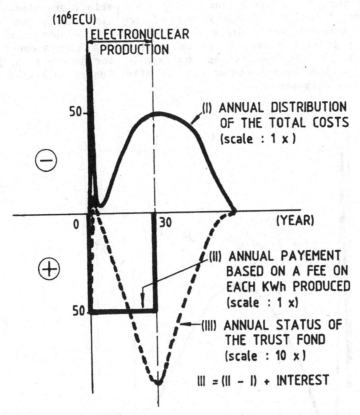

FIGURE 11 : PAYMENT OF A FEE PER EACH NUCLEAR KWH PRODUCED AND CONSTITUTION OF A TRUST FUND.

The diagram in the figure is split up into two parts : the upper part correponds to the annual expenses (curve I) and the lower part to the revenue (curve II). The producer, during the electronuclear power production phase, will pay a fee which is here presumed to be constant : it is represented by a straight horizontal line on the figure. Curve III represents the annual status of the fund thus set up : it is the annual difference between the expenditure and the revenue from the fee, supplemented by the interest corresponding to that difference. The value of the fee is calculated in such a way that, on closure of the disposal site, the fund would have been completely exhausted.

The last way of financing considered here is more complex. It chiefly concerns a fee paid by the producer or each unit of waste delivered for geological disposal. The curve of total expenditure as a function of time (Figure 12) has again been plotted. Two phases can be distinguished here :
- the first corresponds to all expenses incurred before the permit to construct the disposal repository is obtained; payment is made as and when the costs arise.

- the second phase corresponds to the construction, operation and site closure. During phase IIa (repository construction), the costs are covered by the loan : the producer, for this period, provides security in the form of contractual committments which bind him to the operator for the subsequent waste disposal. The waste is disposed of during phase IIb : at that point, a fee calculated by taking account of the repayment of the above-mentioned loan and of the interest on it, the disposal cost and the subsequent cost of site closure is levied on each unit of waste disposed of.

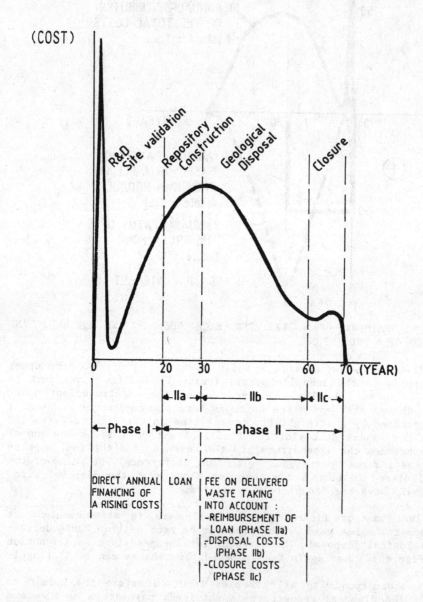

FIGURE 12 : FEE PAID BY UNIT OF WASTE DELIVERED TO GEOLOGICAL DISPOSAL

3.4. Figure 13 shows a characteristic result concerning the nominal total costs, in constant ECU, for each of the geological formations, as a function of the various ways of financing. It concerns once again the reference scenario of 25 GWe over 30 years and simultaneous disposal of waste after 30 years. For the three ways of financing (annual payment of expenses, fee on kWh and fee on disposed waste) and taking account of a real interest rate of 3%, i.e. market rate less inflation rate, following costs in milliards of ECU have been obtained :

- the total cost of disposal with due regard to margins of errors, is perceptibly the same for each of the geological formations concerned.
- over the 70 years which the disposal cycle lasts, the financial cost of disposal to the producer, will amount to 2 milliards ECU in the case of pro rata financing of expenditure, to 1.4 milliard ECU in the case of the payment of a fee on kWh and 2.4 milliard ECU where a fee is made on each unit of waste disposed of.

The difference between the last two ways of payment is considerable (from 1.4 to 2.4 milliard ECU). It can be explained by the fact that in the case of the fee on kWh, a repository trust fund is set up whereas in the case of the fee on buried waste, the interest of the contracted loan should be reimbursed.

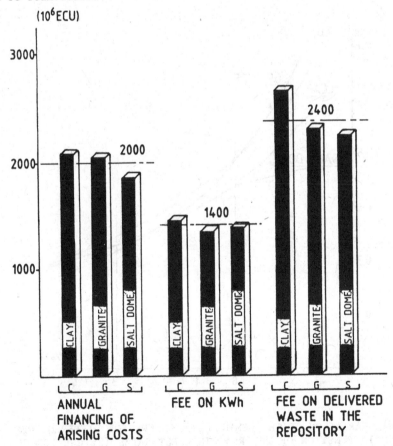

FIGURE 13 : VARIATION OF THE TOTAL FINANCIAL COSTS OF WASTE DISPOSAL (ECU 1984)

It should be noted that the differences which appear between the various ways of financing would certainly be lessened if the total cost of disposal were evaluated, not only from the standpoint of the interest of the waste management body but also as regards the overall cost of nuclear energy considered as a whole.

3.5. In the framework of this study, we varied the interests rates and the inflation rate on the basis of a multi-parametric study. We also assessed by means of simulation the impact of any cost overruns that might arise during disposal and studied the adaptation conditions of the fee as a function of the application date.

By way of Figure 14 shows the variation of the nominal fee in thousandths of ECU (mill), for the reference scenario and the base-case in function of the real interest rates. The real interest rates are indicated in the X-axis and the fee is shown in thousandths of ECU on the Y-axis. This fee comes to about 0.33 mill ECU for a real interest rate of 3%. It increases with the lowering of the interest rate. It can be seen that salt is less sensitive to this variation than the other formations, which is explained by the relatively high level of the expenditure initially committed in the case of salt.

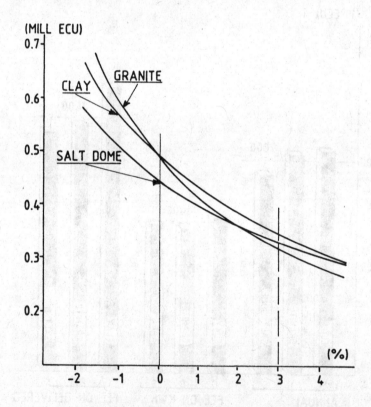

FIGURE 14 : VARIATION OF THE NOMINAL FEE ON KWH FOR DIFFERENT GEOLOGICAL FORMATIONS IN FUNCTION OF THE REAL INTEREST RATES (25 GWE FOR 30 YEARS, ILW 30 - HLW 30)

Figure 15 should be compared with the preceding one. In the case of
the fee per unit of waste delivered for geological storage, the sensitivity of the annual cumulative fee to the real interest rate is being
studied here. In contrast to the preceding case, this fee increases as a
function of that rate. Granite is relatively insensitive to this variation as a result of the better spread of expenditure over the disposal
cycle.

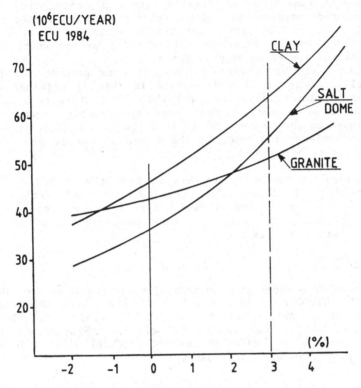

FIGURE 15 : VARIATION OF THE CUMULATED FEE ON DELIVERED WASTE TO THE
REPOSITORY FOR ONE YEAR IN DIFFERENT GEOLOGICAL FORMATIONS IN FUNCTION OF
THE REAL INTEREST RATE (25 GWE FOR 30 YEARS, ILW 30 - HLW 30)

Conclusion

(i) At present, the geological disposal costs can be reasonably assessed.
The amounts at stake, although considerable in terms of absolute
value, are relatively small compared to the total cost of nuclear
energy production. Thus, the total cost of geological storage of
waste coming from a set of 25 nuclear power stations of 1,000 mega-
watts each, as has been previously defined, will be of the same order
of magnitude as the construction cost of two such power stations, or
about two milliard ECU.

(ii) Marked differences between the disposal costs in clay, granite and
salt domes have not been ascertained. Whichever way of financing
chosen, these formations, taking into account the margin of error on
the evaluations made, are quite similar.

(iii) Several ways of financing are possible, and each of them has both advantages and disavantages.

- Under normal economic conditions, the formula of funds capitalised by paying a fee on the kWh is the most advantageous. For a nuclear programme of 25 GWe over 30 years, it is of the order of 0.33 thousandths of ECU per kWh, corresponding to about 1% of the price of the kWh produced. This is the one which offers the largest guarantees as regards future payments since all waste produced corresponds to a pre-emptive taxation of the producers. It is still necessary that the perenniality of such funds be ensured in the long-term, at national or international level, whatever the circumstances may be.

- One of the advantages of the pro rata payment for the waste delivered to geological storage is that it eliminates a great deal of the uncertainty about the effect of costs since, at the moment of payment, these costs are exactly known. What will happen to the payment, which is to be made well after the end of the nuclear programme, if the producer, who is also the payer, ceases to be solvent ?

Therefore, it is to be remembered that the choice of a way of financing can not be based solely on financial considerations but must take into account social and economic factors in the long-term. Be that as it may, solutions do exist, but each authority has to find the one which seems optimal in its case.

Acknowledgements

The authors of this paper wish to express their gratitude to   Mr. MAYLIN of ANDRA, Mr. BRIGGS, Mr. KUNSCH and Mr. VERRAVER of BELGONUCLEAIRE, Mr. MANFROY and Mr. de BRUYNE of CEN/SCK, Mr. PITZ of DBE, Mr. HEREMANS and Mr. ZACCAÏ of ONDRAF/NIRAS, Mr. MAYENCE of TRACTIONEL, Mr. LAFONTAINE of TRANSNUBEL and   Mr. HAIJTINK and Mr. ORLOWSKI of the CEC for their advice and contribution to this study.

DISCUSSION

F. FEATES, DoE London

Are the cost figures discounted?

P. VENET, CEC Brussels

Values given in the curves are expressed in 1984 ECUs. The influence of different economic factors, like inflation and interest rates on financing costs, has been studied.

R. BOSSER, CEA Cadarache

In France, the reference case foresees, for reprocessing of spent fuel from water-cooled reactors, a time of three years after unloading of the spent fuel. What is the base-case used for the calculations and curves, which have been presented?

P. VENET, CEC Brussels

The computations are based on a period of production of electricity by the reference nuclear power plant programme of 30 years. Reprocessing of the fuel is supposed to be carried out 5 years after unloading of the spent fuel elements, with vitrification following five years later. The vitrified waste is supposed to be stored on the reprocessing site for an additional ten years, before being sent to the interim storage or disposal site.

DISPOSAL OF HLW - READY FOR IMPLEMENTATION?

Chairman   : R. Heremans (ONDRAF/NIRAS)
Secretary   : B. Côme   (CEC)
Panel Members: A. Barbreau (CEA-IPSN)
               F. Feates   (UK-DOE)
               F. Girardi   (CEC-JRC)
               K. Kühn   (GSF)
               R. Lyon   (AECL)
               C. McCombie (NAGRA)

Chairman's Introduction

Welcome to this session, in the course of which my colleagues present here, and myself, will be trying, in the light of the statements we have been listening to, and our respective experience, to make a summary of the scientific and technical achievements in the field of the safe and long-term disposal of high-activity and long-lived waste. The properties and characteristics of such waste fully justify the efforts undertaken and the means that are at our disposal for ensuring the protection of the public and of the environment against their potential noxious effects. The final objective of a good management of such waste must therefore be to confine them, thus avoiding any inconsiderate dispersal into our ecosystem. Such a solution, though unique in its principle, may be varied in its applications. A constellation of very different factors, either technico-scientific, e.g. the size of the electro-nuclear programme to take into account, the geological configuration of the countries concerned, or socio-political, e.g. the public's understanding of what the risk is, or quite simply the public's sensitivity to nuclear energy, may considerably influence the final choice of the disposal system in one country or another, or in a given region. The statements we listened to throughout today, as well as the statements from yesterday and the day before, are an image of this range of possibilities which the scientists involved are forging.

In the quest for a balanced solution, suitable for the national context but above all safe, the role played by the multinational bodies is capital. It is, in fact, the regular confrontation of ideas and results of research which will progressively give birth to a consensus on the development of systems acceptable to all.

Ten years after the Commission launched its coordinated programme in the field of radioactive waste management in general, and particularly in its disposal, it is reasonable to ask ourselves the question: "Are we technically ready to answer this latest challenge, which is to implement a final solution for the disposal of radioactive waste, a solution acceptable for all but also valid for the generations to come?"

To start answering this question, the Commission brought around this table some experienced research workers, who for many years, in the

Community and in two countries with which it has preferential links, have been working and directing teams in this matter.

May I introduce to you successively:

- Mr. Barbreau, who will tell us more particularly about reconnaissance and qualification of sites;

- Mr. McCombie, who will tell us about engineered barriers;
- Mr. Kühn, who will tell us about the technological problems and more particularly the mining problems;

- Mr. Girardi, well-known for his work on the safety analysis of geological disposal systems for radioactive waste;

- Mr. Lyon, who will talk about performance assessment of disposal system components.

On tackling these two last subjects, performance assessment and safety analysis, we are anticipating tomorrow's session a little, but we thought these important subjects, where the experimental results crystallize, could not be passed over in silence at this round table session.

So, in order to have a complete round, we asked Mr. Feates to end this series of short statements, talking to us about institutional, legal and financial aspects of geological disposal.

A. Barbreau

Firstly, I would like to recall the interest in geological disposal. It results from the isolation of the waste, and the expected low probability of accessibility in the future, and also from the confinement of radioactivity, obtained by the barrier properties of the geological formation(s) between the respository and the biosphere. Given the importance of the safety of disposal, the qualification of a site will largely depend upon the confinement properties of the geological barrier. According to the situations and the rock types, this confinement may be of three types:

(1)   an absolute confinement, in which no nuclide can escape from the repository; this is an ideal situation;

(2)   a relative confinement vis-à-vis total activity of the waste, resulting from long time for nuclides to travel from the respository to the biosphere, i.e. from radioactive decay;

(3)   a relative confinement obtained when radiological protection standards are met; this can be achieved by dilution and delayed releases.

Transport of radioactivity from a repository up to the biosphere will result from groundwater flow, if and when the integrity of the geological barrier itself is not jeopardized by internal or external factors. The studies to be carried out must therefore be oriented towards acquisition of data about groundwater flow, such as hydrodynamic and hydrodispersive properties of the medium. Rock permeability and porosity, size and shape of aquifers, hydraulic head gradients, groundwater discharges, dispersivity of the medium (combining molecular diffusion and cinematic dispersion) must be known. Moreover, the spatial variability of these parameters has to be taken into account.

These data can then be used in order to define convection and diffusion phenomena in the overall mass transport equation. They must be complemented by the retention capacity of the medium vis à vis radionuclides and, in particular, long-lived emitters, which are the major safety concern in the geological disposal of radioactive waste. This retention capacity is known to play a major role, as was shown by numerous parametric sensitivity studies.

The possibility of constructing deep repositories in the medium considered also has to be assessed.

Finally, the long-term stability of the disposal system and its main possibilities for evolution in the future must be studied; this time factor is probably one of the critical aspects of disposal studies.

In order to qualify a deep disposal site, various studies will have to be performed in several fields of research and with significant investigation techniques: detailed geological and geophysical mapping, deep drilling campaigns, borehole coring; hydrological, thermal and mechanical tests in underground laboratories, which allow investigations to be performed in actual realistic conditions; some basic aspects of deep groundwater flows can also be investigated. Very long time spans can be approached by considering "natural analogues".

Presently, a wide range of adequate <u>techniques</u> exist, which enable us to investigate deep geological formations; new methods of higher sophistication are being developed in several countries, mainly in the fields of predictive geology, of very low permeability measurements, and of actinide speciation. It must be pointed out that underground disposal raises a number of specific problems which were, up to now, rarely considered by "conventional" Earth Sciences, such as long time spans, very low pearmeabilities, and actinide chemistry, as was pointed out yesterday by Mr. Peaudecerf.

It is not the purpose of this talk to consider these various techniques in detail; however, some of them can be listed. Among geophysical methods, mention must be made of remote sensing (satellites, infrared scanning); seismic prospection by reflexion or refraction; gravimetric survey; all of these methods allow the determination of deep rock mass setting. Various geophysical methods, e.g. sonic, electric and neutron logging, can be performed in boreholes.

Hydrodynamic parameters, such as transmissivity, pearmeability and storativity, can be determined by borehole hydrological testing, e.g. improved pumping tests, slug tests, packer tests, pulse tests. Hydrodispersive parameters (effective porosity and dispersivity) result from tracer tests between boreholes with convergent radial flow. Other techniques can complement the above methods (e.g. mapping of isopieometric surfaces) in order to describe the hydrogeological features of deep underground systems. Flow velocities can be estimated by tracer tests; groundwater dating using ^{14}C, $^{18/16}O$, ^{3}H, ^{4}He may prove useful.

Among recently developed technologies, mention should be made of the sinusoidal well testing, the PIEZOFOR probe developed by TELEMAC (F) for low permeability and hydrostatic head measurements in boreholes, the probes for in-situ measurements of Eh and pH. Finally, studies of natural analogues are well under way in several countries.

Let us turn now to deep in-situ experiments in underground laboratories. The research programmes of Stripa, Fanay-Augères, Asse, Mol, Pasquasia, in the USA and Canada, have twofold objectives. On the one hand, they allow (or will allow) better knowledge of deep geological media in representative repository conditions (rock mechanics, heat transfer, hydrogeology, radionuclide migration) to be obtained. On the other hand, they will favour the acquisition of expertise and know-how, both necessary in order to characterize a real repository's environment.

The situation is similar in the field of source-term characterization, and of radionuclide retention by geological formation, mainly for actinides; the behaviour of these elements in realistic geological environments is the subject of intensive research already detailed during this conference: equation of state, Eh and pH stability diagrams for various actinide compounds, complexes with organic and inorganic ligands, and - last but not least - techniques for measuring very low actinide concentrations in solutions.

Finally, a large number of <u>calculation tools</u> are now available, taking into account phenomena such as radionuclide convection, dispersion, retention and radioactive decay; others, such as the CASTOR code, allow the consideration of evolution scenarios and, later, of the impact of a repository on the geosphere and biosphere.

To conclude, a wide variety of investigation techniques, suitable for the study of the geological barrier, are now available; new ones are also being developed. Mention must be made of the deep sea-bed sediments as disposal medium; thanks to the knowledge already acquired, realistic impact studies (Pacific Ocean, North Atlantic) should now become possible.

C. McCombie

As regards the engineered (or technical) barriers, I think that the main question "Are we ready for disposal now" can be subdivided as follows:

(1) do we understand and have we properly defined our requirements on the technical barriers?

(2) have we developed concepts which are adequate to meet these requirements; and here there is an addendum in waste disposal: how do we demonstrate, very convincingly, that they are adequate, more convincingly than other people have to do; and

(3) have we optimized these concepts sufficiently to justify beginning disposal now?

The technical barriers themselves are, of course, system-dependent. Not all host rocks need all technical barriers. They should also not be discussed independently; they should be discussed in their context of integrated near-field. In spite of this, I think I shall now make a few quick remarks on the standard technical barriers, the waste matrix, the container (overpack) and the backfill, as I perceive it, based on the results we had over the past couple of days.

The matrix has undergone a long development; several options are available. The basic options of vitrified waste (glass) or of unreprocessed spent fuel appear adequate. In fact, so much development work was done for unrealistically pessimistic assumptions (e.g. on water flows) that some characteristics may be more than enough. What can still be usefully done is clearing up further "newer" issues seen to be important (e.g. species which can result from glass dissolution in repository environment).

Lifetime requirements for the container/overpack can vary according to the situation. No overpack may be envisaged for disposal in salt; it can be estimated that a 1000 years lifetime allows confinement of activity during most of the "thermal" phase; other concepts envision much longer lifetimes. The justifications for an overpack also vary considerably (e.g. redundancy in the multi-barrier system, or simply chemical conditioning). The work carried out to-date tends to indicate that these performances are achievable now, even for very long lifetimes; recent work has shown that various concepts exist (e.g. corrosion-resistant or "sacrificial" overpacks) which can produce lifetimes in the range of 100-1000 years. The next stage here seems to be to move to full-scale type overpacks.

In the area of backfilling, there are also adequate options. Sophisticated buffer materials may not always be necessary; when they are used, their important purpose is to ensure that radionuclide transport through them (if any) occurs by diffusion rather than by convection. Adequate solutions, primarily based on the use of tight clays of the bentonite type, have been developed. Whether they are optimized or not is another story. We have heard that research is being carried out on less demanding materials: more common clays, cement, spoil materials.

These barriers are combined into an integrated near-field which, by itself, can bring a high degree of isolation with minimum requirements on geology. This near-field can also be "tailored" to enhance performance.

Coming now to conclusions about these technical barriers, I would like to say that we now have concepts which would enable us to begin disposal immediately if we wanted to; we can demonstrate convincingly that these concepts offer adequate safety. However, there are areas where further optimization is possible. We do not have to begin disposal immediately; the critical path does not come from the development of technical barriers in the system, but either from strategy decision (e.g. length of intermediate storage, or from other aspects (e.g. the time needed for site characterization). The time we have should be used carefully, not to generate further options, but to optimize selected options and to validate them so that at time of disposal we have a yet wider body of evidence to convince ourselves and the public that we are implementing a safe disposal methodology.

K. Kühn

I have been asked to say something about the state of the technological
art in the field of high-level waste disposal. There are 3 aspects I
would like to consider: mining questions (repository construction), waste
emplacement, and then backfilling and sealing.

There is a worldwide trend setting up underground laboratories. One of
the first trials here with fully implemented techniques was in the middle
of the 1960's already, at the Lyons salt mine in the USA. It was a
successful demonstration; HLW could be put underground. Again, in the
case of salt, disposal experiments have been made and are going on in the
Asse mine in Germany and in the USA in the Avery Island mine. In granite, a whole series of underground laboratories have been created: at
the Troon site in Cornwall (UK), Fanay-Augères (F). Perhaps more famous
are the underground laboratories in granite at the Stripa mine in Sweden
and at the Grimsel site in Switzerland. At present, the Underground
Research Laboratory in Canada is going ahead; the shaft is 255m deep
already. Last, but not least, we received papers this morning concerning
the underground laboratories in clay at Mol in Belgium, and at Pasquasia
in Italy, the latter being scheduled for this year.

For repository excavation, one of the main questions relates to the
sinking of shafts. The methods we have now make it possible to set up a
repository; in Germany, we have taken this first step and are sinking two
shafts in Gorleben. They are laid out so that, if there is a go-ahead,
then we could certainly turn them into a repository.

For excavations underground, the three formations are to be distinguished. In Germany, we have already decided that we will use point-
attack machines which will make it possible to excavate rooms without
unnecessarily mechanically disturbing the host rock. There is a lot of
experience in hard rocks, particularly in granite, so that there is
nothing new there. In Sweden, very sophisticated methods have been used
for drilling and blasting; in Grimsel, a full-face tunnelling machine has
been used with favourable consequences on safety and cost. Other considerations have been made for excavating large galleries in clay at Mol:
both ground freezing or direct excavation seem possible. Here, of
course, further technological development remains necessary for an
industrial repository.

Transportation and handling of waste are pretty well settled; all countries now have experience. They know how to manage and transport irradiated fuel rods; containers and casks have to be adapted in dimensions
and thermal performances; this should not be difficult. The scene is
different as regards both machine and techniques in the case of HLW being
disposed of underground. There are a number of engineering paper designs, in the US, for example, but none of them have really yet been
tested and I think that this is an important element of the Part B of the
Community's next programme, as mentioned by Mr. Orlowski.

A machinery for transportation and disposal of HLW underground has still
to be tested out and checked for technical reliability. The same goes
for the Mol facility, and in France when a site is selected.

The last aspect is backfilling and sealing. Here, of course, there are differences, which depend on the nature of the voids to be filled in. For the disposal hole for canisters, there are various designs and there should be no problems to put these into technical effect. In his presentation, Mr. Lake gave us an idea of what exists; this is probably a question of optimizing the procedure and finally choosing something, instead of endlessly proceeding with new ideas. An important point which has only been touched upon, is the development and testing of dams in a repository, where you have ground water (e.g. in granite) or where you might have water if something happened (in salt). Within the next EC programme, we are going to test a prototype which would be gas- and brine-tight; this should be available in time for the construction of the Gorleben repository.

As regards the sealing of shafts, this results from mining experience, but, of course, the requirements are stricter than for any coal mine. But procedures are being developed which will successfully enable us to actually fill and seal the holes.

So in technological terms, I would say that there are only a few problems left, and I think that we can put HLW disposal into effect.

F. Girardi

I would like to subdivide the question "Ready for implementation?" into two parts. Have we got the necessary criteria and rules, firstly as regards the protection of the public, and, secondly, as regards the protection of the environment?

Let us consider the protection of the public. Of course, it is difficult to give a yes-or-no answer to the question: " Are we ready for implementing HLW repositories from the viewpoint of long-term safety protection of the public?" The two processes, implementation of disposal, and set-up of laws and codes of practice for protection of the public, proceed in parallel from general concepts towards practical and concrete sets of rules and practice. Presently, the status is the following.

The radiological protection criteria have been established long ago by the International Commission for Radiological Protection (ICRP), and the most recent revision of these rules (ICRP no. 26) has been incorporated in a CEC directive in 1980. They do not apply directly to the waste disposal case, in which there are aspects of protection involving future generations which are not specifically accounted for in the original publication. ICRP is presently drafting a recommendation which interprets the basic rules of ICRP no. 26 and applies them to the waste management case, which will become available in due course. The OECD-NEA has recently convened a group of experts in order to prepare a document in which proposals are made; since the composition of this experts' group also includes ICRP members, one can imagine that the ICRP recommendations will be set on similar guidelines. Mr. Lyon will expand on what the major issues are to meet these guidelines. My personal opinion is that there should be no overwhelming problem in meeting them.

Now, as regards the protection of the environment, a fundamental concept of radioprotection is that if rules are established such as to protect man, they will be more than adequate for protection of the environment, since man is a complex organism, very sensitive to radiation. The problem of environment protection, which is so acute nowadays, is not a radiological one, but essentially a problem of toxic chemical waste. The reasons are that these waste arise in much higher quantities (remember Mr. Orlowski's statement), that they are disposed of in the biosphere, and that they are noxious to specific environmental compartments. In comparison, high-level radioactive waste, which arises in small quantities, will be disposed of outside the biosphere; and there is no special sensitivity to radiation of selected environmental compartments. Therefore, as far as environmental protection is concerned, my answer will then tend to be "yes, we are ready to implement HLW disposal".

R. Lyon

I have been asked to say a few words on the topic of post-closure performance assessment. By performance assessment, we mean the evaluation of how well a disposal system is expected to measure up to its acceptance criteria. For this short presentation, I would first like to suggest that there are a number of challenges facing us in performance assessment. Then I will describe a framework that more or less represents the methodology being developed in various countries to meet those challenges. Finally, I will express an opinion as to how far along we are in establishing the framework.

The challenges that face us relate to: criteria, assimilation, uncertainty and variability, and validation.

Dr. Girardi has given us a concise discussion of criteria and I would just like to add to this that, recently in Canada, the Radiation Protection Association held a workshop on criteria, which included among the attendees representatives of the AECB and our environmental agencies; Messrs. Cadelli and Saltelli also represented the CEC. The conclusions from the workshop, which we are adopting in our programme, were that two basic criteria are appropriate: (a) that there should be a limit expressed in terms of risk (as suggested by the NEA) and (b) that there should be a limit on the probability of individual dose exceeding a particular level.

There is a tremendous amount of research information being generated by the field and laboratory programmes around the world. As so clearly described by Dr. Verkerk on Monday, we face a major challenge in assimilating the detailed information to arrive at conclusions about harm to man. Our objective in performance assessment must be to focus light on the primary objective, which is to evaluate the potential of a disposal facility to cause harm to man or the environment.

The systems to be analysed are highly variable:hydraulic conductivity in one borehole in a geological formation may be quite different to that in a borehole a hundred metres away. Measurements of that hydraulic conductivity will also contain errors of observation and interpretation, resulting in uncertainty in the values used. Further uncertainties result when phenomena are interrelated to evaluate complex processes, such as radionuclide migration, and all of this is compounded by extrapolation into the distant future.

Validation is a particularly difficult challenge for performance assessment. Validation by comparison of predictions with observation obviously cannot be achieved for the total system throughout the period of time that man must be protected from the waste.

The framework that more or less describes the developing methodology in performance assessment is illustrated diagramatically below. Detailed research models interpret directly the field and laboratory research, and produce results which can be cast in the form of simplified submodels and parameter distributions for use in performance assessment calculations. Representation of parameters of the submodels as distributions rather than single values enables account to be taken of uncertainty and variability.

Performance vs. Criteria

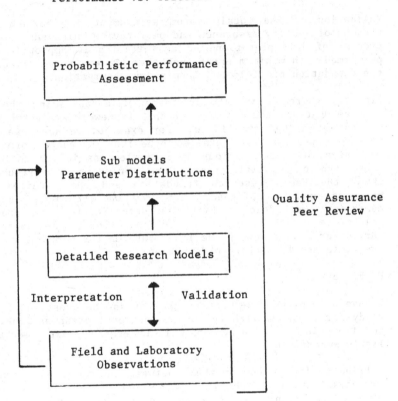

Direct comparison of predictions of the detailed research models with field and laboratory measurement provide for validation. The excellent presentation by Mr. Wallner this morning is an example of where this is being done concerning the thermomechanical behaviour of rock salt.

It is, of course, also possible to interpret directly the field and laboratory observations and use expert opinion, empirical calculations and direct measurements to provide submodels and data distributions. An example of this process of assimilation is the use of a three-dimensional hydrological modelling code to interpret directly the borehole measurements at a site, to infer the total flow field. From the flow field one can then develop distributions of path lengths and water transit times for use in a system model.

Using the submodels and data distributions, the probabilistic performance assessment can then be used to evaluate how well the system measures up against the criteria. The performance assessment can then be used in a sensitivity analysis mode to identify how changes in the subsystems or components of a specific proposed disposal system affect the performance relative to the basic criteria. By this process, the proponent of the disposal system can optimise his site and design to meet the basic criteria. Note that it is not appropriate to define arbitrary criteria for the performance of the subsystems. This would unnecessarily limit the capability of the proponent to design his system to best meet the basic criteria. (I recognise, however, that this may be a contentious statement).

Validation of the total system assessment can be achieved through a process of quality assurance and peer review which ensures that the total process of interpretation and assimilation is appropriate and has been performed with minimum error such that the end result is soundly based on the foundation of field and laboratory observation.

Finally, where do we sit in terms of establishing the framework? I believe that all the elements of this framework are established or in the process of being established. For example, we have heard at this conference papers on the detailed behaviour of waste forms, on container corrosion, on the behaviour of radioelements in the geological environment, and on specific field work including underground experiments. Thus, the basic research foundation and the detailed modelling are becoming well established. Tomorrow we will hear about performance assessment methodology, including deterministic and probabilistic developments. What remains is to consolidate the approaches, and in particular to strengthen the links between the performance assessment and the field and laboratory observation.

F. Feates

I have been given five minutes to address the question of whether we are ready for disposal with respect to national organisations, legal aspects and financing; so I hope you will understand that I will have to be a little superficial.

I tried to find a common theme between the national organisations which now exist, and I thought I had found one. They all seem to spend lot of time finding names which would be a good abbreviation and still be pronounceable. That really is the only common theme we seem to have within the Community. We have organisations which are run by a Board of Management chaired by the atomic energy authority; we have some supervised by Government Departments; some with public sector directors; some without public sector directors; i.e. a total mix of organisations. And this is not unreasonable; I do not think it matters, as long as they are organizations which do work within their own country.

They also seem to have quite dispersed methods of financing. Some have direct invoicing of running cost against waste producers; some have means of pre-financing major systems; others have loaned; some combine all these functions together in one way or another. Again, I do not see that it really matters, so long as we have structures which do the job.

Are we really settling the legal problems which will face disposal?
Well, public relations was not mentioned and will be addressed tomorrow,
but "legal", I think, really means "public acceptability" because if the
public accepts, then the law will allow. We have to recognise that this
is a major problem the nculear industry has to solve. It has, I think,
been tackled very effectively in Germany through a public hearing which
cleared the air immensely and has allowed Germany to progress; it has
been tackled in another way in France, which, again, seems to be very
effective; it has not been tackled very effectively in the United Kingdom
and we made little progress. So one can draw one's own conclusions from
these facts.

There is little left to be said about financing after P. Venet's paper at
the end of the last session, but I would like to comment on it. He did
say that the cost of the disposal facility would only be equivalent to
about two nuclear power stations. He talked about 25 Gigawatts. My
calculations suggest that it is about 10% of the capital cost in the
nuclear power industry; and if you discount the costs in both directions
(I have allowed for old plant which is operating and discounted the
future cost) it could be a lot higher. That does worry me a bit, because
at the time of the INFCE study a few years ago, we were talking about 1
or 2%; we now seem to be talking about 10%. And, in the UK, we are very
conscious of what we like to call the "Concorde Syndrom" (things do not
always cost what you think they will cost); and you can draw very pretty
graphs showing peaks and things like that; but they do not often bear
much relationship towards what is actually spent. Unfortunately, what is
actually spent is so often very much more. And I wonder if these systems
which have been set up will actually provide for the additional cost
which will be needed maybe after the nuclear industry has ceased to
operate, and the organisations may no longer exist. Perhaps that is a
question that we should be addressing.

We also have to recognise that technical problems can crop up: WIPP had
to change direction at a very late stage; I do not know how much it cost,
but it must have cost a bit; when the drilling was done at Gorleben, the
salt dome was found to be not quite what was expected, and I am sure
these problems, if they were problems, will be found on every side.
Maybe our Belgian friends went the other way, and they found that all the
expense that they had incurred freezing could perhaps be saved in a real
repository. But we have to recognise these uncertainties.

There is another uncertainty, and that is the one of the dreadful regula-
tor, which keeps coming along wanting more things doing. It has certainly not helped the American nuclear industry to get off the ground;
and we may find the same sort of problems when we actually get to build
repositories, that in response to public demands, there are additional
regulations coming in at a very late stage and costing the industry a lot
of money which they would claim they would never have thought of when
they made the initial estimates.

So if I could finish, I think that the answer to the question is "some
are more ready than others".

Chairman's Conclusive Talk

I will try to draw a few rapid conclusions, which I think we have reached a consensus on in this panel.

Firstly, as for the geological barrier, we believe we have a clear idea of the essential part it plays in all disposal systems, and we have an arsenal of techniques and instruments which allow us to characterize it in the field. Moreover, our laboratories are now well equipped to proceed with experiments, analyses and tests, all of which are indispensable for validations. The study into natural analogues should allow us to better understand certain physico-chemical phenomena and their evolution in nature.

Secondly, the processing and conditioning techniques for the waste allow us to obtain an end-product whose characteristics meet the main needs of physical and chemical stability. Other engineered barriers, for example the backfilling materials, are already available and ready for implementation.

Thirdly, the intensive research over the past few years into the field of rock and soil mechanics have allowed us to tackle the particular problems linked up with the excavation of galleries and shafts, and their operation. The most striking examples concern the deep dry-drilling experiments in salt in the Asse mine and the excavation of the underground laboratory in clay at Mol.

Fourthly, modelling calculations allow us already to forecast the behaviour of rock masses according to a certain number of parameters relating to the disposal of waste.

Fifthly, as for performance assessment, it is essential to have an in-depth knowledge of the system components, including the disposal site. Performance assessment methodologies have now reached an advanced stage of development. We already have acceptance criteria, but we still have to look into safety in the future and this has to be done at the same time as developing systems.

So to sum up, the development of a high-level waste disposal system could already be envisaged, with of course certain complementary research, and the compulsory passage through a demonstration stage, demonstration aimed at confirming, at the true scale, the conclusions of the research and studies. Now, the time we have left for the construction of industrial disposal facilities allows us to pursue our efforts aimed at clarifying certain incomplete answers to some of our questions, and to optimize the system so that the best economic and safety conditions are created.

The CEC 1985-1989 five year plan is undoubtedly aimed at obtaining these objectives.

SESSION VIII

MODELLING AND PERFORMANCE ANALYSIS

PAGIS, common European methodology for repository performance analysis

Long-term risk assessment of geological disposal : methodology and computer codes

Modelling of radionuclide release and migration

Application of the PAGIS methodology to clay, salt, granite and sub-seabed repositories

PAGIS, COMMON EUROPEAN METHODOLOGY FOR REPOSITORY PERFORMANCE ANALYSIS

N. Cadelli, F. Girardi, S. Orlowski (CEC)

Summary

In 1982 the CEC launched, in the field of risk assessment, the PAGIS action (Performance Assessment of Geologic Isolation Systems) which makes use of a methodology common to the EC countries and of realistic data representative of european sites, repository designs, and waste forms.
PAGIS may therefore provide a commonly accepted basis for future national licensing exercises and contribute, by means of its realism, to the acceptance of disposal solutions by the public.
In PAGIS, the performances of geological media (clay, granite, salt and sub-seabed) and man-made barriers for isolating HLW from man and his environment are being examined on the basis of a well defined normal evolution scenario and several altered scenarios which are superimposed on to the normal one.
Disruptive events will not be the object of consequence analyses, provided that their extremely low probability of occurence can be confirmed.
Best estimates of both the individual and collective doses are being computed and risk distributions for probabilistic events will be evaluated, including the uncertainties affecting the essential parameters.
The sensitivity analysis on the major parameters is expected to show to what extent the retention, retardation and dilution capabilities of the disposal system can be improved by an appropriate selection of the natural and engineered barrier characteristics.
At present, all basic data have been selected and the first preliminary evaluations are being made. The exercise will end with an evaluation of results in comparison with acceptance criteria.

1. Introduction

The pace of development of nuclear energy in Europe depends to a significant extent on the acceptance of solutions for the disposal of radioactive waste.

In order to obtain such an acceptance, by the responsible regulatory authorities on the one hand, and by the public at large on the other hand, a convincing demonstration has to be presented of the capability of selected geological formations to delay the migration of the radio-nuclides until the bulk of the radioactivity has decayed and to eventually allow thereafter their release into the biosphere at an extremely low rate with an insignificant risk to future populations.

Therefore, all countries involved in nuclear energy are making substantial efforts in that direction, with a marked increase since 1977 (1). For its part, the Commission of the European Communities (CEC) started sponsoring R&D work devoted, directly or indirectly to such safety aspects, as early as 1973. It was, however, only at the very beginning of the 80's that the time was judged particularly favourable to launch a major concerted action at Community level: the PAGIS Project

(Performance Assessment of Geological Isolation Systems); one of its objectives is to set up a methodology for evaluating the performances of HLW repositories, enjoying a large european consensus, and which could be regarded as an accepted basis for future national licensing procedures or more comprehensive system analyses.

PAGIS, as it stands, deals with four disposal options for HLW arising from spent fuel reprocessing, namely clay, granite, salt as well as the sub-seabed. It benefits from the considerable amount of information available through the programmes carried out by the Commission of the European Communities and its member states.

PAGIS is also a realistic exercise as it will be explained below. Many safety analyses have already been made in various countries and have shown that, when appropriate measures are taken, waste disposal is feasible and safe. In some cases, however, they are characterized by a generic approach and the lack of concreteness fails to gain the public confidence. In other cases, although based on realistic concepts, systematic conservative assumptions are made in order to cover the uncertainties inherent in long term risk assessment involving, for example, such elements as a model for the radionuclide migration in the host rock. But this approach, sometimes referred to as the deterministic approach, may result in an overdue penalization of the repository design. In order to avoid this drawback, other approaches have been developed which have recourse to probabilistic and statistic methods, and this appears the most promising way to face the problem. In this case, however, there is a risk that the transparency of the assessment, which is granted in the deterministic approaches, could be lost to some extent. PAGIS, on the contrary, makes clear references to real sites, today-repository engineering, and presently existing waste forms.

PAGIS is therefore expected to contribute to the acceptance of solutions for the disposal of radioactive waste.

2. Guidelines adopted

In setting up the methodology for the PAGIS action, four guidelines have been adopted, namely:

a) PAGIS should be a realistic exercise; indeed, it was believed that enough studies had been done on the feasibility and safety "in principle" of disposing radioactive waste. What appeared to be needed was a study on the feasibility and safety "in practice", with a clear reference to real sites having well defined geographical characteristics, using conceptual repository designs in accordance with present engineering standards and referring to waste produced, conditioned and packed with well-proven technologies.

b) PAGIS should be carried out with a methodology and model studies which not only should be made understandable for the scientific community but which also could be "translated" into comprehensible and convincing terms for the information of the public. The objective of PAGIS is, in fact, also a practical one: to increase the confidence of the public on the safety of waste disposal. The difficulty in implementing this guideline is, however, much greater than that expected to arise from the first one (point a. above).

c) The approach and methods adopted for PAGIS should be <u>flexible</u> enough both to accommodate the knowledge becoming available during the exercise and to adapt itself to the various possible radiological protection criteria which could be adopted in the future. It should also be able to accommodate minor changes in the waste management policies which could occur during the exercise.

d) Finally, PAGIS should not only represent a scientific approach but, it should also have a relevant <u>political content</u> at the E.C. level; the degree of development of nuclear energy differs considerably within the European Communities, ranging from the case where countries have implemented the entire fuel cycle, including reprocessing and refabrication of fuel, to the case where countries, with nuclear power installed capacities, will be customers for the cycle services, including waste solidification, and eventually will need to dispose their waste within their own territory. There are also countries with no nuclear programme at all. It is essential for the european policy that the PAGIS exercise helps in creating within the European Community a harmonized view on the question of waste disposal and contributes in reducing the gaps existing between its countries.

3. <u>Aspects related to realism and credibility</u>

a) <u>Realism</u>

Assuring the need for realism and concreteness contained in the first guideline implies a very large spectrum of aspects related to:

- the selection of the site itself and its characterization;

- the engineering of the repository, its parameters and the materials to be used;

- waste arisings, waste forms and conditioning.

No real difficulty was found in selecting repository designs appropriate for the various options, nor for the waste form. However, various packagings, modes and backfilling materials are still under consideration and they might have a non negligible influence on the retardation effect of radionuclide migration during the first post-closure period, when the radioactivity still has high levels.

b) <u>Credibility</u>

Much more difficult is trying to be understandable and convincing. The most challenging problem which had to be faced is related to the scentific approach to the various phenomena which have to be taken into account. The evolution of the various barriers and their eventual degradation is normally reproduced by models which are assessed and validated at laboratory scale and, to a lesser extent, through in-situ experiments. But everybody is aware of the conflicting situation arising between the need for long term predictions and the predictive capability of these models based on results obtained after tests carried out during several years at most. This is typically the case of the models developed for the waste form degradation and the corrosion of the waste packaging. But another difficulty is associated with the migration of the radionuclides through the backfilling, the geosphere and the biosphere, up to man. Various phenomena occur which need modelling: transport of radionuclides with the underground water movements and their dispersion, diffusion following concentration gradients and retardation effects due to several processes, such as ion exchange, sorption/desorption phenomena, filtration, etc. For each of these phenomena, global or, as is more generally the case, individual models are to be made available, and where individual models are used, their coupling results in a heavy mathematical complexity preventing analytical solutions. The mathematical burden, which results in exceedingly long computation times, is jeopardized by the fact that the coefficients pertaining to each model need to reflect the properties of the medium through which the radionuclides are supposed to migrate and these properties are varying in space and time, or are even subject to discontinuities, as is the case for the fissured granite. Moreover, if the variability range of a given property in a given formation can be determined through appropriate in-situ borehole experiments, there is no hope of reproducing in a model the real distribution of the rock properties throughout the formation.

Another kind of uncertainty was related to acceptance criteria for underground disposal of radioactive waste. Basic recommendations have been formulated by the ICRP and they have been adopted throughout the world. Some difficulties, however, are encountered when applying them to the case of underground waste disposal and criteria for practical application are being derived from these recommendations in various countries as well as within international organisations such as the IAEA and the NEA and the ICRP itself (2,3). For the moment, however, the absence of agreed, detailed, quantitative criteria makes the assessment an open ended problem, although progress is being made in this field and targets are expected to be fixed before the end of the exercise.

Finally, in addition to the difficulties mentioned above, which are common to all safety assessments in the field of underground waste disposal, a specific one had to be faced in PAGIS, coming from the intrinsic differences between the various options. As an example, repositories in granite formations are likely to show an extremely low but continuous leak also in the case of a normal evolution scenario, whereas the absence of water in salt formations prevents almost any movement of the radionuclides unless a perturbation occurs. These formation-specific differences may suggest different answers to the same question (such as "what is safe?") from teams used to studying different options. Such differences are not easy to reconcile without a substantial good-will of all assessors involved.

4. Setting up the methodology

Due to the nature of the problem and the existence of conflicting requirements, the solutions adopted for implementing the guidelines outlined in Section 2 cannot always be ideal, but we believe that, on the one hand, they should satisfy the scientific community and, on the other hand, they should be sufficiently transparent so as to be appreciated by the members of the public and their representatives.

a) The need for realism to be assured as a basic guideline sugges-ted the choice of real sites, independently from their final destination.   The selection was essentially dictated by the availability of information resulting from in-situ experiments and this was possible in most cases.   A catalogue of the geolo-gical formations in the European Community, recognised as satisfying predetermined criteria for undergound radioactive waste disposal was edited in 1980 (4).   A very limited number of sites, however, has been the object of intensive in-situ inves-tigations such as those made, for instance, on the salt dome of Gorleben or for the Boom clay deposit at Mol.   Moreover, in-situ explorations, even for research purposes, have been constantly delayed due to the resistance of local populations.   Very few sites, therefore, offer enough information to enable a realistic assessment and they do not necessarily represent the location of HLW repositories since no site has yet been definitely chosen by the relevant national authorities.

One reference site has been selected for each one of the conti-nental options: Mol for clay, Gorleben for salt domes and Auriat for granite formations.

While Auriat is essentially a notional site, the activities carried out at Mol and Gorleben are impressive and the evolution of these sites from research to demonstration sites and, final-ly, to real repositories may be expected (if, of course, all the on-going and future studies provide the expected favourable results).

Variants have also been selected but for some sites the limited amount of available experimental results had to be complemented with data expressed in the form of a range of values.   This situation, which could be considered a weakness in the approach, will in fact be used in PAGIS in order to show, through sensi-tivity analyses within each option, the impact of the charac-teristics and properties of the formation on the overall perfor-mances of the disposal system.   A similar situation was also found for the sub-seabed option for which areas were identified rather than sites.   With the selection made (the GME, CV2 and SNAP areas in the Atlantic Ocean) it will be possible to evalua-te the sensitivity of the results to variations in sediment properties and in the ocean dispersion parameters and to two types of submarine floors, plateaux and distal abyssal plains.

The selection of realistic repository design did not create major difficulties since reference designs appropriate to each option could be selected, which are based on proven techniques. For the waste forms, the selection of a borosilicate glass matrix as a reference was obvious, since the production of waste forms using this material has already been proven at industrial scale.

b) The need to be comprehensible and convincing, as shown in Section 3, is much more difficult to fulfill when dealing with extrapolations of situations occurring over periods of hundreds of thousands of years, as is the case for waste disposal systems. Only the final results will show if this guideline was successfully implemented.

Various evolution scenarios can be identified for each disposal site and they differ both in the expected consequences and in their probability of occurrence. In all cases, however, there is a typical evolution which can be described with a good approximation by assuming that no external event, from natural or human origin, would occur. Moreover, it represents the most probable scenario and in PAGIS it is assumed as the base case (normal evolution scenario). Human intrusion into the repository and altered evolutions due to natural phenomena are examined together with the associated probabilities in order to evaluate the corresponding risks. Through these separate evaluations the sensitivity of the results to the various scenarios will be apparent and, eventually, specific measures for the repository design can be suggested.

It is a common practice to select release scenarios on the basis of a list of possible events and this is how the first selection was made in PAGIS, after careful examination of the site specific characteristics. However, this selection will be revised as well as the detailed description of each scenario (including the normal evolution) in order to take advantage of on-going geoprospective studies (5). There are projections into the future of established natural trends which make it possible to compare situations in the context of consistent and plausible scenarios at various points in time and appropriate to each site.

The selection of predictive models describing the release and the migration of the radionuclides from the repository through the various pathways to man brings about a conflicting situation. On the one hand, advantage must be taken of the most advanced results and data, which imply sophisticated subroutines and, on the other hand, sensitivity studies and uncertainty analyses require multiple calculations which may rapidly lead to prohibitive calculation times and costs.

For PAGIS, it has been decided to use two sets of models: the one, with the most sophisticated approaches, will be used to provide best estimates; the other one, with simplified models, will be employed for sensitivity and uncertainty analyses . It is anticipated that, in order to ensure consistency between the two sets of results, a significant amount of adjustment work will be required, but with the approach selected one should have the essential advantage of a high transparency of the whole procedure.

Both the variability of the rock properties and the uncertainties on their values are taken into account in PAGIS by using a probabilistic approach. Distribution functions of the parameter values reflecting the results of appropriate laboratory and in-situ experiments are established whenever possible. A statistical sampling technique, incorporated in a general managing code, will then show for each scenario the spectrum of the possible consequences through a predetermined number of runs. Such type of codes are going to be applied more and more to underground waste disposal safety assessments. These arguments will be developed in more detail in the next paper.

Uncertainties, however, affect the models themselves and it appears advisable, as a first step, to gain experience on the relative importance of the various chemico-physical processes and transport mechanisms modelled with respect to the final radiological consequences. Sensitivity studies are therefore being made in the framework of PAGIS in order to identify the most relevant phenomena for which, eventually, more sophisticated models will be needed.

The procedure outlined above will certainly be essential in order to become familiar with the whole release and migration processes of the radionuclides, but does not entirely solve all the difficulties due to uncertainties since any model, as sophisticated as it may be, cannot be anything more than a simplified approach to the complex reality.

As far as radiological protection criteria are concerned, no single dosimetric quantity appears to be an appropriate criterion for the acceptance of underground radioactive waste disposal facilities and, eventually, a combination of risk and of dose limitations would be a satisfactory means to grant security for present and future generations. In PAGIS, both individual doses and risks are computed; these, along with a wide spectrum of intermediate results, are expected to respond to the most probable formulations of the safety requirements which will have been agreed at the time.

However, no attempt has been made to fully cover the requirements of the ALARA principle recommended by the ICRP, since no cost evaluation is included in the assessment, nor are subjective aspects such as public perception of the risk or the value of health detriments taken into account, since no optimization of the repository is sought in PAGIS. Collective doses, however, are also evaluated in PAGIS as an element of these optimization studies, which will be undertaken in the future.

Moreover, it should be stressed that a comparison among the four options is beyond the scope of PAGIS; the comparison of the results with the acceptability criteria will rather allow to establish, through sensitivity studies within each option, which are the essential parameters and how improvements can be made by appropriate combinations of the natural and engineered barriers.

c)   Until now, the methodology adopted appears sufficiently flexible to be able to incorporate all new knowledge becoming available. The essential means for reducing the errors in a safety assessment is to ensure tight and continuous connections between the safety analysts, the modelers and the experimentators, with a two-way flow of information: on the one hand, for providing the experimentators with guidelines for the R & D work and, on the other hand, for developing models with the corresponding input parameters appropriate for given ranges of applicability. And this is what is happening, both at the level of the national teams and within the CEC programmes, where PAGIS provides the reference for all concerted actions, such as MIRAGE (6), for the migration studies and COSA for the rock mechanics or for the R & D work on the containers, the waste form and the buffer materials.

As the general trend of the R & D work is to proceed to in-situ experiments, it is hoped that PAGIS will be able to incorporate

in time the results of field test campaigns as well as evidence
of the physico-chemical properties and the migration behaviour
of natural analogues, which can provide data for the projections
in the long term, missing in laboratory scale experiments, and
contribute to the validation of the predictive models used.

d) Concerning the political aspects of the PAGIS action, although
it is premature to make any final statement, we can certainly
say that the exercise has already been extremely useful in
increasing the mutual understanding of the various national
teams involved. Many views and opinions which looked diverging
at the beginning, have gradually smoothed out, and the agreement
reached on a common methodology is an evidence of success.
Moreover, PAGIS proved to be an excellent forum where exchange
of information contribute to reduce the gaps between the various
countries.

5. Conclusion

We believe that PAGIS is successfully developing in line with the
guidelines outlined in Section 2. Of course, we do not know how the
results will be appreciated in the end by the international community,
but a big effort has been made in order to be understandable and credible. The highest possible degree of realism has been incorporated in
PAGIS and flexibility did not present a problem. On the other hand, the
harmonization of the approach has been reached and all E.C. countries
involved in nuclear energy are contributing to the exercise so that, in
practice, its political significance and content is assured. The exercise will last for another few years before it ends with an evaluation of
the final results in comparison with acceptance criteria. At present,
all basic data have been selected (7) and the first preliminary evaluations are being made. The calculations are expected to be completed
before the end of next year. As stated in Section 2, PAGIS is intended as
an intermediate step between the generic performance assessments and
those assessments which will be needed for licensing procedures to
construct a disposal facility.

On the one hand, the methodology developed will be a reference for
the activities each country will undertake in the framework of licensing
procedures for a particular repository; on the other hand, PAGIS provides
the basis for setting up derived criteria for engineering use and some of
the essential elements in view of a system optimization satisfying safety
criteria such as ALARA.

The exercise, as it stands at present, is limited to high activity
waste disposal. Its successive extention to other types of waste should
present only minor difficulties and will provide an overall evaluation
for the entire fuel cycle waste.

Cost estimates are not included in the present study, but the studies presently underway in the Community concerning cost quantification
should allow, in due course, this essential parameter to be taken into
consideration for waste management optimization by techniques such as
cost benefit analysis or multi-attribute analyis.

REFERENCES

(1) Risk analysis and geological modelling in relation to the disposal of radioactive waste into geological formation.
Proceedings of the joint OECD-CEC workshop (May 23-27/1977).

(2) Criteria for underground disposal of solid radioactive wastes -
Safety Series N° 60.
IAEA (1983).

(3) Long-term radiation protection objectives for radioactive waste disposal.
NEA/OCDE (1984).

(4) Geological confinement of radioactive waste within the European Community - European Catalogue of Geological Formations.
EUR 6891 (1980).

(5) PEAUDECERF et al. - BRGM.
Etude géoprospective d'un site de stockage.
EUR 9866 (1985).

(6) MIRAGE project - First summary report.
EUR 9543 (1985).

(7) N. CADELLI, G. COTTONE, G. BERTOZZI, F. GIRARDI
PAGIS : Summary report of phase 1 - A common methodological approach based on European data and models.
EUR 9220 FN (1984).

DISCUSSION

R.B. LYON, AECL Whiteshell

I just like to support the statement made by Dr. GIRARDI about the danger
of incorporating very complex codes into a probabilistic performance
assessment. I think the limitation on this is not just in terms of computer power. There are a lot of other factors too. We will not be seeing
the same increase in terms of personal thinking power over the next few
years, as we will see in terms of computer power. The linkage of a complex
three-dimensional hydrological model to, for example, a thermodynamic
geochemical model, and perhaps even kinetic models, is a very complex process which would require linkages that we do not yet understand, and would
demand data that we do not have. I would, however, very strongly support
DeMARSILY's point about the requirement for validation. In the panel yesterday, I did end up with the comment that the major thing that remains
in the performance assessment developments in the future, is to strengthen
the linkage between the performance assessment codes and the field and
laboratory observations.
Another reason for this also is that we want to bring in - besides the
results of the detailed computer models - other aspects into the performance assessment, such as expert judgement, consideration of natural analogues, etc., which can be used to input at the simplified level and bypass
the very detailed analyses, and can indeed be often a stronger support, a
stronger validation than being able to apply a very detailed code.

F. GIRARDI, JRC-Ispra

Thank you for your comments. I think I can entirely agree with what you
said.

GRAMBOW, HMI Berlin

I have a suggestion on how to validate or how to get confidence from the
public for such codes. I think it is not so much a problem of having more
and more processes to be considered. Why is it not possible to predict
what people will obtain in a laboratory? Surely, you cannot predict the
whole model because we would have to wait 1000 years. But is it not possible to predict sub-models? Is it not possible to suggest experiments and
say: please, do this or that experiment, and I will promise you that you
will get this or that result, and show that you get it? I mean, you will
get confidence, because people like that.

F. GIRARDI, JRC-Ispra

This is perhaps the basic feature of the scientific methods, that you
develop a model and then you eventually devise an experiment to check your
model. The difficulty here is the extrapolation time; long experiments
can be done, but there are the natural analogues which are experiments
started by nature or by man, may be thousands of years ago. These analogues will certainly not be able to demonstrate that waste disposal is a

thing which can or cannot be done (I do not think that we have any of
these analogues), but certainly they are able to show whether the model for
a specific part of the assessment code is correct or not.

J. HAMSTRA, AVORA B.V. Bergen

To my opinion PAGIS is evolving towards too much effort in the release
and radionuclide migration part of the total system performance assessment.
I fully agree that the geosphere is the dominant component of the total
system and that we must very carefully evaluate that barrier. But, as a
conceptual designer, I strongly aim at primarily achieving an isolation as
high as possible in capability of the total system. Then, more specifically, the isolation capability in the near field area of the system. We
should be able to assess that in some sort of performance assessment of
that near-field area.
From the point of view of public perception I think that, if you compare
the two mechanisms: the isolation capability versus the effectiveness of
the geosphere from the point of view of radionuclide migration, the isolation capability will win; it will be far more convincing to prove that
the isolation is really high than to assume that there will be a release
- because that is a negative thing - and then prove that that release
will not harm.

F. GIRARDI, JRC-Ispra

We are always talking of the things where we have problems, and this is
perhaps why radionuclide migration, radionuclide release and other things
come up so frequently in discussions on risk assessment. I am convinced
that the normal evolution scenario, which is the one into which we put the
best of our knowledge, will prove that the release is zero in most of the
cases which one can consider, and, if some release occurs, it will occur
under conditions which can be considered as impossible. I think it is a
false impression that we are giving too much weight to release and migration. Of course, we want to be sure that the models we adopt are the best
available.

LONG-TERM RISK ASSESSMENT OF GEOLOGICAL DISPOSAL:
METHODOLOGY AND COMPUTER CODES

G. Bertozzi[1], M.D. Hill[2], J. Lewi[3] and R. Storck[4]

[1]Commission of the European Communities - JRC - Ispra (Italy)
[2]National Radiological Protection Board - Chilton (U.K.)
[3]Commissariat à l'Energie Atomique - IPSN - Fontenay-aux-Roses (France)
[4]Gesellschaft für Strahlen und Umweltforschung - Braunschweig (G.F.R.)

ABSTRACT

The assessment tools required to demonstrate the safety of a final
storage for nuclear waste are computer codes which accommodate the
mathematical models describing the physico-chemical processes which
occur in the whole system. Identification of release scenarios, modelling of radionuclide transport up to the biosphere and man, dose
and risk evaluation are the basic elements of the analysis. The uncertainty which affects the analysis results as a consequence of uncertainties in most of the input data may also be quantified; this is
made with the use of statistical codes, which handle data in the form
of probability distributions. Four computer code systems are being
developed in the European Community, based on these common issues,
which enable to consider them as a harmonized set.

1. INTRODUCTION

It is widely agreed that repositories mined in carefully chosen geological formations can offer an adequate solution for the final disposal of
radioactive wastes generated in the nuclear fuel cycle. This conviction is
based upon the geologist's statement that it is possible, on the basis of
past history and present conditions, to identify formations which are likely to remain unaltered for geological periods of time, thus assuring that
the enclosed wastes will be kept adequately segregated.

Qualitative judgements of experts cannot, however, be considered as
an acceptable demonstration that the disposal of radioactive wastes into a
given site is consistent with present and future public health and welfare.
It is necessary to demonstrate on a sound scientific basis that final disposal of nuclear waste can be effected in the site considered with the
present-day technology, in a manner that meets pre-established requirements
over extremely long time periods.

As the direct demonstration of the safety of a particular disposal
option is impossible, the proof can only be obtained indirectly, through an
evaluation of the system performance, on the basis of a body of scientific
and technical information [1].

Predictive modelling of repository evolution and radionuclide behaviour
in the specific environment can provide the required answer to the demand
for such a demonstration. Quantification of diffusive and advective transport of the various radioelements through the different barriers which are
interposed between the nuclear waste repository and man constitutes, therefore, the focus of the analysis.

The radionuclide's environmental dispersion may be triggered or modified by the occurrence of different events having the potential to perturb and change the hydro-geological situation around the host rock. They are impossible to foresee with any certainty, but their probabilities of occurrence may be quantified, and taken into account, in comprehensive assessments [2]. Of necessity, the assessment tools are computer codes; they consist of a set of mathematical models used to describe the relevant physicochemical processes occurring in the system and the radiological consequences for humans. As our understanding of these processes increases, new models are developed; therefore, the computer codes must be sufficiently flexible, in order to accommodate easily the new or improved models.

Of course, the results obtained will leave room for interpretation and judgement, due to the uncertainties implicit in future situations, and to the consideration of probabilistic events. It will be the role of the competent national authorities to examine the bases of the evidence furnished for the option safety, and to verify their compliance with pre-established objectives.

2. METHODOLOGY

The approaches which can be chosen to make the performance assessment for a given disposal option exhibit a considerable variability, as the methodology depends on both the purpose of the assessment, the type of geological formation and the amount of information available. A fairly simple methodology is appropriate for preliminary assessments, in which only a few release scenarios are considered; full assessments, on the contrary, require that all release scenarios be analysed, and uncertainties in results be quantified.

The possible approaches rely, however, on a set of common issues:

- identification and description of one or more radionuclide release scenarios from the repository;
- modelling of the radionuclide release from the repository and their transport through the various system components, following the scenario considered;
- modelling of the radionuclide dispersion in the environment, and assessment of their concentration in different food chains;
- assessment of the radiological consequences either to individuals or population groups, in terms of radiation exposure or risk;
- identification of the parameters which affect more strongly the model results;
- quantification of the doubt that the uncertainty in input data creates about the model results.

The first item requires the identification of all the events and processes which could either initiate release of radionuclides from the waste, and cause their transport through the geosphere and the biosphere to man, or could influence release and transport rates. They are then grouped in a way which enables to define a number of release and transport scenarios.

For any given type of geologic formation, there will be some processes which are certain to occur. These can be grouped together to define a "normal evolution scenario". The parameters and assumptions used in calculating the consequences of this scenario are based on extrapolations into the future of present and past geological and climatic trends, and it is assumed that the repository is not disturbed by probabilistic external events (natural or human-induced) [3].

A second class of events and processes are those which, if they occurred, would perturb the normal situation, but not change conditions so substantially that a completely different scenario is generated. These phenomena are dealt with by defining "altered evolution scenarios", consisting of sets of parameter values which are outside the normal range, and by taking into account the probabilities that the phenomena will occur [3].

Another type of scenario is that in which the radionuclide release conditions - linked to the occurrence of probabilistic events - are such that it is necessary to use different models to estimate their consequences. An example of such a scenario may be glacial erosion [4] or human intrusion. Again the probabilities of the triggering events have to be included when assessing the risks from such scenarios.

Very unlikely events having the potential to cause abrupt, direct releases of radionuclides into the biosphere may be associated with "disruptive" scenarios: examples are the impact of large meteorites and the occurrence of extrusive magmatic activity.

The process of listing and grouping phenomena leads to qualitative definitions of scenarios. In order to provide a quantitative definition which can be used in a risk assessment it is then necessary to identify all the parameters required to calculate the probabilities and consequences of each scenario and to assign values to them. Even when the data base available is very large, assigning values to parameters is not an automatic, objective process. In almost every case judgement is required and the persons carrying out the assessment will need to consult experts in the appropriate fields. The choice of parameter values will also be influenced by the scope and purpose of the assessment. If it is a full assessment, including an uncertainty analysis, then realistic distributions of ranges of values of all parameters are needed. However, if the assessment is more limited, then conservative choices may be made (for example, assuming that a low probability event is certain to occur), provided that the degree of conservatism is emphasized when the results of the assessment are presented and discussed.

Methods for calculating the risks associated with each scenario are currently the subject of some debate. In the past, two approaches have been used: either only doses have been calculated [5,6] leaving the user of the assessment results to apply the appropriate dose to risk conversion factors, or risks have been calculated as the product of the probability of occurrence of the scenario, the dose (individual or collective) and the dose to risk conversion factor [7].

The present trend is very much towards this "probabilistic" approach, although there are differences in the way risks are being calculated and it has been recognized that there is a need for information about the separate components of risk, as well as the combination of probability and consequence [8].

3.   UNCERTAINTIES

The results of long-term risk assessments always have uncertainties associated with them. Some of these uncertainties can be reduced through further research leading to better knowledge of the system under consideration, while others are inherent and irreducible. The use of "worst case analyses" does not resolve the uncertainty problem, because there will always be arguments about what constitutes a "worst case". There is, therefore, increasing interest in more exhaustive treatments which quantify the degree of uncertainty in analysis results, through the use of probabilistic methods. The procedure involves the use of probability distributions (probability density functions) for the uncertain parameters, to cover their entire range of variability. The probability density functions have, as a rule, a familiar form: uniform, normal or log-normal distributions are the most common, otherwise they may be presented as histograms. Such distributions are then fed into computer codes, which use them to generate the sets of values for the "Monte Carlo" simulations, and produce model outputs which are still in the form of probability distributions (histograms).

The methodology for sampling values of the input variables assumes a particular relevance when the complexity of the model and the running time of the code limit the overall number of runs which can reasonbly be performed. A strategy of sampling may be adopted to allow a faster convergence of the output distribution, thus minimizing the number of runs required, and saving CPU time. Instead of Monte Carlo, analytical codes do exist to handle data distributions, when the system's mathematics is simple; for instance, they are used in Fault Tree Analysis, where the mathematical relationships are elementary. These codes, however, become intractable with increasing model complexity so that they are of no practical use in comprehensive assessments.

In any case, this type of approach for uncertainty treatment leads to an output consisting of a probability distribution of doses, or risks, or any other quantity has been chosen to express the model results. A complementary cumulative distribution can then be derived, which shows the frequency with which consequences greater than a given level are expected[7,8].

The statistical analysis of this output provides then all the required information (mean values, percentiles, probability of exceeding given exposure levels, and so on).

4.   RISK ASSESSMENT CODES

In this paper the word "code" is used to describe computer programs which either solve sets of mathematical equations, manipulate input and output data, or control the running of other codes. The term "model" is used to refer to the set of mathematical equations which are used to represent processes occurring in the repository or the environment.

The common feature of risk assessment codes is that they are all based on models which predict the rates of release of radionuclides from the near-field (i.e. the repository and its immediate geological environment), the rates of radionuclide transport through the far-field, rates and patterns of dispersion of radionuclides in the biosphere, and doses to man (either individuals or populations or both). The models used vary considerably in complexity and level of detail, depending on the purpose for which assessment results are to be used, as well as on the characteristics of the processes represented and the overall structure of the assessment code.

In codes which are designed to produce "best estimates" of risks, the near-field, far-field and biosphere models are usually fairly complex and detailed, and their running times are such that it is not feasible to carry out extensive sensitivity and uncertainty analyses with them. The near-field and far-field models in these codes are often 2- or 3-dimensional, while the biosphere models are usually of the compartment type. In all three cases numerical methods are used to solve the underlying equations. When the assessment code is designed to quantify the uncertainty in dose or risk estimates by utilizing sampling techniques (see Section 3), it is not feasible to use very complex models because the codes must be run many times for different sets of input parameters. Thus, in "statistical" codes such as LISA (see Section 5.1) and the various versions of SYVAC [9], the underlying models have to be the simplest ones which can be used to represent adequately the relevant processes. Such models may be either specially developed for the purpose, or derived from more complex models, and, where it is possible, analytical methods are used to solve the relevant equations because these are less expensive in terms of computer time than numerical methods.

The differences in, and common features of the structures of the various "best estimate" and "statistical" risk assessment codes in use or under development in European Community countries are illustrated in the next section.

5. DESCRIPTION OF CODES

5.1 LISA

The code LISA (Long-term Isolation Safety Assessment) is a statistical code. It has been developed at the JRC-Ispra, in order to analyse different release scenarios from repositories situated in sedimentary formations; it includes the following functions:

- decay chain evolution of the source term;
- radionuclide release from the repository and migration through porous media by diffusion and advection;
- radionuclide dispersion into surface water bodies;
- pathways to man and dosimetry;
- dose rates conversion into risk.

For the uncertain parameters, the procedure adopted in LISA involves the use of probability distributions. They are fed into the code which generates a distribution of the model results following a Monte Carlo simulation technique. The sampling from the input distributions may be performed either with random sampling or with the Latin Hypercube (LH) sampling, which allows a faster convergence of the output. Different distribution types are accepted as an input by LISA: uniform or normal distributions on linear or logarithmic scales are those most currently employed. A multivariate input can be produced by imposing a rank correlation matrix on the LH matrix; otherwise, the input data matrix can be made "correlation-free", thus eliminating all the spurious correlations that the sampling may have introduced among the variables. Once the input distributions have been supplied, LISA runs through the various submodels for the given number of repetitions.

The organization of the code is shown in Fig. 1. The INPUT data set

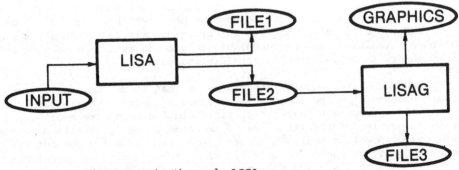

Fig. 1 - Data stream in the code LISA.

contains the data pertaining to the decay chains (nuclide names and inventory, decay constants and specific activities), the parameters of the input data distributions with the requested correlations, the dose factors and transfer coefficients in the biosphere.

The output from the main LISA is divided into two streams. The first, routed to FILE 1, is a record of the simulation characteristics (integration and interpolation data). The second one, routed to FILE 2, contains the actual results; for each run, FILE 2 stores the following:

a) the values sampled for the input parameters;
b) the peak dose rate values with their times of occurrence, and the name of the major contributing nuclides;
c) the curves of dose rate versus time.

FILE 2 constitutes the input for the subroutine LISAG which performs the statistical analysis and sensitivity analysis. These results are stored in FILE 3, while the graphics are routed to the GRAPHIC FILE.

The model results are presented in various formats: average and 95th percentile dose rates as a function of time, histograms of the peak dose rates, cumulative distributions, etc. Further, the doses are converted into a risk figure by combining them with the occurrence probability of the scenario over the time span considered.

The comparison between the input data values associated to the highest doses and the probability distributions assigned to the corresponding parameters permits the identification of the parameters which are more critical in governing the model results, following a procedure already utilized in SYVAC.

A correlation analysis (Spearman correlation) is also performed to identify variables which are strongly associated with the output.

The present version of the code has been implemented for a few cases which include:

- permeable fracture scenario through a clay bed;
- glacial erosion scenario in a clay bed;
- normal evolution scenario in sub-seabed disposal.

The two former scenarios have been analysed in the frame of the collaboration JRC/CEN-SCK for the safety analysis of a potential repository in the Boom formation [4]. The third scenario has been developed in the frame of a feasibility study of the sub-seabed option.

5.2 NRPB codes

The U.K. National Radiological Protection Board (NRPB) has three types of risk assessment code in use or under development. These are:

1) simple codes, which were used in preliminary assessments;
2) more complex codes designed to produce best estimates of risks from normal evolution scenarios and the principal altered evolution scenarios;
3) codes which use sampling techniques to provide estimates of risk and which are to be used for uncertainty analysis.

Each code has a modular structure, so that it can be adapted for use in assessments of disposal in several types of geologic formation by insertion of appropriate near-field, far-field and biosphere models. Fig. 2 shows an outline of the structure of two codes, one for preliminary assessments of disposal in geologic formations on land, and one for detailed assessments of sub-seabed disposal. The latter can also be used for land disposal assessments since the program for predicting radionuclide dispersion in the ocean and doses to man via marine pathways can be replaced by BIOS [10], which predicts dispersion in the terrestrial environment around a repository and subsequently throughout the biosphere, and doses to man via both terrestrial and marine pathways. All the NRPB codes predict doses and risks to members of critical groups, and doses and risks to populations, as a function of time.

While the simple and more complex "best estimate" codes are fully developed, the statistical or sampling type codes are not yet complete. The approach being pursued is to derive simple models for the near-field, far-field and biosphere from the more complex "best estimate" models already available at NRPB. The sampling technique to be used is a stratified one and is based on methods developed for use in risk assessments of nuclear installations other than waste disposal facilities.

5.3 The code MELODIE

Under the responsibility of the "Institut de Protection et de Sûreté Nucléaire"(IPSN) of the Commissariat à l'Energie Atomique (CEA), a risk assessment computer code, called MELODIE, is currently in development in France since the middle of 1984. It is a deterministic code, whose main general features are:

- its large modularity, obtained with a structure which distinguishes the computational modules (near-field, far-field, biosphere, thermo-mechanical effects, etc.) and the management modules (choice of the time step, call of computational modules, input and output processing, etc.) and by gathering all the variables in data files organized independently of the modules. This feature allows a continuous updating of the code along with the progress obtained in the understanding and modelling of the various phenomena involved. Moreover, the modularity gives a great flexibility in the use, depending on the needs, of complex models (best estimate calculations) or simplified models (sensitivity studies). At last, the modularity makes it possible to use the same code to estimate the risk of disposal in the three geological media investigated in France (granite, clay, salt), with models suited for each of them.

- its interactive coupling, at each time step, between the computational modules, in order to take into account the possible physico-chemical interactions between the near-field and the far-field. This way of coupling the modules has another important advantage: at each pre-defined time step, all the variables will be stored, and the job will be stopped. The user will then have the possibility of examining his results, and of making, if necessary, some modifications in the job definition (change of modules, introduction of new events, etc.). Of course, the difficulty of this kind of coupling resides in the determination of a common time step for all the modules.

The present version of the MELODIE code adapted to the granitic formations, is available since the spring of 1985, and is currently being tested. This version which is a very simplified one is constituted by the coupling of the three following modules:

- the phenomena involved in the near-field are modelled by a code called CONDIMENT which is developed by the Waste R&D Department (DRDD) of CEA. It deals with the transport of ground water through the engineered barriers (bentonite), the corrosion of the container, the release of radionuclides from the waste form and their diffusion up to the far-field. For the migration of some fission products, a Kd model is used whereas for the transport of the actinides, a more sophisticated modelling based on thermodynamical equilibria will be used. The code CONDIMENT is monodimensional and uses a totally implicit numerical scheme.

- the groundwater flow rate and the transport of the radionuclides through the geosphere is calculated with the code METIS, developed by the Ecole Nationale Supérieure des Mines de Paris (ENSMP); this code assumes that the geological medium is a continuous porous medium. Two versions exist: the first one is monodimensional and will be used for sensitivity studies. The second one is a two-dimensional finite-element version with an explicit description of the fractures (monodimensional joint elements) and is used for best estimate calculations. This model was tested in some calculations, among which are the INTRACOIN and HYDROCOIN international intercomparisons.

- the transport of the radionuclides in the biosphere and the calculation of the doses to man will be performed by the module ABRICOT developed by the "Département de Protection Technique" (DPT) of CEA.

This first version of the code MELODIE is used to perform the first part of the PAGIS calculations in granite, that is best estimate evaluations of the doses for the reference and variant sites, in the normal scenario. It will be completed by a sensitivity analysis algorithm in order to perform the sensitivity studies foreseen in the PAGIS exercise. At the same time, studies are conducted in order to take into account the thermo-mechanical effects and the impact of future possible geological events (climatological and neo-tectonical events).

5.4 Computer codes for the salt option

Computer codes for assessing radiological consequences of water intrusion into nuclear waste repositories mined in salt formations have been developed and applied by the German project P S E /11/ . For modelling purposes, the disposal system is divided into three subsystems: repository, geosphere and biosphere. For each subsystem either a new computer code has been developed or an existing one developed elsewhere is adopted.

The code EMOS, developed at the Technische Universität of Berlin, calculates radionuclide releases from the repository to the covering layers. A series of barriers are modelled, taking account of the following effects:

- flow resistance of backfill and sealing, during the brine intrusion;
- radionuclides leaching from the conditioned waste forms delayed by the containers;
- solubility limits of the radionuclides;
- radionuclide migration caused by temperature gradients, radiolytic gas transport, diffusion;
- expulsion of the contaminated brine into the covering layers, by convergence.

Due to the modular structure of the program, improvements in the system modelling or inclusion of further submodels would not pose major problems. Changes in the disposal system either for modified disposal concept or due to different release scenarios can also be handled. The main program of the code prepares the input data, organizes the output and calls the subprograms for each barrier, at every time step of the time interval considered for the analysis.

As a result, the release rate of each radionuclide into the overlying layers is given as a function of time.

The "SWIFT" code, by Intera Environmental Consults, models the migration of radionuclides through the geosphere. It simulates transient, three-
dimensional transport processes in heterogeneous geological media, using a
finite difference method. Processes considered are fluid flow, heat transfer and radionuclide transport. The fluid flow and heat transfer provide
the velocity field, which governs the radionuclide transport; the transport processes modelled are advection, hydrodynamic dispersion, sorption
and radioactive decay.

As a result, the concentration of each radionuclide in groundwater is
given as a function of time and space.

The code ECOSYS, developed by the Gesellschaft für Strahlen- und
Umweltforschung, Munich, calculates the nuclide migration through the
biosphere. The following pathways are considered:

- intake of contaminated drinking water;
- intake of vegetables irrigated with contaminated water;
- intake of milk and meat from contaminated cattle.

Finally, as a result of the calculations performed by the three codes,
radiation doses are obtained as a function of time and location for all
radionuclides considered.

6. CONCLUSIONS

In order to demonstrate that a final storage of nuclear waste can be
effected in a given site in a manner consistent with present and future
public health, predictive modelling of the repository evolution and of
possible radionuclide releases and transport must be performed.

The assessment tools are computer codes, which accommodate a set of
mathematical models to describe the relevant physico-chemical processes
which occur - or could occur - in the system, and to evaluate the radiological consequences for humans.

Within EC countries there are a number of computer codes available for
use in risk assessments of geological disposal. These vary considerably in
complexity and level of detail, depending on the purpose and scope of the
assessments in which they are intended to be used, as well as on the cha-
-racteristics of the disposal systems concerned. The lack of internationally
agreed long-term radiological protection criteria for geological disposal
has also led to some divergence in the assessment approaches used in the
various countries. However, it is clear that once such criteria are agreed,
the codes will be available to produce the required form of assessment results, for all the geological formations of interest.

REFERENCES

1. "Geological disposal of radioactive waste - an overview of the current
    status of understanding and development", Report sponsored by CEC and
    OECD, Paris, 1984.
2. M. D'Alessandro, C.N. Murray, G. Bertozzi and F. Girardi, "Probability
    analysis of geological processes: a useful tool for the safety assess-
    ment of radioactive waste disposal", Rad. Waste Management 1, pp.25-42
    (1980).
3. N. Cadelli, F. Girardi and C. Myttenaere, "PAGIS: a European perfor-
    mance assessment of geological isolation systems", Int. Conf. on Rad.
    Waste Manag., Seattle, USA, 16-20 May, 1983.

4.   G. Bertozzi, M. D'Alessandro and A. Saltelli, "Waste disposal into a plastic clay formation: risk analysis for probabilistic scenarios", Report EUR 9641EN (1984).

5.   M.D. Hill and G. Lawson, "An assessment of the radiological consequences of disposal of high-level waste in coastal geologic formations", NRPB-R108 (1980).

6.   "Final storage of spent nuclear fuel - KBS-3", SKBF/KBS, Stockholm (1983).

7.   G. Bertozzi and M. D'Alessandro, "A probabilistic approach to the assessment of the long-term risk linked to the disposal of radioactive waste in geological repositories", Rad. Waste Management and the Nucl. Fuel Cycle, Vol. 3 (2), pp.117-136 (1982).

8.   "Long-term radiation protection objectives for radioactive waste disposal", Report OECD/NEA, Paris (1984).

9.   D.M. Wuschke et al., Environmental and safety assessment studies for nuclear fuel waste management , Vol. 3: Post-closure assessment, Report AECL TR-127-3 (1981).

10. G. Lawson and G.M. Smith, "BIOS - a model to predict radionuclide transfer and doses to man following radionuclide releases from geological repositories for radioactive wastes", NRPB-R169 (EUR-9755EN) (1985).

11. Projekt Sicherheitsstudien Entsorgung, Zusammenfassender Abschlussbericht, Hahn-Meitner-Institut, Berlin, Januar 1985.

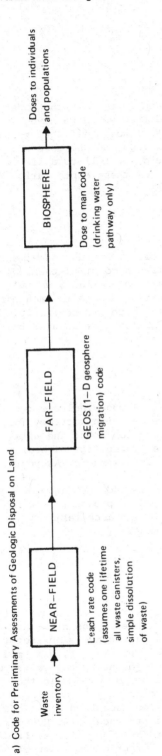

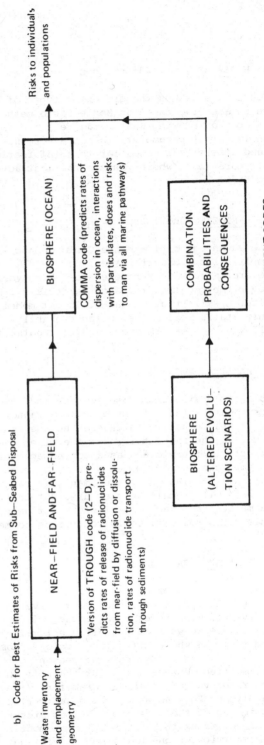

a) Code for Preliminary Assessments of Geologic Disposal on Land

b) Code for Best Estimates of Risks from Sub—Seabed Disposal

FIG. 2 — OUTLINE STRUCTURE OF 2 NRPB RISK ASSESSMENT CODES

DISCUSSION

J. HADERMANN, EIR Würenlingen

You have shown us the different types of codes and especially the codes
which are also based on Monte Carlo methods, and then you have given us an
example of showing what parameters are important. I would say that K_d is
an important parameter is not a very new information and one does not
need statistical codes for this information. Could you tell us a little
bit more about where you see the most advantages of these Monte Carlo
methods?

G. BERTOZZI, JRC-Ispra

The advantage of the Monte Carlo method is very clear. As the assessment
implies large uncertainties, we are obliged to take account of them. When
we are faced with parameters of which value we are not certain, we must
cover their entire range of variability. We cannot associate to each parameter a well-precised value. To get model results which are realistic,
only statistical codes, using probability distributions for input data,
allow us to perform such kind of treatment.

J. HAMSTRA, AVORA B.V. Bergen

Your conversion from dose to risk and also your LISA code includes two
components which - in my opinion - cannot be assessed with accuracy for
future radionuclide releases. The one is the assumption about the uptake
model in the far future; and the other one is the probability of health
effects to individuals as a consequence of radiation exposure. As your
assumptions in your model - in my opinion - are based on the fact that
nothing is going to change, man will remain as he is and the cancerogenous
effects will remain as they are. But that makes your approach a pure mathematical approach and it might not help in achieving acceptability for
this model concept. Could you please comment on that?

G. BERTOZZI, JRC-Ispra

Yes, I think I agree with you. I simply offer mathematical relationships
which link dose rates, dose-risk conversion factors and probability that an
individual will receive that dose. Of course, it is well possible that in
the future we will have a different value for the risk conversion factor
and that the whole situation, socio-economical and demographic, will modify.
But for the moment we have no reason to think that the numerical values will
be modified. However, I think that, if there will be any change in the
numerical values of these parameters, they will not change by orders of
magnitude. For instance, the dose-to-risk conversion factor which is now
estimated of the order of 10^{-2} per Sievert, is rather uncertain. However,
I do not think that it will be modified by orders of magnitude. It may
become twice or one half, but it will not become larger by an order of 10
for instance. In any case, we must perform the assessments with the information which is available at present.

CHAIRMAN

I thank Mr. HAMSTRA for this question. I think that if we have some time
left at the end, this point will be worth a more extensive discussion.
In establishing criteria, undoubtedly this does not belong to us to establish criteria, some weight should perhaps be given to the fact that
scientific progress may alter the dose/effect relationship which we are
presently using.

MODELLING OF RADIONUCLIDE RELEASE AND MIGRATION

G. de MARSILY
Ecole des Mines de Paris, Fontainebleau

T.W. BROYD
Atkins Research and Development, Epsom

B. COME
Commission of the European Communities, Brussels

D. HODGKINSON
UK Atomic Energy Authority, Harwell

M. PUT
Centre d'Etude de l'Energie Nucléaire, Mol

A. SALTELLI
Commission of the European Communities, JRC, Ispra

Summary
One of the key elements in the assessment of the safety of a nuclear
repository is the ability to predict the flow of water and transport
of radionuclides through the geologic barrier. In order to provide
some degree of confidence in its prediction, a model of radionuclide
transport must be shown:
(i) to incorporate all relevant physical and chemical mechanisms which
influence the migration;
(ii) to be mathematically accurate, i.e. to give correct numerical
solutions of the set of equations describing the relevant mechanisms;
(iii) to be validated by in situ experiments, laboratory experiments
and/or interpretation of "analog" studies;
(iv) to be given representative sets of the parameters describing the
medium.
In the last five years, large efforts have been made in the E.C. to
develop or improve calculation tools representing a large number of
physical and chemical mechanisms. In parallel, tests and comparisons
have been set in order to verify the numerical accuracy, and improve
the compatibility and efficiency of these tools.
The validation of these tools in the laboratory is partially under
way, but in situ verifications are still largely to be done.
The development of satisfactory methodologies for measuring and estimating the value of the relevant parameters of a given site is presently receiving much attention.
This paper will review the present status of these activities in the
E.C.

1. Introduction

In the last ten years, a very significant effort has been made in E.C.
countries to develop state of the art mathematical models and computer programs to predict the release and migration of radionuclides from a

repository. In addition to fundamental development work, it is necessary
to maintain an up to date review of relevant computer programs, to ensure
that there is minimal repetition of development and to improve liaison between workers in similar areas of study. The Commission of the European Communities therefore commissioned Atkins R&D (UK) to compile a directory of
computer programs, providing information on the existing models both in
Europe and North America. This work will first be described here, as an overview of the modelling "power" of the E.C. countries.

2. Compilation of a Directory of Computer Programs
A large number of computer programs are available for the prediction
of various aspects of the flow and transport of radionuclides from a disposal facility to the biosphere. Recent work (Broyd et al., 1983-1985) has
led to a directory of programs, and has provided the following:
  a) an overview of the current development and use of appropriate com-
     puter programs in Europe and North America;
  b) a list of all such computer programs in a directory, together with
     detailed reviews of those which satisfied certain selection cri-
     teria.

2.1. Scope of review
It was decided to limit the review to a consideration of postclosure
effects, since this is the region of greatest uncertainty. However, many
of the computer programs listed may be equally applicable to earlier stages in the life of a repository, e.g. the prediction of groundwater flow.
Eight primary subject areas were identified as affecting the transport
of radionuclides from a repository to the biosphere, as follows:
  - nuclide inventory   - stress analysis
  - corrosion   - heat transfer
  - leaching   - groundwater flow
  - geochemistry   - radionuclide transport
Biosphere modelling, surface water flow and risk analysis have not
been covered explicitly, but a section has been included which concerns
those programs which are more relevant to an overall system assessment than
to specific subject areas. Such codes have been considered in a separate
category entitled "total system analysis".

2.2. Methodology
The method used to obtain information included the following:
  - literature searches,
  - postal (and telephone) requests to program authors for publi-
    shed program manuals and relevant reports.
  - visits to relevant organizations in Europe and North America.
In all, more than a hundred different organizations were contacted
during the course of the work, of which more than thirty were visited. The
procedure followed was essentially the same for each program. Draft reviews of relevant computer codes were compiled using the published material
received by post or in meetings. These were then sent to each of the program 'main contacts' to seek their approval of the assessment and also to
request those points of required information not covered in the literature.

2.3. Program Selection Criteria
In compiling the directory, an attempt was made to identify all programs which are applicable to the subject areas listed in Section 2.1.
Due to the large number of programs that were identified, it was not possible
to give the same level of review about each program. In all, four types of

program review were used, as follows:
- detailed review - about 5 pages, covering aspects of author, numerical model, validation studies and applications, program operation, program support, and references;
- summary review - giving program name, summary of its capabilities and methodology, and list of references;
- tabular review - giving program name and area of use;
- total system analysis review - as summary review.

The basic criteria employed when considering which level of review to present were as follows:

a) Detailed reviews - a program should be:
- fully documented;
- directly applicable to radioactive waste repository analysis;
- in use by organizations other than the developing body;
- validated;
- supported for external use.

b) Summary reviews - were presented for those programs which have a definite application in radioactive waste repository analysis but which failed to meet the criteria applied for detailed review status.

c) Tabular reviews - were included for programs which had not been designed for the analysis of nuclear waste repository effects, but which could in principle be used in assessment studies.

d) Total system analysis reviews - as per summary reviews, for those programs designed to handle a large number of individual subject areas. A total of 288 relevant computer programs have been identified and are reviewed in the Directory, of which 63 are presented in detail, 161 in summary and 64 in tabular fashion. A summary of the number of programs reviewed in each subject area combination and of each type of review is given in Table 1.

Table 1: Summary of programs

	DETAILED	REVIEW TYPE SUMMARY	TABULAR
Nuclide Inventory (NI)	3	7	0
Corrosion (C)	2	0	0
Leaching (L)	0	1	0
L + NI	0	1	0
C + L	2	0	0
Geochemistry (G)	7	9	0
Stress Analysis (SA)	1	12	20
Heat Transfer (HT)	4	14	5
SA + HT	6	8	9
Groundwater flow (GF)	3	18	18
GF + HT	6	12	6
GF + SA	0	5	4
GF + HT + SA	1	4	2
Radionuclide Transport (RT)	14	37	0
RT + GF	9	16	0
RT + GF + HT	3	9	0
RT + L	2	0	0
Total System Analysis	0	8	0
TOTALS	63	161	64

3. Model description and status

We will focus here on the modelling of radionuclide release and migration. We will use the following classification:
- modelling of flow and transport in 2 and 3 dimensions,
- simplified one-dimensional modelling of transport,
- models for fractured media,
- models of colloid migration,
- geochemical modelling,
- source term modelling.

3.1. Modelling of flow and transport in 2 and 3 dimensions

The modelling of the regional groundwater flow around a repository will in general require the use of a 3-D model describing a rather large area surrounding the site (e.g. several tens or hundreds of km^2).

For sedimentary rocks, with a succession of pervious and semi-pervious layers, a quasi 3-D multilayered model can however be used: the flow is assumed horizontal in the pervious layers (aquifers) and vertical in the clays and low permeability media. Such a model, NEWSAM, developed at the Paris School of Mines, has been used and applied by the CEN-SCK to represent the flow around and through the Boom Clay at Mol.

For fractured rocks (assuming that they can be represented by an equivalent continuous medium), or for porous rocks, four flow-and-transport codes have been specifically developed within the C.E.C. waste disposal programme. These are:

NAMMU, NAMSOL and NAMTAR at the UKAEA (UK)
TRANSFLOW at the BGS (UK)
METIS at the CEA-Ecole des Mines (F)
METROPOL at the RIVM (NL)

The analysis of potential leakage paths from an underground repository in a well-characterized geological setting required general and flexible two-dimensional or three-dimensional computer codes. They must be capable of handling complicated geometries such as the surface topography, and permeability contrasts arising from large-scale fault zones and sedimentary layers. The finite-element method is particularly suitable for solving such complex problems and thus it has been used as the basis of these four transport models. We will first describe the NAM series codes, and indicate the differences with the other codes.

NAMMU and NAMSOL make use of an extensive library of general finite-element subroutines written over some years at Harwell. This allows the use of several time-stepping methods, numerical integration schemes and choices of element. The partial differential equations themselves are included in a very flexible ways, as are the material properties which can have arbitrary pressure, temperature and saturation dependence.

NAMMU (Rae and Robinson, 1979) calculates the time-dependent groundwater flow and heat transport in a permeable medium with the possibility of incorporating sources. The heat transfer and fluid flow are fully coupled so that buoyancy driven flows arising from the heat output of high-level waste can be treated. It can be used in either two dimensions or three dimensions and its numerical accuracy has been verified by comparison with two dimensional analytical solutions and with three dimensional solutions using an independently written program (Atkinson et al., 1984).

NAMMU has been applied to a wide range of problems such as groundwater flow around a repository for heat-emitting waste in fractured hard rock (Robinson et al., 1980; Rae et al., 1983; Hodgkinson et al., 1983, Wickens and Robinson, 1984), and the pore-water and effective-stress response of clay sediments to a heat source (Wickens, 1984). It has recently been used

to model three-dimensional groundwater flows through three well-characterized sites in Sweden (Atkinson et al., 1984a). An example of the finite-element mesh and water flow paths from a repository at a site which is crossed by intersecting vertical and horizontal fracture zones is shown in fig.
1. This illustrates one of the many post-processing options of NAMMU, namely the ability to draw perspective plots of three-dimensional pathlines.

Considerable effort has been made in order to make NAMMU easily useable by people outside the development group (user's guide, user-friendly'
keyword input, post-processing options).

The time-dependent migration of radionuclides in a steady-state velocity field can be calculated with the program NAMSOL (Dolman and Robinson,
1983). This has many similarities with NAMMU, with the heat transport replaced by solute transport. Thus it can be used in up to three dimensions and
it can handle complicated geometries and physical parameter variations.
Decay chains of up to five radionuclides can be handled and hydrodynamic
dispersion can be described by a general dispersivity tensor. Once again,
a User Manual has been written (Dolman and Robinson, 1983) and the "user-
friendly" input language can be used (Atkinson et al., 1984c).

One problem with using the conventional finite-element method to solve transport problems is that they can become unstable when dispersion
effects are small compared to advection. Two alternatives are being pursued to overcome this problem. The first is to modify NAMSOL to use higher-
order time difference schemes, such as the Taylor-Galerkin method, which
essentially adds in some numerical dispersion in order to stabilize the
solution. The second approach, which is embodied in the program NAMTAR
(Lever and Rae, 1981), uses a new numerical scheme for solving the transport equations. Its basic feature is the use of concentration contours of
marked particles which are tracked in time. In addition to being able to
handle problems where the dispersion is relatively low, it has a number
of other advantages. Complex models for various processes which change the
concentration, such as diffusion into the rock matrix and kinetic sorption,
can be readily incorporated while the explicit time-stepping means that
it should be efficient to run. It should retain these advantages when extended from its current two dimensions to three dimensions. The phenomena
currently included in NAMTAR are advection, linear equilibrium sorption
and anisotropic dispersion. Radioactive decay chains, diffusion into the
rock matrix and more complex geochemistry effects can be added as required.

The other codes quoted above have, with some minor differences, essentially the same characteristics. TRANSFLOW has an extended capacity to represent various forms of sorption. METIS has focused on the representation
of fractured systems, in 2-D, with linear elements to represent selected
major faults and a double-medium approach to include matrix diffusion.
METROPOL is still in the stage of development and is specially written for
representing the behaviour of very dense brines around a salt dome. Many
of these codes have been used in intercomparison studies, in the E.C. (see
paragraph 4) and outside the E.C. (INTRACOIN, HYDROCOIN).

A special note must be given to the code developed at present at Ismes,
Italy, from an existing code, TRAITEME, used in geothermal studies. This
code will compute the coupled mechanical-thermal-hydrological behaviour
of clay at high temperature and pressure. Although it is not exactly a
migration and transport code, this program will give the flow of water inside the clay generated by the heat and mechanical stresses applied to the
clay. This flow can then be introduced into a transport code.

3.2. Simplified one-dimensional modelling of transport

The very complex 2 or 3-D models described above will probably be used to represent the global behaviour of a repository; however, in safety studies or sensitivity analyses, it will often be necessary to represent the transport of radionuclides along one flow path, e.g. the shortest flow path between the repository and an outlet. A one-dimensional model is much simpler and preferable to do so.

Again four transport models have been developed with this purpose in the E.C. countries:

                    MICOF, at the CEN-SCK (B)
                    NAM1D, at the UKAEA (UK)
                    SIMRAP, at the CEA-Ecole des Mines (F)
                    COLUMN, at the RISØ (D)

They generally represent advection, longitudinal dispersion and sorption.

The first one solves the transport equation analytically, using the method of Laplace transforms.

The second is semi-analytical, using a numerical inversion technique of the Laplace transform. The last two are numerical, using simple finite differences. The code COLUMN is now used in the Safety Assessment code LISA at the JRC, Ispra.

In MICOF, the analytical solutions of the migration model have been expanded to the complete range of source parameters, including those causing complex arguments, and to the range of times greater than the time for depletion of the source. The code runs on an IBM VMS system.

Different calculations have been done with the program to evaluate the importance of the migration of the most important radionuclides prevailing in the HSLW. A remarkable result of these computations is given in fig. 2. This figure gives the concentration of the radionuclide Np 237 as a function of time and distance for four different source scenarios, namely:

S1 = a source with a concentration constant until depletion of the source, and equal to the solubility of the radionuclide in the interstitial clay water.

S2 = total linear dissolution of the solidified waste block in a period of 2,000 years.

S3 = total linear dissolution of the solidified waste block in a period of 10,000 years

S4 = instantaneous dissolution of the total amount of radionuclide disposed (no concentration limit).

The figure shows that at a distance of more than about 6 meters from the source, the concentration of Np237 stays always lower than the MPC value of 10^3 Bq/m^3 for drinking water and that after a few thousand years and a distance of a few meters there is no difference between the four source scenarios.

The parameters of the medium assumed in this example are:
                    pore water velocity: 7×10^{-5} m/a
                    retardation factor of Np: 400 (linear Kd)

In comparison with MICOF, NAM1D can handle any complex form of linear interaction between the solute and the medium (e.g. kinetic reactions, matrix diffusion). SIMRAP and COLUMN, using spatial discretization, can handle variability of the properties along the flow-path, and any complex interaction mechanisms. All three can also handle chain decay.

3.3. Models for fractured media

In many rock-masses, the groundwater flow is predominantly through a network of near-planar fractures with separations of many meters. The conditions under which the continuum approach, used for example in NAMMU and NAMSOL, adequatly approximates such a fracture network are not well established. Clearly it can only be valid on a length-scale which is large compared to the fracture dimensions. However, there may also be other limitations on its validity. For instance, the diffusion-like approximation conventionally used for modelling hydrodynamic dispersion due to mixing at fracture intersections may only be valid at very large times (Dieulin, 1984). Also, the retardation mechanisms for fractured and porous media can be different. For example, the time-dependent diffusion of radionuclides from water flowing in fractures into stagnant water in the micropores of the rock should be considered (Rae and Lever, 1980; Glueckauf, 1980; Curtis, 1980). Thus models have been developed which take explicit account of fracture characteristics, as discussed below.

3.3.1. Single fractures

Before the properties of fracture networks can be studied, it is necessary to know some of the physical and chemical characteristics of single fractures. Thus some special models to analyse field experiments in single fractures have been developed. In addition related models of migration in a single fracture have been used to investigate the relative importance of different phenomena (Goblet et al., 1979; 1982, 1984; Hodgkinson, 1984; Glueckauf, 1981; Hodgkinson and Lever, 1983; Lever, 1984).

Radionuclide migration in a single fissure assuming that the solute can diffuse in the matrix has also been modelled (Bradbury et al., 1982; Hemingway et al., 1983; Lever et al., 1983; Lever and Bradbury, 1984; Goblet, 1984).

3.3.2. Fracture networks

A completely deterministic fracture flow model would require a detailed knowledge of the positions, orientations, lengths and apertures of all fractures in the region of interest. Detailed information on these quantities is unlikely to become available. Its collection would probably be prohibitively expensive and destroy the site being investigated. It is also unnecessarily detailed given the large number of uncertainties in modelling radionuclide migration.

The alternative approach being followed at Harwell and elsewhere is to incorporate the essential features of the fracture system into a model in a probabilistic rather than a deterministic way. The data requirements are thereby reduced to manageable proportions (Bourke et al., 1982).

The basic assumptions of the model are that fractures are planar, of finite extent, occur randomly in the rock-mass and have orientations, lengths and hydraulic apertures which are described by probability distributions. In addition, information is required about the number of fractures per unit volume. In principle, further fracture characteristics describing such phenomena as channelling and sorption could be included in the present stochastic approach (Robinson, 1984b; Marsily, 1985; Wilke et al., 1985).

Given this limited statistical information, it is possible to construct a computer model of the fracture system and calculate the flow of water and transport of contaminants through it. Fractures with appropriate orientations, lengths and apertures are generated by random sampling of the probability distributions until the required fracture density is reached. In this way, it is possible to generate a large number of statistically equivalent fracture networks.

One application of this approach is to study the validity of continuum models. Also it can be used to aid the interpretation of experimental data, to simulate field experiments, and to provide permeability values

for use in a regional flow model. It is currently being used to examine
connectivity, water flow and hydrodynamic dispersion using some idealised
probability distributions in two dimensions. In addition, some results on
connectivity in three dimensions have been obtained.

3.4. Models of colloid migration

At the present state of knowledge, a large gap exists between the
models used in the risk assessment studies and those needed to account for
the laboratory experiments. Such a gap appears particularly wide when
colloid migration is considered.

In this respect the work carried out at the JRC of Ispra on Am
colloids has been useful in lightening up that:

a)   in no way colloid migration can be modelled by an "equivalent" or
     "average" Kd;
b)   existing models on colloid migration were not appropriate to describe
     the behaviour of the pseudo-colloids generated by the degradation of
     a glass matrix;

The main finding of the study was in fact that such pseudo-colloids
are distributed in a range of particle sizes, so that the conventional solution of the filtration equations based on a single filter coefficient
was not adequate to describe the outcomes of the experiments. In the following the experimental results are briefly recalled, and the main features
of the model are outlined.

3.4.1. The experiments

In the experimental set up borosilicate glasses, simulating high level
waste (HLW), were leached with an aqueous solution representative of the
water present in the aquifer which overlies the Boom clay formation in Belgium. The porous strata overlying the clay bed is a sand layer containing
glauconite clay mineral. A series of columns were filled with this material,
and the contaminated solutions, generated by the leaching of glasses, were
used as the source term for the column percolation experiments.
Many results were obtained with glasses doped alternatively with ^{238}
Pu, ^{237}Np, ^{99}Tc and ^{241}Am. In parallel with the column percolation experiments, ultrafiltration tests through membrane filters of different porosity have been carried out. The model described here interprets a set of experiments conducted with ^{241}Am doped glass. The experiments are also described in (Avogadro et al.; 1980, 1981; Bidoglio et al., 1983).

The results of the filtration experiments are shown in fig.3 where
the percentage ϕ of americium retained as a function of the filter porosity
is reported. Also the complementary curve $\phi' = 100 - \phi$ is plotted.

In fig. 4, the ^{241}Am concentrations in molar units are reported for
three columns, corresponding to the 28, 80 and 170 days loading experiments.

3.4.2. The model

The large percentage of Am retained on the filter membranes (fig.3)
suggests that the Am transport is largely dominated by the colloidal species. The presence of Am colloids in similar chemical conditions has also
been reported in the literature (Starick and Ginzburg, 1961; Olofson et
al., 1982). It is also evident from the smooth shape of the curve in fig.3
that the colloid size is better described as a distribution rather than
with a single value.

The complementary curve can be thought of as a cumulative distribution, i.e. each point in this curve represents the percentage of colloid
having a size smaller or equal to the value on the horizontal axis. Poly-

dispersion of the Am (pseudo) colloids is also being confirmed by Photon Correlation Particle Analyser tests conducted at the JRC Ispra.

The model which has been developed to account for the three filtration profiles shown in fig.4 originates from the integration - over the size of the particles - of the solution of the monodispersed filtration equation. In practice, a distribution function is derived from fig.3 for the size of the colloids. If λ is a filter coefficient and $f(\lambda)$ is its distribution function, the fraction of particles whose size ranges between λ and $\lambda+d\lambda$ is simply Co f $(\lambda)d\lambda$, where Co is the total colloid concentration at the inlet, and

$$\int_{-\infty}^{+\infty} f(\lambda)d\lambda = 1$$

If $C=C_o g(x,t,\lambda)$ is the solution of the transport problem for the monodispersed colloid, the solution for the polydispersed one can be obtained by writing:

$$d\ Cp = C_o f(\lambda)d\lambda g(x,t,\lambda)$$

for a generical infinitesimal fraction of the polydispersed colloid, and

$$Cp = C_o \int_{-\infty}^{+\infty} g(x,t,\lambda)f(\lambda)d\lambda$$

for the total colloid concentration.

In fig.5, the fit obtained for the 170 day experiment is reported as an example. It is to be mentioned that the model also accounts for the sorption of the colloid, which has to be introduced as a further term in the filtration equation. The sorption of the colloid is described by a Langmuir isotherm, the sorption capacity of the column being saturated already in the 28 day experiment. A complete illustration of the features of the model is given in Saltelli et al., 1984.

In conclusion, the following dominant processes were identified:
a)   Colloid filtration: this is governed by a distribution of values of filter coefficient (i.e. the colloids are polydispersed).
b)   A saturation phenomenon: this appears after a very short time and can be adequately described by a Langmuir isotherm.
c)   Escape of unretained species: after the saturation, a small fraction of the input activity flows through the column with practically no retention. It can not be ascertained at present whether this anionic fraction is given by the lower size microcolloids or by soluble species.

One of the main limitations of the model is the role (if any) of the soluble non-colloidal forms (see point c above). The experimental evidence indicates that in the chemical conditions of the leaching water, the two species Am $(CO_3)_2^-$ and Am $(HCO_3)_2^+$ should predominate (Bidoglio, 1982), in a ratio of 3 to 1 approximately. In the present model, the sorption mechanism has been attributed to the colloids themselves, but it could be due as well to the anionic carbonate complex, generated with some kind of kinetic by the colloids. Furthermore, the model does not include colloid dissolution while a series of scanning tests performed on loaded columns have shown that washing the columns with uncontaminated water slowly removes the activity.

Conversely, incorporating the above phenomena in the model would increase the number of parameters and consequently the validity of the fit would be reduced.

It is evident, in conclusion, that a better fit of these experiments would require more systematic experimental effort. At this stage we can say that a general identification of the main transport mechanisms can be considered as achieved.

3.5. Geochemical modelling

One geochemical speciation code, WHATIF, is being developed in the E.C. at Riso (D). From an initial list of constituants (metal oxyde, CO_2, H_2O), the code calculates which mineral is formed as a function of the progression of the reaction index (i.e. analogous to the progression of the time). It also gives the speciation of the ions in solution (simple or comlex ions). Such calculations are needed to predict the form under which the radionuclides will leave the source and migrate in the medium (or precipitate). If the element remains in solution, its retardation by the medium will indeed be a function of its ionic form (complexation, charges, ...).

If it precipitates, can it form colloids ? Will it be stable indefinitely ? Such geochemical models can also incorporate adsorption.

This kind of modelling is required first close to the source, then in the medium, if the geochemical conditions vary (temperature, pressure, oxydation state, pH, new minerals in the rock, new elements in the solution from a superficial source, etc...).

Other E.C. countries are also testing existing geochemical codes (see paragraph 4). Also attempts are being made to couple these geochemical codes with the transport codes: this is essential to correctly represent the interaction mechanisms, the simple "Kd" approach used mostly so far having been found insufficient.

This task is just at its beginning nowadays and will require considerable efforts of modelling, verification and validation against experiments or analog studies.

3.6. Source term modelling

Many countries are developing source term models for different types of waste. The classical "congruent dissolution" model or the "constant release" model are often no longer considered valid. Solubility-limited release models are being developed: these are closely related to the geochemical models briefly described above since the solubility of a given element will be a function of the composition and speciation of the solution surrounding the waste.

These source term models generally include the simulation of the engineered barriers: corrosion of canister, or transport through a layer of buffer or backfill. Complex transport mechanisms can thus be represented in these models.

Model development is taking place in particular at:
- JRC, Ispra
- HMI, Berlin (code EMOS)
- CEA, Fontenay
- UKAEA, Harwell
- ECN, Petten
- CEN-SCK, Mol
- etc...

but, apart from the code EMOS which has already been applied for the study of a repository in a salt dome, in the FRG, most of these codes are still in the development stage.

4. Computer program intercomparison studies

During the compilation of the Directory (paragraph 2), reliance was placed on information provided by the developers of the computer programs identified. Whilst this was a necessary method of procedure, it did not allow truly objective comments to be made concerning the use of particular programs in realistic situations, the ease of data input and the general adaptability of different codes. Furthermore the intercomparison of computer programs is a valuable technique for instances where verification against field or experimental data is not practicable. Such objective comparisons have recently been made for computer programs in two specific topic areas:

- radionuclide migration (Broyd, 1984)
- geochemistry (Atkins R&D, 1985).

These comparison studies have differed from other recent work such as the INTRACOIN project (SKI, 1984), in that they have attempted to compare both the results and useability of the programs by means of independent observation of program runs during short visits to each participating organisation.

4.1. Methodology

The methodology was similar for both intercomparison studies and proceeded as follows:

- Potential participants in E.C. countries were identified and meetings were held to discuss overall procedures and general aspects of test cases.
- Series of test cases were derived and tested using the study organisers' in-house codes.
- Visits of two days duration were made to each participating organisation during which the running of test cases was observed. Details of test case data had not been provided to participants prior to the visits.
- Results were analysed and discussed during a further plenary meeting of participants.

This methodology worked well for both of the intercomparison studies attempted and enabled the results to be reported within a few months of the start of each piece of work.

4.2. Results

The radionuclide migration and geochemistry studies included five and eight participating organisations respectively, which are listed in Table 2. About ten different types of test case were considered in each study. The type and number of test runs carried out by each organisation provided a rudimentary comparison of user-friendliness and extent of application of the programs used.

It would be erroneous to consider these studies as tests of the accuracy of participating computer programs or their data bases. The result should best be used in deciding in what areas there is general agreement or disagreement between the programs and, where disagreement is apparent, some attempt has been made to ascertain the root cause. Fig. 6 and 7 show typical results from each intercommparison study. A detailed consideration of each study is beyond the scope of this paper and the reader is directed to (Broyd, 1984; Atkins R&D, 1985).

Table 2: Organisations participating in intercomparison studies

Country	Organisation	Migration	Geochemistry
Belgium	SCK/CEN, Mol	X	
Canada	AECL, Whiteshell		X
C.E.C.	JRC Ispra	X	X
Denmark	Riso		X
F.R.G.	GSF, Braunschweig		X
France	BRGM, Orléans EMP, Fontainebleau	 X	X X
U.K.	AERE, Harwell Atkins R&D, Epsom BGS, Keyworth UWIST, Cardiff	X X(*)	 X(*) X X(*)

(*) indicates coordinating organisations.

5. Conclusion

The development, improvement, testing, verification and validation of computer programs are an endless enterprise: as long as the last repository of the last E.C. country will not have been sealed, model improvement and application in waste disposal will have to take place. Why is that so ? Because a model is in the first place a representation of the understanding of the mechanisms taking place in the real world; as scientific knowledge of such mechanisms increases, so must models be improved. On the other hand, the mathematical techniques for solving a given set of equations is constantly evolving as well as the computing power of the standard scientific computers. Therefore a model which is not constantly improved and modified soon becomes obsolete and cannot any longer be of any use.

Modelling efforts in the E.C. countries will therefore have to continue. In this respect, what can be the specific roles of the Commission ? Several can be imagined:

(i)   Help develop, in each country, at least one strong modelling group capable of making a performance assessment, using its own models or models provided by other groups.

(ii) Improve the general capability of the E.C. countries by model comparisons, exchanges of codes, subroutines and methodologies, by modelling of case studies.

(iii) Avoid duplication of efforts and endeavour to make model improvements or developments done in one country available to the other E.C. countries; "clubs" of specific group of modellers could be set up in this view.

(iv) Provide a European independent methodology for validation of the codes which will be used in a final assessment, to improve their credibility.

REFERENCES

1.   ATKINS R&D (1985)
     A comparison of computer programs which model the equilibrium chemis-
     try of aqueous solutions. Progress report to C.E.C. contract 219/1B
     81.7 WAS.UK, Atkins R&D report 20499.20, March 1985.

2.   ATKINSON, R., CHERRILL, T.P., HERBERT, A.W., HODGKINSON, D.P.,
     JACKSON, C.P., RAE, J. and ROBINSON, P.C. (1984a)
     Review of the groundwater flow and radionuclide transport modelling.
     in KBS-3. AERE-R. 11140.

3.   ATKINSON, R., HERBERT, A.W., JACKSON, C.P. and ROBINSON, P.C. (1984c)
     NAMSOL Command reference manual. AERE-R.11365.

4.   AVOGADRO, A., MURRAY, C.N. and DE PLANO, A. (1980)
     Transport through deep aquifers of transuranic nuclides leached from
     vitrified high level wastes. Scientific Basis for Nuclear Waste Mana-
     gement, vol.2; Clyde J.M. Northurp Jr., Ed. Plenum Press.

5.   AVOGADRO, A., MURRAY, C.N., DE PLANO, A. and BIDOGLIO, G. (1981)
     Underground migration of long-lived radionuclides leached from a boro-
     silicate glass matrix. IAEA-SM-257/73 (1982). Proc. of Int. Symp. on
     Migration in Terrestrial Environment of Long-Lived Radionuclides from
     the Nuclear Fuel Cycle, Knoxville, Tenn. (USA), July 27-31, 1981.

6.   BGS. Research carried out in the framework of the MIRAGE project.
     Progress report, January to June 1984. Contract 413-83-7 WAS UK Key-
     worth, 1984.

7.   BIDOGLIO, G., CHATT, A., DE PLANO, A. and ZORN, F. (1983)
     J. Radioanal. Chem. 79, 153-164.

8.   BIDOGLIO, G. (1982)
     Radiochem. Radioanal. Letters 53 (1), 45-60.

9.   BORSETTO, M., CRICCHI, D., HUECKEL, T. and PEANO, A. (1984)
     On numerical models for the analysis of nuclear waste disposal in geo-
     logical clay formations (Ismes, Bergamo). In Numerical Methods for
     Transient and Coupled Problems, edited by R.W. Lewis et al., Pinerid-
     ge Press, Swansea.

10. BOURKE, P.J., EVANS, G.V., HODGKINSON, D.P. and IVANOVICH, M. (1982)
     An approach to prediction of water flow and radionuclide transport
     through fractured rock. Proc. of the OECD-NEA workshop on Geophysical
     Investigations in connection with Geological Disposal of Radioactive
     Waste, Ottawa, September 1982, 189-99.

11. BRADBURY, M.H., LEVER, D.A. and KINSEY, D.V. (1982)
     Aqueous phase diffusion in crystalline rock. Scientific Basis for
     Radioactive Waste Management, vol. V (ed. Lutze, W.), 569-78.

12. BROYD, T.W. (1984)
     A comparison of radionuclide migration computer programs. Part 1:
     preliminary findings. Progress report to C.E.C. contract 219/1B.81.7
     WAS UK, Atkins R&D report 20499.20, May 1984.

13. BROYD, T.W. et al. (1985)
    A directory of computer programs for assessment of radionuclide waste
    disposal in geological formations. Pub. CEC report n° EUR 8669 EN
    July 1983, revised March 1985.

14. CEN-SCK
    R&D programme on radioactive waste disposal into geological formations
    (study of a clay formation), contract WAS-334-83-7-B.
    Semi-annual report n°16, activities for the period 01.7.83-31.12.83
    Semi-annual report n°17, activities for the period 01.1.84 -31.7.84
    Semi-annual report n° 18 (partial draft).

15. CURTIS, A.R. (1980)
    Numerical treatment of the movement of solutes through fissures in
    porous rocks. AERE-R. 9820.

16. DIEULIN, A. (1984)
    L'effet d'échelle en milieu fissuré. Rapport CCE-CEA-EMP LHM/RD/84/17

17. DOLMAN, E.A. and ROBINSON, P.C. (1983)
    NAMSOL: finite element program for migration of radionuclides in
    groundwater. AERE-R 10882.

18. GLUECKAUF, E. (1980)
    The movement of solutes through aqueous fissures in porous rocks.
    AERE-R 9823.

19. GLUECKAUF, E. (1981)
    The movement of solutes through aqueous fissures in micro-porous
    rocks during borehole experiments. AEER- R 10043.

20. GOBLET, P., LEDOUX, E. and MARSILY, G. de (1979)
    Interprétation sur modèle des essais de traçage réalisés sur le massif
    A. Rapport CCE-CEA-Ecole des Mines, LHM/RC/79/8.

21. GOBLET, P. (1982)
    Interprétation d'expériences de traçage en milieu granitique (massif
    B). Rapport CCE-CEA-Ecole des Mines, LHM/RD/82/11.

22. GOBLET, P. (1984)
    Etude hydrodynamique à l'aide de traceurs d'un doublet hydrothermique
    en roches fissurées. Interprétation des traçages aux lanthanides.
    Rapport Ecole des Mines, LHM/RD/84/112.

23. GOBLET, P. (1984)
    Interpretation of tracing experiments in porous and fractured medium
    using the METIS code. INTRACOIN project level 2. Rapport CEA-Ecole
    des Mines, LHM/RD/84/3.

24. GOBLET, P. (1984)
    Etude de modélisation des phénomènes de transfert sur un site de
    stockage aux différents stades de la vie du site. Rapport CCE-CEA-
    Ecole des Mines, DAS 121. Avancement au 15/12/1984.

25. HEMINGWAY, S.J., BRADBURY, M.H. and LEVER, D.A. (1983)
    The effect of dead-end porosity on rock-matrix diffusion. AERE-R 10691

26. HERBERT, A.W., HODGKINSON, D.P., LEVER, D.A., ROBINSON, P.C. and
     RAE, J. (1985)
     Mathematical modelling of fracture hydrology. Final report (draft).
     AERE Harwell, report G.3445.

27. HERBERT, A.W., HODGKINSON, D.P., LEVER, D.A., RAE, J. and ROBINSON,
     P.C. (1984)
     Mathematical modelling of radionuclide migration in groundwater.
     Paper presented at the meeting on "Exotic Uses of Aquifers", Geologi-
     cal Society, London, October, 9, 1984. AERE Harwell TP.1087.

28. HODGKINSON, D.P. and LEVER, D.A. (1983)
     Experiment on the transport of sorbed and non-sorbed tracers through
     a fracture in cystalline rock. Radioact. Waste Manage. Nucl. Fuel
     Cycle, 4, 129-58

29. HODGKINSON, D.P., LEVER, D.A. and RAE, J. (1983)
     Thermal aspects of radioactive waste burial in hard rock. Prog. Nucl.
     Energy, 11, 183-218.

30. HODGKINSON, D.P. (1984)
     Analysis of steady-state hydraulic tests in fractured rock. AERE-R.
     11287.

31. LEVER, D.A. and RAE, J. (1981)
     NAMTAR: A new computer program for the transport of radionuclides in
     groundwater flow. In Proc. of the Intern. Symp. on Migration in the
     Terrestrial Environment of Long-Lived Radionuclides from the Nuclear
     Fuel Cycle, Knoxville (USA), pp. 695-700, IAEA, Vienna.

32. LEVER, D.A., BRADBURY, M.H. and HEMINGWAY, S.J. (1983)
     Modelling the effect of diffusion into the rock matric on radionucli-
     de migration. Prog. Nucl. Energy, 12, 85-117.

33. LEVER, D.A. (1984)
     An analysis of the AECL/CEC field experiment on the transport of [82]Br
     through a single fracture. AERE-R. 11299.

34. LEVER, D.A. and BRADBURY, M.H. (1984)
     Rock-matrix diffusion and its implications for radionuclide migration.
     AERE report TP. 1004, Mineral. Mag. (to be published).

35. MARSILY, G. de (1985)
     Flow and transport in fractured rocks. Connectivity and scale effect.
     Intern. Symp. (IAH) on the Hydrogeology of Rocks of Low Permeability,
     Tucson (USA), January 7-12, 1985.

36. NOY, D.J. (1982)
     Development of computer models for three-dimensional analysis of
     groundwater flow and mass transport: a progress report. IGS, Environ-
     mental Protection Unit report ENPU 82-15, Harwell.

37. OLOFSSON, V., ALLARD, B., ANDERSSON, K. and RORSTENFELT, B. (1982)
     Formation and properties of americium colloids in aqueous systems.
     Scientific Basis for Nuclear Waste Management, vol. 4. S Topp (ed.)
     Elsevier Sci. Publ. Co., N.Y.

38. PATYN, J. (to appear in 1985)
     Docteur-Ingénieur dissertation, CEN-SCK and Ecole des Mines de Paris.

39. PUT, M. (1984)
     Modelling of radionuclide migration in clay. Final report CEN-SCK,

contract 334-WAS(B).

40. PUT, M.J. (1984)
Unidirectional analytical model for the calculation of the migration
of radionuclides in a porous geological medium. Draft submitted for
publication in Radioactive Waste Management and the Nuclear Fuel
Cycle.

41. RAE, J. and ROBINSON, P.C. (1979)
NAMMU: Finite element program for coupled heat and groundwater flow
problems. AERE-R. 9610.

42. RAE, J. and LEVER, D.A. (1980)
Will diffusive retention delay radionuclide movement underground ?
AERE-R. RP.853.

43. RAE, J., HODGKINSON, D.P., ROBINSON, P.C. and LEVER, D.A. (1983-1984)
Mathematical modelling of fracture hydrology. Progress report on CEC
contract n° 378-83-7-WAS-UK(H), January-June 1983 (AERE-G 2896), July-
December 1983 (AERE G) and January-June 1984 (AERE-G 3226) Harwell.

44. RAE, J., ROBINSON, P.C. and WICKENS, L.M. (1983)
Coupled heat and groundwater flow in porous rock. Numerical Methods
in Heat Transfer, vol. II (ed. Lewis, R.W., Morgan, K. and Schrefler,
B.A.) 343-67 (John Wiley and Sons Ltd.).

45. RIBSTEIN, A. (1985)
Couplage des modèles géochimiques avec les modèles de transport.
Rapport final, draft, Ecole des Mines de Paris.

46. RIVM: Transport by groundwater of radionuclides released after flooding of a repository in a salt dome. Interim report, contract WAS-
383-83-7-NL. Period 01.5.83 to 31.10.83, by P. GLASBERGEN and J.M.C.
VAN DER VORST. Period 01.11.83 to 30.4.84, by J.M.C. VAN DER VORST,
P. GLASBERGEN, A. LEIJNSE, N. PRAAGMAN and J. TAAT.

47. RIVM: Metropol-1. A computer package for the simulation of stationary
3-D groundwater flow. Users' manual, by N. PRAAGMAN, F. SAUTER - Draft

48. ROBINSON, P.C. (1984)
Connectivity, flow and transport in network models of fractured media.
Ph.D. thesis, Oxford University. AERE Harwell report TP 1072.

49. ROBINSON, P.C (1984b)
Connectivity, flow and transport in network models of fracture media.
AERE report TP. 1072.

50. ROBINSON, P.C., HODGKINSON, D.P. and RAE, J.
Thermal effects on groundwater flow around a radioactive waste reposi-
tory in hard rock. Subsurface space (ed. M. Bergman) 983-8, Pergamon
Press, Oxford and New York.

51. SALTELLI, A., AVOGADRO, A. and BIDOGLIO, G. (1984)
Americium filtration in glauconitic sand columns. Nuclear Techn., vol.
67, p. 245-254, November 1984.

52. SHARLAND, S.M. and TASKER, P.W. (1984)
Migration processes in a model deep repository for the disposal of
intermediate level waste. AERE Harwell, report R 11240.

53. SKI report 84:3
INTRACOIN - final report level 1, September 1984.

54. SKYTTE-JENSEN, B. (to appear in a later CEC report, 1985)
     Personal communication and document.

55. STARIK, I.E. and GINZBURG, F.L. (1961)
     The colloidal behaviour of americium. Radiokhiya 6, 685.

56. WICKENS, L.M. (1984)
     The pore water and effective stress response of clay sediments to a
     heat source. Ann. Nucl. Energy, 11, 289-304.

57. WICKENS, L.M. and ROBINSON, P.C. (1984)
     Finite element modelling of groundwater flow in hard rock regions
     containing a heat emitting radioactive waste depository. Ann. Nucl.
     Energy, 11, 15-25.

58. WILKE, S., GUYON, E. and MARSILY, G. de (Feb. 1985)
     Water penetration through fractured rocks. A percolation description.
     J. of Int. Assoc. of Math. Geol., 17.

59. HEREMANS, R. and al (1984)
     R & D programme on radioactive waste disposal into a clay formation.
     Final report CEC contract. CEC report EUR 9077, Luxembourg

60. CADELLI, N., COTTONE, G., BERTOZZI, G. and GIRARDI, F. (1984)
     PAGIS: Summary report 7 Phase 1. CEC report EUR 9220, JRC Ispra

61. SALTELLI, A., BERTOZZI, G. and STANNERS, D.A. (1984)
     LISA: a code for safety assessment in nuclear waste disposal –
     Programme description and user's guide. CEC report EUR 9306,
     JRC Ispra

62. RIBSTEIN, A., LEDOUX, E., BOURG, A., OUSTRIERE, P. et SUREAU, J.F.
     (1985)
     Etude du colmatage des fissures en milieu granitique par précipi-
     tation de la silice. Rapport final contrat CCE. Rapport CCE EUR 9476,
     Luxembourg

63. BOURKE, P.J. and al (1983)
     Nuclide migration and mathematical modelling
     CEC report EUR 8656, Harwood Academic Publishers

64. Commission of the European Communities (1984)
     The CEC co-ordinated project MIRAGE. CEC report EUR 9304, Luxembourg

65. Commission of the European Communities (1985)
     MIRAGE project: first summary report covering work period January
     to December 1983. CEC report EUR 9543, Luxembourg

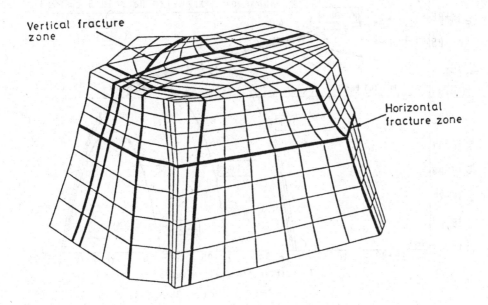

Vertical fracture zone

Horizontal fracture zone

3-D MODELLING OF FLOW AROUND A REPOSITORY. From HERBERT et al., 1984.

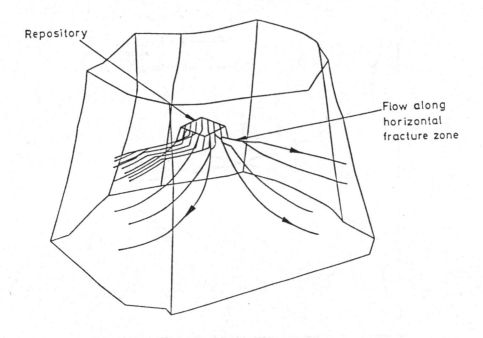

Repository

Flow along horizontal fracture zone

Fig. 1

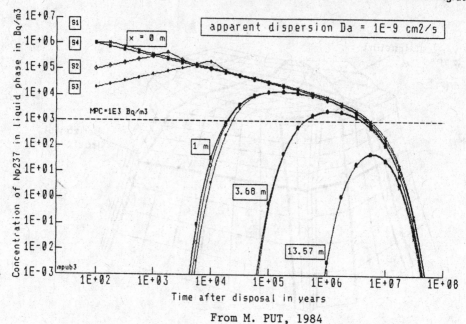

From M. PUT, 1984

Fig. 2   CONCENTRATION OF NP237 AS A FUNCTION OF TIME AND DISTANCE FOR
           4 DIFFERENT SOURCE SCENARIOS

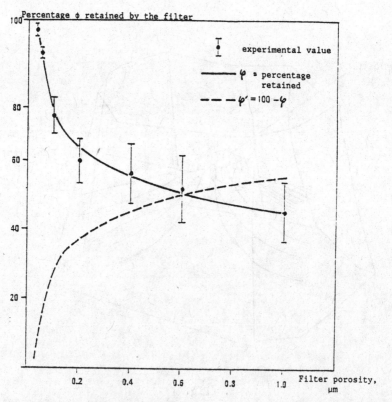

Fig. 3   RETENTION OF COLLOIDAL AM ON FILTER OF DIFFERENT POROSITIES
           From SALTELLI et al., 1984

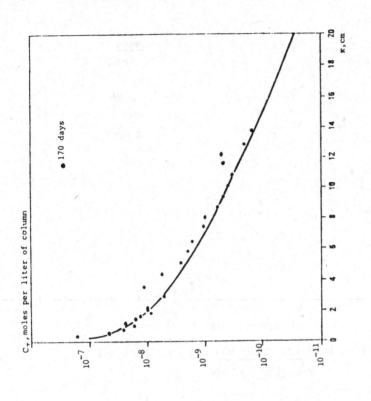

Fig. 5

TOTAL COLLOID CONCENTRATION $C_T = \sigma c c$ (c: concentration
solution; σ: fixed colloid concentration)

From SALTELLI et al., 1984

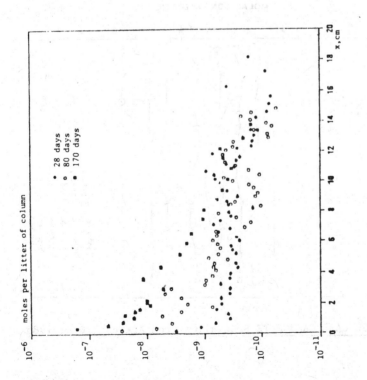

Fig. 4

CONTAMINATION PROFILE OF LEACHED AM^{241} IN
GLAUCONITIC SAND COLUMNS AT VARIOUS TIMES

From SALTELLI et al., 1984

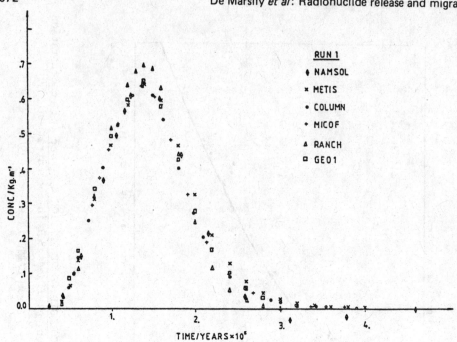

Fig. 6   PILOT STUDY RUN 1 (Baseline case)

From BROYD, 1984

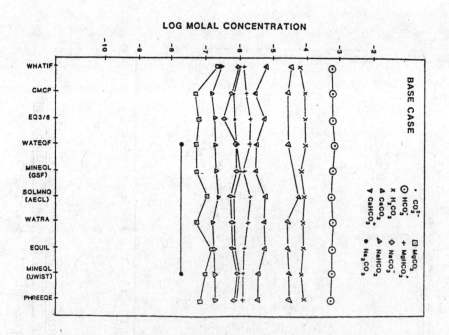

Fig. 7   BASE CASE CARBONATE SPECIATION From Atkins R & D, 1985

DISCUSSION

C. McCOMBIE, NAGRA Baden

I am not entirely in agreement with the conclusion that we need necessarily a constant updating to avoid obsolescence. It is conceivable that we could arrive at a position where we have adequate models for our purposes. I mean: for demonstrating that waste disposal can be safe enough, for example. We can have adequate models which are not necessarily best models. They could be improved for other purposes, for pure research-type purposes, for example.
Besides, referring to the car story: if we only need a car which takes us to work reliably, then we can stop when we have a VW-beetle; we do not have to develop Rolls Royces. And what we have to avoid is what happens with motorcars, where competition that offers the most chromium and the biggest trim, etc., wins.

G. DE MARSILY, Ecole des Mines Fontainebleau

I do not agree with you. I think modelling needs are enormous; improvement possibilities are still gigantic. We heard earlier Mr. BERTOZZI who told us that probabilistic modelling is necessary and that this type of modelling should be carried out with simplified models because it is not reasonable to use complex models. Now, I do not think this is very satisfactory intellectually. Why should we restrict ourselves? For purely economic reasons. It is also a fact that by making a mistake in simplifying approximations, you are making the results less credible. Now, without adding trimmings to the models, I think that evolution in the equipment - computers are evolving very quickly from one year to the next - in order to maintain models which use the expertise and equipment, and in order to be able to use these models in a more and more representative and complex way in probabilistic modelling, there are possibilities of improving things and coming up with Rolls Royces which are not all chromium.

P. PEAUDECERF, BRGM Orléans

I disagree with the judgements issues by Mr. DeMARSILY when he said that at the moment there is no model which allows you to calculate the evolution of a site. Two days ago I presented the geoperspective works in which we integrate models which take into account the constraints and variations in permeability, for example. Now, I am not saying that the problem has been solved; there is still a lot to be done, but I think we are already working in the right direction.

APPLICATION OF THE PAGIS METHODOLOGY TO CLAY, SALT, GRANITE AND SUB-SEABED REPOSITORIES

M.J. Clark[1], A. Bonne[2], F. Van Kote[3] and N. Stelte[4]

[1]NRPB, Chilton, UK.
[2]SCK/CEN, Mol, Belgium.
[3]ANDRA, Paris, France.
[4]GSF, Braunschweig, FRG.

ABSTRACT

The methods adopted for the Performance Assessment of Geologic Isolation systems (PAGIS) in clay, salt, granite and sub-seabed sediments are described, with special reference to the release scenarios adopted and the mathematical models used. The similarities and differences of approach used for the four formations are discussed and some preliminary results from models are presented. Phase 1 of PAGIS was completed in 1984, involving the collection of basic data on HLW inventories, repository designs, sites, release scenarios and radionuclide dispersion models. The current Phase 2 of PAGIS (1984-86) includes the implementation of an agreed methodology for calculating doses and risks using the material collected in Phase 1, and involving both sensitivity and uncertainty analysis.

1. INTRODUCTION

During 1982-4, the Commission of the European Communities (CEC) began the first phase of a Performance Assessment of Geological Isolation Systems (PAGIS) for the disposal of high level radioactive wastes. The objectives of the study have been described in another paper at this conference [1], the major ones being to evaluate different disposal options and to identify research priorities. Several organisations in EC countries are involved in the work, along with the Commission's Joint Research Centre at Ispra, Italy. Disposal in clay formations is being assessed at the Centre d'étude de l'Energie Nucleaire/Studie Centrum voor Kernenergie (CEN/SCK) in Belgium; in salt formations, at the Gesellschaft fur Strahlen und Umweltforschung (GSF) in West Germany; in granite formations, at the Agence Nationale pour la gestion des Déchet Radioactifs and the Commissariat à l'Energie Atomique (ANDRA/CEA) in France; and finally, in sub-seabed sediments, at the National Radiological Protection Board (NRPB) in the UK. The other organisations who have made significant contributions to the work are the Hahn-Meitner Institute (HMI) in West Germany and the Stichting Energiegonderzoek Centrum (ECN) in the Netherlands.

PAGIS is divided into three phases each lasting approximately two years. Phase 1 was completed in 1984 and included the collection of basic data; the choice of sites and repository designs; the choice of release scenarios; and of the methods and models to be used for the assessment of dose and risk. The current Phase 2 of PAGIS (1984-6) is designed to implement the methods and models, and to do calculations for the performance assessment of the various options. A subsequent Phase 3 will review the results and make recommendations on research and development requirements.

2.   DESCRIPTION OF THE PAGIS METHODOLOGY

The CEC have published a summary report of Phase 1 of PAGIS which
contains a general description of the methodology adopted [2]. The basic
methodology is to identify the events and processes which could result in a
radionuclide release from repositories in each formation (a scenario), and
then to use appropriate mathematical models to estimate radionuclide
dispersion in the geologic formation and the environment generally, so that
doses and risks to man can be calculated. The release scenarios can be
classified into those which are certain to occur, representing the normal
evolution of the formation, and those where the repository or its surrounding
geology is perturbed by events of a random (probabilistic) nature which can
alter the normal evolution of the formation. For each formation, the normal
and altered evolution scenarios can be different, and they are classified
into their expected time of occurrence ie, the short, medium or long term.
The scenarios selected for the four geologic formations in PAGIS are shown in
Table 1. They do not include highly disruptive scenarios, such as meteorite
impacts or volcanism, because these are assumed to have a very low
probability, and therefore the radiological impact will not be assessed.

For all the formations, the repository, its surrounding geology and the
general environment are divided into the near-field, far-field and biosphere,
to facilitate the mathematical modelling of radionuclide dispersion and the
calculation of doses and risks to man. The near-field consists of the
vitrified wastes, the containment, any backfilling and sealing material and
the surrounding rock. For three formations, clay, granite and sub-seabed
sediments, a common near-field modelling approach will be used, but for salt
there will be differences because the near-field is assumed to contain the
whole of the host rock. In the far-field, the strata for all formations is
assumed to be a saturated porous medium, which is a good approximation for
most formations. A common approach to biosphere modelling can be used for
the land disposal options, but for sub-seabed disposal, a different model has
been adopted, describing the dispersion in the oceans in detail.

3.   SELECTION OF BASIC DATA, RELEASE SCENARIOS AND RELEASE MODELS.

In order to apply the methodology proposed for PAGIS, it is necessary to
decide on inventories and the form of the high level radioactive waste; to
collect together the relevant data on geologic formations and repository
designs; and to implement radionuclide dispersion models in the near-field
far-field and biosphere, having chosen the release scenarios representing the
expected normal evolution and the possible altered evolution of the geologic
formation. With the exception of the high level waste inventory, where a
common data base can be used (the reference being vitrified high level waste
from Light Water Reactors), many detailed features of the repository design
and modelling are specific to the geologic formations, and these are
discussed in subsequent sections. It should be emphasised that the choices
have been made in PAGIS for the purpose of completing a successful
performance assessment, and are therefore based primarily on the availability
of data. They do not signify any intent, especially in respect of siting
repositories or of repository design.

3.1 Clay

For the purposes of PAGIS a reference site in Belgium has been chosen at
Mol, with two variant sites in Italy (Val d'Era) and the UK (Harwell), as
shown in Figure 1. The Mol site has been chosen as the reference site

because it is well documented and has been the focus of an R & D programme for many years, including repository design, hydrogeology and risk assessment. The top of the clay formation is 160 m below the surface and is about 100 m thick in places. It is a tertiary clay, representative of the open coast type formation, and the reference repository design adopted for PAGIS is shown in Figure 2. The variant sites have a different geology; the Val d'Era site is a representative for the intra-continental deposition environment while Harwell is intermediate between the two.

The normal evolution scenario used for the consequence analysis, involves the natural degradation of the artificial barriers in the repository including corrosion of the waste containers and subsequent leaching of radionuclides from the glass into the surrounding clay. (Figure 3). Dispersion through the clay will be dominated by diffusion rather than advection mechanisms, with radionuclides eventually reaching aquifers above and below the Boom clay, and thereby entering the ground water flow regime. The migration of radionuclides in the clay will be calculated using a code based on an analytical solution of a one dimensional advection/dispersion equation. For dispersion in the aquifers a more complicated three dimensional transport equation has been adopted, where advection rather than diffusion effects dominate, and this is also solved analytically. Radiation exposure is then assessed by assuming that a well is situated in the upper aquifer, immediately above the repository.

3.2 Salt

There are two main types of salt formation in EC countries, the bedded salt formations and salt domes, the latter being extrusions from a lower bedded salt formation, forced by pressure, heat and buoyancy effects. As a reference in PAGIS, a salt dome near Gorleben has been chosen, this site having been the subject of extensive geologic investigations. The variant sites include salt dome and bedded salt formations in the Netherlands and France respectively, as shown in Figure 1.

The normal evolution scenario for the salt dome assumes that direct water intrusion into the repository is prevented by the surrounding salt formation. However water contact is possible due to subrosion effects, where the salt formation and its overlying strata are dissolved, but this is not expected to occur before a million years have elapsed. (Figure 4)

One of the altered evolution scenarios assumed for salt is human intrusion in the future, for the purposes of resource exploration or for salt extraction. If this scenario occurs before the closure of the drift system by convergence it is similar to water intrusion via an associated anhydrite formation. Up to the present this scenario has been treated in a deterministic way, by assuming immediate, post-operational intrusion via the anhydrite, with a minimum flow resistance for water passing through this formation. However, it is intended that the human intrusion scenario will be treated in a probabilistic manner in PAGIS, using available borehole statistics. A near-field code has been written to assess the normal and altered evolution scenarios, which can allow for human intrusion via the main anhydrite formation and a limited brine intrusion via brine pockets in the salt dome itself. In addition to this, codes have been written to calculate radionuclide dispersion and transfer in the far-field and biosphere.

3.3 Granite

Three granite formations have been selected for PAGIS, the reference site being at Auriat in France which is a good example of an outcropping formation. The variant sites are at Barfleur in France, which is a coastal granitic formation, and at a representative site in the North East of England, where the granite is overlaid with sedimentary rocks. (Figure 1) The normal evolution scenario for this formation involves contact of the waste with the small amounts of groundwater in the granite. In the absence of disruptive fissures, this water can move slowly through the near field and eventually contact the waste and lead to radionuclide release into the far field and biosphere. (Figure 5) The altered evolution scenarios include an enhanced water movement in the repository due to geologic faulting, and also intrusion scenarios, such as inadvertent site investigation (drilling). These are expected to occur in the medium to long term: in the short term, radionuclide release due to incomplete backfilling and sealing will be considered.

In order to calculate doses and risks, two main approaches will be used. One will focus on the traditional separation into near-field, far-field and biosphere, considered as individual blocks for modelling purposes. The other will involve a code developed by IPSN which deals with the three above blocks directly in an integrated manner. It describes the joint evolution of the system as a whole, incorporating a 2-D radionuclide transport code which can allow for fracturing, thermal and geochemical effects.

3.4 Sub-seabed sediments

In comparison with the other geologic formations, the investigation of sub-seabed sediments as a suitable medium for a HLW repository is not as well developed, and the extent of basic data for the formation itself and repository design is on a lesser scale. However there is sufficient information for a performance assessment, and recent developments in modelling allow the biosphere to be comprehensively described. Three sites with sufficient data available for the study have been identified, (Figure 6) the reference case being in the Great Meteor East area, which is over 500 km from the nearest fracture zone and is a smooth abyssal plain with scattered elevations. The variant sites are in the Cape Verde Rise and in the South Nares Abyssal plain. These are also relatively flat and geologically stable. Two main repository designs are being considered at present, free-fall penetrator emplacement and drilled emplacement. (Figure 7) For the purposes of PAGIS the penetrator emplacement option is the reference repository design. The normal evolution scenario for the penetrators in sub-seabed sediments allows for thermal effects and the direct contact of the waste container with sediment pore water. The container is assumed to corrode over 500 years leading to direct contact with the vitrified waste and radionuclide release. (Figure 8) This is modelled in the near- and far-field using computer codes which allow for sorption effects on the sediments, in addition to diffusion/advection and decay terms. Eventually some radionuclides will reach the sediment/water column interface and become dispersed in the worlds oceans giving the possibility of exposure via bioaccumulation or by direct irradiation from sediments. The altered evolution scenarios consider the likelihood of incomplete hole closure, and therefore early direct contact of the waste container with sea water, and of thermal transient effects which could enhance radionuclide dispersion in the surrounding sediments. These could both be short term effects and in the linger term the possibility of human intrusion will be considered, as a result of mining or drilling

activities. In all cases, the latest modelling techniques will be used for
the sea biosphere, incorporating the recent work done for the NEA site
suitability review.

3.5 Discussion and preliminary results

Having compiled the methods, models and data during Phase 1 of PAGIS,
the next Phase of the programme is to carry out calculations for radionuclide
release and dispersion, and for the doses and risks to man. At present only
preliminary results are available for some formations, and these will be
briefly described. A fuller description of Phase 2 of PAGIS is given in
Section 4.

For the clay option, the migration of radionuclides in the clay layer
has been calculated assuming a layer thickness of 40m. The results are based
on an analytical solution of the diffusion/advection equation in one
dimension and, as the advective term in clay is much less than the diffusive
term, radionuclides disperse slowly through the clay layer. In order to
estimate dose rates to a maximum exposed individual, a small water well is
assumed to be based in the upper aquifer, just above the repository.
Calculations show that it would take several million years before a maximum
concentration of even a relatively mobile radionuclide like Np-237 is
established in the well-water. Preliminary estimates of exposure from
drinking water extracted from the well, give doses less than 1 μSv from Tc-99
and Np-237 assuming the repository contains 9000 m^3 of vitrified high level
waste and has a total surface area of 1 Km^2.

For salt, some calculations have been carried out for the scenario
involving water intrusion into the repository via an associated anhydrite
formation. The calculations show that convergence effects reduce the water
flow considerably, and if temperature effects are included due to the
presence of high level waste, the convergence effects are enhanced. By means
of a limited sensitivity analysis it has been possible to show that, even if
brine did contact the waste, the effects are limited by the convergence
effect.

For granite, the implementation of dispersion codes has begun and some
calculations of the thermal effects of vitrified waste on the host rock have
been completed. The maximum temperature in the centre of the vitrified waste
blocks will be about 190°C, 10 years after emplacement, for the reference
repository design. Similarly, for the interface between the waste container
and the granite, the maximum temperature of 120°C is reached after 10 years.

For the sub-seabed option, results have been obtained using the sea
biosphere model, which is a three dimensional compartment model describing
water flows in the North Atlantic and the World's Oceans. The sensitivity of
radionuclide concentrations in seawater and doses to man have been
investigated for variations in release rate to the ocean, dispersion in the
ocean, and sedimentation parameters. Calculations have been performed for a
number of activities, including Np-237 and Pu-239, showing that water
concentrations are more sensitive to changes in sedimentation parameters than
are doses to man, while both are equally insensitive to release rate into the
ocean.

4.   PAGIS PHASE 2

The main aim of Phase 2 of PAGIS is to use the data and models collected during Phase 1 to calculate doses and risks to man. The basic framework includes the calculation of radiation exposures to individuals and populations; the calculation of probabilities that doses will be received; a sensitivity analysis for selected parameters; and the evaluation of uncertainties associated with doses and risks. "Best" estimates of the consequences for normal and altered evolution scenarios will be obtained using comprehensive models, while sensitivity analysis and uncertainty analysis may utilise simpler models, because a large number of calculations will be required. In order to place the results of assessments in perspective, the spatial and temporal distribution of individual risks will be calculated, with special emphasis on the time of occurrence of the maximum level of risk in the future. Although there is no intention to compare options in PAGIS, collective doses will be calculated to see whether these quantities are useful in performance assessments. It is expected that the uncertainties in the long term effects will limit the application of collective dose calculations.

In assessing the best estimates of dose and risk, only one dose vs time curve will be produced for each release scenario, utilising best estimates for all parameters and using comprehensive mathematical models. For sensitivity and uncertainty analysis, the use of simpler models will allow sampling of parameters from a distribution of values. Computer codes already exist for such calculations, and have been described in a previous paper in this session [3]. In general they produce a histogram or frequency distribution of dose as a function of time, and provide detailed information on uncertainties and important parameters. Although different codes may be used for each of the geologic formations considered within PAGIS, a consistent approach will be adopted. This will include the method by which radiation exposure is calculated: a hypothetical individual will be assigned to each pathway, as representative of a group which would receive the highest exposures via that pathway. This individual will be an adult, unless it is clear that children or infants would receive higher exposures from the pathway, due to their diet or an age related factor.

5.   CONCLUSIONS

Although it is obviously too early to draw any firm conclusions from PAGIS, because the project is not yet complete, there are some general conclusions which can be made at this stage. There have been virtually no problems in collecting together the data and models necessary for an EC performance assessment of the geologic isolation systems but, inevitably, there are differences in the amount of data available and the types of mathematical model used. It has also been possible to agree a general methodology for the assessment, common to all formations. However, the choice of methodology has been a far more difficult task, complicated by the lack of a wider international agreement on how to assess the long term effects of radioactive waste management, and on criteria for judging the acceptability of various options. This is a rapidly developing subject and it is hoped that some of the problems caused by this lack of agreement will be resolved during the lifetime of PAGIS.

REFERENCES

[1] Cadelli, N, Giradi, F and Orlowski, S. PAGIS. A common European
    methodology for repository performance. 2nd EC Conference on
    Radioactive Waste Management and Disposal (1985).

[2] Commission of the European Communities. PAGIS. Performance Assessment
    of Geological Isolation Systems. Luxembourg, CEC EUR 9220 EN (1984).

[3] Bertozzi, G, Hill, M D, Lewi, J and Stork, R. Long term risk assessment
    of geological disposal: methodology and computer codes. 2nd EC
    Conference on Radioactive Waste Management and Disposal, (1985).

Table 1. Scenario selected for the four options in PAGIS.

OPTION	NORMAL EVOLUTION	ALTERED EVOLUTION		
		Short term	Medium term	Long term
CLAY	· Waste degradation and diffusion · Thermal effects	· Unexpected effects in the near-field	· Human intrusion (water wells)	· Tectonic displacement · Climatic changes
GRANITE	· Waste degradation, diffusion in near-field and transport in fissured rock · Thermal effects	· Convection in near-field · Convection in the shaft	· Change of fracture pattern (far-field) · Human intrusion	· Tectonic displacement · Climatic changes
SALT	· Thermal and convergence effects · Residual uplift · Subrosion	· Water intrusion · Thermal and convergence effects	· Human intrusion (solution mining, drilling)	· Climatic changes
SUB-SEABED	· Waste degradation and diffusion · Thermal effects	· Incomplete hole closure · Thermal transients	· Human intrusion	· Tectonic displacement · Human intrusion · Climatic changes

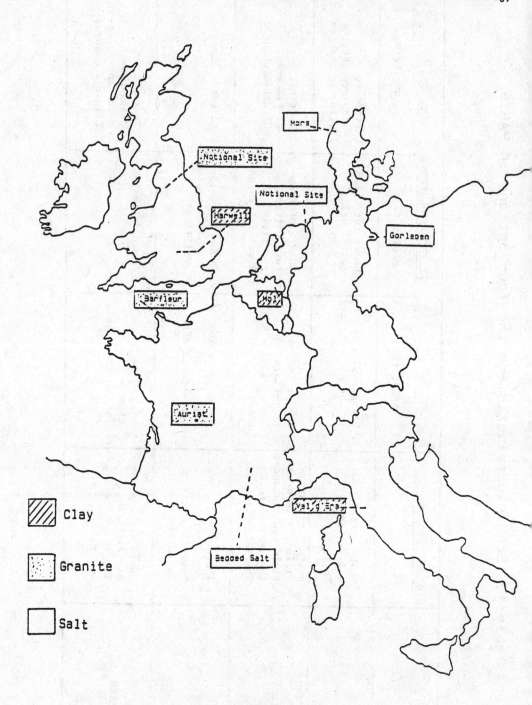

Fig. 1 Formations favorable for waste disposal
        selected in the European Catalogue

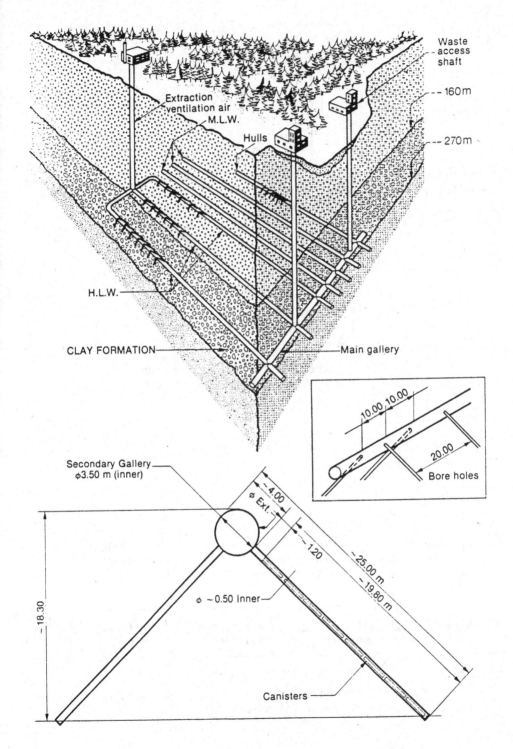

Fig. 2 Reference repository in clay at Mol

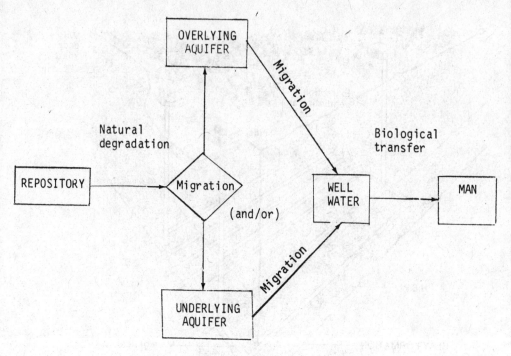

Fig. 3   Scheme of the normal scenario
(natural degradation) in clay.

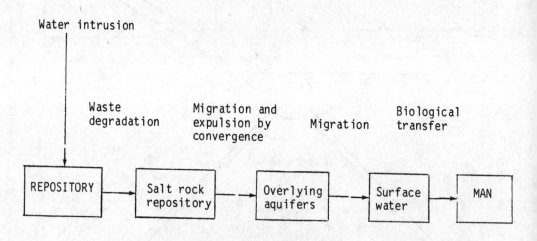

Fig. 4   Scheme of an altered evolution scenario in
salt (water intrusion)

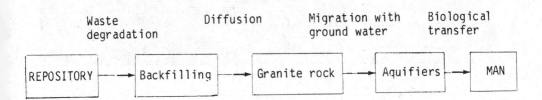

Fig. 5 Scheme of the normal evolution scenario in granite

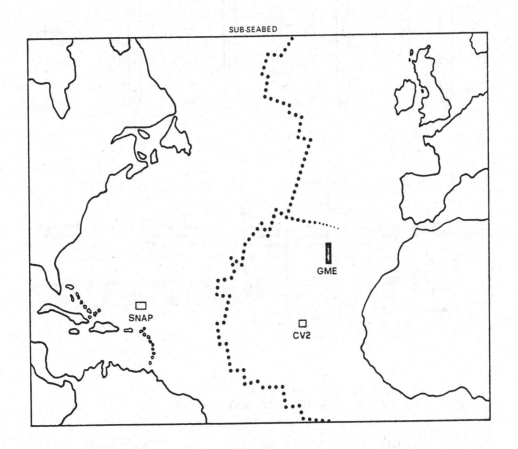

GME   Great Meteor East (reference)
CV2   Cape Verde Rise 2 (variant)
SNAP   South Nares Abissal Plain (variant)

Fig. 6 Sub-seabed disposal study areas

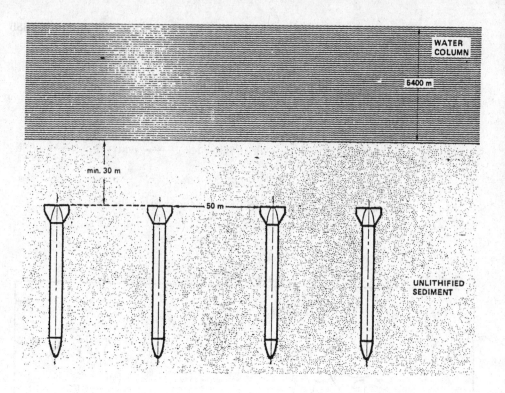

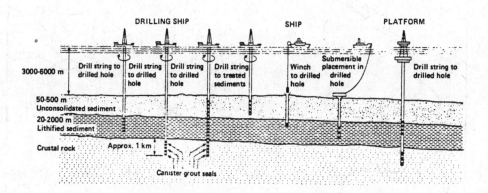

Fig. 7 Emplacement geometry for the reference case (penetrators), and variant case (drilled emplacement).

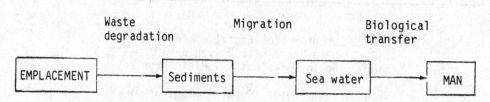

Fig. 8 Scheme of the normal evolution scenario for sub-seabed disposal

DISCUSSION

C. McCOMBIE, NAGRA Baden

I was somewhat surprised to hear you classify glaciation as a disruptive
event with very major consequences you implied. Anyway I realise that it
has major consequences in biosphere modelling. But, on the other hand, for
most places in Europe glaciation should belong to the normal evolution of
the site and must not be a disruptive event by definition.

M. CLARK, NRPB Chilton, Didcot

Yes, perhaps I misled you there. What I meant was exactly that. You can
have something which is disruptive for the biosphere but obviously for
deep repositories and the sub-seabed repositories, glaciation would not be
disruptive for the repository itself.

R. ROMETSCH, NAGRA Baden

In the sub-seabed and variants you have not mentioned the old story about
using the subduction zones. Why is that kind of scenario not taken into
account? Have you special reasons for that? And if so, could you comment
on that?

M. CLARK, NRPB Chilton, Didcot

I think I could comment in two ways. I'd stand to be corrected here, but I
am not sure whether there are many subduction formations - what we would
call - available in Europe. Obviously, the main one, off the coast of
Japan, looks attractive, but it is not something which we would want to
study within the European concept. So, I do not think it is available as
the other options, and we would prefer to concentrate on the two designs
described.
Then, there are the other arguments which I am sure you are aware of,
whether one should put waste near such an active tectonic feature. I think
that the general feeling should be that one should place high-level waste
a fair distance away from such features.

G. De MARSILY, Ecole des Mines Fontainebleau

A few years ago in Luxembourg at the European Conference, Mr. GIRARDI explained the safety concepts which would be used in the future to study repositories, and actually what he said goes along the same lines of what
you presented here today.
The IAEA has also carried out comparative studies of various repositories.
This was published also four or five years ago. Could you stress the difference in the PAGIS approach with respect to these previous approaches?

M. CLARK, NRPB Chilton, Didcot

This project, as I see it, is a potential first within the EC to do this
sort of assessment, to bring the various national authorities and bodies
together, to compare their approaches to this problem and to devise, where
possible, a common approach, while maintaining the necessary differences.
I have already described one, that is for repositories in salt. There is a
genuine difference in the performance assessment there. So I think that it
should be seen in that context, that it is not the first performance
assessment carried out, and we would make no claims for that, but it is
the first to be carried out within Europe; hopefully in a coherent and
cohesive manner.

R. ROMETSCH, NAGRA Baden

You have been rather negative in stating the absence of common criteria of
radioprotection to be compared with the safety assessment results. So, I
would like to ask whether you think it is really so negative or is there
not the beginning of a consensus in this field?

M. CLARK, NRPB Chilton, Didcot

I did not mean to be totally negative. I mean, in the situation now, one
cannot point to a document which sets out criteria which are internationally
agreed, but I did try and say that it is a fast moving field, and there
are documents, I would cite the recent NEA document on long-term radiation
protection objectives, which outlines an approach. That was a report by a
group of experts; I think it does represent an international consensus in
a way, but it is not recommendations to member states.

CHAIRMAN

I am afraid that I have to interrupt the discussion; I will try to reserve
some time at the end for this point, which I think really deserves a more
extensive discussion. On the other hand, some of these points will be
dealt with both in the presentation of Mr. Bertozzi and in mine.

CONCLUDING PANEL

Chairman :

J.A. DINKESPILLER, Deputy Director-General for Science, Research and Development, Commission of the European Communities (Director-General of the Joint Research Centre) Brussels

Panel Members :

P. DEJONGHE,   Deputy Director-General of the SCK/CEN Research Establishment Mol/Belgium

F. FEATES,   Director, UK Department of the Environment, London/UK

S. FINZI,   Director for Nuclear Research and Development, Commission of the European Communities, Brussels

R.E. GREEN,   Acting General Manager of the Whiteshell Nuclear Research Establishment, AECL Pinawa/Canada

K. HÜBENTHAL,   Regierungsrat, Federal Ministry for Research and Technology, BMFT Bonn/Germany

J. LEFÈVRE,   Director, Department of Radioactive Effluents and Waste, CEA Fontenay-aux-Roses/France

R. PURPLE,   Director, Program Integration Div., Office of Civilian Radioactive Waste Management DOE Washington DC/USA

R. ROMETSCH,   President, National Cooperative for the Storage of Radioactive Waste, NAGRA Baden/Switzerland

G. ZORZOLI,   Director, National Agency for Nuclear and Alternative Energies, ENEA Rome/Italy

Secretary :

R.A. SIMON,   Commission of the European Communities, Brussels

Chairman
<u>Chairman</u>

Presents the panel and invites Mr. Heremans and Mr. Hübenthal (on behalf
of Mr. Kroebel) to recall the conclusions of the two preceding technical
panels on HLW-disposal and on Waste Treatment Processes respectively.

<u>R. HEREMANS</u>

I should like to present several conclusions in respect of which a concensus has been reached at the Panel meeting on Geologic Disposal of HLW.

- As regards the geological barrier, we believe that we possess a clear
  idea of the essential role that it plays in any disposal system and we
  have available an arsenal of techniques and instruments which enable
  us to characterize it in the field; our laboratories are nowadays well
  equipped to carry out the indispensable expert assessments, analyses
  and tests. A study of similar situations in nature will allow us to
  achieve better understanding of the occurrence of certain physico-
  chemical phenomena in nature.

- With the waste-processing and conditioning techniques available at
  present, a final product is obtained which possesses characteristics
  that comply with the essential requirements of physical and chemical
  stability. Other engineered barriers, such as filling and plugging
  materials, are now available and ready to be used.

- The intensive research conducted over the last few years in the field
  of rock and soil mechanics have made it possible to identify more
  clearly the special problems associated with the excavation of galleries and, perhaps, with their use and their ultimate operation. The
  most striking examples concern the dry deep-drilling experiments in
  the ground at Asse and the construction of the laboratory in the clay
  at Mol.

  Modelling calculations now make it possible to forecast the behaviour
  of geological formations as a function of certain parameters specific
  to waste burial.

- Where performance assessments are concerned, it is essential to have
  thorough knowledge of the various system components, including the
  disposal site. The methods used for performance assessment have now
  reached a very advanced stage of development. Fundamental acceptance
  criteria already exist, but can be applied to analysis of the safety
  of final waste disposal only in parallel with the development of
  systems.

It is already possible to consider developing a system for the burial of
high-activity waste after certain additional research work has been
carried out and the obligatory demonstration stage, intended to confirm
under actual operating conditions the conclusions reached in the research
work and the studies, has been completed. The time which is still available to us for setting-up industrial-scale underground disposal installations nonetheless allows us to continue our work with a view to obtaining more detailed answers to certain questions and to optimizing the
system so as to ensure that the operation will be conducted under the

best possible economic and safety conditions. The five-year (1985-89)
plan of the Commission of the European Communities is undeniably aimed at
attaining these objectives.

K. HUBENTHAL

For many low and medium active waste streams, a variety of treatment and
conditioning methods for LLW, MLW and gaseous waste are already being
applied on an industrial scale; further processes have qualified for
operation on the more difficult long-lived MLW after extended R and D
efforts.

In fact, a recent independent review of the waste management practice in
France by the Castaing Commission came to the conclusion that waste
management and non-geologic disposal of wastes resulting from fuel
reprocessing had reached industrial maturity.

The disposal of low and medium active waste has also practically been
demonstrated for years in land based facilities. I shall add, that
disposal in the ASSE mine was discontinued for non-technical reasons.

So the present situation is not the lack of technical possibilities.
The questions we face are rather whether and for what waste type we
should select which of the options. The selection must clearly be the
result of an adaptation and optimisation process:
The waste treatment process must be adapted to the process requirements
upstream and the disposal criteria downstream. It must also be optimised
to satisfy the ICRP's ALARA principle.

The optimisation should mainly cover such aspects as:
- volume reduction
- improvement of waste package performance
- separation of long-lived nuclides
in view of identifying strategies, that keep radiological consequences
"as low as reasonably achievable", taking account of economic factors.

The panel welcomed the Commission's initiative to carry out such optimisation studies in its 3rd cost-sharing programme by means of systems
analysis.

The adaptation of processes to acceptance criteria also implies the use
of Quality Assurance methods to assure adequate and consistent performance of the final waste products. The control of the quality of such
final waste products may still be improved by advanced non-destructive
testing methods. Nevertheless, the systematic and comprehensive control
of the production process will remain the most important factor in
achieving satisfactory product quality.

Chairman

May we now discuss these conclusions under the important aspect of
safety?

R. ROMETSCH

Radioactive waste apparently causes more uneasy feelings and downright fears than any other toxic waste. Whatever the reasons for this situation, it has lead, in many countries to an essential social-political requirement, namely, that proof of the feasibility and long-term safety of waste disposal must be established.

The technical-scientific community has accepted this challenge and this conference and the tremendous effort placed in the PAGIS studies are a direct proof of this acceptance.

At least half a dozen safety assessments for waste disposal have been completed in and outside the EC. Some of these are generic, some specific for certain geologic formations and very few practically site-specific or carried out for reference cases. The results of all these studies are formulated in the careful way to which we are accustomed in the technical-scientific community. Results and remaining uncertainties are discussed with a great degree of prudence.

In a number of countries, radiation protection objectives have been fixed pending the establishment of such goals by the ICRP. A certain concensus on such protection goals is emerging on a dose to individuals in the order of 10-100 millirem/year. By comparison the doses resulting from the waste disposal as computed in the safety assessments are generally a number of orders of magnitude lower. This shows the large margin of safety incorporated in the kind of projects studied today. It can be stated with confidence, that the safety of final disposal is confirmed in principle and, in particular cases, the feasibility of its implementation is recognised in spite of the remaining uncertainties. I know of only one case, Sweden, where such a careful scientific statement has been the basis for a clear political decision to justify and permit the continuation of nuclear energy generation.

In Switzerland we are approaching a similar situation, as the project of a safety assessment for final disposal has been submitted to the government and the national and international experts appointed by the government have begun to examine this project. We hope, that in about one year a clear-cut political decision, that the required proof has been established, will be taken.

Generally, however, the safety assessment studies, and this also appears to be the case with PAGIS, will remain statements by the scientific technical community. The requirement, that a safety assessment has to be convincing, has only been applied to the scientific work and not sufficiently in view of convincing the general public. This very important step has yet to be done by all of us, i.e. to convince the public and to gain public and political acceptance for our disposal projects.

R.E. GREEN

Let me say at the outset how much I appreciate the opportunity I've had this week to listen to so many excellent presentations on the Community's Programme on Radioactive Waste Management and Disposal. I have been impressed with both the scope and depth of your activities and also

pleased to discover that our programme in Canada is very much in tune
with yours.

In my remarks to-day, I would like to address the question of public
acceptance, in the context of how we should be planning our future
activities in order to sustain the excellent programmes we have instituted, and to bring them to a successful implementation stage, with the
continuing support of our various publics and their governments.
Why is this question of defining our future programme so important?
Well, I recall that less than 10 years ago, the publics in various countries were quite agitated about the problem of nuclear waste, so much so
that the nuclear industry in some countries was faced with a moratorium
on future growth until the 'problem' could be solved.

Subsequently, many excellent programmes on waste management were put in
place, including your own. Since then, however, the world has changed.
The economic recession has virtually stopped the commitment of new
nuclear power plants, and the public, the utilities and governments seem
to have lost interest in the waste management problem. We don't hear
much of disposal anymore, but rather interim storage, for decades maybe.
Given this new climate, how to keep our programmes in place and progressing towards their logical conclusion, and receive public acceptance
for our waste disposal concepts?

I believe the message is clear from what we have heard this week,
Mr. Chairman. There seems to be a consensus that we are at the point
where we have done enough research to determine the most promising
options, and that we now need to get on with the next phase, which might
be called System Verification or Engineering Development. This does not
mean that further research will not be needed. On the contrary, the
engineering development phase will focus the research activities and new
research needs will appear as the engineering projects proceed.

From the viewpoint of public acceptance, the engineering development
phase will show the public that steady progress is being made towards a
solution of the problem. It may be different to get public acceptance of
disposal based only on mathematical models, so engineering demonstrations
could help us achieve our goal. Furthermore, if we continue with more and
more research, the public could assume that the problem must be much more
difficult than we have said. Some of you will recall how difficult it
was to get geological field research going, because of the public's
suspicion that disposal was about to take place. In this regard an
engineering demonstration phase would be a more reasonable step to
disposal. In any case, as I mentioned earlier, disposal appears to be
much further in the future and my concern is that unless we keep the
momentum in our programmes, the investment we have made in people and
technology will be lost and would have to be regenerated at some future
date.

In summary then, I would recommend that all programme managers plan their
future programmes towards site-related engineering projects, as the most
logical step in gaining public acceptance. In parallel, all of us need
to spend as much time as we can convincing the politicians and the general public that radioactive waste is not all that magic, is not unmanage-
able, and that we have the solution in our hands.

<u>Chairman</u>

I would now like to ask Mr. Dejonghe to comment this from the viewpoint of international cooperation.

<u>P. DEJONGHE</u>, CEN/SCK Mol

The papers presented this week have demonstrated what has been achieved in Europe in the context of international cooperation. That cooperation has intensified over the years and during the five-year plans.

The two five-year plans have made it possible:

- for us to reach an agreement on the major fields to be studied and their respective levels of importance;

- to avoid duplication of effort while ensuring that all essential aspects are covered;

- to create a climate suitable for intensive cooperation between institutes and persons;

- to promote common attitudes in areas which require consent, for example the approaches to safety where discharges and storage are concerned and confidence in the results obtained.

The third period of the joint programme should bring about industrial maturity or, at the very least, the adoption of definitive solutions with regard to storage and processes to be used to prepare the waste for disposal or storage and inspire confidence in broad sectors of the public and in the authorities concerned.

It should also be stressed that such cooperation is not restricted to the Community Member States, but has been extended, in certain forms, to other European countries, North America and the Far East.

Now is perhaps the time to emphasise the role of catalyst played by other international bodies (NEA, IAEA) and certain professional associations.

How can the importance of international cooperation in the field which concerns us, radioactive waste management, be illustrated?

- This is a subject which, by its very nature, goes far beyond the sphere of interest of a single society or of a single country; it concerns entire continents if not, in the long term, humanity as a whole.

- Exchanges of ideas and of methods of approaching the problem are a source of intellectual enrichment, but their chief contribution is that they make it possible to attain objectives more rapidly and, I would say, to have the solutions devised and the conclusions reached to be accepted by a broader spectrum of the scientific and technical community.

- This cooperation will result in the acceptance of common bases for assessment of the safety of nuclear installations which, of course, will have to be applied case by case in the countries concerned.

- The R & D programmes have been stabilized by means of coordination in the European Communities, which serve as a meeting place, not only for heads of programmes, but also for political leaders.

- Work sharing has made it possible to develop a wide range of solutions, each Member State having appropriate access to the results of its neighbour's work.

<u>Future prospects</u>:

- Large scale in-situ experimental installations are an indispensable stage in making a definitive choice in view of the situations specific to each Member State. The programmes are extremely serious, and it is important to obtain the maximum benefit from the results of programmes in neighbouring countries (maintenance, backfilling, etc.) by directly associating a reasonable number of research workers in other countries.

  In order to ensure that such work progresses satisfactorily, it is also important for all such programmes to be coordinated with regard to both subjects and timing.

- As regards the more distant future, and in view of the fact that we consider that the safety problems have been solved or can be solved by means of appropriate application of the methods and systems at our disposal, it will be necessary to expand cooperation up to and including joint operation or any form of cooperation where final storage is concerned. Several ways of doing this can be considered, with regard both to the scenarios and compensation formulae.

- Sight should not be lost of more restricted subjects which could benefit considerably from cooperation:

  - technological development of processes,
  - characterization and interaction with the environment.

Research must continue in order to validate concepts and contribute to their optimization and to promoting a climate of confidence.

<u>Chairman</u>

Before continuing the discussion on these aspects, let me ask Mr. Purple to give us some advance information on recent developments in the United States.

<u>R. PURPLE</u>

In the US, a national decision has been made regarding high-level waste management, which in my terminology essentially comprises spent fuel. Through passage of the Nuclear Waste Policy Act of 1982, the DOE has been authorized and directed to locate, construct, and operate a geologic repository for high-level waste. We have also been directed to locate a 2nd repository, but authorization for its construction has not yet been given.

In December 1984, the DOE issued draft Environmental Assessments on 9 potential repository locations. The documents have been undergoing public review and comment (10000 comments to date). The assessments provide a comparative evaluation of the nine sites and propose three of them for site characterization. The DOE will formally nominate these three sites to the President later this Summer. If he approves, we will initiate a site characterization programme at each of the three sites. This programme will run for 4-5 years and cost about at 700 million dollars per site. In 1990 or 1991 we will then recommend one of them to be the site for construction of the first repository, - which, by law, should be in operation by 1998.

Note that the repository sites are all in the Western part of the US (Texas, Nevada, Washington). The majority of nuclear power plants are in the East. Transport by road and rail therefore become a very important issue.

As a result of extensive systems analysis the DOE has now decided that another major facility should be constructed as part of the overall waste management system for spent fuel. This facility is called Monitored Retrievable Storage (MRS). It will be capable of receiving fuel elements directly from power plants, consolidating the fuel pins, packaging the fuel for ultimate disposal, and providing for temporary above-ground dry storage. The facility would cost about 1 billion dollars and could be ready for operation in 1996. This cost is partially offset by a decreased cost of surface facilities at the repository.

In a public announcement made yesterday, the DOE has released two documents that explain our decision as to the need for an MRS and the preferred locations for it. We have proposed three locations - all in the State of Tennessee, with the preferred location being at the site of the Clinch River Breeder Reactor at the Oak Ridge National Laboratory..
The MRS will significantly reduce the overall transportation of spent fuel, will simplify the design and licensing of the repository, will provide for better and more flexible management of the movement of waste, and will provide higher assurance that the waste management system will be ready to accept waste by the mandated 1998 date.

It is important to note that this new facility is not yet authorized for construction. It is our intent to submit a formal proposal to Congress in January 1986 seeking this authorization. We expect some form of approval action within the next year.

Chairman

I will now call upon the panel members to ask questions or make statements preferably on the topic: what shall we do next? This Conference has clearly established where we stand now, but, although the previous speakers have touched on this subject, I would wish to reach some, at least preliminary conclusions about the way ahead.

J. LEFEVRE

My comments do not exactly deal with the future activities, but rather with the manner in which we should present our results. We have just heard the technical conclusions reached at this conference and they can be said to be very satisfactory. However, all technical solutions put forward are conditioned by the objective to be attained, and this is determined by the radiation-protection criteria.

I should like to be able to share the confidence expressed by Mr. Girardi at the Panel meeting yesterday evening on the future recommendations of ICRP.

Last week, I participated, as did several others here, in the most interesting seminar held in Paris by the NEA of OECD, which brought together (may I dare to say brought into confrontation) radiation-protection specialists, safety specialists and specialists in the management of radioactive waste, and I did not leave it reassured. Those in charge at ICRP seem to me to be far more like theoreticians of radiation protection than concerned by the practical applications of the rules recommended. I shall take as an example the study presented on the "de minimis" (or exemption) value, which, starting from the mean value of natural radioactivity (2 mSv/year) 200 mrem/year resulted, after reductions by factors of 10 dictated by prudence, in 1 mrem/year (10 microSv/year) as the proposed value. That is to say that, in the case of a country like France, where natural radioactivity generally varies from 100 rem to several rem per year according to the locality, only an additional value of 1 mrem/year would be tolerated, or one hundredth of the new recommendation for the population dose, which is 1 mSv/year (instead of 5 mSv/year). Is this reasonable?

Faced with these results, I stated at that seminar that such a recommendation was virtually of no further interest, since it would not even allow general use of certain materials, the natural radioactivity of which gives rise to exposure to doses greater than that threshold: this is only one example.

It is essential to remain realistic with regard to the radioation-protection objectives and not to alarm the public pointlessly by exaggerating the importance of negligible risks, such as doses of 10 mSv, particularly if account is taken of other risks with which we are faced. I am thinking, in particular, of chemical risks and those arising from toxic products, an aspect very competently emphasized by Mr. Orlowski in his talk last Monday.

It therefore appears to me very important that results, particularly when they involve very considerable safety margins, be expressed so that they do not cause the protection objectives to be altered.

Chairman

I am satisfied to note that your criticism is addressed to another meeting and not this Conference.

F. FEATES

I would like to address the questions of what should be done next and the issue of the public perception of our work.
I am in full agreement with the comments made by Mr. Green, that our objective now should be to demonstrate to the public that radioactive waste, at least in the short term, does not present the risks in which many members of the public believe. To do this we must go beyond the underground laboratory stage and construct a facility to accommodate real waste or real fuel in such a way that the public can visit that facility. I therefore hope, that the model of the CLIMAX facility in the US, could be repeated somewhere in Europe. I admit that I would prefer to see such a demonstration located in a major city as a part of a Science Museum, but this may be wishful thinking. But we all recognize, that from the point of view of safety, it could be done.

As to the problem of public perception, I can see a major communications block between the scientific community and the public. I am rather pleased, that there were no members of the public here this morning, because, like Monsieur Lefèvre, I would have been concerned that they would have gone away unconvinced and not assured. The more we do to convince our scientific peers, the less we seem to convince the public. To explain this, I recall Monsieur De Marsily's unnatural analog of the motor car; I feel he is looking at the wrong time frame. He should have taken the situation of 100 years ago, when Daimler wanted to introduce the motor vehicle. Then it was unknown and we went through a long period of public acceptance. The public e.g. feared to be suffocated by high speeds or by the engine fumes. We also have such problems of popular misconceptions now and we must, in the next programme, place a serious effort explaining our projects to the public.

My other concern is the way, in which some problems, which the public would generally not recognize as such, are highlighted. For example, when we discuss the necessity for bigger computers to improve the accuracy of our "forecasting" programmes, the public will, by analogy, believe that risk assessments could be as unreliable as weather forecasts. The scale here is wrong, but we must spend some time explaining to the public, what the scale is in our case.

Then there is the image that nuclear industry projected in the fifties and which emphasized the technology and the skills acquired to handle the risks of nuclear energy. We now all know, that matters are not so complicated and dangerous.

My last point is, that we should pay more attention to the actual concerns of the public. I do not believe that the public really is concerned about a glacier moving over Manchester (in 50000 years) and, our inquiries in the UK confirm this. Glaciation if a worry for the scientist, not the public. A member of the public is concerned about risks he is familiar with, and we do not address these risks. Such risks include fire or flooding in a repository or subsidence. The public is also worried about the quality of the drinking water. It is therefore more important to explain the scale of risk in these areas to the public, in a way the average citizen can understand, than to place all effort on the all-embracing systems analysis, which expresses risks in complex mathematical terms.

Prof. G. ZORZOLI

I agree with the view that the principal result of this conference has been to demonstrate that, henceforth, future programmes will have to be centred on what Mr. Green defined as an "engineering project" and I prefer to call systems experimentation.

Where this subject is concerned, we find ourselves in a situation similar to one in which the principal components of a nuclear power station have been developed, but the power station itself has not yet been constructed.

Everyone knows that the problems of managing a power station are not only more numerous than those to which individual components give rise, but are also of various types and cannot always be predicted through studies and experiments on each component.

For the sake of analogy, it seems to me that this applies to the current problems associated with radioactive waste management.

The transition to systems experimentation accordingly represents a significant step forward and provides confirmation that a technological research and development phase has been in a large measure completed. In the light of present knowledge, however, it would be risky to consider systems experimentation purely and simply as a phase of verification of solutions that have been substantially consolidated. To subscribe to this view would be equivalent to denying the complexity of waste disposal procedures and ignoring the experience in this field which was acquired in other technological programmes.

On the other hand, systems experimentation also represents the most rational instrument for identifying which lines of research are still important and significant and for avoiding as far as possible the risk of academic or abstract research.

Given the present state of knowledge, the most important objectives to be achieved are in fact linked with more rational and safe radioactive waste management and involve optimization of the immobilization matrices and volume-reduction techniques, particularly, in my view, through intensive separation of long-lived actinides and the development of specific techniques for removal.

However, after consistent systems experimentation, these objectives could change in importance. Neither the problems relating to sufficiently consolidated standardization, in particular for the qualification of sites and the subsequent licensing of the repository, nor those associated with the development of adequate quality-assurance procedures can be solved without the practical experience that can be acquired with the management of a radioactive waste disposal system on a significant scale (international cooperation).

Once again the experience acquired with the operation of nuclear power stations is of decisive importance in this connection. It is only as a result of their operation that we have succeeded in developing adequate and reliable standardization. It might be objected that the nuclear plant is a more complex system.

However, a nuclear plant is already a standardized system today and will be so to an even greater extent in the future, while the characteristics of future repositories will always vary considerably from case to case; furthermore, from the standpoint of significance and measurement precision, the experimental data obtained are difficult to compare with those that are typical of a nuclear power station.

I have left to the last that aspect of systems experimentation which is more important in the long run, i.e., its unrivalled capacity to ensure technical and economic optimization from the standpoint of both safety and protection. The "escalation" in cost estimates, as has been confirmed at this conference, is a phenomenon to follow with the closest attention at the moment when there is a transition from programmes focused on individual products and individual research and components to programmes in which there is greater emphasis on systems experimentation. It is specifically in the field of cost analysis that the incentive could be found to revise solutions that today already seem to be consolidated.

It might be objected that this way of interpreting the future phases of our development activities relating to radioactive waste could give rise to a feeling among the public that there was still too much uncertainty.

Well, my personal experience has convinced me of the contrary: the man in the street no longer puts his trust in excessive certainties or in affected optimism, but is more prepared nowadays to accept a description of the situation which points up the progress achieved and the rationality of future programmes which will henceforth be based - and this is the point to be emphasized - on sufficient factual data to make trustworthy management of radioactive waste a realistic objective.

S. FINZI

In attempting to sum up what our conference has to say to the Commission's departments responsible for managing this programme, I think it can be said that the indications given by the technical panels, primarily the one chaired by Mr. Heremans, are very clear.

These conclusions have been accepted as valid by the other members of the panel, and for my part I support this while stressing the need for precise definition of the terms used in order to avoid misunderstandings, the terms I have in mind being "model", "validation" and "demonstration".

But apart from looking at what remains to be done, let us also consider how it is to be done.

I would like, while referring to Mr. Dejonghe's comments, to talk about cooperation as an important element in the success of Community R & D programmes.

We are flattered by, and grateful for, Mr. Dejonghe's remarks on the way in which collaboration has been achieved in the programme hitherto, and I would like to describe this collaboration with a few additional thoughts of my own:

The first aspect of Community cooperation is its depth. In trade union parlance it extends to the "rank and file", involving some 400 Community research workers who have been working together in close and frequent collaboration, often for more than ten years. This represents a degree of coordination of Community action which is quite remarkable. It is, moreover, interesting to note that these researchers come both from countries heavily committed to nuclear energy and from countries which are still undecided.

The second aspect is that this cooperation is very flexible in the forms that it takes, the most rigid being the shared-cost contract and the most flexible the informal meetings between project officers who share the same concerns. The 12-year Community plan of action (1980-1992) adopted by the Council of Ministers in 1980 provides the necessary framework.

The opening up of three projects for experimental underground installations in Germany, Belgium and France to teams from other Community Member States who wish to take part, within the framework of the programme financed by the Commission of the European Communities, is yet another example of this flexibility. Perhaps it could be a source of other ideas on wider international cooperation.

The latter is the subject of my third comment: any activity concerning radioactive waste gains from being opened up to the outside. This certainly applies to Community research, as our bilateral cooperation agreements with Canada, the USA, NAGRA in Switzerland and soon perhaps with Sweden, and our very open relationships with the Agencies in Paris (NEA) and Vienna (IAEA) amply demonstrate.

I feel, therefore, that international cooperation in general in the field of radioactive waste is well developed and we should not be tempted to stimulate it artificially too much and too quickly: especially as far as the final stage of waste management, i.e. the industrial construction of the large underground dumps of the future is concerned.

The creation of a mutual understanding, almost a "common culture" as an important fruit of research is the first stage which opens up plenty of opportunities for more ambitious collaboration in the future.

Chairman

May I now invite questions or comments from the audience.

R.B. LYON, AECL Canada

In respect to Dr. Feates statement about public concerns I wish to add the following: Risk assessment for disposal facilities is normally divided into the pre-closure and the post-closure assessments. The preclosure analysis examines the impact on people "here and now", i.e. the impact of transportation, construction and operation of a facility.

In fact, in Canada in 1981 a probalistic performance assessment and a preclosure assessment were published and widely distributed for comment to the public and the scientific communities.

As a result, essentially all of the comments from the scientists concerned the post-closure assessment, all the comments from the public addressed the pre-closure phase. They were concerned with aspects as transportation and the question of curtailing union activity because of the dangerous nature of work, e.g. restricted right to strike. So I would support Mr. Feates's call to consider what the public wants to know and also stress that assessment of the pre-closure impact is necessary for the implementation of disposal facilities.

T. MARZULLO, ENEL Roma

My comment also addresses the aspects of public opinion. I believe, that it is proper to launch a large research programme to study the impact of disposal, but that we should carefully choose, what should be presented to the general public and how. By investigating the consequences of practically incredible and impossible events, we produce anxieties in the public, because the public does not reason in the same way as we do. Hence, the public cannot appreciate the extremely remote and unlikely. Therefore, if we discuss the migration of radionuclides in geological formations, the public draws the conclusion, that this radioactivity can spread widely, e.g. from Switzerland to Italy.

We should therefore restrain from publicising the scientific work in such areas without presenting the results and explaining these results to the public in a way everyone can understand.

A. SUGIER, CEA Fontenay-aux-Roses

I do not agree with the ideas about the general public implied by the last speaker. The public is not necessarily simple-minded. If the public is confronted by too complex problems, it will look for competent persons outside the circle of people associated - or believed to be associated - with the nuclear industry and its pressure groups. It will ask these other scientists, whether they are convinced (by the safety assessments). It is therefore necessary to have scientifically sound arguments, and if our results can be understood by other scientists not involved in nuclear work, the public will be convinced by the response of these other scientists. We should therefore not seek to simplify too strongly, but continue to do our job in an honest and serious scientific manner.

Chairman

I believe we must first, as Mr. Feates said, understand, what information the public wants and needs. You, however, feel that only the full scientific facts should be issued without simplification, which might obscure the truth.

E. VAN KOTE, ANDRA Paris

It is one of my tasks to prepare, together with my colleagues, the documents intended to convince the licensing authorities and the public of the acceptability of nuclear waste disposal facilities. Performance assessment is certainly an important part of the evidence needed to convince the public, the political bodies and the licensing authorities, but is only one of several parts required. In support of this somewhat abstract and theoretical evidence, one has to prove, that due care has been taken in the original choices of the concepts, the selection and characterization of candidate sites and the realism and soundness of engineering methods to be employed.

Finally, the documents establishing the safety of such facilites must show, that a comprehensive system of quality assurance will warrant the correct implementation of all safty features.

Chairman

I had understood, that we were now embarking on comprehensive experimental facilites because these will be by far more representative of real disposal than studies, on which we have based our arguments until now. As the results of these experimental and demonstration facilites will closely model these in the future large repositories, they should be more convincing. This was the motivation for orientating our research to the in-situ underground laboratories in which disposal activites and conditions can be reproduced authentically.

J.P. PONCELET, Service Programmation de la Politique Scientifique
               Brussels

I note with interest that our scientific community, having convinced itself of the feasibility of these projects, now turns to the problems of public acceptance of these solutions. This will certainly become a major issue in the future, but we should bear in mind that public acceptance depends, to some extent upon irrational factors, which will not be overcome by calculations. We must understand these irrational reactions of the public, since we deal with risks which extend well beyond the life of our generation. Thus acceptance will largely depend on trust, trust in the scientific community and in the institutions.

In this sense, I would like to raise a point we briefly touched at the recent CGC meeting and which essentially is in the competence of the Commission: Mr. De Jonghe has spoken about the expansion of the cooperation in the demonstration projects. Why shouldn't we extend this cooperation to the actual disposal sites and endeavour to create truly Community repositories?

In practice, is it imaginable, that, one day, waste from other countries than Belgium will be emplaced?

I feel, that a truly integrated European solution of these problems could satisfy many people of our generation, who feel they are citizens of the world and expect that the solution to such a problem should extend beoynd national borders.

The Commission, I believe, should be the political advocate of such proposals.

Chairman

Do you suggest we become citizens of this world by disposing our wastes
in someone else's backyard?

S.P. PONCELET, Service Programmation de la Politique Scientifique
                     Brussels

Not exactly, I rather suggested that a common European approach could
stimulate more confidence than solutions merely based on national structures.

P. DEJONGHE

This is an important issue and I will not try to treat it exhaustively. I
only want to make one point. If we categorically refuse to share our
disposal facilities with our neighbours, we undermine our own experiments
about the safety of disposal. This statement may be slightly exagerated,
since in fact there are exchanges of wastes between countries, but these
are far from being optimized.

Another point: one may think lightly of such matters, but I personally
take them rather more seriously. In his very interesting paper, Monsieur
Venet yesterday showed, that although the effect of repository size is
not the most important factor in the cost of repositories, it is a very
important one. We have also seen, that financing conditions and the
actualisation of cost have an important influence on disposal cost and
thus upon the price of the KWh. This contributes to major savings, if it
were optimized on a larger European scale. These implications are currently being investigated by the European Community and the OECD, but we
have to wait for the results of these studies.

R. ROMETSCH

Coming back to my previous statements I repeat that we must gain public
acceptance by explaining our projects to the people of each country. We
have to discuss these problems with the public in its own language and
take account of their concerns.

In order to implement a large international repository, we must first
have the examples of some national repositories. Once we have repositories in, say, ten countries, then we can proceed to optimize on an
international level.

S. ORLOWSKI, CEC Brussels

May I recall, that the European institutions, the Commission and the
European Parliament have always encouraged and called for cooperation in
the field of radioactive waste disposal. In the beginning of the
senventies the European Parliament had made a study of a European approach and submitted it to the governments. The latter felt, that this
study was interesting but very premature. More recently, at the beginning

of 1980, when the Commission presented its 12-year Plan of Action on
Radioactive Waste to the Council, it suggested to work toward a European
network of disposal facilities.

The Plan of Action was approved, but the reference to a European disposal
system was once again qualified as premature and deleted. We shall
continue to bring this approach to the attention of the Council and
progressively work toward such a solution. As Mr. Finzi explained
earlier, such developments should not be forced hastily, but implemented
in cautious steps at the appropriate time.

G. DE MARSILY, Ecole de Mines Paris

May I comment an idea of Mr. Finzi on the definition of terms for the
future programme: I am a bit worried, as a scientist, about the necessity
to repeat experiments, like the CLIMAX facility, with the only objective
to demonstrate that one is capable of handling high-level waste disposal
operation.
I could agree that such a demonstration may be necessary for the sake of
acceptance, but it has no scientific value whatsoever. I therefore
recommend that we should only build one such demonstration in Europe. On
this example at least we could demonstrate European cooperation . But
when it comes to in-situ experiments in suitable geologic formations,
and this should be the real objective of the planned underground
laboratories, these must be built at various sites. Of course, the
scientific work here should be coordinated, but we must draw a clear
distinction between demonstration facilities and underground scientific
laboratories.

B. VERKERK, ECN Petten

Coming back to the subject of public acceptance, we need the help of the
press and the other media as they are capable of translating our technical language to the public. In the past, we have had very little access
to the press, but this is changing and we will nowadays be accepted by
the press.
I therefore suggest that, in our institutions and whenever we can, we be
as open as possible to the press and the other media.

M.J. CLARK, NRPB Chilton

Could I ask the panel a question about what we should present to the
public? I am sympathetic with the argument, that there is no point in
presenting cumulative probabilities or similar data at a public meeting.
These would not be understood by some scientists, let alone the public.
What however, if a witness at a public inquiry claimed that we had not
done as thorough an assessment as those carried out in other countries,
e.g. Canada or the USA, or that we were not capable of doing so. Such
statements would have a strong political effect and therefore we must
carry out the risk assessments seriously and we must try to present their
results understandably.

Chairman

I understood you have also raised a point on international relations, i.e. that one could accuse you of doing less than others in this field. This conference, among others, promotes cross-fertilisation of science, giving you opportunity to exchange views and to learn about progress elsewhere.

G. ZORZOLI

One small comment to this issue of concensus. We should not forget, that there are other scientists, who are part of the anti-nuclear movement. These persons have access to all the information to which we have access and this is certainly a positive fact. None of us wants to place an embargo on information. If, in informing the public, we omit some information, these people will claim that we are trying to hide the one or other element. Since trust in our institutions is the basis of public acceptance, we cannot run this risk. We must expose the problem to the public in a correct' ranking order, explaining what is important and decisive and what is of second priority, but still laying out to the public, that in the vast overall scrutiny of all hypotheses, whe have enhanced safety, and not lost any of it.

R. GREEN

What we obviously need is a collection of messages. On one hand we must satisfy the authorities and fellow scientists with the sophisticated models, but on the other, we need a different kind of message for the general public, which deals with their concerns.
The long-term risks in the post-closure phase, as Mr. Lyon pointed out earlier, are of concern for the authorities, but also for certain anti-nuclear scientists.

In respect to Mr. de Marsily's comment about demonstration, I would agree, that we do not need a CLIMAX type demonstration. What I was suggesting earlier was a demonstration facility, which at a later stage could be turned into a final repository. It should not be a mere show-project, but a first step toward a commercial repository. Then, if something goes wrong, you could still stop the further construction of the commercial repository and start building it somewhere else.

F. FEATES

I agree with Mr. Clark about the necessity for a detailed scientific assessment. My concern is, that we do not present it well enough to the public. As scientists we are trained to identify and highlight the uncertainties of our work. When, however, we present the results to the public we should highlight the certainties without hiding the uncertainties.

Chairman

I must now close this debate. I feel we can go from here as optimists.
From the technical point of view it appears, that the problem is well
under control. Even for long-lived and high level waste, the situation
is better in hand than for non-nuclear waste. This was echoed a number
of times at this Conference.

There is a small danger in this. Probably more attention should be paid
to non-nuclear waste. It would not be such a bad idea to draw the
attention of the public and the authorities to this fact but we should
not believe, that it would distract them from the nuclear problem. The
two problems exist and we in the nuclear field have to deal with ours.
We cannot escape from this task by referring to the non-nuclear waste.

When we talk about our future work, we think of in-situ experiments in
underground laboratories to improve representativity. As to the problem
of acceptance by the public and the relationship with the public, first
of all it has become apparent from the discussion, that public relations
are not a scientist's job. Scientists are not PR specialists, they must
be accurate and objective. Under no circumstances should they leave that
role. Above all, any attempt to select what can be said to the public
from the results, would be very dangerous indeed. So let us stick to our
scientific competence, to our objectivity and to the truth while trying
to understand what the other side wants to know. We must take the
different concerns of the various partners in this exchange into account,
but retain a certain candidness.

I was particularly struck by the international aspect of this meeting. I
had been told that the scientists in this field form a gang, even a mafia
and in more polite terms a scientific community. But these references
were always affectionate; all specialists in this field work closely
together and reach very similar conclusions and, most important, they
give the same advice to the political powers. This is the result of the
close cooperation and a considerable trump-card in the prospect for
political decisions.

Political decisions in this field are mainly taken on a national basis.

As political leaders have the common problem of public acceptance, they
will also tend to look across the fence to their neighbours to see how
they deal with the problem. Here agreement among the international
scientific community is of eminent importance.

I now take the opportunity to greet the large Japanese delegation, which,
for the first time has come to one of our conferences. I hope that they
will show the same interest for our next conference on this subject in
five years.

(The chairman then thanked and congratulated the contributing authors,
speakers and the Conference staff).

LIST OF PARTICIPANTS

AIT, N.
Ingénieur
Centre d'Etudes Nucléaires
de Cadarache
B.P. 1
F - 13115 SAINT PAUL LEZ DURANCE

ALBERS, G.
Ingenieur
RWTH Aachen
Schinkelstr. 6
D - 5100 AACHEN

ANDERSSON, J.E.
Engineer
Swedish Nuclear Power
Inspectorate
Box 27 106
S - 102 52 STOCKHOLM

ANDRE-JEHAN, R.
Ingénieur géologue
Commissariat à l'Energie Atomique
Agence Nationale pour la Gestion
des Déchets Radioactifs
31-33 rue de la Fédération
F - 75752 PARIS CEDEX 15

ANDRIESSEN, H.
Chemieingenieur
KFK
Postfach 3640
D - 7500 KARLSRUHE 1

ANGER, W.
Betriebsdirektor
Kernforschungsanlage Jülich GmbH
Technische Dienste -
Betriebsdirektion
D - 5170 JUELICH

ANTONUCCI, L.
Fisico Sanitario
ENEL - Direzione Produzione
Trasmissione - Settore Nucleare
V.le Regina Margherita 137
I - 00198 ROMA

ARNOULD, M.
Professeur, Directeur du
Centre de Géologie de l'Ingénieur
Ecole Nationale Supérieure
des Mines
60, boulevard St Michel
F - 75272 PARIS CEDEX 06

ATABEK, R.
Ingénieur
Commissariat à l'Energie Atomique
Dép. de Recherche et Dév. Déchets
Serv. d'Etude des Stockages
des Déchets
B.P. 6
F - 92260 FONTENAY AUX ROSES

AVOGADRO, A.
Commission of the European
Communities
Divisione Radiochimica
Joint Research Centre
I - 21020 ISPRA, Varese

BAETSLE, L.
Division Head
Studiecentrum voor Kernenergie
S.C.K./C.E.N.
Boeretang 200
B - 2400 MOL

BAIER, J.
Ingenieur
NOELL GmbH
Bereich Kernenergietechnik (T9)
Postfach 62 60
D - 8700 WUERZBURG

BARBREAU, A.
Ingénieur
Commissariat à l'Energie
Atomique, IPSN/DPT
B.P. 6
F - 92260 FONTENAY AUX ROSES

BAROZZI, M.
Student
Via Trieste 39
I - 20098 SAN GIULIANO MILANESE

BARTHOUX, A.
Adjoint au Directeur
Agence Nationale pour la Gestion
des Déchets Radioactifs
29-31, rue de la Fédération
F - 75752 PARIS CEDEX 15

BATH, A.H.
Geochemist
British Geological Survey
Fluid Processes Research Group
Keyworth
GB - NOTTS NG12 5GG

BAUDIN, G.
Chef de département
Commissariat à l'Energie Atomique
Centre d'Etudes Nucléaires
B.P. 6
F - 92260 FONTENAY AUX ROSES

BAUKAL, W.
Battelle-Institut e.V.
Am Roemerhof 35
D - 6000 FRANKFURT/MAIN 90

BAYAT, I.
Nuclear Waste Management
Atomic Energy Organization of Iran
P.O. Box 1136
Iran - 8486 TEHRAN

BELOT, F.R.
Directeur UNERG - Conseiller UEEB
Union des Exploitations
Electriques en Belgique
4, Galerie Ravenstein, bte 6
B - 1000 BRUXELLES

BERCI, K.
Design Engineer
Power Station and Network Eng. Co.
Szechenyi Rkp. 3
H - 1054 BUDAPEST

BERGER, R.
Ingénieur
Ministère du Redéploiement Industr.
et du Commerce Extérieur
Serv. Central de Sureté des Instal.
Nucléaires, Centre d'Etudes Nucl.
B.P. 6
F - 92260 FONTENAY AUX ROSES

BERNIER, L.
Journaliste
Mc Graw-Hill World News
Nucleonics Week
17, rue Georges-Bizet
F - 75116 PARIS

BERTOZZI, G.
Commission of the European
Communities
Euratom
ED. 46
I - 21020 ISPRA, Varese

BESENECKER, H.
Hydrogeologe
Niedersächsisches Landesamt für
Bodenforschung
Stilleweg 2
Postfach 510153
D - 3000 HANNOVER 51

BEUKEN, G.
Ingénieur Chimiste
Intercom - Centrale Nucléaire
de Tihange
Avenue de l'Industrie
B - 5201 TIHANGE

BEYER, H.
Chemiker
Kraftwerk Union Aktiengesellschaft
Abt. R 451
Hammerbacherstrasse 12 - 14
D - 8520 ERLANGEN

BILLIOTTE, J.
Ingénieur de Recherche
Centre de Géologie de l'Ingénieur
Ecole Nationale Supérieure des
Mines de Paris
60, boulevard St Michel
F - 75272 PARIS CEDEX 06

BILLON, A.
Ingénieur
Commissariat à l'Energie Atomique
Dép. de Recherche et Dév. Déchets
Serv. d'Etude des Stockages
des Déchets, Centre d'Etudes
Nucléaires, B.P. 6
F - 92260 FONTENAY AUX ROSES

BINAS, H.
Physiker
Technischer Ueberwachungs-Verein
Hannover e.V.
Postfach 810740
Am TUEV 1
D - 3000 HANNOVER 81

BLANC, P.L.
Ingénieur
Commissariat à l'Energie Atomique
IPSN/OPT/SEPD
B.P. 6
F - 92260 FONTENAY AUX ROSES

BOGORINSKI, P.
Dipl.-Ing.
Gesellschaft für
Reaktorsicherheit
Schwertnergasse 1
D - 5000 KOELN 1

BOISSON, J.Y.
Ingénieur
CEA - CEN/Saclay
Bâtiment 89
F - 91191 GIF-SUR-YVETTE

BOKELUND, H.
Commission of the European
Communities, JRC, European
Institute for Transuranium Elements
P.O. Box 2266
D - 7500 KARLSRUHE

BONGERS, H.J.G.
Head Mech.Eng. Department
KEMA
Utrechtseweg 310
Postbus 9035
NL - 68ET ARNHEM

BONNE, A.
Projectleader Geological Disposal
Studiecentrum voor Kernenergie
Centre d'Etude de l'Energie
Nucléaire
Boeretang 200
B - 2400 MOL

BOOS, A.
Beamter
Bundesministerium des Innern
Graurheindorfer Str. 198
D - 5300 BONN 1

BORRMANN, H.
Dipl.-Ing.
Kernforschungszentrum Karlsruhe
Projekt Wiederaufarbeitung und
Abfallbehandlung
Postfach 3640
D - 7500 KARLSRUHE 1

BOSSER. R.
Physicien
Commissariat à l'Energie Atomique
CEA/IRDI
SEN/LPA, CEN Cadarache
B.P. 1
F - 13115 SAINT PAUL LEZ DURANCE

BOTHOF, J.
Official
Ministry of Housing, Physical
Planning and Environment
Radiation Protection Division
Directie Stralenbescherming
P.B. 5811
NL - 2280 HV RIJSWIJK

BOURDREZ, J.
Ingénieur
Commissariat à l'Energie Atomique
A.N.D.R.A.
29-33, rue de la Fédération
F - 75752 PARIS CEDEX 15

BOURKE, P.J.
Scientist
United Kingdom Atomic Energy
Authority
B125 CTD AERE, Harwell
GB - OXON OX11 ORA

BOZEC, C.
Ingénieur
Commissariat à l'Energie Atomique
B.P. 6
F - 92260 FONTENAY AUX ROSES

BREWITZ, W.
Abteilungsleiter
Gesellschaft für Strahlen-und
Umweltforschung mbH München
Institut für Tieflagerung
Theodor-Heuss-Strasse 4
D - 3300 BRAUNSCHWEIG

BRODERSEN, K.
M. Sc.
Risoe National Laboratory
Chemistry Department
DK - 4000 ROSKILDE

BRONDI, A.
  Comitato Nazionale per l'Energia
  Nucleare e l'Energie Alternative
  Centro di Studi Nucleari della
  Casaccia
  Casella Postale 2400
  I - 00100 ROMA A.D.

BROYD, T.W.
  Research Engineer
  Atkins Research and Development
  Woodcote Grove
  Ashley Road
  GB - EPSOM, Surrey KT18 5BW

BRUECHER, P.H.
  Dipl.-Ing.
  Institut für Chemische
  Technologie der
  Kernforschungsanlage Jülich GmbH
  Postfach 1913
  D - 5170 JUELICH 1

BRUGGEMAN, A.
  Research
  Studiecentrum voor Kernenergie
  S.C.K./C.E.N.
  Boeretang 200
  B - 2400 MOL

BRUYERE, B.
  Ingénieur
  Commissariat à l'Energie Atomique
  CEN/Cadarache
  DRDD/BECC
  B.P. 1
  F - 13115 SAINT PAUL LEZ DURANCE

BULUT, N.
  US Dep. of Energy
  9800 S.Cass Ave
  USA - ARGONNE, Il, 60439

BUSCA, G.A.
  Commission of the European
  Communities
  Joint Research Centre
  200, rue de la Loi - SDME 1-89
  B - 1049 BRUSSELS

BUTERY, P.
  Directeur Commercial
  Principia Recherche Developpement
  115, rue Saint Dominique
  F - 75007 PARIS

CADELLI, N.
  Commission of the European
  Communities, DG Science, Research
  and Development
  200, rue de la Loi
  B - 1049 BRUSSELS

CAHUZAC, O.
  Ingénieur
  Commissariat à l'Energie Atomique
  Institut de Protection et Sureté
  Nucléaire
  B.P. 6
  F - 92260 FONTENAY AUX ROSES

CALAY, J.C.
  Chimiste
  Laborelec
  B.P. 11
  B - 1640 RHODE SAINT GENESE

CAMARO, S.
  Ingénieur
  Centre d'Etudes Nucléaires
  de Cadarache
  B.P. 1
  DRDD/SEDFMA/SATM Bt 152
  F - 13115 SAINT PAUL LEZ DURANCE

CAMUS, H.
  Ingénieur
  Commissariat à l'Energie Atomique
  Cadarache DERS/SERE
  B.P. 1
  F - 13115 SAINT PAUL LEZ DURANCE

CARLSEN, L.
  Risø National Laboratory
  Chemistry Department
  DK - 4000 ROSKILDE

CATALAYOUD, L.
  Ingénieur
  GOGEMA / Hague
  B.P. 270
  F - 50107 CHERBOURG

CECILLE, L.
  Commission of the European
  Communities
  200, rue de la Loi
  B - 1049 BRUSSELS

CHARLOT, P.
  Ingénieur
  S.T.M.I.
  9, rue Fernand Léger
  F - 91190 GIF SUR YVETTE

CHARO, L.
Ingénieur
Laboratoire de Mécanique des
Solides, Ecole Polytechnique
F - 91128 PALAISEAU CEDEX

CHAUDON, L.
Ingénieur
Commissariat à l'Energie Atomique
Centre d'Etudes Nucleaires
de la Vallée du Rhone, SEIP
B.P. 171
F - 30205 BAGNOLS SUR CEZE CEDEX

CHERADAME, G.
Ingénieur
PEC Engineering
B.P. 205
F - 95523 CERGY PONTOISE CEDEX

CHIRON, A.
Ingénieur
Société BERTIN
B.P. 3
F - 78373 PLAISIR CEDEX

CLARK, M.J.
Scientist
National Radiological Protection
Board, Chilton, Didcot
GB - OXON OX11 ORQ

CLER, M.
Ingénieur
Commissariat à l'Energie Atomique
Centre d'Etudes Nucleaires
de la Vallée du Rhone, SEIP
B.P. 171
F - 30205 BAGNOLS SUR CEZE CEDEX

CODEE, H.
Radiation Protection Officer
Covra
P.O. Box 20
NL - 1755 ZG PETTEN

COME, B.
Commission of the European
Communities, DG Science, Research
and Development
200, rue de la Loi
B - 1049 BRUSSELS

CONRADT, R.
Physiker
Fraunhofer - Institut für
Silicatforschung
Neunerplatz 2
D - 8700 WUERZBURG

COOK, A.J.
Counsellor
Australian Embassy
51-52, ave des Arts
B - 1040 BRUSSELS

COOLEY, C.R.
Chemical Engineer
Department of Energy USA
USA - WASHINGTON DC 20585

COOPER, J.R.
Scientist
National Radiological Protection
Board
Chilton, Didcot
GB - OXON OX11 ORQ

COSTES, J.R.
Ingénieur
Commissariat à l'Energie Atomique
Centre d'Etudes Nucleaires
de la Vallée du Rhone, UDIN
B.P. 171
F - 30205 BAGNOLS SUR CEZE CEDEX

COTTONE, G.
Commission of the European
Communities, DG Science, Research
and Development
200, rue de la Loi
B - 1049 BRUSSELS

COURTOIS, C.
Ingénieur
Centre d'Etudes Nucléaires
de Cadarache
B.P. 1
F - 13115 SAINT PAUL LEZ DURANCE

CRICCHIO, A.
Commission of the European
Communities, DG Science, Research
and Development
200, rue de la Loi
B - 1049 BRUSSELS

CRIPPS, J.
Engineer
OVE ARUP & PARTNERS
13 Fitzroy Street
GB - LONDON W1P 6BQ

DA CONCEICAO SEVERO, A.J.
Assessor
Direcçao General de Energia
Av. Da República, 45 - 5º
P - 1000 LISBON

DAVIES, I.
Taylor Woodrow Construction Ltd
Taywood House
345 Ruislip Road
GB - SOUTHALL, Middx UB1 2QX

DAVIES, D.H.
Deputy Director General
Commission of the European
Communities, DG Science, Research
and Development
200, rue de la Loi
B - 1049 BRUSSELS

DE ANGELIS, G.
Researcher
ENEA
C.R.E. Casaccia
Via Anguillarese 301
I - 00060 S. MARIA DI GALERIA, Roma

DE BATIST, R.
Professeur
Materials Science Department
S.C.K./C.E.N.
B - 2400 MOL

DE BOER, T.C.
Netherlands Energy Research
Foundation ECN
Westerduinweg 3
NL - 1755 ZG PETTEN

DE LAGUERIE, L.
Ingénieur
Geostock
Tour Aurore
F - 92080 PARIS LA DEFENSE CEDEX 5

DE MARSILY, G.
Professeur, Ecole des Mines
de Paris
Centre d'Informatique Géologique
35, rue St Honoré
F - 77305 FONTAINEBLEAU

DE REGGE, P.
Studiecentrum voor Kernenergie
S.C.K./C.E.N.
Boeretang 200
B - 2400 MOL

DE TASSIGNY, C.
Ingénieur
CEA - CEN Grenoble
SPR 85X
F - 38041 GRENOBLE CEDEX

DEBAUCHE, M.
Ingénieur
Belgonucléaire S.A.
25, rue du Champs de Mars
B - 1050 BRUXELLES

DECRESSIN, A.
Commission of the European
Communities, DG Energy
200, rue de la Loi
B - 1049 BRUSSELS

DEJONGHE, P.
Directeur-général-adjoint
Centre d'Etude de l'Energie
Nucléaire   S.C.K./C.E.N.
Boeretang 200
B - 2400 MOL

DELMAS, J.
Ingénieur
Commissariat à l'Energie Atomique
Cadarache DERS/SERE
B.P. 1
F - 13115 SAINT PAUL LEZ DURANCE

DERLICH, S.
Ingénieur
Commissariat à l'Energie Atomique
IPSN/DPT/SEPD
B.P. 6
F - 92260 FONTENAY AUX ROSES

DETILLEUX, E.
Directeur Général
Ondraf
54, boulevard du Régent, bte 5
B - 1000 BRUXELLES

DEVEUGHELE, M.
Maître de recherche
Centre de Géologie de l'Ingénieur
Ecole Nationale Supérieure des
Mines de Paris
60, boulevard St Michel
F - 75272 PARIS CEDEX 06

DI KOMURKA, M.
Wiss. Mitarbeiter
Oesterr. Forschungszentrum
Seibersdorf Ges.m.b.h.
Lenaugasse 10
A - 1082 WIEN

DINKESPILER, J.A.
  Director General of the Joint
  Research Centre of the Commission
  of the European Communities
  200, rue de la Loi
  B - 1049 BRUSSELS

DOERING, L.
  Nationale Genossenschaft für
  die Lagerung radioaktiver Abfälle
  CH - BADEN

DOM, P.
  Directeur Général
  Foraky S.A.
  13, place des Barricades
  B - 1000 BRUXELLES

DONATO, A.
  Chemist
  ENEA
  C.R.E. Cassaccia
  C.P. 2400
  I - 00100 ROMA

DOYEN, J.J.
  Ingénieur
  Framatome
  Tour Fiat
  F - 92084 PARIS LA DEFENSE CEDEX 16

DOZOL, J.F.
  Ingénieur
  Centre d'Etudes Nucléaires
  de Cadarache
  B.P. 1
  F - 13115 SAINT PAUL LEZ DURANCE

DUNCAN, A.
  Department of the Environment
  Romney House
  43, Marsham Street
  GB - LONDON SW1 3PY

DURAND, E.
  Ingénieur Géotechnicien
  Bureau de Recherches Géologiques
  et Minières
  SGN/GEG
  B.P. 6009
  F - 45060 ORLEANS CEDEX

DWORSCHAK, H.
  Commission of the European
  Communities, DG Science, Research
  and Development
  Joint Research Centre
  Div. Radiochemie
  I - ISPRA, Varese

EID, C.
  Commission of the European
  Communities, DG Science,
  Research and Development
  200, rue de la Loi
  B - 1049 BRUSSELS

ENGELMANN, H.J.
  Abteilungsleiter - Dipl.-Ing.
  Deutsche Gesellschaft zum Bau und
  Betrieb von Endlagern für
  Abfallstoffe mbH
  Woltorfer Strasse 74
  D - 3150 PEINE

EWART, F.
  Scientist
  UKAEA Harwell
  GB - OXFORD OX11 ORA

FABER, P.
  Dipl.-Geologe
  Hahn-Meitner-Institut
  für Kernforschung Berlin GmbH
  PSE
  Glienicker Str. 100
  D - 1000 BERLIN 39

FALKE, W.
  Commission of the European
  Communities, DG Science, Research
  and Development
  200, rue de la Loi
  B - 1049 BRUSSELS

FAURE, J.C.
  Chef de la Section d'Assainissement
  Radioactif
  Centre d'Etudes Nucléaires
  de Cadarache
  B.P. 1
  F - 13115 SAINT PAUL LEZ DURANCE

FAYL, G.
  Commission of the European
  Communities, DG Science,
  Research and Development
  200, rue de la Loi
  B - 1049 BRUSSELS

FEATES, F.
  Civil Servant
  Department of the Environment
  Room A5.28, Ronney House
  43 Marshm Street
  GB - LONDON SW1P 3FY

FINDLAY, J.R.
  Chemist
  U.K.A.E.A. AERE HARWELL
  Chemical Technology Division
  B.2220 AERE
  Harwell, Didcot
  GB - OXON OX11 ORA

FINZI, S.
  Director
  Commission of the European
  Communities, DG Science, Research
  and Development
  200, rue de la Loi
  B - 1049 BRUSSELS

FISHER, J.C.
  Physicist
  H.M. Nuclear Installations
  Inspectorate
  Silkhouse Court
  Tithebarn Street
  GB - LIVERPOOL L2 2LZ

FRIDMAN, G.
  Journaliste
  Energie Nucléaire Magazine
  142, rue Montmartre
  F - 75002 PARIS

FUKUYOSHI, T.
  Engineer
  Radioactive Waste Management Centre
  2-8-10, Toranomon
  Minato-ku
  Japan - TOKYO 105

FURRER, J.
  Dipl.-Chemiker
  Kernforschungszentrum Karlsruhe
  KFK-LAF II
  Postfach 3640
  D - 7500 KARLSRUHE

GAHLERT, S.
  Dipl.-Wirtschaftsingenieur
  Kernforschungszentrum Karlsruhe
  IMF 1
  Postfach 3640
  D - 7500 KARLSRUHE

GANDOLFO, J.M.
  Commission of the European
  Communities, DG Science, Research
  and Development
  200, rue de la Loi
  B - 1049 BRUSSELS

GANNON, D.R.
  Senior Engineer
  Central Electricity
  Generating Board
  Walden House
  24 Cathedral Place
  GB - LONDON EC4P 4EB

GANSER, B.
  Chemiker
  NUKEM GmbH
  Postfach 110080
  D - 6450 HANAU 11

GEENS, L.
  Research Engineer
  Studiecentrum voor Kernenergie
  S.C.K./C.E.N.
  Boeretang 200
  B - 2400 MOL

GERONTOPOULOS, P.
  AGIP S.P.A.
  I - 40059 MEDICINA, Bologna

GIACCHETTI, G.
  Commission of the European
  Communities, JRC Karlsruhe
  B.P. 2266
  D - 7500 KARLSRUHE

GIES, H.
  Geologe
  Gesellschaft für Strahlen- und
  Umweltforschung mbH
  Institut für Tieflagerung
  Theodor-Heuss-Strasse 4
  D - 3300 BRAUNSCHWEIG

GINNIFF, M.
  Chief Executive of Nirex
  Nuclear Industry Radioactive
  Waste Executive
  NIREX, Harwell, Didcot
  GB - OXON OX11 ORA

GIRARDI, F.
  Commission of the European
  Communities
  Projects Directorate
  Joint Research Centre
  Bat. 46
  I - 21020 ISPRA, Varese

GLASSER, F.P.
Professor
University of Aberdeen
Dept. Chemistry
Meston Walk
UK - OLD ABERDEEN AB9 2UE

GOBLET, P.
Ingénieur de Recherche
Ecole des Mines de Paris
35, rue St Honoré
F - 77305 FONTAINEBLEAU

GOETHALS, H.
Geologist
Belgian Geological Survey
Belgische Geologische Dienst
Jennerstr 13
B - 1040 BRUSSELS

GOLICHEFF, I.
Ingénieur
Commissariat Energie Atomique
CEA/CEN-FAR
B.P. 6
F - 92260 FONTENAY AUX ROSES

GOMPPER, K.
Dipl.-Chemiker
Kernforschungszentrum Karlsruhe
Institut für Nukleare
Entsorgungstechnik
Postfach 3640
D - 7500 KARLSRUHE 1

GONZE, P.
Directeur Bureau d'Etudes
Foraky S.A.
13, place des Barricades
B - 1000 BRUXELLES

GOOSSENS, W.
Head Chemical Engineering
Studiecentrum voor Kernenergie
S.C.K./C.E.N.
Boeretang 200
B - 2400 MOL

GOUVRAS, G.
Commission of the European
Communities, DG Employment, Social
Affairs and Education
L - 2920 LUXEMBOURG

GRAEFE, V.
Dipl.-Phys, Ing.
Gesellschaft für Strahlen- und
Unweltforschung
Theodor-Heuss-Str. 4
D - 3300 BRAUNSCHWEIG

GRAMBOW, B.
Chemist
Hahn Meitner Institute
Glienicker str. 100
D - 1000 BERLIN 39

GRAS, M.
Ingénieur
Commissariat à l'Energie Atomique
29-33, rue de la Fédération
F - 75752 PARIS CEDEX 15

GREEN, R.E.
Atomic Energy of Canada Ltd.
Whiteshell Nuclear Research
Establishment
Canada - PINAWA, MANITOBA

GRIFFITHS, M.
Taylor Woodrow Management
Engineering Ltd
Alpha House, Westmount Centre
Delamere Road
GB - HAYES, Middx UB4 0HD

GROSSI, G.
Researcher
Energia Nucleare ed Energie
Alternative
CRE Casaccia
B.P. 2400
I - 00100 ROMA

GUE, J.P.
Ingénieur
CEA de Fontenay-aux-Roses
B.P. 6
F - 92260 FONTENAY AUX ROSES

HADERMANN, J.
Physiker
Eidg. Institut für
Reaktorforschung
CH - 5303 WUERENLINGEN

HAHN, V.
Ingenieur
Kraftanlagen Heidelberg
Im Breitspiel 7
D - 6900 HEIDELBERG 1

HAIJTINK, B.
Commission of the European
Communities, DG Science, Research
and Development
200, rue de la Loi
B - 1049 BRUSSELS

HALASZOVICH, S.
Wissenschaftlicher Mitarbeiter
Kernforschungsanlage Jülich GmbH
ICT
Postfach 1913
D - 5170 JUELICH 1

HAMSTRA, J.
Director
AVORA B.V.
Post Box 138
NL - 1860 AC BERGEN (NH)

HARADA, Y.
Manager
Nippon Kokan K.K.
2-1, Suehiro-cho, Tsurumi-ku
Yokohama-shi
Japan - KANAGAWA-KEN 230

HARBECKE, W.
Dipl.-Ing.
Kernkraftwerk Lingen GmbH
Schüttorfer Strasse 100
D - 4450 LINGEN/Ems

HARTJE, B.
Dipl.-Ing.
Projektgruppe Andere
Entsorgungstechniken
Kernforschungszentrum Karlsruhe GmbH
Postfach 3640
D - 7500 KARLSRUHE

HEATH, M.J.
Geologist
Hillcrest
Church Coombe
GB - REDRUTH, Cornwall, TR16 6RT

HEBEL, W.
Commission of the European
Communities, DG Science, Research
and Development
200, rue de la Loi
B - 1049 BRUSSELS

HEIERS, W.
Commission of the European
Communities, DG Information Market
and Innovation
L - 2920 LUXEMBOURG

HEINER, A.
Geologist
SKB AB
Box 5864
S - 10248 STOCKHOLM

HEMMING, C.
Scientist
Department of the Environment
Room A5.28, Romney House
43 Marsham Street
GB - LONDON SW1 3PY

HENRION, P.
Dr.Sc. Chimiques
Centre d'Etude de l'Energie
Nucléaire, C.E.N./S.C.K.
Boeretang 200
B - 2400 MOL

HEREMANS, R.
Chef de Service
Organisme National des Déchets
Radioactifs et des Fissiles
ONDRAF/NIRAS
54, boulevard du Régent, bte 5
B - 1000 BRUXELLES

HERTOG, R.
Journalist
Reformatorisch Dagblad
Postbus 670
NL - 7300 AR APELDOORN

HIDEO, Y.
Chief Researcher
Energy Research Laboratory
Hitachi, Ltd
1168 Moriyama-cho
316 Japan - HITACHI-SHI, Ibaraki-Ken

HIRLING, J.
UN Senior Officer
International Atomic Energy Agency
Wagramerstrasse 5
P.O. Box 100
A - 1400 VIENNA

HISCOCK, S.
   Director
   Lead Development Association
   34 Berkeley Square
   GB - LONDON WIX 6AJ

HOHNEN, P.
   First Secretary
   Australian Embassy
   51-52, ave des Arts
   B - 1040 BRUSSELS

HOLL, F.
   Dipl.-Ing.
   Kraftanlagen AG, Heidelberg
   Postfach 103420
   D - 6900 HEIDELBERG 1

HOLT, G.
   Engineer
   CEGB
   Barnett Way
   GB - BARNWOOD, Gloucester

HUEBENTHAL, K.
   Bundesministerium für
   Forschung und Technologie
   Heinemannstr. 2
   Postfach
   D - 53 BONN 2

HUNT, B.
   Commission of the European
   Communities
   Joint Research Centre
   I - 21020 ISPRA, Varese

IBA, H.
   Senior Researcher
   Energy Research Laboratory
   Hitachi, LTD
   1168, Moriyama-cho, Hitachi-shi
   Japan - IBARAKI-KEN 316

ICHINOSE, H.
   Manager
   Ohbayashi Corporation
   2-3, Kandatsukasa-cho, Chiyoda-ku
   Japan - TOKYO 101

ISHIKAWA, T.
   Manager
   Kobe Steel LTD
   1-8-2, Marunouchi, Chiyoda-ku
   Japan - TOKYO 100

IWABA, K.
   Deputy Director
   Power Reactor and Nuclear Fuel
   Development Corporation
   1-9-13, Akasaka, Minato-ku
   Japan - TOKYO 107

JACQUET FRANCILLON, N.
   Ingénieur Chef de Section
   CEN/VALRHO - SDHA/SEMC
   B.P. 171
   F - 30205 BAGNOLS SUR CEZE CEDEX

JAOUEN, C.
   Ingénieur
   Société Générale pour les
   Techniques Nouvelles (SGN)
   1, rue des Hérons
   Montigny-le-Bretonneux
   F - 78184 ST QUENTIN EN YVELINES CEDEX

JENSEN, B.S.
   M. Sci.
   Risø National Laboratory
   Chemistry Department
   DK - 4000 ROSKILDE

JOCKWER, N.
   Dipl.-Physiker
   Gesellschaft für Strahlen- und
   Umweltforschung mbH
   Institut für Tieflagerung
   Theodor-Heuss-Strasse 4
   D - 3300 BRAUNSCHWEIG

JOURDE, P.
   Ingénieur
   Commissariat à l'Energie Atomique
   Centre de Fontenay aux Roses
   B.P. 6
   F - 92260 FONTENAY AUX ROSES

JUZNIC, K.
   Ing. Chem.
   Josef Stefan Institute
   Jamova 39
   YU - 61000 LJUBLJANA

KANAI, Y.
   Section Manager
   Chiyoda Chemical Engineering &
   Construction Co., Ltd
   2-12-1, Tsurumicho, Tsurumi-ku
   Yokohama-shi
   Japan - KANAGAWA-KEN 230

KAPPEI, G.
  Dipl.-Ing.
  Gesellschaft für Strahlen- und
  Umweltforschung mbH
  Institut für Tieflagerung
  Theodor-Heuss-Strasse 4
  D - 3300 BRAUNSCHWEIG

KASTELEIN, J.
  Abfallbereiter
  N.V. PZEM
  Postfach 48
  NL - 4330 AA MIDDELBURG

KAWANISHI, N.
  Engineer
  Toshiba Corporation
  8, Shin Sugita-cho, Isogo-ku
  Yokohama-shi
  Japan - KANAGAWA-KEN 235

KEEN, N.J.
  Programme Manager
  UKAEA, Harwell, Oxon
  AFPD, Hangar 10
  Harwell, Didcot
  GB - OXON OX11 ORA

KOBAYQSHI, R.
  J G C Corporation
  14-1, Bessho 1- Chome
  Minami-Ku
  Japan - YOKOHAMA 232

KOESTER, R.
  Dipl.-Chem.
  Kernforschungszentrum Karlsruhe
  Postfach 3640
  D - 7500 KARLSRUHE

KOFAHL, A.
  Sachverständiger
  Gesellschaft für
  Reaktorsicherheit mbH
  Schwertnergasse 1
  D - 5000 KOELN 1

KORTHOF, R.M.
  Beleidsmedewerker
  Ministry of Economic Affairs
  P.O. Box 20101
  NL - 2500 EC THE HAGUE

KOWALSKI, E.
  Physiker
  Nagra Nationale Genossenschaft
  für die Lagerung radioaktiver
  Abfälle
  Parkstrasse 23
  CH - 5401 BADEN

KRAEMER, R.
  Projektbevollmächtigter
  Kernforschungszentrum Karlsruhe
  Projekt Wiederaufarbeitung und
  Abfallbehandlung
  Postfach 3640
  D - 7500 KARLSRUHE

KRAMER, J.
  Dipl.-Ing.
  TUEV Rheinland e.V.
  Postfach 10 17 50
  D - 5000 KOELN 1

KRAUSE, H.
  Chemiker
  Kernforschungszentrum Karlsruhe
  Institut für Nukleare
  Entsorgungstechnik
  Postfach 3640
  D - 7500 KARLSRUHE

KRISCHER, W.
  Commission of the European
  Communities, DG Science, Research
  and Development
  200, rue de la Loi
  B - 1049 BRUSSELS

KROEBEL, R.
  Dr.rer.nat./Projektleiter
  Kernforschungszentrum Karlsruhe
  Projekt Wiederaufarbeitung und
  Abfallbehandlung
  Postfach 3640
  D - 7500 KARLSRUHE

KROEGER, H.
  Chemiker
  Technischer Ueberwachungs-Verein
  Hannover e.V.
  Postfach 810740
  Am TUEV 1
  D - 3000 HANNOVER 81

KUEHN, K.
Institutsleiter
Gesellschaft für Strahlen- und
Umweltforschung mbH
Institut für Tieflagerung
Theodor-Heuss-Strasse 4
D - 3300 BRAUNSCHWEIG

KUNZ, W.
Projektleiter
Deutsche Gesellschaft für
Wiederaufarbeitung von
Kernbrennstoffen mbH (DWK)
Abteilung BEE
Postfach 1407
D - 3000 HANNOVER 1

LACHOVIEZ, S.
Ingénieur
Comité Interministériel de la
Sécurité Nucléaire
54, rue de Varenne
F - 75700 PARIS

LAHURE, J.
Secrétaire d'Etat à l'Economie
Ministère de l'Economie et des
Classes Moyennes
19-21, boulevard Royal
L - LUXEMBOURG

LAKE, G.
European Parliament
Information and Public Relations
L - 2929 LUXEMBOURG

LAKE, L.
Mott, Hay & Anderson
20 - 26 Wellesley Rd
GB - CROYDON CR9 2UL

LAMBOTTE, J.M.
Ingénieur Indust. Principal
Ministère de la Santé Publique
Service de Protection contre
les Rayonnements Ionisants
Cité Administrative de l'Etat
2-3/32 Quartier Vésale
B - 1010 BRUXELLES

LANZA, F.
Commission of the European
Communities
Joint Research Centre
I - 21020 ISPRA, Varese

LE MASNE, D.
Géophysicien
BRGM (France)
SGN/GPH
B.P. 6009
F - 45060 ORLEANS CEDEX

LEDEBRINK, F.W.
Chemiker
Alkem GmbH
Rodenbacher Chaussee 6
D - 6450 HANAU 11

LEFEVRE, J.
Directeur
Commissariat à l'Energie Atomique
DED/CEN/FAR
B.P. 6
F - 92260 FONTENAY AUX ROSES

LEFILLATRE, G.
Ingénieur
Commissariat à l'Energie Atomique
Centre d'Etudes Nucleaires
de la Vallée du Rhone, SPI
B.P. 171
F - 30205 BAGNOLS SUR CEZE CEDEX

LEICHSENRING, C.H.
Dipl.-Ing.
Kernforschungszentrum Karlsruhe
Postfach 3640
D - 7500 KARLSRUHE

LEVEQUE, P.
Professeur
Laboratoire de Radiogéologie et de
Mécanique des Roches
Université de Bordeaux I
Avenue des Facultés
F - 33405 TALENCE CEDEX

LEWI, J.
Ingénieur
Commissariat à l'Energie Atomique
IPSN/DAS/SAED
B.P. 6
F - 92260 FONTENAY AUX ROSES

LEY, J.D.
Commission of the European
Communities
Projects Directorate
Joint Research Centre
Bat. 46
I - 21020 ISPRA, Varese

LIESER, K.H.
  Professor
  Technische Hochschule Darmstadt
  Fachbereich Anorganische Chemie
  und Kernchemie
  Technische Hochschule Darmstadt
  D - 6100 DARMSTADT

LI, W.
  Radioactive Waste Treatment Engin.
  Centre d'Etude de l'Energie
  Nucléaire, CEN/SCK
  Boeretang 200
  B - 2400 MOL

LINSTER, R.
  Commission of the European
  Communities, DG Personnel and
  Administration
  L - 2920 LUXEMBOURG

LOIDA, A.
  Wissensch. Mitarbeiter
  Kernforschungszentrum Karlsruhe
  Institut für Nukleare
  Entsorgungstechnik
  Postfach 3640
  D - 7500 KARLSRUHE

LOIZEAU, P.
  Ingénieur
  Electricité de France
  35, rue de Paris
  F - 92110 CLICHY

LOXHAM, M.
  Project Engineer
  Delft Soil Mechanics Laboratory
  P.O. Box 69
  NL - 2600 AB DELFT

LÜEDENBACH, B.
  Dipl.-Ing.
  Interatom GmbH
  Postfach
  D - 5060 BERGISCH GLADBACH

LUTZE, W.
  Scientist
  Hahn-Meitner-Institut für
  Kernforschung Berlin
  Glienicker Strasse 100
  D - 1000 BERLIN 39

LYON, R.B.
  Director, Waste Management Div.
  Atomic Energy of Canada Limited
  Whiteshell Nuclear Research
  Establishment
  Pinawa, Manitoba
  Canada - ROE 1LO

MALOW, G.
  Scientist
  Hahn-Meitner-Institut für
  Kernforschung Berlin
  Glienicker Strasse 100
  D - 1000 BERLIN 39

MANARA, A.
  Commission of the European
  Communities
  Physics Division
  Joint Research Centre
  Bldg. 44
  I - 21020 ISPRA, Varese

MANFROY, P.
  Ingenieur Projet
  Centre d'Etude de l'Energie
  Nucléaire
  C.E.N./S.C.K.
  Boeretang 200
  B - 2400 MOL

MANNONE, F.
  Commission of the European
  Communities
  Divisione Supporto Nucleare
  Joint Research Centre
  I - 21020 ISPRA, Varese

MARCAILLOU, J.H.
  Assistant
  Centre d'Etudes Nucléaires de
  Cadarache
  Déchêts - Serv. de Protec., Prév.
  et Contrôle
  B.P. 1
  F - 13115 SAINT PAUL LEZ DURANCE

MARCUS, F.R.
  Secretary General
  Nordic Liaison Committee for
  Atomic Energy
  P.O.B. 49
  DK - 4000 ROSKILDE

MARPLES, J.A.C.
  Govematerials Scientist
  UKAEA
  B 220.22 AERE Harwell
  GB - DIDCOT, Oxon OX11 ORA

MARSH, G.
  Scientist
  UKAEA
  Materials Development Division
  Building 393
  AERE Harwell
  GB - DIDCOT, Oxon OX11 ORA

MARZULLO, T.
  Manager of radioactive
  waste management
  ENEL - Construction Division
  ENEL DCO
  Via G B Martini 3
  I - 00198 ROMA

MATSUMURA, T.
  Engineer
  Toyo Engineering Corporation
  7-7-1, Hon-cho, Funabashi-shi
  Japan - CHIBA-KEN

MATTEMAN, J.L.
  Maschinenbauer
  N.V. KEMA
  Utrechtseweg 310
  Postfach 9035
  NL - 6800 ET ARNHEM

MATTON, P.
  Chef de Laboratoire
  Centrale Nucléaire
  des Ardennes CHOOZ
  B.P. 160
  F - 08600 GIVET

MATZKE, H.
  Dipl.-Phys.
  Europäisches Institut für
  Transurane GFS
  Postfach 2266
  D - 7500 KARLSRUHE

MAYENCE, M.
  Ingénieur
  Tractionel
  31, rue de la Science
  B - 1040 BRUXELLES

MAYLIN, J.A.
  Ingénieur
  Agence Nationale pour les Déchets
  Radioactifs ANDRA
  31-33, rue de la Fédération
  F - 75015 PARIS

MESTER, W.
  Dipl.-Physiker
  Gesellschaft für
  Reaktorsicherheit (GRS)
  Schwertnergasse 1
  D - 5000 KOELN 1

MIELKE, H.G.
  Sachverständiger
  Gesellschaft für
  Reaktorsicherheit
  Schwertnergasse 1
  D - 5000 KOELN 1

MOHRI, T.
  Manager
  Chubu Environment Engineering Co.
  4-14-16, Osu, Naka-ku
  Nagoya-shi
  Japan - AICHI-KEN 460

MONTAGNINI, A.
  Ingegnere
  C.A.M.E.N.
  Via della Bigattiera
  S. Piero A Grado
  I - 56010 PISA

MONTJOIE, M.
  Ingénieur
  Technicatome
  B.P. 18
  F - 91190 GIF-SUR-YVETTE

MONTSERRAT, W.
  Chemiker
  Deutsche Gesellschaft für
  Wiederaufarbeitung von
  Kernbrennstoffen
  Hamburger Allee 4
  D - 3000 HANNOVER 1

MORIMOTO, H.
  General Manager
  Ishikawajima-Harima
  Heavy Industries Co., Ltd
  1-6-2, Marunouchi, Chiyoda-ku
  Japan - TOKYO 100

MORI, A.
  Ingegnere
  Dipartimento di Costruzioni
  Meccaniche e Nucleari (Universita'
  di Pisa)
  Via Diotisalvi 2
  I - 56100 PISA

MOSAR, N.
  Member of the Commission of the
  European Communities
  200, rue de la Loi
  B - 1049 BRUSSELS

MUKAI, O.
  Engineer
  Hitachi Zosen Corporation
  1-3-40, Sakurajima, Konohana-ku
  Osaka-shi
  Japan - OSAKA-FU 554

MURRAY, N.
  Commission of the European
  Communities
  Divisione Radiochimica
  Joint Research Centre
  I - 21020 ISPRA, Varese

McCOMBIE, C.
  Dr.Phil. Physiker
  Nagra Nationale Genossenschaft
  für die Lagerung radioaktiver
  Abfälle
  Parkstrasse 23
  CH - 5401 BADEN

McEWEN, T.J.
  Geologist
  British Geological Survey
  Keyworth
  GB - NOTTINGHAM NG12 5GG

NAGEL, W.
  Chemiker
  N.V. KEMA
  Utrechtseweg 310
  Postfach 9035
  NL - 6800 ET ARNHEM

NAZARE, S.
  Dipl.-Ing.
  Kernforschungszentrum Karlsruhe
  Institut für Material und
  Festkörperforschung
  Postfach 3640
  D - 7500 KARLSRUHE

NEERDAEL, B.
  Ingénieur Civil
  Centre d'Etude de l'Energie
  Nucléaire C.E.N./S.C.K.
  Boeretang 200
  B - 2400 MOL

NICOLAY, D.
  Commission of the European
  Communities, DG Information
  Market and Innovation
  L - 2920 LUXEMBOURG

NILSSON, K.
  Risø National Laboratory
  Chemistry Department
  DK - 4000 ROSKILDE

NOKHAMZON, J.G.
  Commissariat à l'Energie Atomique
  33, rue de la Fédération
  F - 75752 PARIS CEDEX 15

NOMINE, J.C.
  Ingénieur au DRDD/SESD/CEN.FAR
  Commissariat à l'Energie Atomique
  Centre d'Etudes Nucléaires
  de Saclay
  B.P. 2
  F - 91191 GIF SUR YVETTE

NOWAK, R.
  Dipl.-Chemiker
  Gesellschaft für
  Reaktorsicherheit mbH
  Schwertnergasse 1
  D - 5000 KOELN 1

NOZAWA, S.
  Adviser
  Tobishima Corporation
  2, Sanban-cho, Chiyoda-ku
  Japan - TOKYO 102

OBERSTE-PADTBERG, R.
  Ingénieur
  Etudes et Fabrication Dowell-
  Schlumberger
  B.P. 90
  F - 42003 ST. ETIENNE

ODEN, D.
  Scientist
  Pacific Northwest Laboratory
  P.O. Box 999
  USA - RICHLAND, WA

OFFERMANN, P.
Commission of the European
Communities
Divisione Radiochimica
Joint Research Centre
Ed. 46
I - 21020 ISPRA, Varese

OKAZAWA, M.
Assistant Manager
Japan Atomic Industrial Forum, Inc.
1-1-3 Shimbashi, Minato-ku
Japan - TOKYO 105

OLIVIER, J.P.
Chef de la Division de la
Protection Radiologique et de la
Gestion des Déchets Radioactifs
Agence de l'OCDE pour l'Energie
Nucléaire, AEN/OCDE
38, boulevard Suchet
F - 75016 PARIS

ORLOWSKI, S.
Commission of the European
Communities, DG Science, Research
and Development
200, rue de la Loi
B - 1049 BRUSSELS

OUSTRIERE, P.
Géochimiste
Bureau de Recherches Géologiques
et Minières
SGN/GMX/PG
B.P. 6009
F - 45060 ORLEANS CEDEX

PAGE, R.J.
Waste Management Technologist
British Nuclear Fuels plc
R102
GB - RISLEY, Warrington WA3 6AS

PEAUDECERF, P.
Chef de la Mission Stockages du SGN
Bureau de Recherches Géologiques
et Minières
BRGM
B.P. 6009
F - 45060 ORLEANS CEDEX

PETERS, W.
Editor. Atom
United Kingdom Atomic Energy
Authority
Charles II Street
GB - LONDON SW1Y 4QP

PHILLIPS, D.C.
Scientist
UKAEA
Materials Development Division
Building 47
Aere Harwell
GB - OXON OX11 ORA

PIRS, M.
Dr.Chem.
Institut Jozef Stefan
Janova 39
P.P. 53
YU - 6111 LJUBLJANA

PITZ, W.
Geschäftsführer, Dipl.-Ing.
Deutsche Gesellschaft zum Bau und
Betrieb von Endlagern für
Abfallstoffe mbH
Woltorfer Strasse 74
D - 3150 PEINE

PLUCHINO, G.
Esperto qualificato protezione
nucleare - Direttore tecnico
DE.ME. RAD.
C.P. 2087 - TA/10
I - 74100 TARANTO

POHL, U.
Battelle-Institut e.V.
Am Roemerhof 35
D - 6000 FRANKFURT/MAIN 90

PONCELET, J.P.
Chargé de Mission
Services de Programmation
de la Politique scientifique
8, rue de la Science
B - 1040 BRUXELLES

POPP, M.
Bundesministerium für Forschung
und Technologie
Postfach 200 706
D - 5300 BONN 2

POTEMANS, M.
Directeur EBES - Conseiller UEEB
Union des Exploitations Electriques
en Belgique
4, Galerie Ravenstin, bte 6
B - 1000 BRUXELLES

POTTIER, P.
Adjoint au Chef du DRDD
Commissariat à l'Energie Atomique
CEN/Cadarache
DERDCA/DRDD
B.P. 1
F - 13115 SAINT PAUL LEZ DURANCE

PRICE, M.
UKAEA
GB - WINFRITH, Dorchester, Dorset

PRIJ, J.
Netherlands Energy Research
Foundation ECN
Westerduinweg 3
NL - 1755 ZG PETTEN

PRINS, M.
Soil Scientist
National Institute of Public Health
and Environmental Hygiene
Postbox 1
NL - 3720 BA BILTHOVEN

PUDEWILLS, A.
Dipl.-Geophysiker
Kernforschungszentrum Karlsruhe
Postfach 3640
D - 7500 KARLSRUHE

PUGH, J.
Professional Engineer
National Nuclear Corporation
Limited
Warrington Road
Risley Warrington
GB - CHESHIRE WA3 6BZ

PURPLE, R.
US Department of Energy
USA - WASHINGTON D.C. 20545

PUT, M.
Centre d'Etude de l'Energie
Nucléaire   S.C.K./C.E.N.
Boeretang 200
B - 2400 MOL

REICHARDT, F.
Commission of the European
Communities, DG Science,
Research and Development
200, rue de la Loi
B - 1049 BRUSSELS

RIECH, H.
Dipl.-Ing.
Hochtief AG
Postfach 3289
D - 6000 FRANKFURT AM MAIN 1

RIPPON, S.E.
European Editor
Nuclear News
8, Ruvigny Mansions
Embankment, Putney
GB - LONDON SW15 1LE

ROEST, J.P.A.
Mining Engineer
TH Delft
Afdeling Mijnbouwkunde
Mijnbouwstraat 120
NL - 2628 RX DELFT

ROLANDI, G.
Energia Nucleare ed Energia
Alternative
COM - IRITR
CRE Casaccia
Casella Postale 2400
I - 00100 ROMA A.D.

ROMETSCH, R.
Phys.-Chemiker
Nagra Nationale Genossenschaft
für die Lagerung radioaktiver
Abfälle
Parkstrasse 23
CH - 5401 BADEN

RUDOLPH, G.
Dipl.-Chemiker
Kernforschungszentrum Karlsruhe
Postfach 3640
D - 7500 KARLSRUHE

RUHBAUM, M.
Dipl.- Physiker
Kraftwerk-Union
Postfach 1240
D - 8757 KARLSTEIN

RYHAENEN, V.
Industrial Power Company Ltd
Fredrikinkatu 51-53
SF - 00100 HELSINKI

SAAS, A.
 Chef du BECC/CIDN
 Commissariat à l'Energie Atomique
 CEN/Cadarache
 DRDD/BECC/CIDN
 B.P. 1
 F - 13115 SAINT PAUL LEZ DURANCE

SAKATA, H.
 Senior Advisor
 Sato Kogyo Co., Ltd
 4-8, Nihombashihon-cho, Chuo-ku
 Japan - TOKYO 103

SANTARINI, G.
 Ingénieur
 Commissariat à l'Energie Atomique
 CEN/FAR - SCECF/SECA
 B.P. 6
 F - 92260 FONTENAY AUX ROSES

SCHALLER, K.
 Commission of the European
 Communities, DG Science, Research
 and Development
 200, rue de la Loi
 B - 1049 BRUSSELS

SCHEIBEL, H.G.
 Dipl.-Phys.
 Battelle Institut
 Am Römerhof 35
 Postfach 900160
 D - 6000 FRANKFURT 90

SCHMIDT-HANSBERG, T.
 Dipl.-Chemiker
 NUKEM GmbH
 Postfach 110080
 D - 6450 HANAU 11

SCHMIDT, M.W.
 Dipl.-Geologe
 Gesellschaft für Strahlen- und
 Umweltforschung mbH
 Institut für Tieflagerung
 Theodor-Heuss-Strasse 4
 D - 3300 BRAUNSCHWEIG

SCHMIDT, W.
 Dipl.-Ing.
 DWK Deutsche Gesellschft für
 Wiederaufarbeitung von
 Kernbrennstoffen
 Hamburger Allee 4
 D - 3000 HANNOVER 1

SCHMITT, R.E.
 Dipl.-Chemiker
 Battelle-Institut e.V.
 Am Römerhof 35
 D - 6000 FRANKFURT/MAIN 90

SCHOLZE, H.
 Prof.
 Fraunhofer-Institut für
 Silicatforschung
 Neunerplatz 2
 D - 8700 WUERZBURG

SHIBUYA, M.
 Research Engineer
 JGC Corporation
 C/o Waste Dept.   S.C.K./C.E.N.
 Boeretang 200
 B - 2400 MOL

SIMON, R.
 Commission of the European
 Communities, DG Science, Research
 and Development
 200, rue de la Loi
 B - 1049 BRUSSELS

SMAILOS, E.
 Dipl.-Ing.
 Kernforschungszentrum Karlsruhe
 Postfach 3640
 D - 7500 KARLSRUHE

SOMBRET, C.
 Ingénieur Adjoint Chef de Service
 Commissariat à l'Energie Atomique
 Centre d'Etudes Nucleaires
 de la Vallée du Rhone, SDHP
 B.P. 171
 F - 30205 BAGNOLS SUR CEZE CEDEX

SPIERS, C.J.
 Universiteit Utrecht
 Instituut voor
 Aardwetenschappen
 Budapestlaan 4
 NL - 3508 TA UTRECHT

STALIOS, A.
 Physicien
 Centre d'Etude de l'Energie
 Nucléaire, C.E.N./S.C.K.
 Boeretang 200
 B - 2400 MOL

STARR, M.
Professional Engineer
National Nuclear Corporation
Limited
Cambridge Road
GB - WHETSTONE, Leics LE8 3LH

STELTE, N.
Wiss. Ang.
Gesellschaft für Strahlen- und
Umweltforschung mbH
Institut für Tieflagerung
Theodor-Heuss-Strasse 4
D - 3300 BRAUNSCHWEIG

STILLA, A.
Dipl.-Geologe
Niedersächsisches Landesamt
für Bodenforschung
Stilleweg 2
Postfach 510153
D - 3000 HANNOVER 51

STRAHL, J.F.
Engineer
Weston
2301 Research Blvd
USA - ROCKVILLE, MD 20853

SUGIER, A.
Ingénieur
Commissariat à l'Energie Atomique
Centre de Fontenay aux Roses
B.P. 6
F - 92260 FONTENAY AUX ROSES

SUGIMURA, S.
Director
Ebara Corporation
11-1, Asahi-cho
Haneda, Ota-ku
Japan - TOKYO 144

SWALE, D.J.
Production Manager
British Nuclear Fuels PLC
BNFL, Windscale & Calder Works
Building B 314
GB - SELLAFIELD, Cumbria

TAMAKI, H.
Engineer
Mitsubishi Heavy Industry, Ltd
2-4-1, Shibakoen, Minato-ku
Japan - TOKYO

TAMAKI, H.
Manager
Nippon Nuclear Service Corporation
5-15, Kitahama, Higashi-ku
Osaka-shi
Japan - OSAKA-FU 541

TAORMINA, A.
Nuclear Engineer
Nuclear Assurance Corporation
Weinbergstr. 9
CH - 8001 ZURICH

TASSONI, E.
Ricercatore
ENEA
C.R.E. Casaccia
Via Anguillarese, Km 1,300
I - 00060 S. MARIA DI GALERIA, Roma

THEGERSTROEM, C.
Ms of Science & Nuclear Engineering
Swedish Nuclear Fuel and
Waste Management Co, SKBT
Box 5864
S - 102 48 STOCKHOLM

THEISEN, W.
Dipl.-Ing.
Uhde - GmbH
Friedrich-Uhde-Str. 15
D - 4600 DORTMUND 1

THIMUS, J.F.
Ingénieur Civil - 1e Assistant
Université Catholique de Louvain
Laboratoire du Génie Civil
1, place du Levant
B - 1348 LOUVAIN-LA-NEUVE

THURY, M.
Dr.Phil.
NAGRA, Nationale Genossenschaft
für die Lagerung radioaktiver
Abfälle
Parkstrasse 23
CH - 5401 BADEN

TOMITA, H.
Manager
Nuclear Transport Services Co. Ltd
1-1-3 Shibadaimon, Minato-ku
Japan - TOKYO 105

TRAEGER, S.
Dipl.-Geologe
NUKEM GmbH
Geschäftsbereich Dienstleistung
Postfach 11 00 80
D - 6450 HANAU

TROCELLIER, P.
Ingénieur
Commissariat à l'Energie Atomique
Centre d'Etudes Nucléaires de
Saclay
Laboratoire Pierre Sue
F - 91191 GIF SUR YVETTE

TUNABOYLU, K.
Ingenieur
Motor Columbus
Ingenieur-Unternehmung AG
Parkstr. 27
CH - 5402 BADEN

TURNER, A.
Research Scientist
Atomic Energy Research
Establishment
Materials Development Division
Building 429
AERE Harwell
GB - OXFORDSHIRE OX11 ORA

TURVEY, F.
Principal Scientific Officer
Nuclear Energy Board
20-22 Lower Hatch Street
IRL - DUBLIN 2

UEMOTO,
JGC Corporation
2, place du Palais Royal
F - 75044 PARIS CEDEX 01

UHLENBRUCK, H.
Sachverständiger
Gesellschaft für
Reaktorsicherheit
Schwertnergasse 1
D - 5000 KOELN 1

VAN CAENEGHEM, J.
Commission of the European
Communities, DG Environment,
Consumer Protection and
Nuclear Safety
200, rue de la Loi
B - 1049 BRUSSELS

VAN DE VATE, L.
Geofysicus
Ministerie van Economische
Zaken, RGD
Postbus 157
NL - 2000 AD HAARLEM

VAN DE VOORDE, N.
Ingenieur
Centre d'Etude de l'Energie
Nucléaire   S.C.K./C.E.N.
Boeretang 200
B - 2400 MOL

VAN DONGEN, A.C.
Process Engineer
NUCON b.v.
Weesperzijde 150
NL - 1097 DS AMSTERDAM

VAN ISEGHEM, P.
Ir.
Studiecentrum voor Kernenergie
S.C.K./C.E.N.
Boeretang 200
B - 2400 MOL

VAN KOTE, F.
Ingénieur
C.E.A.
31-33, rue de la Fédération
F - 75015 PARIS

VANDORPE, M.
Chef de la Promotion - Ventes,
Wastes & Systèmes
Belgonucleaire S.A.
25, rue du Champ de Mars
B - 1050 BRUXELLES

VANHELLEMONT, G.
Deputy Manager Fuel
Manufacturing Division
S.A. Belgonucleaire
Europalaan 20
B - 2480 DESSEL

VAUNOIS, P.
Ingénieur
Commissariat à l'Energie Atomique
Cadarache CEN
B.P. 1
F - 13115 SAINT PAUL LEZ DURANCE

VEEL, T.
Ingenieur
NUCON b.v.
Weesperzijae 150
NL - 1097 DS AMSTERDAM

VEJMELKA, P.
Dipl.-Chemiker
Kernforschungszentrum Karlsruhe
D - 7514 LEOPOLDSHAFEN

VENET, P.
Commission of the European
Communities, DG Science, Research
and Development
200, rue de la Loi
B - 1049 BRUSSELS

VERKERK, B.
Radiochemist
Margrietlaan 9
NL - 1862 EC BERGEN (NH)

VIGREUX, B.
Directeur, Secteur Déchets Nucl.
Société Générale pour les
Techniques Nouvelles (SGN)
F - 78184 SAINT QUENTIN YVELINES

VILKAMO, S.
Inspector
Finnish Centre for Radiation
and Nuclear Safety
P.O. Box 268
SF - 00101 HELSINKI

VOLOBRO LARSEN, V.
M.Sc.
Risoe National Laboratory
Chemistry Department
DK - 4000 ROSKILDE

VON BLOTTNITZ, U.
Abgeordnete
Europäisches Parlament
D - 3131 GRABOW

VONS, L.H.
Netherlands Energy Research
Foundation ECN
Westerduinweg 3
NL - 1755 ZG PETTEN

VUORI, S.
Senior Research Scientist
Technical Research Centre
of Finland, Nuclear Engineering
Laboratory
P.O.B. 169
SF - 00181 HELSINKI 18

WALLNER, M.
Ing., Wiss. Angest.
Bundesanstalt für
Geowissenschaften und Rohstoffe
Postfach 510153
D - 3000 HANNOVER 51

WARREN, G.A.
Chartered Engineer
Rolls Royce & Associates Ltd
P.O. Box 31
GB - DERBY DE2 8BJ

WEBER, R.
Wiss. Journalist
Nagra Nationale Genossenschaft
für die Lagerung radioaktiver
Abfälle
Parkstrasse 23
CH - 5401 BADEN

WHITMELL, D.S.
Scientist
Atomic Energy Research
Establishment
Materials Development Division
Building 393
AERE Harwell
GB - DIDCOT, Oxon OX11 ORA

WIECZOREK, H.
Dipl.-Chemiker
Kernforschungszentrum Karlsruhe
Institut für Nukleare
Entsorgungstechnik
Postfach 3640
D - 7500 KARLSRUHE

WILLIAMS, G.
Hydrogeologist
British Geological Survey
BGS, Keyworth
GB - NOTTINGHAM NG12 5GG

WILSON, J.A.
Government Inspector
HM Industrial Pollution
Inspectorate for Scotland
47 Robb's Loan
UK - EDINBURGH EH14 1TY

YAN, J.
  Chemical Engineer
  Studiecentrum voor Kernenergie
  S.C.K./C.E.N.
  Boeretang 200
  B - 2400 MOL

YOKOBORI, Y.
  Editor
  Japan Atomic Industrial Forum, Inc.
  1-13 Shimbashi 1-Chome, Minato-ku
  Japan - TOKYO 105

YONEDA, M.
  Kansai Electric Power
  3, ave Hoche
  F - 75008 PARIS

ZAKORA, M.
  Inspectorate of Nuclear
  Installations
  Risø Huse 11
  P.O. Box 217
  DK - 4000 ROSKILDE

ZAMARRA, F.
  Junta de Energia Nuclear
  Avda Complutense 22
  E - 28040 MADRID

ZAMORANI, E.
  Commission of the European
  Communities
  Joint Research Centre
  I - 21020 ISPRA, Varese

ZIRNGAST, M.
  Geologe
  Bundesanstalt für
  Geowissenschaften und Rohstoffe
  Stilleweg 2
  D - 3000 HANNOVER 51

ZORZOLI, G.
  Energia Nucleare ed Energie
  Alternative - ENEA
  Viale Regina Margherita 125
  I - 00198 ROMA

INDEX OF AUTHORS